Air Bearings

Air Bearings

Theory, Design and Applications

Farid Al-Bender
KU Leuven, Department of Mechanical Engineering
Leuven
Belgium

This edition first published 2021

Registered Offices
John Wiley & Sons, Inc., 111 River Street, Hoboken, NJ 07030, USA
John Wiley & Sons Ltd, The Atrium, Southern Gate, Chichester, West Sussex, PO19 8SQ, UK

Editorial Office
The Atrium, Southern Gate, Chichester, West Sussex, PO19 8SQ, UK

For details of our global editorial offices, customer services, and more information about Wiley products visit us at www.wiley.com.

Wiley also publishes its books in a variety of electronic formats and by print-on-demand. Some content that appears in standard print versions of this book may not be available in other formats.

Library of Congress Cataloging-in-Publication Data applied for

HB ISBN: 9781118511497

Cover image: Courtesy of Farid Al-Bender
Cover design by Wiley

Set in 9.5/12.5pt STIXTwoText by SPi Global, Chennai, India
Printed and bound by CPI Group (UK) Ltd, Croydon, CR0 4YY

10 9 8 7 6 5 4 3 2 1

To the memory of my parents
To Katherine
To the children

Contents

List of Contributors

Name	Affiliation	Contribution
Tobias Waumans	Xeryon, Leuven, Belgium	Chapter 10
Peter Vleugels	ASML BV, Netherlands	Major part of Chapter 12
Steven Cappa	CAPPA precision engineering, Belgium	Major part of Chapter 13
Gorka Aguirre	IDEKO, Spain	Chapter 16
Marius Nabuurs	KU Leuven, Belgium	Chapter 11

List of Figures

List of Tables

Preface

They say you get to know people well only when you have travelled with them. It so happens that I have travelled with each of the contributors of this book many times, in the framework of research and other projects, and discovered their other hidden fine qualities. But my longest journey has undisputedly been with the topic of air bearings itself, spanning several decades. They also say that you don't love anything until you know it by heart. I cannot claim to know this broad topic entirely by heart, but there is still considerable romance in our relation. My passion for air bearings grew more ardent the deeper I went into the subject. Yet it had been almost by a series of sheer accidents that I plunged into this fascinating topic. The journey started at Leeds university, as a Master's student and continued, after a long interruption, at KU Leuven. While it had been a lesser topic of investigation at Leeds, research in air bearings was already well established at Leuven. It had started in the 1970's at the initiative of Eric Blondeel, who undertook a PhD investigation, under the promotion of Prof. Raymond Snoeys. Eric had an enormous passion for the topic, delivering beautiful inventions in it like the famous self-compensating bearing with infinite stiffness. That initiative kicked-off air-bearing research at Leuven, leading afterwards to numerous Master's and PhD investigations often inspired by real industrial needs.

One of these was to be my own, which happened to take place at the point in which research in air bearings worldwide had dwindled down to its lowest level ever. So I had to keep the fire of research and development burning. To make matters worse, my supervisor, Prof. Snoeys passed away unexpectedly handing the torch over to Prof. Hendrick Van Brussels, who luckily accepted the challenge, giving me more impetus but also more responsibility to set the work forth.

Luckily, I was blessed by a mentor and a dear friend, the late Prof. Paul Vanherck, with whom I exchanged ideas, posed problems, sought solutions, designed test rigs,... on almost daily basis. No words can express my gratitude to him.

It took me long to finish the doctoral work, but my dissertation contained, to my mind, the answers to several key questions regarding aerostatic-bearing theory and practice; answers that I had liked to find in the literature, but which had been poorly or scantly treated. I also made, with my own hands as it were, numerous bearings and bearing systems, often in the framework of contract projects, which only went to heighten my attachment to and deepen my love for the topic.

I think it was just a little later, that I first toyed with the idea of making a monograph out of my dissertation, stimulated by the absence of up-to-date books on aerostatic bearings. However, plunging directly into a never-ending series of research projects, academic duties and family engagements meant postponing that project year after year, until an unexpected push finally came. In the meantime, research in air bearings continued through a sporadic series of technical and research projects. And when sometimes I was unable to secure funds to maintain this activity, colleagues Hendrik Van Brussel and Dominiek Reynaerts managed somehow to come up with funded-research opportunities, which kept the ball of air bearings rolling; I am deeply indebted to them for that.

Thus, between the mid 1990's and the present, five PhD's where completed: those of the contributors to this book; I am deeply indebted to them too. The actual push, though, to embark on this project came, again by sheer coincidence, when Sofie Rombouts, of Wiley Chichester, passed a long my office, promoting new books, the one thing led to the other and I told her of my long-postponed book project. Within a couple of weeks, I got an invitation from Eric Willner, then Executive Editor at Wiley, inviting me and encouraging me to submit a book proposal. And here we are now. I am thankful to them both; especially Sofie for the push.

The book draws on all this and previous research knowledge and experience, though obviously does not report on all of it. We have decided, however, to put most of the PhD dissertations of the contributors on the Webpage of the book for the benefit of the reader.

Although an initial Table of Contents had been made for it several years ago, the book project had to be modified several times before settling to its present final form. I needed even to perform quick bits of research in order to fill in some gaps that popped up during the compilation. I trust that it now covers most aspects and facets of the topic adequately, though much more information has since become available, which unfortunately had to be left out.

Many people have contributed directly and indirectly to the completion of this work. Besides the main contributing authors: Gorka, Tobias, Peter, Steven and Marius, I am particularly grateful for the help of several others: To Michael Mayers, of Bosch, Stuttgart, for the revision with valuable comments and discussions on several chapters; to Wim Seymens, of ASML, for revising Chapter 7; to Ron van Ostayen, of Delft university, for revising Chapter 15 as well as approving some material contribution of his research group for it; to my daughter Hanna and my nephew (and namesake) Fareed, who helped me a lot with technical textual issues at the beginning of the journey; to Wim Van de Vijver of LAB Motion Systems and to Byron Knapp of PI, Minnesota, for their interest in the book though perhaps not in the theory *in se*; to Wim Witvrouw who helped and supported me in the air-bearing company affairs. I should also not forget the many family members and friends who encouraged me and egged me on to continue in difficult and dark moments. Thanks are also due to Sonali Melwani, project editor for Wiley, who managed several times to pick up the book project out of a state of confusion and put it back on track again. Last but not least, I am indebted to my Alma Mater, KU Leuven under whose roof this whole story took place.

This book may be regarded as work-in-progress, which serves both as an overview of the state of the art and a capita-selecta of less known topics and aspects. I have endeavoured as far as possible to bring to the fore and consolidate fundamental results, which I trust to be of main interest to the reader: to answer the type of questions I wrestled with at the beginning of my journey and at many other stages later on. The book can also be treated as a textbook that can be read sequentially, or as a reference work to be read selectively whenever the need might arise. The bibliography is restricted to works selected dating to before 2010. I am aware that since then many valuable publications have appeared.

Remains for me only to wish the reader enjoyment, if not benefit reading this book; that it might besides helping them tackle an air-bearing problem, be a source of inspiration to travel some stations further on the journey of air bearings.

Leuven, July 2020 *Farid Al-Bender*

Nomenclature

The most important notations are summarised below. Symbols that occur only locally in one or a few sections are not listed here. Some symbols used in Chapter 11 are not included in this list but explained in the chapter itself. All symbols used, including the ones below, are explained in the text. Boldface letters are used to denote vectors or tensors. Most capital letter denote dimensionless quantities. Many symbols have various meanings.

General symbols

a: parameter/arbitrary constant
a (a_o): sonic speed (at entrance)
A: bearing region, or bearing area
B: bearing width (/length)
B; b: damping (Foil bearings chapter 12)
B_ν: pressure distribution function
c: damping (torque) coefficient
c: journal-bearing radial clearance
C: foil journal-bearing radial clearance (chapter 12)
C: dimensionless damping torque coefficient
C: (arbitrary) constant
C: damping matrix
C_d: coefficient of discharge
c_{ij}; c_{xy}: film damping coefficient
c_p: specific heat at constant pressure
c_v: specific heat at constant volume
D: journal bearing diameter
D: bending stiffness (of foil) (chapter 12)
Ec: Eckert number
f: frequency
f_r: friction factor in turbulent flow
F: bearing axial force (load)
g: velocity profile function
G: normalised velocity profile function
G: modified gap function of foil bearing $= \mathrm{H} + \alpha(\mathrm{P} - 1)$ (chapter 12)
h: film thickness, gap height
h_a: exit (ambient) film thickness
h_o: entrance film thickness
h_v: gap conicity ($h_o - h_a$)

H: normalized film thickness
H_v: dimensionless conicity
I: velocity profile integral function
I: moment of inertia
j: imaginary unit number, $j^2 = -1$
k: thermal conductivity
k: stiffness
k_p: permeability (of porous media) (chapter 13)
K: dimensionless stiffness
$\boldsymbol{K}$: stiffness matrix
Kn: Knudsen number
K_a: dimensionless dynamic axial stiffness
k_{ij}; k_{xy}: film stiffness coefficient
L: flat bearing length
L: journal bearing length
M: dimensionless tilt moment (chapter 6)
M: dimensionless mass (chapter 9)
$\mathbf{M}$: mass matrix
M_c: dimensionless critical mass
M (idem M_0): Mach number (at entrance)
m; n: characteristic values of entrance problem (functions of I) (chapter 5)
m; n: constant and exponent of Reynolds number (in turbulent flow) (chapter 4)
$\dot{m}$: mass flow rate
$\overline{\dot{m}}$: normalized (dimensionless) mass flow rate
n: see m; n
p: static pressure
P: normalized (dimensionless) pressure w.r.t. p_a
P_f: friction power
Pr: Prandtl number
q; Q: velocity amplitude (mid-plane velocity) (chapter 5)
q: volumetric flow rate (chapter 8)
r: radius (coordinate)
r: journal bearing radius
R_f: dimensionless feed-hole radius (chapter 10)
r_f: feed-hole radius (journal bearings)
r_0: feed-hole radius
(r; θ; z): cylindrical coordinates
(r; θ; ϕ): spherical coordinates
R: normalized radius
Re: Reynolds number based on channel width
Re*: reduced Reynolds number
Re_{sq}: Squeeze Reynolds number
Re_ν: Squeeze Reynolds number
s: distance curvilinear coordinate (chapter 12)
s: Laplace variable (chapter 7)
t: time
T: Temperature (chapters 2, 4, 14, 17)

T: Tilting torque (chapter 6)
T: Tension (in foil) (chapter 12)
Ta: Taylor number
$\bar{u}$: main flow average velocity
U: surface velocity
u; v; w: velocity components
U; V; W: normalized velocity components of fluid
V: surface velocity
$\mathbf{v}$: flow velocity vector (u; v;w)
$\mathbf{V}_t$: tangential velocity vector
W; $\bar{W}$: (dimensionless) bearing load
x; y; z: space Cartesian coordinates
X; Y; Z: normalized space coordinates
α: tilt angle (chapter 6)
α: conicity angle (chapter 5)
α: compliance parameter of bump-foil bearing (chapter 12)
α: slip parameter of porous bearing (chapter 13)
β: dimensionless porous feeding parameter
γ: whirl frequency ratio (chapter 12)
γ: dimensionless porosity chapter 13)
Γ: boundary
δ: increment (of perturbation)
Δ: increment (of perturbation)
ε: normalized film oscillation amplitude
ε: normalized tilt angle (chapter 6)
ε: perturbation parameter
ε: journal-bearing eccentricity ratio (chapters 9, 10, 12)
ζ: normalised length coordinate of foil JB (chapter 12)
ζ: overall damping ratio (chapter 10)
ζ_n: damping ratio
θ: angular coordinate
Θ: temperature function
κ: ratio of specific heats of a gas
κ: cross-coupling ratio of JB stiffness coefficients (chapter 10)
λ: separation parameter (eigenvalue) (chapter 3)
λ: mean free path (chapter 4)
λ: eigenvalue (chapter 10)
Λ: bearing number (also called: sliding, Harisson or compressibility number)
$\boldsymbol{\Lambda}$: bearing number (vector form)
Λ_e: entrance number
Λ_f: feed number
Λ_ω: rotation sliding number
μ: dynamic viscosity
μ': bulk viscosity
μ_f: coefficient of friction
ν: kinematic viscosity (chapters 3, 4, 7)
ν: perturbation or oscillation frequency

ν_a: normalized conicity angle (w.r.t. film exit, see Eq. 2.65)
ν_o: normalized conicity angle (w.r.t. film entrance, see Eq. 2.55)
ξ: von Mises-transformation variable, $= x$ or r (chapter 3)
ξ: Galileo-transformation variable, $= x - Ut$ (chapter 4)
ρ: density
ρ: radius of curvature (chapter 12)
σ: squeeze number
τ: normalized time
$\boldsymbol{\tau}$: shear-stress tensor
τ_f: shear stress (of viscous friction)
ϕ: journal-bearing attitude angle (chapters 9, 10, 12)
ϕ: porosity (chapter 13)
Φ: dissipation function (chapter 2)
Φ; Ψ: dimensionless slip-flow parameters (porous bearing)
$\boldsymbol{\Phi}$: flux density
Φ_e: nozzle function
χ: polytropic exponent
ψ: stream function
Ψ: normalized stream function
ω: angular frequency of oscillation (of squeeze film, chapter 15)
ω: angular velocity of journal bearing
ω_d: damped natural frequency
ω_n: natural frequency
Ω_n: whirl ratio
Ω: angular velocity of sliding
$\wp$: normalized density
$\mathfrak{R}$: gas constant
∇: gradient/divergence operator

Subscripts

a: ambient value
a: axial value
c: damping-related
f: friction-related
k: stiffness-related
o: gap entrance value
(0): steady-state value
s: stagnation (or supply) value
r: reference value
thres: threshold value
w: wall value

Abbreviations

b.c.: boundary condition(s)
BL: boundary layer
CCF: circular, centrally fed

EAL: elasto-aerodynamic lubrication
EF: efficiency factor (ch 14)
EHL: elasto-hydrodynamic lubrication
EP: Externally pressurised
FD: finite difference
JB: journal bearing
HGJB: herringbone-groove(d) journal bearing
DN: mean diameter(mm)×rpm; (also called Nd_m)
LHS: left-hand side
o.d.e.: ordinary differential equation
p.d.e.: partial differential equation
NGT: narrow-groove theory
RHS: right-hand side
RE: Reynolds equation
SA: self-acting
TPB: tilting-pad bearing
w.r.t: with respect to

About the Companion Website

This book is accompanied by a companion website:

www.wiley.com/go/AlBender/AirBearings 

The website includes:

- PhD dissertations

Scan this QR code to visit the companion website.

1 Introduction

1.1 Gas Lubrication in Perspective

Gas lubrication is a special branch of fluid film lubrication distinguished by the fact that the lubricating fluid is a gas that is a *compressible* fluid. In fluid film lubrication, the friction between two solid surfaces is reduced through the introduction of a fluid under pressure that flows between them. The pressure flow of the fluid may be induced by shearing the wedge-formed film, caused by the relative motion of the solid surfaces, in which case the lubrication is known as *hydro- (aero-) dynamic*, and the bearing as self-acting. When the flow, on the other hand, is induced by external pressurisation, the lubrication is known as *hydro- (aero-) static*, and the bearing as externally pressurised.

The thickness of the gas film that separates the bearing surfaces may range from sub-micrometre to several tens of micrometres depending on the size of the system and the application at hand. The bearing surfaces might be rigid or flexible, permeable or impermeable, flat or shaped, smooth or grooved. In short, a large variety of sizes, configurations and applications are available, which makes the topic both wide and rich in possibilities.

It would be difficult to imagine today's technology, from the aluminum foil in the kitchen through surgical tools in the health service to the computer disc drives, without gas bearings. Yet gas lubrication remains in relative obscurity when compared with other branches of lubrication: in the mind even of the engineer, the word "bearing" is synonymous with "rolling-element bearing". Although gas bearings are nearly as old as liquid film bearings (which are older than rolling-element bearings), they are still looked upon as a *novelty*. In the following sections, we shall try to situate gas lubrication from a historical point of view, in terms of their utility, from the morphological side, and in regard to their past and present application domains.

1.1.1 Short History

The evolution of gas lubrication cannot be isolated from that of fluid film lubrication in general. However, in the earlier stages, interest in gas lubrication was only marginal; fluid film lubrication being considered mainly *liquid* film lubrication, and the possibility of using a gas looked upon with some suspicion. (Yet, it is ironic that one of the earliest lubrication experiments pointed out that air was acting as a lubricant.) The following historical outline will therefore deal with fluid film lubrication in general while focusing mainly on gas lubrication. It is, further, based on the references (Dowson 1979; Fuller 1990; Gross 1962; Pan 1990) in increasing degree of importance.

Three periods may be identified in the evolution of gas lubrication.

The first is the period of *pioneering*. Evidence of *aerodynamic* lubrication was made known by G. A. Hirn in 1854. He conducted experiments on a "friction balance" (which is a half journal bearing with a means of measuring the

Air Bearings: Theory, Design and Applications, First Edition. Farid Al-Bender.

Companion website: www.wiley.com/go/AlBender/AirBearings

torque) to show that, with a sufficiently high speed of rotation, "non-viscous fluids such as water and air" may be dragged into the bearing; when the speed was reduced to a certain value, the friction force became at once enormous. This is very remarkable since it was one of the earliest pieces of evidence of *fluid* film lubrication.

It was N. Petrov, in 1883, who made the first significant attempt to analyse *theoretically* the friction effect of film lubrication, recognising that it was the film, rather than the bearing materials, that was of prime consideration. However, his unrealistic assumption of *uniform* film thickness, made his results rather limited in importance; falling short of establishing a lubrication theory. He was able, however, to derive the viscous-friction law that bears his name today.

In that same year 1883, B. Tower reported very important experimental results, showing the pressure distribution in a partial journal bearing, and thus pointing out that *pure* fluid film lubrication, that is a fluid film that is able to separate the two relatively sliding surfaces entirely, was possible.

Three years later, in 1886, Osborne Reynolds, unaware of Petrov's theory, was able to explain the results of Tower. He successfully derived the basic differential equation that describes the mechanism of fluid film lubrication that today bears his name. He published at the same time certain solutions that agree well with the results of Tower. Reynolds, however, made no reference to the experiments of Hirn, nor indeed considered the possibility of gas lubrication.

This type of lubrication was not discussed again until A. Kingsbury, unaware of Reynolds' theory, built an air lubricated journal bearing in 1886. (This was based on his earlier rediscovery of air lubrication in a compression piston device of his making.) Having been made aware of Reynolds' theory, he was able to obtain excellent pressure distribution data and publish them in 1887. This was perhaps the first time that air lubrication was demonstrated without any doubts.

In 1913, W. J. Harrison returned to the subject of gas lubrication with a renewed zeal. For the first time, the *pressure dependence of the density* was formally addressed. On the basis of simplified consideration, he introduced the concept of *isothermal state* whereby the density is proportional to the pressure; an assumption that to this day is widely accepted. He also pioneered the idea of numerical computation, laboriously producing solution to the air bearing problem that compared very well with the experimental data of Kinsbury. He characterised the effect of compressibility in the parameter Λ, known in his honour as the Harrison number.

At this juncture, a word must be mentioned about *hydrostatic* lubrication. As the virtues of fluid film lubrication generated by motion of the bearing surfaces became recognised, attention was naturally diverted to the limitations imposed by excessive loading or low surface speeds. Lord Rayleigh was the first to present an analysis of a hydrostatic thrust bearing, in 1917. He derived expressions for the logarithmic radial decay of pressure from a central supply in the space between parallel circular discs, the resisting torque and the limiting load capacity of the bearing. He pointed out the practical importance of situations that could be accommodated by such bearings, emphasising that with the proper geometrical accuracy of the bearing surfaces and the cleanliness of the lubricant, there should be *no wear* of the solid surfaces, which should never come into contact. (The principle and application of hydrostatic lubrication had already been demonstrated, and patented, by L. D. Girard in 1865.)

Until this time, industrial applications of gas lubrication was hardly in evidence, with the notable exception of high speed spindles for the textile machinery for which patents were awarded in 1906 and 1909. Also, the advantage of high speed grinding, with the aid of air bearings, was recognised in 1909. Otherwise, research in fluid film lubrication concentrated mostly on *viscous* liquid bearings until the end of World War II.

The second period, termed by Pan (1990) as *the Golden Era*, began in the early fifties. Already, in the mid-forties, gas lubrication had played a key role in the preparation of high grade nuclear fuel, focusing attention once more on the possibilities of that technology. Hence, serious concerted effort, including of course material investment, was initiated to develop the potential of gas lubrication and to promote it as a technology base to advance the art of mechanical engineering. Sufficient motivations and numerous possible applications (that will be outlined in the next sections) existed to justify this.

Research groups were formed and programmes were initiated both in the USA and in Europe. Regular meetings, conferences and symposia were held and the information, in the form of minutes, reports or proceedings were disseminated to the involved researchers. The driving force behind this activity came from the areas of: closed-loop process equipment, inertial sensors, tape transports, disc drives and aircraft auxiliary machines.

The first International Symposium on Gas Lubricated Bearings was sponsored by the US Office of Naval Research in 1958, and convened in Washington D. C. The proceedings presented many significant advances in gas lubrication. Later, the University of Southampton (UK) convened regular symposia that attracted many participants from East and West Europe. Initially, attention was focused mainly on commercially viable applications. Ties with the US groups were established and joint conferences took place. This three-way dialogue continued many years until the latter part of the seventies, and was the driving force behind hundreds of scientific papers and publications. This period witnessed also the emergence of the first *comprehensive* text books on gas lubrication by: Gross (1962), Constantinesu (1969) (1962, in Rumanian, and 1969, English translation); Grassam and Powell (1964); MTI's lecture notes: Design of Gas Bearings (Wilcock 1969).

Hydrostatic (and aerostatic) lubrication was for the first time identified as a distinct mode of lubrication, and systematised by D. D. Fuller in 1947; hitherto, there had been only sporadic examples of externally pressurised bearing applications. The principle of external pressurisation was exploited in the field of gas bearings to an even greater extent than self-acting behaviour. They overcome low-speed operating problems and permit satisfactory stiffness while providing low friction, and hence small heat generation and minimal thermal distortion. Hybrid combinations, of self-action with external pressurisation began to come more to the fore, since self-acting gas bearings are otherwise limited in load capacity.

Finally, towards the latter part of the seventies, support for gas lubrication research began to wear thin. Although the accomplishments had been, thus far, quite impressive, there were no tangible signs of return on investment. For this to happen, considerable advances in wear control had to take place first, i.e. the gas lubrication research programme had "outraced itself" (Pan 1990). Thus, with the exception of air lubrication in magnetic storage systems, activities in gas lubrication research slowed down to a snail's pace.

This brings us to the third period of gas lubrication: *the present*. Gas lubrication is now recognised as a fully developed applied science; gas lubrication has become one of the proven principles available to innovators in their perpetual effort to contrive means for higher quality of living. Although gas lubrication research began to slow down considerably at the beginning of the eighties, the ever rising demands of modern technology have begun to breath new life into the subject. Requirements of high precision and high speed in machine tools and metrology, non-contact gas seals, computer storage heads, special test facilities, and many other yet unidentified applications, will continue to press for more advances and gas bearings will have a significant role to play for some time. Perhaps the main feature of the present era is the consolidation of the hitherto achieved results and the further promotion of gas lubrication in technology. "It is an immediate challenge that the practice of gas lubrication be incorporated into the regular mechanical engineering curriculum in our higher education institutions and professional level programs." (Pan 1990).

1.2 Capabilities and Limitations of Gas Lubrication

The capabilities of gas lubrication may be directly related to the physical properties of gases. A gas is a very *stable* lubricant; it does not vaporise, solidify, cavitate or decompose at extreme temperature ranges. Gas bearings can thus be used in turbo-compressors operating at 650 °C, or at the other extreme, for cryogenic units at temperatures nearing absolute zero. Furthermore, gas viscosity is little affected by temperature, as compared to a liquid, retaining satisfactory values over wide temperature ranges. Gases are comparatively free from the adverse effects of radioactivity and are tolerant to hazardous environments. Their low viscosity (which is both an advantage and a disadvantage) results in very low friction coefficients and hence low frictional losses and their associated temperature rise. Furthermore, as in other fluid-film lubrication cases, the absence of surface-to-surface contact leads to negligible wear and implies quasi-infinite bearing life. Gas bearings are particularly suited to precision instruments owing to the form-averaging effect of the fluid film, their low noise during motion and their zero friction in the absence of motion. Air, in particular, is abundantly available, comparatively clean, ecologically sound lubricant; self-acting air bearings such as computer flying-heads are automatically *submerged in air*.

The limitations of gas bearings are mainly due to the compressibility of gases. Thus, by industrial safety as well as by dynamic stability limits, externally pressurised gas bearings are limited in supply pressure to around 10 bar (in industrial environments, often no higher than 6 bar-gauge is available), as compared to hydrostatic bearings, which can be fed with up to a couple of hundred bar. Furthermore, the low viscosity of gases means a higher rate of gas consumption, in comparison to liquid bearings, for the same bearing gap height. This entails the use of very small bearing gaps which, in turn, generally demands a closer control over manufacturing tolerances, surface finish, thermal and elastic distortions, and alignment. As for aerodynamic bearings, unlike hydrodynamic bearings, whose load capacity (or mean pressure) increases proportionally with sliding speed, compressibility in self-acting gas bearings results in saturation of the build up of mean pressure as the sliding speed is increased. Thus, a mean pressure build-up of around twice atmospheric pressure is about the maximum one could reach in realistic situations. Secondly, the combination of compressibility and low viscosity results in gas bearings being *poorly damped*. Pneumatic hammer instability is a well known phenomenon in externally pressurised air bearings. One of its main consequences is to limit the maximum allowable supply pressure, and consequently the load capacity and stiffness. In self-acting bearings, on the other hand, one often encounters self-exited whirl instability.

1.3 When is the Use of Air Bearings Pertinent

In view of the above-mentioned features of gas lubrication, two generic application areas can be identified, namely, high-speed and high-precision applications. The first follows from the low viscosity of gases; the second from the film-averaging effect. A combination of the two is also an emerging industrial need answering demands for mastering, inspection, etc., of CDs, wafers and so on. Besides these, a wide range of particular applications are also known, e.g. in the nuclear industry, magnetic HDDs, etc. Presently, gas-lubricated bearings are found in hundreds of essential applications. These may be classified as follows:

- Machine construction:
  - ultra-high-speed grinding and drilling spindles
  - slideways of (NC) machine tools
  - index tables
  - gyroscopes
  - (miniature) gas turbines and turbo-compressors, etc.
- Precision and metrology:
  - reference spindles and rotary tables for roundness measurement
  - 3D coordinate measuring machines
  - telescope and antenna supports
  - laser plotters for printed circuit boards
  - micro-manipulators, etc.
- Data retrieval systems:
  - transport of magnetic tapes
  - magnetic read/write heads for computers
  - disc supports, etc.

Many other applications fall into these categories such as: zero-gravity test facilities, diamond turning machines for large optics, low-friction tension controllers, hanging robot arms, etc. Generally speaking, air bearings *can* also replace any bearing application that falls within their load capabilities.

1.3.1 What Hinders the Extensive Application of Air Bearings

It has been apparent that the most important drawbacks of gas bearings are their comparatively low load-carrying capacities, high gas consumption (in the case of externally pressurised bearings) and their poor damping. The understanding and partial overcoming of these problems belongs to the realm of design theory. In this respect, there are still a number of unsatisfactorily solved problems, in particular those related to dynamic behaviour, stability and added damping, not to mention other minor problems. Gas bearing research, although great in quantity and quality, remains to a large degree scattered.

On the other hand, we have the problem of manufacturing. If air bearing application is to be widely promoted, low cost air bearing components, of well defined characteristics and, preferably, in somewhat standardised form, have to be available on the market, so that they can be easily purchased and assembled for the particular application. Up to the present, only flat (and some restricted types of radial) aerostatic bearings are available in this form commercially. In most cases, an air bearing application has to begin at the design/drawing stage through the details of manufacture and assembly, thus making the cost too high to be commercially viable. Evidently, the problems of design and of manufacture are strongly interconnected: a clear, thorough and easy-to-use design theory and environment leads to relatively simplified manufacturing requirement, while the facility of manufacture can allow, in its turn, for better design.

The third aspect is the awareness of engineers of the *existence*, practicability and reliability of gas bearings. Air bearings are still treated as a curiosity rather than an established technology. Furthermore the tools to design them do not lie directly at the finger tips of the design engineer. There could be too many design parameters and performance characteristics to be managed, let alone optimised, in a design process.

The author believes that it is high time that these three aspects were seriously taken into consideration in order that a proven engineering art occupy its rightful place in modern technology. This book is a modest step in that direction.

1.4 Situation of the Present Work

1.4.1 Short Overview of Gas-bearing Research at KULeuven/PMA

Gas bearing research began at the KU Leuven, department of Mechanical Engineering, since the early 1970s. It concentrated initially on the development of design methods to improve aerostatic bearing characteristics, namely, to increase the load-carrying capacity and the stiffness while reducing the gas consumption.

One of the basic design ideas explored was the use of a *convergent* gap geometry. This was found to yield bearing characteristics that are far superior to those of the conventional uniform gap bearing. Since the pressure distribution in the bearing depended now not only on the inlet pressure but also on the relative *conicity* of the film, the load capacity was found to increase dramatically with decreasing nominal film thickness. This meant an increase in the stiffness with increasing load, on the one hand, and an increase in the load capacity for a given mass flow rate (or air consumption) for a given mean air gap on the other. Furthermore, the convergent shape of the bearing gap resulted in reduced sensitivity to surface irregularities. At the same time, methods of manufacturing the required bearing profile were suggested, such that the cost of fabrication was kept very low.

This design idea found also an important extension: the compliant gap bearing. By making the bearing surface compliant, e.g. a thin plate, clamped at the inner or the outer edge, the conicity was allowed to vary with the mean bearing pressure. Thus, with the proper choice of design parameters, the bearing could be made *infinitely* stiff over a wide range of loads. A design methodology was developed for this type of bearing, based on the solution of the thin-plate bending problem simultaneously with the bearing pressure distribution problem.

Beside these two major topics, a considerable amount of work was devoted to the investigation of the problem of pneumatic stability of aerostatic bearings and of systems that incorporate them. It so happens that inherently compensated, convergent gap bearings were also less prone to pneumatic instability than their conventional, orifice-compensated pocketed-bearing counterparts.

These and many other related topics resulted in a considerable number of important publications, which are reviewed, in the framework of the development discussed above, in (Snoeys and Al-Bender 1987), which overviews the many outstanding results of two doctoral theses written in the course of this research programme (Blondeel 1975; Plessers 1985). This research was further consolidated and extended, especially in regard to entrance flow, tilt characteristics, and dynamic behaviour, in the thesis (Al-Bender 1992). Furthermore, understanding the dynamic characteristics of the air film led to the development of various types of actively controlled and compensated bearing concepts enabling the realisation of virtually infinite stiff bearings (Al-Bender 2009).

Knowledge of the film flow in pressure-induced turbulent regime led to the design of externally pressurised, over-expansion *hanging* air bearings as a curiosity, which is not, however, without potential applications.

Another type of bearing that has been developed in that same period is that of externally pressurised foil bearings, which also might hold certain advantages regarding manufacturing cost and ease of application.

More recent work has focused on mainly self-acting air bearings for miniature (ultra-) high-speed applications, in particular meso and micro turbines. Both compliant-surface as rigid-surface, self-acting and externally pressurised gas bearings have been investigated in two doctoral theses (Vleugels 2009) and (Waumans 2009) in the framework of ultra-high-speed miniature turbines. This is on the one hand, on the other hand, two other doctoral works have treated ultra-high-precision bearings; one regarding active mechatronics positioning systems (Aguirre 2010), while the other treated the design of nanometre precision rotary tables (Cappa 2014). Recently, another thesis was completed treating tilting-pad journal bearings (Nabuurs 2020), while work is ongoing regadring methods and techniques to of adding damping to high-speed journal-bearing systems.

This book draws on the long and rich experience in gas lubrication at KU Leuven and elsewhere in the world, which forms the basis for compiling it. It is hoped that this book should form an appropriate medium for propagating this knowledge in the research and industry world. In particular, while reviewing salient research findings from the state of the art on the subject, the book endeavours, in the first place, to distil essential knowledge about the fundamental aspects in theory, design and practice of air bearings.

1.5 Classification of Air Bearings for Analysis Purposes

As has been mentioned earlier, there is a broad range of air-bearing configurations, morphologies and working principles, which might make it difficult to deal with in concise and systematic manner. Moreover, there are various ways to classify any collection of items, depending on the specific objectives, the analysis approach or the utility of the results. We shall adopt a classification that makes the build-up and analysis of air bearing systems easier to follow. This classification is also a generic one, being used in other fields of physics and engineering, namely:

- *Dynamics*: relates to the manner in which the bearing force is generated. Here, we have three possibilities:
  - Aerostatic or externally pressurised (EP): in which pressurised air is introduced into the gap from an external means, e.g. a compressor.
  - Aerodynamic or self-acting (SA): in which the pressure in the bearing gap is generated by relative motion of the bearing surface w.r.t. each other. This could be one or a combination of:
    * sliding or tangential motion
    * squeeze or normal motion.

- Hybrid, where both the aerostatic and the aerodynamic effects are present by design. This category is, however, very specific and is mentioned here only for the sake of completeness.

- *Kinematics*: relates to the motion and morphology of the bearing element. Here, we have many possibilities:
  - Motion in a plane: flat bearings
  - Rotation around a line: cylindrical and conical bearings
  - Rotation around a point: spherical bearings.

  These different categories, excepting the squeeze-film type, are depicted in Figure 1.1.

  Parallel to this, it might be also useful to subdivide all the previous types into:
  - Bearings with rigid surfaces, including porous bearings, which constitute the majority of applications.
  - Bearings with compliant surfaces, which comprise:
    * Foil bearings of the bump and tension type, where the compliance is evenly distributed over the bearing surface. The mode of lubrication in that case is elasto-hydrodynamic lubrication (EHL) or, more specifically, elasto-aerodynamic lubrication (EAL), including of course the elasto-aerostatic case.
    * Bearings in which the surfaces are rigid but are flexibly supported to adapt the gap shape to the load dynamics. The most salient example of this type is the tilting-pad bearing (TPB).

    This subdivision is indicated in the last row of Figure 1.1.

  Each of these types of bearing can be further divided into:
  - Single boundary: resulting in bearing pads or shoes whether of the flat, cylindrical, conical or spherical types. These could be combined to achieve the functionality of the next category of bearing morphology.

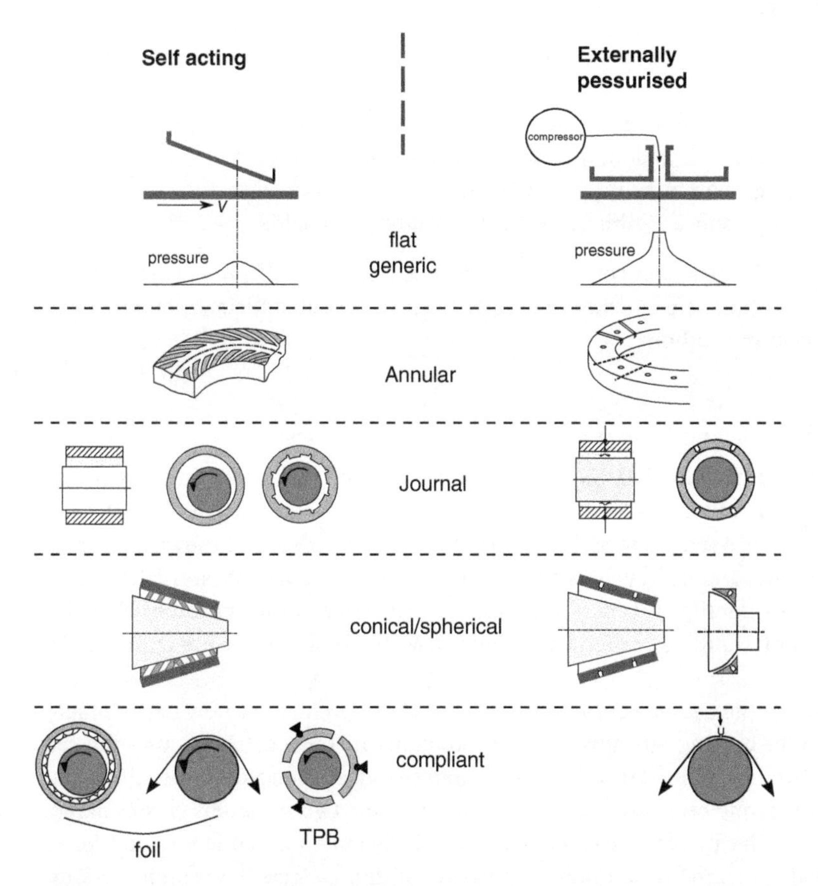

Figure 1.1 Classification of air bearings according to pressure generation (dynamics) and morphology (kinematics). The bottom row depicts compliant-surface bearings.

- Double boundary: resulting in (flat) annular bearings, journal bearings (of both cylindrical and conical type) and spherical ring bearings.

Let us note, however, that not all of these types and bearing configurations are treated in this book. Nevertheless, since specific problems (or situations) exist only as tangible (thus, in a way, unique) manifestations of general ones, the models proposed and the solutions obtained should enjoy some degree of general validity. In particular the results obtained for flat circular bearings should hold also true for *radial* bearings formed by bending the flat surface into a cylindrical one. They should hold at least qualitatively true for *square*- and *sector*-shaped bearing pads. Likewise, knowledge of annular thrust bearings and of journal bearings should put the reader in good stead to deal with conical or spherical bearings.

This book focuses primarily on the underlying theory, since it is that that can form the basis of a good design methodology; no amount of empirical data can replace a proven theory. Yet, theories must be verified by practice and thus particular attention is paid, in this work, to experimental verification. Comparison is carried out with experimental results taken from various sources in the open literature as well as experiments carried out in the laboratories of KU Leuven.

1.6 Structure of the Book

The book has basically five constituent parts:

1. Fundamentals: general modelling, feed flow, Reynolds equation, solution methods, thermal aspects (placed at the end of the book).
2. Flat circular aerostatic bearings: radial flow, basic bearing characteristics, dynamic behaviour and stability.
3. Aerodynamic and hybrid bearings: aerodynamic action, thrust bearings, journal bearings, whirl stability.
4. Other bearing types: tapered-pad bearings, porous bearings, foil bearings, hanging bearings.
5. Active bearings: servo-controlled bearings, squeeze-film bearings, case study: linear slide.

This sequence might be relevant from a scholastic point of view. The advanced reader might prefer to explore the book selectively. There might be some repetition here and there, being dictated by the self-sufficiency of individual chapters or by these coming from contributing authors.

1.6.1 Chapter per Chapter Overview

Chapter 2 formulates the problem, starting from the bearing configurations and the flow-describing equations. Simplifying assumptions are imposed on both of the former in order to facilitate the derivation of practicable working models. Two models are proposed: a refined model for the solution of the feed-and-entrance flow problem in the quasi-steady case, and the conventional Reynolds model with a feed-flow restrictor. Normalisation of the problem is also considered in order to identify the basic dimensionless parameters influencing the behaviour and to systematise subsequent treatment; similitude criteria are deduced. Methods of solution are outlined and discussed.

Chapter 3 proposes methods of solution to the *channel entrance* problem, i.e. the problem of flow development slightly upstream and downstream of the bearing gap entrance. Two situations are treated: (i) pressure-induced flow, such as that obtaining at the entrance of an aerostatic bearing and (ii) shear-induced flow, such as that obtaining at the entrance of an aerodynamic bearing. Also a combination of the two situations is considered. A review of the relevant literature for each case is first carried out from which a method is selected, developed and established as a simple and reliable method of solution for a variety of thin-passage flow configurations. Although nowadays it might be a relatively simple matter to determine the flow field by a purely numerical, e.g.

the CFD method, it is nevertheless rewarding to derive such lumped-parameter methods, as carried out in this chapter. These methods afford us more straightforward means to establish the pressure boundary conditions required in solving the Reynolds equation, which characterises the major part of the bearing film. In that way, not only very fast solution algorithms can be devised to solve the bearing problem, but also useful insights into the design problem are gained.

Chapter 4 is dedicated to the Reynolds equation. First, we give a brief though complete derivation of that equation in its generic form, starting from the viscous laminar flow equations. Thereafter, with the aid of a coordinate transformation scheme, the equation is stated in most coordinate systems pertaining to diverse bearing configurations. Moreover, we consider interpretation of the kinematics for special cases pertaining to situations comprising uneven sliding surfaces having periodic surface features.

Neglected flow effects are categorised as boundary-condition related, mainly slip at the walls, and Reynolds-number related, comprising mainly inertia effects and, more particularly, turbulence. These are reviewed and some appropriate "modified" Reynolds equations have been presented in order to deal with them. Inertia effects are considered at length together with derivation of the momentum and energy integral methods for a flow solution, leading to another type of modified Reynolds equations consisting in a system of simultaneous ordinary differential equations.

Chapter 5 implements the first method developed in Chapter 3, i.e. that pertaining to pressure-fed flow, to provide solutions to the problem of radial channel flow, relevant to circular, centrally fed (CCF) aerostatic bearings. The general trends of the flow are investigated and many comparisons are made of the pressure distributions, obtained by this method, with experimental data from various sources. Finally, owning to the good agreement of the results, the method is used to provide data for the correction coefficient of the restrictor model. This refined restrictor model is used, together with Reynolds model, for the treatment of the remaining problems (Chapters 6 and 7.)

Chapter 6 provides a systematic treatment of the basic characteristics of CCF aerostatic bearings under three categories: (i) static axial characteristics, namely, the load capacity, stiffness and air consumption as a function of the bearing design parameters (the dynamic characteristics are treated in Chapter 7); (ii) the problem of small amplitude tilt. The method of small perturbations is utilised to linearise the problem, which appears sufficient for obtaining fairly accurate results. Both static and dynamic tilt are considered in order to obtain a comprehensive idea about the problem. Experimental verification of the theoretical results is also reported; (iii) the effect of sliding on the otherwise *aerostatic* bearing. The influence of sliding on the load capacity, the stiffness, the flow rate, the centre of force, and the friction force are calculated, presented and discussed. In all these parts, the effect of the conicity, i.e. the shallow pocket around the feed hole, is particularly considered. All these aspects provide a basic understanding of the design problem and furnish a toolbox, as it were, for dealing with the issue of correct bearing design and optimisation

Chapter 7 is devoted to the study of the dynamic behaviour of circular, centrally fed aerostatic bearings and the problem of pneumatic-hammer instability. The various aspects of the problem are examined in conjunction with a literature survey of past treatments. Thereafter, a linearised model, based on the Reynolds/restrictor model is proposed, given certain constraints on the design parameters and the amplitude of oscillation, and its range of validity determined. Such a model is found sufficient to characterise the problem of small amplitude axial oscillations over a wide range of design interest. This methodology and model can be easily applied to other bearing configurations. For a given set of bearing design parameters, the model predicts the dynamic stiffness and damping as a function of the perturbation frequency. A simple equivalent system of springs and dampers (of positive or negative coefficients) is constructed to mimic the air film, which can be used as a design artefact. Stability criteria are deduced, and results are compared with experimental data.

Chapter 8 establishes the fundamentals of the aerodynamic action. Whereas in EP bearings the fluid pressure is provided from an external source, in self-acting bearings, the film wedge combined with viscous shear through the tangential velocity is the mechanism by which fluid is pumped into the gap in order to generate pressure;

this is the principle of the viscous pump, whose mechanical efficiency is shown to be very small. The qualitative difference between incompressible and compressible cases is examined at length, to clarify the limited load capacity of self-acting bearings. Thereafter, a number of generic aerodynamic bearing pads are reviewed to show basic properties, namely, the load capacity, static stiffness and damping, and the friction power dissipated. We consider nominally flat rectangular bearings to demonstrate theory, which could be extended to other situations. A considerable part is devoted to the derivation of the narrow-groove theory (NGT) of herringbone groove bearings and to the discussion of its basic properties. Damping in self-acting bearings can also become negative. This is so especially in spiral and herringbone groove bearings, and we demonstrate this by examining a bearing case.

Chapter 9 treats journal bearings (JBs) and their basic characteristics. We have considered two categories of bearings, namely, self-acting and externally pressurised bearings. The first category is represented by two types: plain and spiral grooved, which significantly differ in their static and dynamic behaviour. The second one comprises various variants of externally pressurised bearings, which are essentially used for low-speed precision applications. As such, they are characterised by high stiffness and, depending on the design, also high damping. Such bearings are also often used for high-speed applications, where they are sometimes referred to as hybrid journal bearings. This type has certain important advantages over the self-acting types, particularly in regard to high stiffness and dynamic stability. When ultra-high speeds are envisaged, it is generally not possible, from this analysis, to select a particular bearing type: the process entails optimisation to avoid the occurrence of whirl instability. In this chapter, we introduce the phenomenon of half-speed whirl and the associated critical mass above which the system will become dynamically unstable. This problem is more systematically treated in Chapter 10.

Chapter 10 treats in a fundamental way the dynamic whirl behaviour of gas lubricated journal bearing systems, which might be an important limiting factor on the performance of these bearing systems. Particular attention is paid to the study of the self-excited whirl phenomenon, which is often destructive, as well as to its remedies. First, the various kinds of rotor whirl are classified into synchronous whirl, being due to imbalance, and self-excited whirl, which sets in at high values of the rotational speed. Hereafter, the cross-coupled stiffness is identified as the root cause or the driving force of self-excited whirling. A simple but effective stability criterion is formulated that relates the maximum allowable cross-coupled stiffness to the other dynamic film parameters. Based on this stability criterion, some stabilising techniques from literature are assessed. Three methods to overcome (or at least postpone the occurrence of) self-excited whirl are proposed. In a first method, the design rules for achieving optimal stability with plain aerostatic bearings are outlined in a dimensionless way. The second stabilising technique tries to eliminate the driving force of self-excited whirl, namely by reducing the cross-coupled stiffness. This approach, in addition to being limited in effect, is not easy to implement in practice. Finally, a third and last strategy works by compensating for the destabilising effects from without the gas film; that is by tuning the dynamic characteristics of the bearing's damped support. This method is most popularly implemented in the form of o-ring supported bearing bush.

Chapter 11 treats foil bearings, which are obtained by making the bearing surface compliant rather than rigid, mostly by disposing a membrane around a shaft. After a review of the state of the art, foil bearings are classified in two major types, namely, (i) *tension* foils, where the surface is wrapped around a shaft by tensioning it, and (ii) *bump* foils, where the foil is not tensioned but rests on a bed of springs. Both cases are treated, although the largest part of the chapter is dedicated to idealised bump foils. Although these are quite complicated systems, it is possible to simulate certain aspects with a modified Reynolds equation where the gap height depends not only on the eccentricity but also on the local pressure via the dimensionless compliance parameter.

The steady-state characteristics, such as the load capacity and stiffness, are treated first, which, for this type of bearing depend not only on the bearing number and eccentricity ratio but also on the bearing surface compliance.

The dynamic characteristics, treated thereafter, are also highly influenced by the surface compliance, as that can be considered as a spring-damper system in series with the air gap stiffness. Furthermore, a two-dimensional rotor-dynamics whirl-stability analysis is carried out and some methods to improve the stability are discussed.

Chapter 12 treats porous bearings, which offer another possibility of introducing the (gas) lubricant into the bearing film, namely through an extremely large number of feeding holes that are of exceedingly small diameters, i.e. the pores. In this way, one achieves a distributed flow all over the bearing surface, which has certain important advantages, but also some drawbacks. This chapter deals with the static and dynamic fluid film modelling of porous bearings. The flow in the porous medium is assumed to be viscous and is governed by Darcy's law. A modified Reynolds equation, describing the flow in the lubricant film, is derived, taking into account the effect of slip flow, and a numerical solution process is developed and explained. First, the static characteristics of porous bearings are presented and discussed. In a second part, the dynamic bearing characteristics are defined and the perturbation method is applied to plain and cylindrical porous bearing configurations in order to obtain them. An expression for the dynamic film coefficients is given and their typical behaviour is discussed. The porous model is also validated with experimental and theoretical data found in the open literature, where good to excellent agreement is found for the static solution over a wide range of bearing parameters.

Chapter 13 treats the theory and practice of tilting-pad air bearings (TPB), in particular of the journal type. This bearing type is famed for its better dynamic, i.e. whirl stability as compared to other-rotary bearing principles. In order to build up the theory, we first look at the plane, infinitely long inclined-pad bearing case in order to establish its basic load, stiffness and damping characteristics. In a TPB, the inclined pad is supported by a pivot so that it can freely adjust its attitude. This requirement introduces additional dynamical aspects in the working of the pad, which need to be carefully formulated and dealt with in order to develop an effective design theory and practice. The greatest part of this chapter is then devoted to the case of journal TPB, where first the static characteristics are addressed followed by the whirl dynamics considerations where a methodology is developed to optimise the design of TPBs, pushing the whirl-onset speed as far as possible beyond the working range of the system. Finally, some fabrication aspects of these bearings are overviewed and discussed, since they are not as evident as for other bearing types.

Chapter 14 looks first at the different possibilities of achieving a bearing that *hangs* from a ceiling rather than *presses* against a floor, as is usual in most applications. It first situates the problem of hanging bearings, showing the advantages and possibilities that they can offer. Thereafter, it presents and evaluates the three possibilities of achieving a hanging bearing, namely, (i) usual thrust bearing combined with a magnet that attracts it against a magnetic-material ceiling, (ii) a vacuum or suction bearing, and (iii) the Bernoulli/over-expansion bearing where the gas flows in a turbulent, supersonic manner through the largest portion of the gap. This latter type has a certain potential as compared to the other realisation possibilities. Moreover, the theory of this type of bearing furnishes a good case study of turbulent-flow lubricating films. A reliable working model is thus derived, and experimentally validated, for use as design and optimisation tool. This new type of developed hanging bearing is very simple and effective, but requires large flow rates. It is easy to implement, requiring relatively large gap heights and thus not especially prepared ceiling surfaces. Suggested applications are: for hanging robot arms in micro-assembly cells; for handling delicate objects such as magnetic discs, chocolate, etc.; flat-wall climbing devices such as window-cleaning robots; or simply as elements to preload other thrust bearings, when light weight is required. The chapter makes also a pros–cons comparison between the three types of realisations mentioned above.

Chapter 15 firstly situates the problem of active compensation or servo control of gas films within the state of the art of air bearings. We see that there are various ways to influence and control the dynamic force in the bearing film, which can be narrowed down to two basic ways: (i) inlet/outlet pressure or flow control, and (ii) film geometry control; the latter having generally a broader frequency range than the former. Then it formulates the general film-force control framework and discusses it in connection with examples pertaining to thrust and journal bearings. Two other special cases are presented. Firstly, control of the traction force on the bearing surfaces and its application in wafer in-plane positioning is presented together with two solutions belonging to inlet/outlet pressure control and to film-geometry control respectively. Secondly, load generation by means of squeeze-film action presents a way of constructing bearings that do not require a pressure supply. The basic theory pertaining to this type of active bearing is over-viewed. From all of this, we see that active gas bearings represent an interesting extension to conventional, passive bearings, all the more so with the increasing utility of the mechatronics engineering methodology.

Chapter 16 describes the development of a linear slide with sub-micrometre accuracy requirement in six degrees of freedom, being chosen as a plausible industrial application for active air bearings, of which a first prototype has been developed. The first challenge is to optimise the design of the active air bearings, considering the coupled interaction between the air–film dynamics, structural flexibility, piezoelectricity and control, for which a multiphysics simulation model is developed and validated as tool for this task. The next step is the design of the active slide, analysing the interaction of several active bearings and their integration into a more complex system, where other effects, such as manufacturing limitations or calibration can determine the design requirements. The basic functionality of the active slide prototype is demonstrated. The implementation of calibration strategies and advanced control techniques should allow for the achievement of sub-micrometer accuracy over very large strokes.

Chapter 17 deals with the thermal aspects that arise during the implementation of air bearings both (i) in high-speed applications, where frictional heat dissipation can be an issue, and (ii) in precision applications, where thermal uniformity might be an issue. Regarding the first issue, we show by means of simple solutions of the thermodynamic energy equation that the heat generated by viscous friction, which may be quite appreciable, can hardly be evacuated by the air itself so that conduction through the bearing surfaces will be the only effective mechanism to accomplish this. It might suffice in certain situations to cool only one bearing member, e.g. the housing, which might simplify the task greatly. Formulas have been provided to quantify the various aspects of the thermal problem based on dimensionless parameters. Regarding the second issue, attention is focused on the thermal behaviour of the flow of centrally fed aerostatic bearings. A method of solving the energy equation simultaneously with the momentum equation is presented. It is based on approximating the convective terms in the energy equation, and using the mean density and viscosity values across the gap, all of which being considered sufficient for obtaining a quantitative idea about the thermal influences on the flow. The results show that the flow becomes isothermal in the viscous region regardless of whether the walls are thermally conductive or not. If the walls are isothermal, as is usually the case in gas bearing surfaces, the mean gas temperature will approach the wall temperature (assumed equal to the stagnation temperature) in the viscous region. For this case, typical behaviour of the polytropic exponent, in relation to the entrance reduced Reynolds number is derived.

References

Aguirre, G., (2010) *Optimized Design of Active Air Bearings for Ultra-Precision Positioning Slides* PhD thesis K.U. Leuven - Dept. Mechanical Engineering.

Al-Bender, F., (1992). *Contributions to the design theory of circular centrally fed aerostatic bearings* PhD thesis Katholieke Universiteit Leuven - Dept. Mechanical Engineering.

Al-Bender, F., (2009). On the modelling of the dynamic characteristics of aerostatic bearing films: from stability analysis to active compensation. *Precision Engineering* 33(2): 117–126.

Blondeel, E., (1975). *Aerostatische Lagers met Lastafhankelijke Spleetkonfiguratie* Doctoral thesis K.U.Leuven, Departement Werktuigkunde. 75D1.

Cappa, S., (2014) *Reducing the error motion of an aerostatic rotary table to the nanometre level* PhD thesis K.U.Leuven, Department of Mechanical Engineering.

Constantinesu, V.N., (1969) *Gas Lubrication*. New York: ASME.
Dowson, D., (1979). *History of Tribology*. London: Longman.
Fuller, D.D., (1990). Hydrodynamic and hydrostatic fluid-film bearings *Achievements In Tribology* (ed L.B. Sibley and F.E. Kennedy), 15–30. The American Society of Mechanical Engineers.
Grassam, N.S., and Powell, J.W., (1964). *Gas Lubricated Bearings*. London: Butterworths.
Gross, W.A., (1962). *Gas Film Lubrication*. John Wiley & Sons.
Nabuurs, M., (2020). Tilting-pad gas bearings for high-speed applications: Analysis, Design and Validation, PhD thesis, KU Leuven, Dept. Mechanical Engineering.
Pan, C.H.T., (1990). Gas lubrication (1915–1990) In: *Achievements in Tribology* (ed. L.B. Sibley and F.E. Kennedy), 31–55, Toronto, Ontario, Canada.
Plessers, P. (1985) *Dynamische instabiliteit van aërostatische gaslagers in mechanische systemen* PhD thesis K.U. Leuven - Dept. Mechanical Engineering.
Snoeys, R. and Al-Bender, F. (1987) Development of improved externally pressurized gas bearings. *KSME Journal* 1(1): 81–88.
Vleugels, P. (2009). *Development of aerodynamic foil bearings for micro turbomachinery* PhD thesis K.U. Leuven - Dept. Mechanical Engineering.
Waumans, T. (2009). On the Design of High-Speed Miniature Air Bearings: Dynamic Stability, Optimisation and Experimental Validation PhD thesis K.U. Leuven, Department of Mechanical Engineering, (2009D16).
Wilcock, D.F. (ed.) (1969) *Design of Gas Bearings*. Mechanical Technology Inc. (MTI), Latham NY.

2

General Formulation and Modelling

2.1 Introduction

Formulation is concerned with defining the configuration of the system to be solved, stating the constitutive equations and specifying the constraints. Modelling, on the other hand, deals with the development of a workable mathematical system that is amenable to treatment in order to yield desired solutions. Contrary to formulation, this step relies on arbitrary, though motivated and substantiated, judgement in order to obtain viable approximations without violating the constitutive equations.

In this chapter, the basic aspects of the gas-bearing problem will be stated, developed and discussed, to the end of constructing a suitable working model. Such a model may be likened to a system that has as input the bearing design parameters, or the bearing configuration: geometry, motion, lubricating fluid, boundary pressures; as output, it gives the bearing characteristics: load carrying capacity, stiffness, damping, mass flow rate, etc. The system consists in the flow regime and its associated pressure field in the bearing film. Adequate characterisation of this system forms therefore the essence of a reliable model.

A model that is both general and practicable is, however, difficult to achieve (Al-Bender 1992). It is, therefore, necessary to affect some simplifications, both in the model and in the input, the two being evidently interconnected. In the first instance, the flow field in the film may be simplified through its division into sub-regions of varying character. In the second, certain constraints may be imposed on the bearing configuration while keeping it, at the same time, within the range of practical interest. In this way, model and input are brought closer together.

Formulation of a reliable model is undoubtedly the more difficult problem of the two. This will involve a general analysis of the flow problem, in order to identify its various aspects, leading to the formulation of sub-models for the various flow regions, or the different bearing configurations. The existing bearing models are, in their greatest part, based upon Reynolds lubrication theory, which adequately characterises the flow in the viscous region of the bearing. It provides no information, however, on the entrance flow, nor is its degree of accuracy known in special cases such as transverse oscillations of the bearing film. A derivation of the flow equations has, therefore, to be given first in order that the background of the working model be known; that the possible errors or uncertainties be traced back to the origin or quantified.

In the next section, the bearing configuration will be described. For this purpose, the flow between two arbitrary surfaces with a feed hole, constituting a bearing, will be considered. For the purpose of concretisation, this general configuration is then restricted to (i) that of an axisymmetric flat bearing with a centrally located simple feed hole and (ii) that of an inclined self-acting slider bearing. While facilitating understanding of the modelling simplifications, this restriction shows how the basic building blocks of a more general numerical model could

Air Bearings: Theory, Design and Applications, First Edition. Farid Al-Bender.

Companion website: www.wiley.com/go/AlBender/AirBearings

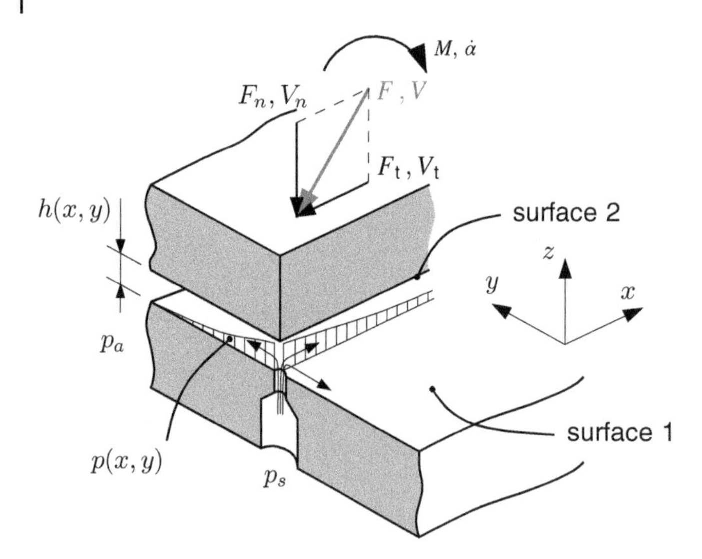

Figure 2.1 General bearing configuration and notation (p_s and p_a denote respectively the supply and atmospheric pressure). Source: Adapted from Al-Bender F 1992.

be formulated. Finally, we can use this as a generic case, to gain insight to the basic properties, in particular the scaling properties, of the solution. Thereafter, the basic equations of the flow are stated and discussed, followed by various proposed simplifications, leading to the formulation of two working models: one for the axisymmetric feed/entrance problem, the other is the conventional Reynolds model with a feed compensator. The basic bearing characteristics are then defined, in relation to the flow problem parameters. A separate section is dedicated to normalisation of the various problems, whereby similarity criteria are deduced that are necessary for a systematic treatment of the problem. Finally, a brief overview of the methods of solutions is given, followed by a summary of the results.

Configuration and Notation

In spite of the wide variety of bearing configurations, let us try to describe the general bearing configuration and its notation. Figure 2.1 depicts two bearing surfaces separated by a thin fluid film. The shape, surface topography and actual position of both bearing surfaces define the film geometry as denoted by $h(x, y)$.

Without loss of generality, assume that one bearing surface is stationary, and that the other bearing member moves according to the vector $\boldsymbol{V}$:

$$\boldsymbol{V} = (\boldsymbol{V}_{\mathrm{t}}, V_{\mathrm{n}}, \dot{\boldsymbol{\alpha}}), \tag{2.1}$$

which holds the following components:

- $\boldsymbol{V}_{\mathrm{t}}$ represents the tangential velocity vector
- V_{n} is the velocity normal to the bearing surface
- $\dot{\boldsymbol{\alpha}}$ is the tilt angular velocity of the bearing member about an arbitrarily given axis.

The latter two components are generally assumed dynamic (i.e. representing an oscillation), while the first component is, in most cases of interest, a steady working condition imposed to the bearing.

The external bearing load can be represented by a single force vector $\boldsymbol{F}$:

$$\boldsymbol{F} = (\boldsymbol{F}_{\mathrm{t}}, F_{\mathrm{n}}, \boldsymbol{M}), \tag{2.2}$$

which can be decomposed into the following forces and moments:

- $\boldsymbol{F}_t$ represents the tangential force vector
- F_n is the normal bearing load
- $\boldsymbol{M}$ is the tilt moment about an arbitrary axis.

Again, the latter two components are in general dynamic, while the first one stands for a stationary friction force due to viscous and hydrodynamic film action.

A pressure distribution $p(x, y, t)$ will be generated as a result of the imposed working conditions. This pressure profile depends of course on the prevalent pressure or flow boundary conditions. Special attention has to be paid to an adequate modelling of the entrance flow effects occurring in the vicinity of feeding sources or at the inlet of self-acting bearings.

The bearing surfaces can have any shape or surface feature such that the combination of film geometry and working conditions results in the generation of a desired pressure profile. The integrated action of the pressure distribution in the film determines the bearing's load carrying capacity and reaction moment that is exerted by the fluid film.

2.1.1 Qualitative Description of the Flow

We shall, for the moment, employ the EP bearing as a generic example. Similar behaviour applies also to aerodynamic bearings, as we shall see later. Figure 2.2 shows a cross section of the flow in which the arrows indicate qualitatively the magnitude and direction of the main flow velocity. In the plenum chamber, the fluid is assumed to be at rest or stagnation, due to the large comparative size of the plenum. There might be also a restrictor, e.g. an orifice, porous insert, etc., placed between this plenum and a second one, as is done in hydrostatic bearings, with the purpose of enhancing the stiffness of the bearing. This type of restriction will be treated further in Chapter 5. We presently will consider the flow from this second plenum, hereafter termed the "feed-hole" till exit of the bearing gap. The flow gradually accelerates through the feed-hole, reaching maximum speed somewhere in or around the annular curtain area of the gap entrance. This will be termed the *feed* region, which is marked by the virtual absence of viscous friction (except perhaps in the immediate vicinity of the gap entrance.) Thereafter, the flow begins to decelerate, at the walls, under the action of viscous friction, and as a whole, owing to the diffuser effect (of increasing flow cross-sectional area.) Initially, the fluid inertia force may be as important as the viscous force, as evidenced by the peculiar pressure dip just downstream of the feed-hole, as we shall see further. In extreme cases, turbulence and supersonic flow are likely to occur. This region will be called the *entrance* region. Further downstream, the flow will have sufficiently slowed down for the viscous forces to become dominant from then on until exit is reached. In a realistically designed bearing, this should constitute the greatest part of the bearing area, which will be termed the *viscous* or *lubrication* region. At exit, the fluid is assumed to be at a pressure equal to that of the ambient medium.

2.2 Basic Equations of the Flow

It is not our purpose to derive the equations that govern the flow. Such a derivation may be found in many fluid mechanics text books such as Lamb (1932 (Re-issue(1974.); Milne-Thomson (1968 (Reprint(1974.); Schlichting (1968). Nor shall we write these equations out in detail since only their simplified forms will be used. They will be merely stated in order to complete the subject, and to facilitate reference. These equations will, further, be given in Cartesian coordinates, enabling their statement in a compact form; their transformation into other coordinate systems may be found in the above mentioned references.

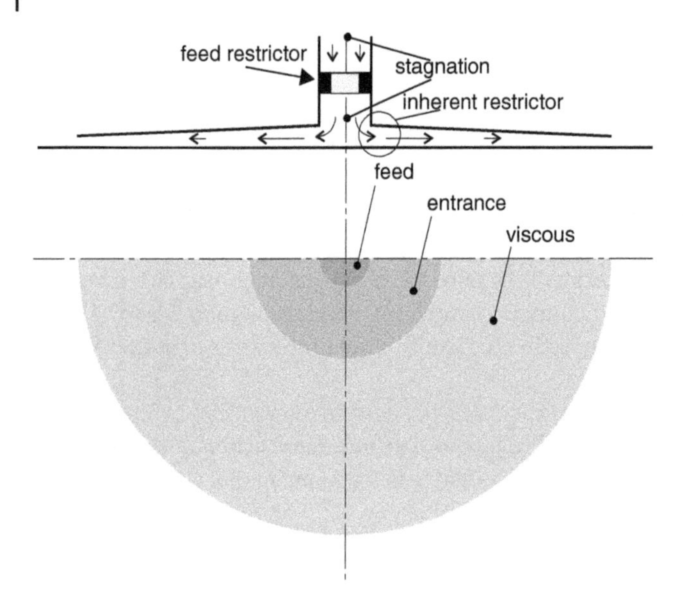

Figure 2.2 Schematic flow configuration of inlet flow to EP bearing. (Not to scale.) Source: Al-Bender F 1992.

2.2.1 Continuity Equation

Denoting the fluid velocity field by $\mathbf{v} = (u, v, w)$, and the density by ρ, the equation of continuity (representing mass conservation) is written:

$$\frac{\partial \rho}{\partial t} + \nabla \cdot (\rho \mathbf{v}) = 0. \tag{2.3}$$

2.2.2 Navier–Stokes Momentum Equation

The equilibrium of forces acting on an infinitesimal element of fluid demands that

$$\rho \left[\frac{\partial \mathbf{v}}{\partial t} + \mathbf{v} \cdot \nabla \mathbf{v} \right] = \boldsymbol{F} - \nabla p + \nabla \cdot \boldsymbol{\tau} \tag{2.4}$$

where p is the static pressure, $\boldsymbol{\tau}$ the shear stress tensor and $\boldsymbol{F}$ the body forces acting on the fluid.

Denoting, further, the dynamic viscosity by μ and the bulk viscosity by μ' (where $3\mu + 2\mu' = 0$ according to Stokes' hypothesis) the shear stress tensor is given by:

$$\boldsymbol{\tau} = \mu(\nabla \mathbf{v} + (\nabla \mathbf{v})^T) + \mu'(\nabla \cdot \mathbf{v})\mathbf{I} \tag{2.5}$$

so that equation of momentum, also referred to as the equation of motion, or the Navier–Stokes equation is written as:

$$\rho[\partial \mathbf{v}/\partial t + (\mathbf{v} \cdot \nabla)\mathbf{v}] = \boldsymbol{F} - \nabla p + \nabla[\mu'(\nabla \cdot \mathbf{v})] + \nabla \cdot [\mu(\nabla \mathbf{v} + (\nabla \mathbf{v})^T)]. \tag{2.6}$$

The bracketed expression on the left-hand side consists of a *local* derivative (first term) and a *convective* derivative (second term), and is often written, in shorthand, as a total or *substantial* derivative:

$$\partial \mathbf{v}/\partial t + (\mathbf{v} \cdot \nabla)\mathbf{v} \equiv \frac{\mathrm{D}\mathbf{v}}{\mathrm{D}t}.$$

2.2.3 The (Thermodynamic) Energy Equation

Denoting the temperature of the fluid by T, its thermal conductivity by k, and its specific heat capacity at constant pressure by c_p, the energy equation is written,

$$\rho \frac{\mathrm{D}}{\mathrm{D}t}(c_p T) = \frac{\mathrm{D}}{\mathrm{D}t}(p) + \nabla \cdot (k \nabla T) + \Phi \tag{2.7}$$

where Φ is the dissipation function, given by:

$$\Phi = \mu'(\nabla \cdot \mathbf{v})^2 + \frac{\mu}{2} \parallel \nabla \mathbf{v} + (\nabla \mathbf{v})^T \parallel^2.$$

2.2.4 Equation of State

For a perfect gas, the pressure, density and temperature are related to one another by

$$p = \rho \Re T \tag{2.8}$$

where $\Re$ is the gas constant equal to the universal gas constant divided by the molecular weight. Further, we have

$$\Re = c_p - c_v \tag{2.9}$$

where c_v is the specific heat at constant volume. The ratio of the specific heats is written,

$$\kappa = c_p / c_v$$

which has the value 5/3 for monatomic gases, and 7/5 for diatomic gases.

2.2.5 Auxiliary Conditions

In general, it is assumed that the values of the flow variables match the corresponding values at the boundaries, thus being continuous (but not necessarily differentiable) functions there. They may be divided into two types of conditions, viz. (i) at the solid walls, and (ii) at the inlet and exit of the fluid control volume. The first are usually referred to as *boundary* conditions; the second, depending on the redundancy of some of them, may be either *initial* or boundary conditions. Also, the number of necessary auxiliary conditions will depend on the order of the derivatives of the respective variable. Therefore, we shall need only two conditions for the pressure (usually at inlet and exit), whereas the velocity and the temperature need to be specified on all boundaries. At this point in the analysis, two important conditions on the velocity at the walls may be in order. These are:

No-slip or that the velocity of the fluid particles along the wall is zero relative to the wall. This is taken to be true only if the gas is not sufficiently rarefied, i.e. when the Knudsen number, $\mathrm{Kn} = \lambda / h$ (where λ is the mean free molecular path) is below about 0.01, a condition that is valid in almost all aerostatic lubricating films.

Impermeability or that no fluid flows through the walls. In other words, we shall exclude the case of porous solid boundaries.

From these two conditions, it follows that the velocity of the fluid at the solid boundaries is identical to that of the boundary.

NB: In Chapter 4, we shall consider cases in which there is a certain degree of departure from these two conditions; namely (i) in case of rarefied fluids, where slip-flow can become appreciable, and (ii) in porous bearings where both slip and permeability are important, treated also in Chapter 13.

2.2.6 Comment on the Solution of the Flow Problem

Assuming the physical properties of the fluid to be given functions of the state variables, Eqs. 2.3–2.8 form a system of four simultaneous equations[1] for the four variables υ, p, ρ, T. This system, together with the boundary conditions mentioned above, should constitute a well-posed problem, if it had a *unique* solution that depended *continuously* on the boundary conditions (Ames 1965). However, no general solution is yet known for the Navier–Stokes equation, nor can we specify correct boundary conditions on the variables at inlet and exit.

A general solution of the flow problem as it stands is, thus, next to impossible. Consequently, simplifications have to be introduced in order to obtain a practically useful solution. In the first instance, if the initial values of the variables are arbitrarily imposed, a general solution may be attempted by numerical means. Such a solution would, however, be necessarily complex and ungainly: impracticable for use as a design model. In the second instance, use may be made of the qualitative differences in the character of the flow in the various regions in order to break the problem down into sub-problems that can be approximated in various degrees. Such a scheme is commonly adopted for the solution of complex problems, and will be the one adopted in this work.

In the next section, the general simplification of flow equations together with the division of the problem into different flow regions will be made, and its problematic discussed.

2.3 Simplification of the Flow Equations

Several levels of approximation are possible. They fall into two main categories: (1) idealisations of the fluid properties, the geometry, or the kinematics, e.g. assumptions of isotropy, streamline layout, surface smoothness, etc. and (2) truncation of the describing equations, based on order-of-magnitude evaluation of the various terms, which will thus vary from region to region. These simplifications will be presented and discussed in the following sections.

2.3.1 Fluid Properties and Body Forces

The physical properties of a fluid are, in general, a function of the temperature and, to a lesser degree, the pressure. If the temperature variation taking place in the fluid is small, and confined to a limited region, as indeed will be later apparent, these properties may be taken as constants (evaluated, e.g., at the average temperature of the fluid.) This constitutes a considerable simplification of the flow equations; in particular the (partial) decoupling of the momentum equation from the energy equation.

It is also generally accepted in aerodynamics that the body forces, generated by the action of gravitation, are negligible on account of the absence of change in free surface and of buoyancy; the change in height, although negligibly small, is included in the pressure. Note, however, that such an assumption would become not valid in the case of magnetohydrodynamics involving strong magnetic fields.

2.3.2 Truncation of the Flow Equations

(I) Entrance Flow into an EP Bearing

Two orders of truncation are usually considered. The lower order is essentially based on geometrical consideration of the flow kinematics; the higher on the predominance of viscous or inertial forces, respectively. In order to facilitate the introduction of such simplifications, the flow will be divided into two main regions, see Figures 2.3 and 2.5: feed flow and film flow (or channel flow); the latter being further subdivided into entrance and viscous regions respectively. Note, however, that since the continuity equation represents conservation of mass, it can under no circumstance be approximated or truncated in any way.

---

1 Six equations, if each velocity component is taken as a variable.

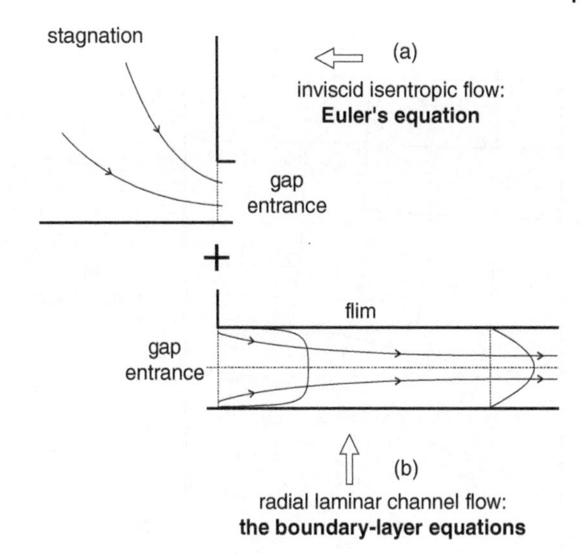

Figure 2.3 (a) Feed flow and its transition to (b) film flow.

Let us remark first that an exact solution of the flow in this region is neither easy nor practically useful; the actual pressure distribution in this region is, in particular, of little importance. Consequently, a lumped parameter approach, viz. that which relates the boundary conditions to one another, should be sufficient for our purpose. As has been mentioned earlier, this is an essentially accelerating flow that may be likened, in the vicinity of the gap entrance, to that taking place in a plane right angle nozzle. This assumption is aided by the fact that the ratio of the gap height to feed radius $h_o/r_o \ll 1$. Figure 2.4(a) shows an incompressible potential flow net for a plane nozzle with $h/L \sim h_o/r_o = 0.1$ that has been obtained by the method of conformal mapping (see Churchill (1948); Vallentine (1967).)

One may notice, first, that the streamlines are nearly uniformly spaced just downstream of the channel entrance, and, second, that the streamlines resemble those corresponding to plane sink flow, i.e. nozzle flow, just upstream of the entrance, as shown in Figure 2.4(b). This latter case is analysed in detail in (Schlichting 1968, p. 153), on the basis of boundary-layer (BL) theory. It is shown that the boundary layers forming at the walls become progressively thinner as the gap entrance (nozzle exit) is approached, in inverse proportion to the square root of the Reynolds number based on the distance from the sink, $\text{Re} = \rho U x/\mu$. By virtue of continuity, this latter is approximately equal to the hydraulic Reynolds number at gap entrance, $\text{Re}_o = \rho_o \bar{u}_o h_o/\mu$, which is typically of the order of 10^2–10^3. Consequently, the BL thickness may, in the first approximation, be taken as negligible in comparison to the potential core of the flow, or that the flow in this region may be regarded as purely *inertial* or *inviscid*.

Thus, if we denote the stream-wise velocity by u, and the distance along a representative streamline by ζ, the momentum equation may then be written (Milne-Thomson 1968):

$$\rho\left(\frac{\partial u}{\partial t} + u\frac{\partial u}{\partial \zeta}\right) = -\frac{\partial p}{\partial \zeta} \tag{2.10}$$

where we may assume additionally that the streamlines become parallel and uniformly spaced at the gap entrance, that $\zeta = 0$ at the gap entrance, and $\zeta = -\infty$ in the plenum chamber. This is seen to be the unsteady compressible one-dimensional Euler equation.

Furthermore, since large velocity changes occur over a very small distance, the flow is assumed to have little time for heat exchange with the walls. Consequently, an adiabatic expansion of the gas may be assumed. For more precision, a polytropic expansion, (given later,) would be more appropriate, with the exponent depending on the magnitude of the entrance velocity (or the Mach number). However, only negligible errors are incurred (on the pressure estimation) by adopting the adiabatic assumption, and since this latter is more applicable for the extreme

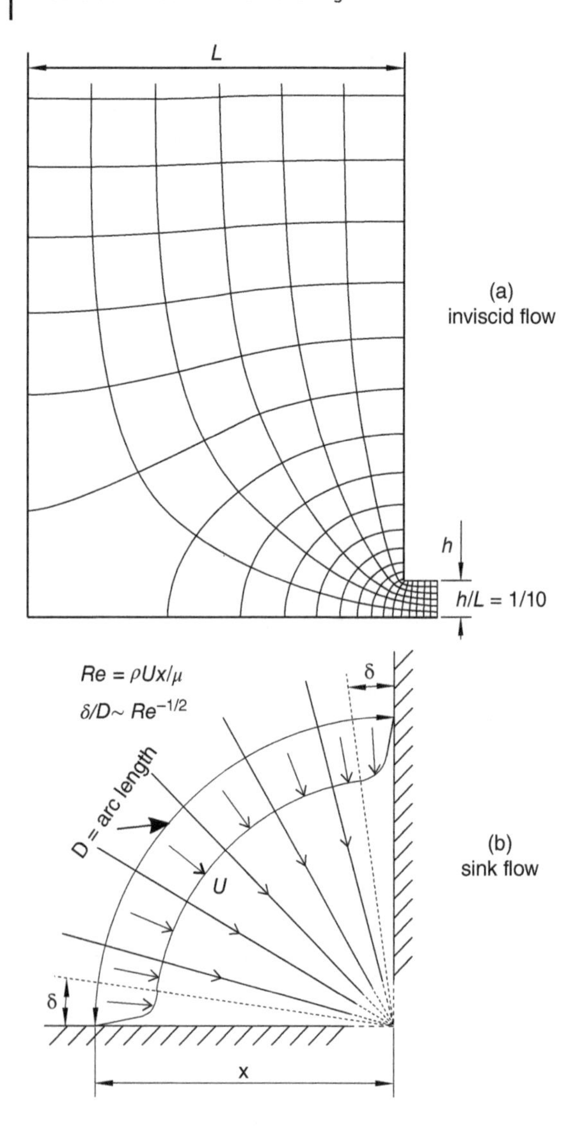

Figure 2.4 Comparison between potential flow in the entrance region and ideal sink flow. Source: Al-Bender F 1992.

case of sonic flow at the entrance, it has been considered more expedient as a general assumption. Finally, since the flow is now both frictionless and adiabatic, it follows that it is *isentropic*.

Further Simplification: the Quasi-steady Euler Equation

Except for the steady state case, Equation 2.10 is difficult to integrate on account of the time derivative term that appears in it. Let us, therefore, deduce conditions whereby this term may be neglected, i.e. where the quasi-steady solution is valid. For this purpose, we need only consider the relative importance of the two inertia terms on the left-hand side of that equation.

We normalise the velocity by the average entrance velocity, $\bar{u}_o$, and the time by $|h/\dot{h}| = 1/\epsilon\omega$, where h is the gap height, $\dot{h}$ its time-derivative and ω is a typical gap oscillation frequency. Since the flow area, as compared to its value at the gap entrance, increases roughly in proportion to the distance from the gap entrance measured by h_o, we have to use this latter as a reference value to normalise the distance ζ, in order that the normalised quantities be all of them of order unity. In this way, we obtain:

$$\frac{\partial u/\partial t}{u\partial u/\partial \zeta} \sim \frac{h_o|\dot{h}_o/h_o|}{\bar{u}_o} = \frac{\epsilon\omega h_o}{\bar{u}_o} \tag{2.11}$$

from which it should be readily apparent that the time derivative is negligible for all but the highest axial oscillation frequencies. As an example, consider that $h_o = O(10^{-4})$ m, $\bar{u}_o = O(10)$ m s^{-1}, $\epsilon = 0.1$, then ω must be at least of order 10^6 rad s^{-1} ($\sim 10^5$ Hz) for the local inertia term to become effective.

Thus, by dropping the time dependent term, the ζ-coordinate becomes also redundant, and we can write:

$$-\frac{dp}{\rho} = udu = d\left(\frac{u^2}{2}\right) \tag{2.12}$$

which is the quasi-steady Euler equation.

(II) Entrance Flow into a Slider Bearing

This problem has been much less treated by researchers than the previous one, i.e. that of an EP bearing. The situation is characterised by the transition from shear-induced flow on a plate (cf. the Sakiadis boundary layer), upstream of the slider bearing, to the Couette–Poiseuille flow in the gap of the slider. This situation is depicted qualitatively in Figure 2.5: (a) sufficiently far away upwards of the lower plate, the flow can be approximated to slow potential (or even viscous) flow that has little effect on the flow near the plate or in the slider-bearing gap; (b) close to the surface of the plate but far (to the left) from the gap entrance, the flow may be represented by classical (Sakiadis) boundary-layer flow. Owing to the lack of a proper initial velocity distribution, this problem, unlike the previous case's flow from a plenum, is somewhat ill defined. As the gap entrance is approached, the BL thickness, see e.g. (Schlichting 1968), will increase in inverse proportion to the square root of the Reynolds number based on the streamwise distance and velocity (U_0). After a transition phase, the lower portion of the streamlines will enter the bearing gap, while the remainder will impinge on the slider face, moving parallel to and away from it as in a typical stagnation corner flow. One might expect the formation of a small eddy owing to reverse flow. If the BL thickness is much larger than the entrance gap height of the bearing (which can be shown to be almost always the case for a practical bearing design), the energy of the flow upstream of the entrance will be greater than that at the gap entrance, which will result in an increase in the static pressure at entrance above its otherwise atmospheric value; the so-called ram pressure. Finally, at (c) just downstream of the entrance, the pressure may still rise owing to entrance effects, before the flow will eventually settle down to Couette–Poiseuille character satisfying Reynolds equation.

2.3.3 Film Flow (or Channel Flow)

A cross section of the film flow configuration pertaining to an EP bearing is depicted in Figure 2.6, together with the associated pressure distribution. the flow is divided into two regions: the entrance region, in which the velocity profile (depicted in the figure) develops from a nearly uniform distribution across the gap; and the viscous region, in which the flow has *settled down* to parabolic velocity distribution. The first-order approximation, that will be discussed below, is applicable to the whole region. The second order approximation applies only to the viscous region.

Similarly, the case of the pressure distribution in a slider bearing is depicted in Figure 2.7, where we see the difference between the actual pressure distribution, with upstream conditions taken into account, and the two approximations based on Reynolds equation.

First Order Approximation: the Boundary Layer (BL) equations

The first order approximation starts from the fact that the problem possesses two length scales, in-plane scale r_r and transverse scale h_r; these being, for the moment, arbitrary reference values for the radius and the film thickness

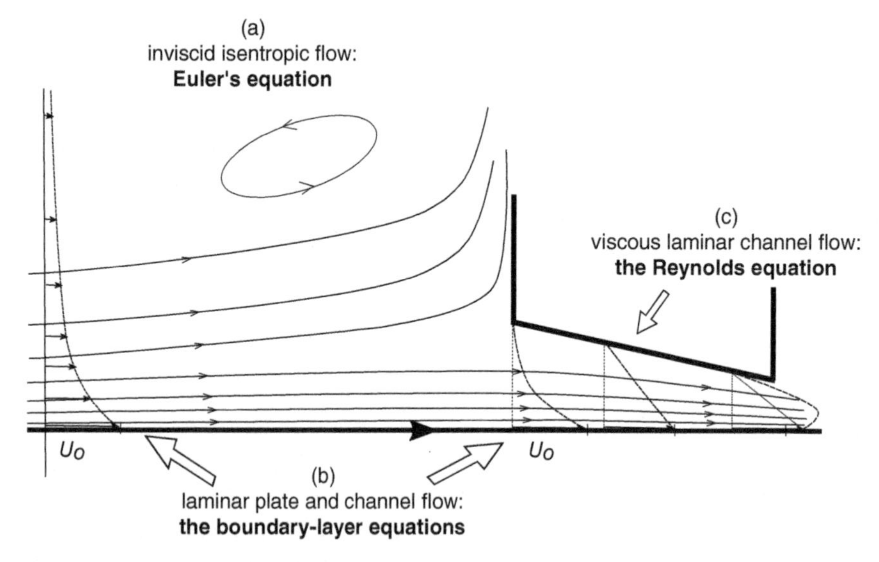

Figure 2.5 Entrance flow into a slider bearing: open shear flow transition to Couette–Poiseuille channel flow.

respectively, and with (h_r/r_r) being in general less than $O(10^{-1})$. The fact that not all the terms in the equations of motion (and energy) have the same order of magnitude is seen if these equations are normalised in such a way that all the (normalised) variables have equal orders of magnitude. Consistent scale factors for the velocities are deduced by asserting that the continuity equation cannot be approximated (otherwise, we have only limiting solutions) and consequently that its different terms must have equal orders of magnitude. Since such reasoning applies equally well to external boundary-layer flow (with h_r replaced by the boundary layer thickness δ, and r_r by the characteristic body length L) as to internal thin-passage flow; the reduced equations are the same, although the flow character for each case is not.

In Cartesian coordinates (x, y, z), the velocity vector is denoted by (u, v, w) in the respective directions. The simplified equations of the flow, obtained by neglecting terms of orders smaller than or equal to $(h_r/x_r)^2$, are then (Constantinescu 1969; Schlichting 1968):

Continuity Equation
(not approximated)

$$\frac{\partial \rho}{\partial t} + \frac{\partial}{\partial x}(\rho u) + \frac{\partial}{\partial y}(\rho v) + \frac{\partial}{\partial z}(\rho w) = 0. \tag{2.13}$$

Reduced Momentum Equations

$$\rho(\frac{\partial u}{\partial t} + u\frac{\partial u}{\partial x} + v\frac{\partial u}{\partial y} + w\frac{\partial u}{\partial z}) = -\frac{\partial p}{\partial x} + \mu\frac{\partial^2 u}{\partial z^2} \tag{2.14}$$

$$\rho(\frac{\partial v}{\partial t} + u\frac{\partial v}{\partial x} + v\frac{\partial v}{\partial y} + w\frac{\partial v}{\partial z}) = -\frac{\partial p}{\partial y} + \mu\frac{\partial^2 v}{\partial z^2} \tag{2.15}$$

$$0 = -\frac{\partial p}{\partial z} \tag{2.16}$$

where the inertia terms have been set on the left-hand side. The last equation of the set means that the pressure is constant across the film.

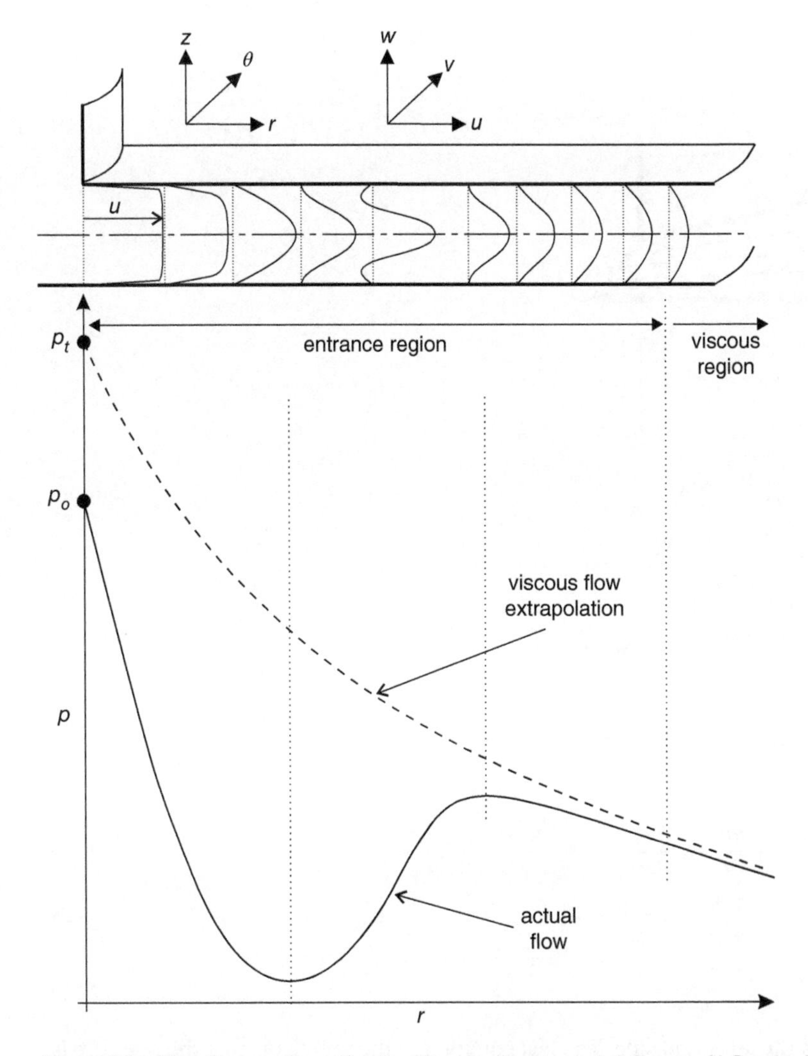

Figure 2.6 Film flow configuration for EP case. (Not to scale.) Source: Al-Bender F 1992.

Reduced Energy Equation

$$\rho c_p\left(\frac{\partial T}{\partial t} + u\frac{\partial T}{\partial x} + v\frac{\partial T}{\partial y} + w\frac{\partial T}{\partial z}\right) = \frac{\partial p}{\partial t} + u\frac{\partial p}{\partial x} + v\frac{\partial p}{\partial y} + k\frac{\partial^2 T}{\partial z^2} + \mu\left[\left(\frac{\partial u}{\partial z}\right)^2 + \left(\frac{\partial v}{\partial z}\right)^2\right]. \tag{2.17}$$

In cylindrical coordinates (r, θ, z), let us again denote the velocity vector, in the respective directions, by (u, v, w). The simplified equations of the flow, obtained by neglecting terms of orders smaller than or equal to $(h_r/r_r)^2$, (written in cylindrical coordinates,) are then:

Continuity Equation

(not approximated)

$$\frac{\partial \rho}{\partial t} + \frac{1}{r}\frac{\partial}{\partial r}(\rho r u) + \frac{1}{r}\frac{\partial}{\partial \theta}(\rho v) + \frac{\partial}{\partial z}(\rho w) = 0. \tag{2.18}$$

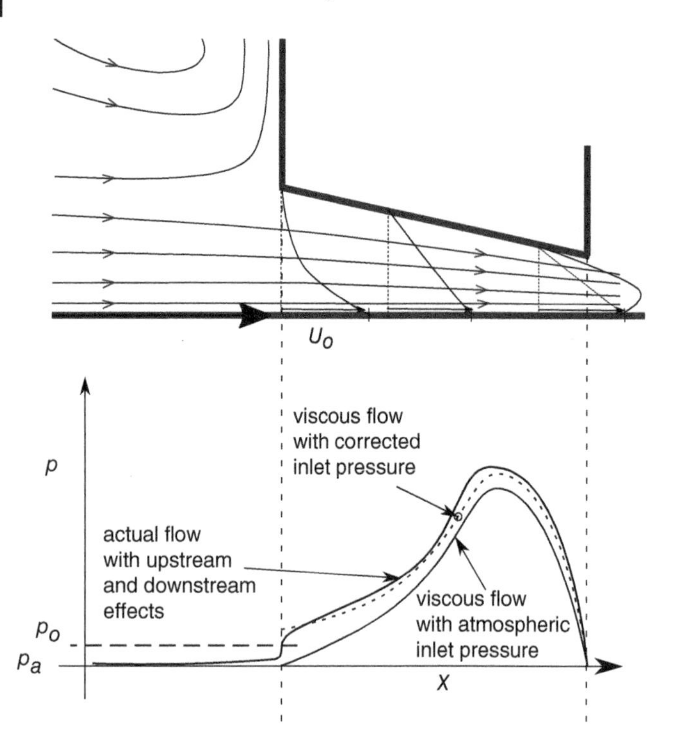

Figure 2.7 Film flow configuration for the slider-bearing case. (Not to scale.)

Reduced Momentum Equations

$$\rho(\frac{\partial u}{\partial t}+u\frac{\partial u}{\partial r}+\frac{v}{r}\frac{\partial u}{\partial \theta}-\frac{v^2}{r}+w\frac{\partial u}{\partial z})=-\frac{\partial p}{\partial r}+\mu\frac{\partial^2 u}{\partial z^2} \tag{2.19}$$

$$\rho(\frac{\partial v}{\partial t}+u\frac{\partial v}{\partial r}+\frac{v}{r}\frac{\partial v}{\partial \theta}+\frac{uv}{r}+w\frac{\partial v}{\partial z})=-\frac{1}{r}\frac{\partial p}{\partial \theta}+\mu\frac{\partial^2 v}{\partial z^2} \tag{2.20}$$

$$0=-\frac{\partial p}{\partial z} \tag{2.21}$$

where the inertia terms have been set on the left-hand side. The last equation of the set means that the pressure is constant across the film.

Reduced Energy Equation

$$\rho c_p(\frac{\partial T}{\partial t}+u\frac{\partial T}{\partial r}+\frac{v}{r}\frac{\partial T}{\partial \theta}+w\frac{\partial T}{\partial z})=\frac{\partial p}{\partial t}+u\frac{\partial p}{\partial r}+\frac{v}{r}\frac{\partial p}{\partial \theta}+k\frac{\partial^2 T}{\partial z^2}$$
$$+\mu[(\frac{\partial u}{\partial z})^2+(\frac{\partial v}{\partial z})^2]. \tag{2.22}$$

This concludes the first order approximated equations, which are exact to the order $(h_r/r_r)^2$; i.e. $\sim 10^{-2}$, at film entrance, and $\sim 10^{-6}$ at exit. They will also be termed the *boundary-layer* (BL) equations.

Second Order Approximation: Viscous Flow

Let us consider, without loss of generality, the cylindrical coordinates case. Normalisation, which will be given later, of Equations 2.20 and 2.22 reveals that the order of magnitude of their left-hand members (representing the inertia terms) depend on the values of two Reynolds numbers: the reduced Reynolds number Re* and the squeeze Reynolds number Re_v. To show this, we need only compare the viscous term on the right-hand side with the two types of inertia terms on the left-hand side; the local and the convective. In addition to the previous radius and film reference values, we shall employ an arbitrary reference main flow average velocity $\bar{u}_r$ (averaged across the film) and a reference density ρ_r. Thus, we have:

$$\frac{\rho u \partial u/\partial r}{\mu \partial^2 u/\partial z^2} \sim \frac{\rho_r \bar{u}_r h_r}{\mu}\frac{h_r}{r_r} = \text{Re}^*$$

and

$$\frac{\rho \partial u/\partial t}{\mu \partial^2 u/\partial z^2} \sim \frac{\rho_r \epsilon \omega {h_r}^2}{\mu} = \epsilon \text{Re}_v.$$

Consequently, a second order approximation of the BL equations is possible, if these numbers (viz. Re^* and ϵRe_v) are negligible in comparison with unity, such as expected to be the case of quasi-steady flow sufficiently downstream of the entrance. This consists then in setting the momentum terms equal to zero, leading to the well known *viscous* flow equations:

$$0 = -\frac{\partial p}{\partial r} + \mu \frac{\partial^2 u}{\partial z^2} \tag{2.23}$$

$$0 = -\frac{1}{r}\frac{\partial p}{\partial \theta} + \mu \frac{\partial^2 v}{\partial z^2} \tag{2.24}$$

$$0 = -\frac{\partial p}{\partial z} \tag{2.25}$$

and the energy equation:

$$0 = \frac{\partial p}{\partial t} + k\frac{\partial^2 T}{\partial z^2} + \mu \frac{\partial^2}{\partial z^2}(\frac{u^2}{2} + \frac{v^2}{2}). \tag{2.26}$$

Remarks

The first set, Equations 2.22–2.22, forms the essence of Reynolds lubrication theory; it may be integrated directly to give the parabolic velocity distribution, which upon substitution in the integrated continuity equation yields the equation for the pressure better known as the Reynolds equation, which will be given in section 2.4.2. and more elaborately in Chapter 4.

The second, Equation 2.26 (which is also readily integrable in the steady state case) may in principle be used to determine the temperature (as a function of the velocity) which, together with the equation of state and the pressure equation, would form a system of three simultaneous equations for the three unknowns p, T and ρ. However, since the temperature variations in a viscous film would hardly be appreciable in most cases (see e.g. (Constantinescu 1969)), the energy equation is usually dropped from the analysis in favour of a simpler approach based on the assumption of polytropy which will be outlined further below.

A more important remark about Equations 2.23 and 2.24 is that they constitute yet another singular perturbation from the previous set. That is to say that the original set of BL equations do *not* approach the viscous solution *uniformly* in the limit $\text{Re}^* \to 0$. In other words, the exact solution cannot be obtained as an expansion in powers of Re^*. Consequently, although the viscous solution constitutes a valid approximation sufficiently far from the feed-hole, it cannot form the starting point for obtaining accurate solutions in the entrance region. This circumstance becomes more apparent if we consider the boundary conditions on the velocity; due to the neglect of the inertial derivatives, the conditions at the solid walls are now sufficient to solve the problem, i.e. the initial conditions at the entrance are made redundant. This will be seen to present a problem of matching the boundary conditions at the entrance with those of the feed flow. This is generally termed a "boundary layer" (not to be confused with flow boundary layer!) problem: the problem of finding a solution near the boundary, satisfying the (otherwise redundant) boundary condition and approaching the singular-perturbation solution (valid) far away from the boundary, see e.g. Ames (1965). It is best illustrated by examining the pressure distribution curves of

Figure 2.6. The theoretical entrance pressure p_t, obtained by extrapolating the viscous pressure distribution, or by assuming that the flow is viscous throughout, does not coincide with the actual entrance pressure p_o corresponding to the exact solution.

Finally, the only unknowns of the problem are now the pressure and the density.

Polytropy

Due to the difficulty, alluded to above, of solving the energy equation and the equation of state simultaneously with the pressure equation, these former two equations are often replaced by a single equation referred to as the equation of polytropy (Constantinescu 1969; Gross 1962):

$$\frac{p}{\rho^{\chi}} = \text{constant} \tag{2.27}$$

in which,

$$1 \leq \chi \leq \kappa$$

is the polytropic exponent. The gas expansion will then be isothermal when $\chi = 1$ and isentropic when $\chi = \kappa$. (Note that if χ is allowed to become infinite, incompressible flow will result.)

However, this representation is limited in usefulness by the fact that χ has to be taken as a constant (or, at most, a pre-assigned function).

2.4 Formulation of Bearing Flow and Pressure Models

In designing a thrust bearing, one is primarily concerned with an axisymmetric flow (and pressure) configuration, since it is in such a disposition that a bearing is intended to perform. The problem would consist then in determining the bearing's static characteristics (load capacity and mass flow rate) for a range of film thickness values. The basic aspect of this problem is the determination of the feed/entrance flow development. This will be the subject of the axisymmetric quasi-static flow model, which may be regarded as a *distributed parameter* model.

In a second instance, other aspects, such as tilt, sliding and dynamic behaviour may be dealt with with the aid of a simpler model based on the Reynolds equation. This is both admissible, since the flow in the bearing film is predominantly viscous, and convenient, since an asymmetric configuration may be easily treated. However, we are then faced with the problem of matching the entrance flow with the viscous flow, or determining the initial value of the pressure needed to solve this problem. This amounts to the formulation of a *lumped parameter* entrance model, that may be obtained, in the vicinity of axial symmetry, from the previous model. A lumped entrance model of this type is usually referred to as a *restrictor* or *compensator*, in analogy with an orifice. Note also that although the problem of determining the axial dynamic characteristics does not violate axial symmetry, it will be shown that the Reynolds/restrictor model is adequate in characterising it in a certain design range.

These two models will be formulated below on the basis of equations derived in the previous section. Solutions to intermediate problems will also be given whenever possible, in order to simplify further treatment.

2.4.1 The Quasi-static Flow Model for an Axisymmetric EP Bearing

This model consists of the feed flow region, characterised by Euler's equation, and the film region, characterised by the BL equations. The feed-flow region might be preceded by another restrictor (e.g. an orifice, a porous insert, etc.), which will be treated later, or it can be completely absent.

Feed Flow Model

This is obtained by substituting the isentropy equation, Equation 2.27 (with $\chi = \kappa$,) into Euler's Eq. 2.12, to yield:

$$-c\frac{\kappa}{(\kappa-1)}d(p^{(\kappa-1)/\kappa}) = udu, \qquad c = \frac{{p_s}^{1/\kappa}}{\rho_s} = \frac{{p_o}^{1/\kappa}}{\rho_o} \tag{2.28}$$

Solution

With boundary conditions,

$$p = p_s \;:\; u = 0, \qquad p = p_o \;:\; u = \bar{u}_o$$

Equation 2.28 is readily integrated to yield:

$$\frac{p_s}{\rho_s}\left(1 - (\frac{p_o}{p_s})^{(\kappa-1)/\kappa}\right) = \frac{(\kappa-1)}{\kappa}\frac{{\bar{u}_o}^2}{2} \tag{2.29}$$

Note that this equation is only valid up to gap entrance velocities which are less than or equal the local speed of sound $a_o = \sqrt{\kappa p_o/\rho_o}$. Equivalently, the inlet pressure is bounded by the relation:

$$\frac{p_o}{p_s} \le \left(\frac{p_o}{p_s}\right)_{\text{critical}} = \left(\frac{2}{\kappa+1}\right)^{\kappa/(\kappa-1)}.$$

In other words, since the flow is converging, a Mach number of unity cannot be exceeded at the throat. When the last condition holds true, we will have:

$$\bar{u}_o = a_o$$

Film Flow Model

Axial symmetry and time-independence will result in greatly simplified describing equations, void of the θ components and the time derivatives, i.e. a 2D problem. Also, since $p = p(r)$, the pressure gradient will be written as a total derivative. Thus, from Equations (2.18–2.22), we have for this region:

Continuity Equation

$$\frac{1}{r}\frac{\partial}{\partial r}(\rho r u) + \frac{\partial}{\partial z}(\rho w) = 0. \tag{2.30}$$

Momentum Equation

$$\rho(u\frac{\partial u}{\partial r} + w\frac{\partial u}{\partial z}) = -\frac{\mathrm{d}p}{\mathrm{d}r} + \mu\frac{\partial^2 u}{\partial z^2}. \tag{2.31}$$

Energy Equation

$$\rho c_p(u\frac{\partial T}{\partial r} + w\frac{\partial T}{\partial z}) = u\frac{\mathrm{d}p}{\mathrm{d}r} + k\frac{\partial^2 T}{\partial z^2} + \mu(\frac{\partial u}{\partial z})^2. \tag{2.32}$$

2.4.2 The Reynolds Plus Restrictor Model

The Reynolds Equation

The Reynolds equation is the differential equation for the pressure distribution in the film obtained by solving the viscous flow equations. Its derivation is quite straightforward, and may, for the general case, be found back in Constantinescu (1969); Gross (1962). It will be briefly derived here for our case of a compressible film with a single moving surface, in order to simplify further referencing, and facilitate derivation of bearing characteristics based thereupon. Its full derivation, forms in different coordinate systems, basic characteristics and extensions to special cases will be dealt within Chapter 4.

Let us denote by $\nabla \equiv \nabla_{xy}$ the two-dimensional gradient/divergence operator in the plane of the bearing. We split the fluid velocity vector into two components, one parallel to the plane of the bearing and one normal to it, by writing:

$$\mathbf{v} = \mathbf{v}_{\mathrm{p}} + w\hat{\mathbf{e}}_z = (u, v) + w\hat{\mathbf{e}}_z.$$

The continuity Equation 2.3 becomes:

$$\frac{\partial \rho}{\partial t} + \nabla \cdot (\rho \mathbf{v}_{\mathrm{p}}) + \frac{\partial (\rho w)}{\partial z} = 0 \tag{2.33}$$

This equation may be integrated across the film, taking care to apply the Leibniz rule, and using the boundary conditions on w at the walls, to give (after rearrangement,)

$$\frac{\partial (\rho h)}{\partial t} + \nabla \cdot (\rho \int_0^h \mathbf{v}_{\mathrm{p}} \mathrm{d}z) = 0 \tag{2.34}$$

where the term under the divergence operator in the last equation is the (macroscopic) *flux density* of the flow $\mathbf{\Phi}$:

$$\mathbf{\Phi} = \rho \int_0^h \mathbf{v}_{\mathrm{p}} \mathrm{d}z,$$

which is determined by integrating the viscous momentum Equations 2.23 and 2.24, which may be combined, for simplicity, in one equation:

$$\frac{\partial^2}{\partial z^2}(\mathbf{v}_{\mathrm{p}}) = \frac{1}{\mu}\nabla p.$$

Integrating this equation across the film, and using the no-slip boundary conditions on $\mathbf{v}_{\mathrm{p}}$ at the walls, we obtain:

$$\mathbf{v}_{\mathrm{p}} = -\frac{1}{2\mu}\nabla p\, z(h - z) + \boldsymbol{V}_t(1 - \frac{z}{h}) \tag{2.35}$$

which is the well known velocity distribution associated with viscous flow. It consists of a *parabolic* component, associated with Poiseuille flow, and a *linear* component, associated with Couette or shear flow. Integrating once more across the gap, yields:

$$\int_0^h \mathbf{v}_{\mathrm{p}} \mathrm{d}z = -\frac{h^3}{12\mu}\nabla p + \frac{h}{2}\boldsymbol{V}_t.$$

Thus, the flux density is given by:

$$\mathbf{\Phi} = -\frac{\rho h^3}{12\mu}\nabla p + \frac{\rho h}{2}\boldsymbol{V}_t \tag{2.36}$$

Substituting for the flux density in Equation 2.34 and rearranging, yields the Reynolds equation:

$$\nabla \cdot \left(\frac{\rho h^3}{12\mu}\nabla p - \frac{1}{2}\rho h \boldsymbol{V}_t \right) = \frac{\partial}{\partial t}(\rho h). \tag{2.37}$$

For more details, see Chapter 4.

Restrictor Model

As has been mentioned earlier, the entrance region usually constitutes only a relatively small part of the bearing area. Errors in the load capacity, committed by assuming the flow to be everywhere viscous, may consequently be negligible. This should mean that the Reynolds equation is sufficient, for practical purposes, to solve the bearing problem. However, we would need, in this case, to determine the first boundary condition p_t of the Reynolds problem. More precisely, we will need to relate p_t to the problem parameters. In this way, the entrance effect is reduced to one point, and characterised by a single relationship, (in the form of entrance head-loss, coefficient of discharge, etc.) This constitutes lumping of the entrance flow which plays an important role in modeling of externally pressurised bearings. Actual formulas, and their associated problems, will be discussed in Chapter 3. It will suffice here to give the general form of such a model, thus:

$$p_t = f(p_s, \text{flow parameters}) \tag{2.38}$$

where f is a function to be determined theoretically or empirically.

This concludes the Reynolds/restrictor model which, with an accurate restrictor formula, should provide a good basis for determining the bearing characteristics.

2.5 The Basic Bearing Characteristics

Here, the definitions of the most basic bearing characteristics will be given. The *bearing region* will be denoted by A:

$$A = \{(r, \theta) \ : 0 < r < r_a \ , \ 0 < \theta < 2\pi\}.$$

2.5.1 The Load Carrying Capacity

This is given as the integral of the net pressure distribution over the bearing area:

$$F = \int_A (p - p_a) \mathrm{d}A. \tag{2.39}$$

Since the feed-hole area is usually negligible in comparison with the total bearing area, A may be taken as the area over which the film extends in the above formula. This has the advantage of simplifying the calculation without greatly impairing the results.

2.5.2 The Axial Stiffness

The axial stiffness, i.e. that which is normal to the tangent plain of the bearing, is defined by:

$$k_a = -\frac{\partial F}{\partial h}. \tag{2.40}$$

In the case of dynamic loading, the stiffness will be complex valued: the imaginary part representing the damping.

2.5.3 The Feed Mass Flow Rate

This is calculated on the basis of the flow issuing through the annular curtain area enclosing the feed hole, using upstream values. Thus,

$$\dot{m}_o = 2\pi r_o h_o \rho_o \bar{u}_o \tag{2.41}$$

substituting for $\bar{u}_o$ from Equation 2.29, and using the equations of isentropy and state, this equation is written:

$$\dot{m}_o = 2\pi r_o h_o \sqrt{\frac{2\kappa}{\kappa - 1}} \frac{p_s}{\sqrt{\Re T_s}} \Phi_e(\bar{p}_o) \tag{2.42}$$

where,

$$\bar{p}_o = \frac{p_o}{p_s}$$

and,

$$\Phi_e(p_o) = \begin{cases} \left(\bar{p}_o^{\,2/\kappa} - \bar{p}_o^{\,(\kappa+1)/\kappa}\right)^{1/2} ; & \bar{p}_o \geq \bar{p}_c = \left(\frac{2}{\kappa+1}\right)^{\kappa/(\kappa-1)} \\ \left(\bar{p}_c^{\,2/\kappa} - \bar{p}_c^{\,(\kappa+1)/\kappa}\right)^{1/2} ; & \bar{p}_o \leq \bar{p}_c. \end{cases} \tag{2.43}$$

2.5.4 The Mass Flow Rate in the Viscous Region

Let Γ be a simple curve in the plane of the viscous region of the bearing. The mass flow rate through the film portion extending over Γ is given by the line integral:

$$\dot{m}_\Gamma = \int_\Gamma \mathbf{\Phi} \cdot \hat{\mathbf{n}} \mathrm{d}\Gamma \tag{2.44}$$

where $\hat{\mathbf{n}}$ is the *unit outward normal* to Γ.

In the special case, of disc bearing, when,

$$\Gamma = \{\ r = \text{constant},\ 0 \leq \theta \leq 2\pi\}$$

(i.e. a circle around the bearing centre,) then

$$\hat{\mathbf{n}} = \hat{\mathbf{e}}_r, \quad \mathrm{d}\Gamma = r\mathrm{d}\theta$$

and the *radial* or *net* mass flow rate out of the bearing is given by:

$$\dot{m}_\mathrm{r} = \int_0^{2\pi} \mathbf{\Phi} \cdot \hat{\mathbf{e}}_r r\mathrm{d}\theta. \tag{2.45}$$

Consequently, we can calculate the mass flow rate of fluid leaving the bearing at any radial location where $\mathbf{\Phi}$ is given. (Evidently, in the steady state, $\dot{m}_\mathrm{r}$ will be independent of the radius.)

In particular, we can simplify the formula for the flow rate, for the case of steady flow at nominal parallelism, by calculating the flow at a circle on which $\rho h =$ constant. This condition is guaranteed, e.g., on $r = r_a$; and, if the film is assumed wholly viscous, also on $r = r_o$. In such case, we have:

$$\int_0^{2\pi} (\rho h \boldsymbol{V}_t \cdot \hat{\mathbf{e}}_r r\mathrm{d}\theta)\,|_{r=r_a} \cdot r_a = 0.$$

It follows then that,

$$\dot{m}_r = \int_0^{2\pi} -\left(\frac{r\rho h^3}{12\mu}\frac{\partial p}{\partial r}\right)_{r=r_o,\ r_a} \mathrm{d}\theta \tag{2.46}$$

which can readily be used to evaluate the mass flow rate leaving the bearing.

2.5.5 The Tangential Resistive "Friction" Force

The tangential resistive, or *frictional* force per unit area $\mathrm{d}\boldsymbol{F}_\mathrm{f}$, at either of the two bearing surfaces is given by:

$$\mathrm{d}\boldsymbol{F}_\mathrm{f} = \left(\mu\left(\frac{\partial \mathbf{v}_\mathrm{p}}{\partial z}\right)_{z=0,h}\right)\mathrm{d}A. \tag{2.47}$$

From Equation 2.35, one obtains, in the viscous region:

$$\mathrm{d}\boldsymbol{F}_\mathrm{f} = \left(\mp\frac{h}{2}\nabla p - \frac{\mu}{h}\boldsymbol{V}_t\right)\mathrm{d}A \tag{2.48}$$

where the $(\mp)$ correspond to the surfaces $z = 0, h$ respectively. It may be noted from Equation 2.48 that both the relative velocity and the pressure gradient contribute to the resisting (friction) force. The term in the velocity represents the linear shearing force associated with Couette flow; this force would be the same even if the bearing is not pressurised. In other words, this represents the true "friction" force in the sense that it is a dissipative force. This is the so-called Petrov's law (Dowson 1979). The term in the pressure, on the other hand, represents the Poiseuille contribution to the velocity gradient at the wall. Consequently, if the pressure distribution in the bearing is axisymmetric, the net (total) friction force due to this term will be zero. In a self-acting slider bearing, this is the force needed to generate the aerodynamic action, i.e. to pump the fluid into the gap. (Note, however, that relative motion will itself cause asymmetry in the pressure distribution).

Finally, if A_v represents the viscous region, the friction force field over that region, is given by:

$$\boldsymbol{F}_\mathrm{f} = \int_{A_\mathrm{v}} (\mp\frac{h}{2}\nabla p - \frac{\mu}{h}\boldsymbol{V}_t)\mathrm{d}A_\mathrm{v}. \tag{2.49}$$

(If the film is assumed wholly viscous then $A_\mathrm{v} \equiv A$.)

Finally, other more specialised bearing characteristics, such as tilt stiffness, centre of force, friction coefficient, etc., will be defined in Chapter 6.

2.6 Normalisation and Similitude

A formal treatment of any dynamic problem is best achieved by way of turning the variables into dimensionless quantities, viz. by *normalisation*. Such should provide not only for a better handling and understanding of the basic aspects of the problem, but should also lead to results that are of a generalised nature. In the first instance, the total number of (absolute) parameters is reduced to a minimum of (dimensionless) parameters; in the second, criteria for similitude are obtained that are essential for deducing the relationship between, e.g., an actual model and an experimental prototype. This may help, in some cases, to scale a test bearing in such a way that some of dimensions, e.g., the film thickness, be easier to measure, as has been done in Mirolyubov and Shashin (1982). Furthermore, a normalisation procedure is not unique (Massey 1968): different reference values may be used as well as different ways of combining them, depending on relevance and convenience. Nor is the knowledge of the governing equations necessary, as long as all the parameters of the problem are included (Parker et al. 1969); simplification of the governing equations may, however, lead to a further reduction in the number of variables (leaving some of the parameters redundant).

It has been apparent that our problem may be addressed at two levels: the first is microscopic, viz. the flow problem, which leads after integration to the second, the macroscopic, or the pressure equation. Further integration of the latter should yield the bearing (global) characteristics. Consequently, the reference dimensions used at each level, as well as the resulting dimensionless parameters, will not, in general, be the same. Although the dimensionless parameters at the first level will, generally speaking, determine the problem as a whole, it is neither convenient nor desirable e.g. to describe the bearing global characteristics in terms of the flow parameters.

Below will be given the normalisation to be adopted for each problem together with an outline of similitude criteria. However, in order to avoid complicating matters unduly, the analysis will, as far as possible, be confined to the cases that find treatment in this work, viz. steady radial entrance flow, and the Reynolds equation.

2.6.1 The Axisymmetric Flow Problem

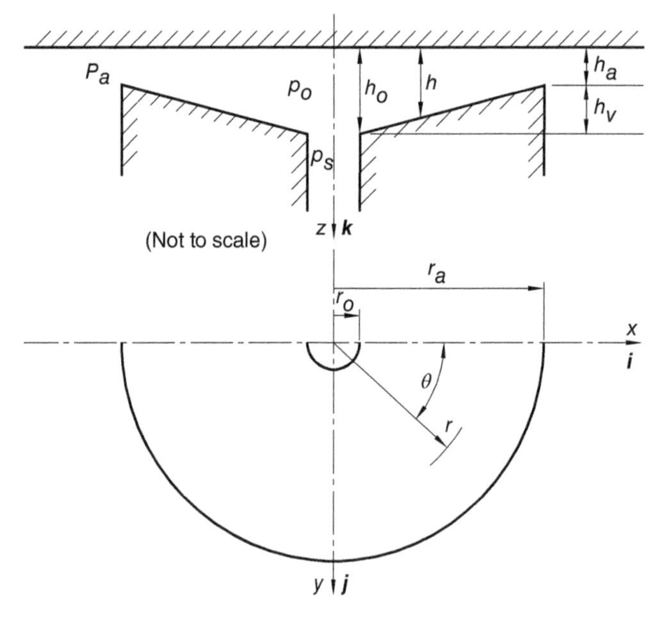

Figure 2.8 Circular centrally fed bearing geometry and notation. Source: Al-Bender F 1992.

2.6.2 Geometry

Figure 2.8 depicts the bearing configuration that will be considered in this chapter. The bearing pad is circular with the pressure source located at its centre. The most important aspect of the geometry (not apparent in the drawing) is the small ratio of the film thickness to the bearing plane dimension. The rest of the specifications are not absolutely necessary for obtaining bearing action, but are rather made as idealisations to facilitate problem solution. The following is assumed to hold true:

- Both of the bearing surfaces are smooth. The platen surface is flat; the pad surface has an axisymmetric profile. A linearly convergent profile will be assumed throughout this work, for simplicity of analysis as well as bearing fabrication, although other profiles are also possible.
- $h/r \ll 1$, in particular, $h_a/r_a \sim O(10^{-3})$ and $h_o/r_o \sim (10^{-1})$ or less. This should ensure also that $|dh/dr| \ll 1$, which is a necessary assumption for simplifying the flow problem.
- $r_o/r_a \ll 1$. This may be seen to be necessary if the bearing is to have reasonable stiffness; typically then r_o/r_a is less than 0.1.

The film thickness function generally consists of an axisymmetric function and a tilt function, thus:

$$h(r,\theta) = h_{\text{ax}}(r) + h_{\text{tilt}}(r,\theta).$$

For the case when the bearing profile is linearly convergent, we have:

$$h_{\text{ax}} = h_o - \frac{h_v}{(r_a - r_o)}(r - r_o) \tag{2.50}$$

$$= h_a + \frac{h_v}{(r_a - r_o)}(r_a - r) \tag{2.51}$$

and,

$$h_{\text{tilt}} = r \sin \alpha \cos \theta \simeq r\alpha \cos \theta$$

since the tilt angle α is very small.

Since we are interested in the film flow, which emanates at the inner radius, it may be handier to normalise relative to the inlet values. We first normalise the coordinates:

$$(R, Z) = (r/r_o, z/h_o). \tag{2.52}$$

Then, by choosing the average inlet radial velocity $\bar{u}_o$ as reference, we define the normalised velocities:

$$(U, W) = \frac{1}{\bar{u}_o}(u, (\frac{r_o}{h_o})w). \tag{2.53}$$

Next the pressure, density and temperature are normalised:

$$(P, \wp, T^*) = (p/p_o, \rho/\rho_o, T/T_o) \tag{2.54}$$

(on the assumption that ρ_o and T_o be uniform across the film.)

The Geometry

Consistent with the above, we choose Equation 2.50 for expressing the film thickness function. We define the normalised film thickness for the flow problem as follows:

$$\begin{aligned} H = \frac{h}{h_o} &= 1 - \frac{(h_v/h_o)}{(R_a - 1)}(R - 1) \\ &= 1 - \nu_o(R - 1) \end{aligned} \tag{2.55}$$

where $\nu_o = (h_v/h_a)/(R_a - 1)$ is the normalised conicity angle.

Feed Flow

Substituting the relevant normalisation forms into Equation 2.12, we obtain the normalised Euler equation:

$$\frac{dP}{\wp} = \kappa \mathrm{M}_o{}^2 U dU \tag{2.56}$$

where M_o is the entrance Mach number.

The equation of isentropy becomes:

$$P = \wp^{\kappa} \qquad \text{or} \qquad \wp = P^{1/\kappa}. \tag{2.57}$$

Substitution of Equation 2.57 into 2.56 and rearrangement yields:

$$d(P^{(\kappa-1)/\kappa}) = (\kappa - 1)\mathrm{M}_o{}^2 U dU. \tag{2.58}$$

The normalised boundary conditions on Equation 2.58, assuming axisymmetric flow at gap entrance, are:

$$U \equiv 0 \ : \ P = P_s, \qquad U \equiv 1 \ : \ P = 1.$$

The solution corresponding to Equation 2.29 is now:

$$P_s = \left(1 + \frac{(\kappa - 1)}{2}\mathrm{M}_o{}^2\right)^{\kappa/(\kappa-1)}. \tag{2.59}$$

Film Flow

Equations 2.31 and 2.32 are written in normalised form, respectively:

$$\frac{1}{R}\frac{\partial}{\partial R}(\wp RU) + \frac{\partial}{\partial Z}(\wp W) = 0, \tag{2.60}$$

$$\wp \mathrm{Re}_o^*(U\frac{\partial U}{\partial R} + W\frac{\partial U}{\partial Z}) = -\frac{\mathrm{Re}_o^*}{\kappa \mathrm{M}_o{}^2}\frac{\mathrm{d}P}{\mathrm{d}R} + \frac{\partial^2 U}{\partial Z^2} \tag{2.61}$$

and

$$\wp \mathrm{Re}_o^*(U\frac{\partial T^*}{\partial R} + W\frac{\partial T^*}{\partial Z}) = \frac{\mathrm{Ec}_o \mathrm{Re}_o^*}{\kappa \mathrm{M}_o{}^2}U\frac{\mathrm{d}P}{\mathrm{d}R} + \frac{1}{\mathrm{Pr}}\frac{\partial^2 T^*}{\partial z^2} + \mathrm{Ec}_o(\frac{\partial U}{\partial Z})^2. \tag{2.62}$$

Auxiliary Conditions

Initial condition

at $R = 1$: $U \equiv 1, P = 1$ and $T^* = 1$.

Boundary conditions

- $Z = 0$: $U = W = 0, T^* = T^*_{w_0}(R)$;
- $Z = H$: $U = W = 0, T^* = T^*_{w_h}(R)$.

2.6.3 Dimensionless Parameters and Similitude

The following dimensionless numbers have been used (evaluated at gap entrance):

Normalised conicity angle

$\nu_o = (h_v/h_o)/(R_a - 1)$, in which $R_a = r_a/r_o$.

Reduced Reynolds number

$\mathrm{Re}_o^* = (\rho_o \bar{u}_o h_o/\mu)(h_o/r_o) = \mathrm{Re}\ (h_o/r_o)$, where Re is the hydraulic Reynolds number.

Mach number[2]

$\mathrm{M}_o = \bar{u}_o/a_o$, where $a_o = (\kappa p_o/\rho_o)$ is the sonic speed at entrance.

Prandtl number

$\mathrm{Pr} = \mu c_p/k$ which is theoretically equal to $4\kappa/(9\kappa - 5)$, (for air, $\mathrm{Pr} = 0.72$ experimentally cf. 0.737 theoretically.)

Eckert number

$\mathrm{Ec}_o = \bar{u}_o{}^2/c_p T_o$, which is equal to $(\kappa - 1)M_o{}^2$ by virtue of the assumption of isentropic feed flow.

Ratio of the specific heats

$\kappa = c_p/c_v$.

Note, first, that due to the neglect of the kinematics of the feed flow, the dimensionless number h_o/r_o did not appear in the above list. We must bear in mind, however, that this has been so only by virtue of the assumption that $h_o/r_o \ll 1$. Secondly, all the dimensionless parameters listed above would, in the general case be independent. However, due to the simplifying assumptions made earlier, as well as the physical properties of the ideal gas, some of these parameters become redundant. The ones left over are:

$$(\nu_o,\ \mathrm{Re}_o^*,\ \mathrm{M}_o,\ \mathrm{Pr},\ \kappa). \tag{2.63}$$

2 In reality only the square of the Mach number is used. This is known as the Cauchy number $\mathrm{Ca} = \mathrm{M}_o{}^2$ which is a measure of the ratio of the inertia force to the elastic force.

Similarity

The axisymmetric flow in two bearing films will be physically similar if the dimensionless parameters 2.63 are the same for both and if, in addition, the dimensionless wall temperatures $T^*_{w_0}$ and $T^*_{w_h}$ are the same for both. This is seen to consist in:

- Geometric similarity, characterised by ν_0, which includes both conicity and radius ratio
- Dynamic similarity, characterised by Re_0^* and $\kappa \mathrm{M}_0{}^2$, the latter accounting for compressibility
- Thermal similarity, characterised by Pr and $\mathrm{Ec}_0 = (\kappa - 1)\mathrm{M}_0{}^2$, as well as the wall normalised temperatures. Note that Pr and κ depend only on the gas properties.
- Kinematic similarity is automatically ensured by geometrical similarity, owing to the simplifying assumptions made on the entrance velocity distribution.

In the case that two films are similar, the dimensionless velocity, temperature and pressure fields (functions of the dimensionless coordinates) will be everywhere identical. Also, scale factors between model and prototype may be deduced on the basis of the dimensionless parameters.

As they stand, these similarity conditions may be very difficult to fulfil in practice, i.e. to make two similar bearings that are not identical; in particular the conditions on the temperature field. Further simplification may be afforded, however, by relaxing some of these conditions. In the first instance, thermal effects would play only a minor role in determining the flow characteristics, owing to the limited temperature variation in the fluid, based on the relatively high thermal conductivity of the bearing surfaces. Secondly, κ, and consequently also Pr, may be taken as constant: indeed, κ is determined only by the number of degrees of freedom of a gas molecule, thus being, e.g., the same for all diatomic gases. When these assumptions are made, the similarity parameters are significantly reduced to:

$$(\nu_0,\ \mathrm{Re}^*,\ \mathrm{M}_0).$$

That is: dimensionless geometry, Reynolds number and Mach number.

Since we shall be interested in the pressure distribution, we can write, in this case:

$$P(R) = f(\nu_0, \mathrm{Re}_0^*, \mathrm{M}_0;\ R).$$

If R_a is an arbitrary radius, for which the flow has become viscous, say the exit radius, then,

$$P_a = P(R_a).$$

Also, by virtue of Equation 2.59, we have,

$$P_s = f(\mathrm{M}_0).$$

We will show later that these similarity conditions can be replaced by equivalent conditions that are easier to apply.

If, finally, the bearing is flat ($\nu_0 = 0$), the flow will be determined completely by Re* and M_0. For incompressible flow, only Re^* is left over.

2.6.4 The Reynolds Equation

Here, it will be more convenient to normalise w.r.t. the exit or ambient conditions, thus we put,

$$(R, \theta, \tau, H) = (r/r_a, \theta, \omega t, h/h_a) \tag{2.64}$$

where, from Equation 2.51, we have:

$$\begin{aligned} H = \frac{h}{h_a} &= 1 + \frac{H_v}{(1 - R_o)}(1 - R) \\ &= 1 + \nu_a(1 - R) \end{aligned} \tag{2.65}$$

in which, $H_v = h_v/h_a$ is the dimensionless conicity. Likewise, we define

$$(P, \wp) = (p/p_a, \rho/\rho_a). \tag{2.66}$$

Substituting Equations 2.64 and 2.66 into Equation 2.37 and rearranging, we obtain the normalised Reynolds equation:

$$\nabla \cdot (H^3 \wp \nabla P - \mathbf{\Lambda} \wp H) = \sigma \frac{\partial}{\partial \tau}(\wp H) \tag{2.67}$$

where the two dimensionless parameters $\mathbf{\Lambda}$ and σ are:

The sliding number

$$\mathbf{\Lambda} = \frac{6\mu \mathbf{V}_t r_a}{p_a {h_a}^2}.$$

The squeeze number

$$\sigma = \frac{12\mu\omega {r_a}^2}{p_a {h_a}^2}.$$

(Both of these numbers are forms of the Harrison number, see Constantinescu (1969).)

One boundary condition on Equation 2.67, at bearing exit, is always available:

$$P(1, \theta, \tau) = 1.$$

If a second boundary condition is specified, (conveniently at film entrance,)

$$P(R_o, \theta, \tau) = P_o(\theta, \tau)$$

together with the function $\partial H(\tau)/\partial \tau = \epsilon(\tau)$, then Equation 2.67 may, in principle, be solved to give:

$$P(R, \theta, \tau) = f(\mathbf{\Lambda}, \sigma, v_a, R_o, P_o, \epsilon(\tau);\ R, \theta, \tau)$$

which sums up the conditions of similarity; it is characterised by the three parameters $\mathbf{\Lambda}$, σ and v_a, in addition to the boundary conditions (R_o, P_o) and $\epsilon(\tau)$.

In the case of steady axisymmetric bearing with no relative sliding, the dimensionless pressure reduces to:

$$P(R) = f(v_a, R_o, P_o;\ R).$$

Note that the condition (R_o, P_o) may be replaced by an equivalent condition on the dimensionless flow rate, as will be apparent later.

The Restrictor Flow

This will be deferred to Chapter 5 where it will be given in conjunction with a proposed formula. For the time being, however, we may use the result of Section 2.6.2.3 to deduce that the extrapolated entrance pressure p_t may be written:

$$p_t/p_o = f(v_o, \mathrm{Re}^*, \mathrm{M}_o).$$

2.6.5 The Bearing Characteristics

The normalisation is effected, most conveniently, w.r.t. the ambient values and external bearing dimensions, which are envisaged to be available at the lowest design level. Let us, in addition, define the dimensionless bearing region as:

$$\bar{A} = \{(R, \theta) :\ 0 \le R \le 1,\ 0 \le \theta < 2\pi\}.$$

The normalised bearing characteristics may thus be written:

Load Capacity

$$W = \frac{F}{\pi r_a{}^2 p_a} = \frac{1}{\pi}\int_{\bar{A}} (P-1)\mathrm{d}\bar{A}.$$

Axial stiffness

$$K_a = \frac{k_a h_a}{A p_a} = -\frac{\partial W}{\partial H}, \qquad (H = h/h_a).$$

Entrance mass flow rate

$$\bar{\dot{m}}_o = \frac{\dot{m}_o}{(\pi p_a \rho_a h_a{}^3/6\mu)} = \Lambda_f R_o H_o P_s \Phi_e(\bar{p}_o)$$

where,

$$\Lambda_f = 12\frac{\mu r_a}{p_a h_a{}^2}\sqrt{\frac{2\kappa \mathfrak{R} T_s}{\kappa - 1}}$$

is the *feed* number.

Mass flow rate in the film

$$\bar{\dot{m}}_r = \frac{\dot{m}_r}{(\pi p_a \rho_a h_a{}^3/6\mu)} = \frac{6\mu \dot{m}_r}{\pi p_a \rho_a h_a{}^3} = -\frac{R_o H_o{}^3 \wp_0}{2\pi}\int_0^{2\pi}\left(\frac{\partial P}{\partial R}\right)_{R=R_o} \mathrm{d}\theta$$

(where we have assumed in the last expression that the film is wholly viscous.)

Friction force

Two normalisations may be adopted here. The first is w.r.t. the bearing load, i.e. the coefficient of friction:

$$\mu_f = \frac{F_f}{F}.$$

This may be normalised w.r.t. the (linear) shearing force of sliding. Thus, if we assume that the sliding is only translational, with speed V, then:

$$\bar{F}_f = \frac{F_f h_a}{\mu V A}.$$

Note that at the limit of zero sliding, the dimensionless friction force tends to a finite limit, as will be shown in Chapter 6 where a normalised coefficient of friction will also be defined.

Finally, other normalised quantities will, when necessary, be given in other parts of the text.

2.6.6 Static Similarity of Two Aerostatic Bearings

In this section, more convenient similarity conditions are deduced and used to deduce scale factors between a model and a prototype.

From the last three sections, it has become apparent (with the neglect of thermal similarity) that two aerostatic bearings will be statically similar if they share the following dimensionless numbers:

$$r_o/r_a, \qquad h_v/h_a, \qquad \mathrm{Re}_o^*, \qquad \mathrm{M}_o.$$

In that case, they will share also the following characteristics:

$$p_s/p_o, \qquad p_o/p_a, \qquad W, \qquad K_a, \qquad \bar{\dot{m}}_o, \qquad \bar{\dot{m}}_r.$$

Consequently, it may be shown, by interchanging some of these dimensionless numbers, that the similarity conditions may be replaced by:

$$r_o/r_a, \qquad h_v/h_a, \qquad \Lambda_f, \qquad p_s/p_o.$$

This last set has the advantage that it may be fully determined from the design values.

In order to simplify presentation of similarity, we shall, without loss of generality, preclude the gas properties, as well as the temperatures, from the analysis. (Note that commonly used gases do not differ appreciably in properties.) In this case, the only absolute parameters at our disposal are (r_a, h_a, p_a). Following in the footsteps of Constantinescu (1969), let us assume now that model and prototype differ by the following α-scale factors:

Model	r_a	h_a	p_a
Prototype	$\alpha_r r_a$	$\alpha_h h_a$	$\alpha_p p_a$

In order that the two bearings be similar, r_o, h_v, p_s must be scaled with the same factors, and, in addition Λ_f must be the same. This last condition yields:

$$\frac{\alpha_r}{\alpha_p \alpha_h^2} = 1. \tag{2.68}$$

Also, the bearing characteristics will be:

Model	F	$\dot{m}$	k_a
Prototype	$\alpha_r^2 \alpha_p F$	$\alpha_h^3 \alpha_p^2 \dot{m}$	$(\alpha_r^2 \alpha_p/\alpha_h) k_a$

The following two possibilities will now be examined:

- If, for the purpose of experimentation, we wish to magnify the film thickness by a factor α_h, then from condition 2.68, we must have:

  $$(\alpha_r/\alpha_p) = \alpha_h^2$$

  which means that we must increase the diameter of the prototype and/or decrease the ambient pressure. As an example, if $\alpha_h = 10$, and $\alpha_r = 5$, then $\alpha_p = 0.05$, or that the experimental ambient pressure must be 0.05 bar. This is how the experiment reported in Mirolyubov and Shashin (1982) was carried out.
- If we assume that the ambient pressure remain unchanged, and wish to examine the influence of the bearing size on the characteristics, then we set $\alpha_p = 1$ to obtain:

  $$\alpha_h^2 = \alpha_r$$

Let us now assume that the prototype bearing has double the area of the model, i.e. $\alpha_r = \sqrt{2}$, then we have:

Model	r_a	h_a	F	$\dot{m}$	k_a
Prototype	$\sqrt{2}r_a$	$2^{1/4}h_a$	$2F$	$2^{3/4}\dot{m}$	$2^{3/4}k_a$

In other words, doubling the bearing size places relatively more stringent tolerance requirement on the gap; it doubles the load capacity, while less than doubling the air consumption; and it less than doubles the stiffness. Thus, in order to make a stiff (though slightly less economic) bearing arrangement, we must use several small bearings rather than one big bearing, which will be at the expense of more air consumption.

Finally, other similarity cases can be considered, and the influence of gas properties can also be included. Similarity in the cases of tilt, sliding, and dynamic behaviour will be considered in the respective chapters.

2.7 Methods of Solution

A mathematical model is, naturally, only useful if it can yield a solution. Therefore, in order to complete our topic, an overview will be given of the methods of solution used in this book, as well as other methods available in the literature.

First, a word should be said about the nature of the problems at hand. Both the flow (B-L) problem and the pressure (Reynolds) equation are, in the general case, nonlinear. However, in certain limiting cases, such as the steady axisymmetric Reynolds equation, analytical solutions are possible. In some other cases, linearisation of the problem may be adequate. However, in general, nonlinearity makes the problem difficult to handle; the more so when the constituent sub-problems, which are of differing nature, have to be matched, or solved simultaneously.

Methods will be classified into three categories: analytic, semi-analytic and purely numerical. By the second category we shall mean that the problem is first reduced, by an analytical procedure, to a simpler one that is then solved numerically.

2.7.1 Analytic Methods

These are only applicable to the steady state axisymmetric or two-dimensional Reynolds equation, yielding the pressure distribution in closed form. Although this is a simple case, it forms the starting point of several important perturbation solutions, thus simplifying the solution considerably. Some compensator flow models, (e.g. the nozzle formula,) are also amenable to analytic solution. However, the complete bearing problem, which involves matching of the two solutions, can be achieved only numerically; usually by a Newton–Raphson or a false position method. Nor can the load capacity (integral) generally be calculated analytically.

2.7.2 Semi-analytic Methods

A considerable number of the solutions, presented in this work, are of this type. The method proposed in Chapter 3, for the solution of the entrance flow problem, is of this type, so are the perturbation and series expansion solutions in the other parts. While offering sufficient accuracy over a wide design range, these methods are easy to derive and to programme. The problem is thereby reduced to a set of first order ordinary differential equations, that can be solved by forward integration techniques. However, the most important feature of these methods is that they enable the study of limiting cases, such as damping at zero frequency, or friction force at zero sliding: in general, a better insight into the problem is afforded than in the purely numerical case.

2.7.3 Purely Numerical Methods

Numerical methods become inevitable for bearing configurations which are more complex than the one considered in this chapter. Air bearings literature includes a large number of numerical solutions to the Reynolds equation. In the first instance, mainly finite difference difference methods, and to a lesser extent, variational methods have been used. These (and others) are reviewed extensively in (Castelli and Pirvics 1968, Gross et al. 1980 ch. 7). More recently, finite element methods have been adapted to the problem, as for example in Booker and Huebner (1972; Reddi and Chu (1970);). The basic difficulty encountered in applying such procedures is the nonlinearity of the compressible Reynolds equation. Thus while the solution of linear equations is somewhat straightforward, with nonlinear equations extra steps are needed to ensure the convergence of the procedure and establish or ascertain uniqueness of the solution. These steps include the various methods of relaxation and over-relaxation, Newton–Raphson, parameter-perturbation, etc.

Those different methods of solution will be presented in various parts of this book in connection with specific modelling and simulation cases.

2.8 Conclusions

The bearing configurations of interest have been defined and idealised with aid of simplifying assumptions. Analysis and various levels of simplification of the flow problem have lead to the formulation of two models for that problem. The first, being based on the BL equations, is proposed for the solution of the steady axisymmetric flow configuration, thus being able to account, more closely, for the entrance inertia effects. The second, being based on the viscous flow assumptions, is none else than the Reynolds equation, which is expected to be sufficiently valid over the greatest portion of the bearing region. This model is however not complete without a compensator or restrictor model, necessary to match the feed/entrance flow with the viscous flow. A general form of the lumped parameter restrictor model has therefore also been proposed. Various ways of normalising the resulting models have been suggested, and used to deduce similitude criteria based on dimensionless model parameters. Finally, solution methods to be used in this work, as well as other methods generally found in the literature, have been briefly reviewed.

References

Al-Bender, F., (1992). *Contributions to the design theory of circular centrally fed aerostatic bearings*. PhD thesis, KU Leuven.

Ames, W.F., (1965) *Nonlinear Partial Differential Equations in Engineering*. Academic Press.

Booker, J.F. and Huebner, K.H, (1972) Application of finite element methods to lubrication: An engineering approach. *ASME - Journal of Lubrication Technology* 313–323.

Castelli, V. and Pirvics, J., (1968). Review of numerical methods in gas bearing film analysis. *ASME - Journal of Lubrication Technology* 777–792.

Churchill, R.V., (1948) *Introduction to Complex Variables and Applications* 1 edn. McGraw-Hill Book Company, Inc.

Constantinescu, V.N., (1969) *Gas Lubrication*. ASME.

Dowson, D., (1979) *History of Tribology*. London: Longman.

Gross, W.A., (1962) *Gas Film Lubrication*. John Wiley & Sons.

Gross, W.A., Matsch, L.A., Castelli, V., et al. (1980). *Fluid Film Lubrication*. New York: John Wiley & Sons.

Lamb, H., (1932). (Re-issue 1974) *Hydrodynamics*. Cambridge University Press.

Massey, B., (1968). *Mechanics of fluids*. London: D. Van Nostrand Company Ltd.

Milne-Thomson, L.M., (1968) (Reprint 1974.) *Theoretical Hydrodynamics* 5th edn Macmillan.

Mirolyubov, I.V. and Shashin, V.M., (1982). Large-scale simulation of flows encountered in gas lubrication. *Fluid Dynamics 17*: 1–6. Translated from Izvestiya Akademii Nauk SSSR, Mekhanika Zhidkosti i Gaza, No. 1, pp. 3–9, January–February, 1982.

Parker, J.D., Boggs, J.H. and Blick, E.F., (1969). *Introduction to Fluid Flow and Heat Transfer*. Addison Wesley Publishing Company, Inc.

Reddi, M.M. and Chu, T.Y., (1970). Finite element solution of the steady state compressible lubrication problem. *ASME - Journal of Lubrication Technology* 495–503.

Schlichting, H., (1968) *Boundary-Layer Theory* 6th edn New York: McGraw-Hill.

Vallentine, H.R., (1967) *Applied Hydrodynamics*. Butterworths, London.

3

Flow into the Bearing Gap

3.1 Introduction

We have seen, in Chapter 2, that the bearing flow problem consists of a feed flow problem and a channel, or film flow problem. A schematic of this situation is depicted in Figure 3.1, where two cases are considered: (a) pressure-fed flow, as in EP bearings, and (b) shear-fed flow, as in self-acting bearings. A combination of those two situations may also arise, such as in "hybrid" bearings. As also indicated on the figure, we propose to treat this problem by considering the flow development (i) upstream of the gap entrance, (ii) just downstream of the gap entrance up to (iii) the attainment of viscous flow, slightly further downstream of gap entrance, where the Reynolds equation becomes applicable. (Obviously, regions (i) and (ii) are characterised by a certain significance or domination of inertia forces in the flow.)

The film problem, more particularly the pressure distribution in the film, which is of main interest to us, is determined by solving the Reynolds equation (see Chapter 4), for which we need to specify two auxiliary conditions: either (i) the pressures at two distinct points along the film (usually at inlet and exit) or alternatively, (ii) the flow rate in addition to the pressure at one point along the film. Usually the flow will, in most cases, have a dominantly viscous character at exit. Thus, to a very good approximation, the pressure there can be taken to be ambient; i.e. the environmental pressure prevailing at the bearing exit, which is a known quantity. Determination of the pressure at another point in the film, or alternatively, the flow rate, however, is not always an easy task, except in simple cases, where the inlet pressure, for instance, is close to the supply value or the ambient value.

Generally then, we need to solve the whole flow problem, from far upstream of the entrance to the exit, in order to determine the inlet pressure or the flow rate, or as is most often the case, a relationship between them: the so-called restrictor model. Here again, we are confronted with the issue of specifying appropriate auxiliary conditions, which might not be easily available. In conclusion, we shall need to develop some appropriate devices to find a solution to this problem.

Let us note here that it is nowadays possible to deal with this problem using direct CFD techniques. However, there are three issues with this approach. They are: (i) that it is time consuming; (ii) that it does not entirely overcome the problem of correct specification of the boundary conditions; and (iii) that it furnishes little insight into the problem. Thus, the semi-analytical modelling approach remains a favourable one.

Let us now, with this understanding, return to Figure 3.1, and consider the particularities and the shared features of cases (a) and (b). In case (a) representing pressure-fed flow, we assume the flow to originate from a plenum stagnating at a pressure p_s (thus, with zero velocity). As discussed in Chapter 2, we may assume that the flow expands from that plenum in accordance with Euler's equation, to reach the gap entrance with a static pressure p_0 and enter with quasi-uniform velocity. In this chapter, we consider how this flow develops downstream of the entrance until it attains a viscous character, not very far from the entrance, as we shall see. Furthermore,

Air Bearings: Theory, Design and Applications, First Edition. Farid Al-Bender.

Companion website: www.wiley.com/go/AlBender/AirBearings

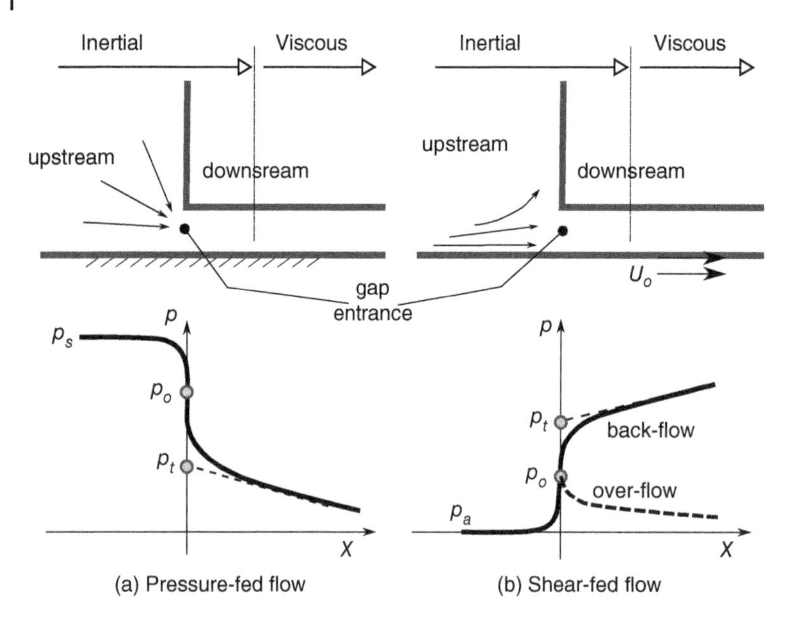

Figure 3.1 Schematic and terminology of the entrance problem. (Not to scale.)

if we extrapolate this viscous-flow pressure curve back to the entrance, we define a fictitious entrance pressure p_t. Our main objective will be to determine the pressure difference $p_s - p_t$, or alternatively, a relationship between the flow rate in function of p_s and p_t, via a so-called "discharge coefficient".

As for case (b), representing shear-fed flow, which more or less obeys boundary-layer equations upstream of the entrance, we see that the pressure builds up, from some ambient value, to the value p_o at the entrance, whence the flow further develops to viscous character, showing further pressure increase (solid line), where we can, in a similar way as in the previous case, correlate the inlet pressure and the flow. The case indicated by a dashed line pertains to the situation when the fluid enters the gap with a velocity distribution that is in excess of the Couette value, as we shall see later.

In the following, we shall treat each of these two cases separately. Let us indicate also here that for both situations, we shall assume parallel bearing, or channel gap, since (i) we are interested in the mechanism of flow development and (ii) for the case of the plain (as opposed to the radial), the flow develops or settles down to viscous character after only a very short distance from the gap entrance so that we can consider the flow rate, rather than the gap variation, as the determining factor.

3.2 Entrance to a Parallel Channel (Gap) with Stationary, Parallel Walls

In the literature, such a problem is commonly referred to as the (channel) "entrance" or "entry" problem. Our first aim will be to develop and establish a general and efficient method for the solution of channel entrance problems, in the pressure-induced flow (EP bearing) situation. (Application of this method actually to circular, centrally fed, bearing problems, by way of validation, will be given in Chapter 5).

Interest in the problem of plane and radial channel flow is, however, not confined to fluid film bearings. It is a problem that has occupied a position of special importance in fluid mechanics both on practical and academic levels. Practical applications include flow intake devices such as turbines, plane ducts, radial diffusers, hydrostatic bearings etc. Theoretical interest includes: the non-linearity of the equation of motion and its associated singular behaviour, and the fact that such flows exhibit most of the diverse aspects of laminar flow such as potential flow with build-up of thin boundary layers, separation and reverse flow, and Poiseuille or "creeping" flow.

Problems of "entrance" of uniform, *pressure-induced* flow into pipes and plane channels have for a very long time been the subject of extensive research, reported in hundreds of papers. Treatments to be found in the open literature fall into two basic categories: (1) semi-analytic solutions, and (2) purely numerical solutions. By the first is meant that the partial differential equations of motion are reduced, by analytic means, to a set of ordinary differential equations. This has, hitherto, involved one simplifying assumption or another, making all those methods approximate. Three approaches may be discerned under this category.

The first is that due to Schlichting (1968) in which one series approximation at the channel entrance (based on the Blasius solution), and another far downstream (based on creeping flow solution), are joined at an appropriate point in the middle to give the complete solution. Although Sclichting's solution has become a standard one for comparison, its validity and accuracy are questionable due to the nature of the approximations made, (practically amounting to linearisation). Van Dyke (1970) indicates a paradox relating to the displacement effect in the upstream assumption. One might add also that the greatest portion of the solution is in fact contributed by the linearised downstream approximation. Roidt and Cess (1962) applied Schlichting's method to magnetohydrodynamic flow, calculating, at the same time, an extra term in the upstream expansion which seems to make a difference of about 10 percent in the entrance length. In short this method is neither handy nor sufficiently accurate.

The second is an integral approach; extended from von Kármán's momentum integral method for external flows (Schlichting 1968), (see also Chapter 4). An assumed velocity function (mostly, a uniform centre core with a parabolic boundary layer), is required to satisfy *macroscopic* mass, momentum and mechanical energy conservation, (obtained by integrating the respective equations over the cross section of the flow). Campbell and Slattery (1963), perhaps the first to employ it with the mechanical energy equation included, reported good results from this method applied to circular pipe flow. Later, Chen (1973) extended it to include approximately the Reynold's number dependence. This approach, which has always claimed good results, is a rough or simplified version of the one that will be developed and adopted in this work.

The third approach is that of local linearisation of the convective terms in the momentum equations. The resulting, locally linearised N-S equations are by no means easy to solve, but they are linear. Han (1960) applied it to the boundary-layer equations for rectangular duct flow, and Narang and Krishnamoorthy (1976) applied it to the complete Navier–Stokes equations for plane channel. This approach is expected to be more accurate than the previous two, at the expense of a more complex solution.

In summary, hitherto available analytic methods, which are preferable to numerical ones, are approximate to an unknown degree. Comparisons of solutions, obtained by the various methods reviewed above, (mostly done against one another—there being very little dependable experimental data), show only qualitative agreement.

Furthermore, these methods have found extensions to the case of radial axisymmetric channel flow. We cite here e.g. (Geiger et al. 1964; Mori and Miyamatsu 1969; Murphy et al. 1978; Savage 1964; Woolard 1957), but shall defer their discussion to Chapter 5. One of these methods, however, advanced by Tang (1968) (and receiving little attention), suggested a new method of attack, but included some anomalies. In the following, we shall rework this idea into an efficient method of solution, using a "separation of variables" technique, or a method of characteristics (Al-Bender and Van Brussel 1992).

3.2.1 Analysis of Flow Development

On the basis of the results of Chapter 2, we shall consider that the BL equations are adequate in characterising a wide class of internal, "narrow" passage, flows, (see also (Constantinescu 1969)). That is not to say, however, that the solution methods commonly used for external flows, for whom boundary-layer theory was devised in the first place, are directly applicable to internal flows. This is so because, although external and internal flows share the same equations, they differ basically in auxiliary conditions. The differences may be summarised as follows:

- In external flows, it is assumed that:
  1. the potential velocity field, far away from the surface, is given;

2. the velocity in the boundary-layer approaches the potential flow velocity at an infinite distance from the boundary (wall).

- In internal flows, it is assumed that:
  1. the integrated continuity equation, which represents the macroscopic conservation of mass, is given;
  2. the velocity distribution is symmetric about the mid-plane between the two walls.

Since the BL equations are non-linear, different auxiliary or boundary conditions lead to different type of solutions. For the purpose of simplicity, the analysis will initially be confined to plane laminar incompressible flow between parallel plates. The results are later easily extended to radial (axisymmetric) polytropically-compressible flow between mildly curved plates.

Referring to Figure 3.2, we have the following steady-state, continuity and momentum, equations respectively:

$$\frac{\partial u}{\partial x} + \frac{\partial v}{\partial y} = 0 \tag{3.1}$$

$$u\frac{\partial u}{\partial x} + v\frac{\partial u}{\partial y} = -\frac{1}{\rho}\frac{dp}{dx} + \nu\frac{\partial^2 u}{\partial y^2} \tag{3.2}$$

where, $u = u(x, y)$, $v = v(x, y)$ and $p = p(x)$ (by virtue of the boundary-layer theory simplification). The density ρ and the kinematic viscosity $\nu = (\mu/\rho)$ are assumed constant.

Auxiliary conditions, at every station x, are:

$$y = 0, h \; : \; u = 0\,, v = 0, \; \text{(impermeability and no slip)},$$
$$y = h/2 \; : \; \partial u/\partial y = 0, \; \text{(symmetry)}.$$

(Further, the flow problem would be well-defined if the velocity distribution were prescribed at any one given section x_0, viz. $u(x_0, y) = f_0(y)$). Introducing the stream function ψ, defined by:

$$\frac{\partial \psi}{\partial y} = u, \qquad \frac{\partial \psi}{\partial x} = -v,$$

(which satisfies the continuity equation identically), and applying the von Mises transformation (Schlichting 1968):

$$(x, y) :\rightarrow (\xi, \psi) = (x, \int_0^y u dy)$$

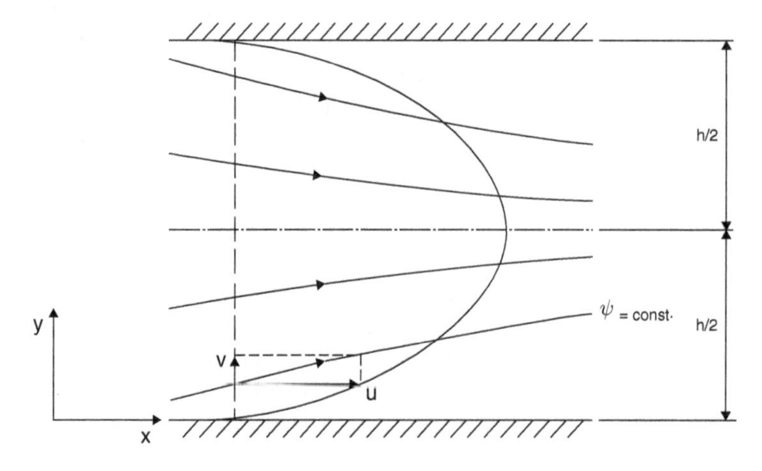

Figure 3.2 A method of "separation of variables" for the solution of laminar boundary layer equations of narrow channel flows. Journal of Tribology 114, 630–636. 1992 by permission of ASME.

which yields

$$\frac{\partial}{\partial x} = \frac{\partial}{\partial \xi} - v\frac{\partial}{\partial \psi}$$

$$\frac{\partial}{\partial y} = u\frac{\partial}{\partial \psi},$$

so that

$$u\frac{\partial u}{\partial x} + v\frac{\partial u}{\partial y} \equiv u\frac{\partial u}{\partial \xi}.$$

With $\psi(x,0) = 0$ and $\psi(x,h) = 2\psi_m$ arbitrarily chosen, Eq. 3.2 is rewritten:

$$u\frac{\partial u}{\partial \xi} = -\frac{1}{\rho}\frac{dp}{d\xi} + \nu u\frac{\partial}{\partial \psi}(u\frac{\partial u}{\partial \psi}) \tag{3.3}$$

with boundary conditions,

$$u(\xi, 0) = 0,$$

$$\frac{\partial u}{\partial \psi}(\xi, \psi_m) = 0.$$

Integrating the continuity equation, Eq. 3.1 over $0 \leq y \leq h$, (noting that $v(x,h) = 0$ by virtue of symmetry), we get:

$$\frac{d}{dx}(\int_0^h u dy) = 0,$$

or

$$\frac{d}{dx}(\psi_m) = \frac{d}{dx}(h\bar{u}/2) = 0,$$

where, $\bar{u} = (1/h)\int_0^h u dy$ is the average velocity of the flow.

The variables are first normalised by defining:

$$X = x/h = \xi/h,\ Y = y/h,\quad \Psi = \psi/\psi_m,$$
$$U = u/\bar{u},\qquad P = p/\rho\bar{u}^2,\ \mathrm{Re} = \bar{u}h/\nu,$$

whereupon, Eq. 3.3 and its boundary conditions are rewritten:

$$\mathrm{Re}(U\frac{\partial U}{\partial X} + \frac{dP}{dX}) = U\frac{\partial}{\partial \Psi}(U\frac{\partial U}{\partial \Psi}),\qquad (0 \leq \Psi \leq 1) \tag{3.4}$$

with,

$$U(X, 0) = 0,\qquad \frac{\partial U}{\partial \Psi}(X, 1/2) = 0. \tag{3.5}$$

We now affect a separation of variables by the assumption:

$$U(X, \Psi) = q(X)g(\Psi) \tag{3.6}$$

where g represents the the normalised profile function of the velocity, while q is its maximum value or amplitude. Applying the conditions 3.5, for $q \neq 0$, gives

$$g(0) = 0,\qquad g'(1/2) = 0,\qquad g(1) = 1 \tag{3.7}$$

Before proceeding further, it would be convenient to rewrite Eq. 3.6 in terms of the space coordinates and to apply the integrated continuity equation to it. Thus,

$$U(X, Y) = U(X, \Psi(X, Y)) = q(X)G(X, Y) \tag{3.8}$$

where,

$$G(X, Y) = g(\Psi) \tag{3.9}$$

(Note that $G(X, 0) = 0, G(X, 1) = 1, G_Y(X, 1/2) = 0$ by virtue of conditions 3.7). Integrating Eq. 3.8 with respect to Y on $0 \leq Y \leq 1$ and applying continuity, we have:

$$q(X)I(X) = 1 \tag{3.10}$$

where,

$$I(X) = \int_0^1 G(X, Y)dY$$

is the normalised profile integral function. We now proceed to substitute the form 3.6 into Eq. 3.4 to obtain:

$$\mathrm{Re}(qq'g^2 + P') = q^3 g(gg')' \tag{3.11}$$

where the primes denote differentiation of the functions with respect to their respective variables. It may be interesting, at this stage, to explain the physical significance of this last equation. Thus, if Eq. 3.11 is transformed back to the space coordinates (X, Y), then:

$$\mathrm{Re}(qq'G^2(X, Y) + P') = q\frac{\partial^2}{\partial Y^2}G(X, Y) \tag{3.12}$$

that is to say that the assumption of separability is equivalent to the assumption that the velocity profile be *locally* invariant in the stream line direction. Equation 3.11 is separable if and only if (Ames 1964):

$$P' = -\alpha qq' \tag{3.13}$$

where α is a parameter that is independent of Ψ. (We shall see later that α would be dependent on the separation parameter.) With the aid of Eq. 3.13, Eq. 3.11 may now be rewritten:

$$(\mathrm{Re}/4)\frac{q'}{q^2}(X) = \lambda = \frac{g(gg')'}{(g^2 - \alpha)}(\Psi) \tag{3.14}$$

where, λ is the separation parameter. Sufficient boundary conditions are available to solve the right-hand member of Eq. 3.14 in order to determine the *eigenvalue* or *characteristic* λ. Thus, we have:

$$g(gg')' = \lambda(g^2 - \alpha) \tag{3.15}$$

with boundary conditions 3.7. We note that Eq. 3.15 has a singularity, at $\Psi = 0$, which, however, may be overcome by considering, in lieu of g, its square:

$$H = g^2 \tag{3.16}$$

which satisfies the same boundary conditions as g. For the moment, we shall consider, without loss of generality, only the case with no reverse flow, i.e. where $\Psi \geq 0$ on $0 \leq Y \leq 1$; and g is thus a single valued monotonic function on $0 \leq \Psi \leq 1$. Substituting Eq. 3.16 into Eq. 3.15 and making use of the relation $f'' = f'df'/df$, we get:

$$\frac{d}{dH}(H')^2 = 4\lambda(\sqrt{H} - \frac{\alpha}{\sqrt{H}})$$

which upon integration once with respect to H, noting that $H' = 0$ when $H = 1$, gives:

$$(H')^2 = 4\lambda[\frac{2}{3}(H^{3/2} - 1) - 2\alpha(H^{1/2} - 1)]$$

or, with $H' \geq 0$, we may write:

$$H' = 2\sqrt{\lambda}\sqrt{\frac{2}{3}(H^{3/2} - 1) - 2\alpha(H^{1/2} - 1)} \tag{3.17}$$

Equation 3.17 may now be rearranged as a definite integral:

$$\int_0^{H(\Psi)} \frac{dH}{2\sqrt{\lambda}\sqrt{\frac{2}{3}(H^{3/2}-1)-2\alpha(H^{1/2}-1)}} = \int_0^{\Psi} d\Psi = \Psi$$

Resorting to relation 3.16,we may equivalently write,

$$\int_0^{g(\Psi)} \frac{gdg}{\sqrt{\lambda}\sqrt{\frac{2}{3}(g^3-1)-2\alpha(g-1)}} = \Psi \tag{3.18}$$

or,

$$\int_0^{g(\Psi)} \frac{tdt}{\sqrt{\lambda}\sqrt{\frac{2}{3}(t^3-1)-2\alpha(t-1)}} = \Psi \tag{3.19}$$

where t is a dummy variable. Equation 3.19 defines Ψ as a function of g and the parameters λ and α. Consequently, the latter, i.e. the eigenvalues (or characteristics) of Eq. 3.15, may be determined if the remaining boundary condition (viz. $\Psi = 1; g = 1$) is substituted in Eq. 3.19:

$$\int_0^1 \frac{tdt}{\sqrt{\lambda} f(t,\alpha)} = 1$$

where we have written for convenience,

$$f(t,\alpha) = \sqrt{\frac{2}{3}(t^3-1)-2\alpha(t-1)}$$

This gives,upon rearrangement:

$$\lambda = [\int_0^1 \frac{tdt}{f(t,\alpha)}]^2 sgn(\lambda) \tag{3.20}$$

which defines a functional relationship between λ and α. Unfortunately, the circumstance is complicated by the fact that the definite integral, appearing in Eq. 3.20, is an elliptic one, such that it cannot be evaluated analytically but rather numerically for any possible value of α (see, for example (Tang 1967)). For our purpose, however, it would suffice to say that it may be shown that Eq. 3.20 defines a one-to-one relationship between λ and α, on $\alpha \in (-\infty, 1/3) \cup (1, \infty)$, which may be obtained, e.g. by numerical integration, in the prescribed interval of α.

We conclude from this analysis that the differential Eq. 3.15 has a continuous set of eigenvalues λ, each of which determining, (or is determined by), a corresponding value of α. Furthermore, each value of λ will yield, through Eq. 3.19 , an eigenfunction $g(\Psi; \lambda)$; and, simultaneously, a solution to the left-hand member of Eq. 3.14. If our original partial differential equation, Eq. 3.4, were linear, we would have been able to superpose these solutions to obtain the sought for general solution. However, since this is not the case, we shall have to pursue a different line of argument which invokes the integrated continuity Eq. 3.10. Thus, returning to Eq. 3.19, and using Eq. 3.9, we can write, at any station X, and for any value of λ,

$$\int_0^{G(X,Y)} \frac{tdt}{\sqrt{\lambda} f(t,\alpha)} = \Psi(X,Y)$$

which, upon differentiating with respect to Y and using Eq. 3.8, yields:

$$\frac{\partial}{\partial Y}\int_0^{G(X,Y)} \frac{tdt}{\sqrt{\lambda} f(t,\alpha)} = \frac{\partial \Psi}{\partial Y} = U = q(X)G(X,Y)$$

or, after performing the differentiation of the left-hand side and cancelling out G from both sides (identically),

$$\frac{1}{\sqrt{\lambda} f(G,\alpha)} \frac{\partial G}{\partial Y} = q \tag{3.21}$$

The velocity profile function G may now be obtained implicitly by rearranging and integrating Eq. 3.21, thus

$$\int_0^{G(X,Y)} \frac{dt}{\sqrt{\lambda f(t,\alpha)}} = qY \tag{3.22}$$

(which may be used to plot G at every station X where q and λ are given). In particular, substituting ($Y = 1; G = 1$) in Eq. 3.22 should provide the missing relation between λ and q or I, thus:

$$\int_0^1 \frac{dt}{\sqrt{\lambda f(t,\alpha)}} = q = \frac{1}{I} \tag{3.23}$$

which means that the velocity amplitude q, or its profile integral function I, are *uniquely* determined by the value of the separation parameter λ. This is, in effect, another statement of the *compatibility* requirement (see (Schlichting 1968, p. 142)). Thus, allowing λ to take values in its domain of definition, a functional (one-to-one) relationship is obtained between I and $\alpha\lambda$ by evaluating the integral on the left-hand side of Eq. 3.23 (in terms of Jacobian elliptic functions as in (Tang 1967), or simply by numerical integration, as was done in our case).

Finally, returning to the left-hand member of Eq. 3.14, Eq. 3.13, the two functions connecting λ, α and I, and Eq. 3.10, we attain the following system of equations:

$$\left.\begin{array}{lcl} \mathrm{Re}(q'/q^2) & = & \lambda \\ P' & = & -\alpha q q' \\ \alpha & = & \alpha(\lambda) \\ I & = & I(\lambda) \\ qI & = & 1 \end{array}\right\} \tag{3.24}$$

in which α, λ and I may, in theory, be eliminated to give:

$$q' = f_1(q), \qquad P' = f_2(q). \tag{3.25}$$

The system 3.24 constitutes a first order initial value problem once the value of any one of its variables (q, q', P', I) is specified at a given section X_0. This is usually done in an arbitrary way; from physical considerations, etc., (see Chapter 5). Consequently, the system may be integrated by a suitable shooting technique starting at X_0.

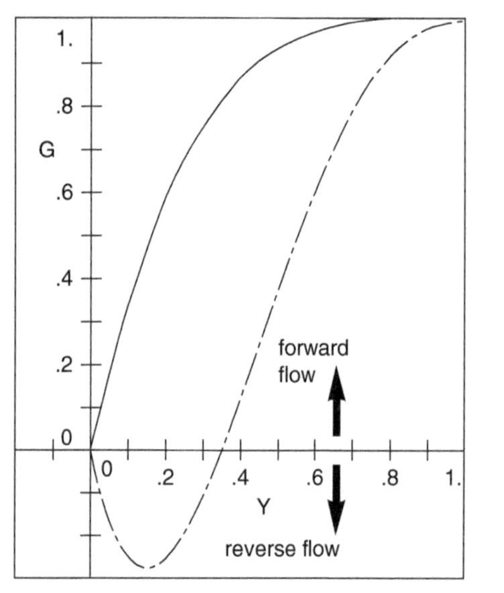

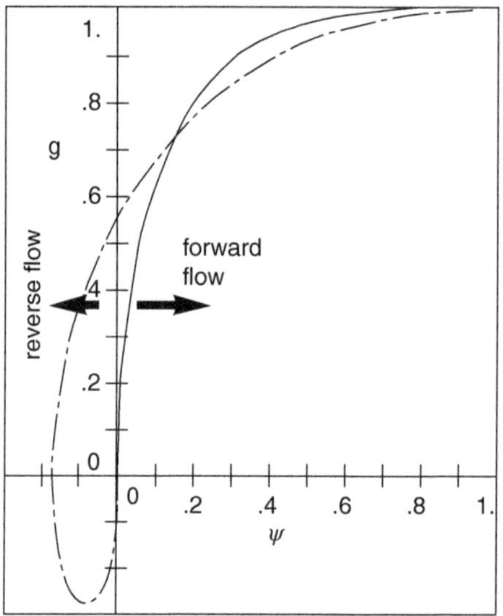

Figure 3.3 Velocity profile functions in forward and reverse flow. Source: Al-Bender and Van Brussel, 1992 by permission of ASME.

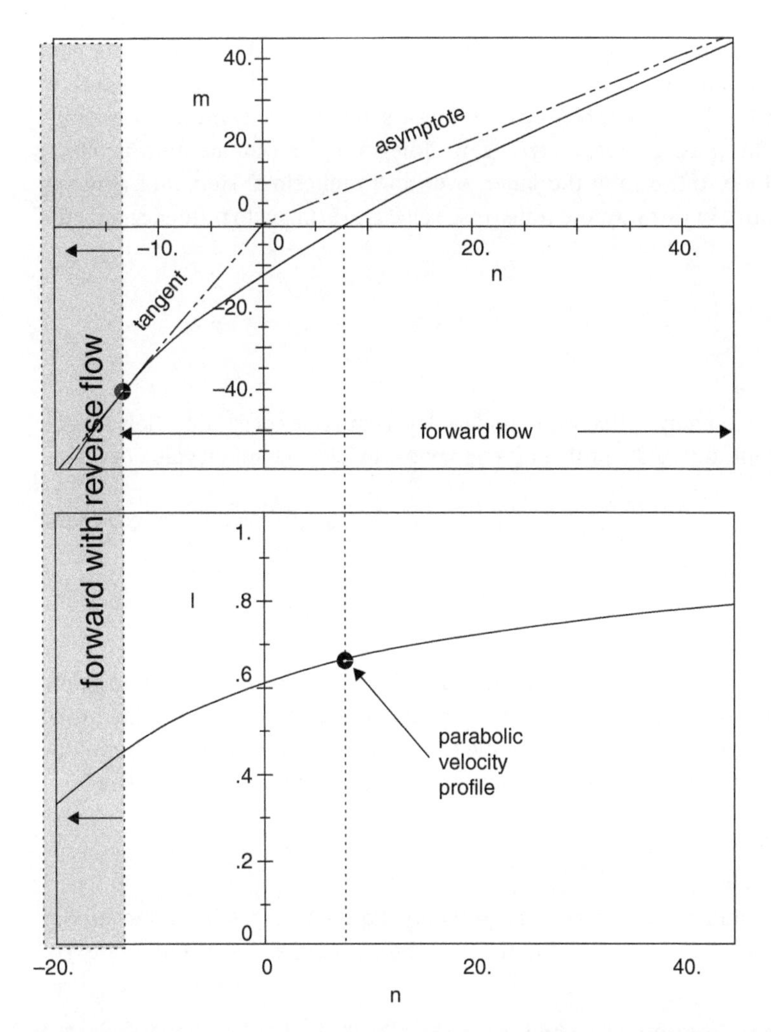

Figure 3.4 $m : \to n : \to I$. Source: Al-Bender and Van Brussel, 1992 by permission of ASME.

Finally, the foregoing analysis may be extended to include the case of reverse flow, viz. when Ψ and g are allowed to take negative values on $0 \leq Y \leq 1$, as shown in Figure 3.3. This represents a branch in the functions $\lambda((\alpha\lambda))$ and $I((\alpha\lambda))$ making them double valued in a certain range. This is an important case when diverging flow, (plane or radial), is considered.

However, $(\alpha\lambda)$ remains single valued when considered as a function of I. It would, therefore, be more convenient, in the general case, to employ the following set of related parameters[1]:

$$I, \qquad m(I) = 4\lambda/I^2, \qquad n(I) = 4\alpha\lambda/I^2 \tag{3.26}$$

which are then connected by the one-to-one relationship shown in Figure 3.4.

The reason for this choice of parameters is made evident if we rearrange Eq. 3.12 and substitute for the coefficients from the system 3.24 and the parameters 3.26 to obtain:

$$\frac{\partial^2}{\partial Y^2} G(X,Y) = m(I)G^2(X,Y) - n(I) \tag{3.27}$$

1 Tang's notation (Tang 1968)

from which the previously deduced relationship between m, n and I may also be achieved by integrating with respect to Y and applying the boundary conditions on G. However, Eq. 3.27 is true only if the procedure of separation of variables is valid. (Note that the solution may be viewed as follows. If the velocity profile is assumed to be *fixed*, at a certain station X, then the transverse velocity is zero. Integrating the momentum equation yields the gradient of the velocity amplitude. Integrating the latter, over an infinitesimal step, and applying continuity, yields the new profile, and so forth. In this way, the transverse velocity v is taken to influence only the profile of u.)

3.3 Results and Discussion

Summarising, it has been shown in the previous analysis that the boundary-layer equations of channel flow may be solved by separating the velocity into an amplitude and a profile function, which yields the initial value problem:

$$\left.\begin{aligned} q' &= (1/\mathrm{Re})m(I) \\ P' &= -(q/\mathrm{Re})n(I) \end{aligned}\right\} \tag{3.28}$$

with,

$$qI = 1, \qquad m :\to n :\to I :\to m$$

The initial condition may be specified, according to the situation, in the form $q(X_0) = q_0$ or equivalent information. However, we should note that there is a unique velocity profile corresponding to each value, i.e. not any initial velocity distribution is permissible. This is owing to the *coupling* of the transverse velocity v with the main flow velocity u, which is inherent in the nature of the BL equations. Some particular cases will now be considered.

3.3.1 Limiting Cases

First, in comparison with the "similarity" solutions of boundary-layer theory, Eqs. 3.27 and 3.28 lead to similar solutions in the following two extreme cases, which correspond to the two singular points ($\lambda \to 0$ and $\lambda \to \infty$) of the solution:

(i) if $m \to 0$, $n \to 8$, the creeping flow solution is obtained: $q = 3/2$, $P' = -12/\mathrm{Re}$, $G(X, Y) = Y(2 - Y)$ (parabola), $I = 2/3$.
(ii) if $m \to n \to \infty$, (see Figure 3.4), then, by putting

$$G(X, Y) = f(\eta), \qquad \eta = \sqrt{\frac{m(X)}{2}}Y$$

we get,

$$f''(\eta) - f^2(\eta) + 1 = 0,$$

with,

$$\eta = 0 \ : f = f' = 0\,; \qquad \eta = \infty \ : f' = 1$$

which is the same equation obtained for sink flow in boundary-layer theory (Schlichting 1968).
This similarity solution enables one also to calculate the function $I(m)$ near the singular limit of $I = 1$ by:

$$I = 1 - \sqrt{\frac{m_r}{m}}(1 - I(m_r)),$$

where, m_r is some sufficiently large reference value of m.

Secondly, if the problem is solved for the case of plane channel flow then:

(1) if the initial velocity profile is taken to be parabolic, corresponding to case (i) above, then it will remain so at any other section. That is to say that Poiseuille flow represents a *stable* limit (or attractor) to the flow;
(2) conversely, if the initial velocity profile is *rounder* than parabolic (i.e. $I > 2/3$), then the parabolic profile will be attained only *asymptotically*.

3.3.2 Method of Solution

In general, the method may be regarded as a step-by-step continuation (Schlichting 1968): with the velocity given at one section X_0, Eqs. 3.28 are used to determine the velocity at the section $X_0 + \Delta X$. As such, the accuracy of the solution will be governed by two factors. The first is the accurate determination of the functions $m(I)$ and $n(I)$, which represent the gradients at each point. Of these, as large a number of values as possible may be calculated and the solution of a channel flow problem. The second factor, being bounded by the first, is the size of integration steps ΔX, which, in addition to general refinement, may be made unequal, e.g. inversely proportional to $max\{|m|, |n|\}$.

3.3.3 Determination of the Entrance Length Into a Plane Channel

This method is now used to solve the "entrance" problem for incompressible plane channel flow, which may be stated as follows: with the main flow velocity assumed to be uniform and parallel at section $X = 0$, it is required to determine the development of the velocity and the pressure downstream of that point. (Physically, this would correspond to the hypothetical situation of channel entrance following a rapid well-rounded contraction(Van Dyke 1970).)

We shall use I as an indication of the profile, (which is, in turn, an indication of the relative influence of the inertial and the viscous forces), the value $I = 2/3$ corresponding to a parabolic profile. This is a more general characterisation of the flow than q when the flow area is not constant. Thus, for this example, the initial value is $I(0) = 1$ (or $q(0) = 1$); for P we may choose any arbitrary reference, e.g. zero. The singularity of Eqs. 3.28 at $I = 1; q = 1$, evidenced by the fact that $m \to n \to \infty$, is easily overcome by taking $I(0)$ close enough to, but smaller than, unity, or by utilising the similarity solution given above. Furthermore, the problem may be integrated with respect to (X/Re) to obtain a general solution for all plane channel cases. Using a fourth order Runge–Kutta procedure, with gradually increasing steps, the system 3.28 was integrated to yield the results shown in Fig 3.5.

It is seen that the profile integral function I is within 1% of the parabolic value, at $X/\text{Re} \simeq .023$, and the corresponding total pressure head loss due to entry is found to be approximately equal to .519, i.e. $\Delta p \approx 1.04(\frac{1}{2}\rho\bar{u}^2)$. These values are comparable to those obtained by the integral methods.

Equivalent coefficient of discharge

Let us recall that the coefficient of discharge C_d, for various constriction types in incompressible flow, is the ratio of actual discharge (= flow rate) to ideal (or theoretical) discharge. For our case of channel entrance, which, it should be noted, is not an actual orifice, we can employ this definition as a useful "expedience" for dealing with bearing problems comprising an external feed.

Let the supply pressure be denoted by p_s, the actual pressure at inlet to the channel by p_o and the theoretical pressure, i.e. the fictitious pressure at the entrance when the flow in the channel is assumed to be viscous throughout, by p_t. This latter is obtained by extrapolating the viscous-flow curve (i.e. the broken straight line in the bottom part of Figure 3.5). Then we have:

$$p_s - p_o = \frac{1}{2}\rho\bar{u}^2,$$

$$p_o - p_t = \alpha\frac{1}{2}\rho\bar{u}^2,$$

where α in our case is equal to 1.04.

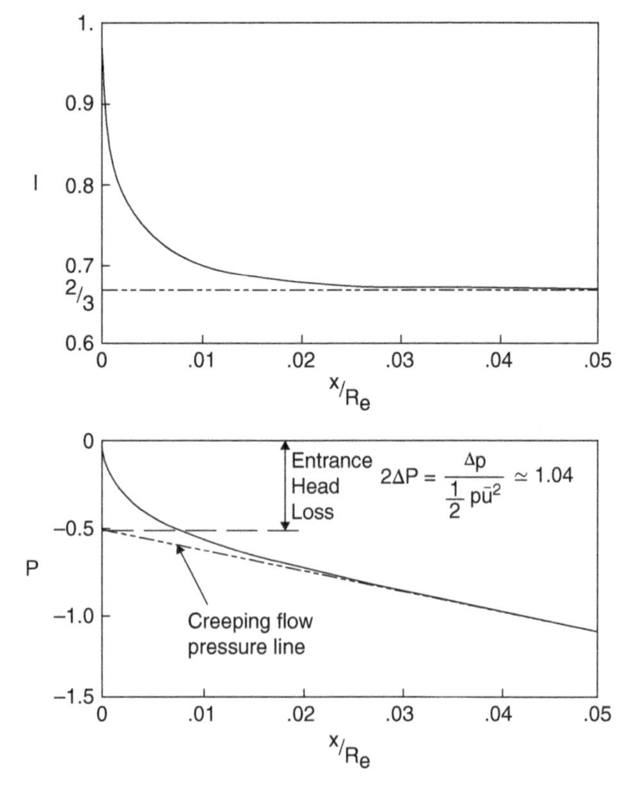

Figure 3.5 Velocity and pressure development in the entrance of a plane channel. Source: Al-Bender and Van Brussel, 1992 by permission of ASME.

Adding the two previous equations results in

$$p_s - p_t = (1 + \alpha)\frac{1}{2}\rho\bar{u}^2,$$

whereas, the ideal pressure drop for the same case (by Euler's equation) is obtained by replacing $(1 + \alpha)$ by unity. Thus the ratio of actual discharge to the theoretical one is

$$C_d = \sqrt{\frac{1}{(1+\alpha)}} \tag{3.29}$$

Substituting for $\alpha \approx 1.04$, we obtain the value $C_d \approx 0.7$ for incompressible flow through a plain inherent orifice. Note that the factor $(1 + \alpha) = 1/C_d^2$ itself is referred to as the flow resistance coefficient k.

The real utility and effectiveness of this method will become apparent when it is applied to compressible and incompressible radial flow, as will be given in Chapter 5. To facilitate this, we shall first extend the method to this latter mentioned case.

3.4 The Case of Radial Flow of a Polytropically Compressible Fluid Between Nominally Parallel Plates

Referring to Figure 3.6, we make the following assumptions:

(i) $h_o/r_o \ll 1$;
(ii) $|dh/dr| \ll h_o/r_o$ (mild film variation);

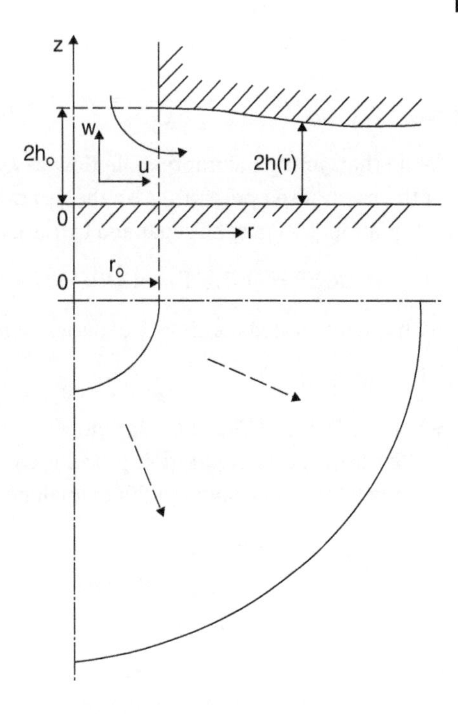

Figure 3.6 Radial channel flow: Notation. Source: Al-Bender and Van Brussel, 1992 by permission of ASME.

(iii) $\rho(r) = \rho(p(r))$; (polytropy);
(iv) $\mu = \mu(r)$ (possibly also).

The equation of motion is of the same form as that of the plane case:

$$\rho(u\frac{\partial u}{\partial r} + w\frac{\partial u}{\partial z}) = -\frac{\mathrm{d}p}{\mathrm{d}r} + \mu\frac{\partial^2 u}{\partial z^2} \tag{3.30}$$

whereas the continuity equation is now:

$$\frac{\partial}{\partial r}(\rho u r) + \frac{\partial}{\partial z}(\rho w r) = 0 \tag{3.31}$$

The stream function ψ may be defined by Constantinescu (1969),

$$\psi_r = -\frac{\rho}{\rho_0} r w, \qquad \psi_z = \frac{\rho}{\rho_0} r u \tag{3.32}$$

where, the subscripts r and z denote partial derivatives, and ρ_o is a reference density. (It can be seen that ψ satisfies the continuity equation identically.)

The von Mises transformation, viz.

$$\xi = r, \qquad \psi = \int_0^z (\rho/\rho_o) r u \mathrm{d}z,$$

may be applied, as before, to yield:

$$\rho u u_\xi = -\frac{\mathrm{d}p}{\mathrm{d}\xi} + \mu(\frac{\rho}{\rho_0})^2 r^2 u(u u_\psi)_\psi.$$

Normalisation is carried out with respect to the entrance and mid-plane values, as follows:

$$R = \xi/r_o, \quad Z = z/h, \quad \bar{h} = h(r)/h_o, \quad \Psi = \psi/\psi(r,h),$$
$$U = u/\bar{u}_o, \quad P = p/p_o, \quad \wp = \rho/\rho_o,$$

with,

$$\mathrm{Re}_0^* = (\rho_o \bar{u}_o r_o/\mu_o)(2h_o/r_o)^2, \quad \text{(the reduced Reynolds number)},$$
$$\mathrm{M}_o = \bar{u}_o/a_o = \bar{u}_o/\sqrt{\kappa p_o/\rho_o} \quad \text{(the entrance Mach number)}.$$

The normalised equation of motion:

$$\frac{\mathrm{Re}^*}{4\wp^2 R^2}(\wp U U_R + \frac{P'}{\kappa \mathrm{M}_0^2}) = U(UU_\Psi)_\Psi \tag{3.33}$$

(Note that, in the incompressible flow case, the term $P/\kappa M_0^2$ becomes $p/\rho\bar{u}_0^2$ which would then be the definition of the normalised pressure. Also then $\wp \equiv 1$.)

Equation 3.33 may be separated by the same assumption as before, viz.

$$U(R, \Psi) = Q(R)g(\Psi) = Q(R)G(R, Z)$$

with the integrated continuity equation being now,

$$\wp RQI\bar{h} = 1,$$

where, $I(R) = \int_0^1 G(R, Z)dZ$ (as before).

We note that the separation procedure developed in the main text may be applied in the same way to Eq. 3.33 leading to an initial value problem analogous to that of Eq.s 3.28, viz.

$$\begin{cases} Q' &= (1/\bar{h}^2 \mathrm{Re}^* \wp)m(I) \\ P' &= (-\kappa \mathrm{M}_0^2 Q/\bar{h}^2 \mathrm{Re}^*)n(I) \\ I &= (1/R\bar{h}^2 \wp Q) \end{cases} \tag{3.34}$$

where, m, n, and I are connected by the same relationship as before. The problem 3.34 may likewise be solved by an appropriate numerical procedure given an initial condition in addition to the functions $\bar{h}(R)$ and $\wp(R) = \wp(P)$. This procedure will find application in Chapter 5.

3.4.1 Conclusions on Pressure-Fed Entrance

The various methods and procedures for the solution of the laminar "channel entrance" problem have been reviewed and evaluated. They have been classified into semi-analytical, and numerical methods. Hitherto, the first type have all been approximate; the second are difficult to implement in practical models.

A method based on the separation of variables has been developed and established. This method, being semi-analytical, is found to be a powerful one capable of extension for a variety of flow configurations: it is both efficient and easy to implement. Comparison with experimental results for radial flow, that will be given in Chapter 5, shows remarkable agreement.

3.5 Narrow Channel Entrance by Shear-Induced Flow

This case pertains to self-acting bearings in which the fluid is "entrained" into, rather than fed to, the narrow gap of the slider bearing by virtue of the motion of the two surfaces, constituting the bearing, relative to each other. This type of shear-induced flow is often referred to as Couette–Poiseuille flow. In contrast to the previous problem of pressure-induced flow, this problem has known only very few treatments, all being in regard to liquid-lubricated slider bearings, where, owing to the high density of liquids, the pressure build-up ahead of gap entrance could become quite appreciable. The earliest treatments are perhaps those of Contantinescu et al. (1975); Pan (1974); Tipei (1978), which were mainly theoretical in nature. Towards the end of the 1980's, with the increased application of high-speed bearings and lubrication in the turbulent regime, the subject received a newly heightened interest reflected in the publications of Hashimoto et al. (1989); Heckelman and Ettles (1988); Kim and Kim (1992); Mori et al. (1990, 1992); Rhim and Tichy (1988, 1989); Rodkiewicz et al. (1990); Tichy and Chen (1985); Tipei (1978, 1982), which addressed the problem analytically, numerically and experimentally.

Our objective is to understand, qualify and quantify entrance flow by entrainment into the gap of a slider bearing both upstream and downstream of the gap entrance.

There are two effects to be distinguished:

1. Pressure build-up just upstream of the gap entrance; the so-called "ram" pressure owing to a sort of stagnation of the BL that begins to develop further upstream of gap entrance;
2. Pressure build-up or build-down just downstream of the gap entrance owing to the development of the entrance velocity distribution under the action of viscosity. Here we have two cases:
   (i) build-up results when the flow rate through the gap is below the Couette value ($\bar{u}_o < U_o/2$);
   (ii) build-down results when the flow rate through the gap is above the Couette value ($\bar{u}_o > U_o/2$). Obviously, no additional entrance pressure effect takes place when the entrance flow is equal to the Couette value.

Generally speaking, there are four aspects that need to be addressed and evaluated for this case of shear-induced flow:

(i) stability of laminar flow at the entrance owing to *adverse* pressure gradient, i.e. when the positive pressure gradient downstream of the entrance becomes sufficiently high that it cause reverse flow;
(ii) the boundary-layer flow development upstream of the gap entrance and the development of the so-called ram pressure;
(iii) the flow development just downstream of the entrance owing to the viscous shear effect;
(iv) the inertia effects on the flow in the bearing film itself.

In the following we shall examine the first three points. The fourth point is treated in Chapter 4.

3.5.1 Stability of Viscous Laminar Flow at the Entrance

At any point along the gap of a slider bearing, the velocity profile across the gap is given by (see Chapter 2):

$$u = U_o\left(1 - \frac{y}{h}\right) - \frac{1}{2\mu}\frac{dp}{dx}\,y(h-y).$$

The variables are first normalised by defining: (N.B. here we use h rather than $2h$ as the channel height to normalise x and y):

$$X = x/h, \qquad Y = y/h, \qquad U = u/U_o,$$
$$P = p/\rho\bar{U}_o^{\,2}, \ \mathrm{Re} = U_o h/\nu$$

Then

$$U = (1-Y) - \frac{\mathrm{Re}}{2}\frac{\mathrm{d}P}{\mathrm{d}X}Y(1-Y). \tag{3.35}$$

Taking the derivative of U with respect to Y at the top surface of the channel $Y = 1$, we get:

$$\left.\frac{dU}{dY}\right|_{Y=1} = -1 + \frac{\mathrm{Re}}{2}P' \qquad (P' \equiv \mathrm{d}P/\mathrm{d}X) \tag{3.36}$$

For forward flow, this derivative should be negative, and vice versa for reverse flow. This yields a condition on the commencement of reverse flow, namely:

$$P' > \frac{2}{\mathrm{Re}}.$$

At the entrance to a convergent gap bearing, be it tapered or step type, i.e. $h_{in}/h_{out} = h(0)/h(L) > 1$, the pressure gradient is positive, increasing with the magnitude of h' or with decreasing flow rate. The Reynolds equation can afford us another relationship between the pressure gradient, the Reynolds number and the mass-flow rate.

Recalling from Chapter 4, the 2D steady-state Reynolds equation, normalised with respect to U_o and h_o and integrated once with respect to X reads:

$$\mathrm{Re}\,\frac{H^3}{12}\frac{\mathrm{d}P}{\mathrm{d}X} - \frac{H}{2} = -\frac{\bar{\dot{m}}}{2},$$

where,

$$\bar{\dot{m}} = \frac{\dot{m}}{\rho_o U_o h_o}.$$

Note that the maximum value that $\bar{\dot{m}}$ can attain by shear-induced flow is 1 (corresponding to pure Couette flow, i.e. zero pressure gradient).

Evaluating the above expression at $H = 1$, i.e. gap entrance, we obtain

$$\frac{\mathrm{d}P}{\mathrm{d}X} = \frac{6}{\mathrm{Re}}(1 - \bar{\dot{m}}) \tag{3.37}$$

N.B. We can arrive at the same result by integrating Eq. 3.35 with respect to Y on $[0, 1]$ and setting $\int_0^1 U\mathrm{d}Y = \bar{\dot{m}}/2$. (This, in fact, is how the Reynolds equation is derived.)

Comparing Eqs. 3.36 and 3.37, we conclude that the condition for reverse flow to occur can be written as:

$$(1 - \bar{\dot{m}}) > \frac{1}{6}$$

Thus, the critical dimensionless mass flow below which reverse flow will ensue is $5/6$ (i.e. an actual flow rate per unit width of the slider equal to $(5/6)\rho U_o h_o/2$.

The flow situations with and without reverse flow are depicted qualitatively in Figure 3.7. In Figure 3.7(a) the flow rate is above critical so that no reverse flow takes place. In Figure 3.7(b), depicted with a larger taper, reverse flow takes place. Let us note that in the reversing part, the flow resembles that taking place in a sharp corner: after the stream lines have reversed against the upper bearing surface they move progressively closer to it as they leave the gap. This means that the boundary layer gets progressively thinner upon leaving the gap: resembling sink flow. We shall make use of this property when setting up the integral method of downstream solution.

3.5.2 Development of the Flow Upstream of a Slider Bearing

Figure 3.8(b) depicts the flow situation under consideration. Panel (a) shows the classical Blasius boundary layer (BL) of a moving fluid on a stationary flat plate (Schlichting 1968). In panel (b), which represents our case, the

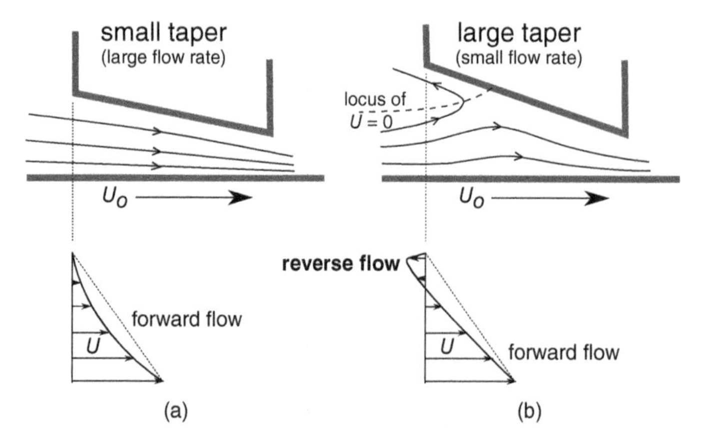

Figure 3.7 Qualitative streamline field for (a) moderate taper (large flow rate) and (b) large taper (small flow rate) sliders and associated entrance velocity profiles.

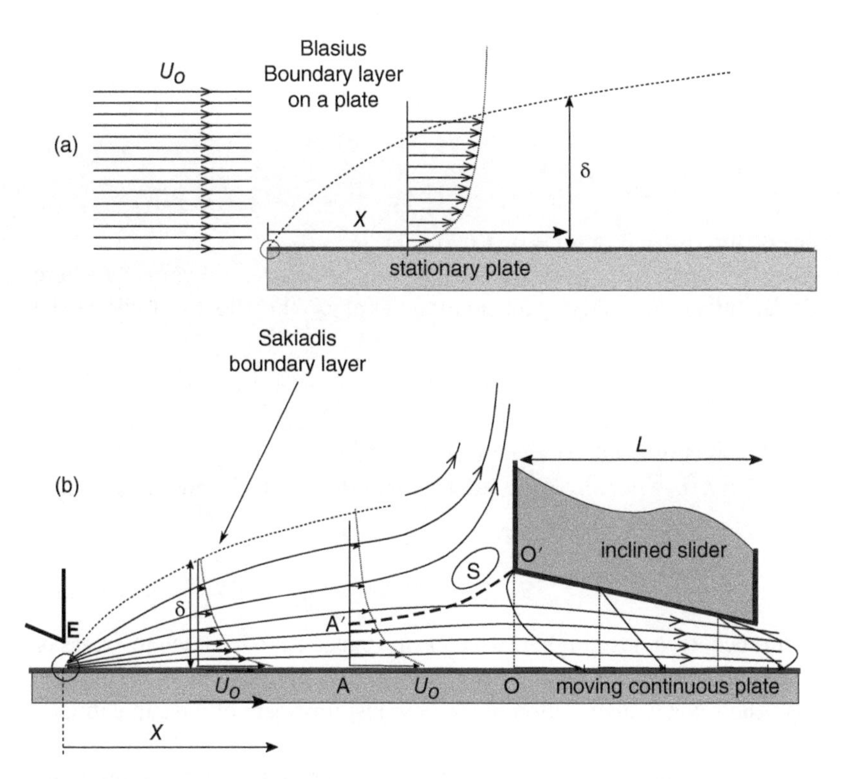

Figure 3.8 Development of the flow upstream of a slider bearing; Blasius BL (a) versus Skiadis BL (b). (Not to scale.)

flow of the entrained fluid upstream of entrance is that of Sakiadis boundary layer, i.e. that of a continuous surface moving in an otherwise undisturbed fluid (Sakiadis 1961).

Now, if we imagine that the entrance of the slider gap has a zero height, then the flow would be that of a BL impinging on a wall (the face of the slider pad), so that we should expect stagnation to take place there, where the dynamic head of the BL is converted to static pressure. This is, roughly speaking, the underlying idea behind "ram" pressure, which will be further quantified in the next section.

In practice, owing to the presence of the slider gap, and depending on the amount of flow of the fluid through it, there will be some interaction between the gap flow and the impinging BL flow so that some swirl flow might take place as indicated on the figure by the encircled S.

The property of the BL that is relevant to our investigation is the displacement thickness δ, which for a Sakiadis BL is given by:

$$\delta \approx 6.33\frac{x}{\sqrt{\mathrm{Re}_x}}$$

where,

$$\mathrm{Re}_x = \frac{U_0 x}{\nu}$$

is the stream-wise Reynolds number (being based on the distance traversed from the source point E).

For the sake of completeness, we briefly indicate the equations describing this type of BL.

The Sakiadis BL is described by the same equations as the similarity solution of Blasius BL, but with different boundary conditions. Thus, if the stream function of the flow is written as

$$\psi = \sqrt{2\nu U_0 x} f(\eta); \qquad \eta = y\sqrt{\frac{U_0}{2\nu x}}$$

then $f(\eta)$ will satisfy the ordinary differential equation

$$f''' + ff'' = 0$$

subject to b.c.'s

$$f(0) = 0;\ f'(0) = 1;\ f'(\infty) = 0$$

(N.B. The b.c.'s corresponding to Blasius BL are $f(0) = 0;\ f'(0) = 0;\ f'(\infty) = 1$).

Solution of the above-mentioned o.d.e. with those b.c.'s yields the velocity profile depicted if Figure 3.9 where we have normalised the ordinates by the BL thickness. Note that for small values of y/δ the velocity profile can be approximated by

$$\frac{u}{U_0} \approx 1 - \frac{1}{C}\frac{y}{\delta},$$

which facilitates the analytic evaluation of momentum and energy integrals of the flow.

If we now normalise $\delta \to \bar{\delta} = \delta/h_o$, $x \to \bar{x} = x/L$, where h_o is the gap height at the entrance and L is the slider width, we obtain

$$\bar{\delta} \approx 6.33\sqrt{\frac{\bar{x}}{\mathrm{Re}_0^*}},$$

where $\mathrm{Re}_0^* = U_o h_o{}^2/\nu L$ is the reduced entrance Reynolds number (which is not expected to be much in excess of 0.1 for the viscous assumption to be valid in the film).

This simple formula can help us assess the velocity distribution of the flow just upstream of the gap entrance (point A on Figure 3.8(b)), as follows.

Let the BL commence at point E (on the same figure), which could be the exit of the previous slider in a row. If the distance x between E and A is about 1/4 of the slider width L (i.e. $\bar{x} = 0.25$) and $\mathrm{Re}_0^* = 0.1$ then the BL thickness at A is $\bar{\delta}_A > 10$. In other words, the BL is more than 10 times thicker than the gap entrance height h_o. In that case, the BL velocity in the region bounded by $A - A'$, which is yet a fraction smaller than $\delta/10$, can be taken to be uniform in the first approximation (since we are now concerned with a rough estimate of the ram pressure magnitude).

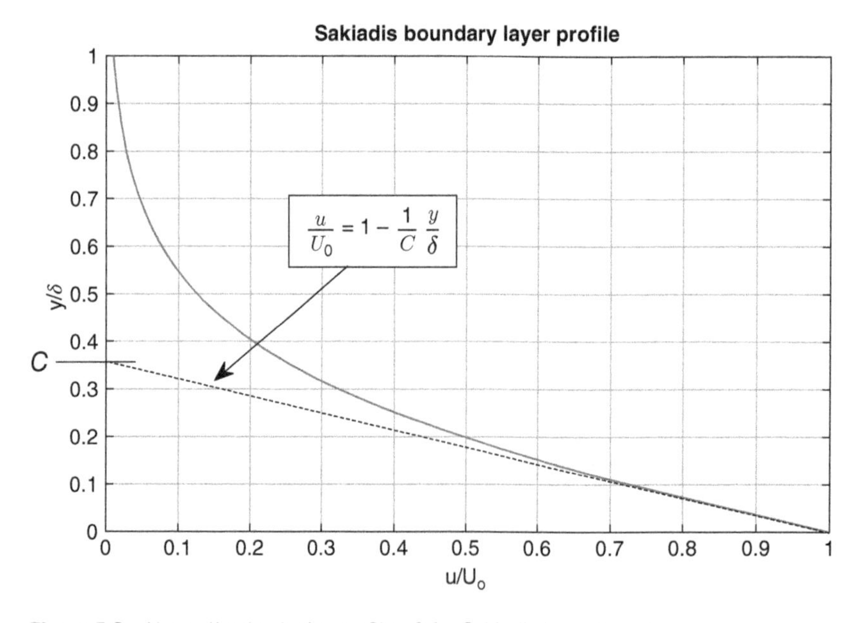

Figure 3.9 Normalised velocity profile of the Sakiadis boundary layer.

Estimation of the "ram" pressure

The analysis, outlined in this section, is based on (Mori et al. 1992), who also performed CFD calculations and comparisons to show that this approximate method agrees well with the more exact numerical solution. Their approximate method is based on the mechanical-energy integral equation. Let us note that Rhim and Tichy (1989) and Hashimoto et al. (1989) use the Bernoulli equation to arrive at similar estimates employing an additional loss coefficient, but we have preferred the former method because it is more elegant.

Referring to Figure 3.8, we consider incompressible flow through the control volume bounded by $A' - A - O - O'$, where the height $A - A'$ is denoted by h^* and of $O - O'$ by h_o respectively, and apply the continuity and mechanical energy integrals, ignoring possible side leakage and viscous effects, yields:

$$\int_0^{h^*} u_A \mathrm{d}y = \int_0^{h_o} u_o \mathrm{d}y, \tag{3.38}$$

$$\int_0^{h^*} \frac{1}{2}\rho(u_A)^3 \mathrm{d}y = \int_0^{h_o} \left(\frac{1}{2}\rho u_o{}^2 + p_o\right) u_o \mathrm{d}y, \tag{3.39}$$

where we have assumed the pressure at station A to be atmospheric $= 0$ so that

$$p_o = p_{\text{ram}}.$$

With u_o, h_o and u_A being known, we can easily calculate p_o.

Remark

in the above derivation, we have neglected viscous effects in, and leakage into or out of the control volume. In regard to the viscous effect, since the BL velocity profile close to the surface is almost linear in y, its second derivative is zero, so that this assumption is justified except perhaps very close to the gap entrance, where the velocity profile will appreciably deviate from linearity. In regard to transverse flow into and out of the control volume, reverse flow from the gap entrance is likely to enhance that situation. Finally, both viscous and leakage effects will result in reducing the value of the ram pressure. Rhim and Tichy (1989) and Hashimoto et al. (1989) find empirically somewhat lower values for the ram pressure in function of the BL head.

Simple example

Without loss of generality, let us approximate the gap-entrance velocity u_o by

$$\frac{u}{U_o} = c - \eta; \qquad \eta = \frac{y}{h_o}; \qquad \frac{1}{2} \le c \le 1$$

By varying c in the indicated range, this velocity distribution covers all possibilities from pure Couette velocity (corresponding to maximum shear-induced flow) to full back-flow (corresponding to zero flow rate). This yields:

$$\int_0^{h_o} u_o \mathrm{d}y = (c - \frac{1}{2})U_o h_o$$

$$\int_0^{h_o} u_o{}^3 \mathrm{d}y = (c - \frac{1}{2})(c^2 - c + \frac{1}{2})U_o{}^3 h_o$$

Solving Eqs. 3.38 and 3.39, using these expressions and assuming $u_A = \alpha U_o$, $\alpha < 1$, yields finally:

$$p_o = \left(\alpha - (c^2 - c + \frac{1}{2})\right) \frac{1}{2}\rho U_o{}^2$$

Assuming $\alpha \approx 1$, the function of c in parentheses will vary between 1/2—3/4 for $c = 1$ —1/2, respectively.

Although more exact calculations are possible, this result provides very good and useful estimation of the expected values of the ram pressure.

Numerical example

Considering a slider with $U_0 = 300$ m s^{-1}, working with air at atmospheric pressure and temperature, yields $\frac{1}{2}\rho U_0^{\,2} \approx 0.5\ 10^5$ Pa (or a half bar). Thus, the ram pressure would vary at most between 1/2 and 3/4 of this value. Most existing bearings to date operate at less than or around $U_0 = 200$ m s^{-1}. This would again, almost halve the value of calculated ram pressure, yielding something around 0.15 bar at the most. This might explain the tepid interest of gas-bearing research in this topic at present. However, we may expect this interest to increase with the ever increasing speed limits of bearings in modern technology machines.

3.5.3 Development of the Flow Downstream of the Gap Entrance

The following analysis is based on notes by the present author. We distinguish between two cases, which yield qualitatively different pressure behaviours.

(1) Over-flow downstream

This case pertains to various instances: supply over pressure (as in a hybrid bearing); high ram pressure with small channel resistance to the flow (as might obtain in divergent gap slider); or a combination of the two. Figure 3.10(a) depicts an assumed known velocity distribution (of the BL) just upstream of the gap entrance. Upon entrance, the no-slip condition on the upper (stationary) bearing, or channel surface will bring the top streamline to a halt, giving rise to the velocity distribution, depicted at the entrance, being composed of the Couette part u_c and an as yet undeveloped Poiseuille part u_p. After a certain distance x_e (the entrance length), u_p will develop into a symmetrical parabolic distribution (offering least resistance to the flow), while u_c remains unchanged.

Figure 3.10(b) depicts the associated pressure field development: at the entrance to the channel, there will be a sharp drop in the pressure followed by development towards a constant slope, equal to $-P_v'$ that is characteristic of viscous flow.

Let the BL velocity distribution just upstream, or at the entrance to a uniform channel be given, we wish to know how the flow will develop downstream and thereby determine the principal entrance parameters: the head loss, the entrance length and the coefficient of discharge.

3.5.4 Method of Solution

Since the method, developed in the first part of this chapter, for the Poiseuille, symmetric velocity profile case, is no longer applicable, owing to the presence of the Couette component and the lack of symmetry, we resort to

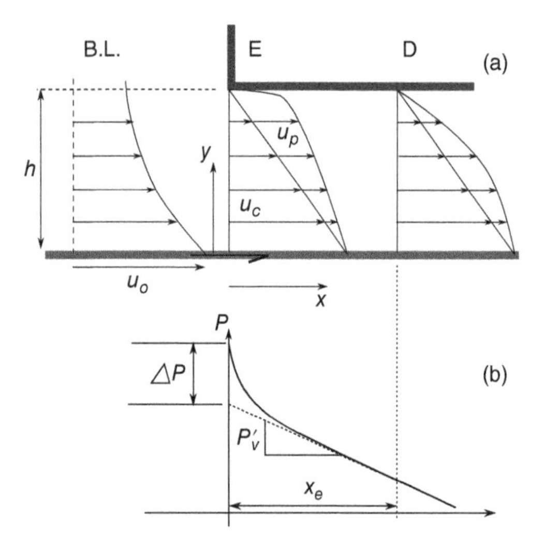

Figure 3.10 Shear-flow entrance into a plane uniform gap channel. Development of (a) velocity profile and (b) pressure (including head loss). (Not to scale)

approximate analysis utilising the momentum and energy-integral method (see Chapter 4) to obtain (at least a proof-of-concept) solution.

Referring to Figure 3.10(a), firstly, we express the velocity profile at entrance as follows:

$$u(y) = \underbrace{(1-y)U_o}_{u_c} + \underbrace{f(y,x)}_{u_p}$$

The variables are first normalised by defining:

$$X = x/h/\mathrm{Re}_p, \quad Y = y/h, \qquad U = u/U_o,$$
$$P = p/\rho\bar{u}_p{}^2, \quad \mathrm{Re}_p = |\bar{u}_p|h/\nu \quad A = \bar{u}_p/U_o$$

(N.B. here we use h rather than $2h$ to normalise x and y.)

In this way, the velocity distribution is written:

$$U = 1 - Y + f(X,Y)$$

We construct a simple but versatile function $f(X,Y) = f(Y; a(X))$ composed of two parabolas (Figure 3.11)

$$U_p = f(Y;a) = \begin{cases} A\left(1-\left(\frac{Y}{a}-1\right)^2\right) & ; 0 \le Y \le a \\ A\left(1-\left(\frac{Y-a}{1-a}\right)^2\right) & ; a \le Y \le 1 \end{cases} \tag{3.40}$$

where, $0 < a < 1$ is a dimensionless parameter, that varies with X, and A is the relative amplitude of the Poiseuille velocity.

The advantage of this velocity profile is that the integral of U_p and all its powers on the interval $0 \le Y \le 1$ are independent of a as could be easily shown. In this way, the continuity equation is automatically satisfied for a constant A and U_o.

Making up the system of integral equations as explained in Chapter 4, one obtains the system of o.d.e.'s:

$$\begin{aligned} -\frac{A}{6}\frac{da}{dX^*} &= -|A|\frac{dP}{dX^*} - \frac{2A}{a(1-a)} \\ -\frac{A(6A-4a+7)}{10}\frac{da}{dX^*} &= -|A|\left(\frac{1}{2}+\frac{2A}{3}\right)\frac{dP}{dX^*} - \frac{2A}{a}\left(1+\frac{2A}{3(1-a)}\right), \end{aligned} \tag{3.41}$$

where $X^* = X/\mathrm{Re}$, and with initial value $a(0) = a_0$.

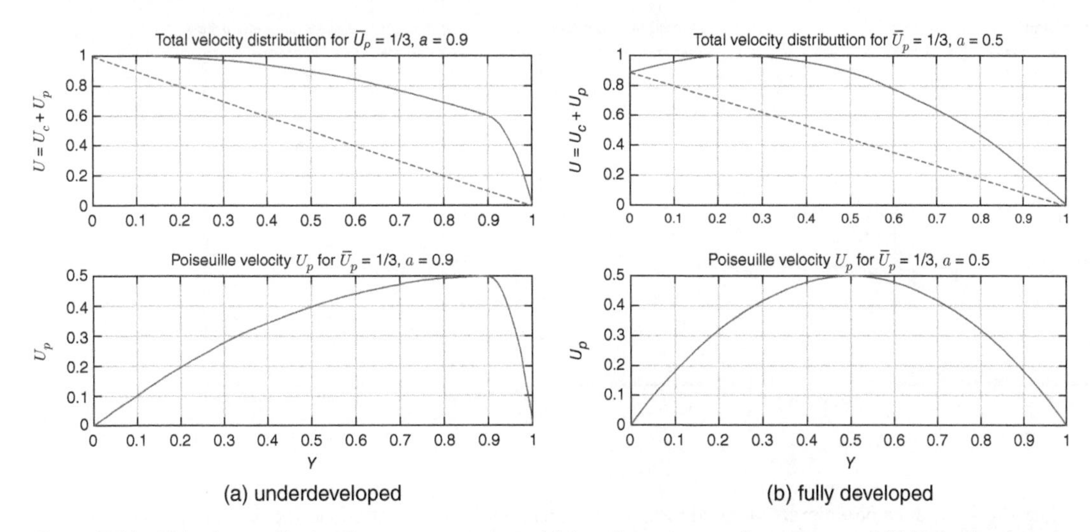

Figure 3.11 Velocity profile and its two components (a) just downstream of entrance and (b) Fully developed.

The value a_0 measures the degree of departure from symmetry of the profile U_p at entrance. Figure 3.11 furnishes an idea about this form for (a) $a = 0.9$ and (b) when the flow is fully developed $a = 0.5$. One can verify that $a = 0.5$ is the only equilibrium point of this system of o.d.e's, with the corresponding value $P' = -\text{sgn}(A)8$.

The solution of Eqs. 3.41 depends not only on the value of a_0 but also on $A = \bar{u}_p/U_o$. This latter is generally less than unity for a BL velocity distribution: its actual value will mainly depend on the flow rate through the channel. Therefore, we need to solve the flow problem as a whole in order to determine the entrance parameters. However, we have solved Eq. 3.41 for values of A ranging from less than 0.1 through about 1000. They have all been obtained for $a_0 = 0.99$, i.e. highly undeveloped flow. The entrance distance X_e is the distance downstream for which $a = 0.505$, i.e. within 1% of fully developed, parabolic profile.

The global results are shown in Figure 3.12 and summarise the main parameters characterising the flow, viz. $\Delta P, X_e/\text{Re}, C_d$, in function of $A = \bar{u}_p/U_o$. The head loss and entrance length decrease logarithmically with A whereas the coefficient of discharge increases. All three attain an asymptotic limit for large A. Let us note that the case of extremely high A actually corresponds to pressure induced flow, but with a skewed entrance velocity distribution owing to the Couette component. Such a situation obtains for instance in an EP bearing with sliding. The results show thus that a head loss is associated with the development of the velocity distribution into a symmetrical profile. This loss is characterised by a coefficient of discharge C_d of approximately 0.78 for high A, associated with high supply pressure values. At the other end, for values of A much smaller than unity, we find values of C_d around 0.4.

Note that the value of C_d has been calculated in the same manner as for an EP gap (Eq. 3.29), i.e.

$$C_d = \sqrt{\frac{1}{(1 + 2\Delta P)}} \tag{3.42}$$

It is important to note that this definition of C_d pertains only to the Poiseuille component of the flow although its value is affected by the Couette component.

In order to understand the behaviour for small ram pressure values, we have plotted the results pertaining to the small $\bar{u}_p/U_o$ range in Figure 3.13, where we have normalised the pressure for better visibility as $P = p/(1/2\rho U_o^2)$.

We note that as $\bar{u}_p/U_o$ approaches zero, the associated head loss also approaches zero, which is consistent with pure Couette flow. As $\bar{u}_p/U_o$ increases, the head loss increases more than linearly with this value.

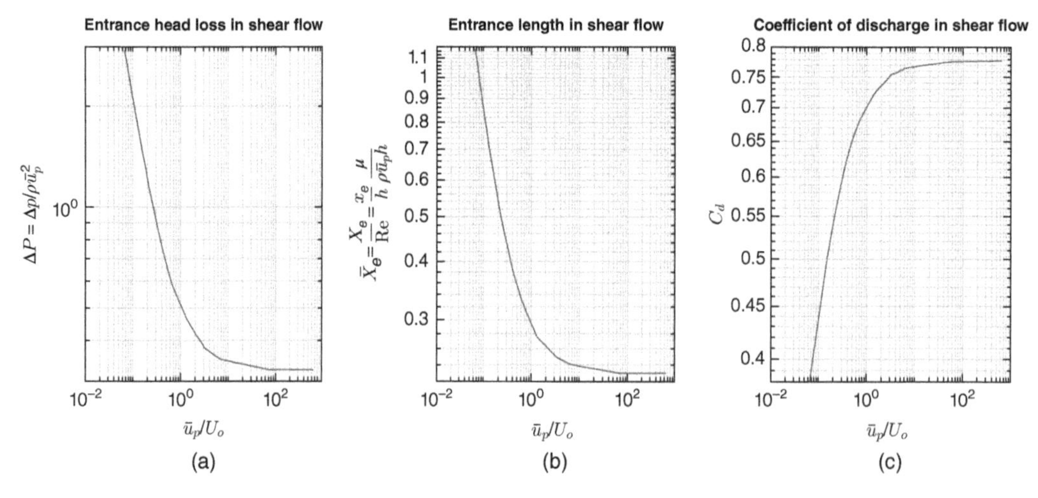

Figure 3.12 Shear-flow entrance parameters (a) head loss, (b) entrance length and (c) coefficient of discharge as a function of $A = \bar{u}_p/U_o$.

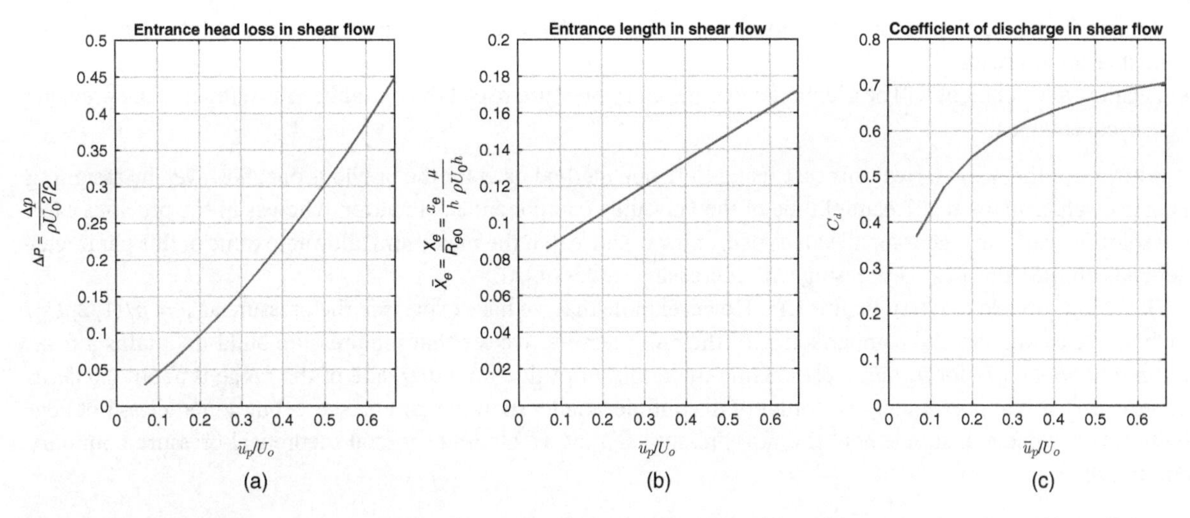

Figure 3.13 Shear-flow entrance parameters for small values of $\bar{u}_p/U_o$ (a) head loss, (b) entrance length and (c) coefficient of discharge as a function of $A = \bar{u}_p/U_o$. Note that the pressure is expressed as $P = p/(1/2\rho U_o^2)$ to facilitate evaluation and comparison.

(2) Under-flow downstream

This situation is more common and likely to be encountered in applications than the foregoing one. In the following, we treat this case using the same method. The situation is depicted in Figure 3.14, where we observe the following differences in comparison to Figure 3.10:

- At the region indicated by "S", there will generally be some sort of stagnation of the flow accompanied by a swirling or formation of eddies. Underneath this region, the flow might accelerate and thus cause some local reduction in static pressure.

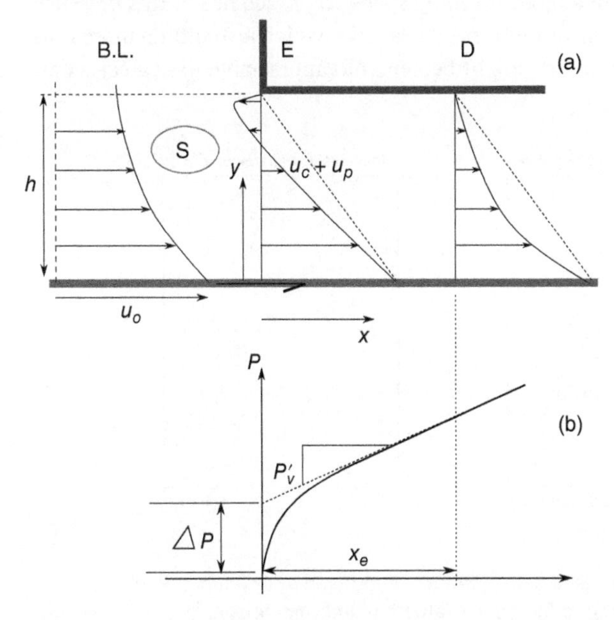

Figure 3.14 Shear-flow entrance into a plane uniform gap channel with reverse flow component. Development of (a) velocity profile and (b) pressure (including head gain). (Not to scale)

- Reverse flow is at its highest at the entrance and, at least for a convergent gap, will develop into forward flow further downstream.
- The flow development will be accompanied by a sharp pressure rise at the entrance, (in contrast to the previous case considered).

It turns out that we can treat this case using the same method used for the previous one, however, by assigning negative values to the relative amplitude of the Poiseuille component A. However, whereas in the previous case, the solution was convergent for all values of a_0, now we have that the maximum allowable value of this parameter depends on the value of A: decreasing with increasing values of $|A|$.

The results are plotted in in Figure 3.15. However, note that we have expressed the pressure as $P = p/(1/2\rho U_0^2)$ to facilitate evaluation and comparison with the ram pressure. We see that the pressure build-up attains a maximum of about 0.075 for $\bar{u}_p/U_o \approx 0.26$, while the minimum value on either side of the range is about 0.035. In other words, it is approximately one order of magnitude smaller than the ram pressure, but nonetheless not negligible. This value will supplement the ram pressure so that we obtain as a total theoretical pressure boundary condition:

$$p_t = (C_{ram} + C_{entrance})\frac{1}{2}\rho U_o{}^2$$

where, $1/2 \leq C_{ram} \leq 3/4$ and $0.035 \leq C_{entrance} \leq 0.075$.

3.5.5 Conclusions Regarding Shear-Induced Entrance Flow

We have seen that entrance by entrainment into a self-acting bearing is generally accompanied by a pressure rise upstream of the gap entrance, the so-called ram pressure, as a consequence of partial stagnation, as well as a further, less pronounced rise downstream thereof, as a consequence of flow development. The global situation is depicted in Figure 3.16. Formulas have been derived, using approximate integral calculations, to quantify those pressure rises, which can be used, through the device of a "discharge coefficient" or otherwise, to determine a theoretical entrance pressure boundary condition p_t leading to a pressure distribution that is different from that obtained assuming ambient entrance pressure. For air with a modest sliding speed, $U_0 \leq 200$ m s^{-1}, this pressure will likewise be modest lying around 0.1 bar. However, for ultra-high speed bearings, which are coming more into use, a pressure increase of around 0.36 bar, for $U_0 \rightarrow 300$ m s^{-1}, might become too appreciable to neglect when modelling and evaluating such bearings.

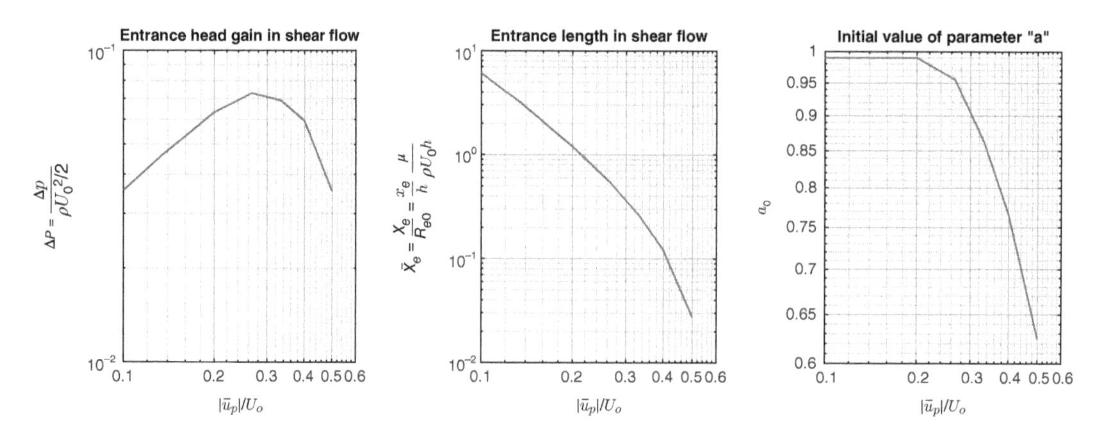

Figure 3.15 Shear-flow entrance parameters from left to right: head gain, entrance length and a_0 as a function of $A = \bar{u}_p/U_o$. Note that the pressure is expressed as $P = p/(1/2\rho U_0^2)$ to facilitate evaluation and comparison.

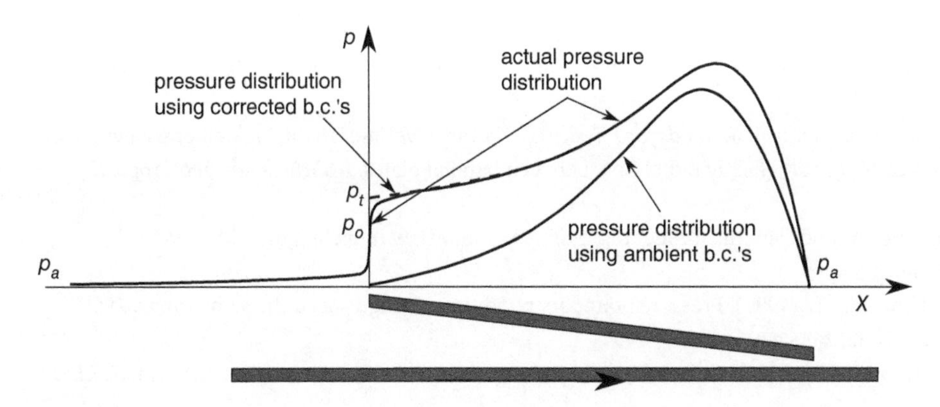

Figure 3.16 Global pressure development, upstream and downstream the gap entrance.

The special case of shear-induced accompanied by pressure-induced entrance flow, as might arise in an EPs slider bearing, may effectively be characterised by a coefficient of discharge lying around 0.78 when the pressure-induced flow is dominant.

3.6 Conclusions

Entrance flow to a bearing film is characterised by flow development upstream and slightly downstream of the gap entrance. We need to characterise this flow development in order to determine either the pressure boundary condition at the inlet to the bearing or, equivalently, a relationship between that and the flow rate through the bearing. This chapter has treated this problem by presenting and discussing methods to evaluate and quantify this flow. Although nowadays it might be a relatively simple matter to determine the flow field by a purely numerical, e.g. CFD method, it is nevertheless rewarding to derive such lumped parameter methods as has been carried out in this chapter. These methods afford us more straightforward means to establish the pressure boundary conditions required in solving the Reynolds equation, which characterises the major part of the bearing film. In that way, not only very fast solution algorithms could be devised to solve the bearing problem, but also useful insights into the design problem are gained.

References

Al-Bender, F. and Van Brussel, H., (1992). A method of "separation of variables" for the solution of laminar boundary-layer equations of narrow channel flows. *Journal of Tribology* 114: 630–636.

Ames, W.F., (1964). *Nonlinear Partial Differential Equations in Engineering*. Academic Press.

Blottner, F.G., (1977) Entry flow in straight and curved channels with slender channel approximations. *ASME - Journal of Fluids Engineering* 666–673.

Bodoia, J.R. and Osterle, J.F., (1961). Finite difference analysis of plane poiseuille and couette flow development. *Applied Scientific Research Section A* 10, 265–276. Section A.

Campbell, W.D. and Slattery, J.C., (1963). Flow in the entrance of a tube. *ASME - Journal of Basic Engineering* 41–46.

Chen, R.Y., (1973). Flow in the entrance region at low reynolds numbers. *ASME - Journal of Fluids Engineering* 153–158.

Constantinescu, V.N., (1969). *Gas Lubrication*. ASME.

Contantinescu, V.N., Galetuse, S. and Jennedy, F., (1975). On the comparison between lubrication theory, including turbulence and inertia forces, and some existing experimental data. *ASME Journal Lubrication Technology* 97, 439–449.

Eilenberger, G., (1981). *Solitons (Mathematical Methods for Physicists)*. Springer-Verlag, Berlin, Heidelberg, New York.

Geiger, D., Fara, H.D. and Street, N., (1964). Steady radial flow between parallel plates. *ASME - Journal of Applied Mechanics* 354–355.

Han, L.S., (1960). Hydrodynamic entrance lengths for incompressible laminar flow in rectangular ducts. *ASME - Journal of Applied Mechanics* 403–409.

Hashimoto, H., Wada, S. and Yoshida, T., (1989). Pressure boundary conditions of high speed thrust bearings. *JSME International Journal, Series III* 32, 269–280.

Heckelman, D.D. and Ettles, C.M.M., (1988). Viscous and inertia pressure effects at the inlet to a bearing film. *STLE Tribology Transactions* 31, 1–5.

Kim, J.S. and Kim, K.W., (1992). Inlet pressure effects on the static and dynamic characteristics of tilting-pad journal bearings. *JSME International Journal, Series III* 35(1): 121–127.

Mori, A., Makino, T. and Mori, H., (1990). Inertia effect in a submerged multi-pad bearing under high Reynolds number with special attention to inlet pressure jump *Proceedings of the Japanese International Tribology Conference, Nagoya*, 911–916.

Mori, A., Makino, T. and Mori, H., (1992). Entry flow and pressure jump in submerged multi-pad bearings and grooved bearings. *ASME. Journal of Tribology* 114(2): 370–377.

Mori, H. and Miyamatsu, Y., (1969). Theoretical flow-models for externally pressurized gas bearings. *ASME - Journal of Lubrication Technology* 181–193.

Morihara, H. and Cheng, R.T., (1973). Numerical solution of the viscous flow in the entrance region of parallel plates. *Journal of Computational Physics* 11, 550–572.

Murphy, H.D., Coxon, M. and McEligot, D.M., (1978). Symmetric sink flow between parallel plates. *ASME - Journal of Fluids Engineering* 477–484.

Narang, B.S. and Krishnamoorthy, G., (1976). Flow in the entrance region of parallel plates. *ASME - Journal of Applied Mechanics* 186–188.

Newell, A.C., (1985). *Solitons in Mathematics and Physics*. Society of Industrial and Applied Mathematics.

Pan, C.H.T., (1974). Calculation of pressure, shear, and flow in lubrication films for high speed bearings. *ASME Journal of Lubrication Technology* 96, 80–94.

Rhim, Y. and Tichy, J.A., (1988). Entry flow of lubricant into a slider bearing –analysis and experiment. *STLE Tribology Transactions* 31, 350–358.

Rhim, Y. and Tichy, J.A., (1989). Entrance and inertia effects in a slider bearing. *STLE Tribology Transactions* 32(4): 469–479.

Rodkiewicz, C.M., Kim, K.W. and Kennedy, J.S, (1990). On the significance of the inlet pressure build-up in the design of tilting-pad bearings. *ASME Journal of Tribology* 112, 17–22.

Roidt, M. and Cess, R.D., (1962). An approximate analysis of laminar magnetohydrodynamic flow in the entrance region of a flat duct. *ASME - Journal of Applied Mechanics* 171–176.

Sakiadis, B., (1961). Boundary layer on continuous solid surface. ii the boundary layer on a continuous flat surface. *AIChE Journal* 7, 221–225.

Savage, S.B., (1964). Laminar radial flow between parallel plates. *ASME - Journal of Applied Mechanics* 594–596.

Schlichting, H., (1968). *Boundary-Layer Theory* 6 edn. McGraw-Hill, New York.

Tang, I.C., (1967). A nonlinear differential equation involving characteristic values. *Z. angew. Math. Mech.* 47, 473–475.

Tang, I.C., (1968). Inertia effects of air in an externally pressurized gas bearing. *Acta Mechanica* 5, 71–82.

Tichy, J.A. and Chen, Sh., (1985). Plane slider bearing load due to fluid inertia-experimental and theory. *ASME Journal of TribolOGY* 107,]32–38.

Tipei, N., (1978). Flow characteristics and pressure head build-up at the inlet of narrow passages. *ASME Journal of Lubrication Technology* 100, 47–55.

Tipei, N., (1982). Flow and pressure head at the inlet of narrow passages, without upstream free surface. *ASME Journal of Lubrication Technology* 104, 196–202.

Van Dyke, M., (1970). Entry flow in a channel. *Journal of Fluid Mechanics* 44(part 4), 813–823.

Woolard, H.W., (1957). A theoretical analysis of the viscous flow in a narrowly spaced radial diffuser. *ASME - Journal of Applied Mechanics* 9–15.

4

Reynolds Equation: Derivation, Forms And interpretations

4.1 Introduction

Although formulated and presented well over a century ago now (Reynolds 1886) and still the cornerstone of lubrication theory, the Reynolds Equation (RE) is to this day a subject of discussions and extensions, and even re-derivations and misinterpretations.

Since this equation makes its appearance in almost every chapter of this book, we have found it appropriate to dedicate a separate chapter to it, at the very least to facilitate easy reference.

We start by providing a compact derivation of the equation accompanied by some rudimentary discussion and explanation. Thereafter the transformation of RE to other coordinate systems is formulated, with some essential cases, pertaining to cylindrical, spherical, conical, etc. configurations being stated for easy (cross)reference. This is followed by interpretation of the equation for cases in which the kinematics of the motion is not straightforward; in particular, moving non-flat bearing surfaces in combination with time variations, (which actually had not been included in Osborne Reynolds original formulation).

In an intermediate part of this chapter, the effects that have been neglected during the derivation of RE are reviewed, systematically discussed and, where possible, included as a modification to the RE.

In this respect, a considerable part of this chapter is devoted to the treatment of inertia effects arising in the flow, which might be a pertinent issue in the mind of design engineers who have to cater for situations of ever increasing demands on speed of motion.

Some numerical estimates and examples of the various effects are provided throughout.

4.2 The Reynolds Equation

The Reynolds equation is the differential equation for the pressure distribution in the film obtained by solving the viscous flow equations. Its derivation is quite straight forward, and may, for the general case, be found back in (Constantinescu 1969; Gross 1962). It will be given here for the general case of a compressible film, with both surfaces moving, in order to simplify further referencing, and facilitate derivation of bearing characteristics based thereupon.

Let us denote by ∇ (nabla) the two-dimensional gradient/divergence operator in the plane of the bearing, i.e.

- for a scalar function $\Phi(x, y)$, we have

$$\text{grad}\Phi \stackrel{\triangle}{=} \nabla_{xy}\Phi \stackrel{\triangle}{=} \frac{\partial \Phi}{\partial x}\hat{\mathbf{e}}_x + \frac{\partial \Phi}{\partial y}\hat{\mathbf{e}}_y,$$

where $\hat{\mathbf{e}}_x$ and $\hat{\mathbf{e}}_y$ are unit vectors in the x and y directions respectively.

Air Bearings: Theory, Design and Applications, First Edition. Farid Al-Bender.

Companion website: www.wiley.com/go/AlBender/AirBearings

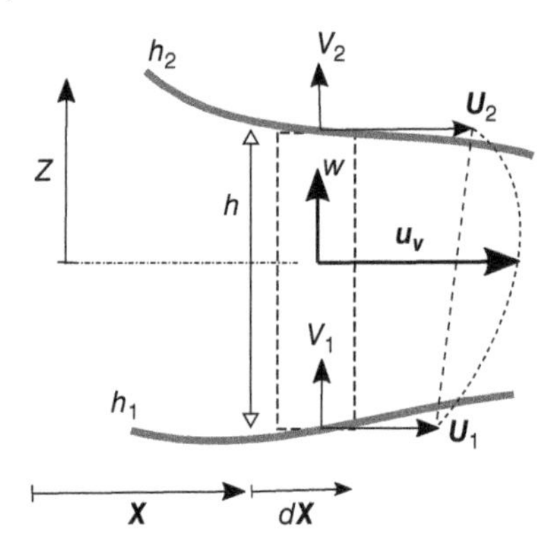

Figure 4.1 Control volume of the Reynolds Equation.

- and for a vector function

$$\mathbf{F}(x,y) = F_x(x,y)\hat{\mathbf{e}}_x + F_y(x,y)\hat{\mathbf{e}}_y,$$

we have

$$\text{div}\boldsymbol{\Phi} \triangleq \nabla_{xy} \cdot \mathbf{F} \triangleq \frac{\partial F_x}{\partial x} + \frac{\partial F_y}{\partial y}.$$

Referring to Figure 4.1, we consider viscous fluid flow having stream-wise velocity field $\mathbf{u}_v$ and one normal to that field $w\mathbf{e}_z$, taking place in a thin gap bounded by two surfaces $h_1(x,y,t)$ and $h_2(x,y,t)$ that are moving at velocities $(\mathbf{U}_1, V_1)$ and $(\mathbf{U}_2, V_2)$ respectively. Thus, we can express the fluid velocity as

$$\mathbf{u} = \mathbf{u}_v + w\hat{\mathbf{e}}_z = (u, v) + w\hat{\mathbf{e}}_z.$$

The continuity equation, $\partial\rho/\partial t + \nabla \cdot (\rho\mathbf{v}) = 0$, becomes:

$$\frac{\partial \rho}{\partial t} + \nabla_{xy} \cdot (\rho \mathbf{u}_v) + \frac{\partial(\rho w)}{\partial z} = 0. \tag{4.1}$$

This equation may be integrated across the film, taking care to apply the Leibniz rule of differentiating under the integral sign, and using the boundary conditions on $\mathbf{u}$ at the walls, namely $\mathbf{u}(h_1) = \mathbf{U}_1 + V_1\hat{\mathbf{e}}_z, \mathbf{u}(h_2) = \mathbf{U}_2 + V_2\hat{\mathbf{e}}_z$, to give

$$\begin{aligned} \frac{\partial}{\partial t}\int_0^h \rho \text{dy} - \underline{\rho\frac{\partial(h_2 - h_1)}{\partial t}} + \nabla_{xy} \cdot (\rho \int_{h_1}^{h_2} \mathbf{u}_v \text{dz}) - \rho(\mathbf{U}_2 \cdot \nabla h_2 - \mathbf{U}_1 \cdot \nabla h_1) \\ + \underline{\rho(V_2 - V_1)} = 0, \end{aligned} \tag{4.2}$$

where the underlined terms cancel out identically by virtue of $V_{1,2}$ being the total derivative of $h_{1,2}$, i.e.

$$V_{1,2} = \mathbf{U}_{1,2} \cdot \nabla h_{1,2} + \frac{\partial h_{1,2}}{\partial t},$$

so that we eventually have (after rearrangement),

$$\frac{\partial(\rho h)}{\partial t} + \nabla_{xy} \cdot (\rho \int_{h_1}^{h_2} \mathbf{u}_v \text{dz}) = 0, \tag{4.3}$$

where the term under the divergence operator in the last equation is the (macroscopic) *flux density* of the flow $\boldsymbol{\Phi}$:

$$\boldsymbol{\Phi} \triangleq \rho \int_{h_1}^{h_2} \mathbf{u}_v \text{dz}.$$

We see thus that the (precursor to the) Reynolds equation is none other than the macroscopic continuity equation, stating that the total mass flux through a control volume at any station (x, y) that is fixed in space (see the dashed box in Figure 4.1) bounded by the bearing surfaces is null. $\mathbf{\Phi}$ is determined by integrating the viscous momentum equation (see, Chapter 2) combined, for simplicity, in one equation:

$$0 = -\nabla p + \mu \frac{\partial^2 \mathbf{u}_v}{\partial z^2}. \tag{4.4}$$

Integrating this equation across the film, and using the no-slip boundary conditions on $\mathbf{u}_v$ at the walls, we obtain:

$$\mathbf{u}_v = -\frac{1}{2\mu} \nabla p \, z(h - z) + \left(\mathbf{U}_1 + (\mathbf{U}_2 - \mathbf{U}_1)\frac{z}{h} \right). \tag{4.5}$$

This is the well known velocity distribution associated with viscous flow. It consists of a *parabolic* component (first term of R.H.S.), associated with Poiseuille flow, and a *linear* component (second term of R.H.S.), associated with Couette or shear flow. Integrating once more across the gap, yields:

$$\int_{h_1}^{h_2} \mathbf{u}_v dz = -\frac{h^3}{12\mu} \nabla p + h \frac{(\mathbf{U}_1 + \mathbf{U}_2)}{2}.$$

Thus, the flux density is given by:

$$\mathbf{\Phi} = -\frac{\rho h^3}{12\mu} \nabla p + \frac{\rho h}{2} \mathbf{V}_t, \qquad \mathbf{V}_t = \mathbf{U}_1 + \mathbf{U}_2. \tag{4.6}$$

Substituting for the flux density in Eq. 4.3 and rearranging, yields the Reynolds equation:

$$\nabla \cdot \left(\frac{\rho h^3}{12\mu} \nabla p - \frac{1}{2} \rho h \mathbf{V}_t \right) = \frac{\partial}{\partial t} (\rho h) \tag{4.7}$$

Finally, normalising by

$$X \rightarrow x/L, Y \rightarrow y/L, H \rightarrow h/h_0, P \rightarrow p/p_0, \wp \rightarrow \rho/\rho_0, \tau \rightarrow \omega t,$$

where $L, h_0, p_0, \rho_0, \omega$ are some suitable reference values for the length (size), gap, pressure, density and pulsation perpendicular to the plane of the film, the Reynolds equations is written in dimensionless form as:

$$\nabla \cdot (H^3 \wp \nabla P - \mathbf{\Lambda} \wp H) = \sigma \frac{\partial}{\partial \tau} (\wp H) \tag{4.8}$$

where the two dimensionless parameters $\mathbf{\Lambda}$ and σ are:

The sliding number also called: the bearing number, the Harrison number or the compressibility number.

$$\mathbf{\Lambda} = \frac{6\mu \mathbf{V}_t L}{p_a {h_a}^2}$$

The squeeze number which has the same form as the previous number, i.e. a compressibility number in the squeeze direction.

$$\sigma = \frac{12\mu\omega L^2}{p_a {h_a}^2} = \frac{12\mu(\omega h_a)L}{p_a {h_a}^2} \left(\frac{L}{h_a} \right)$$

Note, that although we have developed Eq. 4.8 in Cartesian coordinates, we have written a general form (i.e. $\nabla \cdot$) that is actually valid for other coordinate systems, as we shall see further below.

Remarks

As alluded to above, the Reynolds equation may be seen as a macroscopic continuity equation. It comprises three basic terms (from left to right): the pressure (Poiseuille) term, the velocity (Couette, or "wedge") term, and the squeeze term, respectively. Important to notice also is that $\mathbf{V}_t$ is the sum of the surface tangential velocities and not the difference. This point, which might cause some confusion in more complex applications, will be discussed further below with examples.

It is worthy of mention that the Reynolds equation has been also adapted for or extended to situations in which there is a degree of departure (that might be appreciable) from the basic assumptions on which it is based. The most important of these may be the effect of inertia, which could become relevant at high sliding or squeeze velocities being characteristic of modern-day applications. This will be treated in the second part of this chapter in conjunction with the integral methods for the solution of channel flow. The other effects will also be reviewed in sections 4.5 through 4.8 (inclusive), in which the Reynolds equation is stated in various coordinates followed by interpretations pertaining to complex kinematics.

4.3 The Reynolds Equation for Various Film/Bearing Arrangements and Coordinate Systems

We shall adopt the form of the Reynolds equation 4.7 with the general grad and div operators and, for simplicity, with only one sliding surface, that is furthermore assumed to be flat. This form is applicable to all cases, if we take care to use the correct expression for the grad and div operators. In the following, we shall work out which of the cases one is most likely to encounter in practice. Other, more exceptional cases my be then easily dealt with by the reader.

First, we state the general system for obtaining grad and div operators in orthogonal curvilinear coordinates Constantinescu (1969):

Let us transform a system from the reference coordinates (x, y, z) to another orthogonal curvilinear coordinate system (q_1, q_2, q_3) such that:

$$q_i = q_i(x, y, z); i = 1, 2, 3.$$

We calculate the Lamé coefficients:

$$\xi_i = \left(\left(\frac{\partial x}{\partial q_i}\right)^2 + \left(\frac{\partial y}{\partial q_i}\right)^2 + \left(\frac{\partial z}{\partial q_i}\right)^2 \right)^{1/2},$$

(obviously all positive). Then,

$$\text{grad } \Phi = \frac{1}{\xi_1}\frac{\partial \Phi}{\partial q_1}\hat{\mathbf{e}}_1 + \frac{1}{\xi_2}\frac{\partial \Phi}{\partial q_2}\hat{\mathbf{e}}_2 + \frac{1}{\xi_3}\frac{\partial \Phi}{\partial q_3}\hat{\mathbf{e}}_3$$

and

$$\text{div } \mathbf{F} = \text{div}(F_1, F_2, F_3) = \frac{1}{\xi_1\xi_2\xi_3}\left[\frac{\partial}{\partial q_1}(F_1\xi_2\xi_3) + \frac{\partial}{\partial q_2}(F_2\xi_1\xi_3) + \frac{\partial}{\partial q_3}(F_3\xi_1\xi_2)\right].$$

N.B. The Reynolds equation contains only two spatial coordinates, which can be adapted from the previous equations by reduction of dimension.

For cylindrical and spherical coordinates, we have the following table:

	q_1	q_2	q_3	ξ_1	ξ_2	ξ_3
Cartesian coordinates	x	y	z	1	1	1
Cylindrical coordinates	r	θ	z	1	r	1
Spherical coordinates	r	θ	ϕ	1	$r \sin \phi$	r

4.3.1 Cartesian Coordinates (x, y)

This is the generic case, employed for flat bearings and journal bearings (when putting $y \mapsto z$ and $x \mapsto R\theta$), for which:

$$\text{grad} = \text{div} = \left(\frac{\partial}{\partial x}, \frac{\partial}{\partial y}\right),$$

where we have used parentheses to denote the nabla operator. This results in the equation:

$$\frac{\partial}{\partial x}\left(\frac{h^3}{12\mu}\rho\frac{\partial p}{\partial x}-\rho\frac{V_x h}{2}\right)+\frac{\partial}{\partial y}\left(\frac{h^3}{12\mu}\rho\frac{\partial p}{\partial y}-\rho\frac{V_y h}{2}\right)=\frac{\partial}{\partial t}(\rho h). \tag{4.9}$$

Note that often the sliding velocity is chosen to be in the direction of one of the coordinate axes, so that the Reynolds equation will contain only one sliding term. This is, however, not always possible or desirable as in the cases of (i) a square bearing moving along its diagonal, (ii) a journal bearing rotating around its axis and reciprocating along it, (iii) a spherical bearing executing precession motion (rotation and wobble), etc.

4.3.2 Plain Polar Coordinates (r, θ)

A typical example for this application is the flat circular bearing.

In Figure 4.2, we put $r_1 = r$. The grad operator is then:

$$\nabla_{r\theta}\Phi \triangleq \frac{\partial\Phi}{\partial r}\hat{\mathbf{e}}_r+\frac{1}{r}\frac{\partial\Phi}{\partial\theta}\hat{\mathbf{e}}_\theta.$$

The div operator is:

$$\nabla_{r\theta}\cdot\mathbf{F}\triangleq\frac{1}{r}\left(\frac{\partial(rF_r)}{\partial r}+\frac{\partial F_\theta}{\partial\theta}\right).$$

The Reynolds equation becomes:

$$\frac{1}{r}\frac{\partial}{\partial r}\left(r\left[\frac{h^3}{12\mu}\rho\frac{\partial p}{\partial r}-\rho\frac{V_r h}{2}\right]\right)+\frac{1}{r}\frac{\partial}{\partial\theta}\left(\frac{1}{r}\frac{h^3}{12\mu}\rho\frac{\partial p}{\partial\theta}-\rho\frac{V_\theta h}{2}\right)=\frac{\partial}{\partial t}(\rho h). \tag{4.10}$$

4.3.3 Cylindrical Coordinates (z, θ) with Constant R

A typical example for this application is a journal bearing of radius R rotating about the z-axis with angular velocity ω.

In Figure 4.2, we put $r_1 = R$.

The grad operator is:

$$\nabla_{z\theta}\Phi\triangleq\frac{\partial\Phi}{\partial z}\hat{\mathbf{e}}_z+\frac{1}{R}\frac{\partial\Phi}{\partial\theta}\hat{\mathbf{e}}_\theta.$$

The div operator is:

$$\nabla_{z\theta}\cdot\mathbf{F}\triangleq\frac{1}{R}\left(\frac{\partial(RF_z)}{\partial z}+\frac{\partial F_\theta}{\partial\theta}\right).$$

The Reynolds equation becomes:

$$\frac{\partial}{\partial z}\left(\frac{h^3}{12\mu}\rho\frac{\partial p}{\partial z}-\rho\frac{V_z h}{2}\right)+\frac{1}{R}\frac{\partial}{\partial\theta}\left(\frac{h^3}{12\mu}\rho\frac{1}{R}\frac{\partial p}{\partial\theta}-\rho\frac{R\omega h}{2}\right)=\frac{\partial}{\partial t}(\rho h), \tag{4.11}$$

which is equivalent to Eq. 4.9, if we put $z = x$ and $R\theta = y$, etc.

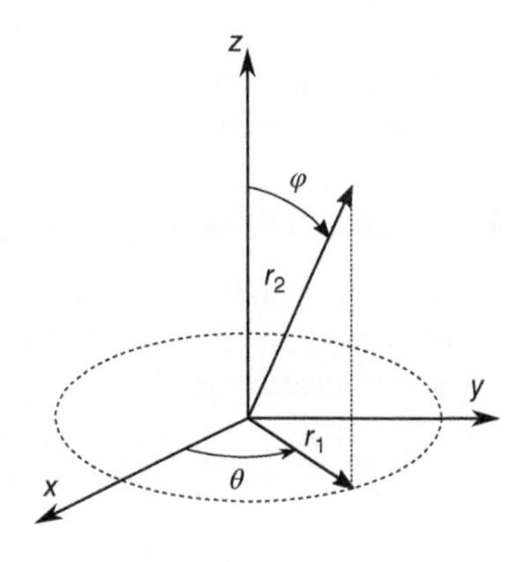

Figure 4.2 Polar and spherical coordinates.

4.3.4 Conical Coordinates (r, θ) $(\phi = \alpha = \text{Constant})$

A typical example for this application is the conical bearing. In Figure 4.2, we put $r_2 = r$ and $\phi = \alpha = $ constant.

The grad operator is:

$$\nabla_{r\theta}\Phi \triangleq \frac{\partial \Phi}{\partial r}\hat{\mathbf{e}}_r + \frac{1}{r\sin\alpha}\frac{\partial \Phi}{\partial \theta}\hat{\mathbf{e}}_\theta.$$

The div operator is:

$$\nabla_{r\theta}\cdot\mathbf{F} \triangleq \frac{1}{r\sin\alpha}\left(\frac{\partial(r\sin\alpha F_r)}{\partial r} + \frac{\partial F_\theta}{\partial \theta}\right).$$

The Reynolds equation becomes:

$$\frac{1}{r}\frac{\partial}{\partial r}\left(r\left[\frac{h^3}{12\mu}\rho\frac{\partial p}{\partial r} - \rho\frac{V_r h}{2}\right]\right) + \frac{\partial}{\partial \theta}\left(\frac{1}{(r\sin\alpha)^2}\frac{h^3}{12\mu}\rho\frac{\partial p}{\partial \theta} - \rho\frac{\omega h}{2}\right) = \frac{\partial}{\partial t}(\rho h), \tag{4.12}$$

where we have used $V_\theta = r\omega$.

The only difference between these equations and the ones belonging to plain polar coordinates is the appearance of the multiplicator $\sin\alpha$ to the r terms. Thus, assigning the value $\sin\alpha = 1$ $(\alpha = \pi/2)$ retrieves the plain polar, or rather the **annular bearing** case:

$$\frac{1}{r}\frac{\partial}{\partial r}\left(r\left[\frac{h^3}{12\mu}\rho\frac{\partial p}{\partial r} - \rho\frac{V_r h}{2}\right]\right) + \frac{\partial}{\partial \theta}\left(\frac{1}{r^2}\frac{h^3}{12\mu}\rho\frac{\partial p}{\partial \theta} - \rho\frac{\omega h}{2}\right) = \frac{\partial}{\partial t}(\rho h). \tag{4.13}$$

On the other hand, taking the limit $(\sin\alpha) \to \alpha \to 0$ and mapping $r\sin\alpha \mapsto R = $ const. and $r \mapsto z$, retrieves the cylindrical journal bearing case.

Furthermore, to facilitate approximate calculation or numerical solution, it might be useful to perform a further coordinate transformation as follows.

First, we multiply Eq. 4.12 by $(r\sin\alpha)^2$ to yield:

$$\begin{aligned} r\sin\alpha\frac{\partial}{\partial r}\left(r\sin\alpha\frac{h^3}{12\mu}\rho\frac{\partial p}{\partial r} - \rho\frac{V_r h}{2}\right) + \frac{\partial}{\partial \theta}\left(\frac{h^3}{12\mu}\rho\frac{\partial p}{\partial \theta} - \rho\frac{(r\sin\alpha)^2\omega h}{2}\right) \\ = (r\sin\alpha)^2\frac{\partial}{\partial t}(\rho h). \end{aligned} \tag{4.14}$$

Then we affect the transformation

$$r\frac{\partial}{\partial r} = \frac{\partial}{\partial \xi},$$

so that

$$\xi = \log\left(\frac{r}{r_o}\right),$$

which yields:

$$\begin{aligned} \sin\alpha\frac{\partial}{\partial \xi}\left(\sin\alpha\frac{h^3}{12\mu}\rho\frac{\partial p}{\partial \xi} - \rho\frac{V_r h}{2}\right) + \frac{\partial}{\partial \theta}\left(\frac{h^3}{12\mu}\rho\frac{\partial p}{\partial \theta} - \rho\frac{(e^\xi\sin\alpha)^2\omega h}{2}\right) \\ = (e^\xi\sin\alpha)^2\frac{\partial}{\partial t}(\rho h). \end{aligned} \tag{4.15}$$

The advantage of this formulation is that for small aspect ratio of the cone (or annulus), i.e. $r = r_o + \Delta r$ with $\Delta r/r_o \ll 1$, (so that $\xi \approx \Delta r/r_o$ and $e^\xi \approx 1$), we can approximate the solution by that of a flat rectangular bearing.

4.3.5 Spherical Coordinates (θ, ϕ) $(r = R = \text{Constant})$

A typical application is the spherical bearing with radius R.

In Figure 4.2, we put $r_2 = R = $ const.

The grad operator is:

$$\nabla_{\theta\phi}\Phi \triangleq \frac{1}{R\sin\phi}\frac{\partial \Phi}{\partial \theta}\hat{\mathbf{e}}_\theta + \frac{1}{R}\frac{\partial \Phi}{\partial \phi}\hat{\mathbf{e}}_\phi.$$

The div operator is:

$$\nabla_{\theta\phi} \cdot \mathbf{F} \triangleq \frac{1}{R\sin\phi}\left(\frac{\partial F_\theta}{\partial\theta} + \frac{\partial(R\sin\phi F_\phi)}{\partial\phi}\right).$$

With rotation around the z-axis with velocity $V = R\omega$, the Reynolds equation becomes:

$$\frac{1}{R^2\sin\phi}\left[\frac{\partial}{\partial\theta}\left(\frac{1}{\sin\phi}\frac{h^3}{12\mu}\rho\frac{\partial p}{\partial\theta} - \rho\frac{R^2\omega h\sin\phi}{2}\right) + \frac{\partial}{\partial\phi}\left(\sin\phi\frac{h^3}{12\mu}\rho\frac{\partial p}{\partial\phi}\right)\right] = \frac{\partial}{\partial t}(\rho h). \tag{4.16}$$

Note that here too, if we put $\sin\phi \approx 1$, we retrieve the journal-bearing equation. Whereas, if we put $\sin\phi \to 0$ and $R\sin\phi \mapsto r$, we retrieve the plane polar or annular bearing equation.

4.4 Interpretation of the Reynolds Equation When Both Surfaces are Moving and Not Flat

In many interesting design applications, it is not obvious how to apply the Reynolds Equation 4.7 to the situation at hand so as to formulate the problem correctly either for analytical, approximate, or, more particularly, a numerical solution. Such situations often involve simultaneous motion of both bearing surfaces in and normal to the film direction, rolling with or without sliding, and all of these combined with features present on the otherwise smooth surfaces, such as grooves and so forth. This has particular relevance to journal bearings and grooved bearings. This section will deal with these situations systematically, by considering, without loss of generality, the plane, incompressible Reynolds Equation, written as:

$$\frac{\partial}{\partial x}\left(\frac{h^{*3}}{12\mu}\frac{\partial p}{\partial x} - \frac{1}{2}Uh^*\right) = \frac{\partial h^*}{\partial t}, \tag{4.17}$$

where we have used U to denote the effective velocity to be used in the equation. As has been shown above, this equation is in fact the integrated continuity equation, which we can rearrange for more clarity as:

$$\frac{\partial}{\partial x}\left(\frac{h^{*3}}{12\mu}\frac{\partial p}{\partial x}\right) = \frac{U}{2}\frac{\partial h^*}{\partial x} + \frac{\partial h^*}{\partial t}, \tag{4.18}$$

where $U = U_1 + U_2$ is the effective sliding velocity and h^* is the general gap-height function. We have, as is customarily done, put the pressure (or Poiseuille) induced flow term on the L.H.S. and the shear (or Couette) and the squeeze terms on the R.H.S. of the equation.

Generally, the gap function may be expressed as:

$$h^*(x,t) = H(x) + h_1(x - U_1 t) + h_2(x - U_2 t) + \delta(t),$$

where

- H may be viewed as a stationary, nominal-gap function,
- h_1 and h_2 are due to features on either or both of the bearing surfaces, which move at the velocity of the respective surface,
- $\delta(t)$ is the space-independent time variation of the gap, e.g. owing to vibration of either or both (rigid) bearing members.

The problem of correct formulation lies to a large extent with the choice of a suitable spatial coordinate system to facilitate solution. Our task is thus to define a reference point in the space that enables defining the correct expressions for the film thickness function h and the effective velocity U from the problem specification. In hydrodynamic lubrication theory, the procedure adopted was as follows (Halling 1978):

- first, ignore the $\delta(t)$ term, i.e. $\partial h^*/\partial t$ arising from purely temporal variation of the film height (by putting $U_{1,2} = 0$)

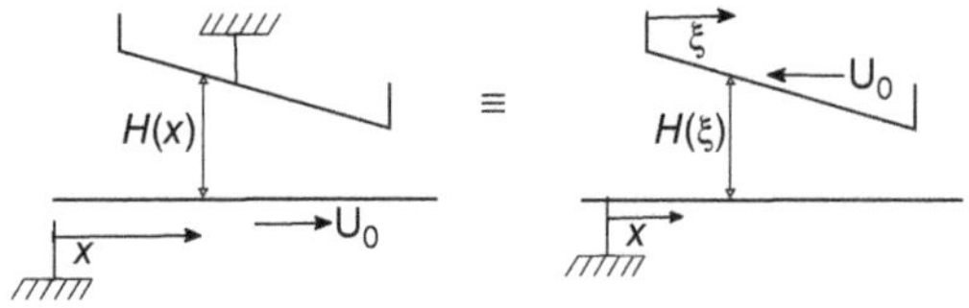

Figure 4.3 Fixed inclined upper surface and moving flat lower one.

- then, whenever possible, define the space coordinate system ξ such that the film thickness function h^* is stationary, that is, $h^*(\xi)$ is constant at point ξ,
- if and when this is accomplished, enter the temporal term back again.

The purpose is to *separate* the kinematics into purely *spatial* and purely *temporal* ones.

Although this might look sufficiently clear, there are situations in which it is not so obvious, straightforward or even possible to apply correctly. Let us thus go through the following examples as illustration. Here, we shall denote the stationary gap height by $H(x)$ (or $H(\xi)$), the (periodic) features on the gap by $h(x)$ (or $h(\xi)$) and the time variation of the gap by $\delta(t)$:

4.4.1 Stationary Inclined Upper Surface, Sliding Lower Member

For the situation depicted in Figure 4.3, we have for the left panel

$$\begin{aligned} h^* &= H(x) \\ U &= U_0 \\ \delta(t) &= \frac{\partial h^*}{\partial t} = 0 \\ \Rightarrow \frac{\partial}{\partial x}\left(\frac{H^3}{12\mu}\frac{\partial p}{\partial x}\right) &= \frac{U_0}{2}\frac{\partial H}{\partial x}. \end{aligned}$$

This situation is equivalent to the right figure. This most classical example can be shown to be true by affecting the (Galilean) transformation:

$$(x, t) \rightarrow (\xi, t)$$

$$\xi = x + U_0 t, \tag{4.19}$$

under which $U \rightarrow -U_0$ in Eq. 4.18 and

$$\begin{aligned} \frac{\partial}{\partial x} &\rightarrow \frac{\partial}{\partial \xi} \\ \frac{\partial}{\partial t} &\rightarrow \frac{\partial}{\partial t} + U_0\frac{\partial}{\partial \xi} \end{aligned}$$

to yield

$$\frac{\partial}{\partial \xi}\left(\frac{H^3}{12\mu}\frac{\partial p}{\partial \xi}\right) = \frac{U_0}{2}\frac{\partial H}{\partial \xi},$$

which is identical to the equation describing the L.H.S. figure but with ξ replacing x. That is, with the motion being observed from the moving slider.

Finally, we can impose a pure time variation, $\delta(t)$ normal to the gap at either or both the upper or the lower member.

4.4.2 Pure Surface Motion

When the tangential motion takes place in the surface only, such as in plane journal bearings, we have the *sliding-rolling* situation as illustrated in Figure 4.4. In this case the effective velocity is the so-called *entrainment*

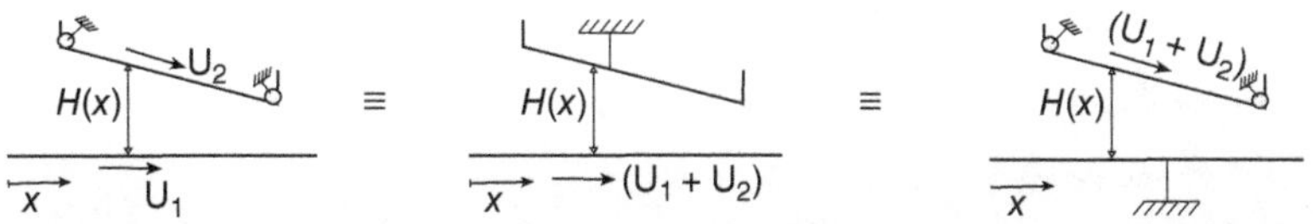

Figure 4.4 Pure surface motion with various surface velocities.

velocity, which is the sum of the surface velocities $U_1 + U_2$. Note that in the left-hand and the right-hand panels, we have made use of the fact that $\partial H/\partial x \ll 1$, which is the case in most practical bearing applications, so that

$$U = U_{\text{surface}} \cos(\arctan \partial H/\partial x) \approx U_{\text{surface}}$$

Obviously, this scheme will lose its validity if there are discontinuities (such as grooves) present on the bearing surfaces (see section 4.6).

4.4.3 Inclined Moving Upper Surface With Features

For the situation depicted in Figure 4.5 (left-hand panel) we have

$$U \approx U_0$$
$$\frac{\partial h^*}{\partial t} = \dot{\delta}(t).$$

Thus,

$$\frac{\partial}{\partial x}\left(\frac{H^3}{12\mu}\frac{\partial p}{\partial x}\right) = \frac{U_0}{2}\frac{\partial H}{\partial x} + \dot{\delta}(t).$$

If we now transform $(x, t) \rightarrow (\xi, t)$ by

$$\xi = x - U_0 t, \tag{4.20}$$

so that,

$$\frac{\partial}{\partial x} \rightarrow \frac{\partial}{\partial \xi}$$
$$\frac{\partial}{\partial t} \rightarrow \frac{\partial}{\partial t} - U_0 \frac{\partial}{\partial \xi},$$

then, transforming for this case first with $\frac{\partial h}{\partial t} \equiv 0$ and then plugging in $\dot{\delta}(t)$, we obtain the middle panel of Figure 4.5.

$$\frac{\partial}{\partial \xi}\left(\frac{H^3}{12\mu}\frac{\partial p}{\partial \xi}\right) = \frac{U_0}{2}\frac{\partial H}{\partial \xi} + \dot{\delta}(t).$$

Finally, placing $\dot{\delta}(t)$ on the top member, instead of the lower one, we get the right-hand panel of Figure 4.5.

4.4.4 Moving Periodic Feature on One or Both Surfaces

This constitutes the most difficult situation to deal with since it is not always evident or even possible to obtain a stationary gap by transformation. Here, the gap function is composed of a stationary as well as a periodic function.

We begin with the simpler special case with parallel nominal (average) gap.

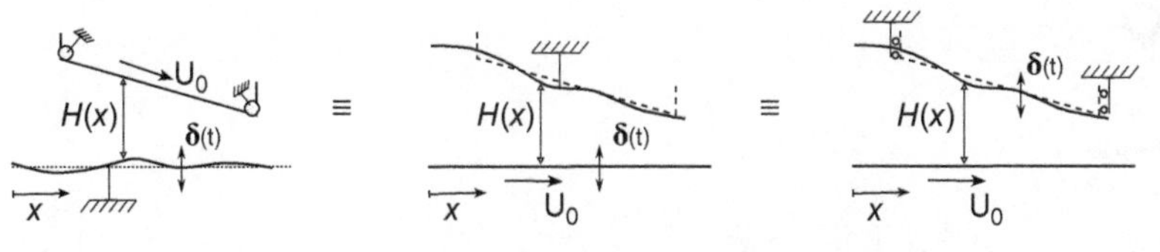

Figure 4.5 Inclined moving upper surface (with possible features).

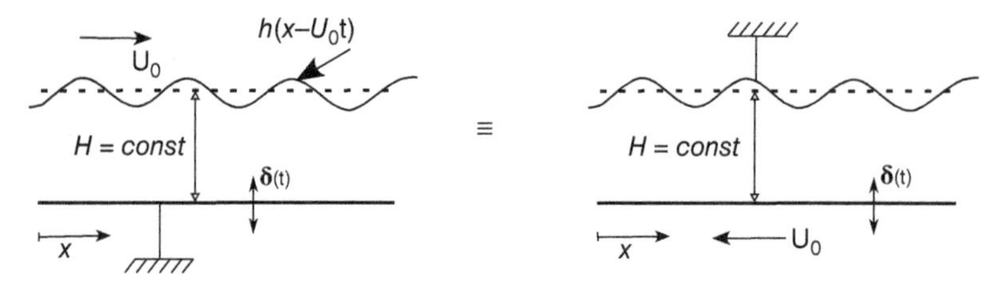

Figure 4.6 Moving and stationary grooves.

Moving and Stationary Grooves With Parallel Nominal Gap

Consider the case of two parallel surfaces one of which is grooved, as depicted in Figure 4.6. The appropriate Reynolds equation is

$$\frac{\partial}{\partial \xi}\left(\frac{h^{*3}}{12\mu}\frac{\partial p}{\partial \xi}\right) = \frac{U_0}{2}\frac{\partial h}{\partial \xi} - U_0\frac{\partial h}{\partial \xi} = -\frac{U_0}{2}\frac{\partial h}{\partial \xi}$$

with $h^* = H + h$, which can be obtained from Eq. 4.18 using the transformation Eq. (4.20). N.B. if either of the surfaces is, in addition, moving with $\delta(t)$ normal to the flow, then

$$\frac{\partial}{\partial \xi}\left(\frac{(H+h+\delta)^3}{12\mu}\frac{\partial p}{\partial \xi}\right) = -\frac{U_0}{2}\frac{\partial h}{\partial \xi} + \dot{\delta}(t).$$

Moving and Stationary Grooves With Sliding Surface

Now we are ready to consider the case most relevant to our inquiry, owing to its complexity of solution, i.e. pattern on the upper sliding surface with a non-constant $H = H(x)$, as depicted in Figure 4.7. Applying the transformation Eq. (4.20) to the relevant terms of the R.H.S. of Eq. 4.18, without the $\dot{\delta}(t)$ term for the time being, we get

$$\frac{\partial}{\partial x}\left(\frac{(H(x)+h(\xi))^3}{12\mu}\frac{\partial p}{\partial x}\right) = \frac{U_0}{2}\frac{\partial H}{\partial x} + \frac{U_0}{2}\frac{\partial h}{\partial \xi} - U_0\frac{\partial h}{\partial \xi} = \frac{U_0}{2}\left(\frac{\partial H}{\partial x} - \frac{\partial h}{\partial \xi}\right). \tag{4.21}$$

N.B. When $H =$ const. we retrieve the case of the previous section.

Thus we have a wedge component arising from the nominal gap H together with the moving feature (grooves, etc.) component, which is difficult to interpret geometrically, owing to non-linearity of the L.H.S. Let us note that now the time t appears only as a parameter in the equation within $\xi = x - U_0 t$. Thus, for any value $t = t_0$ in the interval, or feature period, $0 \leq t < \lambda_h/U_0$ (where λ_h is the wavelength of the periodic feature h), we can solve Eq. 4.21 as an ordinary differential equation in x to obtain the pressure $p(x, t_0)$. In that way, we obtain a periodic solution for the pressure. Thus,

- The greatest difficulty in obtaining a solution arises from the fact that even when $\delta(t)$ is set equal to zero, the L.H.S. is time dependent. Therefore, this equation admits only a periodic solution, (for periodic features).

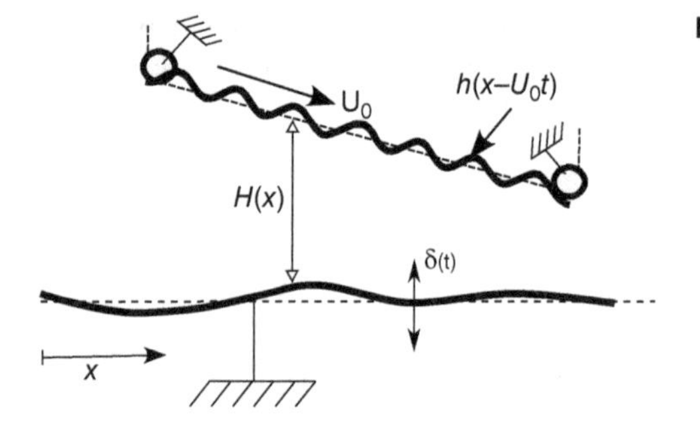

Figure 4.7 Moving and stationary grooves with sliding surface.

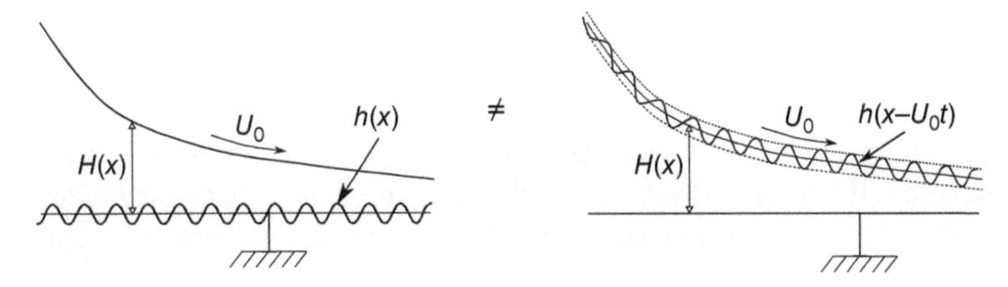

Figure 4.8 Moving and stationary grooves with sliding surface: the two situations are not equivalent.

- We can consider a time-averaged solution over one period. Then $\partial h/\partial \xi$ disappears from the R.H.S., but this does not make the solution much easier.
- Of great interest is the possible effect of rotating features on the film pressure, especially with regard to the dynamic stability aspects. This situation is depicted in (the grooved rotational bearing application example of), Figure 4.8, where it is clear that the left panel is not equivalent to the right panel. This aspect is also treated in Chapter 8.

Pure rolling motion

When both surfaces of Figure 4.8 move with identical speeds $U_1 = U_2 = U_0$, pure rolling motion results. Following the same procedure above, we can show that:

$$\frac{\partial}{\partial x}\left(\frac{(H(x)+h(\xi))^3}{12\mu}\frac{\partial p}{\partial x}\right) = U_0\frac{\partial H}{\partial x} + U_0\frac{\partial h}{\partial \xi} - U_0\frac{\partial h}{\partial \xi} = U_0\frac{\partial H}{\partial x}. \tag{4.22}$$

Thus we see that the Couette term (R.H.S.) is now independent of the surface feature, and thus of the time, which is intuitively obvious, since the entrainment is now without slip; but the Poiseuille term (R.H.S.) remains moving-feature dependent as in the previous case.

Rough bearing surfaces

In particular, when the surface features are random processes, as one assumes in many roughness models, we obtain the situation corresponding to lubrication with rough surfaces. This will be worked out further in Section 4.6.1 below.

4.5 Neglected Flow Effects

This section addresses the flow effects that are neglected in the usual formulation of the Reynolds equation and offers a means of assessing these in real situations and, where possible, alternative forms of the Reynolds equation to account for them.

The neglected effects are by necessity related to the basic assumptions adopted in deriving the Reynolds equation; namely (i) the conditions at the boundaries (or walls) of the film, or the degree of departure from smoothness and non-slip conditions, i.e. the wall boundary-condition related; (ii) the inertia effects, or the degree of departure from viscous flow, i.e. the Reynolds-number related, and (iii) fluid-rheology related, or when the viscosity departs from the Newtonian character.

Thus,

1. The boundary-related effects comprise (i) the smoothness of the boundary (with absence of slip) and (ii) the slip effects (with or without smooth boundaries).
2. The Reynolds-number-related effects comprise (i) the inertia effects *in se* arising from the convective terms in the momentum equation being no longer negligible (but which leave other fluid properties such as viscosity

and thermal conduction unaffected) and (ii) the turbulence effects (that affect the basic nature of the flow), which might occur while the convective terms are still relatively negligible. It is possible for one of the two effects to arise without the other, as we shall see further below.

3. The fluid-rheology related effects comprise the possibility of lubrication by a conducting gas under the influence of a magnetic field, as given in Constantinescu (1967), which leads to a similar pressure equation to the Reynolds equation, or lubrication with non-Newtonian fluids, etc. as reviewed in (Szeri 1987). This case will not be treated further in this chapter.

Regarding the wall boundary-related effects, the smoothness effects are gauged by the ratio of the size of the surface features to the gap height, whereas the slip effects are gauged either by the Knudsen number or by the porosity parameters of the surface. As for the inertia effects, namely convective inertia and turbulence, these are gauged by hydraulic and reduced Reynolds numbers respectively.

In the following, we shall treat each of those four situations in a separate section.

4.6 Wall Smoothness Effects

Departure from smoothness may be purposefully introduced, such as in various surface features that are added to a surface with the purpose of enhancing lubrication as in grooves, etc. or it might be adventitious such as surface roughness or damage incurred by repeated starting and stopping cycles of a self-acting bearing.

Film Height Discontinuities

At film height discontinuities, the assumptions leading to the Reynolds equation can become questionable. Burton (1968) performed an experimental study on the turbulent flow effects in bearings with a spiral groove configuration, while van der Stegen (1997) indicates that the effect of inertial flow terms may no longer be negligible at the transition from land to groove (Figure 4.9). In the literature, this transition is often assumed to be discontinuous (i.e. sudden). However, as this is impossible to manufacture, the transition is considered continuous over a length l. Now, with air as lubricant, $h = 10\ \mu$m, $d = 10\ \mu$m, $l = 100\ \mu$m and with an in-plane flow velocity $u = 10$ m s^{-1} and transversal velocity $w = u\ d/l$, the order-of-magnitude of the derivatives in the Navier–Stokes momentum equation is:

$$\underbrace{\rho u\frac{\partial u}{\partial x}}_{O(10^6)} + \underbrace{\rho w\frac{\partial u}{\partial z}}_{O(10^6)} = -\frac{\partial p}{\partial x} + \underbrace{\mu\frac{\partial^2 u}{\partial x^2}}_{O(10^4)} + \underbrace{\mu\frac{\partial^2 u}{\partial z^2}}_{O(10^6)}. \tag{4.23}$$

The Reynolds equation is based only on $\frac{\partial p}{\partial x}$ and $\mu\frac{\partial^2 u}{\partial z^2}$, while some other derivatives are of equal magnitude. When the in-plane velocity u is increased or when the length l is reduced, these inertial terms become even more important. Locally, the assumptions leading to the Reynolds equation will therefore become invalid. The overall effect on the global flow pattern may however be limited, in as much as the number and occurrence remain small. Arghir et al. (2002) use finite volume analysis to investigate the effect of surface discontinuities, including associated inertia effects. They show that the film discontinuities should always be taken into account in a consistent physical manner in order to be clear about their impact on global results of technological interest.

As far as the author is aware, there is no specific form of an equivalent Reynolds equation to deal with this type of situation.

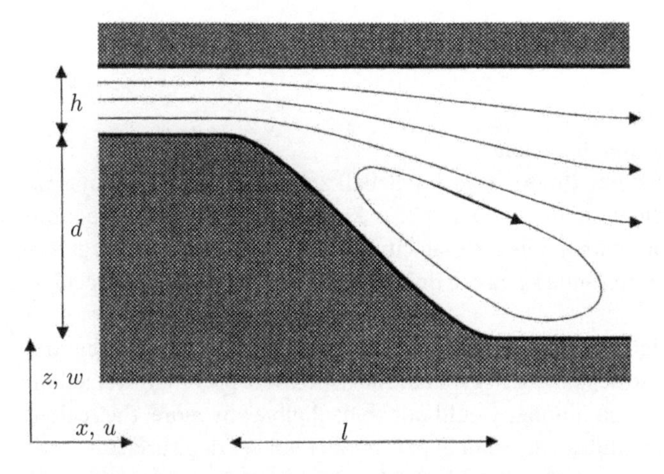

Figure 4.9 Probable flow pattern at the transition from land to groove. Source: Van der Stegen RHM 1997.

4.6.1 Effect of Surface Roughness

As mentioned earlier, if we assume roughness to be a random process with zero mean and σ standard deviation, and further assume that $\sigma \ll H$ and mean roughness slope $\langle |\mathrm{d}h/\mathrm{d}x| \rangle \ll 1$, then we can apply this to Eq. 4.21 with h representing the roughness profile and H the nominal gap. In this way, we have

$$H - h \approx H$$

and

$$(H + h)^3 \approx H^3(1 + 3\sigma/H)$$

so that the modified Reynolds equation for this case (now written with h as the nominal film height) becomes:

$$\nabla \cdot \left(\frac{\rho h^3(1 + 3\sigma/h)}{12\mu} \nabla p - \frac{1}{2}\rho h \mathbf{V}_t \right) = \frac{\partial}{\partial t}(\rho h) \tag{4.24}$$

The various effects of roughness on lubrication have been dealt with in a dedicated symposium (*Surface Roughness Effects in Lubrication* 1978). Berthe and Godet (1973) dedicated their paper to (re)derive the Reynolds equation so as to deal more systematically with moving rough surfaces. de Kraker et al. (2010) developed a texture-averaged Reynolds equation taking account of the dynamics of the fluid entrapped inside the surface features. Let us note that usual surface roughness, as distinguished from textured surfaces, is characterised by only very small roughness slopes so that Eq. 4.24 should suffice to deal with them.

4.7 Slip at the Walls

Molecular Effects and Rarefaction

In the case of very thin fluid films, the lubricant gas may not be considered as a continuum. The validity of the continuum mechanics formulation is indicated by the Knudsen number. This dimensionless number expresses the ratio between the mean distance a molecule travels between collisions, known as the mean free path λ, and the characteristic film thickness value h_m

$$\mathrm{Kn} = \frac{\lambda}{h_\mathrm{m}}, \tag{4.25}$$

with (Bird 1994)

$$\lambda = \frac{kT}{p\pi\sqrt{2}\sigma^2}, \tag{4.26}$$

and where k is Boltzmann's constant and σ the molecular diameter.

Based on the Knudsen number, flows may be divided into three categories (Schaff and Chambre 1958, chap. H). If Kn < 0.01, the Reynolds equation may be used without modification. If $0.01 <$ Kn < 0.1, the Reynolds equation must be redeveloped from the Navier–Stokes equations with a non-zero velocity, known as the slip velocity, at the wall (see further below). If Kn > 0.1, the continuum representation of the fluid is usually replaced by its molecular counterpart.

Air at standard conditions has a mean free path length $\lambda = 0.064$ μm. This means that the basic assumptions are valid down to a gap height of 5 μm. However, the inverse dependence of λ on the local pressure prevents rarefaction from being an issue in this case. And even if the Knudsen number would not be negligible any more, the region for which rarefaction becomes important, usually constitutes only a small part of the total bearing surface.

At any rate, it is possible to estimate, with high accuracy the slip velocity (in Couette flow) as a function of the mean free path. From this, the general velocity distribution as a function of the pressure gradient is derived, leading to the modified Reynolds equation (Burgdorfer 1959):

$$\nabla \cdot \left(\frac{\rho h^3}{12\mu} \left(1 + 6\frac{\lambda}{h}\right) \nabla p - \frac{1}{2}\rho h \mathbf{V}_t \right) = \frac{\partial}{\partial t}(\rho h) \tag{4.27}$$

We note that the difference between this equation and Eq. 4.7, is the term $1 + 6\frac{\lambda}{h}$ (which is approximately equal to unity for Kn < 0.01 as has been mentioned above).

Other Slip Effects

There are two more effects that give rise to virtual slip at the boundary. The first is the presence of surface texture such as "riblets", cavities and roughness. This is a relatively new area of investigation with application to open boundary-layer and turbulent flows, which could be eventually of interest to gas bearing applications. The interested reader can find an overview in (Lee et al. 2014; Martin and Bhushan 2016). The second is the situation in which fluid issues from the surface as in porous bearings, see Chapter 13.

Inertia Effects in Aerodynamic Bearings

The effect of inertia in fluid bearings, and more particularly gas bearings, forms an old topic that, however, continues to be investigated up to today owing to its importance to design theory (Arghir et al. 2002; Belforte et al. 1999; Contantinescu et al. 1975; de Kraker et al. 2010; Dousti et al. 2016; Dowson 1961; Elkouh 1984a,b; Frêne and Constantinescu 1996; Frêne et al. 2006; Greenberg and Weger 1960; Heckelman and Ettles 1988; Kwan and Corbett 1998; Livesey 1960; Mori and Mori 1986; Mori et al. 1980a,c, 1990; Rhim and Tichy 1989; Stolarski and Chai 2008; Tanaka et al. 2007; Tang 1968; Tichy and Chen 1985; Tichy and Winer 1970a, to cite but a few.

One speaks of inertia effects when the convective terms in the flow equations (Navier–Stokes or boundary-layer) become appreciable in comparison with the viscous ones. In that case, the viscous flow formulation, namely, the Reynolds equation, might yield results that are inaccurate in proportion with the increase in inertia effects. These effects may be classified into (i) quantitative, i.e. departure from the viscous regime while the flow remains laminar and (ii) qualitative, when the flow departs from the laminar regime altogether. We note the following in this regard. Recently, the terms "non-laminar" or "super laminar" flow have been coined to characterise the occurrence of either or both of those conditions. A more appropriate term would, in our opinion, be "non-viscous" flow, since the flow could be inertial and laminar, as has just been mentioned. Let us note that:

- Transition to turbulence is gauged by the usual (hydraulic) Reynolds number based on the film thickness: $\mathrm{Re}_h = \frac{\rho U_0 h}{\mu}$, where U_0 is usually the mean flow velocity, for Poiseuille flow, and the slider surface speed in self-acting bearings. Usually, transition to turbulence takes place roughly in the range $1000 < \mathrm{Re}_h < 2000$ (Rhim and Tichy 1989) (for a slider, and even lower values for Poiseuille flow).
- The relative importance of inertial (convective) forces with respect to the viscous forces, whether laminar or turbulent, is gauged by the *reduced* Reynolds number Re*. This is the aforementioned mentioned Reynolds number that is further reduced by the aspect ratio of the channel h/L, where h and L are the characteristic height and length of the channel, respectively. Thus, $\mathrm{Re}^* = \mathrm{Re}_h \frac{h}{L} = \frac{\rho U_0 h}{\mu} \frac{h}{L}$. Usually, inertia effects are considered negligible when $\mathrm{Re}^* \ll 1$. Thus, when this value exceeds $O(10^{-1})$, inertia effects can become appreciable.

Let us consider as an example, a self-acting (aerodynamic) journal bearing, having a mean gap height h_m and sliding speed equal to the circumferential rotor speed $u = \omega R \Rightarrow \bar{u} = \omega R/2$, where R is the shaft radius. For the aspect ratio, if we use h_m/R, we obtain a reduced Reynolds number

$$\mathrm{Re}^* = \frac{\rho u h_\mathrm{m}}{\mu} \frac{h_\mathrm{m}}{2R} = \frac{\rho u h_\mathrm{m}^2}{2\mu R} = \frac{\rho \omega h_\mathrm{m}^2}{2\mu}. \tag{4.28}$$

Taking air as a lubricant, for the following case: $h_\mathrm{m} = 10\ \mu\mathrm{m}$, $R = 10$ mm, and $u = 250\ \mathrm{m\ s^{-1}}$ (corresponding to an Ndm of ca. 5×10^6 mm×rpm), the reduced Reynolds number $\mathrm{Re}^* = 0.095$. This example shows that in these extreme operating conditions, the inertial effects may become appreciable.

In the following, the effect of turbulence and of the convective inertia terms will be treated separately.

4.8 Turbulence

Turbulence is the departure from (layered) laminar flow, with its well-behaved pressure and velocity fields, to a chaotic behaviour (in those characteristics) that is superposed on the so-called "main" (or more appropriately, "mean") pressure and velocity fields. This turbulence is induced by the increase of kinetic energy as compared to viscous dissipation. In that respect, low-viscosity fluids, in particular gases, water, … are more susceptible to turbulence than the more viscous fluids, such as oils. The susceptibility is gauged by the kinematic viscosity $\nu = \mu/\rho$. In actual fact no fluid is exempt from turbulence when the kinetic energy has reached a certain critical level, as gauged by the Reynolds number (in one of its various forms that characterise different flow configurations and kinematics). Here too there has been much written, in regard to applications in fluid-film bearings, which continues to this day, e.g. (Burton 1968; Constantinescu 1962; Contantinescu et al. 1975; Dousti et al. 2016; Elrod and Ng 1967; Frêne and Constantinescu 1996; Frêne et al. 2006; Hélène et al. 2003; Hirs, 1973; King and Taylor, 1975; Martin and Bhushan 2016) being a representative pick, the last one of which providing yet more references on the topic.

4.8.1 Formulation

Turbulent flow may often be adequately described by constitutive relationships derived by considering that the "fluctuating" quantities to be much smaller in magnitude than the "main" quantities. Thus writing (Schlichting 1968):

$$\mathbf{u} = \bar{\mathbf{u}} + \mathbf{u}'$$

$$p = \bar{p} + p'$$

$$\rho = \bar{\rho} + \rho'$$

$$T = \bar{T} + T',$$

where, we have on the L.H.S. the quantity in question being set equal to the sum of the "main" value and the "fluctuating" value, respectively, on the R.H.S.

This reduces the flow problem to a "perturbation" system in which the equations relating the main quantities have the same *form* as the usual Navier–Stokes, boundary-layer, laminar case, but with modified "viscosity" and "conduction" coefficients. Namely, considering the main flow to be in the x-direction with velocity u, (Hirs 1973)

$$\tau_y = -\rho v' u' \triangleq \mu_{y-\text{turb}} \frac{\partial \bar{u}}{\partial y}$$

and

$$q_y = v' \rho C_p T' \triangleq -k_{y-\text{turb}} \frac{\partial \bar{T}}{\partial y}$$

where, τ_y is the "viscous" shear stress, $\mu_{y-\text{turb}}$ is the equivalent "viscosity", q_y is the heat flow and $k_{y-\text{turb}}$ is the equivalent "thermal conductivity coefficient", all being in the y-direction and in regard to the main flow in the x-direction.

Two pertinent remarks are in order here regarding this formulation,

(i) These "effective" coefficient values depend on the perturbation values and thus on the *local* Reynolds number. Usually, some power-law relation is assumed, e.g. (Hirs 1973)

$$\frac{\tau_i}{1/2\rho\bar{u}^2} = n_i \mathrm{Re}^{m_i} \tag{4.29}$$

where, $\mathrm{Re} = \rho\bar{u}h/\mu$, and these values are different in the different flow directions (as indicated by the subscript i), since turbulence is *anisotropic.*

(ii) These effective coefficient values (viscosity, ...) are usually much larger than their laminar counterparts. Thus, Gross et al. (1980) states that film turbulence leads to an apparent increase in viscosity, which results in an increase in load capacity and in frictional losses.

The above analysis and derivation is often referred to as the "bulk flow" model for turbulent (channel) flow (Hirs 1973).

Applying the integral forms of the continuity and momentum equations as in section 4.9.2, we obtain

$$\frac{\partial}{\partial t}\int_0^h \bar{\rho} f(\bar{\mathbf{v}})\mathrm{d}y + \frac{\partial}{\partial x}\int_0^h \bar{\rho}\bar{\mathbf{v}} f(\bar{\mathbf{v}})\mathrm{d}y = \int_0^h (-\frac{\partial \bar{p}}{\partial x} + \frac{\partial \tau_i}{\partial y}) f'(\bar{\mathbf{v}})\mathrm{d}y. \tag{4.30}$$

In particular, taking $f \equiv \bar{\mathbf{v}}$, we obtain the integral momentum equation

$$\frac{\partial}{\partial t}\int_0^h \bar{\rho}\bar{\mathbf{v}}\mathrm{d}y + \frac{\partial}{\partial x}\int_0^h \bar{\rho}\bar{\mathbf{v}}^2\mathrm{d}y = -\frac{\partial \bar{p}}{\partial x}h + 2\tau_{w_i}, \tag{4.31}$$

where, τ_{w_i} is the shear stress at the wall (in the respective flow direction).

4.8.2 Turbulent Reynolds equation without and with inertial effects

(i) **Without inertia effects** The general Reynolds equation may be modified in a simple way to account for the effect of film turbulence. Elrod and Ng (1967) propose a method using pressure dependent turbulence factors. However, the method that is widely adopted nowadays is based on (Hirs 1973), in which RE is written as:

$$\frac{\partial}{\partial x}\left(\frac{h^3}{k_x\mu}\rho\frac{\partial p}{\partial x} - \rho\frac{V_x h}{2}\right) + \frac{\partial}{\partial y}\left(\frac{h^3}{k_y\mu}\rho\frac{\partial p}{\partial y}\right) = \frac{\partial}{\partial t}(\rho h) \tag{4.32}$$

where,

$$k_x = n_x\left(\frac{\rho V_x h}{\mu}\right)^{m_x} \triangleq n_x(\mathrm{Re}_x)^{m_x}$$

$$k_y = n_y\left(\frac{\rho V_y h}{\mu}\right)^{m_y} \triangleq n_y(\mathrm{Re}_y)^{m_y}$$

Typical values of the n's and m's, (which should be greater than 12), are empirically determined. Thus, some authors prefer to correlate the k's as follows:

$$k_x = 12 + n'_x(\mathrm{Re}_x)^{m'_x}$$

$$k_y = 12 + n'_y(\mathrm{Re}_y)^{m'_y}$$

Some authors again prefer to use the notation $G_x = 1/k_x$ and $G_y = 1/k_y$ instead.

(ii) **With inertia effects** This is essentially the same method used for dealing with inertia effects in general, see section 4.9.2.

4.8.3 Taylor Vortices

These constitute a distinct type of turbulence in Couette annulus flows. Before the onset of fully-turbulent flow in journal bearings, specific types of vortices may appear in the film, viz. the Taylor vortices. For concentric cylinders, these large-scale features occur in the flow field if the Taylor number Ta exceeds a certain value (Taylor 1923):

$$\mathrm{Ta} = \frac{\rho^2 u^2 h_\mathrm{m}^3}{\mu^2 R} \geq 1700. \tag{4.33}$$

For the case outlined above, with $h_\mathrm{m} = 10\ \mu\mathrm{m}$, $R = 10$ mm and $u = 250\ \mathrm{m\ s^{-1}}$, the Taylor number Ta = 35.6. This value remains far below the onset value of 1700, which indicates that Taylor vortices will not occur for this case.

4.9 Approximate Methods for Incorporating the Convective Terms in Integral Flow Formulations and the Modified Reynolds Equation

4.9.1 Introduction

Approximate integral methods still enjoy some popularity in fluid mechanics because of their simplicity and easy application. In this section, the boundary-layer equations, that are also used to describe narrow-passage flow, are expressed as a single conservation equation with an arbitrary weighting function of the main flow velocity. Such a formulation leads to a generalised integral flow conservation equation which forms the basis of the integral solution methods. These solutions are obtained by inserting an arbitrarily assumed (main flow) velocity profile into the integral (conservation) equation, thus reducing it to a set of ordinary differential equations.

One of the main claims of this method is that the solution is somewhat insensitive to the velocity profile chosen, as long as that would correspond qualitatively to the actual situation. The velocity profile may, in the general case, be assumed to vary in the flow direction in which case an approximate solution may be obtained for the entrance problem. This will be the first case considered. However, when the flow does not greatly deviate from the viscous regime, this method may be used, with a fixed (parabolic) velocity profile, to account for the inertia effects in the flow. This, sometimes known as the method of "averaged inertia", will be the second case considered. In particular, this last formulation has been used by some authors to derive a "modified" Reynolds equation that might account more closely for the inertial effects in the case of small amplitude transverse oscillation. Some illustrative examples will be given.

Although nowadays CFD solutions are capable of dealing with very complex and varied flow situations, it remains nevertheless interesting to have approximate solutions at our disposal, which provide for some semi-analytical insights into a problem while facilitating quick simulations.

The subsequent analysis is based on (Al-Bender 1992).

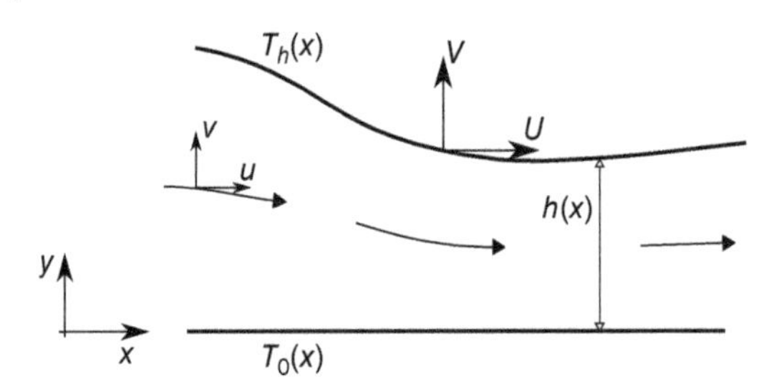

Figure 4.10 Schematic of the 2-D flow configuration.

4.9.2 Analysis

For simplicity, the derivation will be confined to plane flow; the results being easily generalised for the other cases, and will be given subsequently.

Some necessary remarks

- Since we are dealing with plane flow, it is more customary to use (x, y), instead of (x, z), to denote the flow space as shown in Figure 4.10.
- We shall initially, consider also the (thermodynamic) energy equation together with the flow equations since it completes the systematic energy-conservation analysis. Solutions to the thermal problem will, however, be dealt with separately in Chapter 17.

Consider the flow configuration shown in Figure 4.10. The equations of continuity, momentum and energy, in boundary-layer approximated form (see Chapter 2), are:

$$\frac{\partial \rho}{\partial t} + \frac{\partial (\rho u)}{\partial x} + \frac{\partial (\rho v)}{\partial y} = 0, \tag{4.34}$$

$$\rho(\frac{\partial u}{\partial t} + u\frac{\partial u}{\partial x} + v\frac{\partial u}{\partial y}) = -\frac{\partial p}{\partial x} + \mu\frac{\partial^2 u}{\partial y^2}, \tag{4.35}$$

and

$$C_p\rho(\frac{\partial T}{\partial t} + u\frac{\partial T}{\partial x} + v\frac{\partial T}{\partial y}) = \frac{\partial p}{\partial t} + u\frac{\partial p}{\partial x} + k\frac{\partial^2 T}{\partial y^2} + \mu(\frac{\partial u}{\partial y})^2, \tag{4.36}$$

where it is assumed that u, v and T are functions of (x, y, t); p, ρ and h functions of (x, t).

The boundary conditions that are immediately available on the velocity and temperature fields are:

$$y = 0 : u = 0, \quad v = 0, \quad T = T_0$$
$$y = h : u = U, \quad v = V = \partial h/\partial t, \quad T = T_h.$$

Condition of Compatibility at the Wall

Since, in Eq. 4.35, the pressure is independent of y whereas the velocity is not, a certain relationship is imposed between the pressure and the velocity fields, see (Schlichting 1968). In its lowest order, such a relationship is obtained by evaluating Eq. 4.35 at the wall ($y = 0$). Thus, using ($y = 0 : \ u = v = \partial u/\partial t = 0$) we obtain:

$$\frac{\partial p}{\partial x} = \mu\frac{\partial^2}{\partial y^2}u(x, 0, t). \tag{4.37}$$

Conservation Form of the Flow Equations

To derive the conservation form, we first define an arbitrary function $f(u)$ that is continuous and possesses a continuous first derivative for all values of u. Denoting by f' the first derivative of f w.r.t. u, (therefore, e.g., $\partial f(u)/\partial x = f'(u)\partial u/\partial x$), we multiply Eq. 4.34 by f and Eq. 4.35 by f' and add them to obtain (after rearrangement),

$$\frac{\partial}{\partial t}(\rho f(u)) + \frac{\partial}{\partial x}(\rho u f(u)) + \frac{\partial}{\partial y}(\rho v f(u)) = (-\frac{\partial p}{\partial x} + \mu\frac{\partial^2 u}{\partial y^2})f'(u). \tag{4.38}$$

In a similar manner, by defining an arbitrary function $g(T)$, we may write the energy equation in conservation form as:

$$c_p[\frac{\partial}{\partial t}(\rho g(T)) + \frac{\partial}{\partial x}(\rho u g(T)) + \frac{\partial}{\partial y}(\rho v g(T))]$$
$$= [\frac{\partial p}{\partial t} + u\frac{\partial p}{\partial x} + k\frac{\partial^2 T}{\partial y^2} + \mu(\frac{\partial u}{\partial y})^2]g'(T). \tag{4.39}$$

Equations 4.38 and 4.39 represent a hierarchy of denumerable equations whose left-hand sides are in conservation form. They are dimensionally sound, providing that $f(u)$ and $g(T)$ are expressible as convergent polynomial functions. We shall call them, for convenience, *pseudo*-conservation equations. Note that, dimensionally consistent, linear combinations of these equations are also possible. Let us consider the special cases corresponding to the known conservation equations, (N.B. we shall use subscripts to indicate partial derivatives):

- $f(u) \equiv 1, g(T) \equiv 0$ yields the conservation of mass or Eq. 4.34
- $f(u) = u, g(T) \equiv 0$ yields the conservation of momentum:

$$\frac{\partial}{\partial t}(\rho u) + \frac{\partial}{\partial x}(u\rho u + p) + \frac{\partial}{\partial y}(v\rho u - \mu\frac{\partial u}{\partial y}) = 0$$

- $f(u) = u^2/2, g(T) \equiv T$ yields the conservation of energy:

$$\frac{\partial}{\partial t}(\rho H) + \frac{\partial}{\partial x}(u\rho H) + \frac{\partial}{\partial y}(v\rho H - kT_y - \mu\frac{\partial}{\partial y}(u^2/2)) = 0,$$

where H is the total enthalpy:

$$H = c_p T + u^2/2.$$

One of the most important features of Eq. 4.38 (also Eq. 4.39), with regard to channel (or internal) flow, is that it can be readily integrated across the film, taking advantage of the available boundary conditions, thereby eliminating the velocity component v. Integrating Eq. 4.38 thus, on $0 \leq y \leq h$, and making use of the Leibniz rule of differentiation under the integral sign, we obtain:

$$\frac{\partial}{\partial t}\int_0^h \rho f(u)\mathrm{d}y - \underline{\rho(h)f(U)\frac{\partial h}{\partial t}} + \frac{\partial}{\partial x}\int_0^h \rho u f(u)\mathrm{d}y + \underline{\rho(h)Vf(U)}$$
$$= \int_0^h (-\frac{\partial p}{\partial x} + \mu\frac{\partial^2 u}{\partial y^2})f'(u)\mathrm{d}y,$$

where the underlined terms cancel each other out identically, yielding

$$\frac{\partial}{\partial t}\int_0^h \rho f(u)\mathrm{d}y + \frac{\partial}{\partial x}\int_0^h \rho u f(u)\mathrm{d}y = \int_0^h (-\frac{\partial p}{\partial x} + \mu\frac{\partial^2 u}{\partial y^2})f'(u)\mathrm{d}y. \tag{4.40}$$

Equation 4.40 is a *macroscopic* form of Eq. 4.38. The function $f(u)$ (and $g(T)$ in case of the energy equation) is often viewed as a *weighting* function that determines the nature of the conservation considered. We have seen that $f(u) = 1,\ u$ correspond to actual conservation laws. Some writers regard the case $f = u^2/2$ as representing the macroscopic conservation of *mechanical* energy. In that case, there is no reason why higher orders of $f(u)$ may not be considered as representing other (unknown) conservation laws. (Note that Eq. 4.40 is valid for all f's satisfying the above mentioned conditions.)

It is noteworthy also that Eq. 4.40 can be written entirely in u and its derivatives by eliminating the pressure gradient with the aid of Eq. 4.37. Thus, if u is to be specified by a number n of unknown parameters (functions of x and t), Eq. 4.40 can be used to determine these through the solution of an equivalent number of equations of the hierarchy. In the case of steady state flow, this would be a set of n non-linear first order ordinary differential equations. In the limit of infinite n, this would be equivalent to the solution obtained from series expansion.

For the case of radial flow, it may be easily shown that the corresponding version of Eq. 4.40 may be written as:

$$\frac{\partial}{\partial t}\int_0^h \rho f(u)dz + \frac{1}{r}\frac{\partial}{\partial r}\int_0^h r\rho u f(u)dz = \int_0^h (-\frac{\partial p}{\partial r} + \mu\frac{\partial^2 u}{\partial z^2})f'(u)dz, \tag{4.41}$$

where (r, z) are the radial and transverse coordinates respectively, and u is the main flow velocity. Other versions, corresponding to pipe flow or the general case of three-dimensional flow, may be similarly obtained.

The utility of Eq. 4.40 has been hitherto limited to obtaining approximate solutions to the problem, through specifying arbitrary velocity profiles. But before discussing this, it would be handier to normalise the transverse coordinate w.r.t. the film thickness. (The other variables may also be normalised if need be.) Thus by setting:

$$\eta = y/h\ ,$$

Eq. 4.40 becomes:

$$\frac{\partial}{\partial t}\int_0^1 \rho h f(u)d\eta + \frac{\partial}{\partial x}\int_0^1 \rho h u f(u)d\eta = \int_0^1 (-\frac{\partial p}{\partial x} + \mu\frac{1}{h^2}\frac{\partial^2 u}{\partial \eta^2})hf'(u)d\eta. \tag{4.42}$$

(The version corresponding to radial flow is similarly obtained.)

4.9.3 Limiting Solution: the Reynolds Equation

Equation 4.42 admits an exact solution, which is obtained by setting its left-hand side identically equal to zero, with $f'(u) \equiv 1$, or $f(u) = u$. This corresponds to viscous flow in which the inertia forces are negligible. It may be shown, in such a case, that this is equivalent to a parabolic velocity distribution across the gap, viz.

$$u = u_p + u_c = q(x,t)\cdot 4\eta(1-\eta) + (1-\eta)U, \tag{4.43}$$

where the first term u_p corresponds to Poiseuille, or pressure induced flow and the second u_c to simple Couette, or shear flow. The equivalence is ensured by the compatibility condition 4.37 [1], which further gives the relation

$$q = -\frac{h^2}{8\mu}\frac{\partial p}{\partial x}. \tag{4.44}$$

Substituting Eqs. 4.43 and 4.44 in Eq. 4.42 with $f \equiv 1$, i.e. the continuity equation:

$$\frac{\partial}{\partial t}(\rho h) + \frac{\partial}{\partial x}\left(\rho h\left(\frac{2}{3}q + \frac{1}{2}U\right)\right) = 0 \tag{4.45}$$

and rearranging yields then the celebrated Reynolds equation for plane flow:

$$\frac{\partial}{\partial x}(\frac{\rho h^3}{12\mu}\frac{\partial p}{\partial x}) = \frac{\partial}{\partial t}(\rho h) + \frac{\partial}{\partial x}(\rho U h/2). \tag{4.46}$$

Note that the flow problem would thus be completely solvable by specifying appropriate boundary conditions on p. That is to say that the remaining higher orders of Eq. 4.42, (for $f = u, u^2, \ldots$), are now redundant (or automatically satisfied).

4.9.4 Approximate Solutions to Steady Channel Entrance Problems

If $U = 0$ and $\partial/\partial t \equiv 0$, the problem becomes that of steady plane channel flow whose limiting viscous solution is also given by Eq. 4.46 above. However, the fluid is often assumed to enter the channel with a uniform velocity

---

1 If u is parabolic in η then the R.H.S. of Eq. 4.40 vanishes. Conversely if the R.H.S. vanishes for all functions $f'(u)$ then the terms inside the parentheses must vanish identically yielding thus a parabolic profile.

Figure 4.11 The boundary-layer velocity profile.

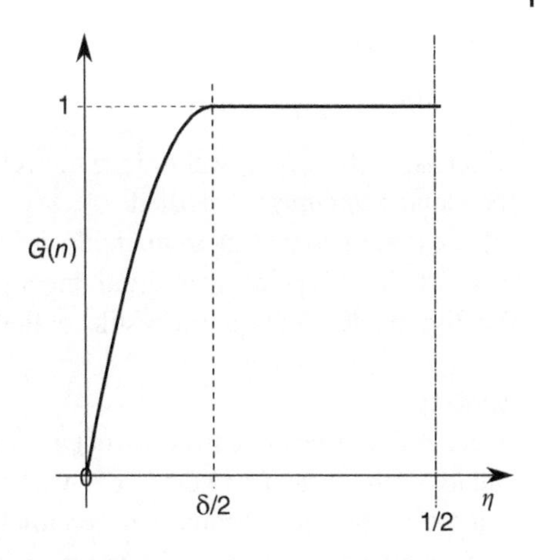

distribution which then develops asymptotically into the parabolic profile. This development region, assumed for practical purposes to be finite, is termed the entrance length. Here, the viscous approximation is no longer valid and a special solution becomes necessary, (mainly to determine the access pressure loss due to entry). An approximate solution to this problem may usually be afforded by constructing an arbitrary velocity profile and requiring that it satisfy Eq. 4.42. In general, one may then write:

$$u = q(x)G(\eta, \delta)$$

with δ a parameter depending on q.

The simplest and most commonly used profile is that shown in Figure 4.11.

The velocity is assumed to consist of a boundary layer, of thickness $\delta/2$, and a uniform core, thus:

$$G = \begin{cases} 1 - (1 - \frac{\eta}{\delta/2})^2 & ; \ \eta \le \delta/2 \\ 1 & ; \ \eta > \delta/2, \end{cases} \tag{4.47}$$

where only the lower half of the profile is considered, the whole being assumed symmetric. This assumption gives:

$$\frac{\partial u}{\partial \eta}(x, 0) = \frac{q}{\delta}$$

$$\int_0^1 (\frac{\partial u}{\partial \eta})^2 d\eta = \frac{4q^2}{3\delta}$$

$$\int_0^1 u d\eta = q(1 - \frac{\delta}{3})$$

$$\int_0^1 u^2 d\eta = q^2(1 - \frac{7\delta}{15})$$

$$\int_0^1 u^3 d\eta = q^3(1 - \frac{19\delta}{35})$$

$$\vdots \quad = \quad \vdots$$

Substituting these expressions into the steady state version of Eq. 4.42, with $f(u) = 1, \ u, \ u^2$ respectively, yields then a system of ordinary differential equations for the three unknowns (q, δ, p), (the density may be assumed a given function of the pressure, i.e. we have ignored the thermodynamic energy equation). This system represents an initial value problem that can be solved by forward integration. An example of applying this method to pipe flow may be found in (Campbell and Slattery 1963).

In its present formulation, this method is commonly known as the *energy integral* method. A simpler, and older, approach ignores the (mechanical) energy equation replacing it by the potential (inviscid) flow at the mid-plane, viz.,

$$dp/dx = -\rho q dq/dx,$$

which is, in theory, also valid since the velocity is uniform at the mid-plane. In that case, the method is known as the *momentum integral* method.

Note that in both of these methods, the compatibility condition at the wall is usually ignored; it cannot, in fact, be satisfied. This points to a certain inconsistency in this method that makes it of doubtful accuracy. In particular, the discrepancy between the results of the two above mentioned methods can be unrealistically big.

Remarks

In general, it is possible to construct a continuous, higher order polynomial representation of the velocity profile and using the system 4.42 to solve. This, being equivalent to a series expansion of the velocity, would however lead to a set of non-linear ordinary differential equations which is quite difficult to handle. Rather, the advantage of the integral approach lies in its assumed "insensitivity" to the actual profile.

In the case of radial flow, the profile prescribed by Eq. 4.47 is not adequate, as it stands, to characterise the velocity, since it does not provide for the possibility of reverse flow (see Chapter 3). After the two boundary layers have joined, at the mid-plane, a new, supplementary, profile has to be constructed in order to account for the further velocity development, namely the possibility of a point of inflexion, or even a turning point, in the profile. This complicates the otherwise simple method by requiring a smooth transition between the two types of profile.

Since the method remains, in any case, approximate, we may suggest a more realistic, and more general, profile function that accounts for all the profile development possibilities. Thus, in place of Eq. 4.47, let us propose:

$$G(\eta; \beta) = \frac{\cos \beta(\eta - 1/2) - \cos(\beta/2)}{1 - \cos(\beta/2)} \tag{4.48}$$

where β is a parameter that may be positive real or positive imaginary. (Complex values may also yield interesting profile possibilities). This function satisfies the boundary conditions:

$$\eta = 0 \ : \ G = 0, \qquad \eta = 1/2 \ : \ G = 1, \ \partial G/d\eta = 0$$

and has the following properties, Figure 4.12:

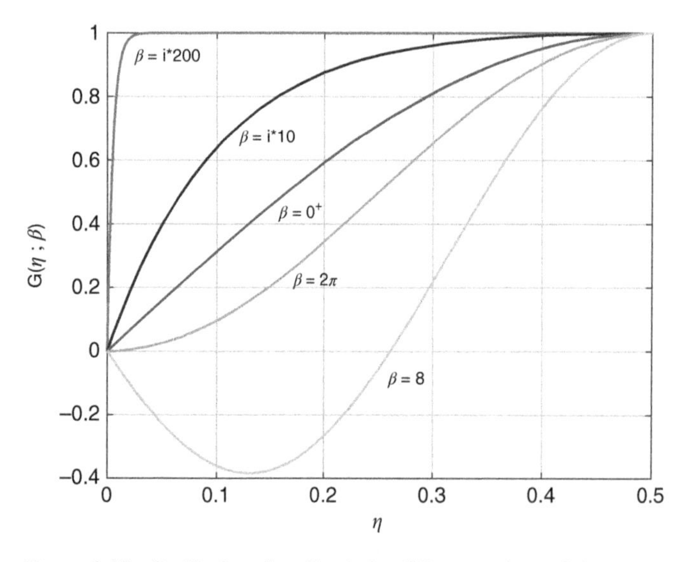

Figure 4.12 Profile function $G(\eta; \beta)$ for different values of β.

- $\beta \to i\infty$: G is uniform, i.e. $G \equiv 1$ for $\eta \neq 0, 1$,
- $\beta \to 0$: $G \to 4\eta(1-\eta)$; a parabola,
- $\beta \geq 2\pi$: $\partial G/\mathrm{d}\eta(\eta = 0) \leq 0$.

However, this profile assumption leads to integral equations that cannot be solved analytically.

4.9.5 Approximation of Convective Terms by Averaging: the Modified Reynolds Equation

When the entrance effects are considered to be of little importance, an alternative simpler approach is possible. This consists of assuming (again) a parabolic velocity profile, Eq. 4.43, but ignoring the compatibility condition, Eq. 4.37. In this way, the profile is used only as an *expedience* to estimate the relative magnitude of the different terms appearing in Eq. 4.42. That is to say that the actual velocity profile may depart from parabolic so long as the integrals of the velocity and one of its first two powers remain nearly the same. The accuracy of this procedure will thus depend on the degree of departure from parabolic of the actual velocity profile. Surprisingly good agreement with experiment has been reported by this method for the case of steady radial flow, sufficiently downstream of the entrance. Note, however, that for the case of steady incompressible plane channel flow (with uniform channel height) this method leads only to the viscous flow solution given above. This shows the real nature (and limitation) of this method: it accounts only for inertia effects arising from, or associated with, the *variable flow cross-section* or of the *density*.

Analysis

If the velocity profile in the gap is assumed to be that of Eq. 4.43, then

$$\int_0^1 u\mathrm{d}\eta = \frac{2}{3}q + \frac{1}{2}U$$

and

$$\int_0^1 u^2\mathrm{d}\eta = \frac{8}{15}q^2 + \frac{2}{3}qU + \frac{1}{3}U^2$$

Then the first two equations of the system 4.42 (continuity and momentum, respectively) are:

$$\frac{\partial}{\partial t}(\rho h) + \frac{\partial}{\partial x}(\rho h(2/3q + U/2)) = 0 \tag{4.49}$$

$$\frac{\partial}{\partial t}(\rho h(2/3q + U/2)) + \frac{\partial}{\partial x}(\rho h(\frac{8}{15}q^2 + \frac{2}{3}qU + \frac{1}{3}U^2)) = -h(\partial p/\partial x + 8\mu q/h^2). \tag{4.50}$$

These two equations *together* constitute the "modified Reynolds" equation. Let us note the following:

- We have now two equations, which need to be solved simultaneously, rather than a single compact equation.
- In some cases, iterative approaches will be necessary in order to solve this set of equations.
- Equation 4.49 is the continuity equation, which must not/cannot be further approximated.
- Equation 4.50 is the momentum equation, the L.H.S. of which could be further approximated depending on the value of the Reynolds number (see further). Notably, if the L.H.S. is set equal to zero, then we can substitute for q form this equation into the first one to retrieve the classical Reynolds equation as has been shown in a previous section.

Normalising by $X \to x/L, H \to h/h_o, P \to p/p_o, \wp \to \rho/\rho_o, \tau \to \omega t, Q \to q/U$, the equations become:

$$\Omega^* \frac{\partial}{\partial \tau}(\wp H) + \frac{\partial}{\partial X}\left(\wp H\left(\frac{2}{3}Q + \frac{1}{2}\right)\right) = 0 \tag{4.51}$$

$$\begin{aligned} \mathrm{Re}_{\mathrm{sq}} \frac{\partial}{\partial \tau}\left(\wp H\left(\frac{2}{3}Q + \frac{1}{2}\right)\right) + \mathrm{Re}^* \left[\frac{\partial}{\partial X}\left(\wp H\left(\frac{8}{15}Q^2 + \frac{2}{3}Q + \frac{1}{3}\right)\right)\right] \\ = -H\left(\frac{6}{\Lambda}\frac{\partial P}{\partial X} + \frac{8Q}{H^2}\right), \end{aligned} \tag{4.52}$$

where

$\Omega^* = \frac{L\omega}{U}$ is the reduced squeeze pulsation

$\mathrm{Re}_{sq} = \frac{\rho_o \omega h_o^2}{\mu}$ is the squeeze Reynolds number

$\mathrm{Re}^* = \frac{\rho_o U h_o^2}{\mu L}$ is the reduced (sliding) Reynolds number

$\Lambda = \frac{6\mu UL}{p_o h_o^2}$ is the sliding number.

- N.B.1: if $U = 0$ we can normalise by another reference velocity, e.g. mean flow velocity.
- N.B.2: $2\Omega^* \cdot \Lambda =$ the squeeze number.

Example

Applying this scheme, in the steady-state case (i.e. with $\partial/\partial\tau \equiv 0$) and isothermal conditions ($\wp = P$), to an inclined slider with $h_1/h_2 = 4$, $\Lambda = 6$, we obtain the pressure distributions, for $\mathrm{Re}^* = 0, 0.1, 0.2, 0.5, 1, 5$ (as well as the asymptote $1/H - 1$ for $\Lambda \to \infty$) depicted in Figure 4.13. We see that the effect of inertia on the pressure distribution, as gauged by the value of Re^*, is (i) to increase the pressure (i.e. move it closer to the asymptotic value) and (ii) that this effect remains very limited even for very high values of Re^*. These conclusions have also been arrived at by Constantinescu (1969). This is owing to both the density and the acceleration of the fluid remaining very small.

We can take this discussion a step further by considering the effect of turbulence on this pressure distribution. We have seen that the main effect of turbulence is to increase the apparent viscosity of the fluid. Since, at high values of Λ, the viscosity plays little or no role in the process, we must conclude that the pressure distribution will not be much effected, as long as the flow remains subsonic. The only issue left is that of heat generation in the film, which will be much higher in inertial and, in particular, turbulent flow than in the viscous laminar case.

Furthermore, we can now assert with confidence that, apart from the film-entrance, in which fluid inertia my be taken to affect the pressure boundary condition, see Chapter 3, we should not expect fluid inertia to play any appreciable role in the pressure distribution further downstream of the entrance.

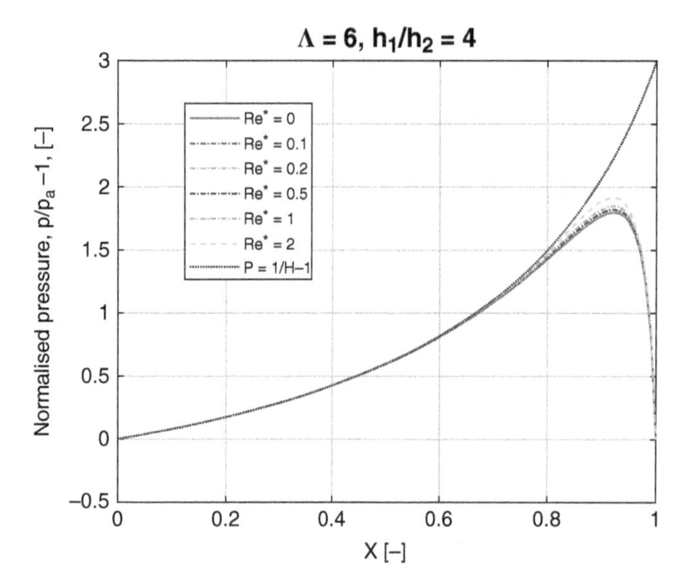

Figure 4.13 Pressure distribution in an inclined slider with and without inertia effects; large Λ.

Centrally fed bearings

When we apply the integral formulation method to channel flow, without the sliding term ($U = 0$), it is possible to write the hierarchy of all the equations. Thus, with the velocity profile fixed (as parabolic), the velocity may be written as:

$$u(x, \eta, t) = q(x, t)\, 4\eta(1 - \eta), \qquad \eta = z/h(x, t)$$

This gives:

$$u_{\eta\eta} = -8q$$

and, for any positive integer n,

$$\int_0^1 u^n \mathrm{d}\eta = \frac{2}{3} \cdot \frac{4}{5} \cdot \frac{6}{7} \cdots \frac{(2n)}{(2n+1)} q^n = \frac{(n!2^n)^2}{(2n+1)!} q^n \equiv I_n q^n$$

Obviously, $I_0 = 1$.

Substituting these expressions into Eq. 4.42, with $f(u) = u^n$, $n \geq 0$, we obtain the following hierarchy of equations:

$$I_n \frac{\partial}{\partial t}(\rho h q^n) + I_{n+1} \frac{\partial}{\partial x}(\rho h q^{n+1}) = \begin{cases} -nI_{n-1} q^{n-1} h(\partial p/\partial x + 8\mu q/h^2) & ;\ n > 0 \\ 0 & ;\ n = 0 \end{cases} \tag{4.53}$$

For radial flow, we can similarly substitute into the radial-flow version of Eq. 4.42, to obtain:

$$I_n \frac{\partial}{\partial t}(\rho h q^n) + I_{n+1} \frac{1}{r} \frac{\partial}{\partial r}(r\rho h q^{n+1}) = \begin{cases} -nI_{n-1} q^{n-1} h(\partial p/\partial r + 8\mu q/h^2) & ;\ n > 0 \\ 0 & ;\ n = 0 \end{cases} \tag{4.54}$$

The equations corresponding to $n = 0, 1, 2$ represent the continuity, momentum and mechanical energy, respectively. However, we have only two unknowns q and p, (the density being assumed a given function of the pressure), and therefore we cannot require all of these equations to be satisfied. In other words, we have to choose (arbitrarily) only two of these equations. Usually, the continuity equation, $n = 0$, is invariably chosen, to avoid violating the basic physics of the problem. Thereafter, the method is that of *averaged momentum* or *averaged energy*, depending on whether $n = 1$ or 2, respectively. The latter case is thought to yield more accurate results on account of the fact that it constitutes *minimisation* of the energy of the flow, (for that specific velocity profile only, it must be noted).

Equation 4.53 (or 4.54 for radial flow) is proposed by some authors as a "modified" Reynolds equation that accounts for the convective inertia effects (while flow is laminar). However, the limitations, mentioned above, of the accuracy of this method should be borne in mind. Furthermore, we now have a set of simultaneous equations rather than one equation. This makes the solution more difficult.

Example 1: steady incompressible radial flow

In this case, we set $\rho = \rho_0 = \text{const.}$, $h = h(r)$, and $\partial/\partial t \equiv 0$ to obtain:

$$n = 0 : I_1 \frac{1}{r} \frac{\mathrm{d}}{\mathrm{d}r}(r\rho_0 h q) = 0 \tag{4.55}$$

$$n = 1 : I_2 \frac{1}{r} \frac{\mathrm{d}}{\mathrm{d}r}(r\rho_0 h q^2) = -I_0 h(\frac{\mathrm{d}p}{\mathrm{d}r} + \frac{8\mu}{h^2} q) \tag{4.56}$$

$$n = 2 : I_3 \frac{1}{r} \frac{\mathrm{d}}{\mathrm{d}r}(r\rho_0 h q^3) = -2I_1 q h(\frac{\mathrm{d}p}{\mathrm{d}r} + \frac{8\mu}{h^2} q). \tag{4.57}$$

The first of these equations yields upon integration:

$$q = \frac{c}{r\rho_0 hI_1}, \qquad c = \frac{\dot{m}}{2\pi},$$

where $\dot{m}$ is mass flow rate.

The second and third equations yield respectively:

$$\frac{dp}{dr} = -[\frac{8\mu q}{h^2} + \frac{I_2}{rhI_0}\frac{d}{dr}(r\rho_0 hq^2)]$$

and

$$\frac{dp}{dr} = -[\frac{8\mu q}{h^2} + \frac{I_3}{2rhqI_1}\frac{d}{dr}(r\rho_0 hq^3)],$$

which may, with q and h given, be integrated numerically for the pressure. Note that the viscous term is the same in both equations, as should be expected, but the inertia terms are different. For the case when h is constant, an analytic solution is possible in the form:

$$\frac{p - p_a}{\frac{1}{2}\rho_0 \bar{u}_0^2} = \pm\frac{24}{\mathrm{Re}_0^*}\log(\frac{R_a}{R}) + K_c(\frac{1}{R_a^2} - \frac{1}{R^2}) \tag{4.58}$$

where, $\bar{u}_0$ is mean gap-entrance velocity, $R = r/r_0$ and K_c is somewhere between the momentum correction factor $= 6/5$ or the mechanical energy correction factor $= 54/35$.

The difference in the inertia terms between the energy and the momentum equations is the factor:

$$\frac{I_3 I_0}{I_2^2} = \frac{9}{7}.$$

That is to say that the energy integral over estimates and the momentum integral under estimates the inertia effects. The difference is seen to be quite appreciable, about 28.5 percent.

Integration of Eq. 4.54 in the time domain may prove to be quite difficult. Linearisation, by perturbation around an equilibrium position, may be carried out to simplify application of this method to dynamic cases (Chapter 7). It might appear from the derivation, however, that this method cannot be sufficiently accurate for large amplitude oscillations of film thickness or the mass flow rate or both. Nevertheless, for the plain squeeze-film problem, one can expect that the velocity profile across the film would be close to parabolic over the largest portion of the bearing width and so we might use this method to assess inertia effects in this situation as in the following.

Example 2: plain squeeze film

Here too we consider the incompressible case since then the analysis is greatly simplified without loss of generalisation. The configuration is depicted schematically in Figure 4.14.

We know that entrance (exit) effects are restricted to a negligibly small portion of the bearing width (Al-Bender and Van Brussel (1992) and Chapter 3) If we write the continuity equation of the flow at any station x (or Eq.(4.53) with $n = 0$ integrated between 0 and x), we have:

$$Vx = -\bar{u}h = -\frac{2}{3}qh \Rightarrow q = -\frac{3}{2}\left(\frac{Vx}{h}\right). \tag{4.59}$$

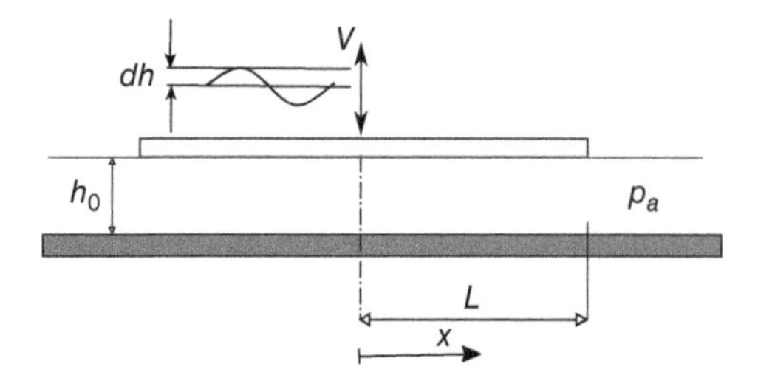

Figure 4.14 Schematic of squeeze film bearing configuration.

Substituting for q in Eq.(4.53) with $n = 1$, we get:

$$\rho\left(\ddot{h}+12\frac{\dot{h}^2}{h}\right)=-\frac{\mathrm{d}p}{\mathrm{d}x}\frac{h}{x}+12\mu\frac{\dot{h}}{h^2} \tag{4.60}$$

Normalising by $X \to x/L, H \to h/h_o, P \to p/p_a, \tau \to \omega t, \dot{()} \to d/d\tau$ and arranging, we obtain:

$$I_{sq}\left(\ddot{H}+12\frac{\dot{H}^2}{H}\right)=-\frac{\mathrm{d}P}{\mathrm{d}X}\frac{H}{X}+\sigma\frac{\dot{H}}{H^2} \tag{4.61}$$

where, we have defined the inertia number,

$$\frac{\rho L^2\omega^2}{p_a}\stackrel{\Delta}{=}I_{sq},$$

and the squeeze number

$$\frac{12\mu\omega L^2}{p_a h_o{}^2}\stackrel{\Delta}{=}\sigma.$$

The L.H.S. of this equation contains the inertia terms, which are seen to be of order I_{sq}, while the R.H.S. comprises the usual Reynolds equation terms.

N.B. we can further assume that the pressure distribution in the bearing is parabolic in form (as it is in the inertia-less case):

$$P = P_o(X^2 - 1),$$

with $P_o(t)$ being the centre pressure, and so obtain a closed-form solution to the whole problem (ignoring cavitation), as:

$$P_o(t)=\frac{1}{2}\left(\sigma\frac{\dot{H}}{H^2}-I_{sq}\left(\ddot{H}+12\frac{\dot{H}^2}{H}\right)\right) \tag{4.62}$$

$$=\frac{1}{2}\sigma\left(\frac{\dot{H}}{H^2}-\mathrm{Re}_{sq}\left(\ddot{H}+12\frac{\dot{H}^2}{H}\right)\right). \tag{4.63}$$

where we see that the squeeze Reynolds number is the ratio of the inertia number to the squeeze number:

$$\mathrm{Re}_{sq}=\frac{I_{sq}}{\sigma}=\frac{\rho\omega h_o{}^2}{\mu},$$

so that the inertia term can be neglected when the $\mathrm{Re}_{sq} \ll 1$.

4.9.6 Approximation of Convective Terms by Averaging in Turbulent Flow

In regard to approximating the convective terms in turbulent flow, this case may be treated in exactly the same way as for the laminar case, as given e.g. in (Dousti et al. 2016) (and the references cited therein). However, for pure Poiseuille flow, as in externally-pressurised bearings, the velocity profile is assumed to be approximately uniform except for a thin boundary layer close to the surface. For more details the reader may refer to the case study of the hanging, over-expansion bearing in Chapter 14.

4.9.7 Summary

A general formulation has been given of the integral methods used for the solution of internal flows. It is shown that the flow equations can be expressed by a single integral pseudo-conservation equation with an arbitrary weighting function of the main flow velocity. This is a hierarchy of denumerable equations which must be satisfied by the main velocity function in order to obtain an exact solution. In practice, however, the utility of the integral method lies in assuming an arbitrary velocity profile and requiring that the integral equations be satisfied only approximately, i.e. by neglecting all but a few of them. Special cases of application, pertaining to the so-called "modified Reynolds Equation", have been given that show the potential but also the limitations of this method.

4.10 Conclusions

In this chapter, we have derived the Reynolds equation and stated it in different coordinate systems, useful for dealing with different bearing configurations. Moreover, we have attempted interpretation of the kinematics for special cases pertaining to situations comprising uneven sliding surfaces having periodic surface features. Neglected flow effects have been categorised as boundary-condition related, mainly slip at the walls, and Reynolds-number related, mainly inertia effects and, more particularly, turbulence. These are reviewed and some appropriate ("modified") RE's have been presented in order to deal with them. Inertia effects have been considered at length together with derivation of the momentum and energy integral methods for flow solution, leading to another type of "modified" RE consisting in a system of simultaneous ordinary differential equations.

References

Al-Bender, F., (1992). Contributions to the Design Theory of Circular Centrally Fed Aerostatic Bearings PhD thesis Leuven, Belgium.

Al-Bender, F., and Van Brussel, H., (1992). A method of "separation of variables" for the solution of laminar boundary-layer equations of narrow-channel flows. *Transactions of the ASME - Journal of Tribology* 114(7): 623–629.

Arghir, M., Alsayed, A. and Nicolas, D., (2002). The finite volume solution of the reynolds equation of lubrication with film discontinuities. *International Journal of Mechanical Sciences* 44(10): 2119–2132.

Belforte, G., Raparelli, T. and Viktorov, V., (1999). Theoretical investigation of fluid inertia effects and stability of self-acting gas journal bearings. *Transactions of the ASME - Journal of Tribology* 121(4): 836–843.

Berthe, D. and Godet, M., (1973). A more general form of Reynolds equation-application to rough surfaces. *Wear* 27, 345–357.

Bird, G.A., (1994). *Molecular Gas Dynamics and the Direct Simulation of Gas Flows*. Oxford Engineering Science. Oxford University Press, New York, NY.

Burgdorfer, A., (1959). The influence of molecular mean free path on the performance of hydrodynamic gas-lubricated bearings. *ASME - Journal of Basic Engineering* 80, 94–100.

Burton, R.A., (1968). An experimental study of turbulent flow in a spiral-groove configuration. *Transactions of the ASME - Journal of Lubrication Technology* 90, 443–449.

Campbell, W.D. and Slattery, J.C., (1963). Flow in the entrance of a tube. *ASME - Journal of Basic Engineering* 41–46.

Constantinescu, V., (1962). Analysis of bearings operating in the turbulent flow regime. *ASME Journal of Lubrication Technology* 82, 139–151.

Constantinescu, V.N., (1967). On the possibilities of magnetogasodynamic lubrication. *ASME - Journal of Lubrication Technology* 314–322.

Constantinescu, V.N., (1969). *Gas Lubrication*. ASME.

Contantinescu, V.N., Galetuse, S. and Jennedy, F., (1975). On the comparison between lubrication theory, including turbulence and inertia forces, and some existing experimental data. *ASME Journal of Lubrication Technology* 97, 439–449.

de Kraker, A., van Ostayen, R. and Rixen, D., (2010). Development of a texture averaged Reynolds equation. *Tribology International* 43(11): 2100–2109.

Dousti, S., Allaire, P., Dimond, T. and Cao, J., (2016). An extended Reynold equation applicable to high reduced reynolds number operation of journal bearings. *Tribology International* 102, 182–197.

Dowson, D., (1961). Inertia effects in hydrostatic thrust bearings. *ASME - Journal of Basic Engineering* 227–234.

Halling, J., (ed.) (1978). *Principles of Tribology*. McMillan.

Elkouh, A.F., (1984a). Fluid inertia effects in a squeeze film between two plane annuli. *ASME - Journal of Tribology* 106, 223–227.

Elkouh, A.F., (1984b). Fluid inertia effects in a squeeze film between two plane annuli. *Transactions of the ASME - Journal of Tribology* 106(22): 223–227.

Elrod, H.G. and Ng, C.W., (1967). A theory for turbulent fluid films and its application to bearings. *Transactions of the ASME - Journal of Lubrication Technology* 89(3): 346–362.

Frêne, J. and Constantinescu, V., (1996). Non-laminar flow in hydrodynamic lubrication In *The Third Body Concept Interpretation of Tribological Phenomena* (eds. D. Dowson, C. Taylor, T. Childs, et al.) vol. 31 of *Tribology Series* Elsevier 319–333.

Frêne, J., Arghir, M. and Constantinescu, V., (2006). Combined thin-film and navier-stokes analysis in high Reynolds number lubrication. *Tribology International* 39(8): 734–747. 4th AIMETA International Tribology Conference.

Greenberg, D.B. and Weger, E., (1960). An investigation of the viscous and inertial coefficients for the flow of gases through porous sintered metals with high pressure gradients. *Chemical Engineering Science* 12(1): 8–19.

Gross, W.A., (1962). *Gas Film Lubrication*. John Wiley & Sons.

Gross, W.A., Matsch, L.A., Castelli, V., et al., (1980). *Fluid Film Lubrication*. John Wiley & Sons, New York.

Heckelman, D.D. and Ettles, C.M.M., (1988). Viscous and inertia pressure effects at the inlet to a bearing film. *STLE Tribology Transactions* 31, 1–5.

Hélène, M., Arghir, M. and Frêne, J., (2003). Numerical three-dimensional pressure patterns in a recess of a turbulent and compressible hybrid journal bearing. *Transactions of the ASME - Journal of Tribology* 125(2): 301–308.

Hirs, G., (1973). A bulk-flow theory for turbulence in lubricant films. *ASME, Journal of Lubrication Technology* 94, 137–146.

King, K.F. and Taylor, C.M., (1975). An experimental investigation of a single pad thrust bearing capable of operating in the turbulent lubricant regime *The 2nd Leeds–Lyon Symposium on Tribology*.

Kwan, Y. and Corbett, J., (1998). A simplified method for the correction of velocity slip and inertia effects in porous aerostatic thrust bearings. *Tibology International* 31(12): 779–786.

Lee, T., Charrault, E. and Neto, C., (2014). Interfacial slip on rough, patterned and soft surfaces: A review of experiments and simulations. *Advances in Colloid and Interface Science* 210, 21–38. Thin liquid films in wetting, spreading and surface interactions: a collection of papers presented at 6th Australian Colloid & Interface Symposium.

Livesey, J.L., (1960). Inertia effects in viscous flows. *International Journal of Mechanical Science* 1, 84–88.

Martin, S. and Bhushan, B., (2016). Modeling and optimization of shark-inspired riblet geometries for low drag applications. *Journal of Colloid and Interface Science* 474, 206–215.

Mori, A. and Mori, H., (1986). Inlet boundary condition for submerged multi-pad bearing relative to fluid inertia forces *The 13th Leeds-Lyon Symposium on Tribology*.

Mori, A., Aoyama, K. and Mori, H., (1980a) Influence of the gas-film inertia forces on the dynamic characteristics of externally pressurized, gas lubricated journal bearings—Part I: Proposal of governing equations. *Bulletin of the JSME* 23(178): 582–586.

Mori, A., Aoyama, K. and Mori, H., (1980b) Influence of the gas-film inertia forces on the dynamic characteristics of externally pressurized, gas lubricated journal bearings—Part I: Proposal of governing equations. *Bulletin of the JSME* 23(178): 582–586.

Mori, A., Aoyama, K. and Mori, H., (1980c). Influence of the gas-film inertia forces on the dynamic characteristics of externally pressurized, gas lubricated journal bearings—Part II: Analysis of whirl instability and plane vibrations. *Bulletin of the JSME* 23(180): 953–960.

Mori, A., Aoyama, K. and Mori, H., (1980d). Influence of the gas-film inertia forces on the dynamic characteristics of externally pressurized, gas lubricated journal bearings—Part II: Analysis of whirl instability and plane vibrations. *Bulletin of the JSME* 23(180): 953–960.

Mori, A., Makino, T. and Mori, H., (1990). Inertia effect in a submerged multi-pad bearing under high Reynolds number with special attention to inlet pressure jump *Proceedings of the Japanese International Tribology Conference, Nagoya*, 911–916.

Reynolds, O., (1886). On the theory of lubrication and its application to Mr.Beauchamp tower's experiments including an experimental determination of the viscosity of olive oil. *Philosophical Transactions of the Royal Society A.* 177, 157–234.

Rhim, Y. and Tichy, J.A., (1989). Entrance and inertia effects in a slider bearing. *STLE Tribology Transactions* 32(4): 469–479.

Schaff, S.A. and Chambre, P.L., (1958). *Fundamentals of Gas Dynamics.* Princeton University Press, Princeton, NJ.

Schlichting, H., (1968). *Boundary-Layer Theory* 6th edn. McGraw-Hill, New York.

Stolarski, T. and Chai, W., (2008). Inertia effect in squeeze film air contact. *Tribology International* 41(8): 716–723.

Szeri, A., (1987). Some extensions of the lubrication theory of Osborne Reynolds. *ASME - Journal of Tribology* 109, 21–36.

Tanaka, S., Esashi, M., Isomura, K., Hikichi, K., Endo, Y. and Togo, S., (2007). Hydroinertia gas bearing system to achieve 470 m/s tip speed of 10 mm-diameter impeller. *Transactions of the ASME - Journal of Tribology* 129(3): 655–659.

Tang, I.C., (1968). Inertia effects of air in an externally pressurized gas bearing. *Acta Mechanica* 5(1): 71–82.

Taylor, G., (1923). Stability of a viscous liquid contained between two rotating cylinders. *Philosophical Transactions of the Royal Society of London - Series A* 223, 289–343.

Surface Roughness Effects in Lubrication

Dowson, D., Taylor, C.M., Godet, M. and Berthe, D. (ed.) (1978). *Surface Roughness Effects in Lubrication* In: *Proceedings of the 4th Leeds–Lyon Symposium on Tribology, September, 1977* Mechanical Engineering Publications Ltd.

Tichy, J.A. and Chen, Sh., (1985). Plane slider bearing load due to fluid inertia-experimental and theory. *ASME Journal of Tribol.* 107,]32–38.

Tichy, J.A. and Winer, W.O., (1970a). Inertial considerations in parallel circular squeeze film bearings. *ASME - Journal of Lubrication Technology* 588–592.

Tichy, J.A. and Winer, W.O., (1970b). Inertial considerations in parallel circular squeeze film bearings. *Transactions of the ASME - Journal of Lubrication Technology* 92(4): 588–592.

van der Stegen, R.H.M., (1997). *Numerical modelling of self-acting gas lubricated bearings with experimental verification* PhD thesis Faculteit Werktuigbouwkunde, Universiteit Twente, Enschede.

5

Modelling of Radial Flow in Externally Pressurised Bearings

5.1 Introduction

An externally pressurised (EP) bearing is usually fed from a pressure source having a hydraulic resistance that is very small compared to that of the bearing. Thus, to treat a bearing modelling problem, one is often justified in assuming that the bearing is fed from an infinitely large reservoir maintained at a fixed supply pressure p_s. This supply is then led through one or a series of *restrictions* before it enters (the viscous part of) the bearing film. The restriction can be of different types, see Figure 5.1:

(a) The curtain area between a simple feed hole and the film entrance, also known as *inherent* restriction. As implied by its name, this restriction is present in all EP bearings unless the the feed-hole corners are rounded off. The flow development in this restriction type has been treated in Chapter 3 for the 2D case.
(b) Orifice issuing into a pocket (or recess), preceding the inherent restrictor.
(c) A porous insert issuing into a (shallow) pocket, preceding the inherent restrictor.
(d) In some exceptional cases, a flow valve is used in order to perform active compensation.

This chapter will deal with the characterisation of flow through a "generalised" inherent restrictor, i.e. the problem of flow development between gap entrance and the attainment of viscous flow downstream thereof.

For the sake of completeness, let us first overview briefly the formulas used for the other types of restrictors, which might precede the inherent restriction.

Orifice Formula

Generally speaking, an orifice is placed between two plenum chambers: the supply reservoir and the bearing pocket or recess chamber. The flow through an orifice depends on the geometry upstream and downstream of the orifice, the geometry of the orifice itself, the Reynolds and Mach numbers, and the prevailing heat-transfer characteristics. This could be quite complex to solve, the more so owing to lack of reliable estimates of the aforementioned parameters. The best way to determine the pressure-flow relationship is experimentally (see e.g. (Belforte et al. 2007; Neale 1973)).

The lumped-parameter formula usually used to determine the mass flow through an orifice is (see Figure 5.1 (b) for notation):

$$\dot{m}_o = C_d \pi \frac{d_o^2}{4} \sqrt{\frac{2\kappa}{\kappa - 1}} \frac{p_s}{\sqrt{\Re T_s}} \Phi_e(\bar{p}_o) \tag{5.1}$$

Air Bearings: Theory, Design and Applications, First Edition. Farid Al-Bender.

Companion website: www.wiley.com/go/AlBender/AirBearings

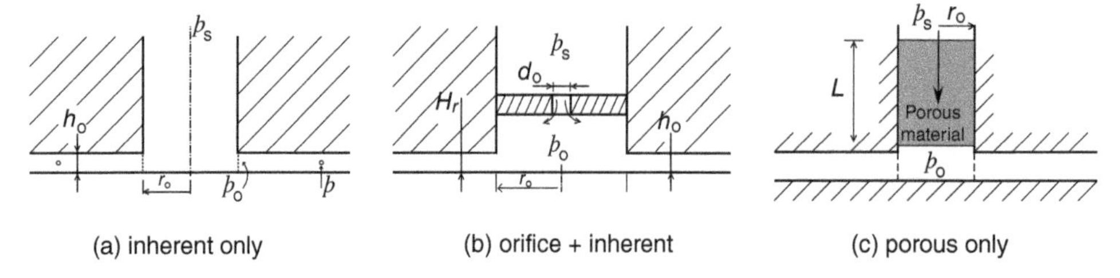

Figure 5.1 Various restrictor types.

where,

$$\Phi_e(\bar{p}_o) = \begin{cases} \sqrt{\bar{p}_o^{\;2/\kappa} - \bar{p}_o^{\;(\kappa+1)/\kappa}} & \text{for } \bar{p}_o \geq \bar{p}c \\ \sqrt{\bar{p}c^{\;2/\kappa} - \bar{p}c^{\;(\kappa+1)/\kappa}} & \text{for } \bar{p}_o \leq \bar{p}c \end{cases} \tag{5.2}$$

with,

$$\bar{p}_o = \frac{p_o}{p_s} \tag{5.3a}$$

$$\bar{p}c = \left(\frac{2}{\kappa+1}\right)^{\kappa/(\kappa-1)} \tag{5.3b}$$

The most critical part of Eq. 5.1 is the *coefficient of discharge* C_d that depends generally on the Reynolds and Mach numbers of the flow as well as the aspect ratio and other geometrical details of the orifice. The value of C_d can be taken to be repeatable if the following two conditions are satisfied (Figure 5.1):

$$\pi \frac{d_o^2}{4} \ll 2\pi r_o h_o$$

(i.e. the orifice cross-sectional area should be much smaller that the inherent-restrictor's curtain area,) and

$$h_o \ll H_r$$

In that case, C_d can be determined, for large Reynolds numbers $4\dot{m}/\pi\mu d_o$, to have a value ranging between 0.6 and 0.7 for incompressible flow (Neale 1973) and between 0.5 and 0.9 for compressible flow (Belforte et al. 2007) depending on the geometry, the Reynolds and the Mach numbers of the flow.

Porous-Insert Formula

A porous insert is sometimes used to compensate a flat, pocket-less bearing pad as an alternative to other laminar-flow restrictors often used in oil-lubricated bearings, such as capillary tubes or narrow slits, which become impractically long or narrow when used with air, owing to the latter's low viscosity. Laminar restrictors differ from orifices in that the flow does not depend on the inertia, but qualitatively resembles Poiseuille flow in a duct, or the current through an electrical conductor, so that for incompressible flow, we have (Darcy's law):

$$dQ \propto \Delta p$$

where the constant of proportionality is directly proportional to some area factor and inversely to the viscosity times the length of the element (which does not appear in the orifice formula).

For compressible fluids, we need to calculate the mass-flow rate by:

$$d\dot{m} = \rho dQ \propto \rho dp \Rightarrow \dot{m} \propto (\rho_s p_s - \rho_o p_o)$$

For an element of porous material the flow, which three-dimensional, will depend on the form of that element. However, the element is slender, the flow will be approximately one dimensional, and the constant of proportionality is given by:

$$\frac{\kappa A}{\mu L},$$

where, κ is the permeability of the porous medium, A is the cross-sectional areal of the flow, μ is the dynamic viscosity and L is the length of the element. More details pertaining to flow through porous media can be found in Chapter 13.

Inherent Restriction: the Entrance Into an Axisymmetrical Gap

Inherent restriction is the name commonly given to the resistance encountered by the flow upon the entrance to a narrow channel (See Chapter 3). In that sense, it is present in all bearing feed systems; whence the relevance of studying its behaviour. Thus, below, we shall examine the flow development in a centrally fed air gap, making use of the results obtained in Chapter 3. Nevertheless, these results enjoy a certain degree of generality, being *mutatis mutandis* applicable to other, single or multiple feed-hole configurations.

Interest in the problem of entrance flow in aerostatic bearings dates back many decades. A basic understanding of the effects and a correct prediction of the relevant parameters has proven to be essential in the successful application of aerostatic bearings. As no general formula exists to quantify the entrance pressure (or mass flow), a range of mostly empirical methods have been developed to allow the solution of the complete flow problem. A comprehensive overview of these methods will be given in sections 5.3 and 5.4. The value of the results produced by empirical methods is, however, always limited to the underlying experimental conditions that produced them. In a wider context, the stated problem of entrance flow in the neighbourhood of feeding sources can be regarded as an example of a narrow-channel internal flow configuration. This broad class of flow phenomena is of both practical and theoretical importance in fluid mechanics: flow intake devices, plane ducts, radial diffusers etc. The solution method proposed here is easily adaptable to describe the flow effects encountered in these applications.

The first part of this chapter outlines the specific problem configuration and its notation; thereafter, a brief overview of existing solution methods for the stated problem is provided. Further on, the method of "separation of variables", developed in Chapter 3, is applied to the laminar boundary-layer equations describing narrow-channel flow, resulting in an initial value problem. This set of equations is then solved for different bearing configurations and feeding systems. Comparison of the obtained results with experimental data from various sources shows good agreement. Finally, the formulation of an expedient coefficient of discharge leads to a lumped parameter formulation of the problem, suitable for practical bearing design.

5.2 Radial Flow in the Gap and its Modelling

Not being confined to hydrostatic bearings, interest in the problem of radial flow between closely spaced plates dates back many decades. Engineering applications include radial diffusers, flapper valves, geothermal reservoirs, viscometry etc. Moreover, the favourable, axisymmetrical, configuration of the flow facilitates experimentation, thus making the problem a suitable one for comparison of solutions.

The basic aim of the present study, which is largely based on (Al-Bender and Van Brussel 1992; Al-Bender 1992) is to provide a solution for the axisymmetric aerostatic bearing problem, formulated in Chapter 2. Nevertheless, all the important aspects of the radial flow problem will be treated, in order to show the versatility of the method of solution.

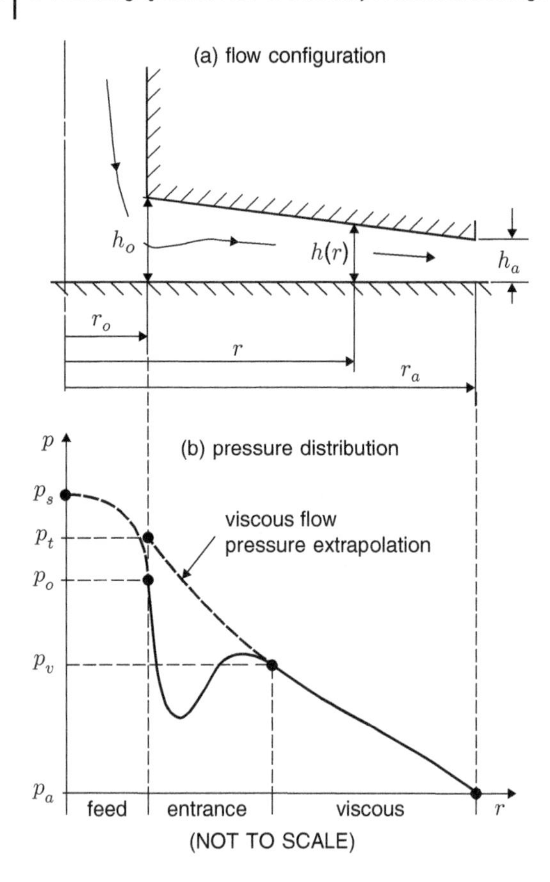

Figure 5.2 Entrance region and notation. (Source: Al-Bender and Van Brussel 1992 by permission of ASME.)

Figure 5.2 shows the bearing configuration and the corresponding pressure distribution, (typical for radial diverging flow.) As has been shown in Chapter 2, three flow regions may be distinguished:

- The feed-hole region, in which the flow quickly accelerates from stagnation, (pressure p_s,) to reach gap entrance (with total pressure $p_o + \rho_o \bar{u}_o^2/2$),
- The (channel) entrance region, in which both inertial and viscous forces are of equal importance, the pressure curve undergoes a characteristic trough and a subsequent crest, see (Snoeys and Al-Bender 1987), before the flow settles down to,
- The fully developed flow region, in which viscous forces dominate and the velocity profile is sufficiently close to parabolic. This region constitutes the greatest portion of a realistically designed bearing, and extends to the outer radius where ambient conditions prevail.

The flow development in the converging case is adequately discussed in Murphy et al. (1978), and will be also described later.

The pressure curve of the developed region may be extrapolated back to the channel entrance, in analogy to the plane channel case, to obtain the theoretical inlet pressure p_t, (Figure 5.2), which corresponds to the case of viscous flow throughout the channel and is needed to assess the entrance head loss[1]. In the first approximation, the bearing designer may assume the flow to be fully developed over the entire bearing area without incurring significant errors on the load capacity. However, an accurate prediction of p_t, (or, equivalently, the mass flow rate,) would still be needed as a necessary and sufficient condition for solving the viscous flow problem or Reynolds equation. (This is usually referred to as the *inherent* compensator or restrictor.) It will be seen in the following that

1 This may, in fact, be a loss or a *gain*, i.e. *recovery* ($p_t \geq p_o$,) depending on the flow parameters.

no effective formula exists yet for estimating either; indeed, a general formula would be necessarily a very complex one. One of the objectives of this study will be to assemble the solution data for use as a correction coefficient for the ideal feed flow formula. Furthermore, a precise knowledge of the actual flow development in the entrance region would also be needed if a better design of that region is to be contemplated or as a basis for unsteady flow solution, etc.

In the following section, the most important empirical formulas will be discussed; thereafter, a short review of other solution methods will be given. The method of "separation of variables", developed in Chapter 3, is then applied to the problem and the results discussed and evaluated. Thereafter, the feasibility of the lumped parameter approach is discussed, and a correction coefficient function is proposed.

5.3 Lumped Parameter Models

As has been mentioned above, in order to solve the complete flow problem assuming that the flow is viscous throughout, it is necessary to have a good estimate of either the pressure p_t or the mass flow rate $\dot{m}$. Formulas that predict these, represent an expedience which has played, and still does, an important role in studying the static characteristics of aerostatic bearings. Two basic approaches have been known: the orifice/nozzle formula and Vohr's correlation formula.

5.3.1 The Orifice/Nozzle Formula

Designers have been tempted from a very early stage to regard the feed-hole curtain area as an orifice or a nozzle (it being the smallest flow area.)[2] Further, the pressures, upstream and downstream of this orifice, were conveniently taken to be p_s and p_t respectively, and the ideal flow rate $\dot{m}_{id}$ was calculated using Euler's formula. The actual flow rate $\dot{m}$ is then obtained as fraction of the ideal one through the multiplication by a coefficient of discharge C_d[3], viz.,

$$\dot{m} = C_d \dot{m}_{id} \tag{5.4}$$

where $\dot{m}_{id}$ is given, from Chapter 2, (cf. also Eq. 5.1 above) as:

$$\dot{m}_{id} = 2\pi r_o h_o \sqrt{\frac{2\kappa}{\kappa - 1}} \frac{p_s}{\sqrt{\Re T_s}} \Phi_e \left(\frac{p_t}{p_s} \right), \tag{5.5}$$

where $2\pi r_o h_o$ represents the throat area, i.e. the "curtain" area of the orifice.

This would be a very powerful formula, if C_d were a *constant* or even a simple function of the entrance Reynolds and Mach numbers, as in the case of a round orifice in turbulent flow. However, due to the qualitative differences between the actual situation and the model, C_d will have widely ranging values, depending on the flow parameters. Consequently, it is difficult to seek to determine these values experimentally. As an example, we give here one such formula deduced empirically by Dudgeon (1970):

$$C_d = f\left(r_o, \frac{r_o}{r_a}\right) \left(\frac{p_t}{p_s}\right) \left(\frac{p_t}{p_s} - 1\right)^{0.2} \left(\frac{h}{r_o}\right)^{0.56} \tag{5.6}$$

where the function f is given graphically. Beside the fact that it is dimensionally unsound, the validity of this formula is restricted to a limited design range. This example was given in order to show the difficulty of the problem as well as to demonstrate a basic issue with *ad hoc* empirical solutions.

---

2 This has been further encouraged by a questionable hypothesis that turbulence and separation would occur downstream of the feed-hole, presumably caused by the sharp turn.

3 Note that C_d may, in this case, be bigger or smaller than unity.

However, the nozzle analogy seems to become valid in the extreme case of a fully developed turbulent supersonic flow downstream of the feed-hole. In this case, one may deduce from the results of Moller (1966), that beyond a certain supply pressure limit, there exists a nearly constant coefficient of discharge, approximately equal to 0.86 for a square edge feed-hole. For more details pertaining to this case, the reader is referred to Chapter 14.

The other extreme case, viz. that of very small inlet Reynolds numbers, seems to yield better to viscous laminar flow analysis as is given, e.g., by Strauss (1976) who shows that the entrance pressure loss is then approximately equal to that obtaining in viscous flow over a length of the channel equal to 0.339 times its height.

5.3.2 Vohr's Correlation Formula

This is perhaps one of the earliest attempts to quantify the entrance "losses" associated with inherent compensation. Instead of the orifice analogy, the entrance region is treated as a diffuser in which the dynamic head, resulting at the gap entrance, would ideally be fully recovered (into static pressure). However, due to the postulated entrance phenomena, a fraction of that is, in practice, lost. The formula endeavours to relate this loss to the entrance Reynolds number $\mathrm{Re} = 2\rho_0 \bar{u}_0 h_0 / \mu$. The pressure loss is taken to be equal to $(p_s - p_t)$, the dynamic head at entrance to be $(p_s - p_o)$, being based upon the incompressible flow assumption, and the following formula is then proposed (Vohr 1966):

$$(p_s - p_t) = K(\mathrm{Re})(p_s - p_o) \tag{5.7}$$

where the loss coefficient K, is determined experimentally to be an increasing function of Re. Apart from the questionable accuracy of evaluating K, especially at low Reynolds numbers[4], one may assert with some confidence that the large scatter in the experimentally obtained data points is due to the fact that such a correlation cannot, in reality, exist. (It is noteworthy that experiments conducted by Lowe (1970), to check the validity of this formula, failed to show the existence of any pattern.) Indeed, the subsequent analysis, or even only a careful statement of the basic equations characterising the problem, should show that if such a loss coefficient existed, it would be a function of the reduced Reynolds number Re_0^* and the inlet Mach number M_o. That is to say that the basic problem of this correlation is that it is based on the assumption of plane flow which is fundamentally different from radial flow. (An attempt to explain Vohr's correlation theoretically was later given by Hagerup (1974) who, again using plane incompressible flow analysis, attributed the losses to the effect of oblique fluid entry into the gap and obtained K values lying some 40 percent above those of Vohr.)

In summary, lumped parameter empirical methods, though they may be a useful expedience to the designer, are very restrictive in their application: satisfactory empirical formulas may prove too complicated or anomalous; hardly valid outside the limited experimental conditions that produced them.

5.4 Short Review of Other Methods

Short review of other methods As has been mentioned in Chapter 3, the various methods used for the treatment of the problem of plane channel entrance have found extensions or parallels for the case of a radial channel. It may, therefore, be interesting to outline and discuss these briefly:

5.4.1 Approximation of the Inertia (or Convective) Terms

This is by far the oldest and most effective method. The convective terms in the equation of motion are approximated in various ways. The simplest of these is replacing them by their average values (over the film thickness,)

4 One would expect, for instance, that as Re becomes small, p_t would approach p_o, i.e. that K should approach unity. However, $(p_s - p_o)$ is then a very small quantity and, consequently, K is evaluated as a ratio of two minute numbers.

assuming a parabolic velocity profile throughout, see Chapter 4. Applied to incompressible, parallel channel, radial flow, the resulting pressure distribution is then (Eq. 4.58),

$$\frac{p - p_a}{\frac{1}{2}\rho_o \bar{u}_o^2} = \pm \frac{24}{\mathrm{Re}_o^*} \log\left(\frac{R_a}{R}\right) + K_c\left(\frac{1}{R_a^2} - \frac{1}{R^2}\right) \tag{5.8}$$

which is a linear combination of the pressure distribution corresponding to viscous flow and that corresponding to ideal (inviscid) flow. The latter term is multiplied by a coefficient K_c, which is taken either as the mechanical energy correction factor (= 54/35) as in McGinn (1955); Savage (1964), or as the momentum correction factor (= 6/5), yielding the better results, as is given in Gross (1962) or by Livesey (1960). The positive or negative sign corresponds to diverging or converging flow respectively. Despite the simplicity of this method, the results it provides (in terms of the pressure distribution) seem to be adequate for engineering purposes except in the immediate vicinity of the entrance where rapid changes in the velocity profile are to be expected. As such, it cannot be used to predict the flow rate in diverging flow with large Reynolds numbers since significant pressure variations occur near the feed hole, but is rather more suited to the case of converging flow where entrance pressure losses are usually negligible. The apparent accuracy of the results is discussed and shown to be fortuitous by Murphy et al. (1978). Later, Lee and Lin (1985) published a method for converging flow based on linearising the convective terms (and discarding the normal velocity,) leading to a simple differential equation for the pressure. Beside the fact that their assumptions are open to question, the results they obtain are not appreciably different from those cited above.

Further, this method may be extended to the case of compressible unsteady channel flow. Some authors have suggested using this method to formulate a "modified" Reynolds equation that accounts for inertia effects. Our own formulation, and discussion of the various possibilities of this method have been given in Chapter 4.

5.4.2 The Momentum Integral Method

Here, an arbitrary velocity profile is constructed (see Chapter 4) The flow is divided into two regions: the entrance region and the filled region[5]. In the first of these, the main flow is assumed to consist of a uniform inviscid core flanked at the walls by thin boundary layers (usually parabolic in shape) whose thickness increases progressively with the radius. The filled region is reached when the two boundary layers join together at the mid-plane. In the plane channel case (with parallel walls,) since the average velocity is constant by virtue of continuity, this would represent the onset of creeping flow where the velocity distribution reaches a state of equilibrium and should remain subsequently parabolic throughout. For radial (diverging) case, however, the mean flow velocity is decreasing and, as a result, the fluid nearer the wall is expected to be further retarded relative to the mid-plane flow. It is, therefore, necessary to make an additional velocity profile assumption for the filled region that would allow for further reduction of the velocity profile gradients at the wall (including negative values corresponding to back flow). This is conveniently done by using the Pohlhausen fifth degree polynomial velocity distribution (see (Schlichting 1968).) Such a scheme provides a good *qualitative* representation of the flow development. The accuracy of the solution, obtained by inserting these assumptions into the integrated equations of motion and continuity[6], is, however, difficult to judge. This method has been applied to the problem of a radial diffuser e.g. by Woolard (1957) who made, in addition, a number of heuristic and very questionable assumptions in order further to linearise the problem. (Application of our method to the one example he gives, reveals, contrary to his results, that the flow must be dominated by turbulence.) Mori (1969) applied it to the aerostatic bearing problem (overlooking the filled region assumption, and thus obtaining very bad comparison with experimental data). While this method is not very easy to apply (cf. the preceding one), the results obtained can be very approximate indeed. This method is discussed also in Chapter 4.

5 Such a division is necessary in the case of diverging flow. In converging flow, as will be shown later, the channel is theoretically never filled.

6 Thus, reducing them to ordinary differential equations for the mid-plane velocity and the pressure.

5.4.3 Series Expansion

This is only applicable to diverging flow since it takes advantage of the fact that the flow approaches creeping character at the limit of very large radii[7]. The stream function may then be expressed as a series in powers of $(1/r^2)$ whose coefficients are functions of the normal coordinate z, thus:

$$\psi(r,z) = f_0(z) + \sum_{n=1}^{\infty} f_n(z)\left(\frac{1}{r}\right)^{2n}$$

with,

$$\partial\psi/\partial z = ru, \qquad \partial\psi/\partial r = -rw$$

Such a scheme may be used to solve the complete Navier–Stokes equations of incompressible radial channel flow as has been shown independently by Geiger et al. (1964) and Savage (1964). It results in a system of ordinary differential equations for the coefficient functions which must be solved recursively, (the first equation corresponding to creeping flow.) However, it is neither easy to calculate new higher order terms nor possible to establish convergence of the series. This is a problem shared by most series solutions, in particular those representing solutions to non-linear equations, making them mainly of academic interest. It may thus be impossible to express the uniform flow, expected near the entrance, by a functional series. (It is noteworthy, however, that using only the first two terms of the series leads to a pressure distribution function identical to that given by equation 5.8, with $K_c = 54/35$.) This method is, further, not suitable for compressible fluid flow or variable hight channel.

5.4.4 Pure Numerical Solutions

These are usually carried out by a suitable finite difference formulation. Murphy et al. (1978) used a forward marching technique to solve the boundary-layer incompressible convergent flow problem. They had to devise a means of ensuring that there be always sufficient grid points in the boundary-layer, (a difficulty that is always encountered in such numerical solutions.) Hayashi et al. (1975) solved the complete Navier–Stokes equations for the divergent flow problem of a flapper valve. They used their solution to deduce a simpler approximate one. As has been pointed out in Chapter 3, such schemes are complicated and often fraught with problems of convergence which make them impracticable and unhandy for general application.

5.5 Application of the Method of "Separation of Variables"

Application of "separation of variables" The underlying theory for this method has been given in Chapter 3. We recall briefly here that, for the case of radial compressible flow with a variable gap, if the velocity is separated into amplitude and profile functions:

$$U(R,Z) = Q(R)G(R,Z)$$

with,

$$I = \int_0^1 G dZ$$

7 In converging flow, the velocity field approaches the limit of *similar* profiles for very small radii, see e.g. (Murphy et al. 1978), which have no explicit solution.

then, the following initial value problem is obtained:

$$\begin{cases} Q' &= (1/H^2\mathrm{Re}_0^*\wp)m(I) \\ P' &= (-\kappa M_0^2 Q/H^2\mathrm{Re}_0^*)n(I) \\ I &= (1/RH\wp Q) \end{cases} \tag{5.9}$$

where, the primes denote derivatives w.r.t. R, and m and n are given functions of I. (Note that for the incompressible case, $\wp = 1$ and κM_0^2 is replaced by $\rho_o \bar{u}_0^2/p_o$ in the last set of equations.)

The film thickness function H, for the general case of a convergent gap, is given from Chapter 2, as:

$$H = 1 - \nu_0(R-1),$$

where, $\nu_o = (h_v/h_o)/(r_a/r_o - 1)$ is the normalised conicity angle.

The velocity profile integral function I may take values between 0 and 1, with the value 1 corresponding to uniform velocity distribution; 2/3 to parabolic profile; less than about 0.457 to the onset of reverse flow. The functions $m(I)$ and $n(I)$ were calculated, by numerical integration, at intervals $\Delta I = 0.01$ and tabulated in a data file so that intermediate values could be obtained, (sufficiently accurately,) by second order interpolation.

In order to solve the system 5.9, for a given geometry, Reynolds and Mach numbers and density (polytropy) function, we need to specify an initial value for Q or I (or equivalent information.) To solve the complete flow problem, we must, in addition, solve for the flow from stagnation up to gap entrance, and subsequently match the two solutions. Let us then briefly discuss the various possibilities, conditions, and constraints on the range of applicability of this method.

5.5.1 Boundary Conditions on *I*

Strictly speaking, the problem constitutes one continuous whole in which stagnation, gap entrance and exit conditions influence one another and are, thus, almost impossible to specify. However, it has become customary, under certain considerations, to make some simplifying assumptions whereby one or more of the boundary conditions are made redundant. Such considerations relate to the *stability* of the flow, viz. that it should, firstly, remain laminar and, secondly, *settle* down to a regular pattern well before the exit is reached. The latter condition requires that the channel be sufficiently long, which would then mean that the exit conditions would have negligible effect on the rest of the flow and may consequently be ignored. The former, however, depends, on whether the flow is converging or diverging.

Figure 5.3 shows the development of the velocity profile integral function I for both cases, for incompressible flow starting from uniform velocity profile at entry.

We note first, that converging flow is unconditionally stable, i.e. the velocity profiles remain always rounder than parabolic[8]. As for diverging flow, stability depends on the value of the reduced Reynolds number Re_0^*. It may be noted first that, in this flow configuration, the function I always undershoots its limiting parabolic value, to approach it asymptotically from below. The value of the undershoot increases, however, with Re_0^*: back-flow becomes inevitable beyond a certain value. (Although back-flow does not mean automatically the onset of permanent turbulence, since the flow may re-attach further downstream, it does lead to, or further accentuates, phenomena such as asymmetry, by virtue of the separation bubble tending to form on one side of the channel as is shown (Vallentine 1967), that would undermine the basic assumptions of the method. Solutions which display a large region of back-flow should, thus, be taken only with reservation.)

Having been able to neglect exit conditions, we may turn our attention to the entrance conditions. Here, a uniform velocity profile is usually (conveniently) specified, ($I_o = 1$). This assumption may seem very arbitrary, but is

8 The parabolic profile is, in theory, never attained in converging radial flow.

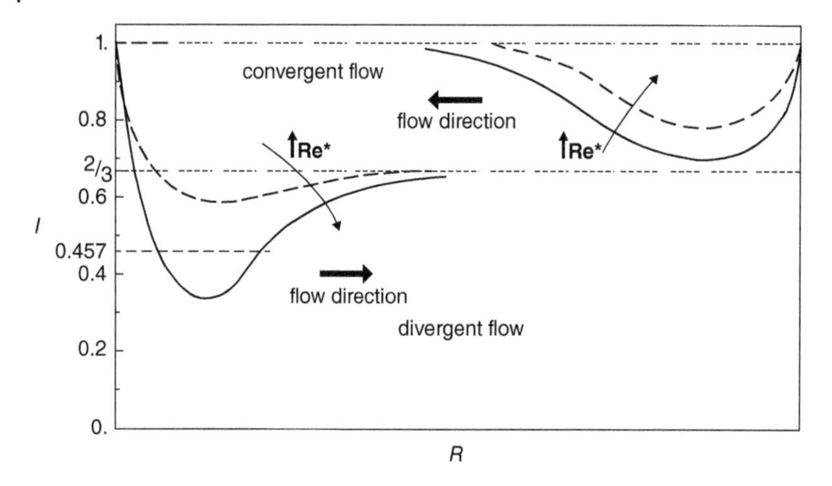

Figure 5.3 Qualitative development of the velocity profile integral function in radial channel flow. Source: Al-Bender and Van Brussel 1992 by permission of ASME..

in fact justifiable to a certain extent, for small h_0/r_0 values, by the following reasoning. The flow from stagnation to gap entrance will be a converging, accelerating one, which may, just upstream of the feed-hole, be regarded as plane sink flow. This means that the boundary-layer at the walls would be of a progressively decreasing thickness, which at the gap entrance would be inversely proportional to the square root of the Reynolds number based on the film thickness Re, (see (Schlichting 1968, p. 153)), which is, typically, of the order of 10^3, leading to a boundary-layer to gap height ratio of less than 0.1 which corresponds, in turn, to a value of I_0 which is greater than about 0.9. If we keep in mind that velocity profiles are similar for high I values (see Chapter 3,) the pressure loss to viscous friction in the feed-hole region would be roughly equal to that which would occur in the gap if the flow were to start from a uniform velocity distribution. The only error would be a radial shift of the whole pressure curve. However, since a uniform velocity profile decays over a very short radial distance, that shift error would likewise be small. It will therefore be subsequently assumed throughout that initial velocity distribution is uniform unless otherwise stated.

5.5.2 Flow From Stagnation to Gap Entrance

From Chapter 2, this flow may be determined from Euler's formula. Thus, we have:

$$\frac{p_s}{p_o} = 1 + \frac{1}{2}\frac{\rho_o \bar{u}_o^2}{p_o} \qquad \text{(incompressible)} \tag{5.10}$$

$$\frac{p_s}{p_o} = \left(1 + \frac{(\kappa - 1)}{2} M_0^2\right)^{\kappa/(\kappa-1)} \qquad \text{(compressible)} \tag{5.11}$$

5.5.3 The Density Function in the Gap

It should suffice, for practical purposes, to use the polytropic equation of state with a conveniently chosen exponent. For simplicity, the solutions presented in this paper are all based on the isothermal assumption:

$$\wp = P \tag{5.12}$$

where a step change in the temperature, to its ambient value, is imposed at gap entrance. This, somewhat artificial, assumption is acceptable (see, e.g., (Gross 1962, p. 287)): it ensures continuity of the pressure function, and compatibility with the assumed isothermal exit conditions.

In a theoretical investigation that is reported in Chapter 17, the energy equation is solved simultaneously with the equation of motion in order to assess the thermal characteristics of the flow. The results show that, for isothermal channel walls, no significant errors are incurred on the overall solution by an isothermal flow assumption, but small local differences are observed in the entrance region.

5.5.4 Solution Procedure

Given the geometry, the supply and ambient pressures, and the fluid's physical properties, the overall problem is split into two, viz. stagnation to gap entrance ($s \rightarrow o$) and gap entrance to ambient ($o \rightarrow a$), which are matched through equating the end conditions of the first with initial conditions of the second. The first problem is solved analytically by equation 5.10 or 5.11, the second, governed by equations 5.9 with 5.12, must be solved numerically, e.g. by a Runge–Kutta method (of the fourth order as was done in our case). Matching of the two solutions is done by the method of false position *(regula falsi,)* through equating the parameter p_0. (However, due to the non-linearity of the problem, some care must be taken in choosing a starting value that will insure convergence.)

5.6 Results and Discussion

This method has been applied extensively, for the purpose of evaluation, to a wide class of bearing problems. Comparisons have been made with experimental results obtained by other authors as well as those obtained in our laboratory. Since it was considered not very relevant to compare with approximate solutions, comparison has been made in one instance only with the numerical solution of Murphy et al. (1978) for incompressible converging flow.

5.6.1 Qualitative Trends

It may be worthwhile first to indicate the general trend of the solution in the gap, starting from a uniform velocity profile. In addition to the velocity development characteristics already discussed in Figure 5.3, which may qualitatively be deduced from a momentum integral method, our solution shows the following additional characteristics:

- The mid-plane flow is always initially accelerating, i.e. $Q'(1) > 0$, whatever the value of the reduced Reynolds number and the entrance Mach number, (this result is already known for converging and parallel channels).
- Consequently, an initial pressure trough is inevitable in diverging flow. Many researchers and analysts have, in the past, attributed this depression to an assumed *vena contracta*, caused e.g. by a separation bubble, just downstream of the feed-hole. However, the results of this analysis show that the pressure depression may be very large without the occurrence of separation[9]. The relationship between the pressure behaviour and the velocity profile is as depicted in Figure 2.6 Chapter 2): it can be directly obtained from the behaviour of the sign of n.
- Further, in the compressible diverging case, the average flow is initially accelerating, i.e. $d\bar{U}(1)/dR > 0$, except for very small entrance Mach numbers. This means, among other things, that if the flow should become choked, it would do so only some distance downstream of the throat (the feed-hole curtain area,) a phenomenon that is already known in the case of de Laval nozzle.

It will also be noted, in the subsequent comparisons, that the agreement is only fair near the feed-hole. This is due primarily to two factors. First, this method assumes uniform pressure across the gap, and would as such predict the average pressure, whereas the experimental results give the wall pressure (at one of the two walls), which

9 Conversely, when separation does occur, it does so only some distance downstream of the feed-hole, i.e. not necessarily as a consequence of the sharp edge or the oblique entry; the latter affects rather the symmetry of the flow.

may be, to a varying degree, different from the average pressure in the entrance region (even if the flow were symmetric). Secondly, and more importantly, the oblique entry of the flow will invariably cause asymmetry in the initial region, as has been shown by the experiments of McGinn (1955), that will in turn cause asymmetry in the pressure distribution across the gap, thus enhancing the difference between wall pressure and the average pressure. However, approximate analysis, which is outside the scope of this work, shows that the development of the average flow and pressure is not appreciably affected by mild asymmetry, so that the average pressure should lie between the two wall pressures. Moreover, since the centre stream line will experience a turn, the wall pressure curve (on either of the two walls), must cross the average pressure curve at a point downstream of the entrance. This will be seen to be the case in most of the comparisons.

5.6.2 Comparison with Experiments

In Figure 5.4, the present solution is compared with the data of McGinn (1955) for a water bearing, (where the pressure was measured on the orifice side). The solutions used as input the indicated values of supply pressure, film thickness and temperature (for correct viscosity). However, the inner and outer diameters of the disc were estimated from a figure. The top part represents diverging flow. It is noted that the agreement improves with the decreasing reduced Reynolds number values; with the maximum error in the predicted flow rate value remaining, however, within 3 percent. This is a remarkably good agreement in view of the fact that the error in the flow rate is three times as big as that in the film thickness[10] which is not expected to have been measured with an accuracy better than 1 percent.

In the lower part, which represents converging flow, the agreement is slightly poorer, although again the flow rates are within the same accuracy as before. Beside the fact that there are less data points here, it is expected that exit conditions should play a more significant role than in the diverging case.

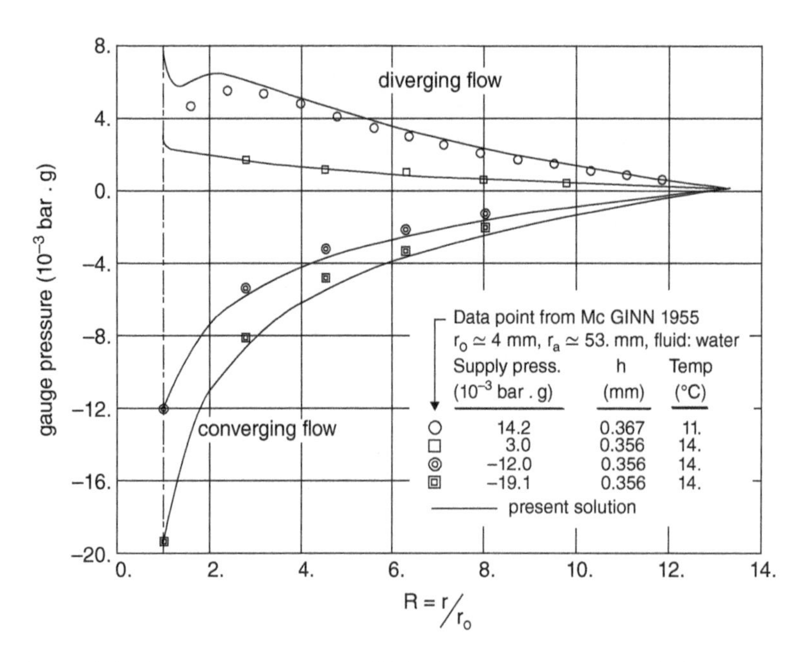

Figure 5.4 Comparison of pressure distributions for converging and diverging incompressible radial flow. Experimental data points adapted from McGinn (1955). Source: Al-Bender F and Van Brussel H 1992 by permission of ASME.

10 Since, for a given pressure gradient, the flow rate is proportional to the cube of the film thickness.

Further, the converging flow solution was compared with the numerical solution of Murphy et al. (1978). The pressure distribution results, (Figures 3.3 and 3.4 of Murphy et al. (1978),) were found to be identical, so were the velocity profile development values, (Figure 3.5 thereof), for values of I less than about 0.85; for values higher than those, a small discrepancy was found which may be attributed either to the inaccuracy of their numerical scheme for thin boundary-layers, or to the use of the similarity solution mentioned therein.

Figures 5.5 and 5.6 are comparisons with results of gas bearings with large radius ratios ($R_a > 30$). The pressure was measured (as in all of the subsequent examples) on the side opposite to that of the feed-hole. The qualitative and, in view of earlier general remarks, also quantitative agreement near the inlet is significant. The mass flow rates are again correct to within 3 percent. The lowest of the two curves of Figure 5.6 represents an extreme case of high Re^* and M_0. The solution indicates the occurrence of reverse-flow in the region $3 < R < 6$, i.e. around the point where there is a small jump in the data, and the rise (and subsequent fall) of the average Mach number to above unity in the film. Mori (1969) treated the case using turbulent supersonic flow assumptions, obtaining questionable results. This is not justified, in our opinion, since the present laminar flow method seems to be adequate even for more extreme cases than this one.

Figures 5.7 and 5.8 are comparisons for small radius ratios, ($R_a \leq 10$). Although the data points of Figure 5.7 do not themselves seem to be accurate or complete, nor is the solution sufficiently close, this example has been given to show the presence, albeit minute, of the pressure depression even in this case of very slow flow and to indicate the qualitative agreement which cannot be predicted by other approximate solutions. Figure 5.8 typifies how the solution crosses the data points near the inlet (discussed above) to approach the exact distribution in the viscous downstream region.

Figure 5.9 shows, beside the close agreement of theory and experiment, the superior pressure distribution characteristics of a convergent gap bearing. (The measurement was carried out in our laboratory.)

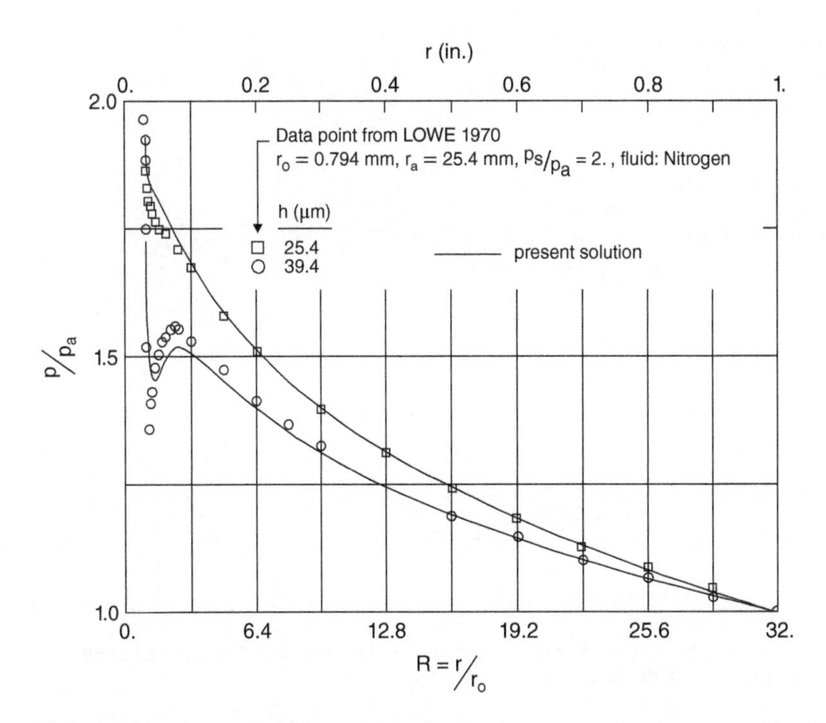

Figure 5.5 Comparison of pressure distributions for a bearing with large radius ratio. Experimental data points adapted from Lowe (1970). Source: Al-Bender F and Van Brussel H 1992 by permission of ASME.

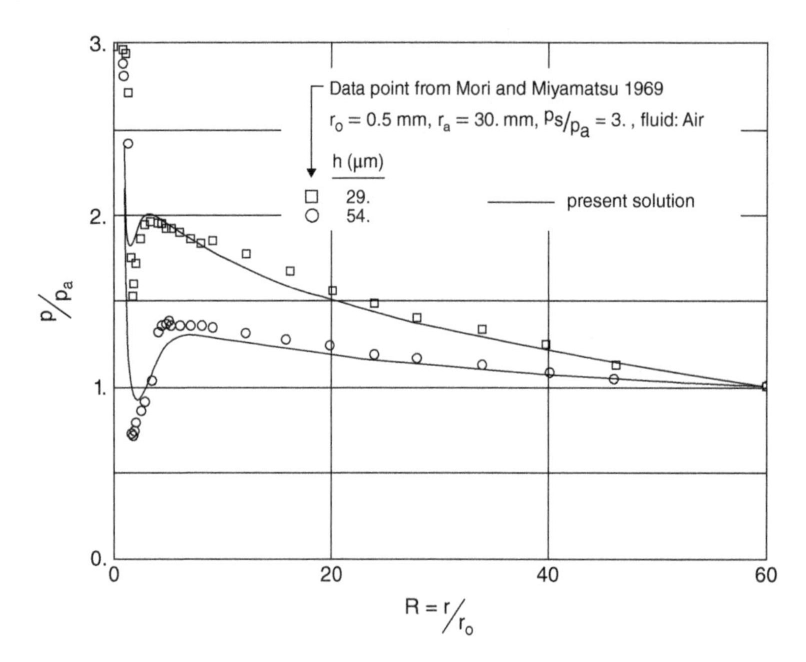

Figure 5.6 Comparison of pressure distributions for a bearing with large radius ratio. Experimental data points adapted from Mori (1969). Source: Al-Bender F and Van Brussel H 1992 by permission of ASME.

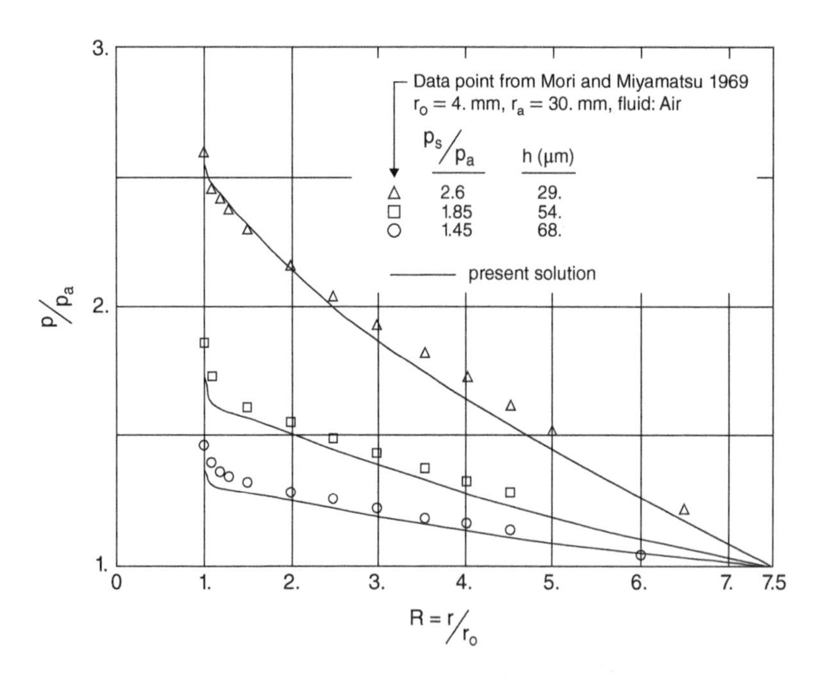

Figure 5.7 Comparison of pressure distributions for a bearing with small radius ratio. Experimental data points adapted from Mori (1969). Source: Al-Bender F and Van Brussel H 1992 by permission of ASME.

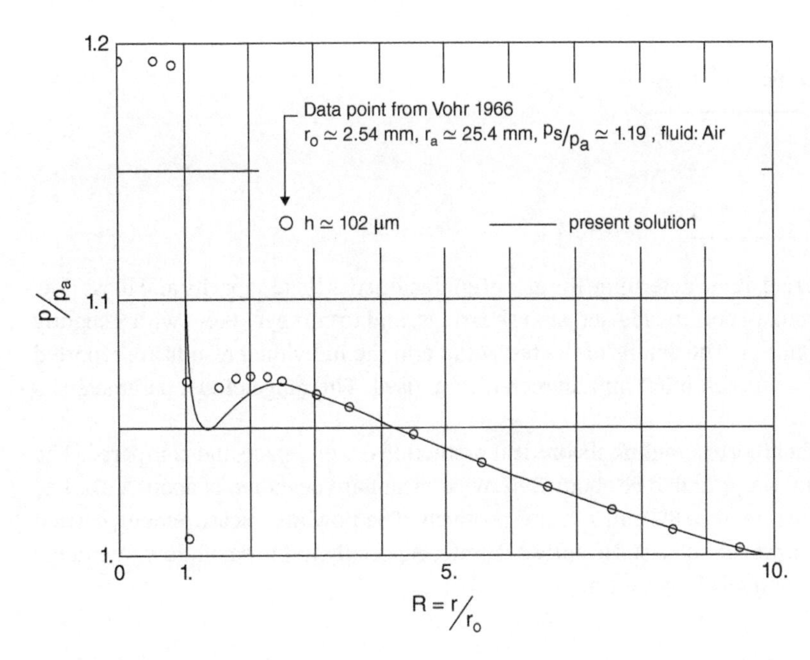

Figure 5.8 Comparison of pressure distributions for a bearing with small radius ratio. Experimental data points adapted from Vohr (1966). Source: Al-Bender F and Van Brussel H 1992 by permission of ASME.

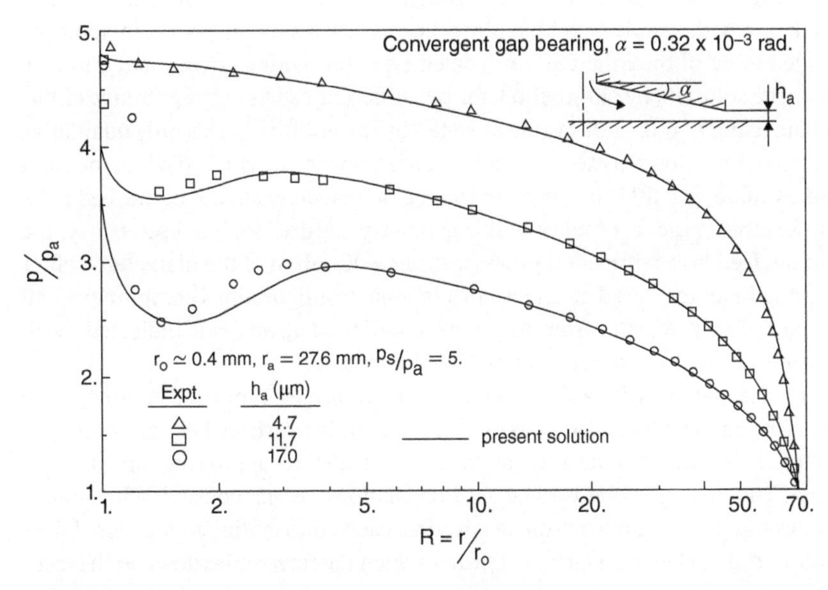

Figure 5.9 Comparison of pressure distributions for a convergent gap air bearing. Source: Al-Bender F and Van Brussel H 1992 by permission of ASME.

Table 5.1 Range of test bearing parameters.

Parameter	Range	Parameter	Range
R_a	20–111	p_s/p_a	2.5–7.5
h_o/r_o	0.01–0.1	Re^*	2–220
h_o/h_a	1–3	M_o	0.1–0.8

Furthermore, an experiment was carried out to determine the global characteristics (load capacity and mass flow versus film thickness) of a range of circular centrally fed aerostatic bearings, and to compare them with a slightly more refined version of the present solution. The details of the test setup and the individual results are reported in (Lambrechts 1989). Eight bearing pads, each of 60 mm diameter, were used. The range of test parameters is summarised in table 5.1.

A total of 170 measurements (and their corresponding theoretical predictions) were made and compared. The average discrepancy in the load capacity was found to be about 2.9%, with a standard deviation of about 2.2%. The corresponding results for the mass flow rate were 8% and 7.9%, respectively. The flow rate measurement, carried out by a rotameter, was not expected, however, to be sufficiently accurate, especially in the small flow rate range which was mainly responsible for the relatively larger errors[11].

5.7 Other Comparisons

These comparisons have been carried out by Waumans et al. (2008) against more recent data from Belforte et al. (2007). Two bearing cases are studied: one with simple feed hole; the other having a shallow pocket in addition. In the first, the test bearing is regarded to be of the inherently restricted type, but with a fairly small entrance radius $r_o = 0.1$ mm. Figure 5.10 shows the solution, by our method, for two different values of the entrance radius and of the gap height. The solution data using $r_o = 0.1$ mm (value as stated by the authors) yields only qualitative agreement with their experimental data points. Good to excellent agreement is, however, achieved when the solution is calculated for an entrance radius value $r_o = 0.15$ mm, even in the region just downstream of the feed hole, as is shown in detail in Figure 5.11. Regarding the fact that only this geometry features such a discrepancy, the cause is most likely an inaccuracy in the feed hole fabrication process, either a deviation of the diameter itself or a chamfer at the hole exit, both resulting in an increased entrance curtain area. Configurations combining small entrance radii and thin air gaps are more likely to suffer from the above mentioned effect (as is indicated by all other comparisons having either larger entrance radii or an increased gap height).

We now consider the second bearing case, namely with a shallow pocket of radius $r_p = 2$ mm and depth $d_p = 20$ μm. Although the solution method was originally intended to be applied to inherently restricted bearing configurations, it can also handle feeding systems which incorporate a pocket/recess. Two distinct approaches are outlined and compared with experimental data. The first approach is to be used on shallow feeding pockets. When entering the bearing gap, the flow passes through two separate entrance curtains, each contributing to the global flow pattern. A first restriction occurs at the curtain given by $2\pi r_o(h_o + d_p)$ after which the flow settles down in the feeding recess. From this point, the flow undergoes a second (minor) restriction characterised by $2\pi r_p h_o$. Even further

11 This was ascertained, for example, by the fact that in many instances where the flow was nearly a slow viscous one over the whole bearing, and consequently, the flow rate would be uniquely determined from the load capacity, the measured flow rate was nearly double that value, which is physically impossible.

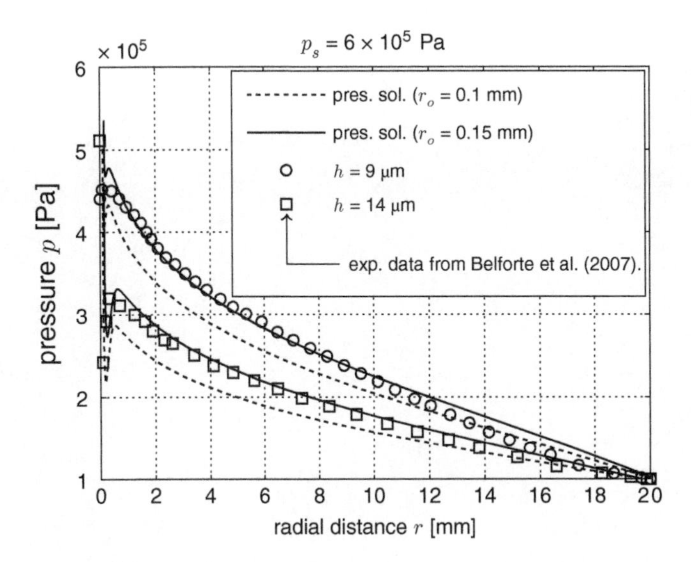

Figure 5.10 Comparison of pressure profile for an inherently restricted bearing (detail view provided in Figure 5.11). $\mathrm{Re}_0^* = 27.1$, $M_0 = 0.407$ for $h = 9$ μm; $\mathrm{Re}_0^* = 86.9$, $M_0 = 0.627$ for $h = 14$ μm. Experimental data from Belforte et al. (2007). Source: Waumans T, Al-Bender F and Reynaerts D 2008 by permission of ASME.

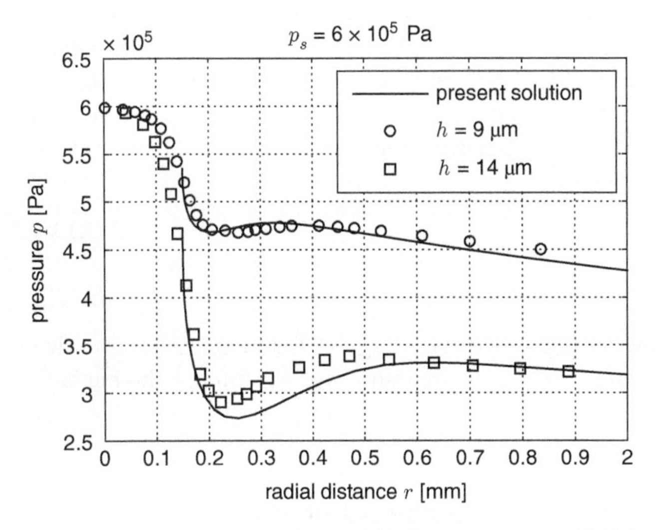

Figure 5.11 Detail view of Figure 5.10. Source: Waumans T, Al-Bender F and Reynaerts D 2008 by permission of ASME.

downstream, viscous flow sets in up to ambient conditions at gap exit. To model this flow pattern, it is sufficient to introduce a step in the gap height function $h(r)$ at $r = r_p$ (a gradual step can prevent numerical problems when integrating the initial-value value problem). The outlined approach is adopted to the geometry of Figure 5.12, demonstrating good overall agreement with measurements performed by Belforte et al. (2007). The discrepancy for $r < 1$ mm is caused by the fact that the conditions in this region are not completely in line with the method's assumptions. Comparison of the restriction surfaces given by respectively $2\pi r_0(h_0 + d_p)$ and πr_0^2, indicates that the first entrance geometry does not entirely act as an inherent restrictor ($2h_0 \ll r_0$).

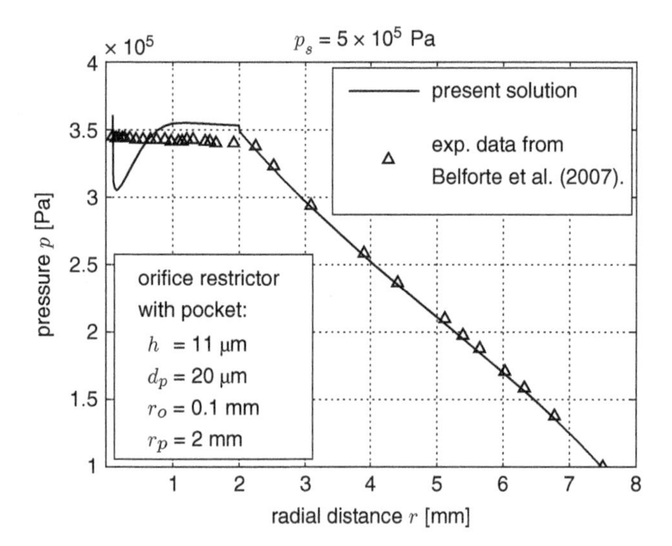

Figure 5.12 Comparison of pressure profile for an orifice restricted bearing with a shallow feeding pocket. Experimental data from Belforte et al. (2007). Source: Waumans T, Al-Bender F and Reynaerts D 2008 by permission of ASME.

For geometries with much deeper pockets, the entrance conditions may be such that severe reverse flow takes place in the recess volume involving the onset of turbulence. At this point the basic assumptions of the solution method are violated and we have to refer to another solution. For this purpose, a feeding geometry with a deep recess is regarded as a combination of an orifice flow characterised by r_o and a certain orifice discharge coefficient $C_{d,\text{ orifice}}$ followed by an entrance phenomenon at the curtain denoted by $2\pi r_p h_o$. The orifice flow is assumed to lose all its dynamic pressure before leaving the recess volume. An iterative solution strategy yields the value of the intermediate recess pressure p_p by matching of both flow rates:

$$\dot{m}_1 = f(p_s, p_p, C_{d,\text{ orifice}}) \tag{5.13a}$$

$$\dot{m}_2 = f(p_p, p_o). \tag{5.13b}$$

Figure 5.13 shows the results obtained in this way. A discharge coefficient $C_{d,\text{ orifice}} = 0.6$ is assumed for the orifice flow. No information regarding the actual pressure distribution before gap entrance is gained through this method. The comparison should therefore only be made for $r > r_p$.

5.8 Formulation of a Lumped-Parameter Inherent Compensator Model

Formulation of a compensator model In the foregoing, the accuracy of the method of separation of variables has been amply demonstrated. Although the method is not difficult to apply for a given bearing configuration, it would, nevertheless, be handier to use the method to deduce a lumped-parameter entrance flow formula. Such a formula can then be easily incorporated into the Reynolds model from which bearing characteristics can then readily be obtained.

Both the nozzle formula and the entrance-loss formula (cf. Vohr's correlation) may be used as a basis for such a model. They can, in fact, be shown to be equivalent; i.e. different formulation for the same phenomenon. The basic aspect of the problem lies quantifying the coefficient of discharge C_d, or the loss coefficient K, respectively.

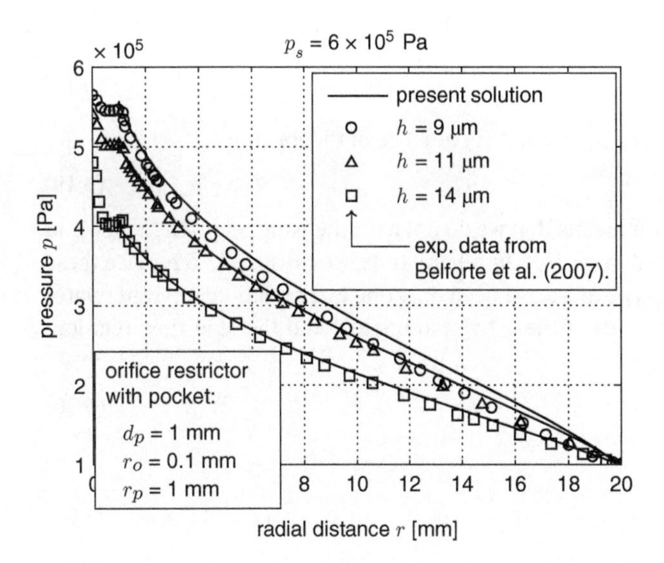

Figure 5.13 Comparison of pressure profile for an orifice restricted bearing with a feeding pocket. Experimental data from Belforte et al. (2007). Source: Waumans T, Al-Bender F and Reynaerts D 2008 by permission of ASME.

The nozzle formula is, however, easier to implement, since it does not involve the intermediate parameter p_o. It will, therefore, be used as the basis for the lumped-parameter model.

5.8.1 The Entrance Coefficient of Discharge

First, let us restate the ideal entrance mass flow formula, (from Chapter 2):

$$\bar{m}_o = A_o \sqrt{\frac{2\kappa}{\kappa - 1}} \frac{p_s}{\sqrt{\Re T_s}} \phi_e \left(\frac{p_o}{p_s}\right) \tag{5.14}$$

(where $A_o = 2\pi r_o h_o$ for simplicity.)

We propose to replace (p_o/p_s) by (p_t/p_s) and correct the flow rate by a coefficient C_d, thus:

$$\bar{m}_o = C_d A_o \sqrt{\frac{2\kappa}{\kappa - 1}} \frac{p_s}{\sqrt{\Re T_s}} \phi_e(\bar{p}_t) \tag{5.15}$$

where,

$$\bar{p}_t = \frac{p_t}{p_s}$$

The problem is to ascertain whether or not this presentation is valid, and if so, to determine the form of C_d. This we will do in the following way.

The basic results from the previous analysis, expressed in a general way, are:

$$\frac{p_t}{p_o} = f_1(\mathrm{Re}_o^*, \mathrm{M}_o, \nu_o) \tag{5.16}$$

$$\frac{p_o}{p_s} = f_2(\mathrm{M}_o) \tag{5.17}$$

Eliminating p_o between equations 5.16 and 5.17 yields:

$$\bar{p}_t = \frac{p_t}{p_s} = f_3(\mathrm{Re}_o^*, \mathrm{M}_o, \nu_o) \tag{5.18}$$

This means, initially, that the correction coefficient C_d, in equation 5.15, must be of the form:

$$C_d = C_d(\mathrm{Re}_o^*, \mathrm{M}_o, \nu_o) \tag{5.19}$$

However, having eliminated (p_o/p_s) from the calculation means that we do not have the parameters Re_o^* and M_o at our disposal, since these are directly or indirectly related to (p_o/p_s). In other words, we must seek to replace these parameters by new ones to fit the new problem. For this purpose, we need only one extra relationship that relates Re_o^* with M_o, which is easily obtained from the definitions of these two parameters and the feed flow relation. Thus, after manipulation and simplification, we obtain:

$$\mathrm{Re}_o^* \Lambda_e = f_4(\mathrm{M}_o) \tag{5.20}$$

where,

$$\Lambda_e = \frac{\mu r_o}{p_s h_o^2}\sqrt{\frac{\Re T_s}{\kappa}}$$

is the *entrance* number.

Now, Re_o^* and M_o may be eliminated between relations 5.18 and 5.20, to give:

$$(\mathrm{Re}_o^*, \mathrm{M}_o) = f_5(\Lambda_e, \bar{p}_t, \nu_o) \tag{5.21}$$

Finally, we may write, for C_d:

$$C_d = f_6(\Lambda_e, \bar{p}_t, \nu_o) \tag{5.22}$$

which is the sought after representation.

The question of validity of this representation is determined by its uniqueness, i.e. whether the function f_6 is single valued over its domain of definition. This is not obvious from the analysis, since we do not have an analytic expression for the function f_1 given above. There is only one way to settle this question, and that is to calculate the function f_6. This is, however, also one of our primary objectives.

5.8.2 Calculation of C_d

The calculation of C_d amounts to the representation of the entrance flow characteristics in condensed form. That is to say that the entrance problem has to be solved over the range of interest of the parameters $(\mathrm{Re}_o^*, \mathrm{M}_o, \nu_o)$; determining each time the values $(\bar{p}_t, C_d)$. A sample of such results is depicted in Figure 5.14 (Al-Bender 1992), for $\nu_o = 0$, i.e. a flat bearing.

The range of the results is bounded from below by the onset of turbulence, and from above by approaching viscous character. It is evident from this result that C_d is single valued in $\bar{p}_t$ for each value of Λ_e; this holds true also for other values of the normalised conicity ν_o. It is noteworthy also that for $\bar{p}_t$ below about 0.97, the C_d curves may be approximated by a small degree polynomial. Beyond that value, on the other hand, as marked by the grey-shaded region, the actual value of C_d is not that important for determining the flow balance and consequently the value of p_t in an actual bearing problem: the flow will be viscous on almost the entire extent of the film.

In order to facilitate the use of the nozzle formula, an approximate representation for C_d had to be sought. This was done in the following way. Each C_d–$\bar{p}_t$ curve was represented by three points through which a parabola may be drawn, namely:

$$C_d = a_0 + a_1\bar{p}_t + a_2(\bar{p}_t)^2 \tag{5.23}$$

where the coefficients a_i are functions of (Λ_e, ν_o).

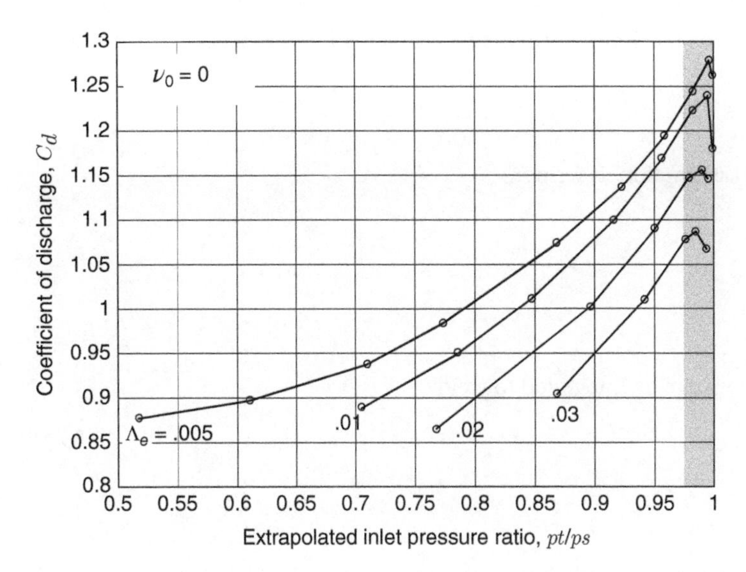

Figure 5.14 Calculated values of the Coefficient of discharge C_d. Source: Al-Bender F 1992.

Next, these coefficients where calculated for five values of Λ_e and three values of v_o, thus:

$$\Lambda_e = 0.003,\ 0.005,\ 0.01,\ 0.02,\ 0.03 \qquad v_o = 0,\ 0.03,\ 0.06$$

(the reason for taking fewer points in v_o is because C_d is less sensitive to v_o.)

The data points were, finally, set in an array from which the appropriate values for C_d were retrieved through Lagrangian interpolation. When the value of Λ_e lies outside the range given above, C_d is assigned the values (1 : $\bar{p}_t > 0.97$) or (0.86 : $\bar{p}_t < 5.28$). This would result in only negligible errors.

Remark

In a design problem, both Λ_e and v_o are directly available from the design parameters, namely the geometry and the supply pressure. With these values given, the coefficients a_i of equation 5.23 are determined, and C_d is given as an analytic function in $\bar{p}_t$. This can then be used to solve the Reynolds problem, i.e. for matching the the viscous mass flow rate with the inlet flow rate, by a Newton–Raphson method.

5.8.3 The Normalised Inlet Flow Rate

Equation 5.15 may be normalised, as given in Chapter 2, through its division by $\pi r_a \rho_a h_a^3/6\mu$, to give:

$$\bar{m}_o = C_d \Lambda_f \left(\frac{r_o}{r_a}\right)\left(\frac{h_o}{h_a}\right)\left(\frac{p_s}{p_a}\right)\Phi_e(\bar{p}_t) \tag{5.24}$$

where Φ_e is defined in Chapter 2.

5.8.4 Solution of the Static Axisymmetric Bearing Problem by the Reynolds/Compensator Model

Let the bearing geometry, supply and ambient pressures and temperatures be given. Normalisation w.r.t. ambient (exit) values yields:

$$H(R) = 1 + v_a(1 - R), \qquad H = h/h_a,\ R = r/r_a, R_o = r_o/r_a$$

$$P = p/p_a.$$

Also, Λ_f, Λ_e and v_o are determined. Further, let us for simplicity denote p_t by p_o. (This notation will henceforth be adopted for Reynolds/compensator problems.)

Film solution

The Reynolds equation, with the isothermal assumption, is written:

$$\frac{1}{R}\frac{\mathrm{d}}{\mathrm{d}R}(RH^3\frac{\mathrm{d}P^2}{\mathrm{d}R}) = 0$$

with boundary conditions:

$$P(R_o) = P_o, \qquad P(1) = 1$$

With the film function $H(R)$ given, this problem can be integrated to yield:

$$P^2 = C_1 B_v(R) + 1 \tag{5.25}$$

where,

$$\begin{aligned} B_v(R) &= \int \frac{\mathrm{d}R}{RH^3} + \text{const.} \\ &= \frac{1}{a^3}\left(\log(\frac{R}{H}) + \frac{(a+H)^2 - (1+a)^2H^2}{2H^2}\right), \qquad a = 1 + v \end{aligned}$$

(note that $B_v(1) = 0$, and that when $v = 0$, then $B_0(R) = \log(R)$) and,

$$C_1 = \frac{P_o^2 - 1}{B_v(R_o)}$$

The normalised mass flow rate in the film is given from Chapter 2, as:

$$\bar{m}_r = -R_o H_o^3 P_o \frac{\mathrm{d}P_o}{\mathrm{d}R} = -\frac{C_1}{2} \tag{5.26}$$

Feed flow solution

This is given from equation 5.24, as

$$\bar{m}_o = C_d \Lambda_f R_o H_o P_s \Phi_e(\bar{p}_o) \tag{5.27}$$

In order to obtain the complete solution, we have to solve equations 5.26 and 5.27 simultaneously, thus determining the only unknown P_o. Let us note that, with C_d determined from equation 5.23, both of the two flow equations are determined only by, and are analytic in, P_o. Thus, a Newton–Raphson procedure may be used to find the solution. This has the advantage of quick convergence. (Usually, an accuracy of 0.1 percent is obtained in about four iterations.)

It is seen, from the foregoing, that the Reynolds/restrictor model is much easier to apply than that with the exact method.

5.9 Conclusions

This chapter has treated the radial flow into an axisymmetric bearing gap. Prior to the flow settling to viscous character sufficiently far downstream the feed hole, where the Reynolds equation will become valid further downstream, the flow has to develop from almost uniform transverse velocity distribution, corresponding to high inertia contribution, at gap entrance to parabolic distribution, corresponding to viscous flow, over a relatively short entrance interval. In this way, the entrance region resembles in its behaviour a restrictor: the so-called "inherent restrictor", which, when compared to the standard orifice or capillary restrictors, might show a head loss or a head gain with respect to viscous-flow pressure when that is extrapolated from the downstream viscous solution back to gap entrance. After reviewing and evaluating the state of the art in regard to solutions to this problem, the method of "separation of variables", developed in Chapter 3, is applied to this radial flow configuration, yielding satisfactory results.

Furthermore, the general trends of the flow are investigated and many comparisons are made of the pressure distributions, obtained by this method, with experimental data from various sources, showing good agreement. Finally, owing to this good agreement of the results, the method is used to provide data for the correction coefficient of the restrictor model, i.e. the generalised "coefficient of discharge", which is shown to depend on the film geometry and the entrance Reynolds and Mach numbers.

Let us note that this model is based entirely on the laminar-flow assumption, which seems thus to remain valid for very high values of the entrance Reynolds and Mach numbers. The case of supersonic, turbulent entrance differs considerably from the ones considered here. That case is treated in Chapter 14.

The results of this refined restrictor model are used, together with Reynolds model, for the treatment of the remaining problems (Chapters 6 and 7).

A final remark may be in order. Although the considered solutions pertain to the case of axisymmetric radial channel flow, as exemplified by circular, centrally-fed (CCF) aerostatic bearings, the results may be adapted, with approximation, to other situations in which there might be a certain departure from those conditions, e.g. for multi-feed-hole bearings, where the pressure field might not be symmetric, journal bearings, where the gap function around the feed-hole might depart from symmetry, etc.

References

Al-Bender, F. and Van Brussel, H., (1992). Symmetric radial laminar channel flow with particular reference to aerostatic bearings. *Trans. ASME - Journal of Tribology* 114(7): 630–636.

Al-Bender, F., (1992) *Contributions to the design theory of circular centrally fed aerostatic bearings*. PhD, KU Leuven.

Belforte, G., Raparelli, T., Viktorov, V. and Trivella, A., (2007). Discharge coefficients of orifice-type restrictor for aerostatic bearings. *Tribology International* 40(3): 512–521.

Dudgeon, E.H., (1970) Performance factors for circular, hydrostatic, gas-lubricated thrust bearings—Part I. Single inlet thrust bearings. Mechanical engineering report mt-62, National Research Council of Canada. 32 pp.

Geiger D., Fara, H.D. and Street, N., (1964). Steady radial flow between parallel plates. *ASME - Journal of Applied Mechanics* 354–355.

Gross, W.A., (1962). *Gas Film Lubrication*. John Wiley & Sons.

Hagerup, H.J., (1974). On the fluid mechanics of the inherent restrictor. *ASME - Journal of Fluids Engineering* 341–347.

Hayashi, S., Matsui, T. and Ito, T., (1975). Study of flow and thrust in nozzle-flapper valves. *ASME - Journal of Fluids Engineering* 39–50.

Lambrechts, C., (1989). *Feed flow into an aerostatic bearing* Engineering thesis Department of Mechanical Engineering, K.U. Leuven.

Lee, P.M. and Lin, S., (1985). Pressure distribution for radial inflow between narrowly spaced disks. *ASME - Journal of Fluids Engineering* 107, 338–341.

Livesey, J.L., (1960). Inertia effects in viscous flows. *International Journal of Mechanical Science* 1, 84–88.

Lowe, I.R.G., (1970). A study of flow phenomena in externally pressurized gas thrust bearings. Mechanical engineering report mt-61, National Research Council of Canada. 45 pp.

McGinn, J.R., (1955). Observations on the radial flow of water between fixed parallel plates. *Applied Science Research Section A* 5, 255–263.

Moller, P.S., (1966). Radial flow without swirl between parallel disks having both supersonic and subsonic regions. *ASME - Journal of Basic Engineering* 147–154.

Mori, H. and Miyamatsu, Y., (1969). Theoretical flow-models for externally pressurized gas bearings. *ASME - Journal of Lubrication Technology* 181–193.

Murphy, H.D., Coxon, M., and McEligot, D.M., (1978). Symmetric sink flow between parallel plates. *ASME - Journal of Fluids Engineering* 477–484.

Neale, M.J., (1973). *Tribology Handbook*. Butterworths, London.

Savage, S.B., (1964). Laminar radial flow between parallel plates. *ASME - Journal of Applied Mechanics* 594–596.

Schlichting, H. (1968). *Boundary-Layer Theory* 6th edn. McGraw-Hill, New York.

Snoeys, R. and Al-Bender, F., (1987). Development of improved externally pressurized gas bearings. *KSME Journal* 1(1): 81–88.

Strauss, K., (1976). Die langsame strömung einer newtonschen flüssigkeit durch einen endlichen spalt. *Acta Mechanica* 24, 305–311.

Vallentine, H.R., (1967). *Applied Hydrodynamics*. Butterworths, London.

Vohr, J.H., (1966). An experimental study of flow phenomena in the feeding region of an externally pressurized gas bearing. Report mti-65tr47, Mechanical Technology Inc.

Waumans, T., Al-Bender, F. and Reynaerts, D., (2008). A semi-analytical method for the solution of entrance flow effects in inherently restricted aerostatic bearings (GT2008-50499) *Proceedings of the ASME Turbo Expo 2008: Power for Land, Sea and Air*, Berlin, Germany.

Woolard, H.W., (1957). A theoretical analysis of the viscous flow in a narrowly spaced radial diffuser. *ASME - Journal of Applied Mechanics* 9–15.

6

Basic Characteristics of Circular Centrally Fed Aerostatic Bearings

6.1 Introduction

As mentioned in previous chapters, owing to its relative simplicity, the case of axisymmetric flow furnishes a good generic example to study basic behaviour of a flat bearing. In this chapter, we take a look at this behaviour in order to distill some salient properties that could help the bearing designer gain better insight into the problem.

The knowledge accumulated in the foregoing chapters puts us in good stead to derive the basic static characteristics of CCF (circular centrally-fed) bearings. In the first place, we would like to form an idea about load-carrying and stiffness characteristics as well as typical air consumption values as related to the bearing design parameters. A second category of information relates to the effect of misalignment and relative speed on the characteristics mentioned above as well as the performance of a bearing as a whole. With this information, we can go about designing air bearing systems from a good starting point, being at least able to estimate and evaluate the feasibility of various solutions. Let us mention again that, although the results that will be developed in this chapter relate to CCF bearings, in essence they are applicable, *mutatis mutandis* to non-circular bearings; however, at least for the time being, not to bearings with multiple feed holes.

We consider first the axial characteristics of the bearing, i.e. assuming a perfectly parallel (nominal) gap and with zero relative sliding velocity. Thereafter, we consider tilt and relative speed. Throughout this treatment, isothermal state of the gas is assumed. The solutions for the general polytropic case could be easily derived in a similar way. Let us note however that the characteristics obtained are very similar for all cases.

Bearing characteristics may be dealt with under three classes, namely (i) axial characteristics, which include load capacity, stiffness, air consumption and their inter-relationships, (ii) tilt behaviour, which pertains to the situation in which the air film deviates from nominal parallelism, and (iii) the effect of relative sliding of the bearing surfaces on load and flow characteristics.

6.2 Axial Characteristics: Load, Stiffness and Flow

For easy reference, let us first us recall the bearing geometry and notation as depicted in Figure 6.1.

The main dimensionless design parameters are:

- the dimensionless conicity: $H_v = h_v/h_a$
- the dimensionless feed-hole radius: $R_o = r_o/r_a$
- the dimensionless Supply pressure: $P_s = p_s/p_a$
- the feed number: $\Lambda_f = 12\frac{\mu r_a}{p_a h_a^2}\sqrt{\frac{2\kappa \Re T_s}{\kappa - 1}}$.

Air Bearings: Theory, Design and Applications, First Edition. Farid Al-Bender.

Companion website: www.wiley.com/go/AlBender/AirBearings

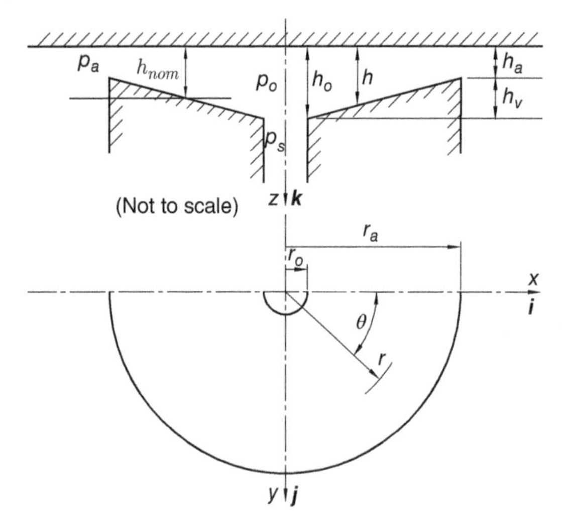

Figure 6.1 Geometry and notation. Source: Al-Bender F 1992.

6.2.1 Determination of the Pressure Distribution

Using the restrictor–viscous film model (chapter 5), one may construct the solution to all possible configurations arising from the different combinations of supply pressure, ambient pressure, bearing size, gap height, conicity and feed-hole size.

In order to encompass all the design space without redundancy of solutions, we shall continue to adopt the dimensionless formulation of design parameters and characteristics, while providing some dimensional design examples by way of illustration. Let us briefly recall the relevant equations (Chapter 5):

Assuming an axisymmetric gap geometry with a height distribution given by:

$$H(R) = 1 + \nu_o(1 - R), \tag{6.1}$$

with ν_o the normalised conicity angle ($\nu_o h_o / r_o = \tan(\alpha) \simeq \alpha$, where $\tan(\alpha) = h_v/(r_a - r_o)$). In this case, the Reynolds equation is written as (without hydrodynamic or squeeze contribution):

$$\frac{1}{R}\frac{d}{dR}(RH^3\frac{dP^2}{dR}) = 0. \tag{6.2}$$

with boundary conditions:

$$P(R_o) = P_o, \qquad P(1) = 1.$$

Integration results in:

$$P^2 = C_1 B_\nu(R) + C_2, \tag{6.3}$$

where, ($a = 1 + \nu_o$)

$$B_\nu(R) = \frac{1}{a^3}\left(\log(\frac{R}{H}) + \frac{(a+H)^2 - (1+a)^2H^2}{2H^2}\right). \tag{6.4}$$

(note that $B_\nu(1) = 0$, and that when $\nu = 0$, then $B_0(R) = \log(R)$,) and, with boundary conditions:

$$P(R_o) = P_o, \qquad P(1) = 1,$$

we get

$$C_1 = \frac{P_o^2 - 1}{B_\nu(R_o)}, \tag{6.5}$$

$$C_2 = 1.$$

The normalised mass flow rate in the film is given from Chapter 5, as:

$$\bar{m}_r = -R_o H_o^3 P_o \frac{dP_o}{dr} = -\frac{C_1}{2} \tag{6.6}$$

Note that when the supply pressure is specified, the entrance pressure P_o can be determined from equating the entrance flow with the film flow.

6.2.2 Typical Results

PressurE Distribution for a Given P_0

Firstly, we consider **flat-surface bearing** ($H_v = 0$)

In this case, the pressure distribution is independent of the gap height, but is determined completely by P_0 and R_0:

$$P^2 - 1 = A \log R$$

where

$$A = (P_0^2 - 1)/\log R_0$$

Let us note that the function $\log R/\log R_0$ represents the form of the pressure distribution of incompressible flow through the gap. Figure 6.2 plots this form for various values of R_0.

We notice that as $R_0 \to 0$, the pressure gradient close to the feed hole becomes unbounded resulting in very small values of the pressure downstream of the feed hole. This points to an important design principle, namely not to make the feed hole too small, in order not to starve the bearing, in a manner of speaking. This effect has also implications when one employs numerical modeling in which a feed hole is represented by a single node: this could lead to a gross misrepresentation of the real situation.

Figure 6.3 plots P for $P_0 = 5$ (as an example) and three values of R_0, for compressible flow. The qualitative difference between this and Figure 6.2 is the more "convex" shape of the curves, in the middle and high R_0 range, owing to taking the square root of the incompressible case. This "convexity" increases with increasing P_0. In other

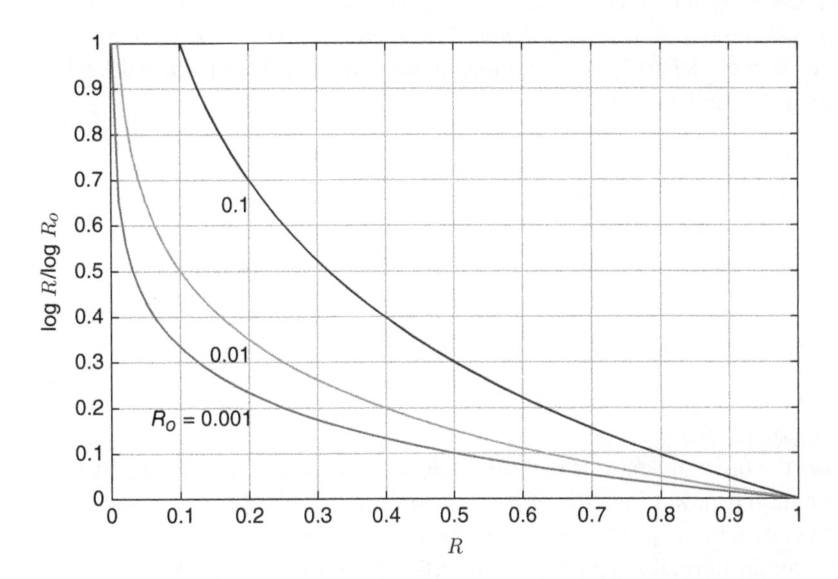

Figure 6.2 Normalised pressure distribution corresponding to incompressible fluid flow for $R_0 = 0.001, 0.01, 0.1$.

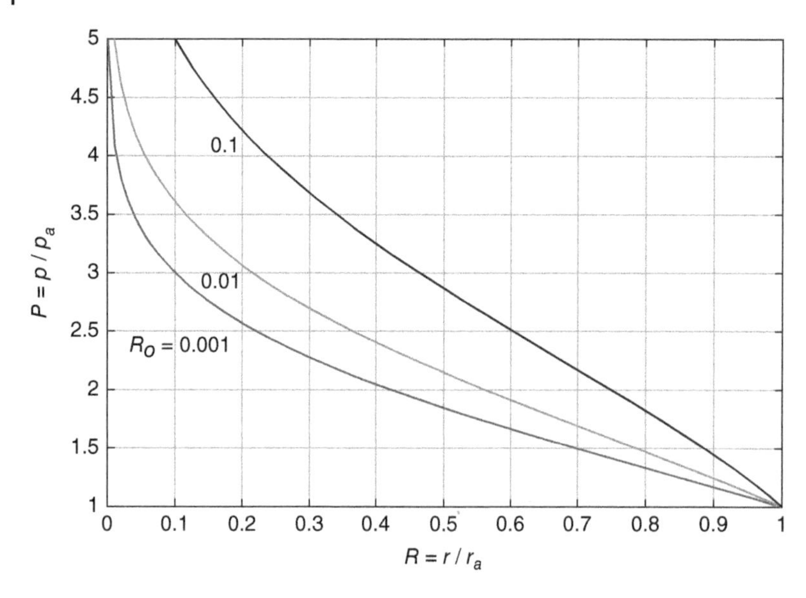

Figure 6.3 Pressure distributions for $P_o = 5$ and three values of R_o, for compressible flow.

words, the effect of compressibility becomes more manifest as the film pressure increases, which is logical. Thus, when $P_o >> 1$, we get

$$P \sim \sqrt{(\log(R))}.$$

Conversely, when P_o is very close to unity, we can write $P = 1 + \Pi$, with $\Pi << 1$, and substitute in the Reynolds equation (6.2) to find out that

$$\Pi \sim \log(R),$$

which is the gauge pressure in the incompressible case.

The effect of varying the entrance pressure P_o for a given value of R_o is depicted in Figure 6.4. The value of P_o, especially when large, roughly scales P as could also be readily deduced from Eqs. 6.3 and 6.5. Thus, Figure 6.5 furnishes a clearer presentation of the effect of higher P_o, i.e. of compressibility, which is to increase “convexity” of the pressure distribution (by enhancing the inflexion point).

Behaviour of Mass Flow

Referring to Eq. 6.6, we note that the flow (which equals the gas consumption) increases with $(P_o^2 - 1)/\log(R_o)$ and so increases with increasing R_o $(0 < R_o < 1)$. Furthermore, if, for not too small values of R_o, we assume the load generated in the film, $R_o < R < 1$, to be roughly proportional to P_o, then the ratio of load to flow is

$$\sim \frac{P_o \log(R_o)}{(P_o^2 - 1)},$$

which increases with decreasing R_o.

Second, a bearing with **convergent gap**, $H_v \neq 0$

Equation 6.3 shows that for this case, the function $\log(R)$ is replaced by B_v, which involves the film function.

This means that the pressure distribution is now a function of P_o, R_o as well as v_o (whereas for a parallel-gap film, the pressure distribution is independent of the gap height, for a constant P_o, as we saw above). To see this, and without loss of generality, we consider the normalised pressure distribution for one selected value of $R_o = 0.05$ and a range of conicity values.

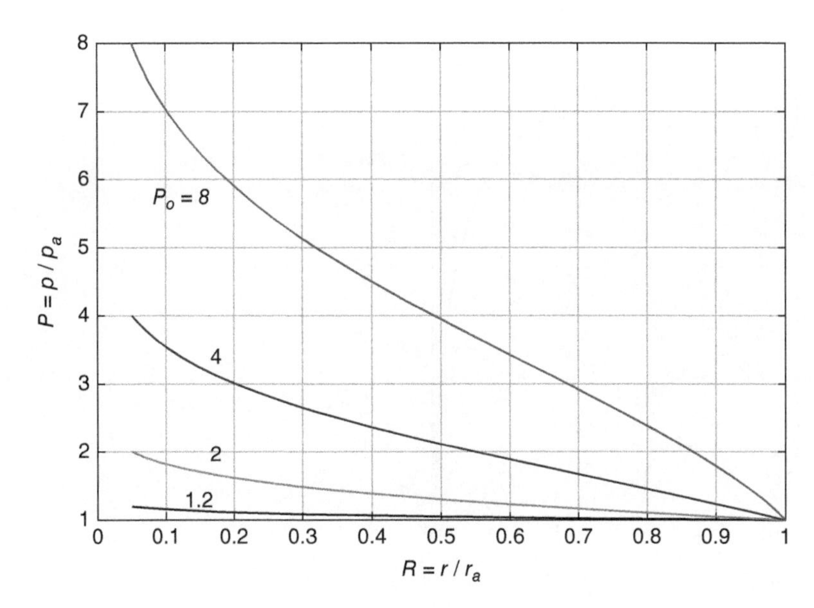

Figure 6.4 Pressure distributions for $R_o = 0.05$ and, from bottom to top, $P_o = 1.2, 2, 4, 8$.

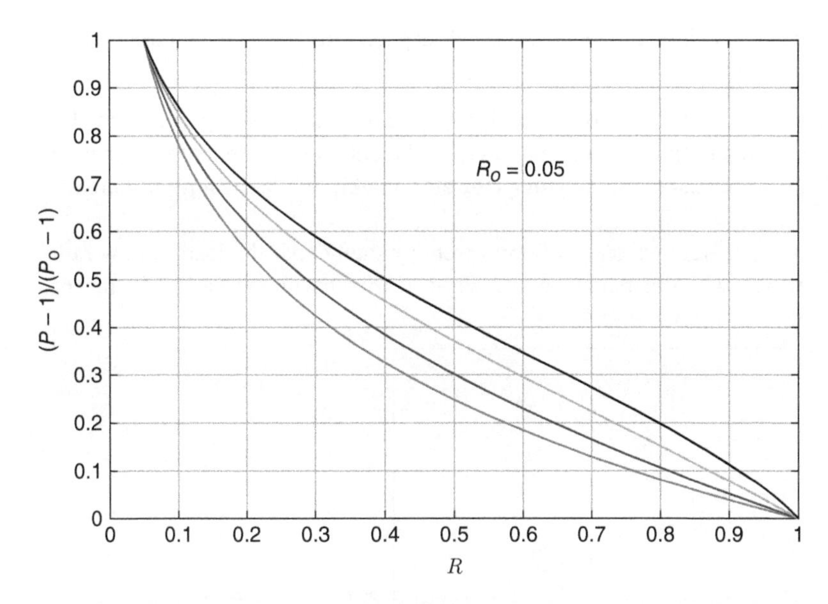

Figure 6.5 Clearer presentation: the effect of higher P_o. Curves from bottom to top are for $P_o = 1.2, 2, 4, 8$.

Figure 6.6 shows that as the conicity $H_v = h_v/h_o$ increases, the area under the pressure curve increases until it eventually fills the whole space at $H_v \to \infty$. The mass flow likewise increases with increasing conicity.

Flow efficiency

The question arises whether the bearing becomes more flow-efficient with increasing conicity. If we characterise this efficiency by the ratio of load to flow, then the outcome will depend on the choice of the value of the nominal gap height. This latter is equal to h_a = const., for a flat bearing (i.e. without conicity). For a bearing with conicity, we can define a value for the nominal gap, see Figure 6.1, following various design and construction criteria. Perhaps

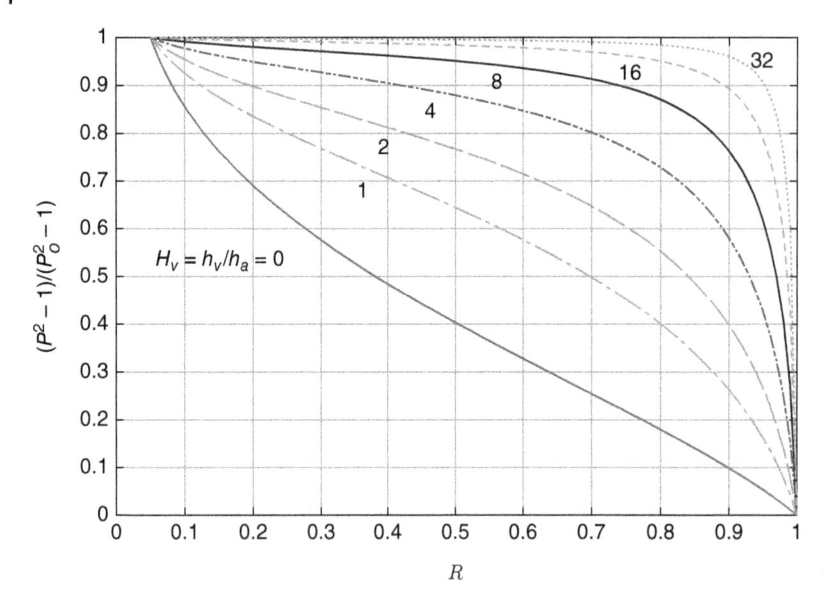

Figure 6.6 Pressure distribution for different conicities, from bottom to top, H_v = [0.0; 1; 2; 4; 8; 16; 32].

the most reasonable is to take the 2D mean gap height $h_a + h_v/2$ as a nominal gap. Thus we have two cases, which produce different results in regard to flow efficiency:

case 1 Nominal gap = minimum gap = h_a. Here, we see, as depicted in Figure 6.7, that this ratio (i.e. of load to flow) decreases with increasing H_v. That is, the bearing becomes less efficient. The reason for this is that the flow rate increases faster in H_v than the load does; the latter saturates quickly, while the former increases almost linearly, see Figure 6.9.

case 2 Nominal gap = mean gap = $h_a + h_v/2$. Here we see a different trend; namely, that the load-to-flow ratio increases with conicity. In other words, with the increase of conicity, a bearing becomes more load efficient,

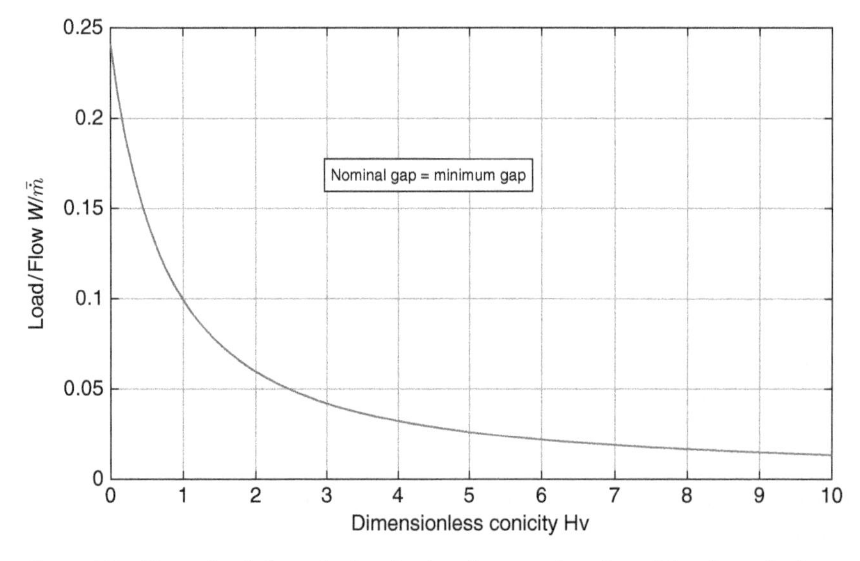

Figure 6.7 The ratio of dimensionless load-to-flow as a function of the dimensionless conicity, when the minimum gap h_a is taken as the nominal gap for calculating the flow rate.

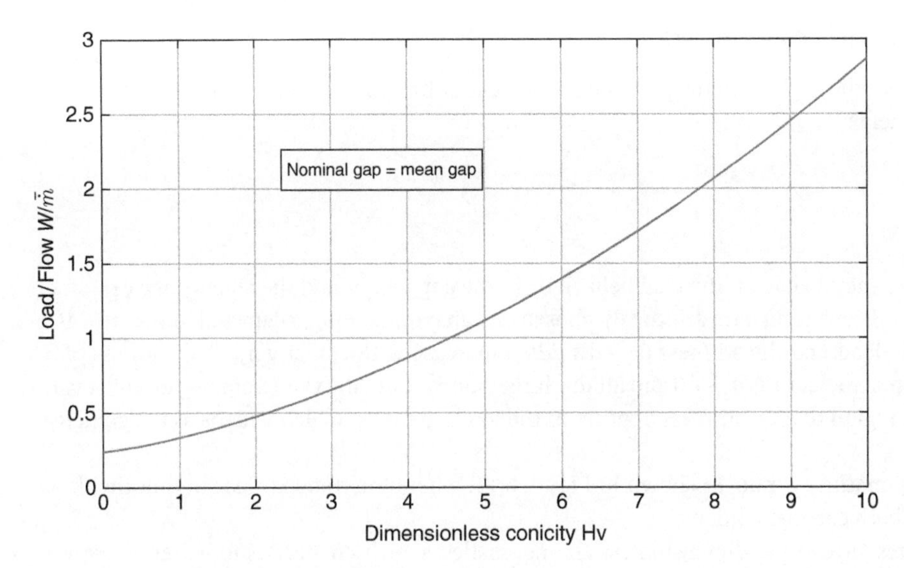

Figure 6.8 The ratio of dimensionless load-to-flow as a function of the dimensionless conicity, when the mean gap $h_a + h_v/2$ is taken as the nominal gap for calculating the flow rate.

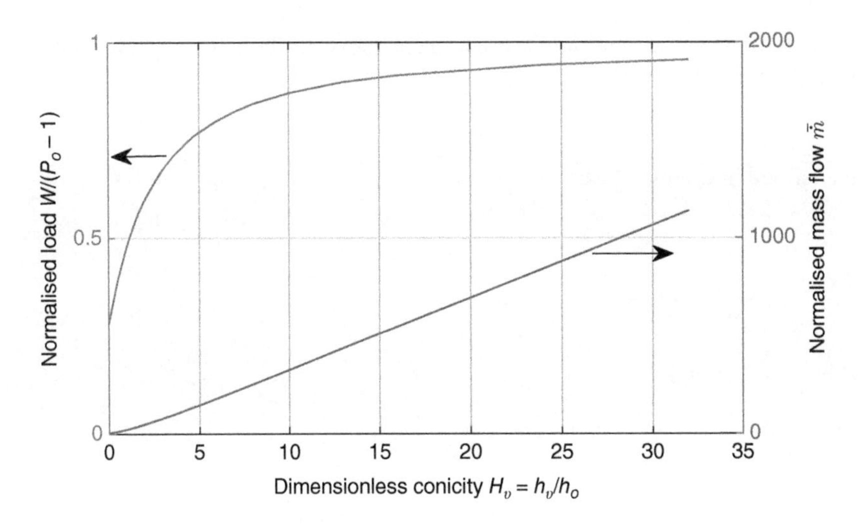

Figure 6.9 The dimensionless load and flow as a function of the dimensionless conicity.

see Figure 6.8. Let us note however, that this mean-gap criterion is reasonable only for small values of the dimensionless conicity H_v since the ratio of minimum gap to nominal gap might otherwise become unrealistically small. This ratio is given by:

$$\frac{\text{minimum gap}}{\text{mean gap}} = \frac{h_a}{h_a + h_v/2} = \frac{1}{1 + H_v/2}.$$

Thus, e.g. for $H_v = 1$, we have a ratio of 2/3, which might be acceptable. However, for $H_v = 3$, we get a ratio of 2/5, which might be unacceptably small.

6.2.3 Characteristics with Given Supply Pressure

These characteristics follow from those given in the previous sections after first solving the flow balance between feed and film. This generally yields:

$$W = f_{\mathrm{w}}(R_{\mathrm{o}}, P_{\mathrm{s}}, \Lambda_{\mathrm{f}}, H_{\mathrm{v}})$$

$$\bar{\dot{m}} = f_m(R_{\mathrm{o}}, P_{\mathrm{s}}, \Lambda_{\mathrm{f}}, H_{\mathrm{v}})$$

Firstly, in order to provide a general view on the load behaviour in the gap height and the concity, we examine the dimensional characteristics of a bearing with arbitrarily-chosen design parameters, as depicted in Figure 6.10.

The general trend is that the load and the stiffness ($= -\mathrm{d}W/\mathrm{d}h_{\mathrm{a}}$) increase with conicity h_{v}. The stiffness of a bearing with zero conicity approaches zero as $h_{\mathrm{a}} \rightarrow 0$ and attains large values at middle values of the gap. However, when the conicity increases, a region of high stiffness appears in the low gap range, which widens as the conicity further increases.

In a similar way, we plot the mass flow rate, as shown in Figure 6.11, where the trend is clear in that the flow increases monotonically with both gap and conicity.

Let us turn our attention presently to the dimensionless characteristics, which enables us to better condense the results than in the dimensional case. In these presentations the dimensionless bearing characteristic is plotted as a function of the dimensionless feed number Λ_{f} and the dimensionless conicity H_{v}. In order to facilitate the interpretation of the results, however, the dependence on the feed number is expressed as that on $1/\sqrt{\Lambda_{\mathrm{f}}}$, which is proportional to h_{a}.

Figures 6.12 through 6.14 correspond to previous sections and are self-explanatory. In the following, we shall discuss the stiffness and the influence of the feed hole size.

The Dimensionless Stiffness K

The dimensionless stiffness has been defined in Chapter 2 as:

$$K = \frac{kh_{\mathrm{a}}}{Ap_{\mathrm{a}}},$$

where A is the bearing area.

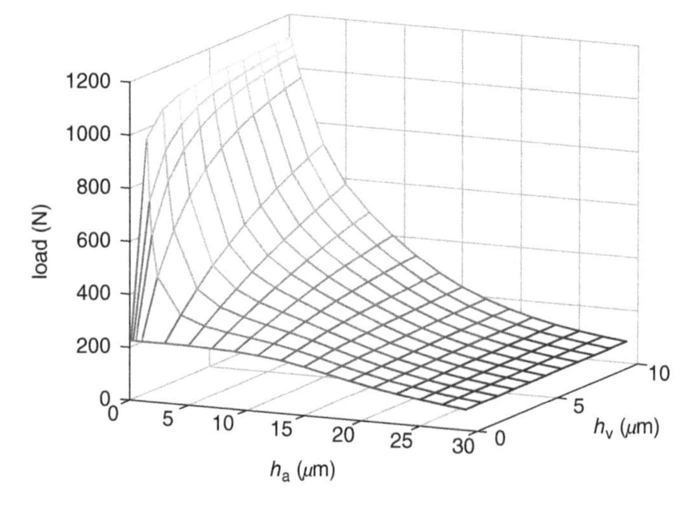

Figure 6.10 Dimensional load as a function of the gap-height and concity for bearing with $P_{\mathrm{s}} = 5$ bar; $P_{\mathrm{a}} = 1$ bar; $r_{\mathrm{o}} = 0.3$ mm; $r_{\mathrm{a}} = 30$ mm.

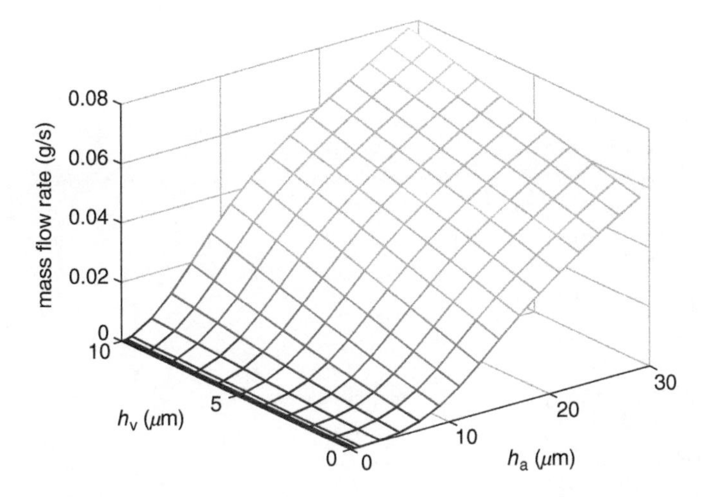

Figure 6.11 Dimensional flow rate as a function of the gap-height and conicity for bearing with $P_s = 5$ bar; $P_a = 1$ bar; $r_o = 0.3$ mm; $r_a = 30$ mm.

Figure 6.16 shows that for any given value of the conicity H_v, there is a value of the feed number Λ_f that yields maximum stiffness. This value shifts towards higher Λ_f (lower $1/\sqrt{\Lambda_f}$) values as H_v increases.

Effect of R_o

The load and the flow rate both increase with increasing feed-hole radius r_0, or size generally. One wonders what happens to the stiffness. To find out, we consider a bearing with constant $P_s = 5$ and $\Lambda_f = 400$ and vary $R_o = r_o/r_a$ and H_v, so as to obtain the bearing characteristics depicted in Figures 6.17 through 6.22.

With regard to Figures 6.21 and 6.22, we make the important observation that: there is an optimum of the stiffness in R_o, for which the optimum value of R_o increases with increasing H_v, which could be a useful design guideline.

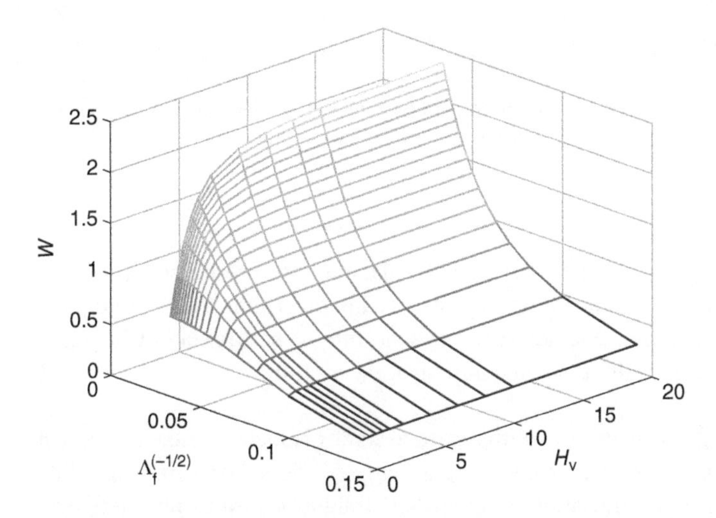

Figure 6.12 The dimensionless bearing load as a function of the dimensionless feed number Λ_f (or $1/\sqrt{\Lambda_f}$) and the dimensionless conicity H_v.

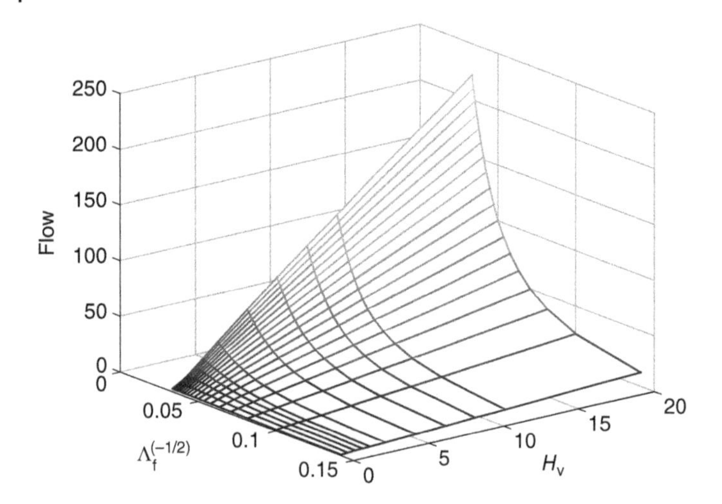

Figure 6.13 The dimensionless bearing flow as a function of the dimensionless feed number Λ_f (or $1/\sqrt{\Lambda_f}$ and the dimensionless conicity H_v.

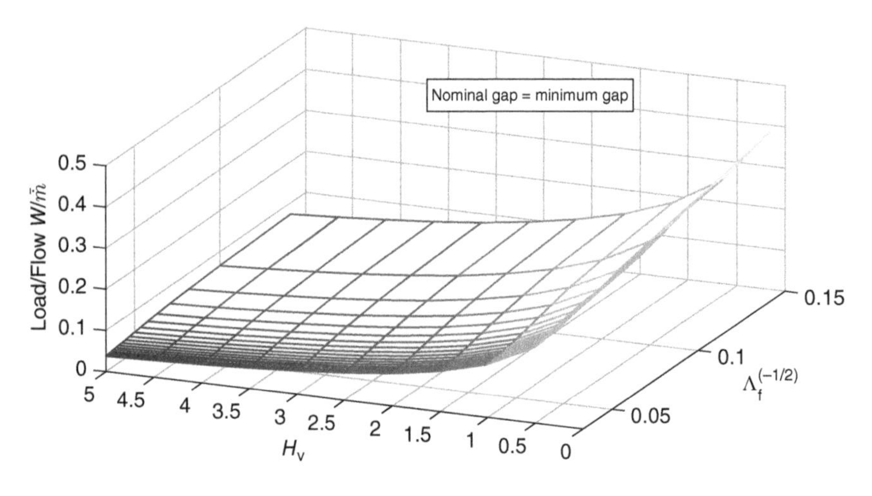

Figure 6.14 The dimensionless bearing load-to-flow ratio as a function of the dimensionless feed number Λ_f (or $1/\sqrt{\Lambda_f}$) and the dimensionless conicity H_v. The ratio increases with $1/\sqrt{\Lambda_f}$ and decreases with H_v, when the minimum gap h_a is taken as the nominal gap for calculating the flow rate.

- **Remark on utility**
  When the damping in the gas film is not relevant; e.g. when sufficient external damping is present in the system, then these results are significant for general optimisation. Otherwise, we need to determine the damping and include it in the optimisation procedure. The dynamic characteristics of a gas film is the subject of Chapter 7, which comprises the determination of the dynamic stiffness and damping of the air film.
- **Note on optimisation**
  Let us assume that we wish optimise a bearing of a given size, with regard to static characteristics, with respect to a certain objective function comprising load capacity, air consumption and stiffness. For a fixed nominal gap h_a and given supply and ambient pressures, optimisation is somewhat straightforward using the figures above since the behaviour in the feed-hole size and the conicity is monotonic except for the stiffness. Seeking optimum behaviour over a range of gap values (which is the more interesting case in practice) will prove more

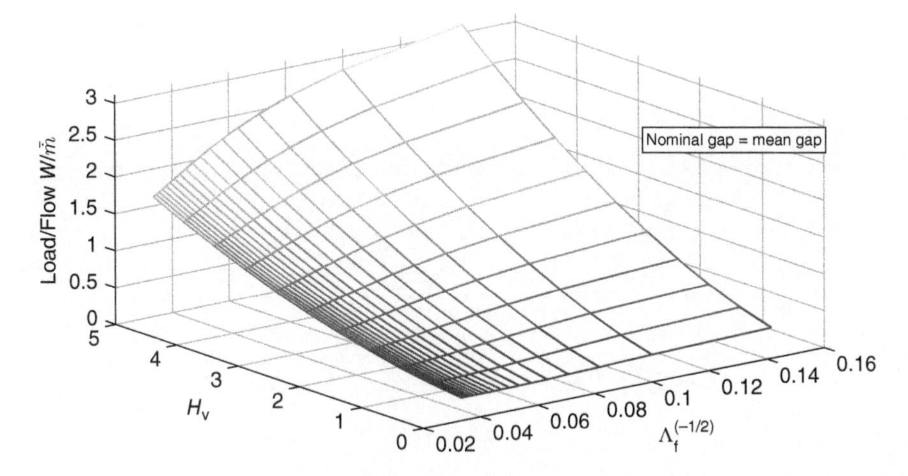

Figure 6.15 The dimensionless bearing load-to-flow ratio as a function of the dimensionless feed number Λ_f (or $1/\sqrt{\Lambda_f}$) and the dimensionless conicity H_v. The ratio increases with $1/\sqrt{\Lambda_f}$ and also increases with H_v, when the mean gap $h_a + h_v/2$ is taken as the nominal gap for calculating the flow rate.

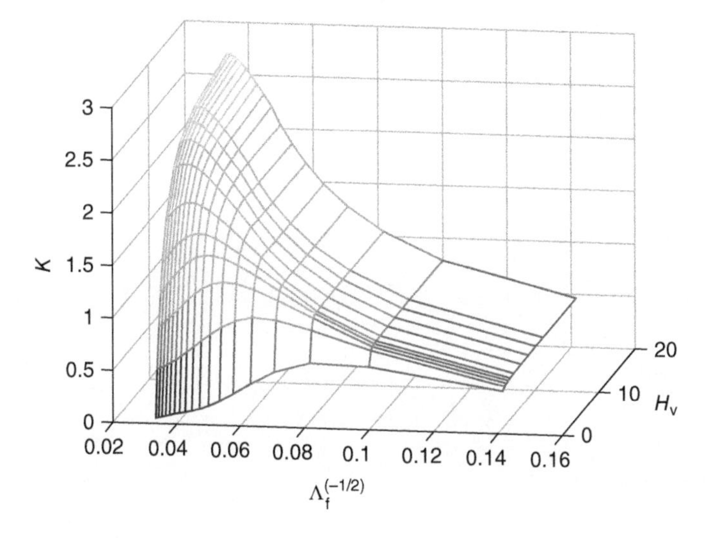

Figure 6.16 The dimensionless bearing stiffness as a function of the dimensionless feed number Λ_f (or $1/\sqrt{\Lambda_f}$) and the dimensionless conicity H_v.

complex as the range of h_a increases. The task becomes yet more difficult if we add to it the size and the supply pressure. These optimisation issues, whose ingredients have been indicated here, are, however, outside the scope of this book.

An important design parameter in this process is the feed number Λ_f, which is proportional to r_a/h_a^2. Let us note here that from a fabrication point of view, this value is related to the manufacturing tolerance of bearing surfaces for a given bearing size (outer radius). Since relatively tighter tolerances could be achieved with larger parts (tolerance scales approximately with the cube root of size, as observed by E.T. Fortini as cited in (Nigam and Turner 1995)), optimality ($\sim$ large r_a/h_a^2) favours larger bearing sizes. It is indeed much easier to manufacture a bearing with a radius of 50 mm operating with a gap height of 10 micrometers or less, than a bearing radius 10 mm operating with a gap of 5 micrometer. Furthermore, the larger the bearing size is, the lower will be the relative air consumption.

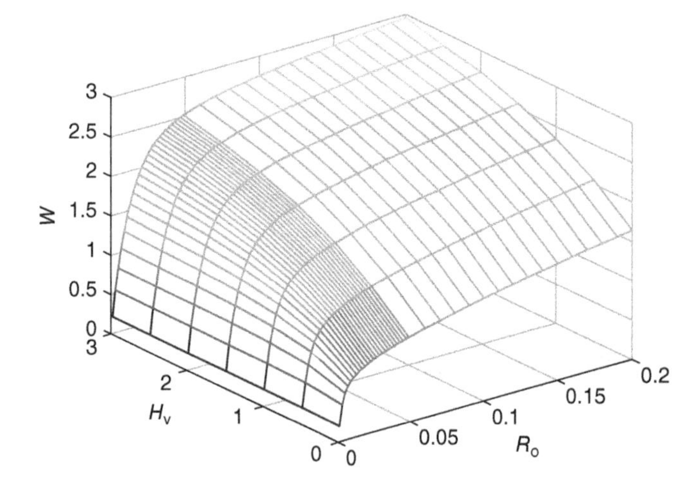

Figure 6.17 The load increases monotonically with both R_o and H_v.

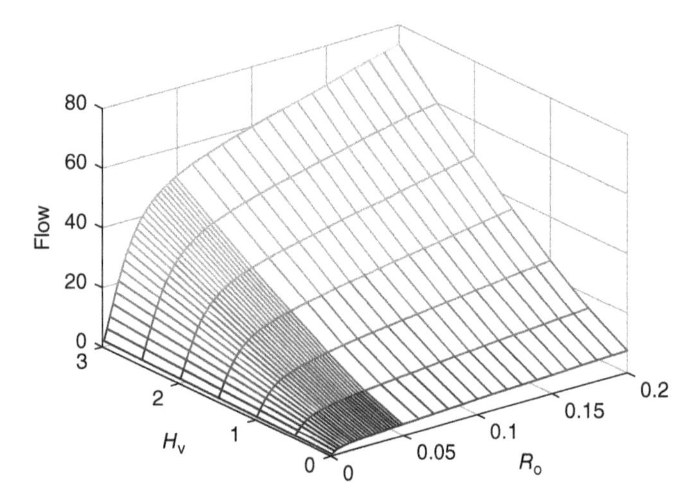

Figure 6.18 The flow increases monotonically with both R_o and H_v.

- **Note on estimating load capacity and stiffness**

  It is relatively easy to obtain an estimate of the load capacity, especially in the small gap range.
  Firstly, we have an upper limit to the load capacity

$$\frac{W}{P_s - 1} < 1. \tag{6.7}$$

  That is, the bearing force $F_{max} = (p_s - p_a)A$.
  This value of the bearing force is approached for a bearing having a convergent gap that approaches zero at the outer bearing edge. In that case, the air consumption and the stiffness both approach zero.
  Thus, when we wish to design a bearing having some stiffness, some part of the load capacity must be sacrificed. In a classical, flat-surface bearing, without pocket, the maximum load capacity is around 0.2 of the theoretical maximum (of Eq. 6.7). When the bearing is designed for maximum stiffness, approximately only half of this value remains, i.e.

$$W \simeq 0.1(P_s - 1)$$

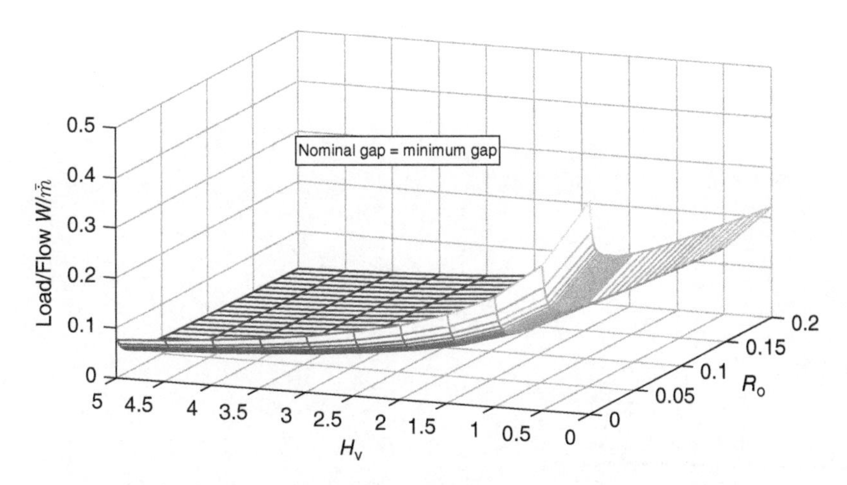

Figure 6.19 The load-to-flow ratio decreases with R_o (especially for small R_o). It also decreases with H_v, when the minimum gap h_a is taken as the nominal gap for calculating the flow rate.

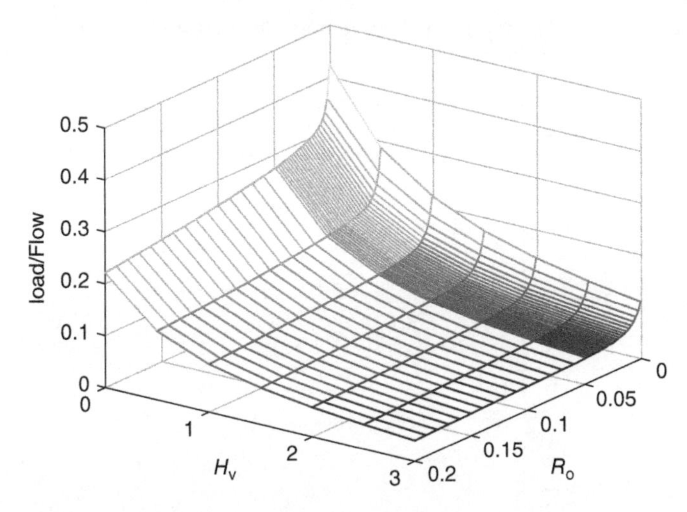

Figure 6.20 The load-to-flow ratio decreases with R_o (especially for small R_o). It increases with H_v, when the mean gap $h_a + h_v/2$ is taken as the nominal gap for calculating the flow rate.

Often, the stiffness of the bearing at that point is approximated as:

$$K \simeq \frac{1}{3}W,$$

or,

$$k \simeq \frac{2F_a}{3h_a}$$

Adding conicity to the bearing surface, i.e. making the gap convergent, can increase both load capacity and stiffness considerably, however at the expense of increasing the air consumption and reducing the damping. Referring to Figures 6.12 and 6.16, both load and stiffness can be increased by factors of up to 4 or more.

- **Note on the conicity**

Nowadays, it should not prove that difficult to produce a convergent surface (a slope of between 0.15–1 10^{-3}, or 0.15–1 mm m^{-1}) by conventional machining processes, ranging from turning to grinding and EDM. However, one might still prefer an alternative way to achieve the same effect of the conicity.

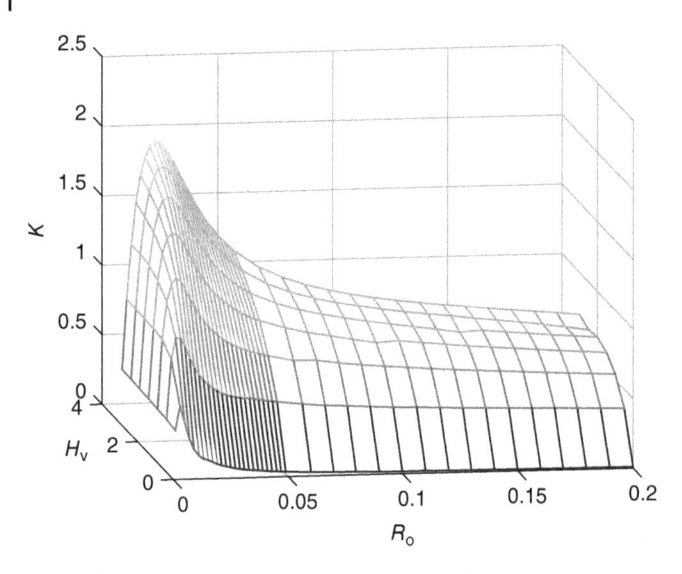

Figure 6.21 The dimensionless stiffness as a function of R_o and H_v. For any given H_v, there is a unique value of R_o that yields maximum stiffness. As H_v increases, the value of the optimum R_o increases.

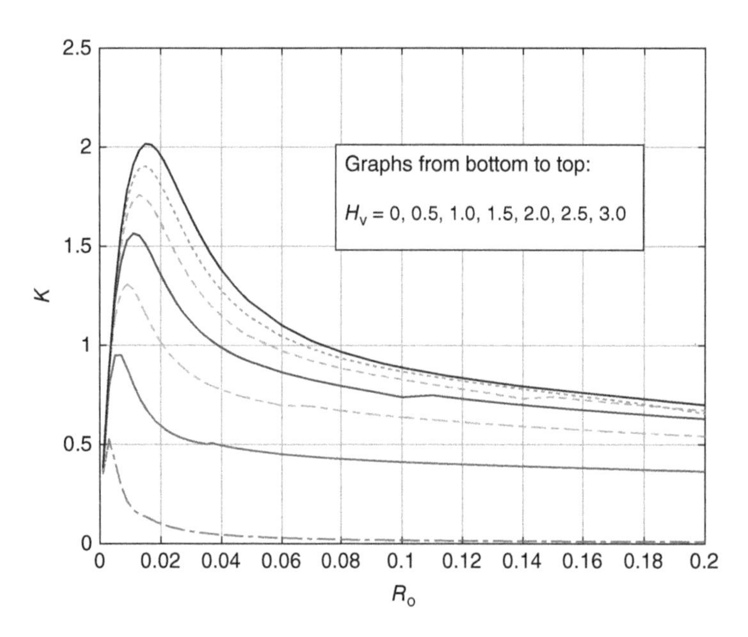

Figure 6.22 Parametrised graphs corresponding to the previous figure to facilitate reading the values.

This is possible in a variety of ways. The simplest is replacing the conicity with a step (i.e. a shallow pocket) or a series of steps (= nested shallow pockets). Other shapes are also possible, however, one needs to be able to model the resulting bearing in order to perform the design correctly. For other production methods, see (Snoeys and Al-Bender 1987).

- **Note on air consumption**

  Air or gas consumption might be one of the pertinent issues when designing an air bearing system, for the following reasons:

- There could be a limitation on the amount of compressor power available to supply the needed gas. It is often necessary to keep the total required flow rate well below this limit in order to avoid undulations in the supply pressure.
- Using too much lubricating fluid (or gas in this case) is not desirable from economic and ecologic point of view.
- Large flows issuing from the bearing's outlets are liable to induce undesired turbulences that could have adverse effects on the accuracy of high-precision machines (in which gas bearings are often used).
- Undue contamination of clean rooms.

For these reasons, it is often desirable to minimise the gas consumption for a given set of load, stiffness,... requirements, and so to optimise the application. Figures 6.13 and 6.18 furnish some idea about the trends of bearing flow w.r.t. its design parameters. Gas flow obviously increases with increasing P_s. However, for a given P_s, maximum load to air consumption is afforded by a bearing with a flat surface. An optimisation process with a more complicated objective function, might require a dedicated computer program.

It might be also pertinent to estimate the compressor power necessary to achieve a certain mass flow rate. This depends on the thermodynamic process considered. denoting the required compressor power by P_{comp}, we have that

$$dP_{\text{comp}} = dp.Q = \frac{\dot{m}}{\rho}dp.$$

Assuming a polytropic process, then

$$p/p_o = (\rho/\rho_o)^{\chi},$$

or,

$$(p/p_o)^{(1/\chi)} = \rho/\rho_o.$$

Substituting for ρ from this in the first equation, we obtain

$$dP_{\text{comp}} = p_o/\rho_o d\bar{p}/(\bar{p})^{(1/\chi)}; \quad \bar{p} = p/p_o.$$

Integrating this between outlet (o) and supply (s), putting $p_o/\rho_o = R_g T_o$, we obtain:

$$P_{\text{comp}} = \dot{m} R_g T_o \frac{\chi}{(\chi - 1)} (P_s^{((\chi-1)/\chi)} - 1), \chi \neq 1$$

For $\chi = 1$ (isothermal case), integrating in the same way, yields

$$P_{\text{comp}} = \dot{m} R_g T_o (\log(P_s) - 1)$$

which can also be obtained by taking the limit of the previous result as χ approaches unity.

N.B. In order to obtain the net total power, (i) p_s should be taken as the compressor pressure, and (ii) the result should be divided by the compressor efficiency.

6.2.4 Conclusions on Axial Characteristics

In the foregoing, we have examined the basic axial, static characteristics of centrally fed bearings. We saw that the bearing characteristics of load, stiffness and flow rate or air consumption depend on the bearing size, feed-hole size ratio and conicity, or gap convergence ratio. Increasing this latter leads to increased load and flow simultaneously, but with decreasing load to flow ratio, that is, a lower flow efficiency. Increasing the conicity leads also to increased stiffness (but reduced damping as we shall see in Chapter 7). The stiffness attains a maximum for a combination of optimum conicity and feed-number values. For a given feed number, the stiffness will be maximum for an optimum value of the feed-hole ratio, for each given value of the conicity. Thus, we need to formulate an elaborate optimisation problem, if we wish to achieve effective and energy-efficient bearing behaviour, in particular over a range of gap heights or loads. Finally, let us emphasise that the stiffness characteristics considered here concern the static values. Increasing the stiffness will generally reduce the damping, sometimes to a negative level, which thus forms another optimisation constraint. The dynamic bearing characteristics are treated in Chapter 7.

6.3 Tilt and Misalignment Characteristics (Al-Bender 1992; Al-Bender and Van Brussel 1992)

By tilt we mean the deviation of the bearing gap from the (design) position of nominal parallelism (or axial symmetry.) Such misalignment, usually ignored by the designer in the first approximation, may arise in a variety of ways.

- Statically, it may be the result of geometrical inaccuracies of fabrication and assembly, both in the rigidly fixed and the centrally pivoted cases; in the latter case, it may also be due to friction in the pivot contact.
- Dynamically, it may manifest itself as angular vibrations about the bearing diameter when either or both of the bearing parts have a tilt degree of freedom.

In view of the small order of magnitude of the bearing film thickness, in comparison with the rest of the dimensions, and of the limitations of manufacture, it may be concluded that tilt is, to a varying degree, inevitable in most bearing systems. Consequently, the prediction of tilt characteristics is of considerable importance in the design of these components. This comprises two interconnected aspects, namely, (i) the influence of tilt on the axial characteristics, in particular, the load capacity, and (ii) the magnitude of the restoring torque per unit tilt angle or the tilt stiffness (which also includes damping in the dynamic case). Moreover, the problem is interesting from a theoretical point of view for two reasons: the first is its susceptibility to linearisation for relatively large tilt values, as will be shown later, and the second is its qualitative analogy with the concept of "passive" stiffness, or pure "squeeze film", viz. that arising from film thickness variation at constant boundary pressure. (see also Chapter 7.)

In comparison with the axial bearing characteristics, the tilt problem has received relatively little attention. In the context of journal bearings, the problem of misalignment was first treated by Ausman (1960) using first-order perturbation, as alluded to in Chapter 9. Gross (1962) mentions briefly, on the basis of a simple argument, that circular aerostatic bearings would be expected to have a tilt stiffness. The first theoretical treatment, perhaps, was given, in a Brief Note, by Licht and Kaul (1964) using a first- and second-order perturbation method. Their analysis was confined, however, to uniform gap geometry, and considered mainly the static characteristics. Later, Ausman (1967) treated the case of tilt squeeze film of a disc using approximate analysis. Safar (1981) treated the case of tilt combined with rotation (about the axis) of a flat circular centrally fed aerostatic bearing. Pande (1985) extended the solution to the case of convergent gap bearings. They both employed finite difference methods to solve the resulting two-dimensional, non-linear Reynolds equation, calculating only static characteristics, and excluding the tilt stiffness. Apart from some discrepancies and inconsistencies, the results they obtained showed that the effect of tilt with rotation about the axis of the bearing is significant only at extreme values of those two parameters (unlikely to arise in practical application). Furthermore, no experimental verification, either for the static or the dynamic tilt cases, has, to our knowledge, been reported in the open literature.

The objective of this section is to give a more comprehensive treatment of the problem of small order pure tilt, both theoretically and experimentally, including the case of convergent gap bearings. As has been mentioned in the first part of this chapter (see also (Snoeys and Al-Bender 1987),) the axial characteristics of the convergent-gap bearings are far superior to those of the uniform gap type; however, it has been expected that the tilt stiffness of such bearings might be further impaired by the increasing order of gap convergence. Simple design guidelines will, thus, be deduced for the general case.

In the following section, a theoretical model is proposed on the basis of some simplifying assumptions. A solution of the problem is then given, followed by typical results. Thereafter, the test apparatus is described together with the range of experiments carried out. The results are then compared and discussed, with some general formulas and tendencies given, followed by appropriate conclusions.

6.3.1 Analysis

The bearing arrangement is schematically shown in Figure 6.23. The fluid enters the bearing gap at r_o and discharges to atmosphere at r_a. In the position of nominal parallelism, the pressure field is axially symmetric; when the bearing is tilted by an angle α, about the diameter through $\theta = \pi/2$, only the pressure profile through $\theta = \pi/2$ remains symmetric (about the diameter through $\theta = 0$). The following inter-related assumptions are now made and discussed:

1. We assume the Reynolds-restrictor model of Chapter 2 to be valid; i.e. we ignore entrance effects. Also, for simplicity, the inlet pressure will be denoted by p_o (not p_t).
2. The change in the inlet film thickness due to tilt is negligible, i.e. $\alpha \ll h_0/r_0$, such that the inlet pressure, p_o may be taken to be independent of θ. The exit pressure is taken to be the ambient pressure p_a. Furthermore, since α is small, then $\tan\alpha \approx \alpha$.
3. The radius ratio $R_o = r_0/r_a$ is much smaller than unity, such that the contribution of the feed-hole to the load capacity would be negligible, (as is usually the case in inherently compensated bearings). This assumption is not necessary for the solution of the problem, but it makes the final results approximately independent of R_o, thus considerably simplifying the treatment of the results.

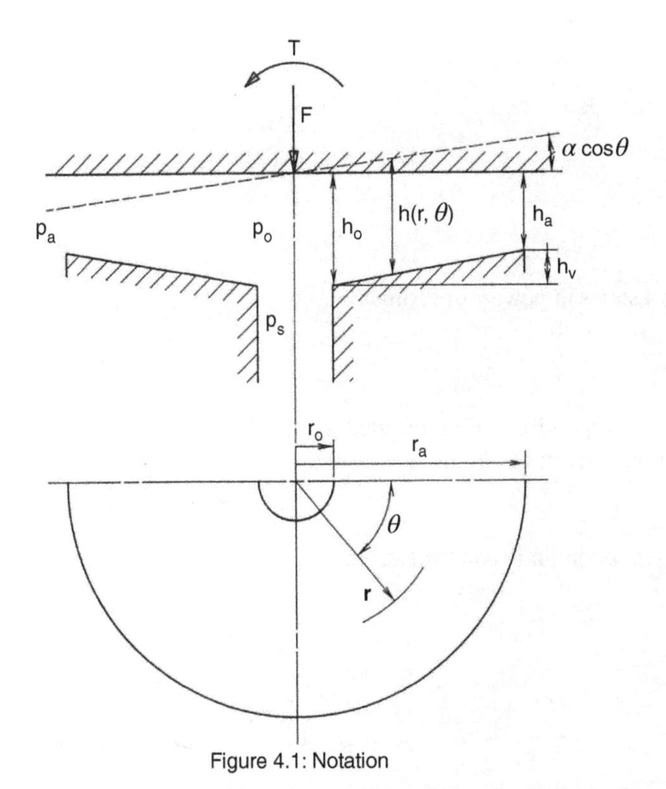

Figure 6.23 Tilt notation. Source: Republished with permission from Elsevier, from Al-Bender F and Van Brussel H 1992; permission conveyed through Copyright Clearance Center, Inc.

In view of these assumptions, the (time dependent) Reynolds equation of lubrication (Chapter 4) may be applied:

$$\frac{1}{R}\frac{\partial}{\partial R}(RH^3\frac{\partial}{\partial R}P^2)+\frac{1}{R^2}\frac{\partial}{\partial\theta}(H^3\frac{\partial}{\partial\theta}P^2)=2\sigma\frac{\partial}{\partial\tau}(PH) \quad (6.8)$$

where the variables have been normalised w.r.t. the exit parameters and a characteristic angular frequency:

$$P=p/p_a,\ R=r/r_a,\ H=h/h_a,\ (H_v=h_v/h_a,)\ \tau=\nu t,$$

and

$$\sigma=\frac{12\mu\nu}{p_a}\left(\frac{r_a}{h_a}\right)^2 \quad \text{is the squeeze number.}$$

The boundary conditions (following from the second assumption) are:

$$P(R_o,\theta,\tau)=P_o;\quad P(1,\theta,\tau)=1 \quad (6.9)$$

Since an approximate solution is to be sought, and its bounds of accuracy evaluated, it would be more convenient to consider the cases of static tilt and dynamic tilt separately, starting with the former.

Static Tilt

Here, σ is set equal to zero, and we have to a good order of approximation:

$$h(r,\theta)=h_{(0)}(r)+\alpha r\cos\theta\ ,$$

which, upon normalisation, becomes:

$$H(R,\theta)=H_{(0)}(R)+\epsilon R\cos\theta \quad (6.10)$$

in which,

$$H_{(0)}(R)=1+\nu_a(1-R),\qquad \text{with,}\quad \nu_a=H_v/(1-R_o)$$

is the untilted normalised film thickness function, and

$$\epsilon=\alpha\frac{r_a}{h_a},\qquad |\epsilon|\lesssim 1$$

is the normalised tilt angle.

Correspondingly, the pressure may be expanded as a series in powers of ϵ, thus:

$$P(R,\theta)=P_{(0)}(R)+\sum_{n=1}^{\infty}\epsilon^n P_{(n)}(R,\theta) \quad (6.11)$$

where $P_{(0)}$ is the axisymmetric pressure distribution corresponding to the untilted gap configuration. Since the boundary conditions 6.9 must hold true for any arbitrary ϵ, it follows from Eq. 6.11 that:

$$P_{(0)}(R_o)=P_o\ ;\qquad P_{(0)}(1)=1 \quad (6.12)$$

and that all the other pressure terms satisfy homogeneous boundary conditions, viz.,

$$P_{(n)}(R_o,\theta)=P_{(n)}(1,\theta)=0,\qquad n=1,2,\ldots \quad (6.13)$$

The basic bearing characteristics may be listed:

- The dimensionless load capacity:

$$W\equiv\frac{F}{\pi r_a^2 p_a}=\frac{1}{\pi r_a^2 p_a}\int_0^{2\pi}\int_0^{r_a}(p-p_a)r\,dr\,d\theta$$

$$\approx 2\int_{R_o}^{1} P_{(0)}RdR - 1 + \sum_{n=1}^{\infty}\epsilon^n \frac{1}{\pi}\int_0^{2\pi}\int_{R_o}^{1} P_{(n)}RdRd\theta$$

$$= W_{(0)} + \sum_{n=1}^{\infty}\epsilon^n W_{(n)}$$

- The dimensionless tilt moment:

$$M \equiv \frac{T}{\pi {r_a}^3 p_a}$$

$$= \frac{1}{\pi {r_a}^3 p_a}\int_0^{2\pi}\int_{r_o}^{r_a} pr\cos\theta rdrd\theta$$

$$= \sum_{n=1}^{\infty}\epsilon^n \frac{1}{\pi}\int_0^{2\pi}\int_{R_o}^{1} P_{(n)}R\cos\theta RdRd\theta$$

and the dimensionless tilt stiffness is likewise:

$$K \equiv \frac{k_\alpha}{\pi {r_a}^3 p_a}\frac{h_a}{r_a}$$

$$= -\frac{\partial}{\partial\epsilon}M$$

(where, in the last expressions, $k_\alpha = -\partial T/\partial\alpha$).

Further, the square of the pressure function, needed for subsequent calculations, may be expressed as follows:

$$P^2 = \sum_{n=0}^{\infty}\epsilon^n \phi_{(n)}$$

where,

$$\phi_0 = P_{(0)}^2$$
$$\phi_1 = 2P_{(0)}P_{(1)}$$
$$\phi_2 = 2P_{(0)}P_{(2)} + P_{(1)}^2$$
$$\phi_3 = 2P_{(0)}P_{(3)} + 2P_{(1)}P_{(2)}$$
$$\vdots \quad \vdots$$

Substituting Eqs. 6.10 and 6.11 into Eq. 6.8, (with $\sigma = 0$,) and equating the terms in equal powers of ϵ, yields the following system of linear differential equations for the component pressures (written, for convenience, in shorthand notation):

$$\nabla\cdot({H_{(0)}}^3\nabla\phi_0) = 0 \tag{6.14}$$

$$\nabla\cdot({H_{(0)}}^3\nabla\phi_1 + 3{H_{(0)}}^2R\cos\theta\nabla\phi_0) = 0 \tag{6.15}$$

$$\nabla\cdot({H_{(0)}}^3\nabla\phi_2 + 3{H_{(0)}}^2R\cos\theta\nabla\phi_1 + 3H_{(0)}(R\cos\theta)^2\nabla\phi_0) = 0 \tag{6.16}$$

$$\vdots \quad \vdots$$

The first of these equations corresponds to the untilted situation and has an analytic solution subject to the boundary conditions 6.12. Consequently, the system may be solved recursively, with each step being determined by the previous one. Thus, the solution of Eq. 6.14 is found to be (Eq. 6.3):

$$\phi_0 = {P_{(0)}}^2 = C_1 B_{\nu_a}(R) + 1 \tag{6.17}$$

where,

$$B_v(R) = \int \frac{\mathrm{d}R}{RH_{(0)}{}^3} + \text{const.}$$
$$= \frac{1}{a^3}\left(\log(\frac{R}{H_{(0)}}) + \frac{(a+H_{(0)})^2 - (1+a)^2H_{(0)}{}^2}{2H_{(0)}{}^2}\right), \quad a = 1 + v$$

and

$$C_1 = \frac{P_o{}^2 - 1}{B_v(R_o)}$$

The first-order tilt pressure Eq. 6.15 is rewritten as:

$$\frac{\partial}{\partial R}(RH_{(0)}{}^3\frac{\partial \phi_1}{\partial R}) + \frac{H_{(0)}{}^3}{R}\frac{\partial^2 \phi_1}{\partial \theta^2} = -3\frac{\mathrm{d}}{\mathrm{d}R}(R^2H_{(0)}{}^2\frac{\mathrm{d}\phi_0}{\mathrm{d}R})\cos\theta$$
$$= -3C_1\frac{\mathrm{d}}{\mathrm{d}R}(\frac{R}{H_{(0)}})\cos\theta \tag{6.18}$$

It may be shown that, due to its periodicity in θ, ϕ_1 can be expressed in separated form as:

$$\phi_1(R,\theta) = F_1(R)\cos\theta,$$

thus reducing Eq. 6.18 to an ordinary differential equation for F_1, (which, likewise, satisfies homogeneous boundary conditions):

$$\frac{\mathrm{d}}{\mathrm{d}R}(RH_{(0)}{}^3\frac{\mathrm{d}}{\mathrm{d}R}F_1) - \frac{H_{(0)}{}^3}{R}F_1 = -3C_1\frac{\mathrm{d}}{\mathrm{d}R}(\frac{R}{H_{(0)}}) \tag{6.19}$$

This is a boundary value problem which may, by virtue of its linearity, be easily converted into an initial value problem, e.g. by the method of superposition, (see (Na 1979)) and consequently solved by an appropriate forward numerical integration technique.

Similarly, it may be shown that the second pressure approximation can be expressed in the form:

$$\phi_2(R,\theta) = G_2(R) + F_2(R)\cos 2\theta, \tag{6.20}$$

which, upon substitution in Eq. 6.16, with ϕ_0 and ϕ_1 already known, should yield the required solution; and so on, for the higher-order terms.

However, since the process becomes more laborious the more terms are calculated, while it is not possible to establish convergence of the series expansion, it is considered advantageous, first to determine the range of validity of first-order perturbation, which is our main concern. For this purpose, two cases, amenable to analytical solutions were considered. The first is the case of plane parallel incompressible fluid bearing, which yields closed form expressions for the bearing tilt characteristics. The second is the case of uniform gap, circular, aerostatic bearing, reported in (Licht and Kaul 1964) (although with typographic errors,) which admits an analytic solution, up to and including the second-order perturbation term, thus enabling an easy assessment of the relative increase in the load capacity and flow rate. The principal results may be summarised as follows:

- both the load capacity and the flow rate increase with tilt, initially in proportion with the square of the tilt angle;
- first-order perturbation provides an accuracy approximately in proportion with the square of the tilt angle, being within 1 percent for 20 percent tilt (i.e. $\epsilon = 0.2$).

The first result means that the load capacity and the flow rate will be hardly affected by small order tilt. As the latter increases, the load capacity would increase if the inlet pressure should remain constant, (e.g. if the inlet restrictor flow is choked). However, since the flow increases simultaneously, this will lead generally to a reduction in the inlet pressure, and consequently to a reduction in the untilted load capacity. The net effect may be determined, if the restrictor flow-pressure characteristics are known. Being considered of secondary importance, this problem is presently dropped from the analysis. However, simplified calculations, as well as our experiments, have shown that the net effect is, in most cases, an increase in the overall load capacity. (The results of Safar (1981) show no variation of the load capacity with tilt, even at large tilt angles.)

The second result is of primary importance to our investigation since it shows that first order perturbation is sufficient in characterising the problem even for relatively large tilt angles. This is further corroborated by the results of Pande (1985), and by our experimental results.

In conclusion, the foregoing discussion provides sufficient reason to consider only first-order perturbation, which will be adopted then directly for the more general case of dynamic tilt.

Dynamic First-Order Tilt

Confining ourselves thus to first-order tilt, which amounts to linearisation of the problem, we may further assume, without loss of generality, harmonically varying tilt. Adopting the more convenient complex notation, we may then write:

$$\epsilon(\tau) = \epsilon_0 e^{j\tau},$$

and, consequently, we have:

$$H(R,\theta,\tau) = H_{(0)}(R) + \epsilon_0 e^{j\tau} R\cos\theta, \tag{6.21}$$

and,

$$P(R,\theta,\tau) \approx P_{(0)}(R) + \epsilon_0 e^{j\tau} P_{(1)} \tag{6.22}$$

where, $P_{(0)}$ is the same as before, and,

$$P_{(1)} \equiv -(P_k(R) + j\sigma P_c(R))\cos\theta$$

where, P_k is the pressure function in-phase with the tilt, i.e. that responsible for the *stiffness* torque, and P_c is the out-of-phase pressure function, yielding the *damping* torque coefficient. (The utility of the form $P_k + j\sigma P_c$ becomes more evident when the case $\sigma \to 0$ is considered.)

Substituting Eqs. 6.21 and 6.22 into Eq. 6.8, and simplifying as before, we obtain the same expression for $P_{(0)}$ as given by equations 6.14 *et seq.*; but now, the dynamic, first-order tilt pressure is given by:

$$\begin{aligned}\frac{\partial}{\partial R}\left(RH_{(0)}{}^3\frac{\partial\phi_1}{\partial R}\right) + \frac{H_{(0)}{}^3}{R}\frac{\partial^2\phi_1}{\partial\theta^2} &= -3C_1\frac{\mathrm{d}}{\mathrm{d}R}\left(\frac{R}{H_{(0)}}\right)\cos\theta \\ &\quad + j2\sigma\frac{R}{P_{(0)}}(R\phi_0\cos\theta + H_{(0)}\phi_1/2)\end{aligned} \tag{6.23}$$

where, the ϕ's denote the same quantities as before. This is the same as equation 6.18 with the addition of the term in the squeeze number σ. Since only harmonic oscillation is considered, this equation should be valid for any bounded value of σ. When σ becomes infinite, the equation reduces to a singular perturbation problem for which the solution, inside the film, may be simply obtained by setting the term in σ equal to zero to give: $P_{(1)} = -(R/H_{(0)})P_{(0)}\cos\theta$, $R_o < R < 1$.

Equation 6.23 is solved by the same procedure outlined for the static case. However, since $P_{(1)}$ is now complex, the integration may be carried out either by using complex variables, or by splitting Eq. 6.23 into two equations, for the real and the imaginary pressure components respectively.

6.3.2 Theoretical Results

It can be clearly seen from the analysis that the results depend on four bearing parameters, viz. P_o, R_o, H_v and σ. It may, further, be shown that the general characteristics are not appreciably affected by the value of R_o as long as that

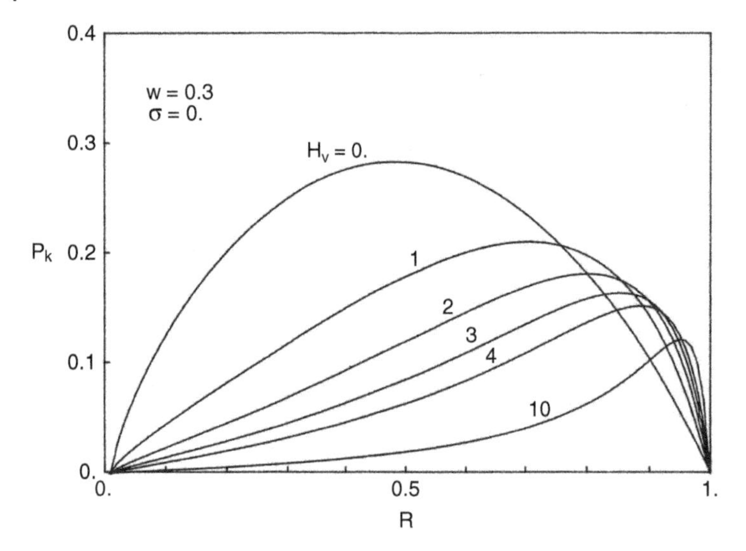

Figure 6.24 Static tilt stiffness-pressure distribution. Source: Republished with permission from Elsevier, from Al-Bender F and Van Brussel H 1992; permission conveyed through Copyright Clearance Center, Inc.

remains small. Therefore, in order to simplify presentation, the results will be restricted to the case of $R_o = 0.01$, being approximately valid for $0 < R_o < 0.1$. The pressure distribution results will be first given followed by the global characteristics.

Figures 6.24 and 6.25 show the static first-order tilt pressure distributions (P_k and P_c) for various values of the conicity H_v, for the same dimensionless load W. One may note first that P_k is about one order of magnitude bigger than P_c; a result that holds true for other values of W. Secondly, increasing conicity has the effect of reducing the the magnitude of both of these pressures and shifting their peak values toward the outer radius, until they

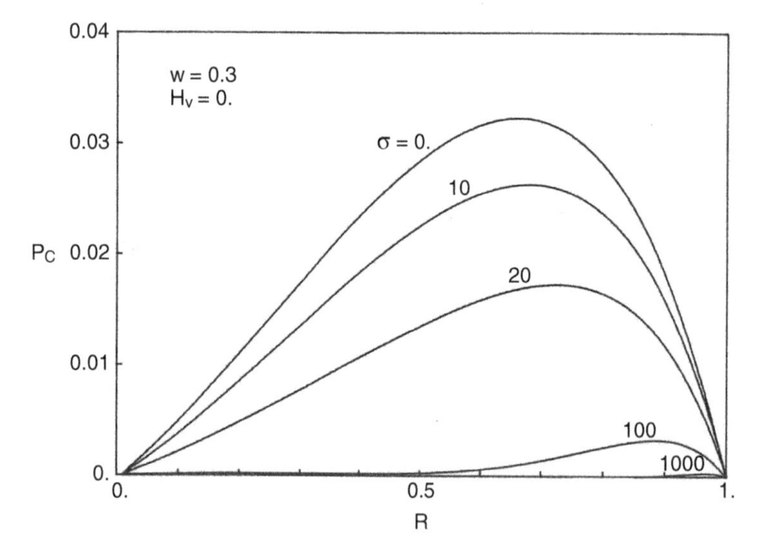

Figure 6.25 Static tilt damping-pressure distribution. Source: Republished with permission from Elsevier, from Al-Bender F and Van Brussel H 1992; permission conveyed through Copyright Clearance Center, Inc.

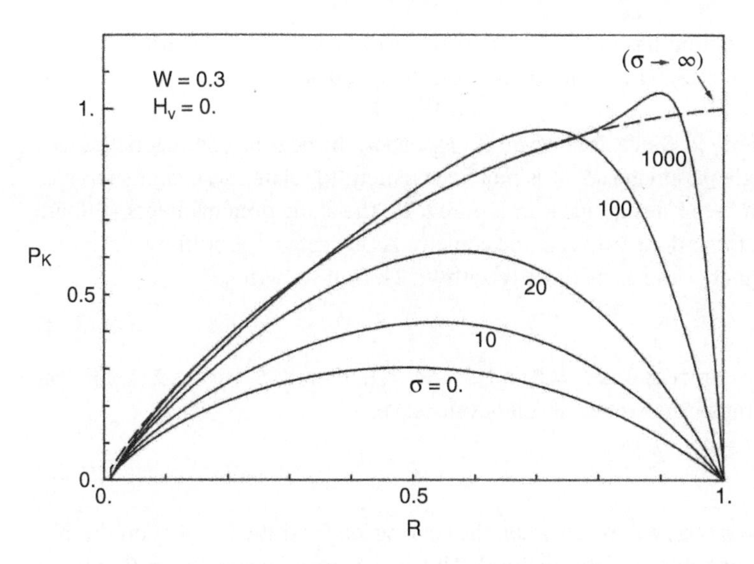

Figure 6.26 Dynamic tilt stiffness-pressure distribution. Source: Republished with permission from Elsevier, from Al-Bender F and Van Brussel H 1992; permission conveyed through Copyright Clearance Center, Inc.

are reduced to tiny bulges near the bearing exit. This situation may be physically interpreted as that in which the bearing film becomes more like a gas pocket. Furthermore, increasing conicity has a more pronounced effect on P_c than on P_k.

Figures 6.26 and 6.27 show the behaviour of the same pressure components for increasing values of the squeeze number σ, for a flat bearing, $H_v = 0$. The real part P_k increases with σ to reach the maximum asymptotic limit alluded to previously; while the imaginary part, P_c decreases to zero. If we recall that the dynamic stiffness and the damping arise from the incremental viscous flow, this extreme situation corresponds then to the case when,

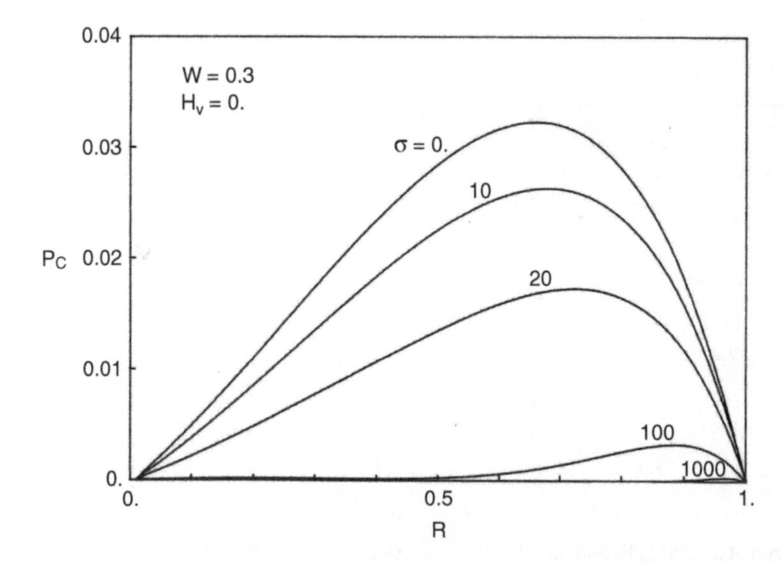

Figure 6.27 Dynamic tilt damping-pressure distribution. Source: Republished with permission from Elsevier, from Al-Bender F and Van Brussel H 1992; permission conveyed through Copyright Clearance Center, Inc.

due to the exceedingly high tilt frequency, the fluid has no time to flow in any direction. Consequently, the local variation in the film thickness (i.e. the gas volume) will induce a pressure change in accordance with perfect gas (isothermal) expansion law, or Boyle's law.

The global tilt characteristics are next examined. Since, for a given R_o, H_v and σ, the bearing characteristics, viz. W, K and C, will be determined by the single parameter P_0, it is more convenient to relate these characteristics directly to one another, which would give a better design idea. In Figure 6.28, the static dimensionless stiffness K is plotted against the dimensionless load, for various values of the conicity. As expected, the stiffness decreases with the conicity; but, more importantly, it increases nearly linearly with W. Thus, if we write,

$$K = \beta_k(W, H_v) \cdot W, \tag{6.24}$$

then, for any given conicity, β_k is nearly a constant; e.g. for $H_v = 0$ and $.2 < W < 1.$, then $.175 < \beta_k < .2$. A physical interpretation of Eq. 6.24 is obtained if we transform back to absolute values, i.e.

$$k_\alpha \frac{h_a}{r_a} = \beta_k F r_a \tag{6.25}$$

In other words, by extrapolation, full tilt ($\alpha = h_a/r_a$) will result when the bearing load is shifted by $\beta_k r_a$ off the axis of symmetry, (on the assumption of linear tilt behaviour throughout). This may immediately be seen to impose rather *stringent* requirements on the geometrical accuracy of bearing manufacture and assembly. As an example, if tilt is to be kept to within 10 percent, (i.e. $\epsilon = .1$,) then the load must be applied with an eccentricity not exceeding $.02\ r_a$.

Figure 6.29 shows that the static dimensionless damping coefficient C decreases with the dimensionless load W, and we may, in a similar manner, write

$$C = \beta_c(W, H_v) \tag{6.26}$$

where β_c, may again, for design purposes, be taken as a constant in W. Resorting to absolute dimensions, we have:

$$c = \beta_c 12\pi\mu r_a^3 (r_a/h_a)^3 \tag{6.27}$$

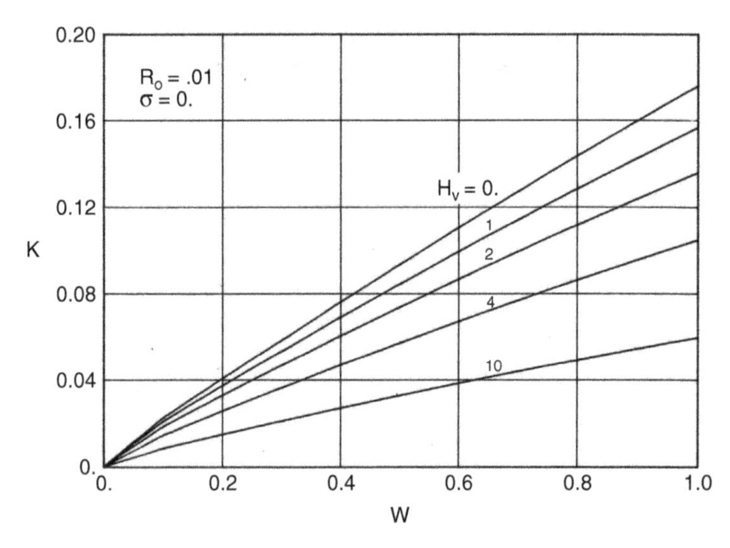

Figure 6.28 Static tilt stiffness versus load capacity. Source: Republished with permission from Elsevier, from Al-Bender F and Van Brussel H 1992; permission conveyed through Copyright Clearance Center, Inc.

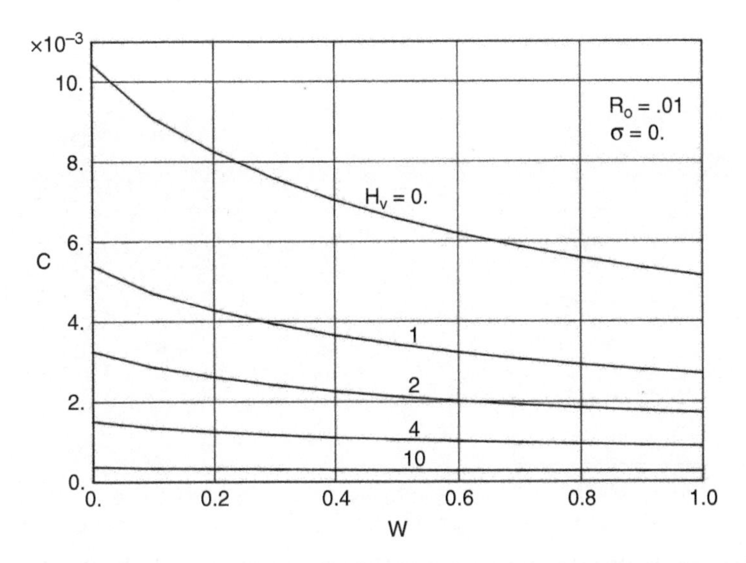

Figure 6.29 Static tilt damping coefficient versus load capacity. Source: Republished with permission from Elsevier, from Al-Bender F and Van Brussel H 1992; permission conveyed through Copyright Clearance Center, Inc.

which physically means that, for a given dimensionless load and conicity, the damping coefficient c increases in direct proportion to r_a^6, (or the cube of the bearing area,) and in inverse proportion to the mass flow rate, (since the latter is proportional to h_a^3/μ). We may also note, upon comparing Figure 6.29 to Figure 6.28, that the damping is more strongly influenced by the conicity, than the stiffness. The behaviour of both the stiffness and the damping with the conicity is then summarised in Figure 6.30.

All of the tilt characteristics given above have been for the static case, $\sigma = 0$. Their utility extends, however, a little further if we note (a) that the results are not significantly affected by σ, as long as that remains small,

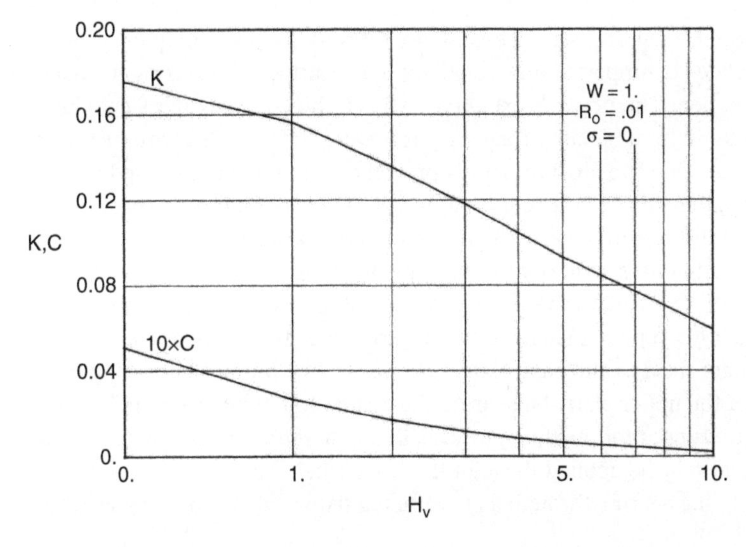

Figure 6.30 Influence of the conicity on the static stiffness and damping. Source: Republished with permission from Elsevier, from Al-Bender F and Van Brussel H 1992; permission conveyed through Copyright Clearance Center, Inc.

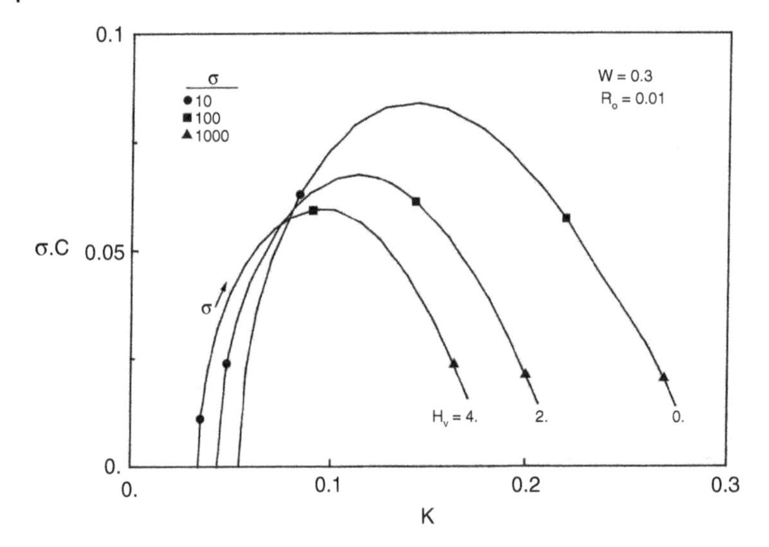

Figure 6.31 Dynamic tilt Nyquist diagram. Source: Republished with permission from Elsevier, from Al-Bender F and Van Brussel H 1992; permission conveyed through Copyright Clearance Center, Inc.

(cf. Figures 6.26 and 6.27,) and (b) that σ is small in most practical applications, e.g. for an air bearing with r_a =30 mm, h_a =30 μm, p_a = 1 bar, tilt frequency = 70 Hz, then $\sigma < 1$. For higher squeeze numbers, it is more convenient to provide the tilt Nyquist diagram, as shown in Figure 6.31, which then summarises dynamic tilt behaviour: it roughly resembles a "phase lead" system. This means that the damping torque is always positive, and consequently, tilt instability should never arise if the bearing is assembled in a passive mechanical system.

6.3.3 Experimental Investigation

The experimental apparatus, depicted in Figure 6.32, was designed essentially for static, and low frequency, resonance tests. It comprises a heavy, rigid base plate (1) supported on two pillars. The bearing pad (2) is rigidly fixed in the middle of the base plate, by means of a bolt. The upper bearing part (3) is a thick rectangular steel plate, (260 ×100 ×25 mm^3, such that its fundamental natural bending frequency is greater than 2 kHz,) with its lower surface chrome-plated and ground flat. It rests upon the air film, and is prevented from moving in the bearing plane by four long, adjustable elastic chords (4) that attach its corners to four posts at the corners of the base plate. This arrangement, reckoned to offer negligible stiffness in the vertical direction, serves also for coarse adjustment of the parallelism of the bearing gap. Fine adjustment is achieved through the displacement of small compensation masses along the four screw-threaded staffs (5) fixed to the bearing plate. These staffs, being oriented with the diameters of the plate, serve also as levers for applying small static tilt moments. Two pairs of contactless, eddy current type, displacement sensors (6) are fixed to the base plate, symmetrically around the bearing centre and approximately facing the diameters of the upper plate. Measuring the displacement between the bearing pad and the bearing surface, these sensors can be used for the dual purpose, first of ascertaining parallelism, and later of measuring tilt. Additional bearing load may be applied through the lever/steel-chords arrangement (7). Pressurised, dried and filtered air is supplied to the bearing through a pressure regulator; provision for measuring relative air flow is also available.

Two types of tests were carried out, as shown in Figure 6.33:

1. **Static tests** Small and equal mass loads, of negligible weight in comparison with the bearing load, are progressively hanged at the extreme end of one of the threaded staffs in the longitudinal direction. At the same

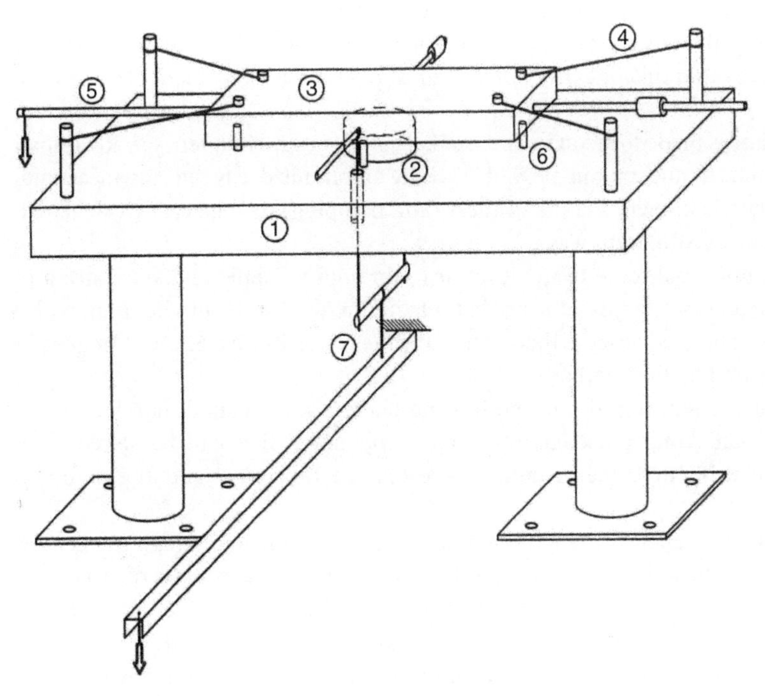

Figure 6.32 Test apparatus. Source: Republished with permission from Elsevier, from Al-Bender F and Van Brussel H 1992; permission conveyed through Copyright Clearance Center, Inc.

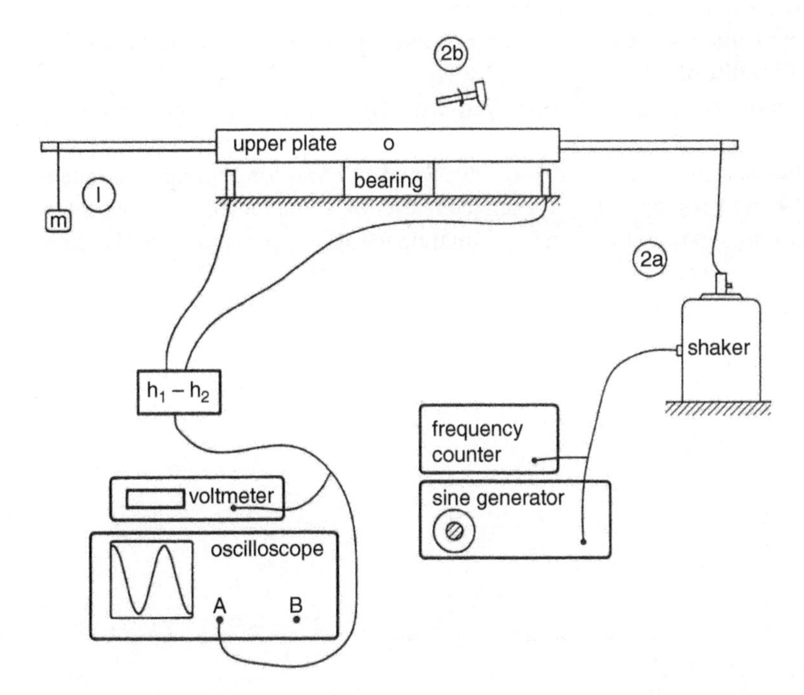

Figure 6.33 The different test types. Source: Republished with permission from Elsevier, from Al-Bender F and Van Brussel H 1992; permission conveyed through Copyright Clearance Center, Inc.

time, the difference between the two displacement pick-ups is recorded. Thus, the tilt moment is obtained as a function of the tilt angle. This test has the advantage of enabling the verification of the linearity of the tilt moment in the tilt angle. It was observed that this linearity holds in most cases up to 30 percent of maximum tilt angle.

2. **Resonance tests** Here, the bearing plate is made to resonate (i.e. oscillate at its natural frequency) with a small amplitude in the tilt mode. The resonance frequency, and possibly also the amplitude decay rate, are measured. Since the moment of inertia of the plate is known, the tilt stiffness (and damping) may be easily calculated. Excitation to resonance was achieved by two different ways:
   (a) A shaker transmits a very weak sinusoidal force to the lever arm, through a supple and *slack* string of negligible mass. The excitation frequency is adjusted such that maximum amplitude tilt vibration, corresponding to resonance, is achieved. This frequency is then accurately measured by an electronic frequency counter. Damping cannot be measured by this method.
   (b) Resonance is induced by applying a small impulse to the bearing plate, e.g. by a small hammer. After the transient has died out, a typical damped-oscillations curve is obtained, that can be stored on a memory-oscilloscope. Analysis of this curve yields both the resonance frequency and the damping factor.

   It should be noted, however, that these frequencies where limited to about 50 Hz, due to the high moment of inertia of the bearing plate, which was in turn dictated by its rigidity requirements. The axial (vertical) movement of the bearing plate was verified to be negligible by an electronic micro-comparator set at its centre. Nevertheless, these minute axial displacements seemed to give rise to a small amplitude, small frequency modulation of the damped-oscillations curve. For this reason, the measurement of the damping coefficient was, in general, poorer than that of the stiffness; more so when the former was of a large magnitude, since then it had to be estimated from a limited range of data.

As had been expected, the (stiffness) measurements of the different tests agreed very well with one another; the discrepancies, remaining within a few percent, could well be attributed to systematic errors. Furthermore, this was a confirmation of the excellent repeatability of the tests.

Altogether seven bearing pads, of equal diameter, $2r_a = 60.$ mm, were used for the tests. These are detailed in Table 6.1 for subsequent reference.

The experiments were carried out at two bearing loads; approximately 53 and 273 N, (corresponding to dimensionless loads of approximately $W = 0.19$ and 1 respectively). At each load, tests were conducted at regular intervals of film thickness in the range $10\ \mu\text{m} < h_a < 50\ \mu\text{m}$, by means of regulating the supply pressure to the bearing.

Table 6.1 Test bearing geometry.

Bearing ref. no.	r_o (mm)	h_v (μm)
1	0.30	0
2	0.76	0
3	1.02	0
1c	0.51	15
2c	0.63	22
3c	0.35	27
4c	0.78	40

a) $r_a = 30$ mm for all bearing pads.

6.3.4 Results, Comparison and Discussion

Attention is initially focussed on the flat bearings since here the influence of the conicity is excluded. Use was made of the Eq. 6.25 to check the validity of the theoretically linear relation between the tilt stiffness k, and the reciprocal of the maximum tilt angle h_a/r_a. The results are shown in Figure 6.34. We may remark first that, as had been expected, the feed-hole radius (in the range $0.01 < R_o < 0.033$) seems to have no influence on the results. Secondly, the data points, for both of the loads appear to conform approximately to one and the same pattern, suggesting perhaps that, in practice, the tilt stiffness is independent of the dimensionless load. Further, this pattern, although it shows the same qualitative tendency as the theoretically predicted line, crosses the latter somewhere in the middle of film thickness range such that the theory presents an underestimation of the stiffness at the high film thickness range, and vice versa at the other end. (This will be seen to be the case also for the other results.) Since the validity of the linearisation of the problem was confirmed by the static tests, the reason for this discrepancy must lie in the nature of the two basic assumptions that were made: in reality, neither is the flow viscous near the feed-hole, (see Chapter 3), nor does the inlet pressure, p_o, remain constant during tilt.

Figure 6.35 plots Eq. 6.27 for a comparison of the damping coefficient values. The same remarks, made above in regard to the stiffness, apply also here, with the addition that measurement accuracy becomes poorer with the increasing magnitude of the damping, as evidenced by the large scatter of the data in that region. (The reason for this has already been given in the previous section.)

Figures 6.36 and 6.37 compare the predicted and the experimental influence of the conicity, at a constant load, (see Eqs. 6.24, 6.26, and Figure 6.30.) The same tendency discussed above is evident here.

In general, there seems to be a fair agreement between the theoretical model and the experiment; one could, indeed, not have awaited an exact agreement, in view of the limitations of the model. While the excellent repeatability of the experimental results makes them of considerable practical value, the theory can provide a reasonable estimate of the tilt characteristics. The present theoretical results may thus provide good design guidelines, with the empirical results showing the limits of accuracy.

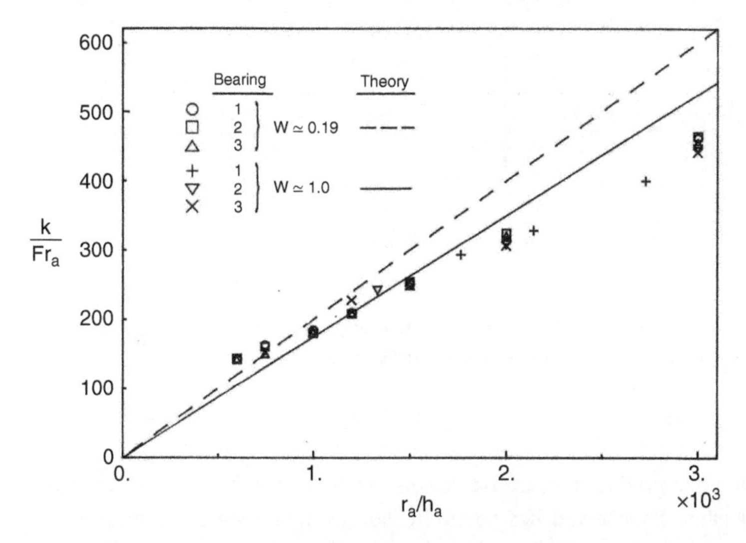

Figure 6.34 Comparison of stiffness values (flat bearings.) Source: Republished with permission from Elsevier, from Al-Bender F and Van Brussel H 1992; permission conveyed through Copyright Clearance Center, Inc.

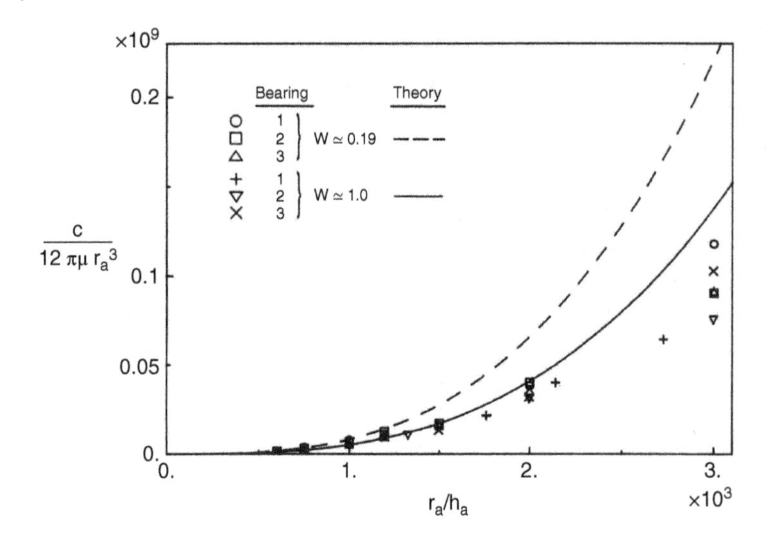

Figure 6.35 Comparison of damping values (flat bearings.) Source: Republished with permission from Elsevier, from Al-Bender F and Van Brussel H 1992; permission conveyed through Copyright Clearance Center, Inc.

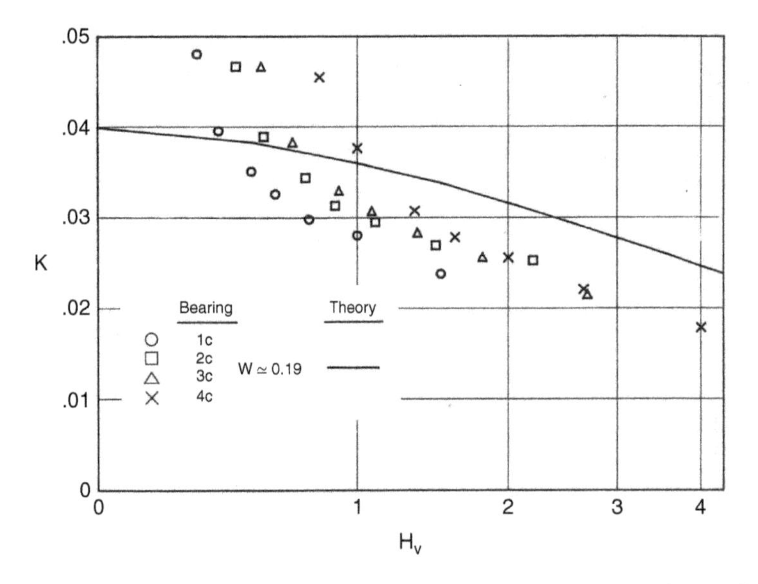

Figure 6.36 Comparison of stiffness values (convergent bearings). Source: Republished with permission from Elsevier, from Al-Bender F and Van Brussel H 1992; permission conveyed through Copyright Clearance Center, Inc.

6.3.5 Conclusions on Tilt

A simple theoretical model has been constructed to predict the various aspects of small-order tilt in circular aerostatic bearings, with uniform or convergent gaps. This model has been checked against a set of accurate tests carefully designed and carried out for this purpose, and covering a broad range of practical interest. The tests show a consistent trend of results that agreed fairly well with the theory; the discrepancies being attributed to the idealising assumptions made in the latter. For a centrally-fed bearing, the main tilt characteristics may be summarised as follows:

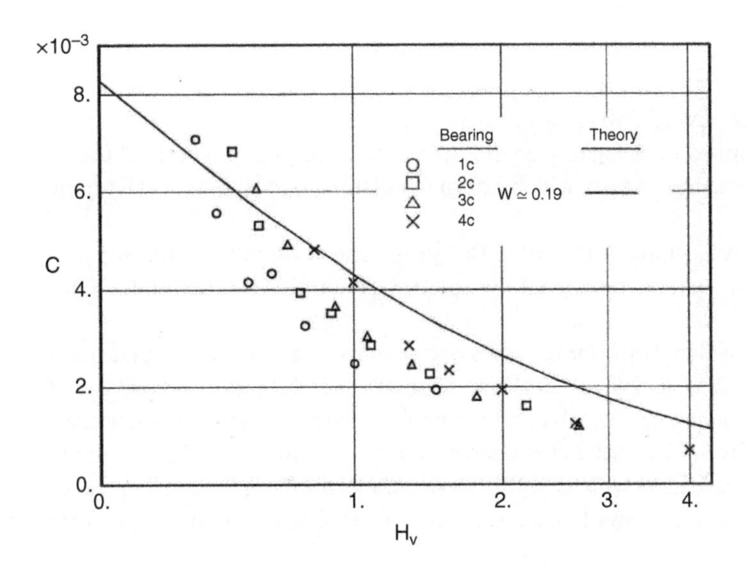

Figure 6.37 Comparison of damping values (convergent bearings). Source: Republished with permission from Elsevier, from Al-Bender F and Van Brussel H 1992; permission conveyed through Copyright Clearance Center, Inc.

- The dimensionless tilt stiffness increases approximately in proportion with the dimensionless load; the proportionality factor decreases with increasing dimensionless conicity. The value of the tilt stiffness is generally small in comparison with the friction moment expected in a conventional pivot or ball bearing support.
- The dimensionless tilt damping coefficient decreases with increasing dimensionless load; it decreases with the dimensionless conicity, for a given load, more drastically than the stiffness.
- The (dynamic) stiffness increases (up to an asymptotic limit,) and the (dynamic) damping decreases (down to zero,) with increasing squeeze number.

6.4 The Influence of Relative Sliding Velocity on Aerostatic Bearing Characteristics (Al-Bender 1992)

Motivation

Although compact, the term *aerostatic* may be a little misleading in that the bearing surfaces are always intended to slide relative to one another. Sliding is taken to mean motion that is tangential to the "plane" of the bearing. As such, it may include both translation of the bearing plate, and rotation about the geometrical axis of the bearing. The relative speed of sliding may be negligibly small, as in slider bearings of measuring machines or machine tools; or relatively big, as in thrust collars or high speed spindles. Since aerostatic bearings are usually designed on the assumption of zero sliding, it is interesting to gain a qualitative as well as a quantitative idea about the influence of sliding on the bearing characteristics. Such an understanding is particularly necessary, if a standardised application of aerostatic bearings is to be envisaged.

In the "stationary" case, the flow field is induced entirely by external pressurisation, i.e. pressure flow. Relative sliding motion will induce shear flow, that will influence the resulting pressure distribution and, consequently, the bearing characteristics. This influence may, in the first analysis, be ignored if the sliding speed is small in comparison with the average pressure flow speed of the fluid. When this is not the case, however, a solution to the bearing problem has to be sought with this effect included. Generally, a numerical solution should provide a satisfactory result. However, our purpose is to gain some insight through adopting a perturbation approach.

Brief Literature Survey

It must be mentioned first that, in contrast to incompressible fluid lubrication, the sliding (aerodynamic) influence cannot be isolated from the aerostatic influence, owing to the non-linearity of the lubrication equation in this case. Sliding has thus been treated either in the context of self acting bearings, in which the pressure on the bearing boundaries is ambient, or in the context of hybrid bearings, in which external pressurisation is combined with the sliding action.

Self acting bearings constitute a specialised subject that is treated in Chapters 8 and 9. Although valuable information may be gained about the behaviour of such bearings, that information is not directly applicable to our problem.

Hybrid configurations have usually been considered only for the cases of full journal bearings (Elrod and Glanfield 1971; Kazimierski and Trojnarski 1980; Pink and Stout 1981), see also Chapter 9, and tilting pad radial bearings (Nemoto and Ono 1981), to mention but a few. In such configurations, the aerodynamic influence gives a positive, although limited, contribution to the load capacity characteristics of the bearing. This is owing to the favorable "wedge" shape of the film profile. Situations in which the film has a parallel profile are thus excluded. In all of these treatments, numerical solutions where employed: usually, finite difference schemes with some form of relaxation.

Linear sliding, of an otherwise aerostatic bearing, outside the above mentioned cases, has to our knowledge hardly been considered. Constantinescu (1969) considers the case of plane parallel gas bearing with a constant inlet pressure, which has an analytic solution. He shows that the load capacity approaches an asymptotic limit, and the flow rate increases without limit, as the relative speed between the bearing surfaces increases. In other words, sliding has, in this case, an unfavourable effect on bearing performance. Shih and Yang (1990) analysed the effect of sliding on a plane parallel gap porous bearing by means of regular, and matched singular, perturbations, in addition to a numerical solution. They reached qualitatively similar results to those that will be given in this section.

The other situation of practical interest, viz. that of pure rotation of one of the bearing surfaces about the axis of (geometrical) symmetry, does not present any difficulty when the bearing surfaces remain nominally parallel, and will be discussed in the main text. The case of rotation combined with tilt is analysed by Safar (1981) for the case of flat bearing, and by Pande (1985) for a convergent gap bearing, both using finite difference solutions. It is found that the effect of rotation becomes only significant at high tilt values.

Objectives and Scope

The aim of this part of the chapter is to study the effect of relative velocity for a nominally parallel bearing gap. In particular, the effect of sliding will be considered in some detail, in order to gain an initial quantitative idea about it. The analysis will be purely theoretical, since an attempted experimental verification (whose issues will be discussed) proved difficult.

In the following section, the problem will be posed, formulated, and discussed. A solution by series expansion is then proposed and evaluated. Thereafter, typical results are given and discussed. A practical application of high speed sliding on aerostatic thrust bearings is then briefly sketched; the problem of measurement is exposed, with a measurement principle deduced. Relevant conclusions are finally drawn.

6.4.1 Formulation of the Problem

Basic Assumptions

The bearing configuration is sketched in Figure 6.38. The problem will be analysed on the basis of the Reynolds equation, with a lumped parameter entrance flow, given by the nozzle formula. This is strictly speaking not correct,

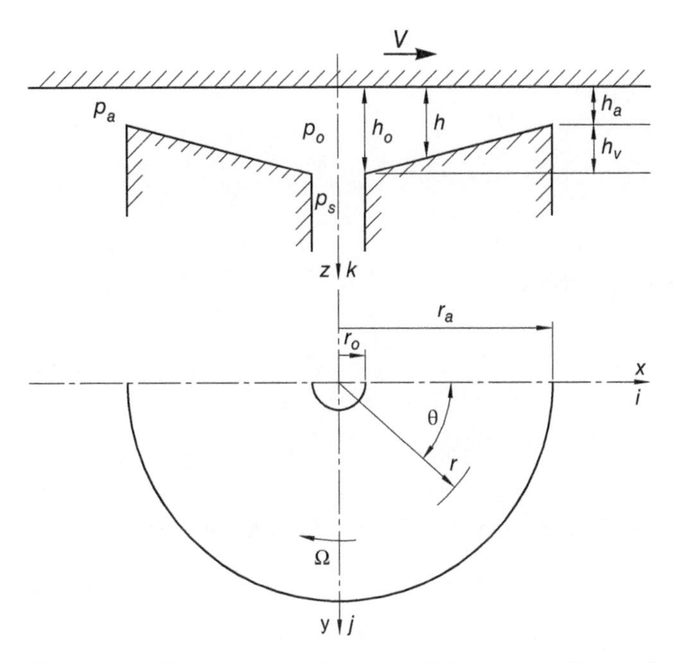

Figure 6.38 Plane bearing with relative sliding velocity. Source: Al-Bender F 1992.

as shown in Chapters 3 and 5, but is considered sufficient for establishing the basic nature of the problem. An exact solution of the entrance flow, which includes the effect of relative speed, was considered to form only a minor, or secondary aspect of the problem, and was thus dropped from the analysis. Consistent with the latter assumption, the pressure at gap entrance ($r = r_o$) is taken to be independent of the angular position θ. In other words, we consider a point or a line source that is also consistent with the assumption of a small feed-hole ($R_o \ll 1$). The relative motion will be assumed steady; the film remaining nominally parallel; the moving bearing surface flat, smooth and rigid (i.e. no stretch). Other assumptions will be made as and when the sliding problem is considered. Furthermore, the dynamic aspects will not be considered.

Basic Equations

With the assumptions made above, the equation governing the pressure distribution in the bearing film is the steady Reynolds equation (see Chapter 4):

$$\nabla \cdot (H^3 \wp \nabla P - \Lambda_f \wp H) = 0 \tag{6.28}$$

in which,

$$\Lambda_f = \frac{6\mu \mathbf{V}_t r_a}{p_a {h_a}^2} \tag{6.29}$$

is the *sliding number*.

Referring to Figure 6.38 for notation, the velocity $\mathbf{V}_t$ has the general form

$$\mathbf{V}_t(r, \theta) = V\mathbf{i} + r\Omega\hat{\mathbf{e}}_\theta \tag{6.30}$$

where V is a constant sliding velocity of the rigid bearing plate, in a given direction, while Ω is the angular velocity of rotation about the centre of the bearing. Also, $\mathbf{i}$ is a unit vector in the x-direction, such that

$$\mathbf{i} = \cos\theta\hat{\mathbf{e}}_r - \sin\theta\hat{\mathbf{e}}_\theta \tag{6.31}$$

where $\hat{\mathbf{e}}_r$ and $\hat{\mathbf{e}}_\theta$ are unit vectors in the curvilinear (polar) r and θ coordinate directions.

Substituting Eqs. 6.29, 6.30 and 6.31 into equation 6.28, rearranging and simplifying the R.H.S., yields:

$$\nabla \cdot (H^3 \wp \nabla P) = \mathbf{\Lambda} \left(\frac{\partial}{\partial R}(\wp H) \cos\theta - \frac{1}{R}\frac{\partial}{\partial \theta}(\wp H) \sin\theta \right) + \Lambda_v \frac{\partial}{\partial \theta}(\wp H) \tag{6.32}$$

where,

$$\mathbf{\Lambda} = \frac{6\mu V r_a}{p_a {h_a}^2} \qquad \text{and} \qquad \Lambda_\omega = \frac{6\mu\Omega {r_a}^2}{p_a {h_a}^2}$$

are the translation and the rotation bearing numbers respectively.

The boundary conditions on Eq. 6.32 are:

$$P(R_o, \theta) = P_o \qquad \text{and} \qquad P(1, \theta) = 1 \tag{6.33}$$

Nominal parallelism will mean that

$$H = f(R)$$

In addition to the Reynolds equation, we need the inlet flow, or compensator formula (see Chapter 5):

$$\bar{m}_o = C_d \Lambda_f H_o R_o P_s \Phi_e(\bar{p}_o) \tag{6.34}$$

from which, the inlet pressure may be determined through equating the film flow with the inlet flow.

6.4.2 Qualitative Considerations of the Influence of Relative Velocity

Conditions Under Which the Relative Velocity Does Not Influence the Pressure Distribution

Following Constantinescu (1969), let us first consider the conditions by which the relative velocity should have no influence on the pressure distribution. This is so when the wedge term vanishes identically:

$$\nabla \cdot (\Lambda_f \wp H) \equiv 0 \tag{6.35}$$

The assumption of rigidity of the moving surface is equivalent to the fact that the sliding velocity field $\mathbf{V}_t$ is solenoidal, viz.

$$\nabla \cdot \mathbf{V}_t \equiv 0$$

In this case, Eq. 6.35 becomes:

$$\Lambda_f \cdot \nabla(\wp H) \equiv 0 \tag{6.36}$$

This would be so when either of the two multiplied vectors vanishes, i.e. when the velocity vector is normal to the gradient of $(\wp H)$.

The first condition is satisfied, besides in the trivial case of zero sliding, also in the case when the product of the density with the film thickness is constant ($\wp H \equiv 1$,) e.g. in the case of incompressible lubricant in a uniform gap bearing. We shall later consider the case of incompressibility, in order to further qualify the origin of the effect of sliding.

The second condition, in which neither of the two terms vanishes, may be satisfied when

$$\wp H = f(R) \qquad \text{and} \qquad \Lambda_f = g(R)\hat{\mathbf{e}}_\theta,$$

where f and g are arbitrary functions, i.e. when the relative velocity is purely rotational while the film geometry remains axisymmetrical. In other words, rotation about the axis of the bearing should not influence its load and flow behaviour.

Remark

Let us note that this result has been possible only by virtue of the assumption of viscous flow on which the Reynolds equation is based and by which the inertia effects are neglected. As such, this result will become less valid the higher the rotational speed is. Dowson (1961), who included the centrifugal inertia terms, has shown, for the case of incompressible lubricant, that high speed rotation may have a detrimental effect on the load capacity. The basic aspect of this problem is similar to the case of "rotation close to a stationary wall" treated in (Schlichting 1968, pp. 213–218.): at sufficiently high rotational speed, a *secondary* flow is created which pumps the fluid towards the rotating surface and the centre of the bearing. Safar (1983) carried out a similar analysis that included, in addition, the case of tilt.

Conditions Under Which the Relative Velocity Does Not Influence the Load Capacity

To this class belongs only the case of incompressible lubricant with axisymmetric film profile in the absence of cavitation, i.e.

$$\wp \equiv 1 \qquad \text{and} \qquad H = f(R)$$

In this case, Eq. 6.32 becomes:

$$\nabla \cdot (H^3 \nabla P) = \Lambda \frac{1}{R} \frac{dH}{dR} \cos\theta \tag{6.37}$$

which is a linear equation in the pressure. Consequently, it may be easily shown that its general solution may be written as:

$$P(R,\theta) = P_{(0)}(R) + \Lambda P_{(1)}(R) \cos\theta \tag{6.38}$$

in which, $P_{(0)}$ is the solution to the homogeneous problem, satisfying the conditions 6.33, and $P_{(1)} \cos\theta$ is the particular integral that satisfies homogeneous boundary conditions (i.e. vanishes at the boundaries). It can be directly seen that, by virtue of its circular (i.e. sinusoidal) form in θ, this particular integral does not contribute anything to the load capacity, nor to the flow rate.

This analysis has shown, in addition, an important difference between compressible and incompressible lubrication, viz. that, in the incompressible case, the hydrodynamic effect is linearly independent from the hydrostatic effect: the net effect is a simple superposition of the two effects. The hydrodynamic effect is, furthermore, directly proportional to the sliding number Λ. This constitutes a considerable simplification to the problem, which is unfortunately not at our disposal for compressible lubricants, except perhaps in the limit of small bearing pressures (Chapter 8). In this latter case, the aerodynamic effect is non-linearly dependent on the sliding number[1].

6.4.3 Solution Method

First, we make the additional assumption that the flow is isothermal, i.e.

$$\wp = P.$$

A polytropy assumption would have also been possible, with a fixed exponent, but this would make only a secondary difference; not affecting the basic nature of the problem.

Secondly, without loss in generality, we shall consider only the case of purely translational sliding: the other case, with the rotational component included, can be solved by the same method. Thus, with these assumptions, equation 6.32 becomes, after rearrangement:

$$\nabla \cdot (H^3 \nabla P^2) = 2\Lambda \left(\frac{\partial}{\partial R}(PH) \cos\theta - \frac{1}{R} \frac{\partial}{\partial \theta}(PH) \sin\theta \right) \tag{6.39}$$

subject to the same boundary conditions 6.33. We shall use two types of solution:

1 It is noteworthy that some researchers attempted to derive a superposition principle for hybrid journal air bearings, based on some arbitrary averaging assumptions, see (Pink and Stout 1981).

1. Small-Λ functional series solution, which is valid for low to moderate values of Λ and has the advantage of being semi-analytical, so that it can generate a range of results for each solution run. Details of the solution can be found in (Al-Bender 1992).
2. The purely numerical F.D. solution, which is possible for any value of Λ, in particular to show the behaviour at large values of this parameter. A rectangular grid is used since this is the most suitable in view of the sliding velocity being unidirectional.

Calculation of the Bearing Characteristics

Once the pressure distribution $P(R, \theta)$ is determined, we can calculate the following bearing characteristics, the definitions and basic formulas of some which are taken from Chapter 2.

The Load Capacity

The normalised bearing load is given, from Chapter 2, by:

$$W = \frac{1}{\pi} \int_{\bar{A}} (P - 1) d\bar{A}.$$

The Flow Rate in the Film

The normalised flow rate, evaluated at R_o, is given by:

$$\bar{m}_r = -R_o H_o^{\ 3} P_o \frac{1}{2\pi} \int_0^{2\pi} \left(\frac{\partial P}{\partial R} \right)_{R_o} d\theta$$

The Centre of Force

The centre of force x_f is defined by:

$$x_f = \frac{\int_A (p - p_a) x dA}{\int_A (p - p_a) dA}$$

In normalised form, this may be written:

$$\bar{X}_f = \frac{x_f}{r_a} = \frac{\frac{1}{\pi} \int_{\bar{A}} (P - 1) R \cos\theta d\bar{A}}{W}$$

The Traction Force

Calculated at the moving surface, and expressed, for convenience, in a direction *opposite* to the motion, the normalised friction force is given by:

$$\bar{F}_f = \frac{1}{\pi} \int_{\bar{A}} (\frac{3}{\Lambda} H \nabla P \cdot \mathbf{i} + \frac{1}{H}) d\bar{A}$$

where,

$$\nabla P \cdot \mathbf{i} \equiv \partial P / \partial X.$$

As has been mentioned in Chapter 2, the traction force consists of a pressure (Poiseuille flow) contribution—the first term, and a velocity (linear shear, or pure Couette) contribution, or viscous friction—the second term. It may be interesting, at this stage, to examine their relative merit, while at the same time simplifying the first term.

Let us note that, by virtue of the assumptions made, the product PH is constant on the boundaries $R = R_o$ and $R = 1$. Thus, through integration by parts, we have the following result:

$$\int_{\bar{A}} H \partial P / \partial X d\bar{A} = - \int_{\bar{A}} P \partial H / \partial X d\bar{A}$$

Now, for the case of a flat bearing, $H \equiv 1$, this term vanishes, and we have the simple, but very useful result:

$$\bar{F}_{\mathrm{f}} = 1.$$

This means that, for this case, the resistance to the motion is due entirely to viscous shear, and is independent of the pressure distribution. In absolute dimensions, the friction force will then be given by:

$$F_{\mathrm{f}} = \frac{\mu VA}{h_a}.$$

That is to say that the friction force is directly proportional to the viscosity, sliding speed and bearing area, and inversely proportional to the film thickness. This is often called Petrov's law Dowson (1979).

For the case of a convergent-gap bearing, $H = H(R)$, we have:

$$\partial H/\partial X = \partial H/\partial R \cos\theta = -v_{\mathrm{a}} \cos\theta$$

So that we have:

$$\int_{\bar{A}} H \partial P/\partial X \mathrm{d}\bar{A} = v_{\mathrm{a}} \int_{\bar{A}} P \cos\theta \mathrm{d}\bar{A}$$

It can be seen that, just as in the case of the centre of force, only the terms in $\cos\theta$ will contribute to the pressure friction. In particular, in the limit $\epsilon = 2\Lambda \to 0$, only the first perturbation pressure $F_1^{(1)}$ will remain, and the normalised aerodynamic traction force will have a finite value.

The Coefficient of Friction

Expressed in terms of dimensionless numbers, the coefficient of friction may be written as:

$$\mu_{\mathrm{f}} = \frac{\bar{F}_{\mathrm{f}}}{W} \frac{\Lambda}{6} \frac{h_a}{r_a}.$$

It may be noted from this expression that similar bearings do not have the same coefficient of friction, but rather the same value for a *normalised* coefficient of friction, given by:

$$\bar{\mu}_{\mathrm{f}} = \mu_{\mathrm{f}} \frac{r_a}{h_a}.$$

6.4.4 Results and Discussion

There are two levels of treating the problem of sliding. The first is its effect on the film characteristics assuming a constant inlet pressure P_0. This we shall term the *passive* effect. The second is its effect on the global bearing performance, i.e. the load capacity and stiffness, assuming a constant supply pressure P_s. This will be termed the *active*, or the global, effect. The passive effect furnishes the basic idea about the problem, and constitutes the necessary first step for determining the active effect. This latter is characterised by the fact that sliding influences (increases) also the flow rate of the fed gas, so that, for a constant supply pressure, the inlet pressure will drop with increasing sliding speed, thus further reducing the load capacity. This problem must, in general, be solved by iteration, starting from the passive solution.

The Passive Characteristics

The problem is governed by the parameters (R_0, H_v, P_0, Λ). However, the characteristics are not equally sensitive to all these parameters; in particular, R_0 and P_0 may be fixed at typical design values, in order to facilitate representation of the results. Thus, the bearing characteristics will be investigated as a function of H_v and Λ, the qualitative influence of R_0 and P_0 being indicated. The following results are given for the values of $R_0 = 0.05$ and $P_0 = 4$. (Note also that the range of the given results corresponds approximately to the limits of accuracy of the solution method.)

Pressure Distribution

First, typical pressure distributions are given along the diameter of sliding, $y = 0$, to give an initial idea about the problem. These results are obtained using the series expansion method.

In Figure 6.39, such pressure distributions are plotted for different values of Λ; (a) for a flat bearing, and (b) for a bearing with conicity $H_v = 1$. It is seen that the pressure distribution becomes asymmetric the more the sliding number Λ increases. The pressure decrease on the "trailing" side is relatively larger than the pressure increase on the "leading" side. This behaviour is qualitatively similar to that of an aerodynamic step bearing as given in Chapter 8. Comparison of the top and bottom panels of Figure 6.39 shows also that the convergent gap

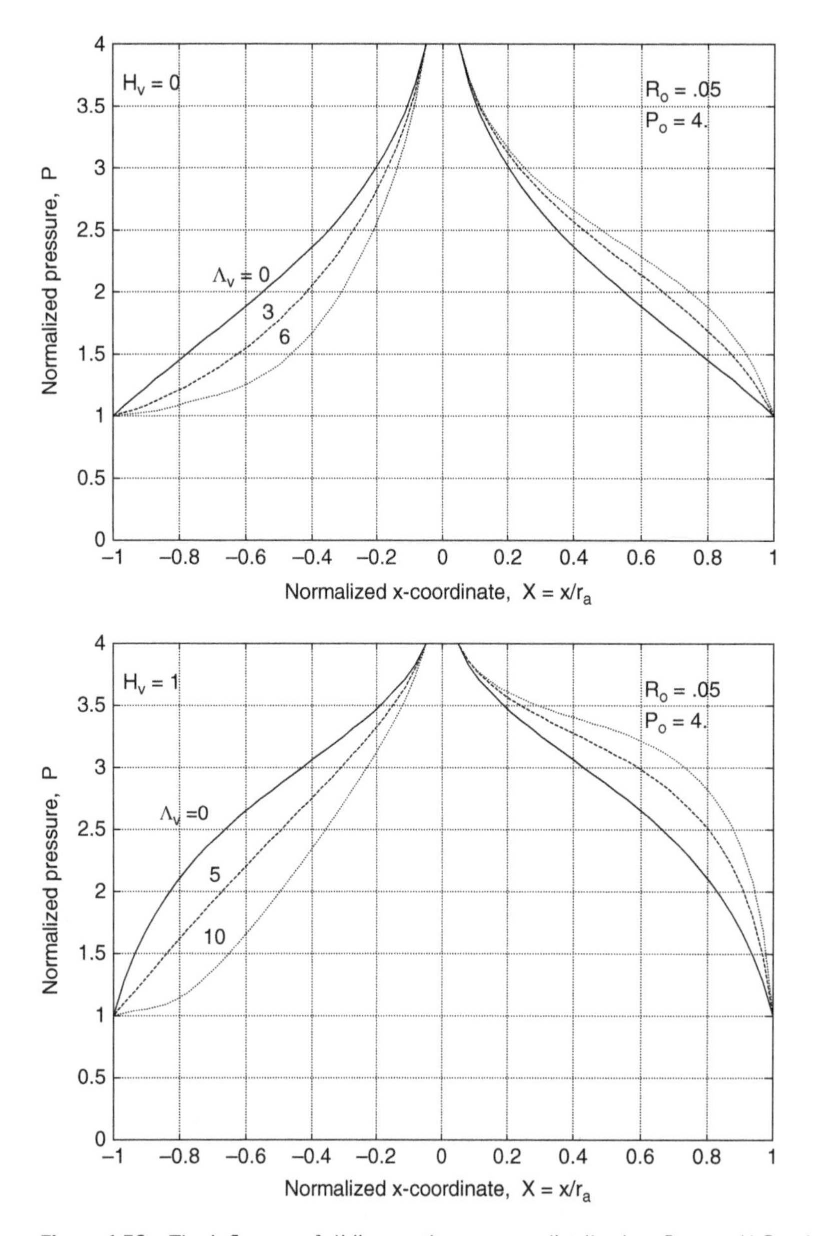

Figure 6.39 The influence of sliding on the pressure distribution. Source: Al-Bender F 1992.

bearing is less influenced by sliding (at the same value of Λ). If the bearing were two dimensional, i.e. plane infinite, the net effect on the load capacity would have been negligible at these low sliding numbers, as is shown in (Constantinescu 1969, pp. 183–186). However, the pressure on the sides, $x = 0$, i.e. the left and the right halves of the bearing decreases also making the net effect much more pronounced in the case of a finite width bearing. This should be clear from the 3D-plots of Figures 6.40 and 6.41, which pertain to the same two bearing cases of Figure 6.39. In panel (b) of each figure, this effect is not yet appreciable, owing to the the small sliding number. In panel (c) of both figures, it becomes already noticeable. Finally, as Λ reaches very high values (as 100 in panel (d)), positive pressure greatly disappears except on a thin strip in the middle of the bearing. This corresponds to the singular solution of Eq. 6.39, as discussed in Chapter 8, so that the term in parentheses on the R.H.S. approaches zero. This yields the singular solution:

$$PH = \text{Const.}$$

Load and flow rate

Figure 6.42 shows the relative decrease in the load capacity as a function of Λ, for various values of the conicity. The loss in load capacity may become appreciable for high values of Λ. However, this loss decreases significantly as

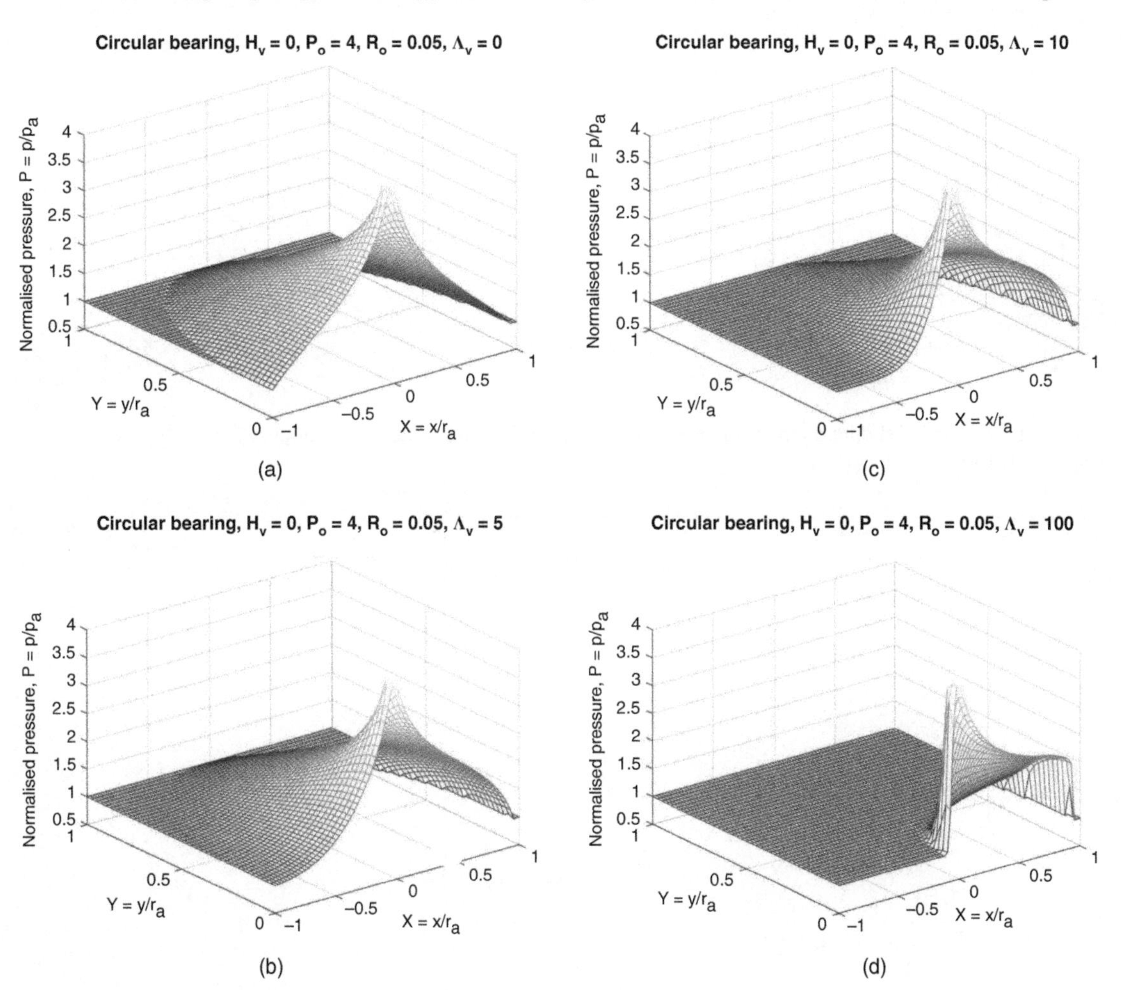

Figure 6.40 The influence of sliding on the pressure distribution obtained from the F.D. solution. $R_o = 05, H_v = 0, P_o = 4, \Lambda = 0, 5, 10, 100$.

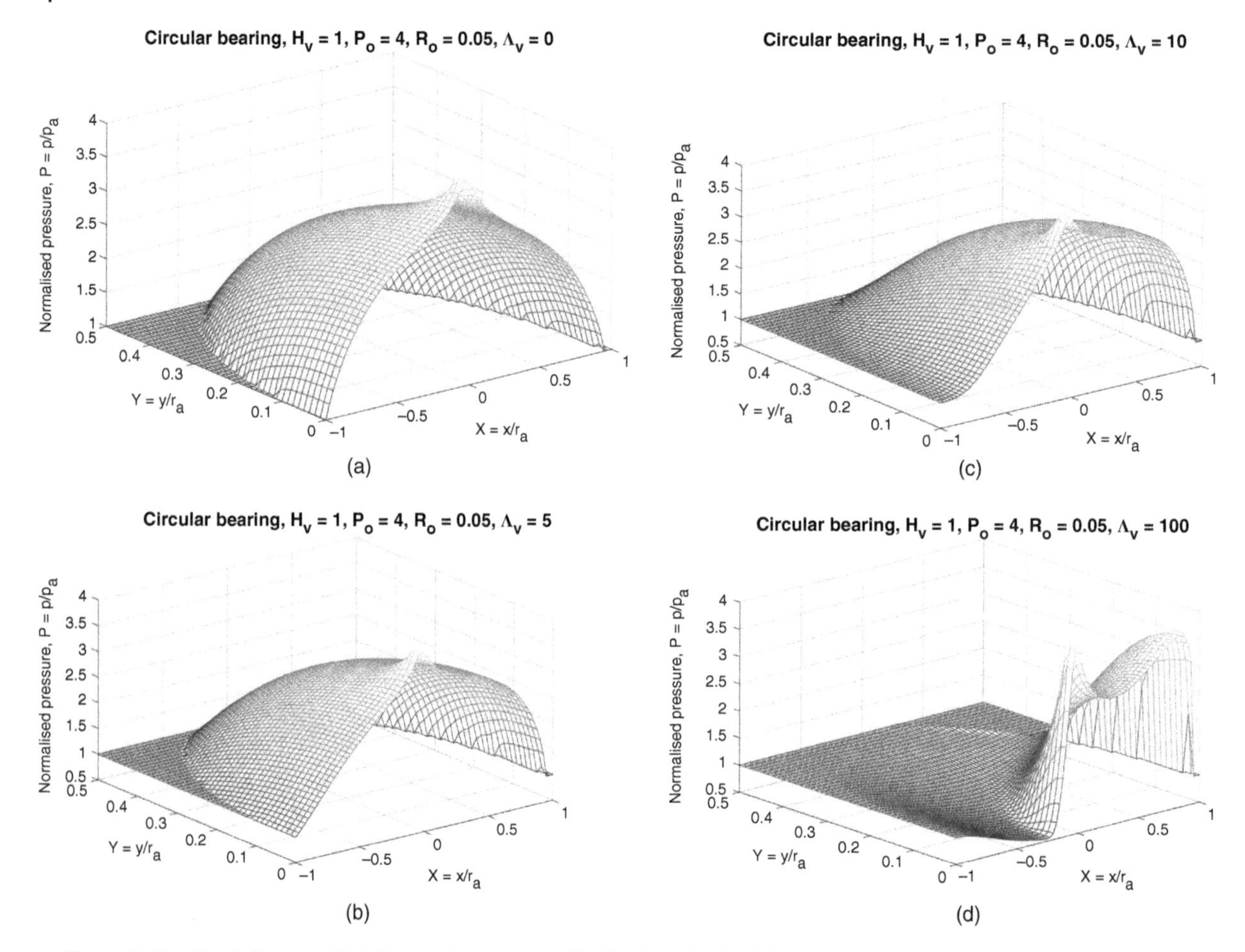

Figure 6.41 The influence of sliding on the pressure distribution obtained from the F.D. solution. $R_o = 05, H_v = 1, P_o = 4, \Lambda = 0, 5, 10,100$.

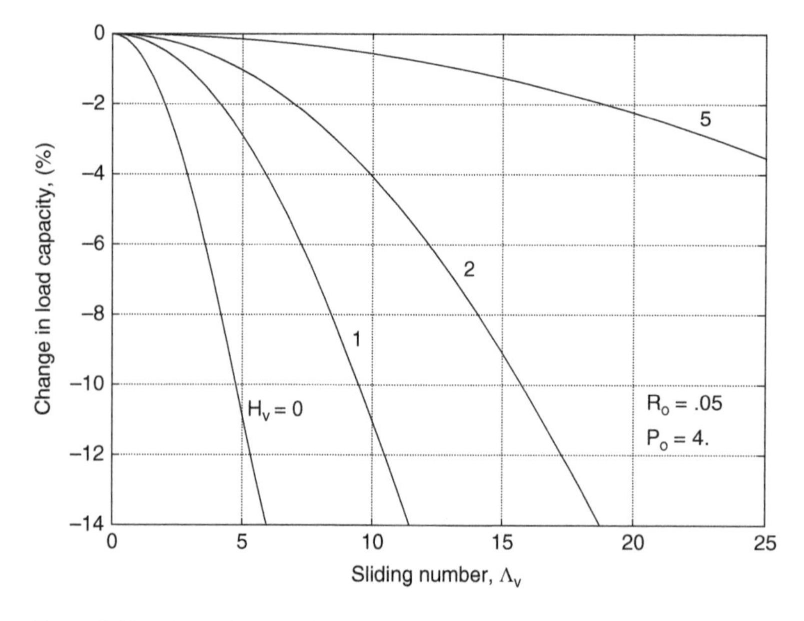

Figure 6.42 The influence of sliding on the load carrying capacity. Source: Al-Bender F 1992.

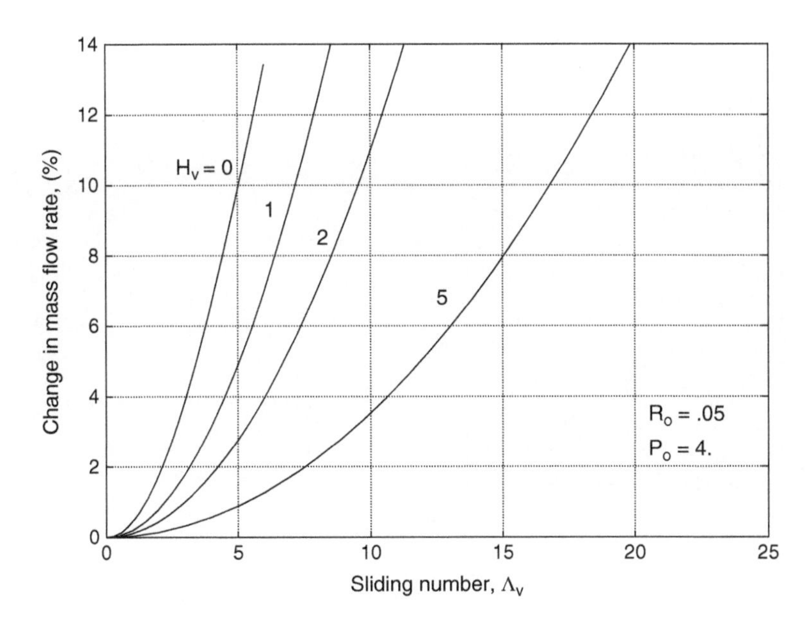

Figure 6.43 The influence of sliding on the mass flow rate. Source: Al-Bender F 1992.

the conicity increases. Similar behaviour is shown in regard to the relative increase in the mass flow rate, depicted in Figure 6.43. These two results show that, increasing conicity yields a more favourable sliding behaviour. A word about the influence of the feed-hole radius and the inlet pressure may be mentioned here. An increase of R_o or P_o, in the vicinity of the specified values, makes the influence of sliding less pronounced, i.e. the relative change in the load capacity and the flow rate is reduced.

Centre of force

Figure 6.44 gives the values of the associated shift in the centre of force. This is approximately linear in $\boldsymbol{\Lambda}$. For a flat bearing, at $\boldsymbol{\Lambda} = 5.$, the centre of force is shifted by approximately $0.15r_a$. This means that if the bearing is centrally pivoted, it will tilt about its pivot point in order to attain equilibrium. Since the shift is in the direction of motion, the bearing gap will increase on the leading side, and vice versa, thus further reducing the load capacity, and increasing the flow rate. Thus, a rigidly fixed bearing would be expected to perform better than a centrally pivoted one.

Traction force

Figure 6.45 shows the behaviour of the normalised friction force as a function of the conicity, at a low value of $\boldsymbol{\Lambda}$. This result is, however, little affected by $\boldsymbol{\Lambda}$, R_o or P_o, and can thus provide general design values. Both the total traction force as and its shear (friction) and pressure (aerodynamic) components are plotted, in order to show their relative influences. It is seen that, while the pressure component attains a maximum value somewhere around $H_v = 7.$, the shear component decreases monotonically, so that their orders of magnitude become comparable at high conicities. Hence, the total traction force decreases substantially with the conicity.

The normalised coefficient of friction is plotted in Figure 6.46. As has been evident from the friction force characteristics, the coefficient of friction is initially linear in $\boldsymbol{\Lambda}$. However, as the latter increases, the load capacity decreases, while the friction force remains nearly constant, so that $\bar{\mu}_f$ will lie above its linear value at high sliding numbers. Also, since the conicity increases the load capacity and decreases the friction force simultaneously, the resulting effect on the coefficient of friction becomes very pronounced. It may be seen, from Figure 6.46, that the

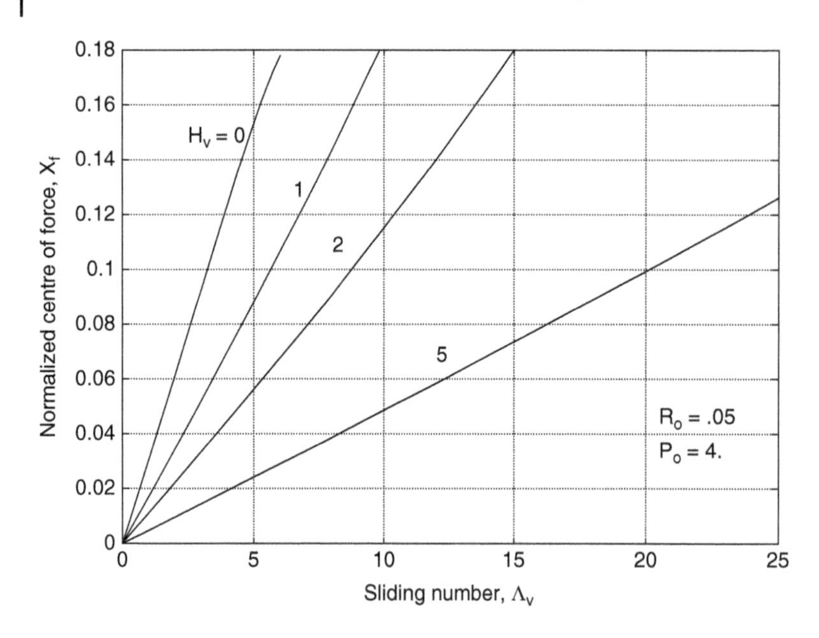

Figure 6.44 The influence of sliding on the centre of force. Source: Al-Bender F 1992.

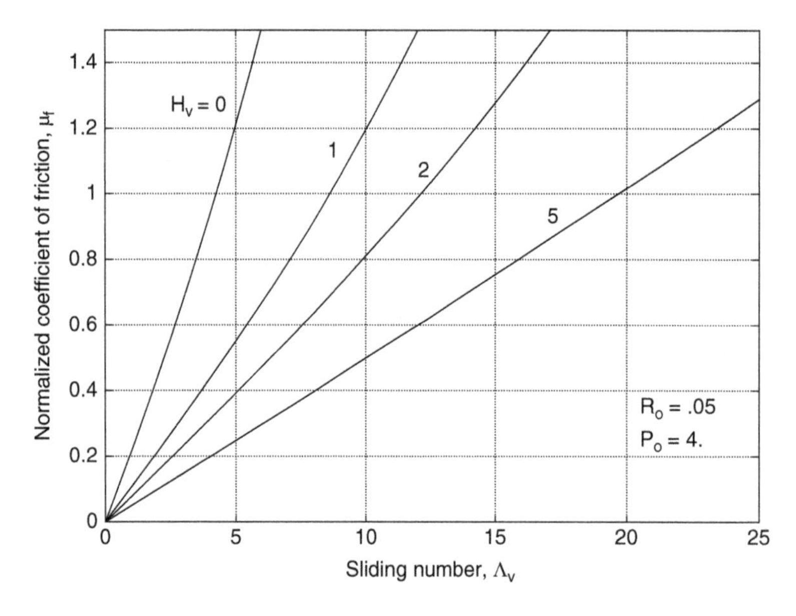

Figure 6.45 The friction force as a function of the conicity. Source: Al-Bender F 1992.

coefficient of friction is nearly halved for a bearing with $H_v = 1$, in comparison with a flat bearing ($H_v = 0$.) This may be regarded as one of the important advantages of convergent gap bearings.

Using the flat bearing as a reference, an idea about the order of magnitude of the coefficient of friction may be obtained. Thus, assuming linear behaviour in $\mathbf{\Lambda}$, we have the following result, (for $H_v = 0$):

$$\bar{\mu}_{\mathrm{f}} \simeq 0.24\mathbf{\Lambda}.$$

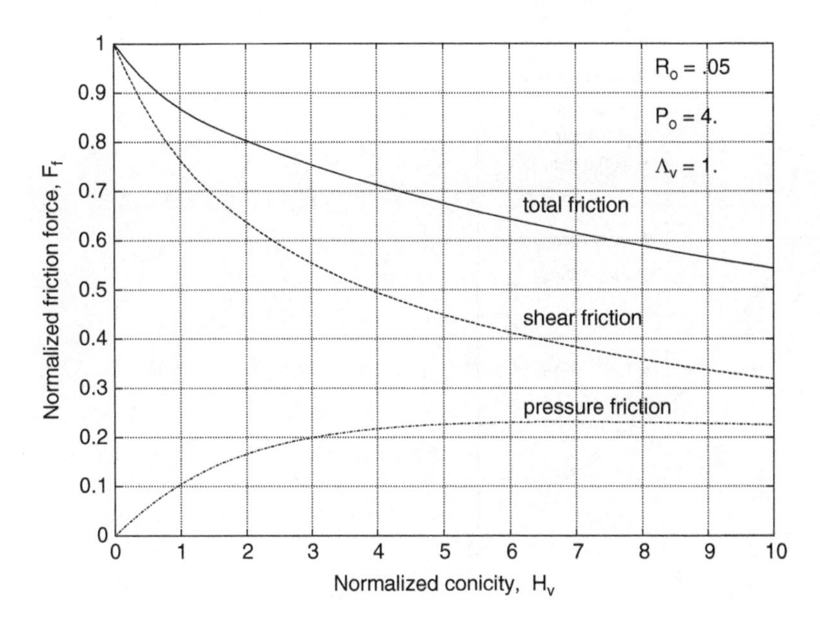

Figure 6.46 The normalised coefficient of friction as a function of the sliding number. Source: Al-Bender F 1992.

Substituting for $\bar{\mu}_{\mathrm{f}}$ and $\boldsymbol{\Lambda}$, with the respective absolute dimensions, we obtain:

$$\mu_{\mathrm{f}} \simeq 1.44 \frac{\mu V}{p_a h_a}.$$

Note that the coefficient of friction is, in this case, independent of the bearing size r_a, and is determined by the dimensionless number $\mu V / p_a h_a$.

As an example, let the lubricating gas be air, the ambient pressure be 1 bar, the sliding velocity be 100 m s^{-1}, and the film thickness 10 μm, then the coefficient of friction is about 0.0026, (or 2.6 10^{-5} per m s^{-1}).

Global Characteristics

As has been seen in the previous section, sliding will result in a decreased load capacity and an increased flow rate. The latter fact means that, in a bearing fed with a constant supply pressure, sliding should result in a reduction in the inlet pressure p_0. Although this reduction is generally small, it nevertheless plays a role in the further reduction of the load capacity under sliding. This means also that the passive solution cannot be used directly to determine the global characteristics, but constitutes rather the first step for obtaining that solution. The solution is obtained, by iteration, as follows.

The passive solution provides all the relevant sliding characteristics. What we are interested in here is the load capacity characteristics over a range film thickness values, under the influence of constant velocity sliding, which should also provide an idea about the stiffness under these conditions. These characteristics are best illustrated by practical examples. Two bearings will be considered for this purpose; a flat one, and a convergent gap one. They have the same external, and feed hole radii, and the same supply pressure; the convergent bearing has a conicity of 10 μm.

Figure 6.47 shows the load capacity characteristics of the flat bearing, at the different sliding speeds indicated. Let us bear in mind that the loss in load capacity depends on the sliding number $\boldsymbol{\Lambda}$ which is proportional directly to the speed, and inversely to the square of the film thickness. Thus, even when the sliding speed is moderate, e.g. 10 m s^{-1}, the load capacity begins to decrease drastically, as smaller gap heights are approached. Although this

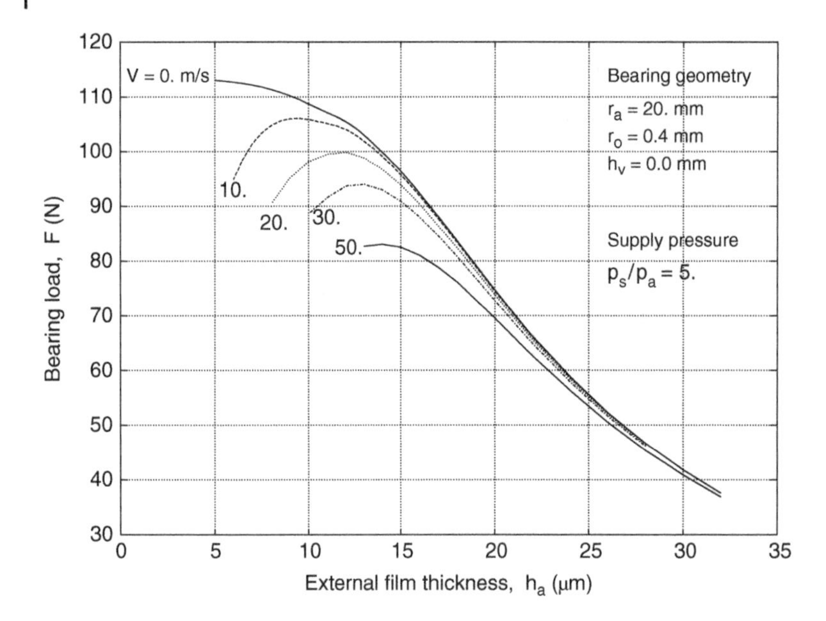

Figure 6.47 The influence of sliding on the load capacity and stiffness, for a flat bearing. Source: Al-Bender F 1992.

loss remains bounded, the shape of the load curve is qualitatively changed. It has now a turning point at a film thickness of about 10 μm. To the right of this point, the stiffness of the bearing is positive; to the left, it is *negative*. This means that at this point, the bearing will become statically unstable: it will collapse under its own weight, despite the fact that the load capacity remains positive.

Returning to Figure 6.47, we note that the point of static instability shifts to the right, as the sliding speed increases, such that it takes place at nearly equal values of the sliding number $\boldsymbol{\Lambda}$.

Figure 6.48 shows the same characteristics for the convergent gap bearing. The behaviour is remarkably more favourable, as has been evident also in the previous section. Static instability does not occur for small speeds. When it does take place, it is shifted more toward the small film thickness range. It must be noted also that instability does not take place now at equal sliding numbers. This is owing to the fact that the normalised conicity $H_v = h_v/h_a$ is not a constant, but varies rather with the film thickness: the lower the latter, the higher the former, and thus the better the characteristic. Thus, we can assert that gap convergence, or the conicity, contributes to static stability during sliding.

Finally, a remark about normalisation and similarity may be in order. It has been evident that two bearings are physically similar, when sliding is included, if they share the following set of parameters:

$$R_o, \quad H_v, \quad P_s, \quad \Lambda_f, \quad \boldsymbol{\Lambda}$$

In a systematic representation of load characteristics, the normalised load capacity W is plotted as a function of the bearing number Λ_f, for a given normalised geometry. In such cases, a constant sliding speed will result in a variable sliding number. In order to circumvent this difficulty, a new sliding parameter has to be defined. Let us note, for this purpose, that the the sliding number and the bearing number are of the same form; they differ by a dimensionless factor:

$$\frac{\boldsymbol{\Lambda}}{\Lambda_f} = \frac{1}{2}\sqrt{\frac{\kappa-1}{2}}\frac{V}{\sqrt{\kappa \Re T_{\mathrm{a}}}} = \frac{1}{2}\sqrt{\frac{\kappa-1}{2}}\mathrm{M_v}$$

in which $\mathrm{M_v}$ is a Mach number of the sliding velocity relative to ambient sonic speed. This Mach number can thus be used instead of the sliding number, as a similarity parameter.

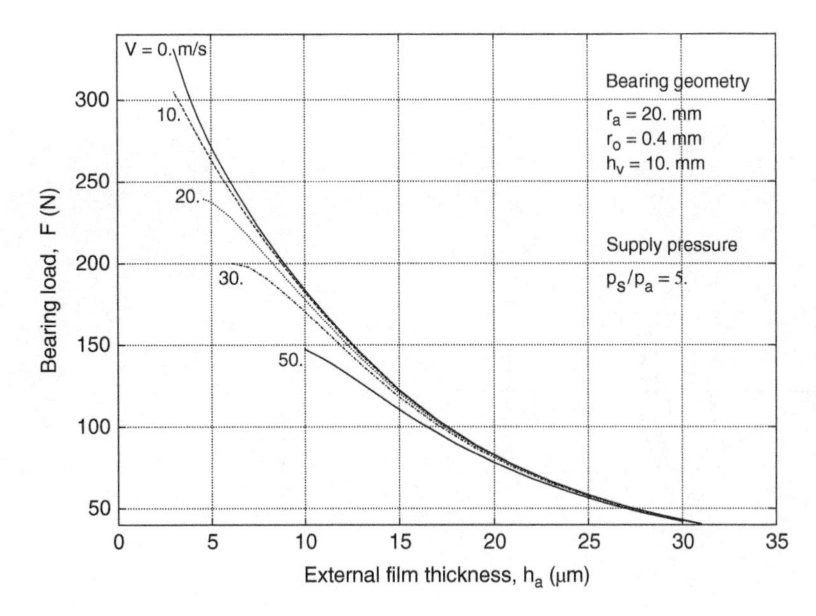

Figure 6.48 The influence of sliding on the load capacity and stiffness, for a convergent gap bearing. Source: Al-Bender F 1992.

On the Question of Experimental Verification

One of the motivations of this study was the design of a diamond polishing disc, so-called "scaife", supported horizontally on three aerostatic thrust bearings, as reported in (Al-Bender et al. 1989) and depicted in Figure 6.49.

The disc rotates at 3000 r.p.m. resulting in a sliding speed of about 40 m s^{-1}, at the bearing circle. It was, therefore, important to know whether the system would maintain sufficient stiffness at that speed. This was also a good opportunity to verify the proposed theory. However, the problem turned out to be more complex than was initially envisaged. This was primarily due to the fact that it was difficult to achieve a sufficiently flat disc surface. As it was, the disc surface had a waviness of about 5 μm peak-to-peak, measured in a circumferential direction. It was estimated that the influence of sliding would become appreciable only at small bearing gaps, e.g. $\simeq$ 10 μm. At this limit, however, surface waviness becomes comparable in amplitude to the film thickness, such that not only the surfaces can no longer be regarded as parallel, but also other dynamic influences arise, rendering meaningful measurement very difficult. It should be noted that measurement of the aerodynamic influence has to be carried out in the positive stiffness range. For a constant load, this means only a variation of a few micrometers, (cf. Figure 6.48). To be able to detect such variation requires a highly smooth running, extremely well balanced and geometrically accurate system.

Nevertheless, the configuration as it stood attracted no less interest, by virtue of being an *actual* situation more likely to be encountered in practice than the idealised model. The two interacting dynamics of the air film, viz. the axial stiffness and the aerodynamic influence, together with gyroscopic motion of the disc, presented an interesting problem of "gyro-dynamics". Such a problem could be treated in a semi-empirical manner, i.e. through qualitative theoretical results coupled with experimentation.

6.4.5 Conclusions on Relative Sliding

The origin of the influence of relative sliding has been outlined by examining the conditions that bring it forth. The Reynolds equation, including the sliding (wedge) term, has been solved, for the case of circular nominally parallel bearings, by the method of regular perturbations. Such a solution, although limited to moderate sliding

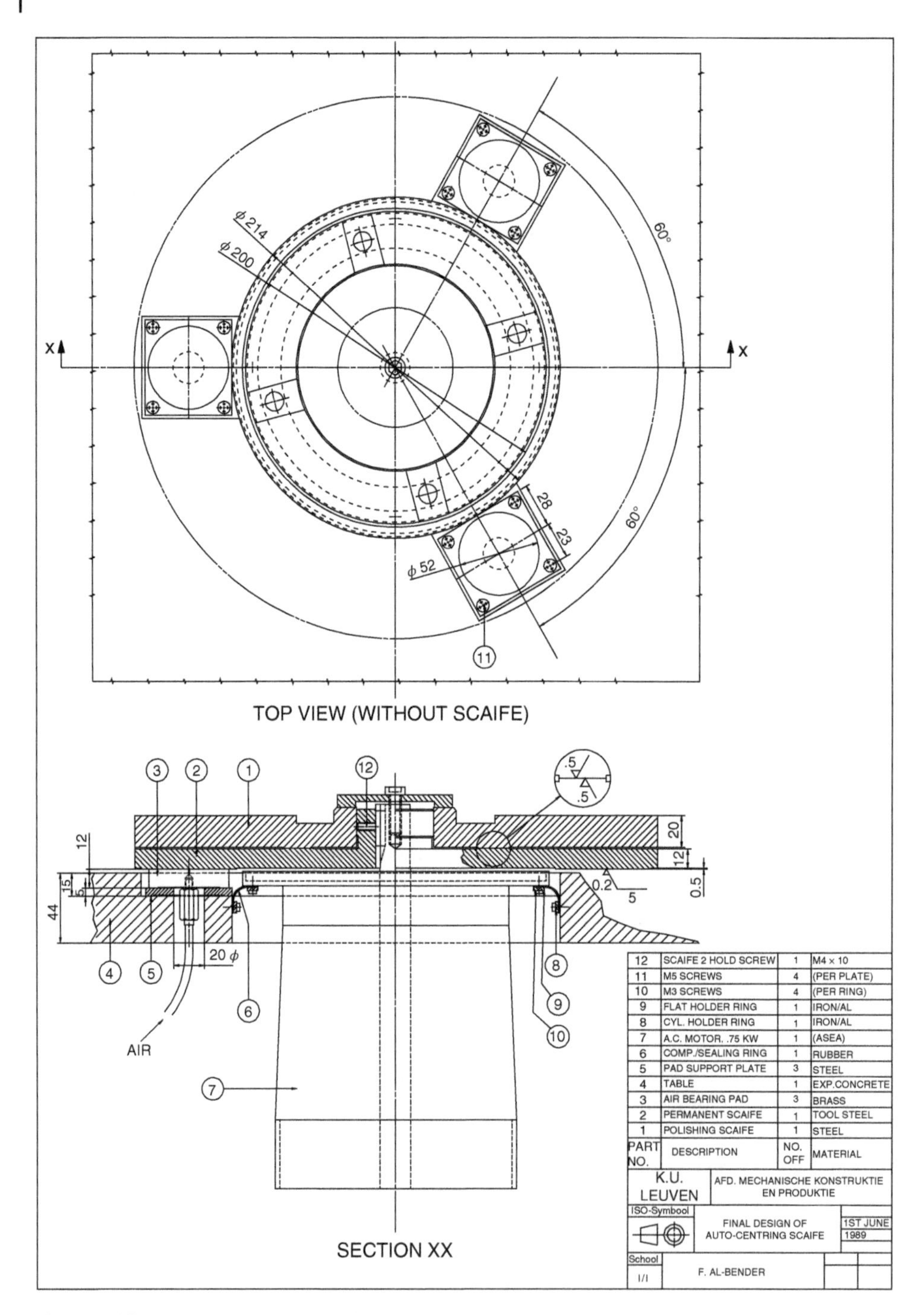

Figure 6.49 Auto-centering scaife, for diamond polishing, supported on three circular aerostatic pads. Source: Al-Bender F, Vanherck P and Van Brussel H 1989.

numbers, has been found to be sufficient to give a general idea about the problem in a semi-analytical manner. It has the advantage of simplicity and easy implementation, as compared to a purely numerical solution, of which some examples have been given.

The basic characteristics of bearings under the influence of sliding have been determined. The load capacity is found to decrease and the mass flow rate to increase with increasing sliding number. The combined effect of these two factors reduces the stiffness of the bearing. It may lead, in extreme situations of very small film thickness or very high sliding speeds, to making the bearing statically unstable. Bearings with convergent gap geometry are less adversely prone to be affected by sliding than flat ones.

The asymmetric pressure distribution, caused by the aerodynamic effect, means, moreover, that the centre of force is shifted away from the geometric centre of the bearing during sliding. This shift takes place in the direction of sliding. If the bearing is centrally pivoted, it will tilt, in order to attain equilibrium, in a way that will further reduce the load capacity. The shift in the centre of force decreases with the increasing conicity.

The normalised friction force, opposed to sliding, has a constant value of unity for a flat bearing. It decreases with increasing conicity, and remains insensitive to the design parameters. The coefficient of friction varies linearly with the speed, at low sliding numbers; it lies above its linear value at high sliding numbers. Since a convergent gap bearing gives a higher load capacity and smaller friction force than a flat bearing, the coefficient of friction is remarkably reduced with increasing conicity.

Summing up, it is seen that a convergent gap bearing has much better sliding characteristics than a flat bearing.

An application, wherein three flat air-bearing pads are used to support a diamond polishing disc rotating at a surface speed of approximately 50 m s^{-1} relative to the bearing-pad surfaces, has been presented.

6.5 Conclusions

This chapter has presented the basic characteristics of circular, centrally-fed aerostatic bearings. These characteristics have been put in three categories: (i) axial static characteristics comprising: load, stiffness and flow rate; (ii) tilt static and dynamic characteristics comprising tilt stiffness and damping; and (iii) the effect of relative speed on the axial characteristics.

In the first part, we saw that the bearing characteristics of load, stiffness and flow rate (or air consumption) depend on the bearing size, feed-hole size ratio and conicity, or gap convergence ratio. Increasing this latter leads to increased load and flow simultaneously, with (a) decreasing load to flow ratio, when the minimum film thickness is taken as reference and (b) increasing load to flow ratio, when the average film thickness is taken as reference. Increasing the conicity leads also to increased stiffness. The stiffness attains a maximum for a combination of optimum conicity and feed-number values. Thus, we need to formulate an elaborate optimisation problem, if we wish to achieve effective and energy-efficient bearing behaviour, in particular over a range of gap heights or loads. Finally, the stiffness characteristics considered here concern the static values. Increasing the stiffness will generally reduce the damping, sometimes to a negative level, which thus forms another optimisation constraint. The dynamic bearing characteristics are treated in Chapter 7.

In the second part, a simple theoretical model has been constructed to predict the various aspects of small-order tilt in circular aerostatic bearings, with uniform or convergent gaps. This model is, furthermore, verified by experiments covering a broad range of practical interest. The tests show a consistent trend of results that agreed fairly well with the theory. For centrally-fed bearing, the main tilt characteristics are (i) the dimensionless tilt stiffness increases approximately in proportion with the dimensionless load; the proportionality factor decreases with increasing dimensionless conicity; (ii) the dimensionless tilt damping coefficient decreases with increasing dimensionless load and dimensionless conicity, (iii) the dynamic stiffness increases and the damping decreases with increasing squeeze number.

In the third part, the basic characteristics of bearings under the influence of sliding have been determined. The load capacity is found to decrease and the mass flow rate to increase with increasing sliding number. The combined effect of these two factors reduces the stiffness of the bearing. It may lead, in extreme situations of very small film thickness or very high sliding speeds, to making the bearing statically unstable. Bearings with convergent gap geometry are less adversely prone to be affected by sliding than flat ones. The centre of force and the friction force are also determined showing also favourable behaviour of bearings with conicity over flat bearings.

Finally, knowledge of the totality of these bearing characteristics is essential when designing and optimising systems that employ aerostatic bearings of the circular, centrally-fed type. Furthermore, they are almost directly applicable to bearings that are square or oblong forms and they are fundamental, *mutatis mutandis*, in dealing with aerostatic bearings of other configurations, owing to their genericness. The dynamic axial bearing characteristics, which are also necessary in the design optimisation process, have not been considered in this chapter since this needs dedicated treatment. This is done in Chapter 7.

References

Al-Bender, F., (1992). *Contributions to the Design Theory of Circular Centrally Fed Aerostatic Bearings* PhD thesis Leuven.

Al-Bender, F. and Van Brussel, H., (1992). Tilt characteristics of circular centrally fed aerostatic bearings. *Tribology International* 25(3: 189–197.

Al-Bender, F., Vanherck, P. and Van Brussel, H., (1989). Autocentring scaife for diamond polishing machine. Report on development of prototype 89 R 22, KULeuven.

Ausman, J.S., (1960). Torque produced by misalignment of hydrodynamic, gas lubricated journal bearings. *ASME - Journal of Basic Engineering* 335–341.

Ausman, J.S., (1967). Gas squeeze film stiffness and damping torques on a circular disk oscillating about its diameter. *ASME - Journal of Lubrication Technology* 219–221.

Constantinescu, V.N., (1969). *Gas Lubrication.*

Dowson, D., (1961). Inertia effects in hydrostatic thrust bearings. *ASME - Journal of Basic Engineering* 227–234.

Dowson, D., (1979). *History of Tribology*. Longman, London.

Elrod, H.G. and Glanfield, G.A., (1971). Computer procedures for the design of flexibly mounted, externally pressurized gas lubricated bearings *Proceedings of the 5th Gas Bearing Symposium* number 22 University of Southampton. BHRA Fluid Engineering, Cranfield, U.K.

Gross, W.A., (1962). *Gas Film Lubrication*. John Wiley & Sons.

Kazimierski, Z. and Trojnarski, J., (1980). Investigations of externally pressurized gas bearings with different feeding systems. *ASME - Journal of Lubrication Technology* 102, 59–64.

Licht, L. and Kaul, R.K., (1964). Effect of misalignment on a circular, externally pressurized, gas-lubricated bearing. *ASME - Journal of Applied Mechanics* 141–143.

Na, T.Y., (1979). *Computational Methods in Engineering Boundary Value Problems*. Academic Press.

Nemoto, M. and Ono, S., (1981). Static characteristics of hybrid tilting pad gas journal bearings *Proceedings of the 8th Gas Bearing Symposium*, 199–132 number 8 Leicester Polytechnic. BHRA Fluid Engineering, Cranfield, U.K.

Nigam, S.D. and Turner, J.U., (1995). Review of statistical approaches to tolerance analysis. *Computer-Aided Design* 27(1): 6–15.

Pande, S.S., (1985). Analysis of tapered land aerostatic thrust bearing under the conditions of tilt and rotation. *Wear* 104(4): 297–308.

Pink, E.G. and Stout, K.J., (1981). Characteristics of orifice compensated hybrid journal bearings *Proceedings of the 8th Gas Bearing Symposium*, 29–44 number 3 Leicester Polytechnic. BHRA Fluid Engineering, Cranfield, U.K.

Safar, Z.S., (1981). Tilting effects on the performance of hydrostatic and hybrid gas thrust bearings. *Tribology International* 199–202.

Safar, Z.S., (1983). Centrifugal effects in misaligned hydrostatic thrust bearings. *ASME - Journal of Lubrication Technology* 621–624.

Schlichting, H., (1968). *Boundary-Layer Theory* 6th edn. McGraw-Hill, New York.

Shih, Y.D. and Yang, J.Y, (1990). Analysis of narrow rectangular aerostatic porous thrust gas bearings moving with a uniform transverse velocity. *Wear* 141, 125–136.

Snoeys, R. and Al-Bender, F., (1987). Development of improved externally pressurized gas bearings. *KSME Journal* 1(1), 81–88.

7

Dynamic Characteristics of Circular Centrally Fed Aerostatic Bearing Films, and the Problem of Pneumatic Stability

7.1 Introduction

The prediction of the static characteristics of air bearings represents the necessary first condition for a bearing design: it provides the general domain of feasibility of a bearing application. Such a domain is bounded by maximum allowable bearing number Λ_f and supply pressure ratio P_s.

Knowledge of the dynamic behaviour of such components is, however, needed to complete or modify this picture. This may be either to limit or to extend the domain of feasibility.

In the first instance, air bearings are prone to a type of self-excited vibration known as the "pneumatic hammer", the likelihood of which seems to increase the more one tries to enhance the static characteristics. In other words, dynamic stability represents one of the limits for bearing design optimisation. However, in understanding the problem of stability, we hope not only to be able to predict its likelihood, in order to avoid it, but eventually also to be able to overcome it by modifying either bearing or structure or both.

In the second instance, since the air film characteristics depend, among other things, on the supply pressure (or flow compensation), it may be possible, by manipulating the latter, to influence the dynamic stiffness of the air film, so as to compensate for the externally applied disturbances on the supported object, or even to suppress pneumatic hammering. This should represent an extension of the range of application of bearings.

A third aspect is the use of a fluid film as a *squeeze* bearing, damper, or both. Here, the fluid flow in the film is caused solely by the reciprocating motion of the bearing part(s). In this way, a load carrying capacity may be generated (when the oscillation amplitude is sufficiently large, see Chapter 15) and viscous damping may be introduced to an oscillating system.

Evidently, all of these aspects belong to (and are special cases of) the same problem, namely the dynamic behaviour of the air bearing film. Let us, therefore, first examine these aspects more closely. This chapter is for the greatest part based on (Al-Bender 1992).

7.1.1 Pneumatic Instability

This is a type of self-excited vibration which manifests itself as "pneumatic hammer" in aerostatic bearings, and as "half-speed" (or, more generally, sub-synchronous or asynchronous) whirling in journal bearings, both self-acting and hybrid, see Chapters 9 and 10. In the case of aerostatic bearings, this phenomenon finds its origin in the lubricant compressibility (Constantinescu 1969; Gross 1962) which should provide the possibility of storing energy in the lubricating film and releasing it in phase with the motion of the bearing system. This constitutes the essence of the phenomenon of *negative* damping.

Air Bearings: Theory, Design and Applications, First Edition. Farid Al-Bender.

Companion website: www.wiley.com/go/AlBender/AirBearings

It follows then that if an externally-pressurised film is either (a) *passive*, or (b) *incompressible*, no danger of instability can arise. A typical instance of (a) is the dynamic, small order, tilt behaviour analysed in Chapter 6: since the boundary pressures (and consequently also the excess mass flow rate) remain constant during tilt, no external energy is supplied to the system. In consequence, as it has been apparent (Chapter 6), the damping remained positive. (One must note, however, that although its boundary pressure is constant, a self-acting bearing is *not* passive, since external energy is supplied through film shearing caused by the relative motion of the bearing surfaces.) As for the second point (b), it has been observed (Gross 1962) that externally-pressurised liquid bearings do not, in general, have problems of stability. However, liquids are only relatively incompressible: their bulk modulus is very high as compared to gases. Moreover, liquid compressibility increases often due to the gases that get dissolved in them. (Also, the possible elastic deformation of the bearing surfaces, the supply lines, etc. must be taken into account as compressibility factor.) Consequently, the effect of compressibility on the dynamic behaviour of externally-pressurised bearings is not restricted to gas bearing applications, but may extend also to hydrostatic bearings, especially when those have large recess and supply line volumes. In this respect, the reader is referred to (Rohde and Ezzat 1967; San Andres 1991). In the remainder of this chapter, the treatment will be confined to compressible-fluid films.

7.1.2 Squeeze Film

This constitutes a special case of the passive bearing in which the (dynamic) flow is induced by the motion of the bearing surface(s) in a normal direction to the "plane" of the film. Confined mostly to liquid lubricants, such an arrangement can provide an important means of applying viscous damping to mechanical systems, and has thus been the subject of extensive research. In gas lubrication, on the other hand, squeeze film dampers are less effective in practice owing to the small viscosity of gases. Rather, this phenomenon has been used as a load generating mechanism: by virtue of the non-linearity of the flow problem, a net (or "D.C.", in the language of electricity) bearing force is generated when the amplitude of oscillation is sufficiently large. This will be treated in more detail in Chapter 15. Moreover, squeeze film solutions, which are easier to formulate and solve than the general case comprising pressure feed, are often used in constructing approximate (externally-pressurised) active-bearing solutions.

7.1.3 Active Compensation

Active compensation applies to all kinds of externally-pressurised as well as self-acting bearings. The problem has hitherto, however, attracted much less attention than the previous two. This may be attributed to the difficulty of realising an external compensation mechanism in practice. Thus, although *static* active load compensation dates back to the early 1960s (Wunsch 1964), the possibility of active *dynamic* compensation has been explored only as recently as the late 1980s (Sato and Harada 1988; Shimokohbe et al. 1986), through restrictor control and supply pressure control, respectively. Horikawa et al. (1991) successfully employed active displacement compensation, viz. through controlling the bearing mounting displacement. This, however, belongs rather to structure compensation than to film compensation. This area of research will be treated in more detail in Chapter 15.

7.1.4 Objectives and Layout of this Study

This chapter will be concerned with the construction of an adequate model for the axial dynamic behaviour of circular, centrally fed aerostatic bearings. In the following section, a short review of the previous treatments of the problem will be given, followed by an evaluation of special aspects pertaining to the adequacy of theoretical models. A linearised model is then proposed and its range of validity evaluated. Thereafter, typical theoretical results are given together with some comparisons with experiments. Design charts for unconditional stability are briefly sketched. Appropriate conclusions are finally drawn.

7.2 Review of Past Treatments

Interest in the subject of dynamic behaviour of fluid film bearings spans a long period of time; beginning in the early 1950s and continuing to the present day. Research involves, further, nearly all types and configurations of bearings: numerous papers, covering the various aspects of the problem are to be found in the open literature, and it would be not only ungainly but well-nigh impossible to list all of them.

Since the basic aim of this review is to situate, and provide the necessary background for the present work, it will, therefore, be confined to externally pressurised (mainly gas) bearings. For self acting bearings, the reader is referred to (Constantinescu 1969; Dai et al. 1992; Gross 1962; Marsh 1980; Tan and Mote 1990; White and Chan 1992) to mention but a few. But, even confined to our area of interest, the survey will include only the milestones; the more salient, representative or relevant works, to our purpose.

The survey will comprise the following topics:

- models for the dynamic bearing problem, and their underlying theory,
- system analysis tools and stability criteria,
- methods of stabilisation,
- discussion and evaluation.

In each topic, a chronological sequence will be adhered to as far as possible.

7.2.1 Models and Theory

Models may in general be divided into two types: empirical, and analytical. The former do not require full knowledge of the governing equations of the system, but rather the parameters influencing the behaviour. A relationship between those parameters is then sought experimentally (and obtained if the problem is well-posed, and the experiment is sufficiently accurate and comprehensive). Needless to say, dimensional analysis is imperative in this approach, otherwise the results cannot be generalised. A typical (and nearly singular) example of the empirical treatment is to be found in (Turnblade 1963) where a normalised reciprocal of the average lubricant velocity, termed "the molecular transit time" is correlated to a bearing Reynolds number, for a given bearing geometry. The method is amply discussed in (Blondeel 1975; Plessers 1985). Generally, if a problem is correctly normalised, an empirical solution can always be attempted. However, since experimental verification will, in any case, be needed, it would be more advantageous first to construct a theoretical model, and then test its validity. Therefore, in the rest of this section, only analytical models will be treated.

In discussing analytical models, let us also distinguish, right from the start, the basic methods of solution from the stability criteria proposed with them: these latter will be discussed in a separate section.

Models to be found in the literature may be divided into three main categories:

- lumped parameter methods: based on a pneumatic (quasi-static) model,
- distributed parameter methods: based on the time-dependent Reynolds equation,
- refined methods: including (or accounting for) inertia effects.

Lumped Parameter Models

These constitute the very earliest attempts to assess the stability of externally pressurised air bearings. Such a model has been presented in two papers, appearing in 1958. The first and most cited work is that of Licht et al. (1958); the second is by Richardson (1958). They differ only by the stability criterion they employ, and the fact that the second reference considers the case of inherent compensation in more detail.

The model is based on the assumption of quasi-static, linear, pressure distribution in the bearing coupled with small amplitude dynamic variation in the film thickness (and consequently also the other flow parameters). Consideration of the dynamic mass rate of flow balance through the bearing yields, in closed form, the dynamic stiffness in terms of the bearing's basic design parameters: geometry, pressures, and compensation flow characteristics.

As a testimony to its importance, it should be mentioned that this model has been re-presented and discussed in (Blondeel 1975; Constantinescu 1969; Gross 1962; Plessers 1985; Roblee 1985), to which the interested reader is referred. For our purpose, it will suffice to give the basic features and results of this model.

The lumped parameter model contains nearly all the essential aspects of the problem (though in an approximate way). The only aspect not included in the analysis is the squeeze-film dynamic stiffness of the viscous region. As such, it yields only a conservative estimate of the damping. The result obtained by this model shows that the dynamic stiffness of the bearing may be represented by a one-pole, one-zero system, i.e. either a phase-lead or a phase-lag. In other words, the damping is either always positive or always negative over the entire frequency range. Consequently, the stability of a mass-bearing system will be independent of the system's natural frequency (or of the *mass*, since that is one of the factors determining the natural frequency).

The factors influencing the nature of the dynamic stiffness (or its two time constants) are the flow sensitivities to film height and to inlet pressure variations, and the bearing film geometry; in particular the ratio of the recess volume (if any) to the viscous film volume. Richardson (1958) shows that inherently compensated bearings are always well damped. It also shows that measures taken to increase the static stiffness through adjusting the flow sensitivities (i.e. through flow compensation) lead inevitably to a reduction in the damping: the only "safe" way is achieved by increasing the inlet pressure or decreasing the film thickness.

One last remark about the lumped parameter method must be made in order to situate its underlying theory more precisely. Although the pressure is assumed linear along the viscous film, for simplicity, an exact pressure distribution would not have altered the qualitative nature of this method, nor indeed its results, as long as the pressure is assumed quasi-static. That is to say that this method is indirectly based on the quasi-static Reynolds equation (i.e. that in which the squeeze term is neglected), and on linearisation of the problem in the time domain by considering small amplitude perturbation from static equilibrium.

Distributed Parameter Methods

These include by far the greatest part of treatments of the problem to be found in the literature. Owing to the limitations of the lumped parameter method, it was natural for researchers to seek more refined models, accounting for the missing aspects in that model, in particular the squeeze film damping. In view of the stringent requirements for the achievement of stable bearing behaviour, every contribution, however small, had to be included. The distributed parameter model is then based on the time-dependent Reynolds equation. However, solutions differ in their degree of approximation. They may be divided into two classes:

1. perturbation solutions: based on time-linearisation of Reynolds' equation,
2. solutions including the amplitude effects, i.e. not linearised.

Perturbation Method

One of the oldest papers pertaining to this method is that by Licht and Elrod (1960). Having realised the shortcomings of the lumped parameter approach, they formulated a new model based on the Reynolds equation with small amplitude perturbation from equilibrium. An explicit solution of the problem was, however, no longer possible by analytical means. Instead, the authors attempted an implicit solution through the Laplace transformation of the perturbed Reynolds equation. They were able then to assess the influence of the various bearing parameters on stability. Their main results showed that the bearing mass played also a role in stability, (whereas it cancelled out in the lumped parameter approach). In other words, the dynamic stiffness of the bearing film is not the simple one-pole/one-zero, but rather, a higher order system, in which the damping is either positive over the whole frequency range; or it may be negative at low frequencies, but becomes positive at higher frequencies. Qualitatively, such a representation summarises an aerostatic film dynamic behaviour: subsequent research preoccupied itself with facilitation and refinement of this model. This included: devising methods of solution of the linearised Reynolds equation; refinement of the entrance flow model; and amplitude effects.

The first derivation of the time-dependent Reynolds equation appears to have been given by N. Tipei, in 1954, as cited in ref. Gross (1962). In a monumental paper, published in 1962, Langlois (1962) re-derived that equation with a careful consideration of the approximations made, giving also solutions to a number of elementary or limiting cases of the pure squeeze film problem. As has been given in Chapter 2 and 4, and what will later be indicated, the accuracy of the Reynolds equation is governed by the relative order of magnitude of the inertia terms, represented by the squeeze and the reduced Reynolds numbers.

Various methods of linearisation, and subsequent solution, of the Reynolds equation have been proposed to suit the different bearing configurations considered. (See, e.g., the relevant journals of the Transactions of the ASME, 1960–the present.) Some references will be cited, pertaining to thrust aerostatic bearings.

A method of approximate solution has been given, among others, by Lund (1969), for a uniform gap bearing. The approximation is effected through turning the coefficients of the linearised equation into constants, by taking their averaged values. The resulting equation has then an analytic, closed form, solution that seems to be sufficiently satisfactory. Considering both orifice compensated and inherently compensated bearings, Lund confirms that the latter type are always well damped. A finite difference solution was later given by Chiang and Pan (1969), for an inherently compensated bearing. Blondeel (1975); Blondeel et al. (1980) attempted to strike a match between the perturbation solution and Turnblade's (Turnblade 1963) idea of a "molecular transit time", obtaining questionable results. Still later, Holster et al. (1991) applied even a finite element procedure to solve the linearised Reynolds equation for a convergent gap bearing.

Amplitude Effects

In order to estimate the validity of linearisation of the Reynolds (and the feed flow) equations, the effect of large amplitude oscillations had to be considered. Only a few instances of this, however, are to be found in the literature, not only because of the difficulty of the problem, but also owing to the contradictory nature of the resulting model that will be indicated later.

Stiffler and Tapia (1979) considered amplitude effects on a plane parallel gap inherently compensated aerostatic bearing. They considered harmonic film thickness variation, and solved the resulting equations by finite difference techniques. Their main conclusions are that the stiffness is hardly affected by the amplitude, while the damping may be appreciably affected by large amplitudes: at any rate, linearisation, or even lumped parameter methods, would yield good results for an amplitude ratio (or film thickness variation) of 10 percent. Plessers (1985) Plessers and Snoeys (1988a) used a predictor–corrector method to solve the Reynolds equation for a circular convergent gap aerostatic bearing. Their conclusions regarding the validity of linearisation are similar to those stated above.

These results, in fact, only go to confirm the heuristically made linearisation assumption, viz. that it is valid for perturbations one order smaller than the perturbed variables. In other words, such solutions have a mathematical rather than a physical value. This is so because large amplitude oscillations would mean an increased influence of the fluid inertia forces that would render the Reynolds equation not valid. It is of little use, therefore, to consider amplitude effects outside the context of inertia (and possibly also thermal) effects. This remark has been also made in (Langlois 1962) with regard to the classical plane squeeze film problem.

Inertia Effects

As has been mentioned above, consideration of the effects of inertia entails in the first place derivation of models that incorporate these effects: the conventional Reynolds equation is no longer sufficient for this purpose. The model established in Chapters 3 and 5 is one instance of this. However, it is valid only for steady state (or quasi-static) flow. This is, however, sufficient for our present purpose as will be apparent later.

Two of the methods reviewed in Chapters 4 and 5, namely the integral, (or averaged inertia), method and the series expansion method, have found extensions for the dynamic case. Both of the methods are approximate, although in qualitatively different ways. The first, by far the most popular on account of its simplicity, has been used by Mori et al. (1980a,b) to derive a so-called modified Reynolds equation. This is, in effect, a set of three simultaneous non-linear differential equations in the pressure and its two derivatives in the bearing "plane". Without indicating a method of solution in the static case, the authors proceed to linearise that equation, in the time domain, in order to apply it to externally pressurised journal bearings. Making, at the same time, other assumptions regarding the feed flow, they show that inertia forces can have a significant effect on the dynamic performance of those bearings. Their results, which seem to agree fairly well with experimental data, show that the effect is most pronounced on the damping; the stiffness being hardly affected. This is especially so, when the supply pressure, the film thickness, the flow rate, and the feed hole diameters are large. They indicate also that the results are only valid for small vibration amplitudes and small squeeze numbers. This somewhat paradoxical result is attributed to the possibility of the local inertia effects being "magnified" by the convective inertia that is induced by pressure flow. Although neither sufficiently rigorous nor, indeed, easy to apply, this method indicates nevertheless the deficiency of the conventional models.

The second method, namely that of series expansion, has been applied only for the case of incompressible axisymmetric flows, namely those taking place between parallel flat discs. We recall that the velocities and the pressure were expressed as power series in $(1/r)$, in the steady state case. It so happens that the squeeze induced flow comes as a term in r, thus simply complementing the series. Naturally, the coefficients of the series are now functions of the time in addition to the transverse coordinate, which necessitates their expansion, in turn, by power series (in the time, or more conveniently, in the time derivatives of the film thickness).

One of the first applications of this method to the problem of pure squeeze between parallel flat discs appears to have been made by Ishizawa (1966). Several other papers followed including both theoretical and experimental evaluation of the results, see e.g. (Elkouh 1984); Jones and Wilson 1975; Tichy and Winer 1970; the last one applying the method to the case of flat annuli. Finally, and more recently, Ishizawa et al. (1987) and Wang et al. (1990) extended the method to the case of two parallel discs with a central fluid source, when one or both of the film thickness and the feed flow vary harmonically, i.e. the dynamic behaviour of a circular hydrostatic bearing. Their results compares, moreover, very favourably with experiment.

One of the essential advantages of this method is that it is, to a certain degree, analytical, thus providing a very suitable means of assessing and/or approximating the inertia effects. The parameters governing the problem are clearly brought to the fore and evaluated: these appear to be, beside the reduced Reynolds number of the main steady state flow Re^*, a set of squeeze Reynolds numbers, $h\dot{h}/(\mu/\rho)$, $h^3\ddot{h}/(\mu/\rho)^2$, etc., that will be referred to later. Furthermore, it seems to be possible, within a certain range of these parameters, to affect a superposition of the pure squeeze flow onto the quasi-steady flow (caused by external pressurisation). Although the method is not directly applicable to compressible flows, nor can it account for the entrance flow development, its results can provide a good overall idea about the problem.

7.2.2 System Analysis Tools and Stability Criteria

In its simplest guises, a bearing system may appear as a lubricating film supporting a pure mass load; this is indeed the most common application of a bearing. Consequently, the stability of such a system may be ascertained directly by writing the equation of motion and applying Routh's stability method, as has been done in the early analyses of Licht and Elrod (1960), Licht et al. (1958). It is possible, in this case, to determine the value of the *critical mass* beyond which the system becomes unstable. Other stability criteria have also been applied to such a system, e.g. by the response to a step jump (Elrod et al. 1967) and by Lyaponov's direct method (Gorez 1973).

However, such a representation is often inadequate in describing a bearing system, especially when the latter is more complex, e.g. involving more than one bearing, and possibly also additional external springs, dampers and masses. Therefore, rather than determining a critical mass (or such other condition), researchers have been more inclined, from the start, to determine the dynamic stiffness of the bearing film over the frequency range of interest, see e.g. (Lund 1968; Richardson 1958). This involves, of course, more effort in comparison with the previous method, since a range of data (rather than one point) has to be calculated. However, once this data is available, the dynamic behaviour of a given mechanical structure comprising the bearing film(s) may be readily determined.[1]

In particular, a bearing system may be represented as a servomechanism. This idea was perhaps first suggested by Wilcock (1967), who gave an extensive analysis of an arbitrary (compensated) bearing film on this basis, showing the hydrodynamic analog of the elementary electrical units. He concluded that much better insight into the dynamic behaviour can be gained by this method than by the conventional Routh criterion. Later, Blondeel (1975), Blondeel et al. (1980) employed this method to analyse the stability of an orifice fed, recessed aerostatic bearing. Three frequency response functions of the bearing were defined for this purpose: the squeeze film (or passive) stiffness; the film force response to inlet pressure variation; and the inlet pressure response to film thickness variation (i.e. the compensator characteristic). The product of the latter two functions represents the active stiffness. The bearing force was taken as input to the system; the film thickness as output. The direct loop consisted of the passive stiffness (to which may be added also other external passive elements); the feedback loop of the active stiffness. Stability could be ascertained by applying the Nyquist criterion. Plessers (1985), Plessers and Snoeys (1988b), with an eye to the practical application of air bearings in mechanical structures, adopted a global version of this representation: they simply set the structure's transfer function in the direct loop; and the bearing film(s) dynamic stiffness in the feedback loop. They were able then to analyse the stability of single as well as multiple degree of freedom systems, with the aid of Nyquist's criterion. The problem boiled down to the determination (theoretical and/or experimental) of the bearing film dynamic stiffness. In order to facilitate the subsequent analysis, they sought also to express the film dynamic stiffness, in a closed form, as a multi-pole/multi-zero system. This they did by means of curve-fitting the obtaining stiffness data, concluding that a two-pole/two-zero representation is an adequate approximation.

7.2.3 Methods of Stabilisation

In view of the difficulty of achieving damped bearing behaviour at relatively extreme loads and stiffnesses, a considerable part of research was preoccupied with the problem of stabilisation of bearing systems. Two approaches may be discerned: introduction of external damping, and introduction of internal damping.

The oldest and simplest approach was that of introducing damping to the structure, see e.g. (Boffey 1978; Boffey and Desai 1980; Kazimierski and Jarzecki 1979; Lund 1965). In nearly all cases, rubber bearing support elements were considered. Theoretically, it was feasible to achieve stability through the proper choice of the mass, stiffness and damping of the supports. In practice, however, considerable problems arose, owing to the difficulty of isolating the three aforementioned quantities. Nor was it indeed easy to measure the frequency-dependent, and often also load- and amplitude-dependent, dynamic stiffness of rubber elements. (In this respect, see (Boffey and Desai 1980; Devis 1985; Kazimierski and Jarzecki 1979) which also contain dynamic stiffness test results for various types of rubber "O" rings.) Another important drawback is the reduction of the bearing stiffness upon the addition of a flexible element in series with it. In short, although external (rubber) stabilisation may be an effective way of enhancing damped behaviour, it has also important shortcomings.

The second approach, not so popular on account of its difficulty, consists of introducing damping to the bearing film or to the flow (pneumatic) network. The damping element comprises a restrictor (orifice or capillary) connected to a capacitance (volume). It may be attached to the bearing recess, (i.e. in parallel with supply), in series with pressure supply, or at the bearing's exhaust. An instance of the first two cases is given in (Mori and Mori 1967), where it is shown, theoretically and experimentally, that an originally unstable bearing may be so stabilised. (However, an originally stable bearing may then become unstable.) In the third case, (see (Mori and

1 This excludes the possible transient behaviour.

Mori 1973)), since the stabilising element lies at the film exit, vacuum has to be applied to force the flow through it. Needless to say, such methods are less handy and more expensive to realise than the previous.

7.2.4 Discussion and Evaluation

First, in regard to the modelling of the dynamic characteristics, it appears that distributed parameter methods have to be opted for. A linearised viscous model may be quite adequate, if the oscillation amplitudes are small and the inertia effects are negligible. The latter may be assessed by the consideration of certain flow parameters (Reynolds numbers), as will be shown later. If only the stability of a bearing system is of interest, small amplitude perturbations would be quite adequate to analyse the problem: when the film is well damped for small amplitudes, there is no cause for small disturbances to grow in amplitude, i.e. self-excited vibrations cannot arise. When inertia effects are not negligible, a more complex model has to be considered: initially an inertial squeeze film solution may be superposed on the inertialess solution. We will show, however, that the linearised viscous model is sufficient for our purpose.

Second, assessment of the stability of bearing systems may be easily carried out once the dynamic characteristics of the structure and, more importantly, of the bearing film are known. The determination of the latter is of prime relevance to the design theory of air bearings; the stability analysis (of a bearing system) belonging rather to the field of structural dynamics. Therefore, the rest of this study will concentrate only on the dynamic characteristics of the bearing film (including the determination of conditions of unconditional stability).

Third, although methods of stabilisation have been suggested in the literature, they are not very practicable if standardised bearing components are envisaged. It would be more interesting to be able to design unconditionally stable, i.e. well damped, bearings (over a given performance range) than to try to stabilise them afterwards. The latter procedure may always be resorted to in extreme situations.

7.3 Formulation of the Linearised Model

The static bearing solution has been discussed in Chapters 5 and 6. We recall that the flow was divided into three distinct regions: feed; entrance; viscous; in order to facilitate its analysis. Each region yielded a boundary value problem of varying difficulty. In general, the temporal solution for each of these regions would be needed, in order that a complete dynamic solution be obtained. This latter would then be constructed by joining (matching) these solutions at the common boundary points. A comprehensive solution in the time domain (in which all the time dependent terms in the flow equations are retained) was considered, however, it is outside the scope of this work. Such an undertaking would, indeed, be necessarily involved to the extent that it would constitute a research topic on its own. Moreover, such a solution is not immediately necessary in order to understand the problem, given especially that the capability of experimental verification of dynamic behaviour is still lacking in many respects. It might thus be considered sufficient for our purpose to adopt a line of approximate analysis, namely that which is based on the linearised Reynolds equation. However, an essential feature of the proposed model, in comparison with previous ones, is the refined entrance solution, or the compensation model.

7.3.1 Basic Assumptions

The analysis will be based on the following assumptions:

- The flow is viscous throughout the film, i.e. the entrance section is ignored (since a temporal solution for it has not been envisaged yet). The load carrying capacity contribution of this section is usually negligible in comparison with the total value. However, on account of the large pressure gradients likely to take place in this section, it would be difficult to judge what its dynamic characteristics are likely to be. Instead, the refined nozzle feed flow formula will be used to correlate the inlet pressure to mass flow rate.

- Linearisation of the problem, in the time domain, through the use of the method of small perturbations from the steady state equilibrium, provides sufficient accuracy (for a range of oscillation amplitudes to be determined).
- The flow in the film is isothermal. This is consistent with the assumption of viscous flow: a polytropic assumption (with fixed exponent) does not present any added difficulty, but it will complicate the presentation unduly.
- The film remains nominally parallel (or axisymmetric), i.e. tilting is assumed negligible.
- There is no relative motion of the bearing surfaces in the plane of the bearing; or that its effect is negligible.

7.3.2 Basic Equations

In view of the previous assumptions, the film pressure will be given by the normalised, one-dimensional (axisymmetric), time-dependent Reynolds equation (see Chapter 2):

$$\frac{1}{R}\frac{\partial}{\partial R}\left(RH^3P\frac{\partial P}{\partial R}\right) = \sigma\frac{\partial}{\partial \tau}(PH) \tag{7.1}$$

where the variables have been normalised w.r.t. exit values $r_\mathrm{a}, h_\mathrm{a}, p_\mathrm{a}$ and the *perturbation angular frequency* ν; i.e.

$$R = r/r_\mathrm{a}, H = h/h_\mathrm{a}, P = p/p_\mathrm{a}, \tau = \nu t$$

and

$$\sigma = \frac{12\mu\nu}{p_a}\left(\frac{r_a}{h_a}\right)^2$$

is the squeeze number.

The boundary conditions are:

$$P(1,\tau) = 1 \tag{7.2}$$

and,

$$P(R_\mathrm{o},\tau) = P_\mathrm{o}(\tau) \qquad\qquad P_\mathrm{o} = p_\mathrm{o}/p_\mathrm{a}, \tag{7.3}$$

where P_o is obtained by equating the feed flow with the film flow.

The normalised film flow is given by:

$$\bar{m}_\mathrm{r} = -RH^3P\partial P/\partial R \tag{7.4}$$

Evaluated at R_o, this yields:

$$\bar{m}_\mathrm{r} = -R_\mathrm{o}H_\mathrm{o}{}^3P_\mathrm{o}\partial P_\mathrm{o}/\partial R \tag{7.5}$$

which is needed in obtaining the flow balance solution.

The feed flow is given by the nozzle formula:

$$\bar{m}_\mathrm{o} = C_\mathrm{d}\Lambda_f H_\mathrm{o}R_\mathrm{o}P_s\Phi_e(\bar{p}_\mathrm{o}) \tag{7.6}$$

in which,

$$\Lambda_f = 12\sqrt{\frac{2\kappa\Re T_\mathrm{s}}{\kappa-1}}\frac{\mu r_a}{h_a{}^2 p_a}, \tag{7.7}$$

$$C_d = C_d(\bar{p}_\mathrm{o}, \Lambda_\mathrm{e}, \nu_\mathrm{o}), \tag{7.8}$$

$$\bar{p}_\mathrm{o} = \frac{p_\mathrm{o}}{p_s} = \frac{P_\mathrm{o}}{P_s}, \tag{7.9}$$

$$\Lambda_e = \frac{\mu r_o}{h_o^2 p_s}\sqrt{\frac{\Re T_s}{\kappa}}, \tag{7.10}$$

$$v_o = \frac{h_v}{h_o}\frac{1}{(R_a - 1)}, \tag{7.11}$$

$$\Phi_e(\bar{p}_o) = [\bar{p}_o^{\,2/\kappa} - \bar{p}_o^{\,(\kappa+1)/\kappa)}]^{1/2}. \tag{7.12}$$

Finally, flow equilibrium is obtained by equating 7.5 with 7.6 to yield, after rearrangement:

$$\Lambda_f C_d P_s \Phi_e(\bar{p}_o) = -H_o^2 P_o \partial P_o/\partial R \tag{7.13}$$

which provides the required boundary condition on P_o.

7.3.3 The Perturbation Procedure

Denoting the static value of each variable, say X, by $X_{(0)}$, and assuming small order harmonic variation in the dynamic regime, we may write:

$$X(R,\tau) = X_{(0)}(R) + \Delta X(R)e^{j\tau} \tag{7.14}$$

where ΔX is complex valued in the general case.

Substituting this form into equation 7.1, cancelling the static terms out identically, and discarding second order terms (i.e. retaining only the terms in $e^{j\tau}$, we obtain:

$$\frac{1}{R}\frac{\mathrm{d}}{\mathrm{d}R}\left\{R\left[3H_{(0)}^{\ 2}P_{(0)}\frac{\mathrm{d}P_{(0)}}{\mathrm{d}R}\Delta H + H_{(0)}^{\ 3}\left(\frac{\mathrm{d}P_{(0)}}{\mathrm{d}R}\Delta P + P_{(0)}\frac{\mathrm{d}\Delta P}{\mathrm{d}R}\right)\right]\right\} = j\sigma(P_{(0)}\Delta H + H_{(0)}\Delta P) \tag{7.15}$$

(Note that, due to the assumed rigidity of the bearing surfaces and the absence of tilt, ΔH will be independent of R.)

Boundary Conditions:

These follow from applying the perturbation forms to equations 7.2 and 7.13 respectively.

First, by virtue of the fact that $P_{(0)}(1) = 1$, Eq. 7.2 will give:

$$\Delta P(1) = 0 \tag{7.16}$$

Second, by applying the perturbation procedure to Eq. 7.13 as before, we obtain for the general case:

$$A\Delta P_o + Bd(\Delta P_o)/\mathrm{d}R + C\Delta H + D\Delta P_s = 0 \tag{7.17}$$

where the coefficients are given by:

$$A = \frac{1}{P_o}\left\{1 - \left(\frac{1}{C_d}\frac{\partial C_d}{\partial \bar{p}_o} + \frac{\Phi_e'}{\Phi_e}\right)\frac{P_o}{P_s}\right\}$$

$$B = \frac{1}{P_o'}$$

$$C = \frac{1}{H_o}\left\{2 + \frac{2\Lambda_e}{C_d}\frac{\partial C_d}{\partial \Lambda_e} + \frac{v_o}{C_d}\frac{\partial C_d}{\partial v_o}\right\}$$

$$D = \frac{1}{P_s}\left\{-1 + \frac{\Lambda_e}{C_d}\frac{\partial C_d}{\partial \Lambda_e} + \left(\frac{1}{C_d}\frac{\partial C_d}{\partial \bar{p}_o} + \frac{\Phi_e'}{\Phi_e}\right)\frac{P_o}{P_s}\right\}$$

(where the primes have been used, for simplicity, to denote derivation of the functions w.r.t. their respective variables).

In summary, the linearised model consists of Eq. 7.15 with the boundary conditions 7.16 and 7.17.

7.3.4 Range of Validity of the Proposed Model

In the first instance, since the steady state model represents an initial condition to the dynamic model, it follows that the range of validity of the latter lies within the range of validity of the former, i.e. the proposed dynamic model cannot, in general, be valid if the assumptions made in the static case are violated. The most important of these assumptions is that the bearing film be, in its greatest part, viscous, or that the entrance region be of a negligible area.

Now, let us assume that the steady state assumptions are fulfilled and ask what extra constraints have to be imposed in the dynamic case?

We assume, as before, harmonic variation of the film thickness h, i.e.

$$h(t) = h_{(0)}(1 + \epsilon \sin \nu t)$$

where $\epsilon \sim \Delta H$ is the relative amplitude of oscillation.

The first condition of validity is that of the linearisation itself, namely that,

$$\epsilon^2 \ll 1 \qquad \text{Condition 1} \tag{7.18}$$

That means that the oscillation amplitude may be allowed to take values up to approximately $\epsilon = 0.3$ or 30% of the nominal gap value. Obviously, the smaller ϵ is, the better is the approximation. Thus, we should expect excellent results when $\epsilon \sim O(10^{-1})$, if the other conditions below are also fulfilled.

The second condition has to do with the validity of the Reynolds equation, i.e. the orders of magnitude of the inertia terms in the equations of motion. Let us, for this purpose, recall the continuity and the momentum equations (see Chapter 2), where we have underlined the temporal and inertia terms:

$$\underline{\epsilon \nu^* \frac{\partial \wp}{\partial \tau}} + \frac{1}{R}\frac{\partial}{\partial R}(\wp R U) + \frac{\partial}{\partial Z}(\wp W) = 0, \tag{7.19}$$

$$\wp\left(\underline{\epsilon \mathrm{Re}_\nu \frac{\partial U}{\partial \tau}} + \underline{\mathrm{Re}^* \left[U\frac{\partial U}{\partial R} + W\frac{\partial U}{\partial Z} \right]} \right) = -\frac{\mathrm{Re}^*}{\kappa \mathrm{M}_0{}^2}\frac{\partial P}{\partial R} + \frac{\partial^2 U}{\partial Z^2}. \tag{7.20}$$

where the normalisation is assumed to be w.r.t. general reference dimensions, velocity, indicated by (*) and time, w.r.t. $1/\nu$:

$$(R, Z, \tau, U, W, \wp) = (r/r^*, z/h^*, \epsilon \nu t, u/u^*, (r^*/h^*)w/u^*, \rho/\rho^*),$$

ν^* is the normalised squeeze pulsation, Re_ν is the squeeze Reynolds number, (cf. Ω^*, Re_{sq} in section 4.9.5) and Re^* is the reduced Reynolds number; see formal definitions further below.

In this way, the normalised variables are all of unit order of magnitude. (Note that $|\dot{h}/h| = \epsilon\nu$.)

The evaluation is concerned with the underlined terms in these equations which are thus of relative order of magnitude equal to their respective coefficients:

$$\epsilon \nu^* = \epsilon \nu r^*/u^*, \quad \epsilon \mathrm{Re}_\nu = \epsilon \rho^* h^{*2} \nu/\mu, \quad \mathrm{Re}^* = \rho^* u^* h^{*2}/\mu r^* .$$

Note, particularly, that

$$\mathrm{Re}_\nu/\mathrm{Re}^* = \nu^* \tag{7.21}$$

We proceed to evaluate the two basic idealising assumptions of the proposed model. Here, we extend the analysis of section 4.9.5 in a more systematic way by involving also the temporal inertia term of the momentum equation (being proportional to $\epsilon \mathrm{Re}_\nu$), which has not been considered before. We shall show that this term is negligible except in very extreme situations.

Let us recall that the film has been subdivided into (i) the entrance region in which the convective inertia term (being proportional to Re^*) is not negligible and (ii) the viscous region in which that term is neglected, which is a condition for the validity of the Reynolds equation there.

Entrance Region

Here, we have used the quasi-static entrance-flow model, or the inherent-restrictor model, in which the first term of each of Eqs. 7.19 and 7.19 is neglected. We would like to ascertain under which conditions this is justifiable. Here, the degree of approximation will depend on the relative order of magnitude of the neglected terms, namely ϵv^* in both equations (in Eq. 7.19 by virtue of relationship 7.21). We may then assume that the quasi-static entrance solution is valid when,

$$\epsilon v^* = \epsilon v r^*/u^* \ll 1 \qquad \text{Condition 2} \qquad (7.22)$$

As an example, if we assume that, in the vicinity of the feed hole, we have $r^* = 2$ mm, $u^* = 20$ m s^{-1}, $\epsilon = 0.1$, then for oscillation frequencies up to $v = 10^4$ rad s^{-1} ($\sim 10^3$ Hz), the model would be free from significant errors.

Viscous Region

Here, the first term of the continuity equation is retained, which leads to the squeeze term in the Reynolds equation, but the two inertia terms (temporal and convective) of the momentum equation are both neglected. In this case, since the convective inertia term must be neglected, we need only consider the influence of the first term. We may, for this purpose, make use of the results of the references cited in section 7.2.1. Those results show that the error committed by the use of Reynolds equation depends on the the following set of squeeze Reynolds numbers:

$$\mathrm{Re}_1 = h\dot{h}/(\mu/\rho), \qquad \mathrm{Re}_2 = h^3\ddot{h}/(\mu/\rho)^2, \ldots \mathrm{Re}_n = h^{(2n-1)}(\mathrm{d}^n h/\mathrm{d}t^n)/(\mu/\rho)^n$$

(where (μ/ρ) is the kinematic viscosity).

The error is equal, in the first approximation, to the order of magnitude of the first Reynolds number; and thereafter on the relative order of magnitude of the higher order numbers. In other words, if the error is to remain of second order, we must have:

$$\mathrm{Re}_1 \ll 1, \qquad \mathrm{Re}_{n+1}/\mathrm{Re}_n \leqslant 1, \quad n > 0.$$

The first Reynolds number Re_1 is equal to $\epsilon\mathrm{Re}_v$, while

$$|\mathrm{d}^n h/\mathrm{d}t^n| = \epsilon h_{(0)} v^n,$$

so that

$$|\mathrm{Re}_{n+1}/\mathrm{Re}_n| = \mathrm{Re}_v, \quad \text{for all } n$$

Thus, the above condition becomes:

$$\epsilon\mathrm{Re}_v \ll 1, \qquad \mathrm{Re}_v \leqslant 1 \qquad \text{Condition 3} \qquad (7.23)$$

This last result provides a sufficient condition for the accuracy of the Reynolds equation. As an example, let us assume the lubricating fluid to be air which, at some point in the film, has a density $\rho = 3$ kg m^{-13}, $\mu = 1.82 \times 10^{-5}$ N sm^{-12}, film thickness $h_{(0)} = 30$ μm, oscillation amplitude $\epsilon = 0.1$, oscillation frequency $f = v/2\pi = 10^3$ Hz, then

$$\mathrm{Re}_v \simeq 1, \qquad \epsilon\mathrm{Re}_v \simeq 0.1$$

and the inertia effects are negligible. If the film thickness is doubled, however, then the errors might remain small but they are not quite negligible. The same thing may be said about doubling the oscillation amplitude.

In summary, the proposed model is expected to be sufficiently accurate if the three deduced conditions, Eqs. 7.18, 7.22 and 7.23 are simultaneously satisfied. The most crucial of these conditions is the first one, namely that the oscillation amplitude should be less than about 30% of the nominal gap height. It may be further concluded, therefore, that the model would be adequate for most aerostatic bearing applications, and over reasonably broad frequency ranges, as testified by the examples given above. In any case, a means of ascertaining accuracy has been provided.

N.B. Similar criteria, that are not reported here, may show that the isothermal assumption is likewise adequate in characterising the problem in the range of conditions considered.

7.3.5 Special and Limiting Cases

As has been mentioned above, the general problem of dynamic behaviour consists in equation 7.15 with boundary conditions 7.16 and 7.17. Such a formulation corresponds to the case of harmonically varying film thickness and supply pressure, relevant to the problem of active compensation for stability, that has scarcely been treated in the literature. Although its solution should not present any difficulty, the practical importance of this problem depends on the feasibility of supply pressure control, which is hitherto still problematic.

The case most commonly considered is that pertaining to stability at constant supply pressure. However, in order to physically understand the various aspects of the problem, it is often convenient to consider the following additional special cases.

The Squeeze Film Problem

The squeeze film problem is that in which the pressure at the film inlet is assumed constant. Thus, the second boundary condition 7.17 is replaced by:

$$\Delta P_o = 0 \tag{7.24}$$

In this case, as has been mentioned earlier, the film is passive, since no energy is input to it during oscillation. Consequently, the damping is always positive, and it is of interest to gauge its order of magnitude for a given film configuration. The most important feature of this problem is that it requires no feed-flow (or compensator) model.

This case may be seen also as a limiting case of the problem of stability, at constant supply pressure, as the film thickness approaches zero. In this case, since the inlet pressure ratio $\bar{p}_\mathrm{o}$ will approach unity, the function Φ_e will approach zero; and $\Phi_\mathrm{e}'(= \mathrm{d}\Phi_\mathrm{e}/\mathrm{d}\bar{p}_\mathrm{o})$ infinity. Thus, in Eq. 7.17, A will approach infinity, while B and C remain bounded (and $D = 0$), which reduces that equation to Eq. 7.24. In other words, we are able to deduce that the damping will be positive for this case without having to solve the problem.

Active Compensation: Response to Supply Pressure

We assume here that the film thickness remains constant in time, while the supply pressure, and consequently also the bearing force, vary harmonically. In this case, we set $\Delta H \equiv 0$ in Eq. 7.15 and boundary condition 7.17.

Although, in the actual problem of active compensation, the supply pressure would be controlled to follow the external disturbance (force), with the film thickness variation ΔH representing the error function, consideration of the above situation, i.e. where $\Delta H \equiv 0$ may serve to show the feasibility of achieving this.

Here also, we may deduce from physical considerations that the bearing force will always lag behind the supply pressure. This is mainly owing to the compressibility of the fluid, so that the volume of gas in the film will have the effect of a (nearly) first-order delay. The time constant t_c is proportional to the squeeze number divided by the perturbation frequency:

$$t_\mathrm{c} \propto \frac{12\mu}{p_a}\left(\frac{r_a}{h_a}\right)^2$$

which will increase with increasing viscosity and bearing size, and decreasing gap height and ambient pressure.

7.4 Solution

7.4.1 Integration of the Linearised Reynolds Equation

Equation 7.15 is a linear differential equation with variable coefficients that depend on the steady state pressure distribution, geometry and squeeze number. The coefficients of its second boundary condition 7.17 are also determined from the static solution. Once this latter is known, therefore, the problem may be solved by integration.

Since the problem and its boundary conditions are linear, a solution can be obtained through forward numerical integration using the method of superposition (Na 1979). The method consist in finding two integrals subject to certain prescribed initial conditions and constructing a linear combination of them (superposition) in such way that both boundary conditions are satisfied. This makes it possible to use forward integration techniques which are both efficient and easy to program. A fourth order Runge–Kutta method has been used throughout for obtaining the solutions presented further below.

Another advantage of the superposition method, in comparison, e.g., to finite difference, or finite element methods, is its flexibility: only one of the two boundary conditions is initially needed to find the general solution. Although this flexibility was not utilised in our work, since only solutions for our particular bearing configuration were sought, it may prove very handy in studying the influence of the compensation function on the dynamic behaviour. This can be done in the following manner.

If the film geometry, and inlet pressure P_o are given, the static solution can be found. Consequently, Eq. 7.15 is obtained with a single boundary condition, viz. at $R = 1$ (Eq. 7.16). This problem can be solved by the method of superposition starting at $R = 1$ to yield a general solution that depends on one undetermined parameter. We can then pose the question as to what compensation characteristic (boundary condition at $R = R_o$ or Eq. 7.17) is necessary in order that the dynamic stiffness should have a certain desired value (or range of values). In particular, if the compensation model is given, as in our case, we can ask, e.g., beyond what value of the supply pressure would the bearing develop negative damping?

7.4.2 Bearing Dynamic Characteristics

The basic characteristic that we are concerned with is the dynamic stiffness (at constant supply pressure), since it is this that determines stability of a bearing in a mechanical structure. In the second place, the bearing force response to supply pressure variation will also be considered.

The Dynamic Stiffness

As defined in Chapter 2, this is given by:

$$K_a(\sigma) = -\left(\frac{\partial F}{\partial H}\right)_{\Delta P_s=0} = \int_{\bar{A}} \left(\frac{\Delta P}{\Delta H}\right)_{\Delta P_s=0} d\bar{A}$$

Since the Reynolds equation has been linearised, the solution for this case may be obtained by assigning ΔH any arbitrary finite non-vanishing value. For simplicity of analysis, we may choose this value to be real (say unity). In that case, the dynamic stiffness, which is generally complex valued, will have a real part (thus in phase with ΔH) which corresponds to the spring stiffness; and an imaginary part, corresponding to the damping force. We may thus write:

$$K_a = K + j\sigma C$$

where K is the dimensionless spring stiffness, and C is the dimensionless damping coefficient, (σC is the dimensionless damping force). Note that C usually attains its maximum absolute value as σ tends to zero, and therefore care has to be exercised in calculating the damping at this limit. (One may, e.g., substitute a similar form for the dynamic pressure in the linearised Reynolds equation, in order to split it into two equations; for the stiffness and the damping pressures respectively. Alternatively, and more simply, a small value may be assigned to σ in the solution.)

The Supply Pressure Response

This is defined as

$$K_p = \left(\frac{\partial F}{\partial P_s}\right)_{\Delta H=0} = \int_{\bar{A}} \left(\frac{\Delta P}{\Delta P_s}\right)_{\Delta H=0} d\bar{A}$$

It may be split, as before, into in-phase and out-of-phase components.

Other Characteristics

The passive (or squeeze film) stiffness, and the response to the inlet pressure may be similarly defined and calculated.

7.5 Results and Discussion

7.5.1 General Characteristics and Similitude

We have seen that the solution of the dynamic problem requires the static solution as initial value. In addition to that, the squeeze number σ determines the solution completely. In mathematical form, any one of the dynamic characteristics given above may be expressed as a function:

$$K_a(\sigma) = f(R_o, H_v, P_s, \Lambda_f;\ \sigma)$$

where the first four parameters belong to the static solution.

This relation summarises the conditions of similitude of bearing films in the dynamic regime:

two films are dynamically similar *at a given squeeze number* if they are statically similar.

This results is, of course, only true by virtue of linearisation. When linearisation is not valid, the amplitude of oscillation has to be added to this condition. Note also that this condition is not, in general, exclusive, i.e. it is sufficient but not always necessary, unless we consider similarity over a wide range of σ values, when it would be unlikely for two statically not similar bearings to yield the same dynamic characteristics.

As has been emphasised in (Al-Bender 1991), knowledge of the similitude parameters is essential for carrying out effective tests; in particular for scaling the test model and deciding the frequency range.

On the other hand, the fact that the dynamic characteristics are functions of five distinct parameters makes the presentation of general results in compact form next to impossible; the more so since the characteristics themselves are not single valued (i.e. scalar) functions. Since the problem of the actual design optimisation of bearing systems is beyond the scope of this work, the presentation will be confined to the general trends of behaviour, and some comparisons with available experimental results. Of main concern, here too, will be the dynamic stiffness (at constant supply pressure), since it is this that determines design optimisation with regard to stability.

The Dynamic Stiffness K_a

First, *typical dynamic pressure distributions* are given. In order to facilitate presentation, the dynamic pressure will be expressed in the following form

$$-\Delta P/\Delta H = P_k + j\sigma P_c$$

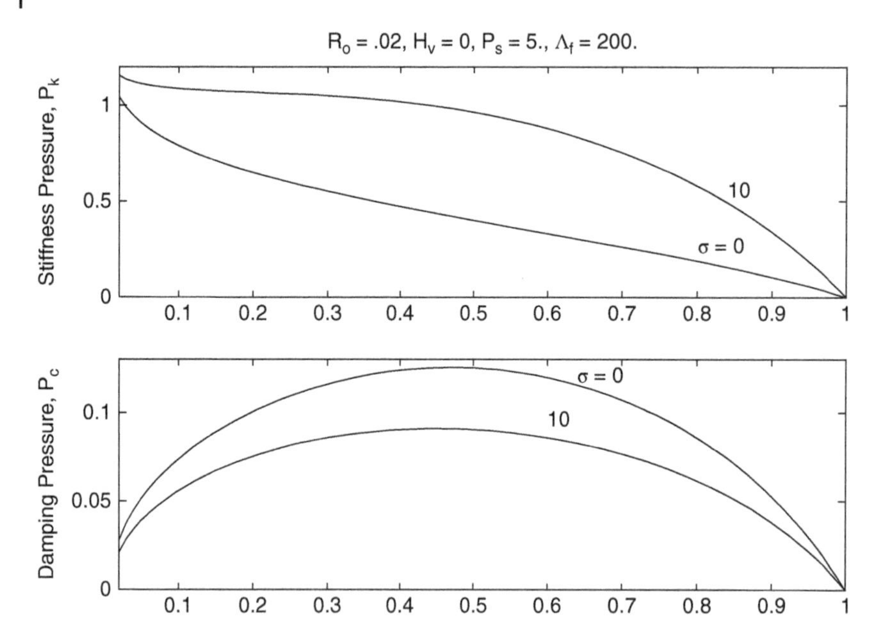

Figure 7.1 Dynamic pressure distributions pertaining to stiffness P_k and damping P_c at small squeeze numbers. $R_o = 0.02, H_v = 0, P_s = 5, \Lambda_f = 200$. Source: Al-Bender F 1992.

where P_k and P_c represent the components of the dynamic pressure responsible for the stiffness and the damping coefficient respectively.

Starting with a uniform gap bearing, i.e. $H_v = 0$, *the influence of the squeeze number* on the *pressure distribution* is considered. Figure 7.1 shows the pressure distributions for small values of the squeeze number σ.

At $\sigma = 0$, the P_k distribution has nearly the same shape as the static pressure $P_{(0)}$, whereas P_c has a "bow" shape, with the maximum somewhere in the middle radius range. As σ increases, the initial value of P_k increases and its shape changes with the increasing slope at the boundaries; the initial value of P_c decreases progressively together with its maximum value, and it flattens out in the middle. Figure 7.2 shows the pressure distributions for large values of the squeeze number σ. At those very large values, the pressure distributions approaches the so-called "trapped gas" case, i.e. where the film gas does not have time to flow any more, but is rather purely compressed by the changing volume caused by gap oscillation. Disregarding the "mathematical boundary-layers" at entrance and exit, the solution (inside the film) for this case may be obtained, by singular perturbation, simply by setting the right-hand side of equation 7.15 to zero. Thus, one obtains (the algebraic equation):

$$(P_{(0)}\Delta H + H_{(0)}\Delta P) = 0, \qquad \sigma \to \infty, \ R_o < R < 1.$$

which has the solution:

$$-\Delta P/\Delta H = P_{(0)}/H_{(0)} \tag{7.25}$$

(cf. $dp/p = -dv/v$: Boyle's law, see also (Langlois 1962)).

This solution is shown plotted by a *dotted* line in Figure 7.2. Note that since $H_{(0)} \equiv 1$ for this case, $(P_k)_{\sigma\to\infty}$ is identical to $P_{(0)}$. On the other hand, P_c will tend to zero. Note also how the solution for $\sigma = 1000$ agrees with this limiting case on the domain $0.2 < R < 0.9$. In conclusion, the film acts in this case as a pure spring. Behaviour for other bearing-film geometries is qualitatively similar, except for the case when the damping at $\sigma = 0$ is negative,

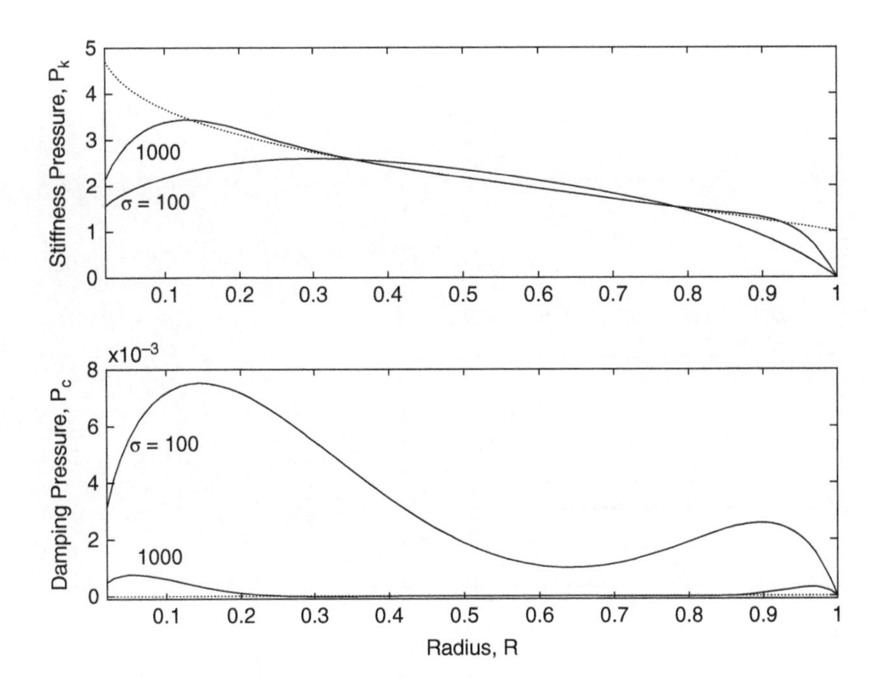

Figure 7.2 Dynamic pressure distributions pertaining to stiffness P_k and damping P_c at large squeeze numbers. $R_o = 0.02, H_v = 0, P_s = 5, \Lambda_f = 200$. Source: Al-Bender F 1992.

in which case the stiffness will decrease and the damping will increase up to the point where it becomes positive. From then on the stiffness begins to increase, and the damping to decrease, as before.

The influence of the conicity (being the most important geometrical aspect) *on the pressure distribution*, at $\sigma = 0$, is shown in Figure 7.3. Note how the mean value of P_k increases, and that of P_c decreases with increasing conicity: for this particular set of design parameters, P_c is already negative when H_v reaches the value of 2. The following set of theoretical results will be concerned with the influence of the four bearing static design parameters on the stiffness and the damping coefficient in the range $0 < \sigma < 1000$. In each instance, three of the said parameters will be held constant, at arbitrary (typical) values, while the fourth is discretely varied.

The behaviour of the dynamic stiffness is now considered.

Let us, for clarity, recall that:

$$K_{\text{dyn}} = K(\sigma) + j\sigma C(\sigma),$$

where,

$$K = \frac{kh_a}{p_a \pi r_a^2}$$

and

$$C = \frac{c}{12\pi\mu r_a}\left(\frac{h_a}{r_a}\right)^3,$$

from which the dimensional values:

$$k_{\text{dyn}} = k(\nu) + j\nu c(\nu)$$

can be easily retrieved.

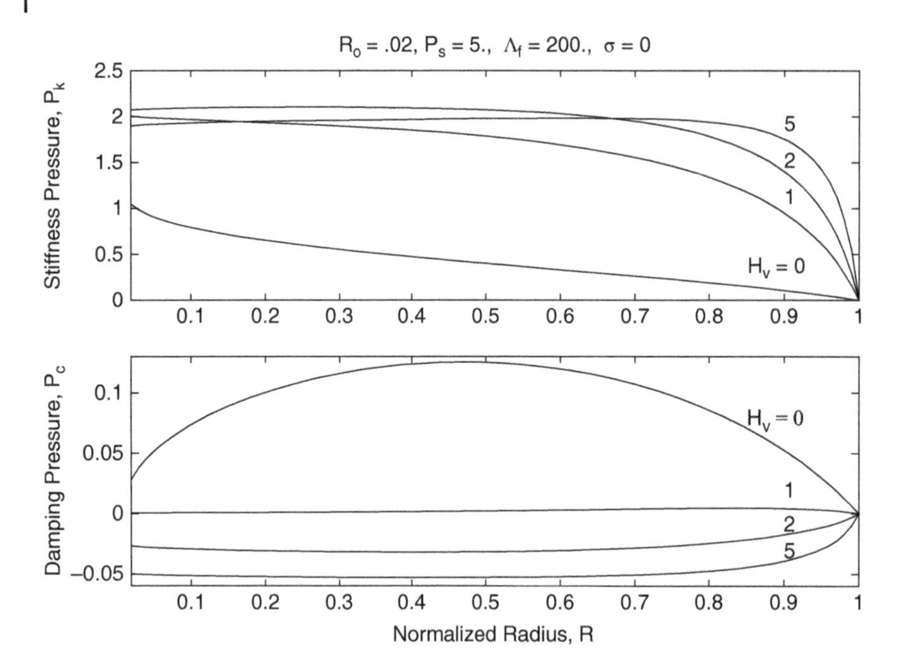

Figure 7.3 Influence of the conicity H_v on the dynamic pressure distribution, at $\sigma = 0$. Source: Al-Bender F 1992.

The *influence of the conicity* on the dynamic stiffness is shown in Figure 7.4. We note that increasing conicity increases the static stiffness, and decreases the static damping; both very strongly: an increase oh H_v from 0 to 1 causes the former to increase, and the latter to decrease, by nearly one order of magnitude. When $H_v = 2.$ the static damping becomes negative. If the conicity is further increased, this behaviour will be further enhanced, (i.e. the static stiffness will further increase; and the damping further decrease). Physically, this has to do with the increasing film volume, and film pressure with increasing conicity. (The stiffness will, however, approach a limiting value at very high conicity.)

Figure 7.5 shows the **influence of the feed hole radius** on the dynamic stiffness. An essential difference between this and the previous figure is that the dynamic stiffness does not vary monotonically with R_o (as it did with H_v). The static damping, e.g., which is positive at $R_o = .01$, first becomes negative and then positive with increasing R_o; i.e. it has a local minimum. The same behaviour holds true for the stiffness, except that it has a local maximum. (These two points do not, in general coincide, but they lie very close to one another.) Furthermore, there seems to be only one extremum point: on either side of that point, the stiffness (or the damping) will then remain monotonic. The physical explanation of this result lies in the influence of the feed hole radius on the flow sensitivity to film thickness variation: the higher the radius, the less is that sensitivity; until the film becomes nearly passive at very high R_o. However, when the radius becomes too small, the film will become "starved", i.e. the inlet pressure ratio $\bar{p}_0$ $(= P_o/P_s)$ will become too small; likewise will be the load capacity and the stiffness.

Figure 7.6 shows the **influence of the supply pressure** P_s. Here, as has been the case with the conicity, the stiffness and the damping vary monotonically with P_s; the stiffness increasing and the damping decreasing, with increasing P_s. This may be attributed simply to the increasing energy supply to the system. (Note that in this example, the extreme values of $H_v = 2$ and the feed number $\Lambda_f = 200$ have been especially chosen to accentuate the influence of the supply pressure on the dynamic stiffness.)

Finally, Figure 7.7 shows the *influence of the bearing number* Λ_f. The behaviour is similar to that obtained for the influence of R_o, and so is the physical explanation. This can also be viewed as the influence of the absolute

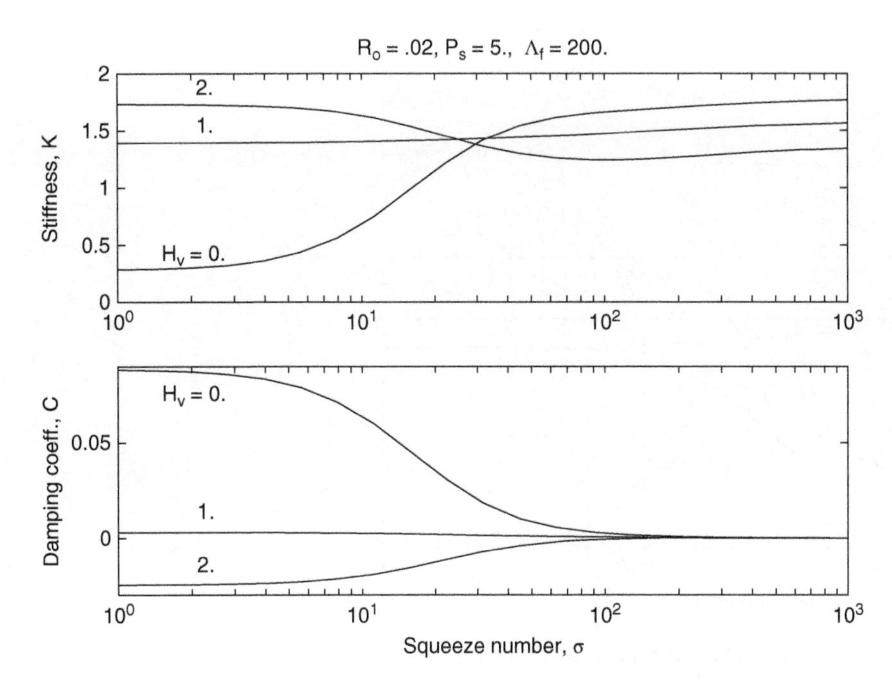

Figure 7.4 Influence of the conicity H_{υ} on the dynamic stiffness. Source: Al-Bender F 1992.

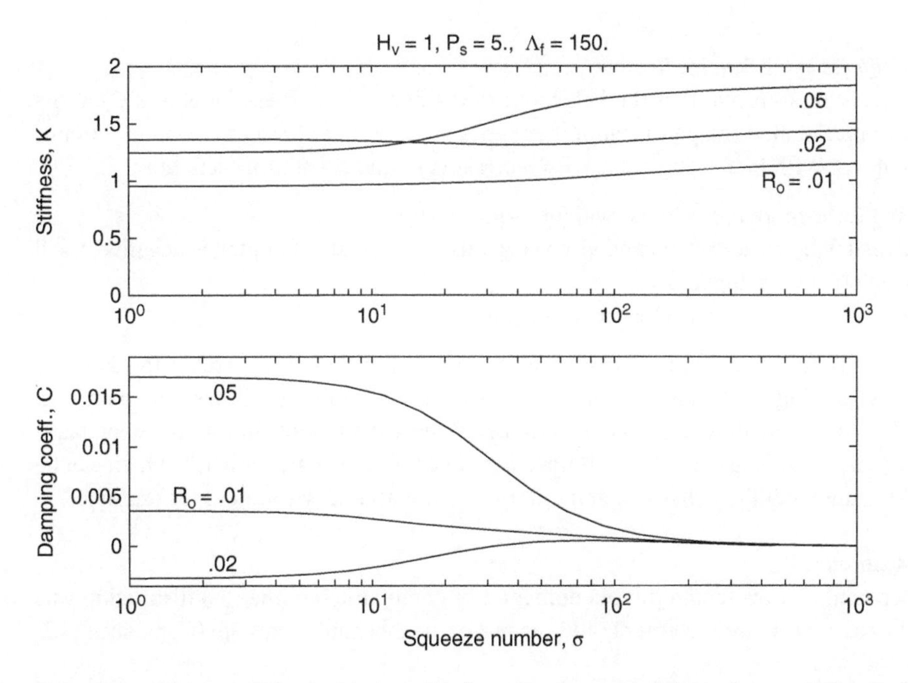

Figure 7.5 Influence of the feed-hole radius R_o on the dynamic stiffness. Source: Al-Bender F 1992.

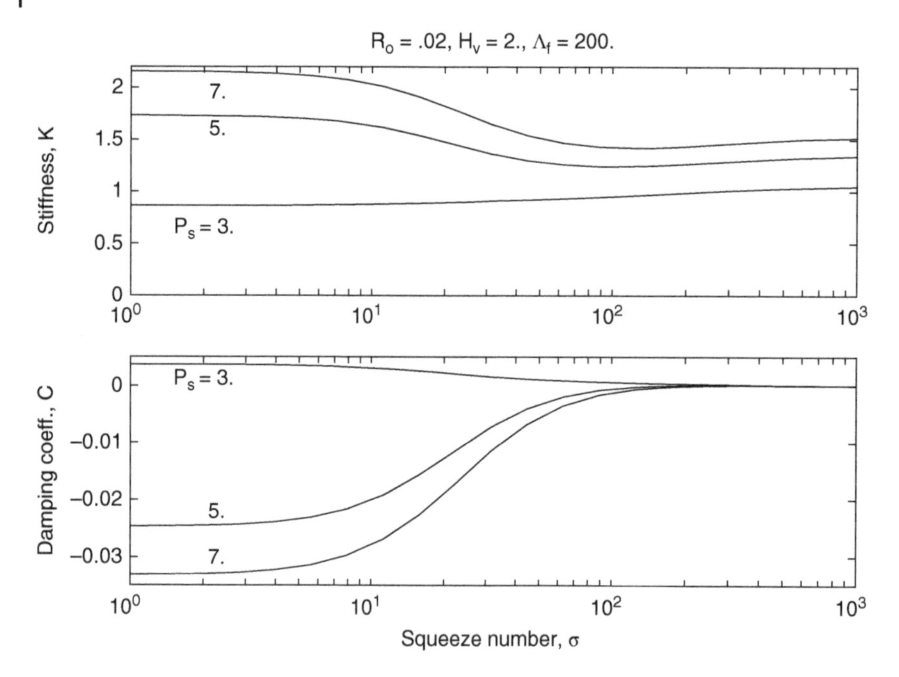

Figure 7.6 Influence of the supply pressure P_s on the dynamic stiffness. Source: Al-Bender F 1992.

film thickness, since Λ_f is inversely proportional to h_a^2 (at constant r_a). It can be seen then that negative damping is a middle-range-film-thickness phenomenon, i.e. when the film thickness is too small or too large, the damping will be positive.

General Trends of the Dynamic Stiffness in the Squeeze Number

Examination of Figures 7.4–7.7, is apt to reveal that the behaviour of the dynamic stiffness (or K and C) in the squeeze number σ, which is proportional to the perturbation frequency ν, seems to follow certain general trends that may be summarised as follows. With increasing σ, i.e. with increasing frequency of film excitation:

- the higher the static damping is the more the stiffness will increase,
- when the damping is positive, it will be decreasing, and vice versa, (also, if the static damping is negative it will increase to a positive value before decreasing to zero),
- when the damping is increasing, the stiffness will be decreasing and vice versa.

These general trends, which are sketched qualitatively in Figure 7.8 are very useful, since they relate the dynamic behaviour at higher frequencies (although only qualitatively) with the static values of the stiffness and damping. Moreover, these trends have been confirmed by a reasonable number of test calculations. In the following paragraph, we shall convert these trends into (sub)models for lumped dynamics system characteristics, which can be useful to derive quantitative dynamic stability behaviour and criteria in simulation environment for example.

Dynamic Stiffness Models and Application

In the early attempts to understand and overcome the phenomenon of pneumatic hammer, no distinction was made between the bearing film (as a mechanical element) and the rest of the mechanical system (Constantinescu

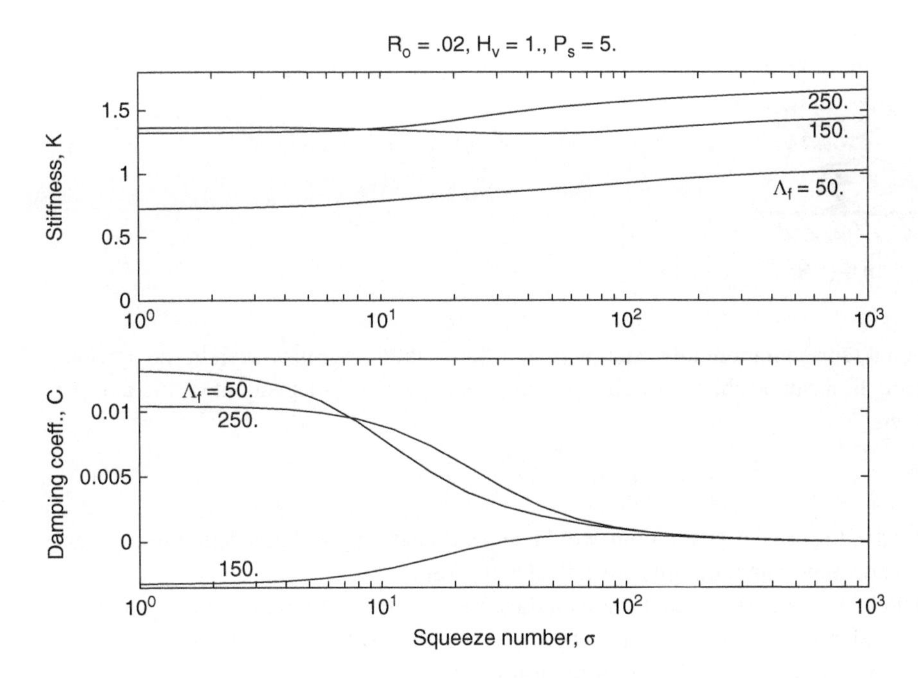

Figure 7.7 Influence of the bearing number Λ_f on the dynamic stiffness. Source: Al-Bender F 1992.

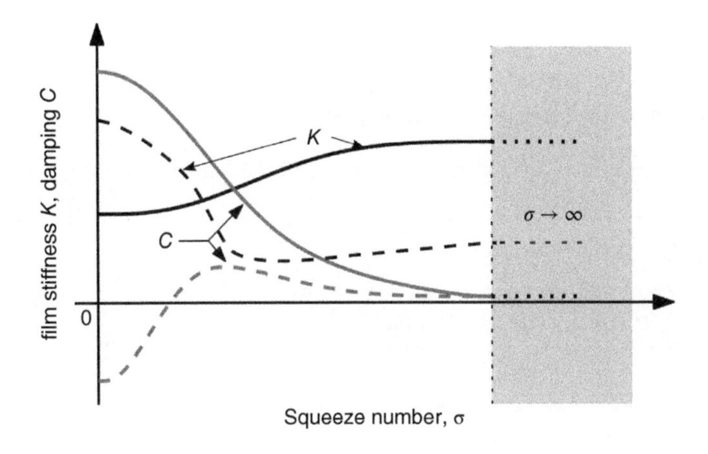

Figure 7.8 General trends of the dynamic stiffness in the squeeze number. Source: Al-Bender F 1992.

1969; Gross 1962), etc. Rather, a dynamic stability criterion was applied to the (simple mass-loaded bearing) system in order to deduce a "critical" mass, supply pressure or film geometry feature, beyond which the system would develop self-excited vibrations.

Such an approach is very restrictive in scope of application, since one generally has often to deal with a multi-body system. A more systematic approach is to quantify the dynamic stiffness of the bearing film, in the frequency range of interest, and then to apply the methods of structural and/or rotor dynamics to analyse the given system that incorporates aerostatic bearing films.

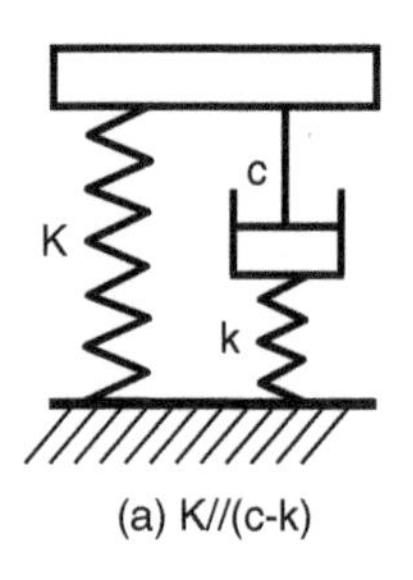

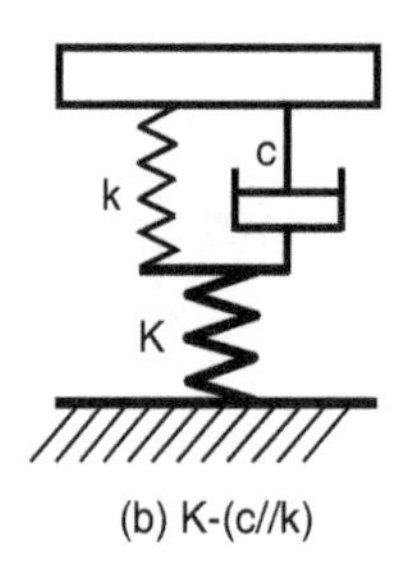

Figure 7.9 Generalised linear scheme for modelling the dynamic stiffness of an air film having positive damping.

Let us revert to the dimensional representation, for easier physical interpretation. We have seen in the previous section that an aerostatic bearing film can be characterised as a non-linear, frequency-dependent spring-damper element, with a dynamic stiffness:

$$k_{\text{dyn}} = k(\nu) + j\nu c(\nu).$$

The trends of k and c, described in the previous paragraph, would help us in deriving an equivalent *linear* system structure that could yield, to a good approximation, this prescribed behaviour.

For the case when $c(0) > 0$, Roblee (1985) shows that these characteristics can be captured in an approximate way by a "phase-lead" system, which may be mechanically expressed as one of stiffness-dashpot combinations depicted in Figure 7.9. Let us adopt the scheme of Figure 7.9(a) for notation, see 7.10. We have:

$$k_{\text{dyn}}(s) = k_0 \frac{1 + \tau_1 s}{1 + \tau_2 s}, \qquad \tau_1 > \tau_2,$$

where s ($= j\nu$ in the periodic case) is the Laplace variable. Obviously, $k(0) = k_0$ and $k(\infty) = k_0\tau_1/\tau_2 > k_0$. The intermediate values could be obtained best in function of the harmonic pulsation ν as:

$$k_{\text{dyn}}(\nu) = k_0 \frac{(1 + \tau_1\tau_2\nu) + j\nu(\tau_1 - \tau_2)}{1 + \tau_2^2\nu}$$

The stiffness is the real part of this expression, viz.

$$k(\nu) = \text{Re}(k_{\text{dyn}}) = k_0 \frac{(1 + \tau_1\tau_2\nu)}{1 + \tau_2^2\nu};$$

the damping is the imaginary part divided by ν:

$$c(\nu) = \text{Im}(k_{\text{dyn}})k_0/\nu = \frac{\tau_1 - \tau_2}{1 + \tau_2^2\nu}.$$

(N.B. the condition $\tau_1 > \tau_2$, together with $k_0 > 0$, implies therefore that $c > 0\ \forall\nu$).

This behaviour is qualitatively depicted in Figure 7.8.

Referring to Figure 7.10, the values of the corresponding spring-dashpot elements values are given by:

$$\tau_1 = \left(1 + \frac{k_0}{k_1}\right)\frac{c}{k_0};$$

$$\tau_2 = \frac{c}{k_1}$$

which provides an alternative way for (element-based) modelling.

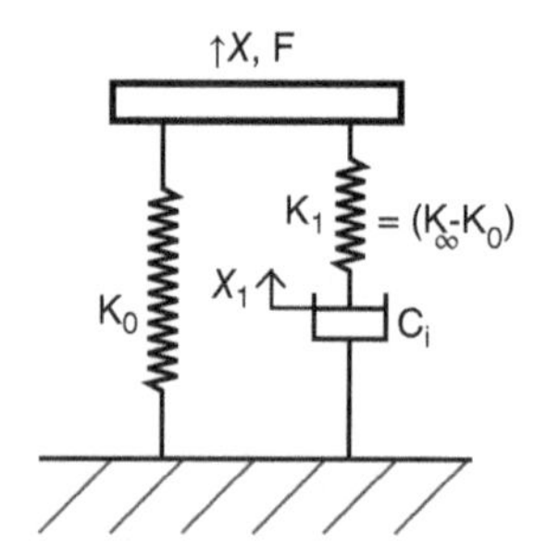

Figure 7.10 Dynamic stiffness and damping components for a positively damped air film.

Remark

This simple model seems to provide a good fit to the actual dynamic-stiffness behaviour not only for this type of bearing but also to other types such as porous bearings. In Chapter 13 an example is provided to show the quality of fit that this model yields as compared to exact numerical simulations over the whole frequency range.

In order to utilise this approximate model, we need to tabulate the values of $k(0), k(\infty)$ and $c(0)$ for each design-parameter combination of an aerostatic bearing. This could be quite a formidable task, if one wishes to cover all the design space. However, we can simplify this endeavour by (i) restricting the treatment to configurations with a phase-lead behaviour, i.e. positively-damped films, (ii) limited supply-pressure range, (iii) interpolate between the results, and (iv) use dimensionless parameters to condense the results; i.e. tabulate $K(0), K(\infty)$ and $C(0)$ in function of R_0 and Λ_f (or rather $1/\sqrt{\Lambda_f}$ to reflect the gap height), for $H_v = 0, 1$ and $P_s = 3, 5$. Such results are depicted in the 3D plots of Figures 7.11–7.14. The trends have already been discussed earlier; in particular, that increasing the stiffness leads to reducing the damping and that small (including negative) damping is a middle-film-height phenomenon.

The Case of Films Having Negative Damping

Although a bearing with negative damping over the whole frequency range has not been reported (being perhaps physically impossible), it is nevertheless expedient to have a model for it, before deducing a model for the general case of a bearing film with a region of negative damping. As a matter of fact, this is obtained by a "phase-lag" system, viz.

$$k_{\text{dyn}}(s) = k_0 \frac{1 + \tau_1 s}{1 + \tau_2 s},$$

as before, but now with

$$\tau_1 < \tau_2,$$

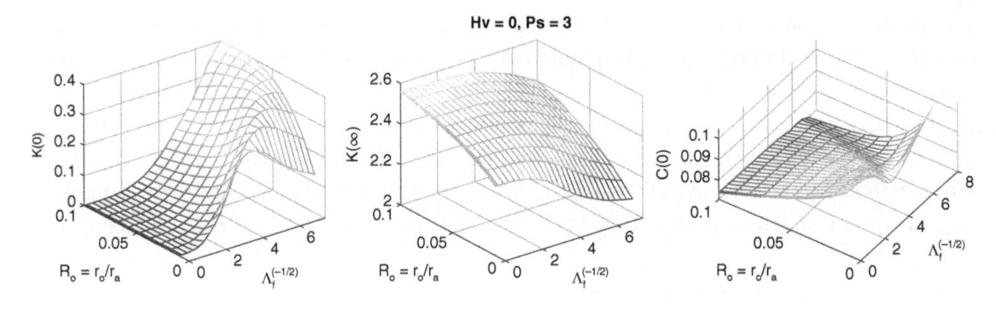

Figure 7.11 The values of $K(0), K(\infty), C(0)$ as a function of $R_o, 1/\sqrt{\Lambda_f}$, for $H_v = 0, P_s = 3$.

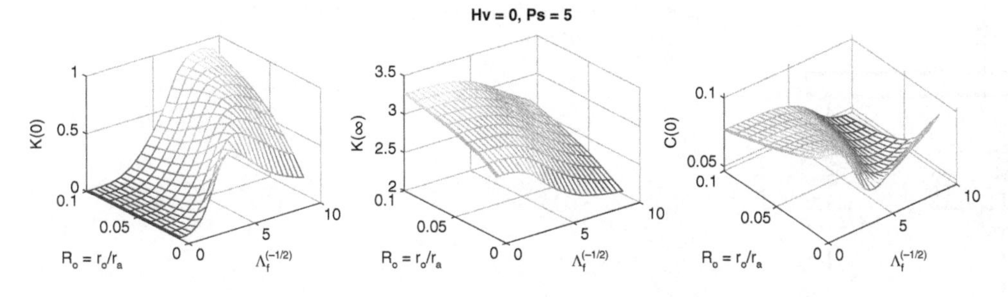

Figure 7.12 The values of $K(0), K(\infty), C(0)$ as a function of $R_o, 1/\sqrt{\Lambda_f}$, for $H_v = 0, P_s = 5$.

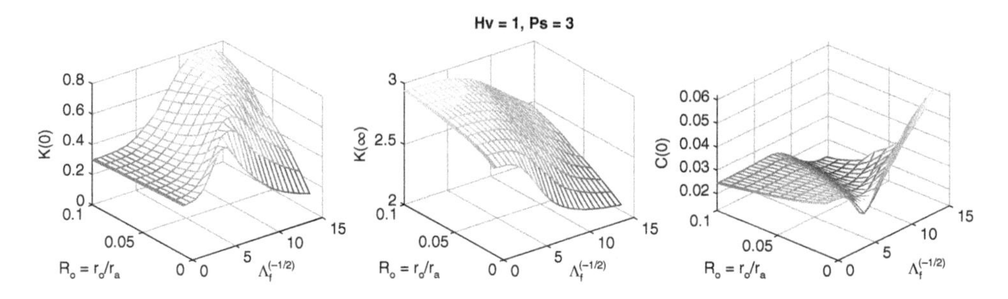

Figure 7.13 The values of $K(0), K(\infty), C(0)$ as a function of $R_o, 1/\sqrt{\Lambda_f}$, for $H_v = 1, P_s = 3$.

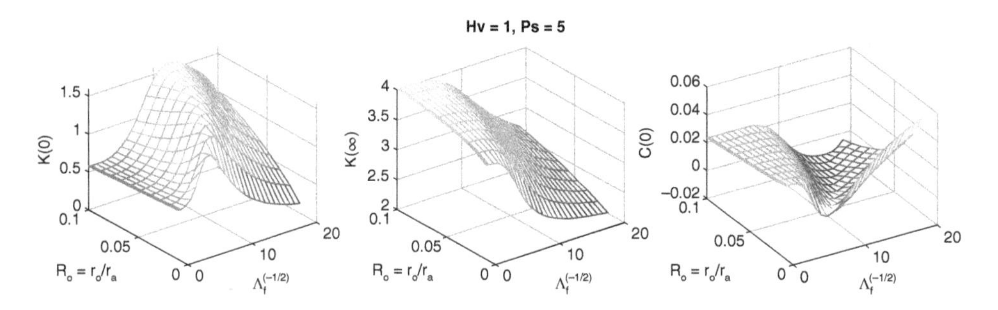

Figure 7.14 The values of $K(0), K(\infty), C(0)$ as a function of $R_o, 1/\sqrt{\Lambda_f}$, for $H_v = 1, P_s = 5$.

i.e. with $c < 0$ assuming $k_1 > 0$.

Finally, we come to the practically encountered case of a bearing with an initial frequency range showing negative damping, but evolving to positive damping at high frequencies. The qualitative behaviour is shown in the dashed-line curves of Figure 7.8.

From consideration of the above two extreme cases, we can conclude that the behaviour can be modelled as a lag-lead system, i.e. a superposition of the two systems outlined above, as depicted in Figure 7.15. It should be noted however that the parameter values (k's and c's) should be chosen such that the damping at zero frequency be negative. It may be argued that this scheme encompasses all the possible combinations of stiffness and damping evolutions in a system of that kind.

To determine the parameter values, we examine the force

$$F = k_0 x + k_1(x - x_1) + k_2(x - x_2)$$

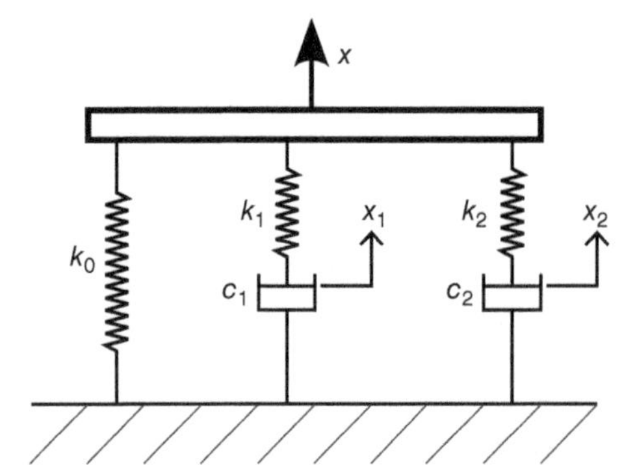

Figure 7.15 Generalised linear scheme for modelling the dynamic stiffness of an air film.

with

$$k_1(x - x_1) = c_1 s x_1 \Rightarrow x_1 = \frac{k_1 x}{k_1 + c_1 s}$$

and

$$k_2(x - x_2) = c_2 s x_2 \Rightarrow x_2 = \frac{k_2 x}{k_2 + c_2 s}$$

This yields,

$$k_{\text{dyn}}(s) = \frac{F}{x} = k_0 + k_1\left(1 - \frac{k_1}{k_1 + c_1 s}\right) + k_2\left(1 - \frac{k_2}{k_2 + c_2 s}\right)$$
$$= k_0 + \frac{c_1 s}{1 + \frac{c_1}{k_1}s} + \frac{c_2 s}{1 + \frac{c_2}{k_2}s}$$

Thus, with positive k's, we have for:

Case 1 positive damping over the whole frequency range $\Rightarrow c_1 \geqslant 0$ and $c_2 \geqslant 0$ as before.
Case 2 negative followed by positive damping $\Rightarrow c_1 > 0 > c_2$ and $c_1 + c_2 < 0$.

In this latter case, we have $k(0) = k_o$, $k(\infty) = k_o + k_1 + k_2$, $c(0) = c_1 + c_2$ and the crossover angular frequency from negative to positive damping is given by

$$v_c^2 = -\frac{c_1 + c_2}{c_1 c_2\left(\frac{c_1}{k_1^2} + \frac{c_2}{k_2^2}\right)}$$

This generalised model, which now has five parameters as compared to three in case of the simple phase-lead system, can be used in dynamic simulation/optimisation purposes for systems incorporating either or both positively and negatively damped bearings. It will obviously yield a better fit for the phase-lead case owing to the higher number of fit parameters.

Finally, to illustrate the stability behaviour of a system with a lag-lead bearing film, let us consider, as an example, the case of a single mass on an aerostatic bearing film and deduce the "critical mass" criterion.

Referring to Figure 7.16, we see that the mass, in the forward part of the system, causes a phase lag of 180 degrees. Thus if the bearing affects a phase lead, the total (closed-loop) systems will be dynamically stable, or positively damped. Conversely, if the bearing gives an additional phase lag, the total (closed-loop) systems will be dynamically unstable, or negatively damped. Note however, that this latter effect is only valid for the region of frequencies equal to and above the resonance frequency of the system = $\sqrt{k(v)/M}$. In other words, for a given k (assumed initially constant) if M is chosen (sufficiently small) such that the natural frequency falls within the positively damped part of the frequency response, then the system will be dynamically stable. The largest mass yielding this property is referred to as the critical mass. For a supported mass that is smaller than the critical mass, the system will be dynamically stable, by virtue of the natural frequency lying in the positively damped frequency region.

Summarising the above, we can see the foregoing problem as follows. If the mass is small enough for the resonance to occur in the positively damped frequency region, then the system will be dissipative and therefore stable. Otherwise, it will be "generative" and thus unstable.

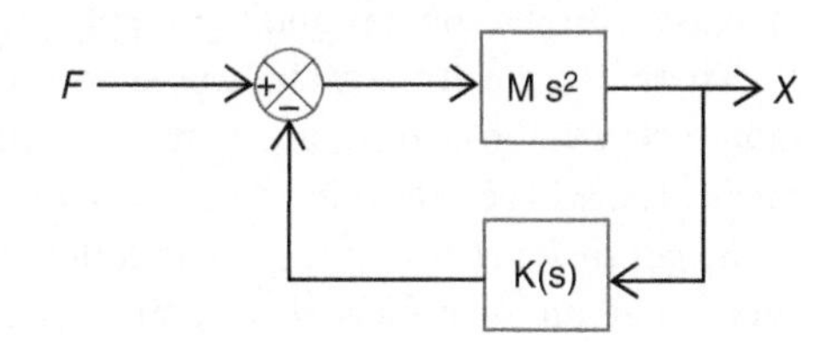

Figure 7.16 Block diagram of a mass on dynamic spring (representing air gap).

From the foregoing, it is obvious that a system that is unconditionally stable (i.e. regardless of the supported mass) should have a positive static damping $C(0)$ or $c(0)$.

Design for Positive Damping

It is hoped that the above results and their accompanying discussion shed sufficient light on the problem of the dynamic stiffness. If the design of *damped* bearings is envisaged, which should furthermore have sufficient load carrying capacity and stiffness over a broad film thickness range, with limited geometrical tolerances—as indeed often is the general design requirement, the foregoing results present an optimisation problem that would involve a *trade-off* between the various design parameters and the characteristics.

If one now wishes to design and optimise a mechanical system incorporating aerostatic bearings, one can proceed as follows:

1. from a static optimisation procedure, decide a range of values of the bearing design parameters,
2. calculate and curve-fit $k_{\text{dyn}}(\nu)$, (equivalently, k_0, τ_1 and τ_2, above, when $c(0) > 0$ for this range of values, and use these for dynamic optimisation.

This could become an elaborate procedure, if it should be carried out on a range of supply pressure, conicity and nominal gap values. In special situations where, for some reason or other, it may be difficult to avoid negative static damping, one can overcome the system's instability by: (i) ensuring that the system's resonance frequencies fall in the positive dynamic damping range (cf. "critical" mass principle mentioned above), (ii) adding external damping to the system (e.g. rubber bearing support, which however leads to a decreased in the total stiffness of the system), or (iii) adding damping to the film (e.g. additional flat squeeze film lands).

Nevertheless, an indication as to the direction of treatment of such a problem may be given in the following.

We have seen that if the static damping is positive, it remained so throughout the frequency range. Let us then try to correlate the bearing design parameters so as to ensure that outcome, i.e. *to define the limits of positive damping*. It has been apparent from the results above that a uniform gap bearing is always well damped; the damping decreasing with increasing H_v. We may, thus, pose the problem as follows. Let P_s, R_o and Λ_f be given, what is the critical value of H_v beyond which the damping becomes negative? Such a value may be found by an iterative solution, (using the *regula falsi*), starting with $H_v = 0$. Figures 7.17 and 7.18 show sample results for supply pressures $P_s = 3$ and 5 respectively. Each curve represents the zero damping line for the particular value of R_o: the values of H_v lying above the curve correspond to negative damping; those below the curve, to positive damping. Thus, the results given above may be seen here more globally.

Before discussing how these diagrams may be used as design charts for stability, let us make the following important observation. In each diagram, there is a lower conicity margin for which the damping is positive, for all values of R_o. This margin becomes smaller with increasing P_s; it takes approximately the values $H_v = 1.7, 0.88$, and 0.6 for $P_s = 3$, 5, and 7 respectively, (and for $\Lambda_f > 100$). This means that the conicity may be dropped out of the stability analysis, when it lies within this margin.

In order to use the diagrams as a design aid for stability, let us note first that they are not complete in themselves, since they do not give the bearing characteristics needed for such design. For this purpose, we would need to construct similar diagrams for the load capacity, stiffness and flow rate. However, we may, for the present, confine the analysis to the general behaviour of the load and the stiffness, as follows. Both the load and the stiffness increase with the conicity and the supply pressure; the load increases and the stiffness decreases with increasing R_o. Therefore, the primary aim would be to maximise the conicity, and then decide upon the size of the feed hole radius. Now if the bearing size r_a and the film thickness range h_a are given (together with the gas properties), the range of Λ_f will be determined. Furthermore, *for every given absolute conicity h_v, the range of H_v will be determined as the parabola*: $H_v^2 \propto \Lambda_f$ (as would follow from the definition of Λ_f). For any given P_s, these parabolas may be superposed on the critical H_v diagram, see Figure 7.17, and required to *lie entirely under* one of the R_o = const.

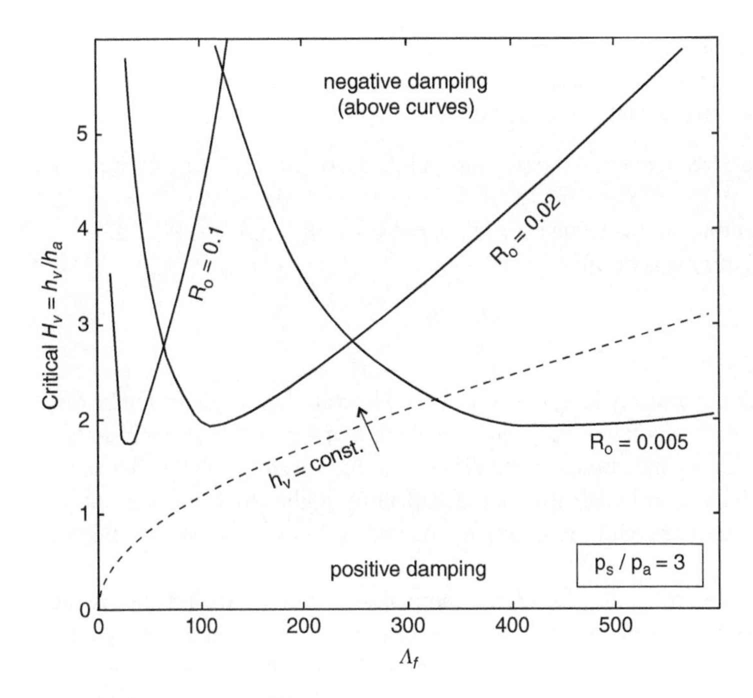

Figure 7.17 Critical conicity H_v at $P_s = 3$.

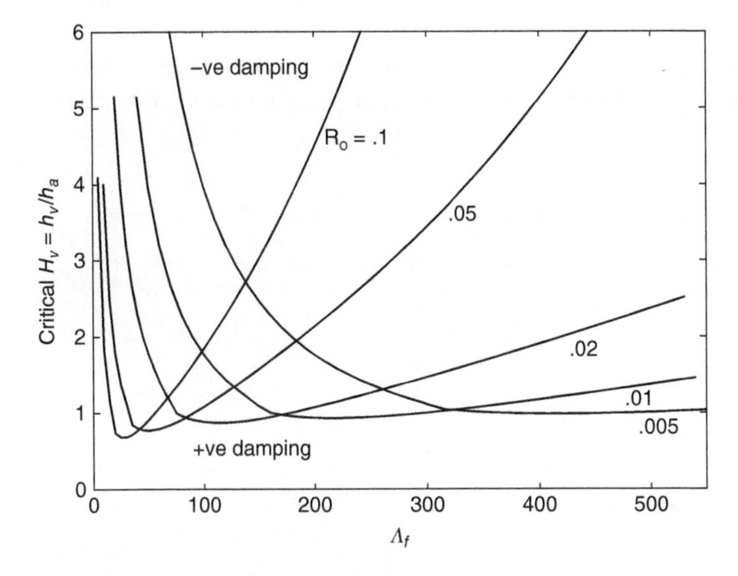

Figure 7.18 Critical conicity H_v at $P_s = 5$.

curves, thus determining a critical R_o value. The rest of the optimisation procedure will consist in deciding the most suitable R_o.

The foregoing should be viewed as a simple method of attack, or just a notion about the problem: optimisation in the "large", or with arbitrarily posed constraints has to be treated more *systematically*, owing to the large number of design parameters, and bearing characteristics to be optimised.

Simple Criterion for Ascertaining Unconditional Stability of a Flat Bearing Pad

If we have at our disposal the static load F and the static stiffness k_0 of a flat $h_v = 0$ bearing at any given gap height h, then we can determine if the damping is positive at that point as follows:

- we know that damping will be positive for all frequencies, if the stiffness at infinite frequency is higher than the static stiffness (Figure 7.8),
- we also know that the dynamic stiffness approaches the value $k_\infty = \int_A (p/h)\mathrm{d}A = F/h + p_a A$ (see Eq. 7.25),
- thus if $k_\infty > k_0$, we can conclude that the damping is positive.

7.5.2 The Supply Pressure Response K_p

Not being of immediate relevance to conventional bearing design, this bearing dynamic characteristic will be discussed in the most general terms. Only one sample result will be given, namely that of Figure 7.19. It represents a Bode plot (with linear amplitude scale) of this response function for different values of the conicity. The static "gain" appears to have a maximum in $0 < H_v < 2$, although it is not significantly higher than the minimum. Exploratory calculations show that this gain increases with increasing R_o, Λ_f and P_s, being very weakly affected by the latter.

What is more important, in the context of controllability (or active compensation) of an air bearing, is the behaviour in σ. In this respect, the frequency response roughly resembles that of a first order (or single lag) system, in the frequency band corresponding approximately to $\sigma < 100$. This "bandwidth" is little affected by P_s or Λ_f, but it broadens with increasing R_o. This means that active compensation would be feasible in this range, providing of course that a means of controlling the supply pressure should be available (over the same range). Also, since σ is directly proportional to the frequency and inversely proportional to h_a^2, for a given bearing, the frequency bandwidth will decrease with decreasing film thickness.

To show typical orders of magnitude of the frequency in relation to σ, we consider the following example. Let the fluid be air, the ambient pressure be 1 bar, and $r_a/h_a = 10^3$, then $\sigma \approx 0.015 f$, where f is the oscillation frequency in Hz.

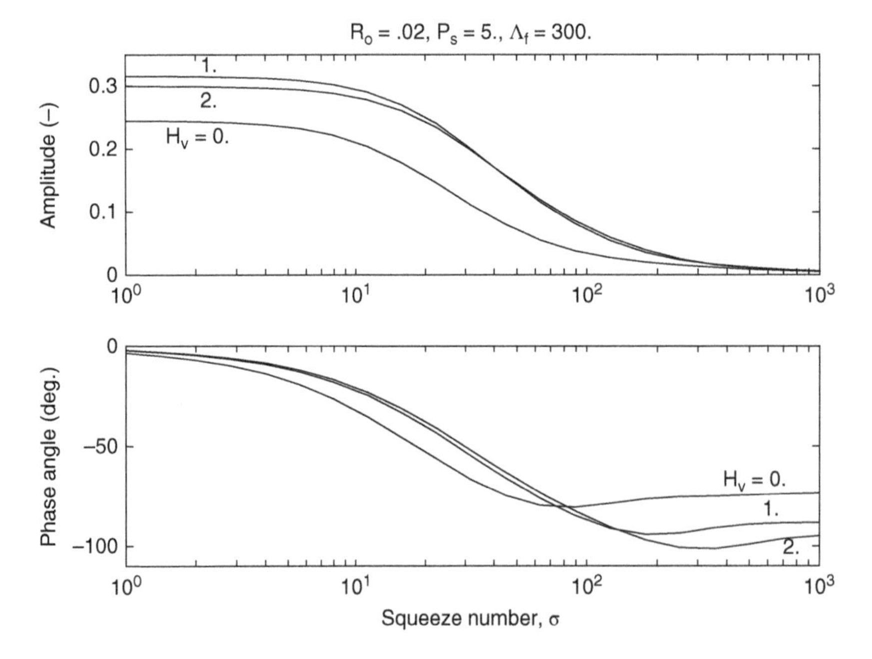

Figure 7.19 Bearing force response to supply pressure, K_p, at different conicities.

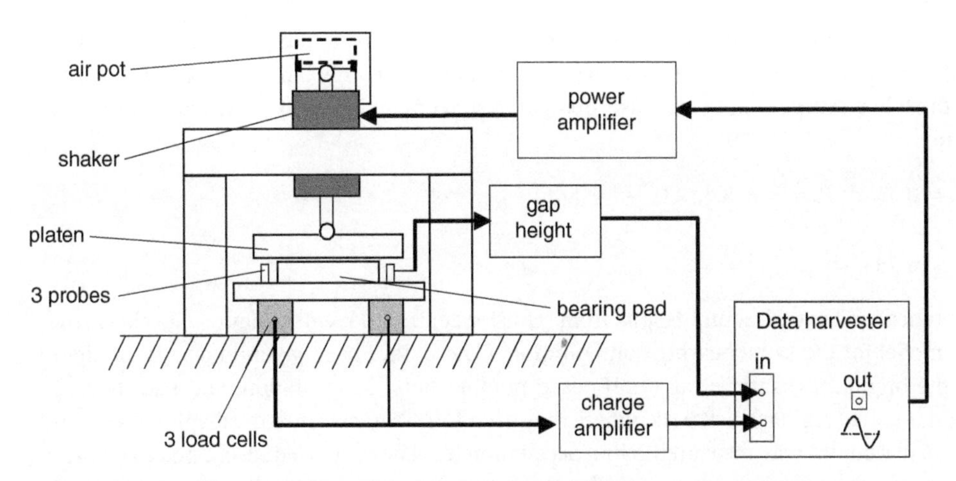

Figure 7.20 Typical test set-up for measuring the dynamic stiffness of the air film.

7.5.3 Comparison with Experiment

The touchstone of the accuracy of any theoretical model is, in the final analysis, its correspondence to reality. Therefore, the importance of (meaningful) experimentation cannot be overstressed. Unfortunately, the measurement of the dynamic stiffness of an air film has, so far, proved to be not an easy task. A typical test set-up is depicted in Figure 7.20, which is self-explanatory. (For a description of test apparatus and the problematic of experimentation, the reader is referred to (Al-Bender 1991; Holster et al. 1991; Plessers 1985) and Chapter 16.)

Figure 7.21 shows a comparison of the theoretically predicted dynamic stiffness with an experiment carried out by the author. The dynamic stiffness of a test bearing is shown, in a *linear* Bode plot, for a number of supply pressure ratios P_s. The data was obtained by stepped-sine excitation at intervals of 50 Hz. The bearing number

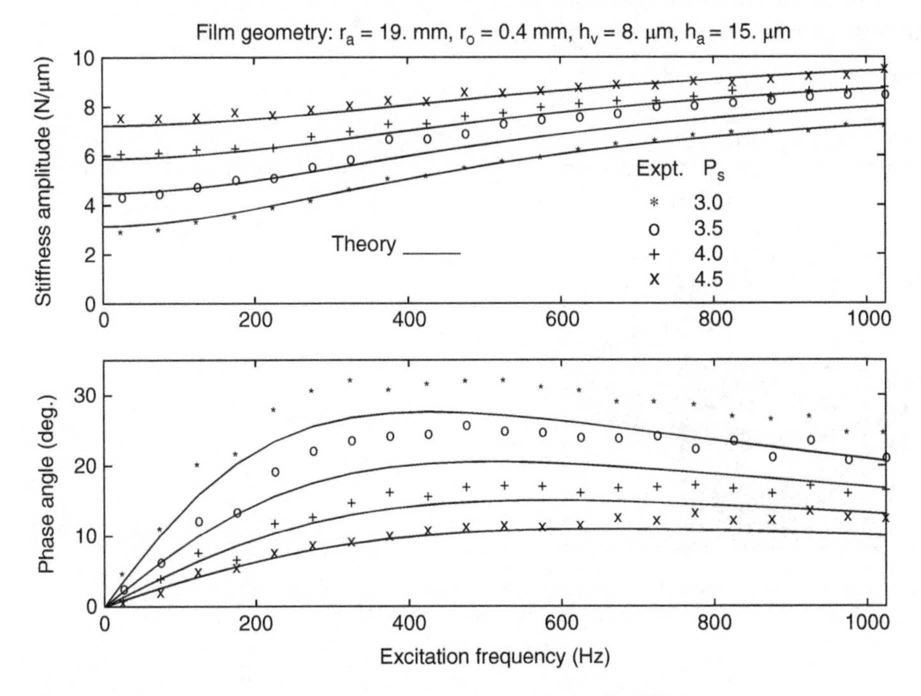

Figure 7.21 Comparison of calculated and measured dynamic stiffness.

Λ_f is equal to about 150, while the maximum value of σ is about 25; $H_v \simeq 0.5$. Note the good agreement of the stiffness amplitude results (except in one case). The phase results show, however, only fair agreement, for reasons that will be discussed in the following section. In any case, qualitative agreement is clearly apparent. For this test, the bearing parameters are:

$$R_o \simeq 0.021, \quad H_v = 0.53, \quad P_s = 3.0,\ 3.5,\ 4.0,\ 4.5$$

$$\Lambda_f \approx 150, \quad (\sigma)_{\max} \approx 25.0$$

Figure 7.22 shows theoretical and experimental results from (Holster et al. 1991), while Figure 7.23 shows theoretical results from our model for the same bearing configuration. (These have been given in a separate figure to avoid overburdening the original.) Holster et al. (1991) used random noise as excitation, and had to subdivide the frequency domain into two regions: 0–400 Hz, where they used LVDTs to measure the displacement; and 400–1200 Hz, where the displacement was measured by an accelerometer. They reported also a possible 0.5 μm error in the film thickness measurement, such can be significant for the accuracy of the results at small film thickness values. Our results (Figure 7.23) compare very favourably both with their theoretical and experimental results.

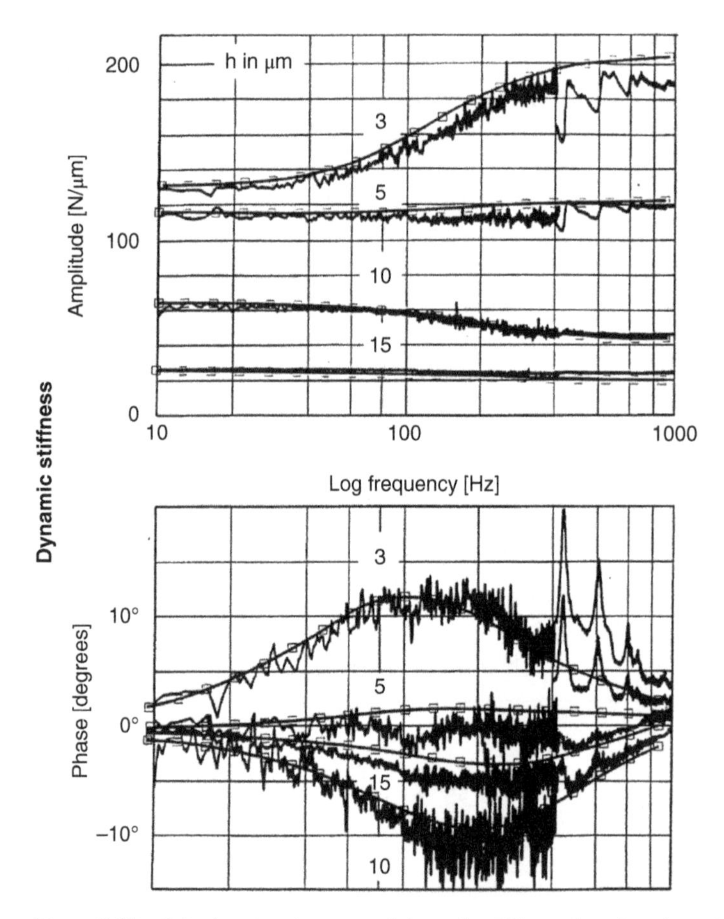

Figure 7.22 Calculated and measured dynamic stiffness, for a bearing with $r_a = 30$ mm, $r_o = 0.265$ mm, $h_v = 13$ μm, $P_s = 7$, from (Holster et al. 1991). Source: Holster P, Jacobs J and Roblee J 1991 by permission of ASME.

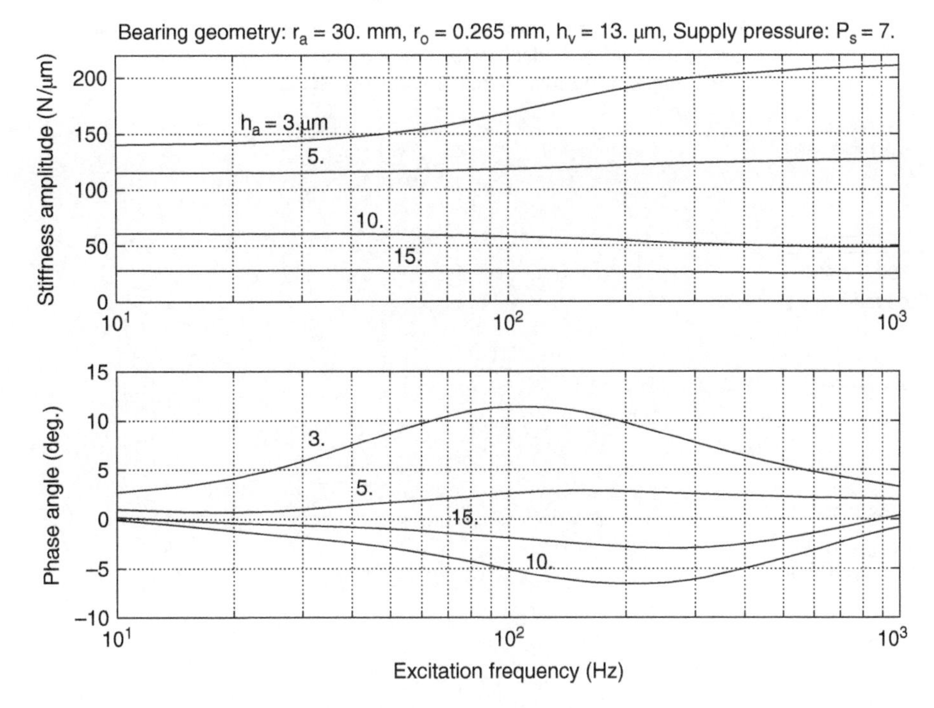

Figure 7.23 Calculated dynamic stiffness, by our model, for comparison with those of Figure 7.22. Source: Al-Bender F 1992.

Note that for this test, the bearing parameters are:

$$R_o \simeq 0.009, \quad 0.87 < H_v < 4.33, \quad P_s = 7$$

$$221 < \Lambda_f < 5524, \quad 524 < (\sigma)_{\max} < 1357.$$

This reference has also been given as a demonstration of the difficulty of conducting accurate tests over a wide frequency bandwidth. In particular, due to the very small order of the damping force, (as compared to the stiffness), it is practically impossible to measure the phase angle with sufficient certainty.

A final comparison is made for the case of film force response to supply pressure. The experimental results have been taken from (Blondeel 1975), where description of the test apparatus may also be found. In this test the bearing gap was held constant, while the supply pressure was varied by means of a rotating valve situated upstream of the feed hole. Figure 7.24 shows a comparison of the theoretical and experimental frequency response function normalised by the static value. Unfortunately, the comparison is not complete since the absolute values (or the static gain) are not given. Nevertheless, except for small discrepancies (that might well have been due to system resonances), the agreement is quite satisfactory. Note that, owing to the large phase angles involved, it is possible to obtain a more fruitful comparison, suggesting that this type of tests may a good starting point for carrying out an experimental verification. This case is further treated in Chapter 15, where additional experimental results are provided.

In addition to the foregoing, the proposed model has been often used by the author for designing unconditionally stable, (i.e. well damped), bearings. So far, the so-designed bearings have proved indeed to be stable in application.

In conclusion, though with a limited number of comparisons as given above, the agreement is found generally acceptable, so that this linearised model may be assumed sufficiently reliable for design purposes.

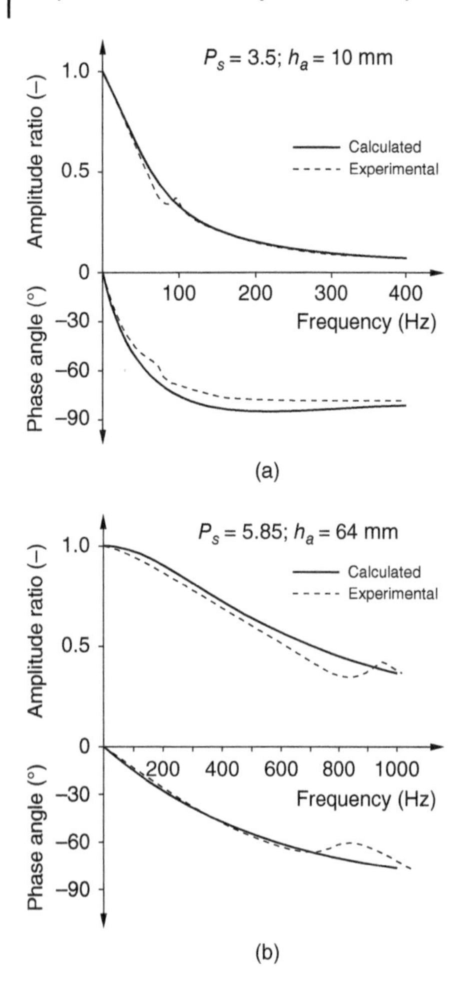

Figure 7.24 Bearing force response to supply pressure: experimental results from ref. (Blondeel 1975). Bearing geometry: r_a = 76.5 mm, $r_o \simeq 1$ mm, h_v = 12 μm. Source: Al-Bender F 1992.

7.6 Conclusions

The various aspects of the problem of the dynamic behaviour of aerostatic bearings have been outlined. The most important one is identified as that of pneumatic stability. (The question of active compensation, which may be also very significant, is treated in Chapter 15.)

A literature survey carried out shows not only the importance of this problem in bearing design, but points out also the various methods of treatment of that problem. It appears that a model based on the linearised Reynolds equation may be sufficient in characterising the problem, given certain constraints on the design parameters and amplitude of oscillation.

Such a model has been constructed for the type of bearing configuration of our concern, with the added feature of an accurate entrance flow model. This methodology and model can be easily applied to other bearing configurations. Limits of the validity of this model have also been carefully outlined, showing, through practical examples, that it may be sufficient for usual bearing design requirements.

If the dynamic stiffness is split into in-phase *stiffness* and out-of-phase *damping*, then the results show that:

- the stiffness increases and the damping decreases with increasing conicity and/or with increasing supply pressure,

- the stiffness has a local maximum and the damping has a local minimum over the feed hole radius range and/or over the feed number range,
- the behaviour in the squeeze number seems to follow certain general trends as given in section 7.5.1.

Furthermore, the problem of designing *unconditionally stable* (or well damped) bearings has been formally addressed. It has been shown that it is possible to find regions in the (Hv, Λ_f, P_s)-space, bounded above by surfaces: R_o = constant, where the damping is positive. These *charts* may be used to design for unconditional stability. The bearing characteristics can then be optimised within these regions of positive damping.

The limited available experimental data show a good agreement with the prediction of the theoretical model both for the dynamic stiffness K_a and for the force response K_p.

References

Al-Bender, F., (1991). Discussion on 'the measurement and finite element analysis… ' (*op. cit.*). *ASME - Journal of Tribology* 113, 774–775.

Al-Bender, F., (1992). Contributions to the design theory of circular centrally fed aerostatic bearings PhD thesis Katholieke Universiteit Leuven - Dept. Mechanical Engineering.

Blondeel, E., (1975). Aerostatische Lagers met Lastafhankelijke Spleetkonfiguratie Doctoral thesis K.U. Leuven, Departement Werktuigkunde. 75D1.

Blondeel. E., Snoeys, R. and Devrieze, L., (1980). Dynamic stability of externally pressurized gas bearings. *ASME - Journal of Lubrication Technology* 102, 511–519.

Boffey, D.A., (1978). A study of the stability of an externally-pressurized gas-lubricated thrust bearing with a flexible damped support. *ASME - Journal of Lubrication Technology* 100, 364–368.

Boffey, D.A. and Desai, D.M., (1980). An experimental investigation into the rubber-stabilization of an externally-pressurized air-lubricated thrust bearing. *Journal of Lubrication Technology* 102, 65–70.

Chiang, T. and Pan, C.H.T., (1969). Refined solution of pneumatic hammer instability of inherently compensated hydrostatic thrust gas bearings. Technical report, Mechanical Technology Incorporated (MTI), Latham, New York. MTI 69TR23.

Constantinescu, V.N., (1969). *Gas Lubrication*. ASME.

Dai, R.X., Dong, Q. and Szeri, A.Z., (1992). Approximations in hydrodynamic lubrication. *ASME - Journal of Tribology* 114, 14–25.

Devis, B., (1985). Ontwerpgerichte Meetprocedures voor de Bepaling van Mechanische Eigenschappen van Rubbermaterialen PhD thesis K.U. Leuven, Departement Werktuigkunde. 85D2.

Elkouh, A.F., (1984). Fluid inertia effects in a squeeze film between two plane annuli. *ASME - Journal of Tribology* 106, 223–227.

Elrod, Jr. H.G., McCabe, J.T. and Chu, T.Y. (1967). Determination of gas-bearing stability by response to a step-jump. *ASME - Journal of Lubrication Technology* 493–498.

Gorez, R., (1973). A study of the stability of externally pressurized gas bearings with porous wall by liapunov's direct method. *ASME - Journal of Lubrication Technology* 204–207.

Gross, W.A., (1962). *Gas Film Lubrication*. John Wiley & Sons.

Holster, P., Jacobs, J. and Roblee, J., (1991). The measurement and finite element analysis of the dynamic stiffness of nonuniform clearance, gas, thrust bearings. *ASME - Journal of Tribology* 113, 768–774.

Horikawa, O., Yasuhara, K., Osada, H. and Shimokohbe, A., (1991) Dynamic stiffness control of active air bearing. *International Journal of the Japanese Society of Precision Engineering* 25(1): 45–50.

Ishizawa, S., (1966). The unsteady laminar flow between two parallel discs with arbitrary varying gap. *Bulletin of the JSME* 9(35), 533–550.

Ishizawa, S., Watanabe, T. and Takahashi, K., (1987). Unsteady viscous flow between parallel disks with a time-varying gap width and a central fluid source. *ASME - Journal of Fluids Engineering* 109, 394–402.

Jones, A.F. and Wilson, S.D.R, (1975). On the failure of lubrication theory in squeezing flows. *ASME - Journal of Lubrication Technology* 101–104.

Kazimierski, Z. and Jarzecki, K., (1979). Stability threshold of flexibly supported hybrid gas journal bearings. *ASME - Journal of Lubrication Technology* 101, 451–457.

Langlois, W., (1962). Isothermal squeeze films. *Quarterly of Applied Mathematics* 20, 131–150.

Licht, L. and Elrod, H., (1960). A study of the stability of eternally pressurized gas bearings. *ASME - Journal of Applied Mechanics* 250–258.

Licht, L., Fuller, D.D. and Sternlicht, B., (1958). Self-excited vibrations of an air-lubricated thrust bearing. *Transactions of the ASME* 411–414.

Lund, J.W., (1965). The stability of an elastic rotor in journal bearings with flexible, damped supports. *ASME - Journal of Applied Mechanics* 87(4): 911–920.

Lund, J.W., (1968). Calculation of stiffness and damping properties of gas bearings. *Journal of Lubrication Technology* 793–803.

Lund, W., (1969). Dynamic performance and stability *DESIGN OF GAS BEARINGS* Mechanical Technology Incorporated (MTI), Latham, New York.

Marsh, H., (1980). Stability and rotordynamics for gas lubricated bearings. *TRIBOLOGY international* 219–221.

Mori, A. and Mori, H., (1973). An application of pneumatic phase shifting to stabilization of externally pressurized journal gas bearings. *ASME - Journal of Lubrication Technology* 33–41.

Mori, A., Aoyama, K. and Mori, H., (1980a). Influence of the gas-film inertia forces on the dynamic characteristics of externally pressurized, gas lubricated journal bearings—part I: Proposal of governing equations. *Bulletin of the JSME* 23(178): 582–586.

Mori, A., Aoyama, K. and Mori, H., (1980b). Influence of the gas-film inertia forces on the dynamic characteristics of externally pressurized, gas lubricated journal bearings—part II: Analysis of whirl instability and plane vibrations. *Bulletin of the JSME* 23(180): 953–960.

Mori, H. and Mori, A., (1967). On the stabilizing methods of externally pressurized thrust gas bearings. *ASME - Journal of Lubrication Technology* 283–290.

Na, T.Y., (1979). *Computational Methods in Engineering Boundary Value Problems*. Academic Press.

Plessers, P., (1985). Dynamische Instabiliteit van AerostatischeGaslagers in Mechanische Systemen Doctoral thesis K.U.Leuven, Departement Werktuigkunde. 85D4.

Plessers, P. and Snoeys, R., (1988a). Dynamic identification of convergent externally pressurized gas bearing gaps. *ASME - Journal of Tribology* 110, 263–270.

Plessers, P. and Snoeys, R., (1988b). Dynamic stability of mechanical structures containing externally pressurized gas lubricated thrust bearings. *ASME - Journal of Tribology* 110, 271–278.

Richardson, H.H., (1958). Static and dynamic characteristics of compensated gas bearings. *Transactions of the ASME* 1503–1509.

Roblee, J., (1985). Design of Externally Pressurized Gas Bearings for Dynamic Applications PhD thesis University of California Berkely, Calf.

Rohde, S.M. and Ezzat, H.A, (1967). On the dynamic behavior of hybrid journal bearings. *ASME - Journal of Lubrication Technology* 98, 90–94.

San Andres, L.A., (1991). Effect of compressibility on the dynamic response of hydrostatic journal bearings. *Wear* 146, 296–283.

Sato, Y., Maruta, K. and Harada, M., (1988). Dynamic characteristics of hydrostatic thrust air bearing with actively controlled restrictor. *ASME - Journal of Tribology* 110, 156–161.

Shimokohbe, A., Aoyama, H. and Watanabe, I., (1986). A high precision straight-motion system. *Precision Engineering* 8(3): 151–156.

Stiffler, A.K. and Tapia, R.R., (1979). Amplitude effects on the dynamic performance of hydrostatic gas thrust bearings. *ASME - Journal of Lubrication Technology* 101, 437–443.

Tan, C.A. and Mote, Jr. C.D., (1990). Analysis of a hydrodynamic bearing under transverse vibration of an axially moving band. *ASME - Journal of Tribology* 112, 514–523.

Tichy, J.A. and Winer, W.O., (1970). Inertial considerations in parallel circular squeeze film bearings. *ASME - Journal of Lubrication Technology* 588–592.

Turnblade, R.C., (1963). The molecular transit time and its correlation with the stability of externally pressurized gas-lubricated bearings. *ASME - Journal of Basic Engineering* 297–303.

Wang, Z., Ishizawa, S. and Takahashi, K., (1990). Unsteady viscous flow between parallel disks with a time-varying gap width and a central fluid source, (the case in which the rate of flow from the source is forcibly varied with time). *JSME International Journal, Series II* 33(3): 446–453.

White, M.F. and Chan, S.H, (1992). The subsynchronous dynamic behavior of tilting-pad journal bearings. *ASME - Journal of Tribology* 114, 167–173.

Wilcock, D.F, (1967). Externally pressurized bearings as servomechanisms. i - the simple thrust bearing. *ASME - Journal of Lubrication Technology* 418–424.

Wunsch, H.L., (1964). GAS LUBRICATED BEARINGS. Butterworths, London. The Application of Externally Pressurized Air Bearings to Measuring Instruments and Machine Tools.

8

Aerodynamic Action: Self-acting Bearing Principles and Configurations

8.1 Introduction

The previous chapters have presented the theory of aerostatic bearings, which are externally fed from a pressure source. They also considered the effect of relative velocity between the bearing surfaces as an exceptional effect on the behaviour of such bearings, which are otherwise stationary. In this chapter, we consider the aerodynamic action as a mechanism that enables in itself the build-up of load-bearing pressure inside the film by virtue of the wedge-effect. This leads to the *self-acting* or aerodynamic bearing, which knows many different configurations and numerous applications. In particular, a single slider bearing is one of the key technologies that led to the success of modern magnetic disk storage devices, for positioning a magnetic head precisely over a high speed rotating recording disk. Miniature turbines, gyroscopes, spindles and many other rotating devices furnish other examples.

Hydrodynamic and aerodynamic lubrication theory is very old and well established, see e.g. (Constantinescu 1969; Grassam and Powell 1964; Gross 1962; Gross et al. 1980; Wilcock 1969) to mention but a few. Moreover, the Reynolds equation has been treated in Chapter 4 and the gap-entrance problem in Chapter 3.

In this chapter, we would like to take a comprehensive look at the aerodynamic action as principle with its merits and limitations. For this purpose, we shall consider firstly, the qualitative difference between incompressible and compressible cases, which will be examined at some length since compressibility represents a limiting factor to pressure generation. As has been indicated in the case of aerostatic bearings, the compressibility of air, or gas generally, plays an important role in the behaviour of the bearing film, which distinguishes it from incompressible bearings in an essential manner. In aerodynamic bearings, compressibility of the fluid imposes an upper limit on the pressure that could be generated in the film at exceedingly high speeds: a saturation in which further build-up of the pressure is converted into density (or elastic energy) rather than potential or pressure.

Secondly, whereas in EP bearings the pressurised fluid is provided from an external source, in self-acting bearings, the film wedge combined with viscous shear through the tangential velocity is the mechanism by which fluid is "pumped" into the gap in order to generate pressure: This is the principle of the viscous pump, whose mechanical efficiency will be shown to be very small, which is an issue of concern when one sees the total energy efficiency of the bearing system.

Thirdly, we shall consider three representative, nominally flat rectangular self-acting bearing configurations and study their typical characteristics to demonstrate theory, which could be extended to other situations. These are: the inclined pad, the stepped pad and the herringbone grooved bearing. Journal bearings will be separately treated in other dedicated chapters (9 and 10).

Relative sliding raises also the issues related to the interaction of the (moving) surface features with the air film, in particular in the case of journal bearings having rotation or stationary features, such as grooves. These issues need careful treatment in order for one to be able to set up the problem correctly even for numerical solution.

Air Bearings: Theory, Design and Applications, First Edition. Farid Al-Bender.

Companion website: www.wiley.com/go/AlBender/AirBearings

In this regard, a considerable part of this chapter is devoted to grooved bearings as a particular type of self-acting bearings characterised by enhanced pumping action. Owing to its importance, a derivation of the narrow-groove theory (NGT) of herringbone grooved bearings is presented and discussed.

Finally, externally-pressurisation may be combined with the aerodynamic action, when the sliding velocity is sufficiently high, leading to the type of bearings often known as *hybrid* bearings. In this type of bearing, depending on the design w.r.t. pressure boundary conditions at the leading and trailing edges, the aerodynamic action may either enhance or mar the load capacity and stiffness of the bearing and should therefore be well quantified, if one considers effective applications of such bearings. This is particularly the case for (ultra) high-speed journal and thrust-collar bearings, which, with the advent of miniaturisation and portable-energy systems, are finding numerous applications. A section of this chapter is devoted to hybrid thrust bearings, while hybrid JB's will be treated in Chapters 9 and 10.

In short, the basic objective of this chapter is to overview the most important aspects of aerodynamic lubrication including dynamic behaviour.

8.2 The Aerodynamic Action and the Effect of Compressibility

Let us start by restating the Reynolds Equation (2.37) in dimensional form from Chapter 4:

$$\nabla \cdot \left(\frac{\rho h^3}{12\mu} \nabla p - \frac{1}{2} \rho h \mathbf{V}_\mathrm{t} \right) = \frac{\partial}{\partial t}(\rho h) \tag{8.1}$$

In order to facilitate treatment of this equation in the general compressible fluid case, we further assume that the relationship between pressure and density is characterised by a (generalised) polytropy equation:

$$\frac{p}{\rho^\chi} = \text{const.} \tag{8.2}$$

where,

- $\chi = 1$ corresponds to a (compressible) isothermal process,
- $\chi = \kappa = c_\mathrm{p}/c_\mathrm{v}$ corresponds to a (compressible) isentropic process,
- $\chi \to \infty$ corresponds to an incompressible fluid.

The first case constitutes the most common assumption in gas bearing problems since it leads to simplified equations without appreciably departing from the second case of adiabatic process. The last case represents incompressible fluid (i.e. $\rho \sim p^{1/\chi} \to$ const.). The intermediate values of χ could simulate different degrees of compressibility. The Reynolds equation for a compressible fluid may thus be written as:

$$\nabla \cdot \left(\frac{h^3}{12\mu} p^{1/\chi} \nabla p - \frac{1}{2} p^{1/\chi} h \mathbf{V}_\mathrm{t} \right) = \frac{\partial}{\partial t}(p^{1/\chi} h) \tag{8.3}$$

Note that χ will vary at different stations of the flow as well as with time, depending on the thermal balance in the film. We can obtain an idea about this by solving the energy equation simultaneously with the flow equations at least for generic cases. This problem is treated in Chapter 17. On the other hand, one can solve the Reynolds equation for the two extreme cases of $\chi = 1$ and κ and so obtain an envelope within which the true solution will lie. The error margin between isothermal and isentropic is however not that large, as shown in Chapter 17.

Without loss of generality, let us examine the steady-state, two-dimensional (infinitely wide) tapered bearing pad case, where the lower bearing surface is flat and moves with velocity V, while the upper surface is stationary, see Figure 8.1. Thus, we have:

$$\frac{\mathrm{d}}{\mathrm{d}x} \left(\frac{\rho h^3}{12\mu} \frac{\mathrm{d}p}{\mathrm{d}x} - \frac{1}{2} \rho h V \right) = 0. \qquad \text{(General)} \tag{8.4}$$

For an isothermally compressible fluid, this becomes

$$\frac{\mathrm{d}}{\mathrm{d}x}\left(\frac{ph^3}{12\mu}\frac{\mathrm{d}p}{\mathrm{d}x}-\frac{1}{2}phV\right)=0, \qquad \text{(Isothermal)} \tag{8.5}$$

whereas, for the incompressible fluid case, we have

$$\frac{\mathrm{d}}{\mathrm{d}x}\left(\frac{h^3}{12\mu}\frac{\mathrm{d}p}{\mathrm{d}x}-\frac{1}{2}hV\right)=0. \qquad \text{(Incompressible)} \tag{8.6}$$

A hydrodynamic bearing operates in actual fact as a viscous pump (or compressor in case of compressile fluid). That is, the motion of the one surface relative to the other displaces the fluid in the direction of motion. If now the passage of the fluid is constricted, namely by the so-called wedge effect, i.e. a decreasing gap in the direction of motion, pressure will build up in the gap, which results in the bearing action.

The term in parentheses represents the negative of the mass flow of the fluid through the pad (i.e. in the direction of V) per unit width B,

$$\left(\frac{\rho h^3}{12\mu}\frac{\mathrm{d}p}{\mathrm{d}x}-\frac{1}{2}\rho hV\right)=-\frac{\dot{m}}{B}. \tag{8.7}$$

We consider the classical case of tapered-pad bearing (Figure 8.1), with smooth gap-height variation, namely,

$$h(x)>0; \qquad \mathrm{d}h/\mathrm{d}x\le 0; \qquad h(0)=h_o; \qquad h(L)=h_a \quad (h_o\ge h_a).$$

In this case, the pressure will build up from ambient (designated (o) at the inlet and (a) at the outlet) reaching a maximum value at station (m) somewhere in between. In the following, we shall show that this maiximum value p_m is bounded in the case of a compressible lubricant, but that it is unbounded for an incompressible one.

Since the pressure slope at the inlet to the bearing is positive, it follows that the maximum flow through the bearing is bounded by the Couette component:

$$\frac{\dot{m}}{B}\le\frac{\rho_o Vh}{2} \tag{8.8}$$

Thus, integrating Eq. 8.4 w.r.t. x, to obtain Eq. 8.7, we can determine the value of p_m by substituting $\mathrm{d}p/\mathrm{d}x=0$ there, to obtain:

$$\frac{\rho_m Vh_m}{2}=\frac{\dot{m}}{B}\le\frac{\rho_o Vh_o}{2} \tag{8.9}$$

$$\Rightarrow \frac{\rho_m}{\rho_o}\le\frac{h_o}{h_m} \tag{8.10}$$

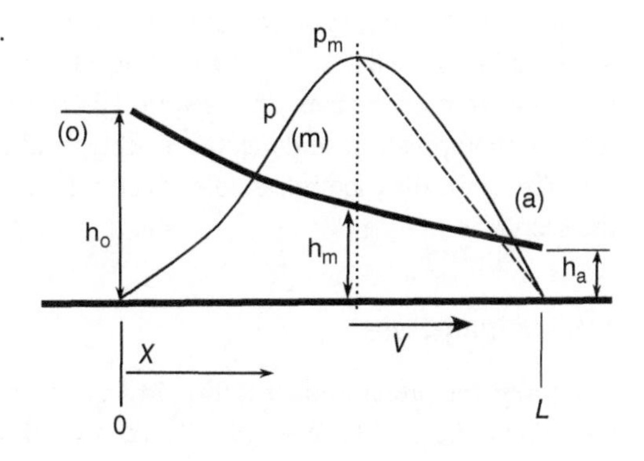

Figure 8.1 The pressure distribution in a generic tapered-pad bearing.

That is to say that the ratio of the maximum density to the ambient value is bounded by the ratio h_o/h_m for all values of the sliding velocity. Applying the polytropy relation, we find that:

$$\frac{p_m}{p_o} \leq \left(\frac{h_o}{h_m}\right)^{\chi} \leq \left(\frac{h_o}{h_a}\right)^{\chi} \tag{8.11}$$

Thus, the smaller χ is, i.e. the higher the compressibility is, the smaller is the pressure build up. Let us note that this result only places an upper bound on the maximum pressure. When $\chi \to \infty$, this upper limit is also infinite. However, we do not know from this if this limit can be actually attained. To do so, we examine the the incompressible-flow Reynolds Equation, in which the density, being constant, drops out:

$$\frac{d}{dx}\left(\frac{h^3}{12\mu}\frac{dp}{dx} - \frac{1}{2}hV\right) = 0 \tag{8.12}$$

Let us examine this equation first as the reference, which is well known from hydrodynamic lubrication theory, see e.g. (Gross et al. 1980), excluding cavitation that actually pertains to compressibility.

Integrating once w.r.t. x, we obtain:

$$\frac{dp}{dx} = \frac{6\mu V}{h^3}(h - h_m) \tag{8.13}$$

From this equation, we see first that for any point in the film different from x_m, i.e. $h - h_m \neq 0$, the pressure gradient increases indefinitely with V. Second, differentiating this equation w.r.t. x, we get:

$$\frac{d^2p}{dx^2} = 6\mu V\frac{h^3 - 3h^2(h - h_m)}{h^6}\frac{dh}{dx} \tag{8.14}$$

Thus, for $x > x_m$, or $h - h_m < 0$, we have that $d^2p/dx^2 < 0$; in other words, the pressure curve is convex and so stands above the chord $m - a$ as depicted in Figure 8.1. (This is then a fundamental property of the pressure curve that does not generally apply in the first half of the bearing $o - m$, as also is depicted in the same figure.) Now applying the *mean-value* theorem to the convex portion of the pressure curve, together with the increasing pressure gradient, results in that p_m increases also indefinitely with V, Q.E.D.

A special, *limiting case* exists for which there is no difference between compressible and incompressible fluid with regard to the resulting pressure distribution, namely that of a *blind* bearing, i.e. for which either $h_a = 0$ or the exit is closed by a step. In that case, the flow is null so that

$$\left(\frac{\rho h^3}{12\mu}\frac{dp}{dx} - \frac{1}{2}\rho hV\right) = 0 \Rightarrow \frac{dp}{dx} = \frac{6\mu V}{h^2} \tag{8.15}$$

(where we assumed realistically that the density is not zero!). In that case, the point of maximum pressure will be situated at the exit of the bearing.

Let us note that this is a singular solution, which represents a discontinuity (or a jump) in the set of solutions, for different flow rates, at the zero-flow value. In practice, this type of situation (i.e. $\dot{m} = 0$) might occur at start-up of the slider-bearing system (plain or journal) from its initial surface-contact situation before lift-off occurs, as depicted in Figure 8.2. In (a), the trailing edge of the bearing is still in contact with the lower surface so that there is no flow though the gap. The pressure distribution for this case, which is linear owing to the uniform gap, is the same for compressible and incompressible fluid with a slope proportional to μV. After lift-off, the pressure profile is different for the incompressible (b) and the compressible (c) cases, being linear for the first and non-linear for the second.

Two Limiting Cases

Generally, the pressure distribution in an aerodynamic bearing will depend on the ambient pressure, as we have seen above (Eq. 8.11 for example). The degree of dependence varies, however, with the level of pressure in the film,

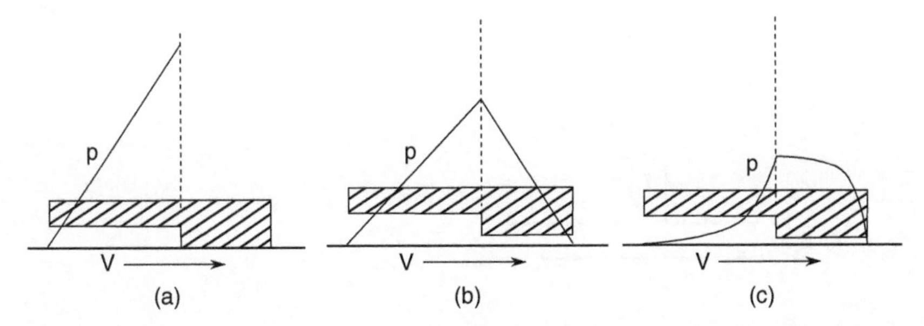

Figure 8.2 Lift-up of a step slider: (a) before lift-off compressible or incompressible, (b) after lift-off, incompressible, (c) after lift-off, compressible.

which increases with μV. More generally than the latter, the sliding, or the bearing number Λ_v will be defined later in the framework of dimensionless formulation, see Sec. 8.4. For low values of the sliding number (or the pressure in the film), we can approximate the pressure by writing:

- The *low pressure* case: For low values of the sliding number (or the pressure in the film), we can approximate the pressure by writing:

  $$p \approx p_o + \Delta p, \qquad \Delta p \ll p_o$$

  Substituting this in Eq. 8.5, ignoring second-order terms, and dividing by p_o, we get

  $$\frac{d}{dx}\left(\frac{h^3}{12\mu}\frac{d\Delta p}{dx} - \frac{1}{2}hV\right) = 0 \tag{8.16}$$

  which is the incompressible Reynolds equation (Eq. 8.6) in Δp. In this (low-pressure) limiting case, the fluid behaves thus as incompressible.
- The *high pressure* case: This limiting case is obtained for exceedingly high values of the sliding term μV. Integrating Eq. 8.5 once w.r.t. x yields

  $$p\frac{dp}{dx} = 6\mu V(ph - C) \tag{8.17}$$

  where C is an arbitrary constant. Now, to the left of the p_m point,in Figure 8.1, which shifts farther to the right as the sliding number increases, the pressure times its gradient (or the L.H.S. of Eq. 8.17) is bounded. Thus, as μV takes on very high values, the term in the parentheses on the R.H.S. of that equation must tend to zero. So, we get (see also (Constantinescu 1969)):

  $$ph \rightarrow C = p_o h_o$$

  This singular limit is a very important results as it provides an easy way to estimate the maximum load capacity of a plane slider bearing.

  N.B. To the right of the p_m point,in Figure 8.1, the pressure will generally reduce smoothly back to exit atmospheric pressure p_a.

8.3 Self-Acting or EP Bearings?

There is a fundamental difference in the working principle between a self-acting and an EP bearing. Referring to Figure 8.3, in the first (a), the wedge effect generates the pressure build-up, i.e. as in viscous pump within the bearing itself; whereas in the second (b), pressurised fluid from an external source, e.g. a compressor, is introduced into the bearing gap. It might be pertinent to wonder whether there be differences in the energy efficiency between the two principles. We take a look at this in the following.

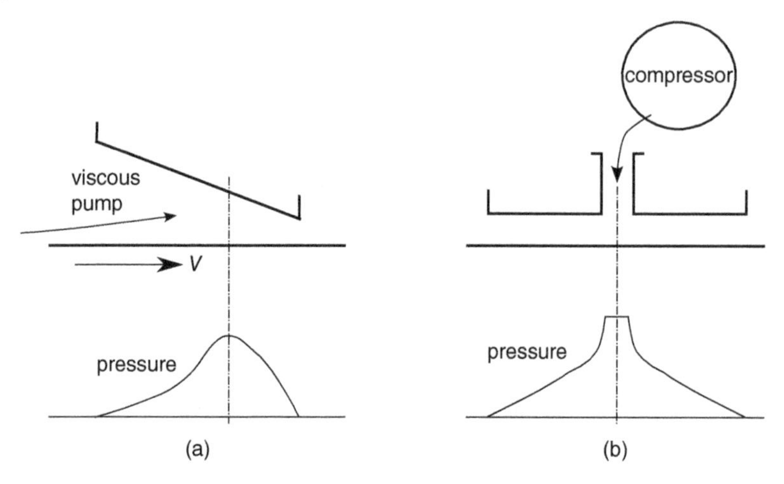

Figure 8.3 Principle of self-acting bearing (a) and externally-pressurised bearing (b).

8.3.1 Energy Efficiency of Self-Acting Bearings

A self-acting bearing generates its own bearing pressure in the gap by viscous shear owing to relative motion of the bearing surfaces. An EP bearing on the hand, uses pressurised fluid, provided by a pump or a compressor, that is introduced into the bearing gap to pressure; i.e. no relative motion is needed. The question that concerns us here is: which of those two principles is the most efficient from energy-consumption point of view? Compressor and pump efficiencies lie approximately in the range 0.65–0.9, depending on type, quality, process, etc. In the following, we shall show that the maximum efficiency of a viscous pump/compressor cannot exceed the theoretically ideal value of $1/3 \approx 0.33$ and, in practical situations, only a portion of this value can be attained.

Let us consider a viscous pump in its simplest guises: two closely spaced parallel surfaces moving relative to one another with a velocity V so that a viscous fluid is pumped from a station at pressure p_a to one that is at $p_m > p_a$, as depicted in Figure 8.4. Without loss of generality, we consider first the case of incompressible fluid, owing to its simplicity.

The volumetric flow rate per unit width of the channel Q', is the negative of the term in the parenthesis of Eq. 8.6

$$Q' = -\frac{h^3}{12\mu}\frac{\mathrm{d}p}{\mathrm{d}x} + \frac{Vh}{2}, \tag{8.18}$$

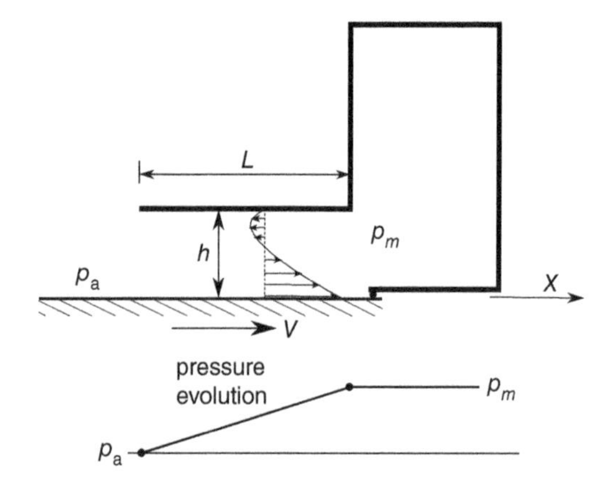

Figure 8.4 Principle of a viscous pump.

whereas the traction stress on the moving surface is given by (Chapter 2):

$$\tau_f = -\left(\frac{h}{2}\frac{dp}{dx} + \mu\frac{V}{h}\right). \tag{8.19}$$

Let us re-emphasise that this traction stress comprises two qualitatively different constituent parts:

- The first term is that due to aerodynamic wedge action owing to the pressure build-up, which increases with increasing film convergence and vanishes when $dh/dx = 0$, (see also Chapter 6). The power lost owing to this component is that needed to pump or compress the fluid in the gap and generate load capacity; i.e. it is the useful power.
- the second term is that due to dissipative friction owing to the viscous action, which is summarised by the so-called Pertrov's law (Dowson 1979): the viscous friction force is directly proportional to the product of the velocity and the viscosity and inversely proportional to film thickness. This friction stress is independent of the hydrodynamic stress, since it depends only on the gap geometry h.

The total traction force on the moving surface, per unit width, is then given by:

$$F'_f = \int_0^L \left(\frac{h}{2}\frac{dp}{dx} + \mu\frac{V}{h}\right) dx \tag{8.20}$$

Since the fluid is incompressible, the pressure distribution is linear in the position, so that we can write:

$$\frac{dp}{dx} = \frac{\Delta p}{L}$$

Substituting this in the equations for the flow and the friction force, we get

$$Q' = -\frac{h^3}{12\mu}\frac{\Delta p}{L} + \frac{Vh}{2} \tag{8.21}$$

and

$$F'_f = \int_0^L \left(\frac{h}{2}\frac{\Delta p}{L} + \mu\frac{V}{h}\right) dx = \frac{h}{2}\Delta p + \mu\frac{VL}{h} \tag{8.22}$$

Now, the energy efficiency μ_e is given by the ratio of the output power $Q'\Delta p$ to the input power $F'_f V$:

$$\eta_e = \frac{Q'\Delta p}{F'_f V} = \frac{\left(\frac{6\mu VL}{\Delta p h^2} - 1\right)}{\left(\frac{2\mu VL}{\Delta p h^2} + 1\right)\frac{6\mu VL}{\Delta p h^2}} \tag{8.23}$$

Putting

$$\frac{6\mu VL}{\Delta p h^2} = \Lambda'$$

which is a sort of sliding number, we have finally:

$$\eta_e = \frac{\Lambda' - 1}{\left(\frac{1}{3}\Lambda' + 1\right)\Lambda'} \tag{8.24}$$

(Note that this equation is valid for $\Lambda' > 1$, which corresponds to positive flow through the pump.)

The maximum value of this function is equal to 1/3 (corresponding to a value of $\Lambda' = 3$) as shown in Figure 8.5.

Let us note here that owing to side leakage in a finite-width aerodynamic bearing, the actual pumping efficiency might be considerably less than this value, depending on the aspect ratio of the bearing.

The previous analysis is valid for any length of channel. Choosing this to be infinitesimal (= dx), the result we arrived at states that maximum pressure-generation efficiency takes place when

$$\Lambda' = \frac{6\mu V}{h^2 dp/dx} = 3$$

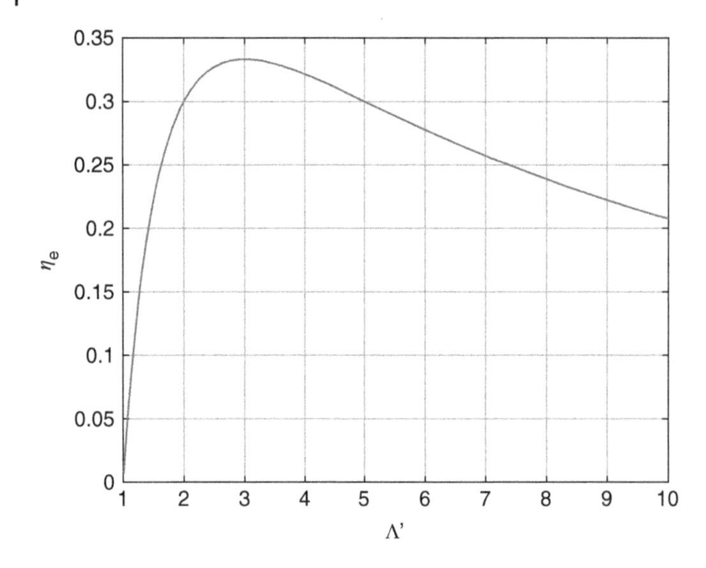

Figure 8.5 Energy efficiency of a viscous pump as a function of the sliding number.

Inserting this requirement, i.e. $\mathrm{d}p/\mathrm{d}x = 2\mu V/h^2$, into the Poissuille term of the incompressible Reynolds equation (Eq. 8.6), we obtain:

$$\frac{\mathrm{d}}{\mathrm{d}x}(Vh) = 0$$

In other words, for any given V, h should be constant and thus the Rayleigh-step bearing would thus emerge as the most energy efficient bearing providing it satisfies the condition on Λ'.

Let us now consider the efficiency in the compressible case.

The results regarding the energy efficiency of the pump apply obviously also to any infinitesimal portion of the channel, where we could assume the density to be (locally) constant. Thus, the same condition on Λ' will hold for maximum efficiency. However, inserting the requirement that $\mathrm{d}p/\mathrm{d}x = 2\mu V/h^2$ now into the compressible Reynolds equation, Eq. 8.4, we obtain:

$$\frac{\mathrm{d}}{\mathrm{d}x}(\rho Vh) = 0,$$

which should determine the most optimal film-height evolution, when the density function is known. As an example, for isothermal flow, the density will be proportional to the pressure so that we get:

$$\frac{\mathrm{d}}{\mathrm{d}x}(pVh) = 0,$$

which, for any given value of V, should lead to a variational problem for determining the gap function $h(x)$. As a special case, we have seen in a previous section that for very high values of V, we have that (Eq. 8.17)

$$ph \to C = p_o h_o$$

on the major part of the bearing, in particular the compression or pressure build-up part. Thus, we need only choose the appropriate value of $p_o h_o$ and $\mathrm{d}h/\mathrm{d}x$ as follows:

$$\mathrm{d}p/\mathrm{d}x = -p_o/h_o \cdot \mathrm{d}h/\mathrm{d}x$$

and

$$\Lambda' = \frac{6\mu V}{h^2 \mathrm{d}p/\mathrm{d}x} = -\frac{6\mu V}{p_o h_o \mathrm{d}h/\mathrm{d}x} = 3$$

which determines $\mathrm{d}h/\mathrm{d}x$ for any given V and $p_o h_o$.

Example

For $p_o = 1$ bar, $h_o = 10\ \mu$m, $\mu = 2\ 10^{-5}$ Pa s, $V = 10$ m s^{-1} $\Rightarrow dh/dx = -0.4\mu$m mm^{-1}.

As a conclusion from the above, we see that the hydrodynamic or the aerodynamic action is not an energy-efficient means of generating pressure for load carrying. The main culprit in this is the viscous-friction dissipation. A significantly more efficient manner is external pressure supply, in particular at low sliding velocities. At high velocities, both types will be dominated by viscous losses. The reason that self-acting bearings are preferred to EP ones has more to do with simplicity of design and construction rather than energy considerations. Recently (Al-Bender 2011) has shown that hydrostatic bearings might be the most energy-efficient of all bearing types even when the cost of the oil-supply ancillary is taken into account.

8.3.2 The Viscous Motor

If in Figure 8.4, we let the lower surface move freely while imposing a pressure difference between inlet and outlet, then the inverse situation to a viscous pump will occur: i.e. a viscous motor. In that case the pressure gradient will drag the plate with a force equal to that calculated for the friction force in the case of the viscous pump. It can be shown that the mechanical efficiency of such a viscous motor is likewise bounded by 1/3.

Although this topic is outside the scope of air bearings as such, this principle has recently been used to advantage in the design of a combined bearing-drive system, where eccentrically fed air to the gap between a stator and a platen results in the controlled motion of that platen (with high dynamics). This is reviewed briefly further in Chapter 15.

8.4 Dimensionless Formulation of the Reynolds Equation

From Chapter 4 (Eq. 4.9), the dimensionless Reynolds equation for a finite-width aerodynamic bearing sliding in the x-direction is given by

$$\frac{\partial}{\partial X}\left(H^3\wp\frac{\partial P}{\partial X} - \Lambda\wp H\right) + \frac{\partial}{\partial Y}\left(H^3\wp\frac{\partial P}{\partial Y}\right) = \sigma\frac{\partial}{\partial \tau}(\wp H) \tag{8.25}$$

where, the normalisation is:

$$X = x/L, H = h/h_a, P = p/p_a, \wp = \rho/\rho_a, \Lambda = \frac{6\mu VL}{p_a h_o^2}, \sigma = \frac{12\mu\nu L}{p_a h_o^2}, \tau = \nu t$$

(ν is the angular frequency of harmonic oscillation or perturbation),

Λ is the sliding number (also called the bearing number, the compressibility number, or the Harrison number) and σ is the squeeze number.

This normalisation of the equation is obviously not unique and other dimensionless formulation may be used to suite other situations.

Boundary Conditions

The formulation of Eq. 8.25 assumes the sliding to be in the x-direction. We can, at any rate, choose the coordinate system so that this becomes the case. The boundary conditions generally correspond to the pressure field prevailing at the boundaries of the bearing pad. However, in most cases of interest, the pressure is assumed to be constant at the ambient value outside the bearing, $p = p_a$.

Let us note however, that at the fluid-entrainment side, or the inlet to the bearing, the flow will generally develop before attaining viscous character downstream of the gap entrance, as shown in Chapter 3. This is associated with an equivalent pressure rise, the "ram" pressure, which should be used as a correction to the ambient value at inlet.

8.5 Some Basic Aerodynamic Bearing Configurations

The previous sections have provided the basic equations and described the resulting behaviour of aerodynamic bearings. Now, we shall consider some well-known slider, i.e. thrust, and journal bearing configurations to gain an idea about general pressure distributions and the trends mentioned above. Let us note that, owing to its strong non-linearity, the Reynolds equation hardly possesses any analytic solutions, even for the simplest of two-dimensional cases, except the two extreme cases mentioned above. We thus have to resort to numerical solutions for most cases.

8.5.1 Slider Bearings

We shall consider the flat rectangular bearing cases, although, being generic, the results can easily be adapted to the annular, radial, etc. cases.

Inclined slider

This is the most common case to consider. It forms the basis for the tilting-pad bearing application.

Adopting the notation of Figure 8.6, the problem is defined by:

- $\Lambda = \frac{6\mu VL}{p_a h_a^2}$, the bearing number or sliding number;
- $\sigma = \frac{12\mu \nu L}{p_a h_a^2}$, the squeeze number;
- h_o/h_a the taper ratio;
- h_a/L film aspect ratio;
- B/L bearing-pad aspect ratio.

This means that two bearings sharing these design parameters will have the same characteristics, see Chapter 2:

- $\bar{W} = \frac{W}{p_a LB}$, the dimensionless bearing load;
- $K = \frac{kh_a}{p_a LB}$, the dimensionless stiffness;
- $C = \frac{c}{12\mu B}\left(\frac{h_a}{L}\right)^3$ the dimensionless damping;
- $\bar{\dot{m}} = \frac{2\dot{m}}{\rho_a h_a BV}$ the dimensionless mass flow rate through the bearing.

Typical pressure distributions are shown in Figure 8.7 for $B/L = 1$ and various values of the sliding number and taper ratio. The peak pressure, and thus also the load capacity, increases with increasing bearing number and taper ratio. Let us note in particular how the pressure distribution approaches the limiting $ph = C$ behaviour as Λ becomes very large in the right most panel.

This bearing principle can be used in fixed or tilting pad configurations, both in thrust annular or journal bearing situations. The latter is treated in detail in Chapter 11.

Figure 8.8 gives us an idea about the load and friction characteristics of this type of bearing. As regards the load, we note that this becomes appreciable only for very small gap heights; in our case, well below 10 μm. We

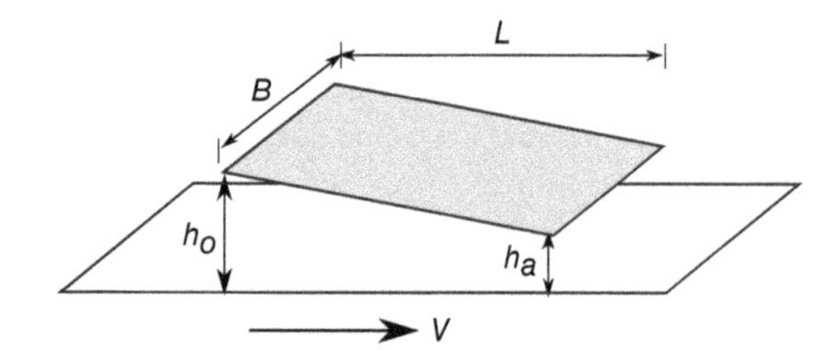

Figure 8.6 Configuration of finite tapered-pad bearing.

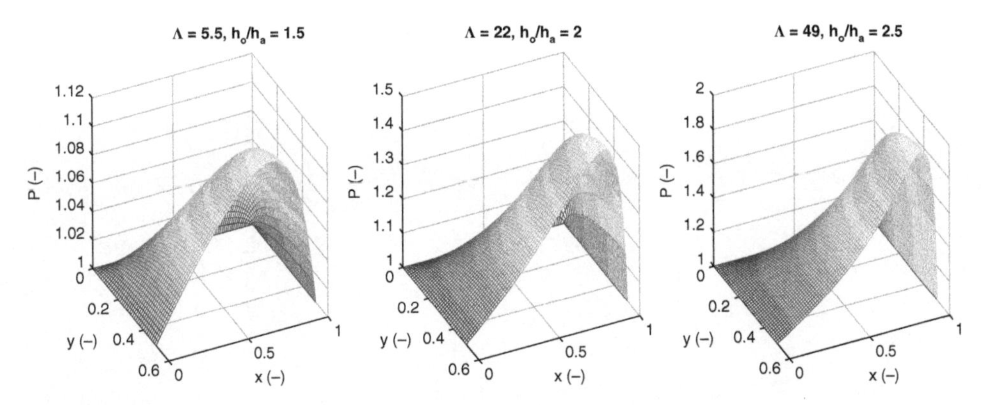

Figure 8.7 The pressure distribution in a finite tapered-pad bearing with $B/L = 1$ for various values of the sliding number and h_o/h_a.

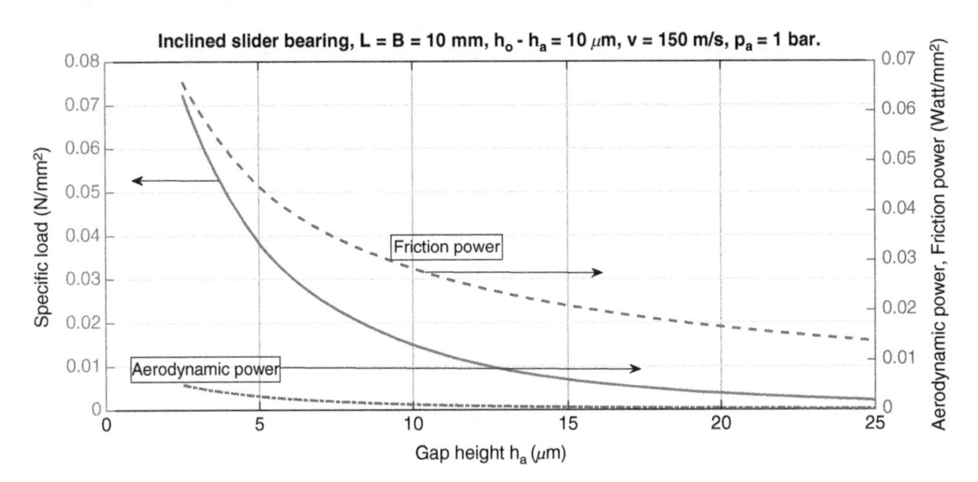

Figure 8.8 The specific load, aerodynamic power and friction power of a finite tapered-pad bearing with $L = B = 10$ mm, $h_o - h_a = 10$ μm, $V = 150$ m s^{-1}, for various values of the nominal gap height h_a.

can verify that, in that region, the approximation $ph =$ const. is valid. This means that while the sliding speed has to be high in order to reach this saturation point, a further increase thereof is pointless and, in actual fact, counter productive as it will only increase the friction losses unnecessarily. On the same figure are plotted the traction power losses, consisting in the (i) viscous-friction part and (ii) the aerodynamic part; cf. Eq. 8.19. We note that the friction power is the most dominant when compared to the aerodynamic power. These friction losses are inevitable and can become quite large; in this example, on the order of 5 Watt cm^{-2}, which can form a serious issue of thermal management in high-speed devices.

Figure 8.9 plots the static stiffness and damping obtained by the method of small-amplitude perturbation, see e.g. Chapter 7. At small gaps, the stiffness may be accurately estimated from the asymptotic pressure behaviour through:

$$ph = C \Rightarrow Wh = C' \Rightarrow \frac{dW}{dh} = -\frac{W}{h} = -k$$

That is, the stiffness at that asymptotic limit is simply the ratio of the bearing load to the gap height.

The static damping (i.e. at zero frequency) is plotted on the same Figure 8.9. Here, there is no direct rule to estimate this, since the result depends on the solution of the dynamic Reynolds equation. In particular, the damping can become negative, as we shall see in the case of grooved bearings below.

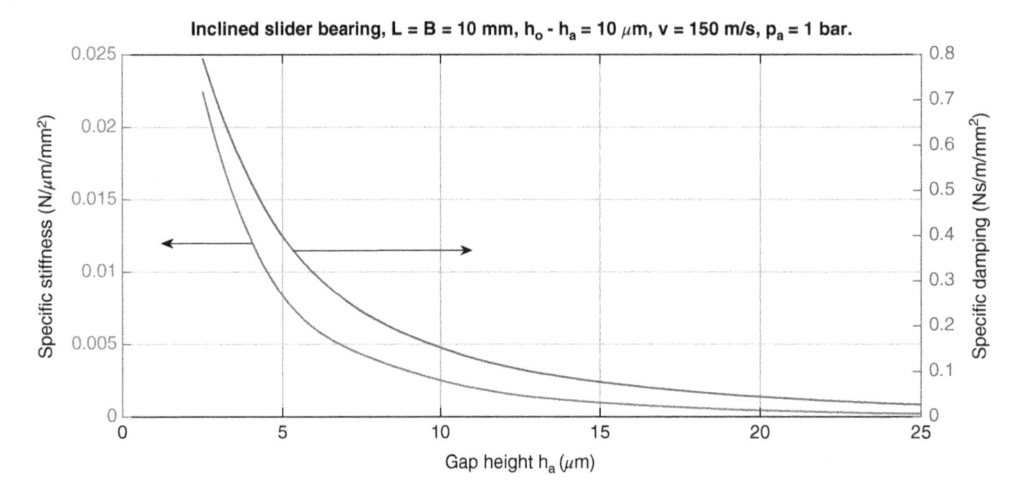

Figure 8.9 The specific static stiffness and damping of a finite tapered-pad bearing with $L = B = 10$ mm, $h_o - h_a = 10$ μm, $V = 150$ m s^{-1}, for various values of the nominal gap height h_a.

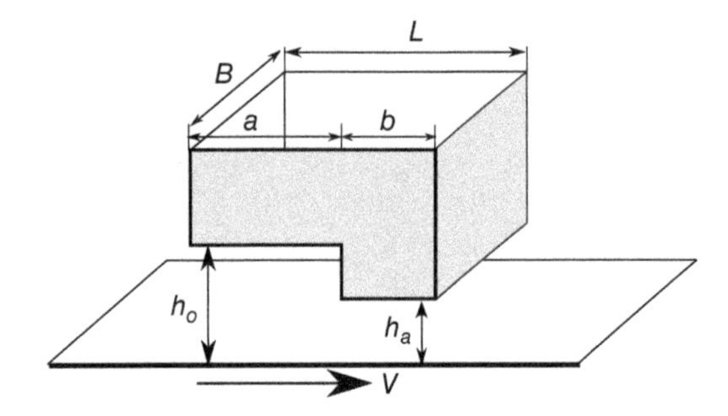

Figure 8.10 Configuration of finite Rayleigh-step bearing.

Rayleigh step

Referring to Figure 8.10, the problem is defined by the same parameters as for the inclined slider in addition to the step-to-length ratio a/L.

Typical pressure distributions are shown in Figure 8.11 for $B/L = 1, a/L = 0.5$ and various values of the sliding number Λ and gap ratio h_o/h_a. As in the previous case, the peak pressure, and thus also the load capacity, increases with increasing bearing number and gap ratio. In addition, owing to the step geometry, i.e. constant gap portions,

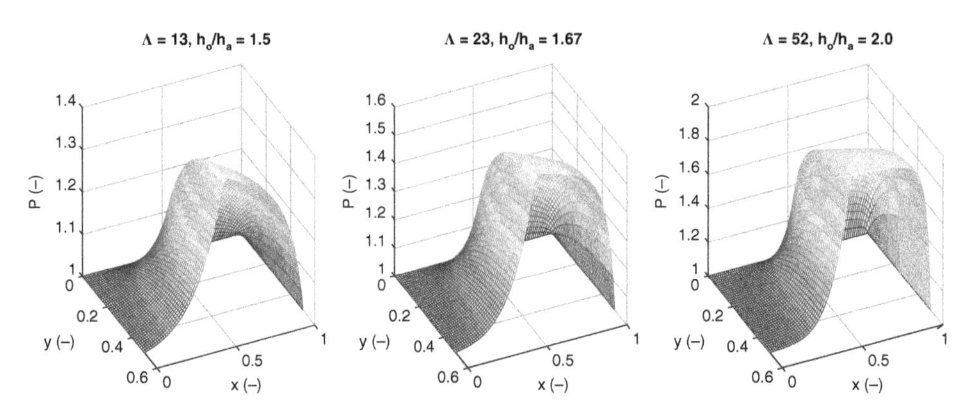

Figure 8.11 The pressure distribution in a finite stepped-pad bearing with $B/L = 1$ and $a/L = 0.5$ for various values of the sliding number and h_o/h_a.

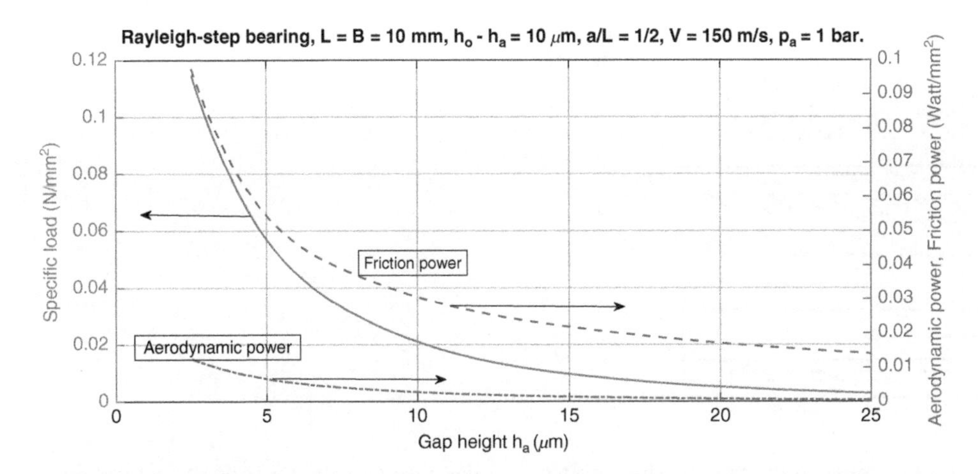

Figure 8.12 The specific load, aerodynamic power and friction power of a finite Rayleigh-step bearing with $L = B = 10$ mm, $h_o - h_a = 10$ µm, $V = 150$ m s^{-1}, for various values of the nominal gap height h_a.

we see more clearly how the pressure distribution approaches the limiting $p = C$ in each potion of the bearing, as Λ becomes very large in the right most panel.

As in the case of inclined-pad bearing, Figure 8.12 plots the load and friction characteristics of the step bearing. The same remarks as for the previous case are applicable. However, two differences can be distinguished: (i) the load capacity is higher for this type than for the previous one and (ii) the friction losses are higher. This latter characteristic is owing to the fact that the friction stress is proportional to the mean of $1/h$, which is larger for a step than for a taper.

Figure 8.13 plots the static stiffness and damping for this step bearing. In comparison with the tapered bearing, we remark that the stiffness is now considerably higher while the damping is considerably lower. This confirms what is appearing to be a general rule that: *increasing the stiffness reduces the damping.*

Herringbone grooved bearing

Figure 8.14 sketches the principle of this type of bearing. The bearing surface has shallow grooves, which can be considered as a series of oblique Rayleigh-steps, creating the form of a herringbone, possibly with a central

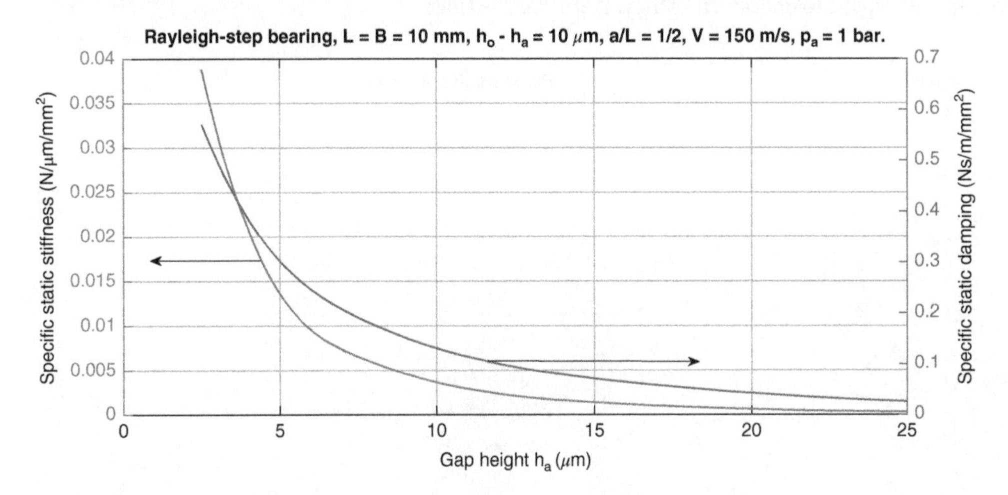

Figure 8.13 The specific static stiffness and damping of a finite Rayleigh-step bearing with $L = B = 10$ mm, $h_o - h_a = 10$ µm, $V = 150$ m s^{-1}, for various values of the nominal gap height h_a.

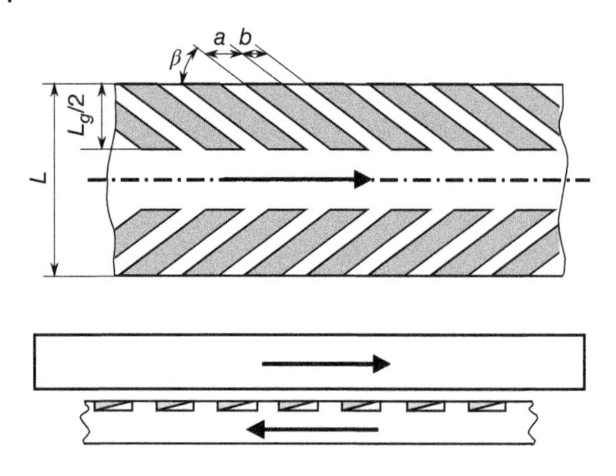

Figure 8.14 Configuration of finite herringbone-grooved bearing.

smooth land. Relative motion between the two bearing surfaces induces viscous flow towards the central land, thus creating an over pressure that lifts up the load.

The behaviour is determined by the same parameters as for a step bearing, in addition to the groove angle β, grooved to smooth part L_g/L and the number of grooves N_g per unit length or per period of repetition of the pattern.

The pressure distribution generated in a nominally parallel gap, arbitrarily chosen grooved bearing arrangement, is depicted in the right panel of Figure 8.15 over the grooved portion, shown in the left panel.

We see that, except for the "corrugation" effect across the grooves, the pressure profile resembles that of an inward build-up of pressure from ambient value to a maximum one at the centre line of the bearing.

Traditionally, owing to the difficulty of constructing a direct numerical solution for a grooved bearing, especially when the number of grooves is so high as to demand a very large number of grid points, an equivalent Reynolds equation, with nominally smooth surfaces, was developed in the 1950's and 1960's. Owing to the importance of this theory, it will be presented in a dedicated section further below.

First, we provide two examples of load and friction characteristics of this type of bearing in order to demonstrate the effect of groove angle, which is responsible for the inward pumping effect that distinguishes this type of bearing from the previously discussed two types. For this, we consider firstly the case of large groove angle depicted in Figure 8.16. Owing to the large angle, the load and friction characteristics are comparable to the step bearing.

In Figure 8.17, we have decreased the groove angle and increased the sliding speed to show the superiority of the herringbone principle in regard to energy efficiency. Here, we see that:

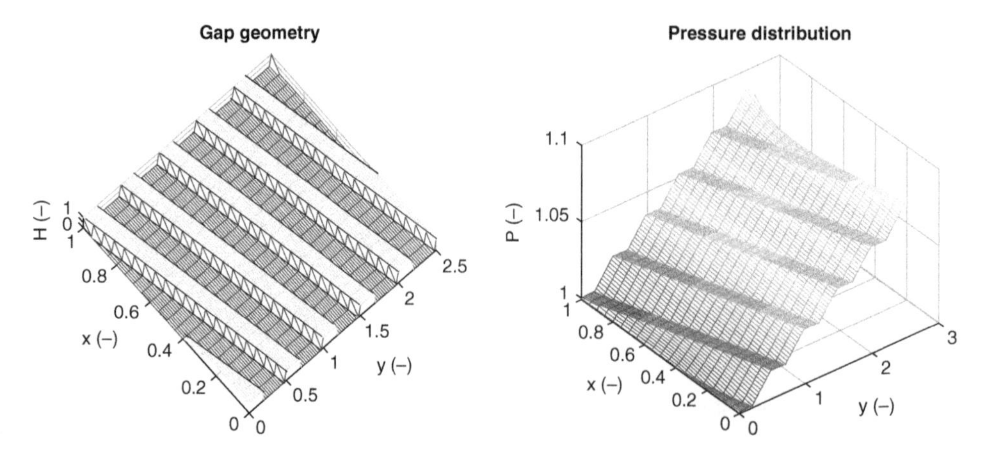

Figure 8.15 (Left) grooved bearing profile (not to scale). (Right) Corresponding pressure distribution

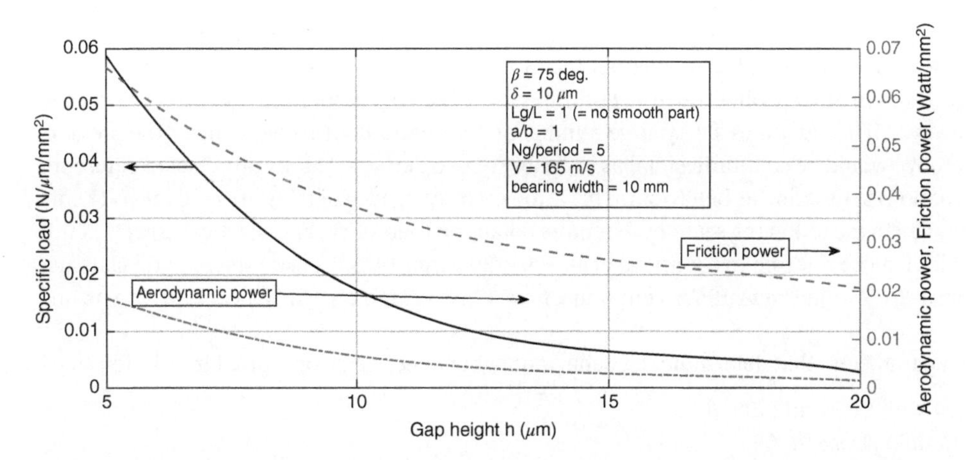

Figure 8.16 The specific load, aerodynamic power and friction power of a herringbone bearing with $L = 10$ mm, $\beta = 75$ deg., $\delta = 10$ μm, $V = 165$ m s^{-1}, number of grooves per period $N_g = 5$, with periodic boundary conditions in the sliding direction, for various values of the nominal gap height h.

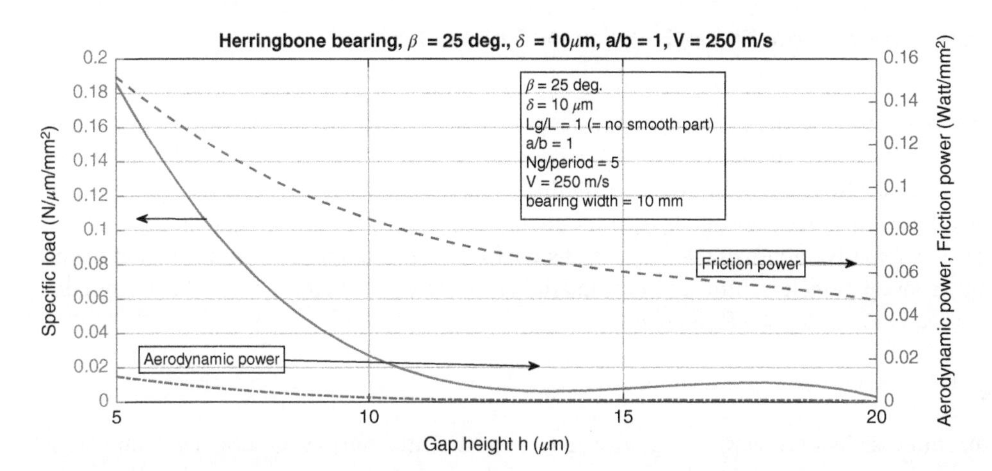

Figure 8.17 The specific load, aerodynamic power and friction power of a herringbone bearing with $L = 10$ mm, $\beta = 25$ deg., $\delta = 10$ μm, $V = 165$ m s^{-1}, number of grooves per period $N_g = 5$, with periodic boundary conditions in the sliding direction, for various values of the nominal gap height h.

- the load does not saturate quickly for high speeds as the previous two types did;
- much higher load is achieved for the same friction power.
- we shall see in a further section that the stiffness is also considerably higher but that the damping suffers as a consequence.

The dynamic stiffness will be discussed in a more elaborate way in section 8.8 below.

8.6 Grooved-Surface Bearings

As mentioned above, grooved self-acting bearings are considered as a special class of bearings having important advantages over other types. Provided that they are properly optimised, they are claimed to produce higher stiffness coupled with good damping. They have shallow grooves, as a cascade of Rayleigh step bearings, that could be generally of various geometries, which are placed on either the moving or the stationary member. Generally, one would need numerical methods in order to obtain an exact solution to the induced pressure, as has been provided in the previous section. However, if one assumes a very large number of grooves, or high groove density, the pressure change from groove to groove would be so infinitesimal as to enable one to derive a "continuum" set of equations (equivalent Reynolds equation) to describe its behaviour. This is the underlying idea of the Narrow-Groove Theory (NGT), which will be pursued in the following sections. For more details on theoretical and practical aspects of this type of bearings, the reader may be referred to (Bonneau et al. 1993; Bootsma 1975; Constantinescu and Galetuse 1987; Malanoski and Pan 1965; Muijderman 1967; Vohr and Chow 1965; Whipple 1958; Wilcock 1972; Xue and Stolarski 1997).

There are many design parameters that determine the behaviour, which need to be optimised in a design step:

- groove angle subtended with the centre line β
- groove depth w.r.t. film thickness δ/h
- groove width/land width a/b
- grooved part width / central land L_g/L
- number of grooves per unit length (or groove density) N_g.

8.6.1 Derivation of the Narrow-Groove Theory Equation for Grooved Bearings

Approach

Figure 8.18 depicts a grooved bearing where the straight grooves are on the stationary upper surface, in a direction normal to the direction of sliding. The grooves are $\delta = h_o - h$ deep and a wide. They are separated by "lands" or "ridges" of width b at a distance h from the sliding surface. Sliding of the lower (flat) bearing surface is associated with flow of the viscous fluid being characterised by volumetric flows Q_x and Q_y across and along the grooves, respectively. Our purpose is to derive approximate expressions for those flow components as a function of the sliding velocities and the groove parameters, film geometry and fluid properties. When these are available, the construction of the Reynolds equation is directly facilitated.

8.6.2 Assumptions

We assume locally incompressible flow across a groove. For a sufficiently narrow groove, i.e. number of grooves $N_g \to \infty$, this should have only negligible effect on the global result since the density variation from groove to groove will be infinitesimal. On the other hand, if this assumption is not made, it becomes very difficult to find a practicable solution to the problem as it becomes strongly non-linear in the velocity and geometry.

Further we assume rigid surfaces, no slip and that the Reynolds equation holds true on each groove-land pair.

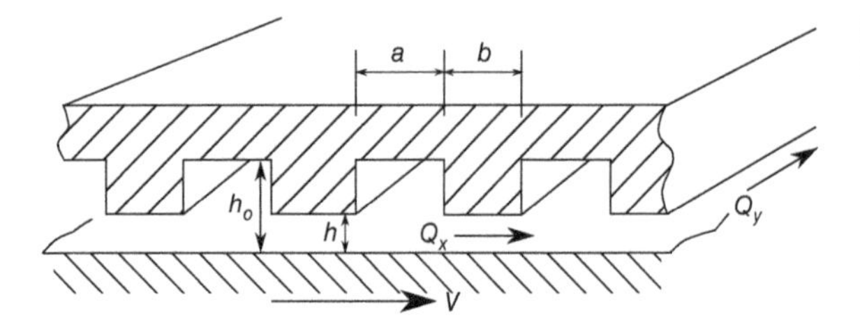

Figure 8.18 Geometry of a generic bearing with grooves in the stationary member (not to scale).

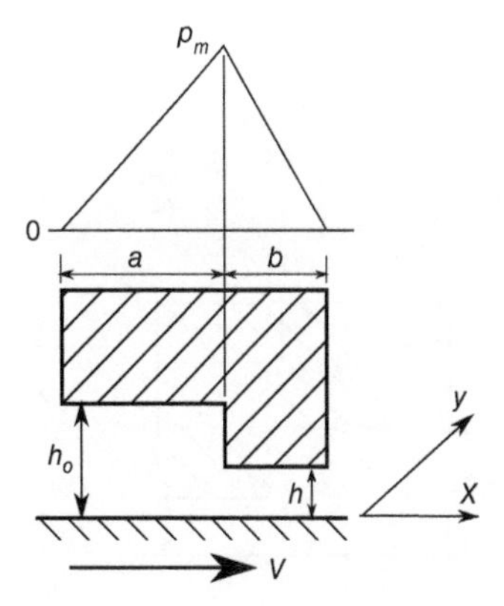

Figure 8.19 Flow across the groove.

8.6.3 Flow in the x-Direction

Shear-Induced Flow

In the following all flows Q are given per unit width in the y-direction. As the flow is assumed incompressible Q must be continuous across the step in Figure 8.19. The shear-induced flow contribution is

$$(Q_x)_S = -\frac{h_o^3 p_m}{12\mu a} + \frac{Vh_o}{2} = \frac{h^3 p_m}{12\mu b} + \frac{Vh}{2},$$

where p_m is the pressure at the step jump, or the maximum pressure.

Thus,

$$\frac{p_m}{12\mu}\left(\frac{h_o^3}{a} + \frac{h^3}{b}\right) = \frac{V(h_o - h)}{2}$$

whence,

$$p_m = \frac{V(h_o - h)}{2} \frac{1}{\left(\frac{h_o^3}{a} + \frac{h^3}{b}\right)} 12\mu := 12C\mu \frac{V(h_o - h)}{2}$$

where,

$$C = \frac{1}{\left(\frac{h_o^3}{a} + \frac{h^3}{b}\right)}.$$

Thus,

$$(Q_x)_S = -h_o^3 \frac{V}{2} \frac{h_o - h}{a} C + \frac{Vh_o}{2} = V\left(-h_o^3 \frac{h_o - h}{2a} C + \frac{h_o}{2}\right) := \frac{V}{2} \cdot B(h_o, h, a, b)$$

with,

$$B = -h_o^3 \frac{h_o - h}{a} B + h_o = \frac{ah_o h^3 + bh_o^2 h}{ah^3 + bh_o^3}.$$

Pressure-Induced Flow

The pressure-induced flow contribution is, (Figure 8.20)

$$(Q_x)_P = \frac{h_o^3(p_1 - p_m^*)}{12\mu a} = \frac{h^3(p_m^* - p_2)}{12\mu b}$$

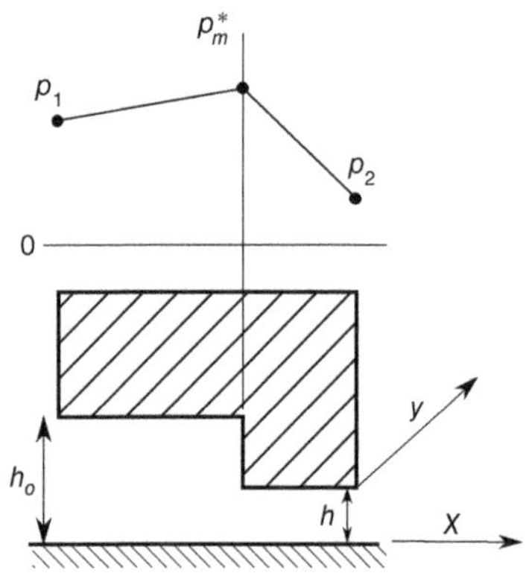

Figure 8.20 Poiseuille flow.

Thus,

$$\bar{\alpha}(p_1 - p_{\mathrm{m}}^*) = p_{\mathrm{m}}^* - p_2$$

with

$$\bar{\alpha} = \frac{h_{\mathrm{o}}^3}{h^3}\frac{b}{a}$$

and so

$$p_{\mathrm{m}}^* = \frac{\bar{\alpha}p_1 + p_2}{1+\bar{\alpha}}$$

is the weighted average of p_1 and p_2 where $p_1 \lesseqgtr p_{\mathrm{m}}^* \lesseqgtr p_2$. Thus,

$$(Q_x)_{\mathrm{P}} = \frac{h_{\mathrm{o}}^3}{12\mu a}\left(p_1 - \frac{\bar{\alpha}p_1 + p_2}{1+\bar{\alpha}}\right) = \frac{h_{\mathrm{o}}^3}{12\mu a}\left(\frac{p_1 - p_2}{1+\bar{\alpha}}\right)$$

Going over to differential formulation by writing

$$\frac{p_2 - p_1}{a+b} \sim \frac{\partial \bar{p}}{\partial x},$$

we finally get the flow (per unit y),

$$(Q_x)_{\mathrm{P}} = -\frac{1}{12\mu}\frac{(a+b)h_{\mathrm{o}}^3}{a + b\frac{h_{\mathrm{o}}^3}{h^3}}\frac{\partial \bar{p}}{\partial x}.$$

8.6.4 Flow in the *y*-Direction

Referring to Figure 8.21 for notation, we write the flow (Poiseuille plus Couette) along the groove and the land respectively, as

$$Q_{\mathrm{h_o}} = -\frac{ah_{\mathrm{o}}^3}{12\mu}\frac{p_{\mathrm{m_1}} - p_{\mathrm{m_2}}}{L} + \frac{aUh_{\mathrm{o}}}{2} \qquad \text{(flow along groove)}$$

$$Q_{\mathrm{h}} = -\frac{bh^3}{12\mu}\frac{p_{\mathrm{m_1}} - p_{\mathrm{m_2}}}{L} + \frac{bUh}{2} \qquad \text{(flow along land),}$$

where, $p_{\mathrm{m_1}}$ and $p_{\mathrm{m_2}}$ are the mean pressures on either side of the channel.

In the limit $L \to 0$, we have $\frac{p_{\mathrm{m_2}} - p_{\mathrm{m_1}}}{L} \to \frac{\partial p_{\mathrm{m}}}{\partial y}$, so that the flow per unit width $(a+b)$ will be,

$$Q_y = -\frac{ah_{\mathrm{o}}^3 + bh^3}{12\mu(a+b)}\frac{\partial p_{\mathrm{m}}}{\partial y} + \frac{U}{2}\left(\frac{ah_{\mathrm{o}} + bh}{a+b}\right)$$

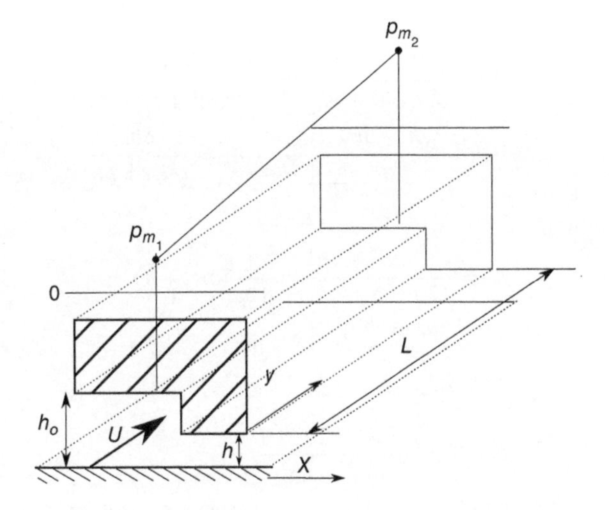

Figure 8.21 Flow along the groove direction (y-direction).

8.6.5 Squeeze Volume

The volume of gas V_g, under a groove-land pair, per unit area $L(a + b)$ is equal to

$$V_g = \frac{ah_o + bh}{a + b}$$

In summary

We have determined the following quantities:

Q_x per unit width y:

$$Q_x = -A\frac{\partial p}{\partial x} + B\frac{V}{2}$$

with

$$A = \frac{1}{12\mu}\frac{(a + b)h^3h_o^3}{ah^3 + bh_o^3}$$

$$B = h_o - h_o^3\frac{h_o - h}{a}\frac{1}{\frac{h_o^3}{a} + \frac{h^3}{b}} = \frac{ah_oh^3 + bh_o^3h}{ah^3 + bh_o^3},$$

and Q_y per unit length x:

$$Q_y = -A'\frac{\partial \bar{p}}{\partial y} + \frac{U}{2}\frac{ah_o + bh}{a + b} := -A'\frac{\partial \bar{p}}{\partial y} + \frac{U}{2}B'$$

with

$$A' = \frac{ah_o^3 + bh^3}{12\mu(a + b)}$$

and

$$B' = \frac{ah_o + bh}{a + b}$$

In order to further simplify notation, let us define new parameters:

$$\bar{a} = \frac{a}{a + b}$$

$$\bar{b} = \frac{b}{a + b}$$

$$\delta = h_o - h$$

(N.B. $\bar{a} + \bar{b} = 1$)

So that

$$A = \frac{h_o^3 h^3}{12\mu} \frac{a + b}{ah^3 + bh_o^3} = \frac{1}{12\mu} \frac{h_o^3 h^3}{\bar{a}h^3 + \bar{b}h_o^3}$$

$$B = -h_o^3 \frac{h_o - h}{a} \frac{1}{\frac{h_o^3}{a} + \frac{h^3}{b}} + h_o = \frac{\delta \bar{b} h_o^3}{(\bar{a}h^3 + \bar{b}h_o^3)}$$

and

$$A' = \frac{1}{12\mu} \frac{ah_o^3 + bh^3}{a + b} = \frac{1}{12\mu}(\bar{a}h_o^3 + \bar{b}h^3)$$

$$B' = \frac{ah_o + bh}{a + b} = \bar{a}h_o + \bar{b}h$$

Also,

$$B - B' = -\frac{\bar{a}\bar{b}\delta\,(h_o^3 - h^3)}{\bar{a}h^3 + \bar{b}h_o^3}.$$

Finally, using the expressions for Q_x, Q_y and the squeeze volume, we can write the Reynolds equation (Eq. 8.1) for this case as:

$$\frac{\partial}{\partial x}\left(\rho\left(A\frac{\partial p}{\partial x} - B\frac{V}{2}\right)\right) + \frac{\partial}{\partial y}\left(\rho\left(A'\frac{\partial p}{\partial y} - B'\frac{U}{2}\right)\right) = \frac{\partial}{\partial t}(\rho V_g) \tag{8.26}$$

8.6.6 Inclined-Grooves Reynolds Equation

We are naturally interested in the general case depicted in Figure 8.22, where the grooves are inclined by an arbitrarily chosen angle β w.r.t. the sliding direction. In order to obtain the Reynolds equation corresponding to this case, we affect the coordinate transformation

$$\xi = x \sin\beta - y\cos\beta$$
$$\eta = x\cos\beta + y\sin\beta$$

$$\frac{\partial}{\partial \xi} = \sin\beta\frac{\partial}{\partial x} - \cos\beta\frac{\partial}{\partial y}$$
$$\frac{\partial}{\partial \eta} = \cos\beta\frac{\partial}{\partial x} + \sin\beta\frac{\partial}{\partial y}$$

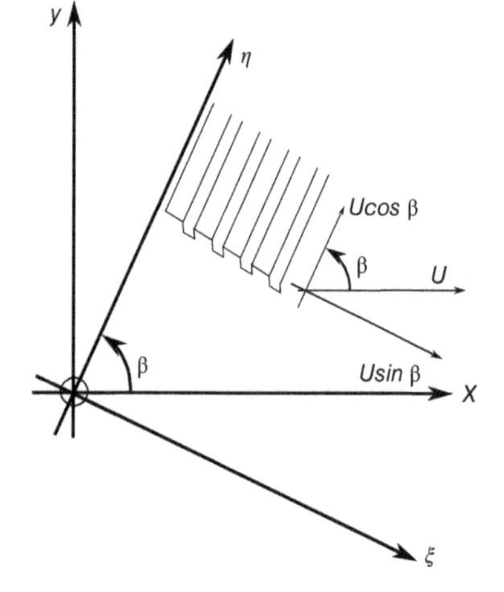

Figure 8.22 Inclined Grooves.

The Reynolds equation in ξ, η-coordinate system is:

$$\frac{\partial}{\partial \xi}\left[\rho\left(A\frac{\partial p}{\partial \xi} - B\frac{U}{2}\sin\beta\right)\right] + \frac{\partial}{\partial \eta}\left[\rho\left(A'\frac{\partial p}{\partial \eta} - B'\frac{U}{2}\cos\beta\right)\right] = \frac{\partial}{\partial t}(\rho V_{\mathrm{g}})$$

where A, B, A' and B' are known from the generic case above. Performing this transformation yields:

$$\begin{aligned}
&\sin\beta\frac{\partial}{\partial x}\left(A\rho\left(\sin\beta\frac{\partial p}{\partial x} - \cos\beta\frac{\partial p}{\partial y}\right) - B\rho\frac{U}{2}\sin\beta\right) - \\
&\cos\beta\frac{\partial}{\partial y}\left(A\rho\left(\sin\beta\frac{\partial p}{\partial x} - \cos\beta\frac{\partial p}{\partial y}\right) - B\rho\frac{U}{2}\sin\beta\right) + \\
&\cos\beta\frac{\partial}{\partial x}\left(A'\rho\left(\cos\beta\frac{\partial p}{\partial x} + \sin\beta\frac{\partial p}{\partial y}\right) - B'\rho\frac{U}{2}\cos\beta\right) + \\
&\sin\beta\frac{\partial}{\partial y}\left(A'\rho\left(\cos\beta\frac{\partial p}{\partial x} + \sin\beta\frac{\partial p}{\partial y}\right) - B'\rho\frac{U}{2}\cos\beta\right) = \frac{\partial}{\partial t}(\rho V_{\mathrm{g}}).
\end{aligned} \tag{8.27}$$

At this point, let us note the following special cases:

- *Smooth bearing*: when $A = A'$ and $B = B'$, we retrieve the case of a smooth bearing:

$$\frac{\partial}{\partial x}\left(\rho\left(A\frac{\partial p}{\partial x} - B\frac{U}{2}\right)\right) + \frac{\partial}{\partial y}\left(\rho\left(A'\frac{\partial p}{\partial y}\right)\right) = \frac{\partial}{\partial t}(\rho V_{\mathrm{g}}), \tag{8.28}$$

where, $A = A' = h^3$ and $B = h$.

- *Inward pumping*: referring to Figure 8.23, for an infinitely long bearing in the x-direction, having a nominally parallel gap, i.e. $\frac{\partial}{\partial x} \equiv 0$, operating in steady state, $\frac{\partial}{\partial t} \equiv 0$, with zero net flow in the y-direction, we have, after integrating once w.r.t. y,

$$(A\cos^2\beta + A'\sin^2\beta)\frac{\partial p}{\partial y} + (B - B')\sin\beta\cos\beta\frac{U}{2} = 0 = \text{flow/unit length,}$$

which after rearrangement yields

$$\begin{aligned}
\frac{\partial p}{\partial y} &= \frac{(B - B')\sin\beta\cos\beta}{(A\cos^2\beta + A'\sin^2\beta)}\frac{U}{2} \\
&= 6\mu U\delta^2\frac{\bar{a}\bar{b}(h_{\mathrm{o}}^2 + h_{\mathrm{o}}h + h^2)\sin\beta\cos\beta}{h_{\mathrm{o}}^3h^3\cos^2\beta + (\bar{a}h^3 + \bar{b}h_{\mathrm{o}}^3)(\bar{a}h_{\mathrm{o}}^3 + \bar{b}h^3)sin^2\beta} \\
&:= 6\mu U\delta^2 G(h, h_{\mathrm{o}}, \bar{a}, \bar{b}, \sin\beta, \cos\beta).
\end{aligned} \tag{8.29}$$

This expression is positive for positive $U\sin\beta\cos\beta$ and vice-versa.

Let us note that, since the flow rate is zero, the pressure gradient is independent of the density, which falls out identically after integration. The maximum value of the pressure gradient, which is constant in the grooved

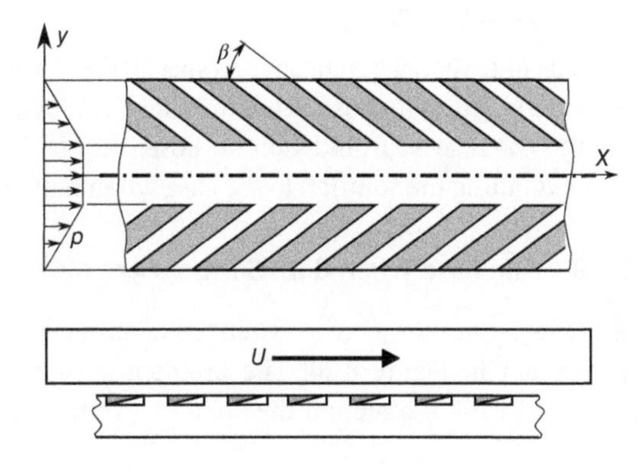

Figure 8.23 Inwardly pumping herringbone bearing with parallel nominal gap.

region, depends then, for any given μU, on h, h_o and β, i.e. the grooved gap geometry. In particular, we note that it vanishes for $\beta = 0, \pi/2$ corresponding to the grooves being parallel or perpendicular to the sliding direction, respectively.

We can integrate Eq. 8.29 w.r.t. y to obtain the pressure rise over the grooved portion as:

$$\Delta p = (p_m - p_a) = 6\mu U\delta^2 G\, L_g,$$

from which, the load capacity per unit width of the bearing may be obtained as

$$\Delta p(2L - L_g).$$

This compact formula may be used to estimate the load capacity of a herringbone-groove flat bearing.

8.6.7 Globally Compressible Reynolds Equation

Inserting the above expressions (for A etc.) in Eq. 8.27 yields the sought for general Reynolds equation with stationary grooves:

$$\frac{\partial}{\partial x}\left(\rho\left(f_x\frac{\partial p}{\partial x} + f_c\frac{\partial p}{\partial y}\right) - g_x\rho\frac{U}{2}\right) + \frac{\partial}{\partial y}\left(\rho\left(F_c\frac{\partial p}{\partial x} + F_y\frac{\partial p}{\partial y}\right) - g_y\rho\frac{U}{2}\right) = \frac{\partial}{\partial t}\left(\rho\frac{ah_o + bh}{a+b}\right), \tag{8.30}$$

with

$$f_x = A\sin^2\beta + A'\cos^2\beta$$

$$f_c = -A\sin\beta\cos\beta + A'\sin\beta\cos\beta = (A' - A)\sin\beta\cos\beta$$

$$g_x = B\sin^2\beta + B'\cos^2\beta = (B - B')\sin^2\beta + B'$$

$$F_c = -A\sin\beta\cos\beta + A'\sin\beta\cos\beta = (A' - A)\sin\beta\cos\beta = f_c$$

$$F_y = A\cos^2\beta + A'\sin^2\beta$$

$$g_y = -B\sin\beta\cos\beta + B'\sin\beta\cos\beta = (B' - B)\sin\beta\cos\beta$$

rearranging and simplifying Eq. 8.30, we obtain:

$$\begin{aligned}&\frac{\partial}{\partial x}\left(\rho\left(f_x\frac{\partial p}{\partial x} + f_c\frac{\partial p}{\partial y}\right)\right) + \frac{\partial}{\partial y}\left(\rho\left(f_c\frac{\partial p}{\partial x} + f_y\frac{\partial p}{\partial y}\right)\right)\\ &+ c_s\left(\sin\beta\frac{\partial}{\partial x} - \cos\beta\frac{\partial}{\partial y}\right)(f_s\rho) = 12\mu\left(\frac{U}{2}\frac{\partial}{\partial x} + \frac{\partial}{\partial t}\right)(\rho f_v),\end{aligned} \tag{8.31}$$

where the new symbols are:

$$c_s = 6\mu U\bar{a}\bar{b}\delta\sin\beta$$

$$f_s = \frac{h_o^3 - h^3}{\bar{b}h_o^3 + \bar{a}h^3}$$

$$f_v = \bar{a}h_o + \bar{b}h$$

Let us note that, when taken as a special case of stationary grooves (with moving smooth surface), these expressions are identical to those found in Wilcock (1969), although the derivation method adopted here is different. This can thus be take as a validation of those results. We shall give the general expressions at the end of this analysis after extending the solution to the case when both surfaces are moving.

8.6.8 The Case When Both Surfaces Are Moving

The extension to the case when both surfaces are moving with velocities U_h (smooth) and U_0 (grooved), is depicted in Figure 8.24. The problem amounts to determining the flow in the ξ-direction normal to the grooves. In the η-direction the solution is obvious. That is, we need to determine the new values of A, A', B and B'.

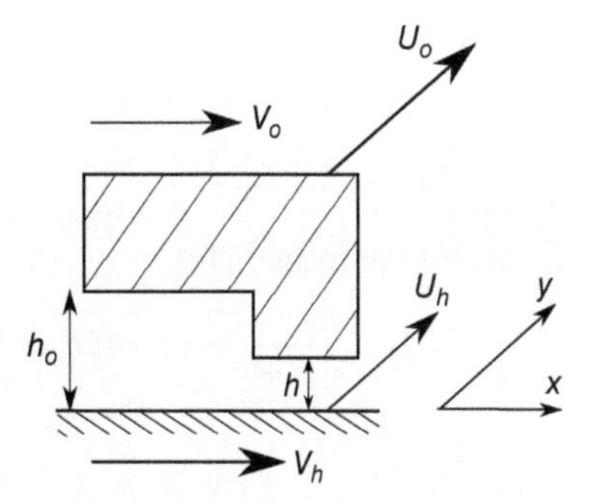

Figure 8.24 New situation sketch with both surfaces moving.

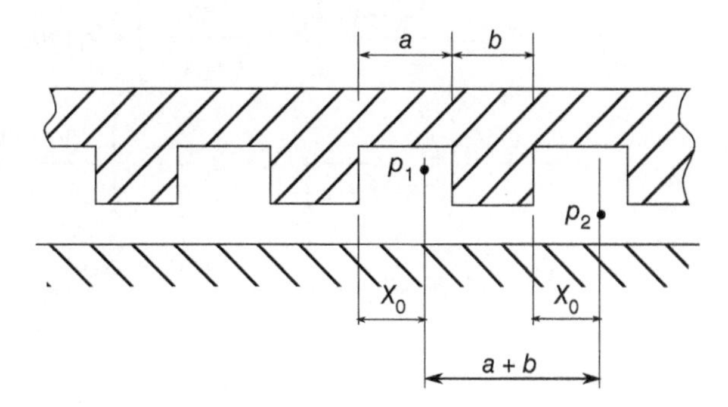

Figure 8.25 Both surfaces moving.

In the y-direction, since the channels are parallel, the flow will be the same as before, if we now substitute for the total entrainment velocity $U = U_0 + U_h$. This affects A' and B'.

In the x-direction the situation is more complex.

- If we determine the flow with respect to the grooved surface U_0, V_0, we can apply the same results as before substituting $V = V_h - V_0$ and $U = U_0 + U_h$.
- A does not include the velocity. We can show that its value remains unchanged, i.e. that the flow between any two points x_1 and x_2 with $x_2 - x_1 = a + b$ is

$$Q = \frac{h_0^3 h^3}{12\mu} \cdot \frac{p_2 - p_1}{ah^3 + bh_0^3}$$

  as before.
- Now it remains only to determine B, or the Couette flow in the x-direction. This value, in actual fact, is time-dependent in a periodic manner when observed from a point x that is fixed in space. We shall seek the time-averaged value (over one step interval) as follows, cf. Figure 8.26.

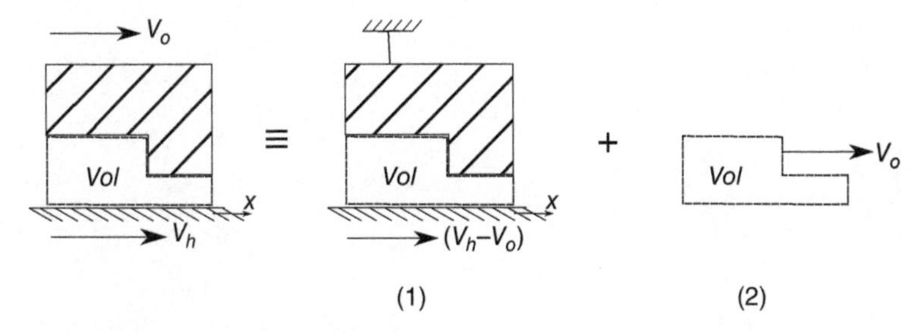

Figure 8.26 Time averaging procedure.

Thus,

1. mass flow of the first part is

$$Q_1 = B\frac{(V_\mathrm{h} - V_0)}{2}$$

2. mass of the second part

$$Q_2 = \frac{ah_\mathrm{o} + bh}{a+b}V_0 = (\bar{a}h_\mathrm{o} + \bar{b}h)V_0$$

Thus the total flow, averaged over one step is

$$\begin{aligned}
Q_x = Q_1 + Q_2 &= \frac{V_\mathrm{h} - V_0}{2}\left(h_\mathrm{o} - \frac{\delta\bar{b}}{\bar{b} + \bar{a}\frac{h^3}{h_\mathrm{o}^3}}\right) + V_0(\bar{a}h_\mathrm{o} + \bar{b}h) \\
&= \frac{V_\mathrm{h} - V_0}{2}\left(h_\mathrm{o} - \frac{\delta\bar{b}}{\bar{b} + \bar{a}\frac{h^3}{h_\mathrm{o}^3}} - (\bar{a}h_\mathrm{o} + \bar{b}h)\right) + \frac{V_\mathrm{h} + V_0}{2}(\bar{a}h_\mathrm{o} + \bar{b}h) \\
&= \frac{V_\mathrm{h} - V_0}{2}\delta\bar{b}\left(\frac{h_\mathrm{o}}{\delta\bar{b}} - \frac{1}{\bar{b} + \bar{a}\frac{h^3}{h_\mathrm{o}^3}} - \frac{(\bar{a}h_\mathrm{o} + \bar{b}h)}{\delta\bar{b}}\right) + \frac{V_\mathrm{h} + V_0}{2}(\bar{a}h_\mathrm{o} + \bar{b}h) \\
&= \ldots \\
&= \frac{V_\mathrm{h} - V_0}{2}\delta\bar{b}\left(\frac{\bar{b}h_\mathrm{o}^3 + \bar{a}h^3 - 1}{\bar{b}h_\mathrm{o}^3 + \bar{a}h^3}\right) + \frac{V_\mathrm{h} + V_0}{2}(\bar{a}h_\mathrm{o} + \bar{b}h)
\end{aligned}$$

which, when $V_0 = 0$ becomes identical to the old result. Finally filling this, instead of the old $BU/2$, into Eq. 8.27, we obtain the generalised case of Eq. 8.31, as:

$$\begin{aligned}
&\frac{\partial}{\partial x}\left(\rho\left(f_x\frac{\partial p}{\partial x} + f_\mathrm{c}\frac{\partial p}{\partial y}\right)\right) + \frac{\partial}{\partial y}\left(\rho\left(f_\mathrm{c}\frac{\partial p}{\partial x} + f_y\frac{\partial p}{\partial y}\right)\right) \\
&\quad + c_\mathrm{s}\left(\sin\beta\frac{\partial}{\partial x} - \cos\beta\frac{\partial}{\partial y}\right)(f_\mathrm{s}\rho) = 12\mu\left(\frac{U_\mathrm{h} + U_0}{2}\frac{\partial}{\partial x} + \frac{\partial}{\partial t}\right)(\rho f_\mathrm{v})
\end{aligned} \tag{8.32}$$

where, U_0, U_h are the velocities of the grooved and the smooth surfaces, respectively, and

$$\begin{aligned}
f_x &= \frac{g_1 + g_2\cos^2\beta}{g_3} \\
f_\mathrm{c} &= \frac{g_2}{g_3}\sin\beta\cos\beta \\
f_y &= \frac{g_1 + g_2\sin^2\beta}{g_3} \\
c_\mathrm{s} &= 6\mu(U_\mathrm{h} - U_0)\bar{a}\bar{b}\delta\sin\beta \\
f_\mathrm{s} &= \frac{h_\mathrm{o}^3 - h^3}{g_3} \\
f_\mathrm{v} &= \bar{a}h_\mathrm{o} + \bar{b}h
\end{aligned}$$

with

$$\begin{aligned}
g_1 &= h_\mathrm{o}^3h^3 \\
g_2 &= \bar{a}\bar{b}(h_\mathrm{o}^3 - h^3)^2 \\
g_3 &= \bar{b}h_\mathrm{o}^3 + \bar{a}h^3 \\
\delta &= h_\mathrm{o} - h
\end{aligned}$$

8.6.9 Discussion and Properties of the Solution

- In the Reynolds equation Eq. 8.32 only the term in c_s is responsible for the groove's action. All other terms correspond to an averagely smooth bearing.
- The case $U_h = U$, $U_0 = 0$, which corresponds to stationary grooves and the case $U_h = 0$, $U_0 = -U$, which corresponds to stationary smooth member and sliding grooved member (in the opposite direction), can be shown to yield the same pressure distribution $p(y)$ for an infinitely long bearing with nominally parallel gap and axisymmetric in the y-direction. This point will be elaborated upon in the following section.
- The groove action—the pumping—disappears if $U_h = U_0$, since then $c_s = 0$. This corresponds to the pure-rolling case (Chapter 4.)
- Considering $\bar{a}h_o + \bar{b}h$ as being the average film height, the term in c_s is consistent with the result of Chapter 4 regarding the 2D consideration on the Reynolds equation with moving and/or stationary features.

8.6.10 The Case of Stationary Grooves Versus that of Moving Grooves

This has perhaps been one of the most unclear issues regarding the design and application of herringbone-grooved and spiral-grooved bearings. Intuitively, one might think that there should be no difference, in regard to the generated pressure field, between the two cases. In Chapter 4, we have shown that there is a difference when the nominal gap is not parallel. In the following, we shall treat this issue more systematically making use of Eq. 8.32. That is, we shall compare the two cases when the direction of pumping (inward or outward) is maintained. Without loss of generality, we consider the case of inward pumping.

Smooth member rotating:
In this case, we have

$$U_h > 0,\ U_0 = 0,\ \beta = \beta_0 : 0 < \beta_0 < \pi/2.$$

Eq. 8.32 now reads:

$$\begin{gathered}\frac{\partial}{\partial x}\left(\rho\left(f_x\frac{\partial p}{\partial x}+f_c\frac{\partial p}{\partial y}\right)\right)+\frac{\partial}{\partial y}\left(\rho\left(f_c\frac{\partial p}{\partial x}+f_y\frac{\partial p}{\partial y}\right)\right)\\ +6\mu U_h\bar{a}\bar{b}\delta\sin\beta\left(\sin\beta\frac{\partial}{\partial x}-\cos\beta\frac{\partial}{\partial y}\right)(f_s\rho)=12\mu\left(\frac{U_h}{2}\frac{\partial}{\partial x}+\frac{\partial}{\partial t}\right)(\rho f_v)\end{gathered} \tag{8.33}$$

Grooved member rotating:
In this case, we have

$$U_0 > 0,\ U_h = 0,\ \beta = -\beta_0 : 0 < \beta_0 < \pi/2.$$

The condition that β should now be negative follows from maintaining inward pumping action. This has effect only on the terms containing $\sin\beta\cos\beta$, inverting there sign. Thus, the Reynolds equation, for this case, now reads:

$$\begin{gathered}\frac{\partial}{\partial x}\left(\rho\left(f_x\frac{\partial p}{\partial x}\underbrace{-f_c\frac{\partial p}{\partial y}}\right)\right)+\frac{\partial}{\partial y}\left(\rho\left(\underbrace{-f_c\frac{\partial p}{\partial x}}+f_y\frac{\partial p}{\partial y}\right)\right)\\ +6\mu U_0\bar{a}\bar{b}\delta\sin\beta\left(\underbrace{-\sin\beta\frac{\partial}{\partial x}}-\cos\beta\frac{\partial}{\partial y}\right)(f_s\rho)=12\mu\left(\frac{U_0}{2}\frac{\partial}{\partial x}+\frac{\partial}{\partial t}\right)(\rho f_v),\end{gathered} \tag{8.34}$$

where we have under-braced the terms that have changed w.r.t. the previous case.

Note that when $\partial p/\partial x = 0$, such as in a steady-state, parallel-nominal-gap bearing (discussed above), or in a concentric journal bearing, these terms vanish, so that Eq. 8.33 and 8.34 become identical.

The direct inward-pumping term, i.e. $-\cos\beta\partial(f_s\rho)/\partial y$ is not affected. Only the terms in the x and the mixed derivatives of p have changed sign.

It is not so straightforward to estimate the magnitude of this difference on the pressure distribution. The effect of the sign change of f_c might be gauged by comparing to f_x and f_y, where we would find out that all three have the same order of magnitude when $\delta/h = h_o/h - 1$ is of order unity. However, we would expect the effect to be small whenever the x derivative of the pressure is appreciably smaller than the y derivative, which is almost always the case in practical applications. The same remark may be made concerning the third under-braced term (in $\sin\beta$), which though it is comparable in magnitude to $\cos\beta$ is also associated with the x derivative of the density (read pressure). There do not seem to be many studies on this relevant issue in grooved-bearing design. However, Bootsma (1975) shows experimentally, that the difference in load capacity between the two cases (of stationary versus rotating grooves) is only slight for most cases of practical interest, namely those for which δ/c, where c is the radial clearance in a journal bearing, is not much larger than unity.

Remarks on NTG

As we have seen, NGT does not lead to easy-to-solve equations, except in generic cases, nor to a model that can entirely replace a finite-number-of-grooves situation, although some correction formulas exist to remedy that, see e.g. Muijderman (1966), Wilcock (1969). Rather, this formulation is interesting for gaining insight to the problem and the role of each design parameter. Furthermore, NTG theory can be used for quick simulation purposes and design optimisation, when the number of grooves is too high for a practical FD or FE solution to be viable; otherwise, calculation by such direct numerical methods is actually easier and exacter. NTG can also help us perform comparisons with direct numerical solutions and to the correct formulation of those, especially when both surfaces are moving.

8.6.11 Grooved Bearing Embodiments

In the foregoing, we have treated, as generic, the case of flat, rectangular herringbone-grooved strip. This same principle can adapted to herringbone and spiral groove, annular and disc thrust bearing bearings, by means of conformal mapping, as depicted in Figure 8.27. When wrapped around a cylindrical shaft or bush, one obtains the herringbone grooved journal bearing (HGJB) treated in Chapter 9. Naturally, other configurations, such as conical or spherical are also possible, as shown in Figure 8.28; see also (Muyderman 1966), which contains in addition useful tables of typical characteristics. For more details about their performance characteristics, the reader is referred to (Bootsma 1975; Muijderman 1966).

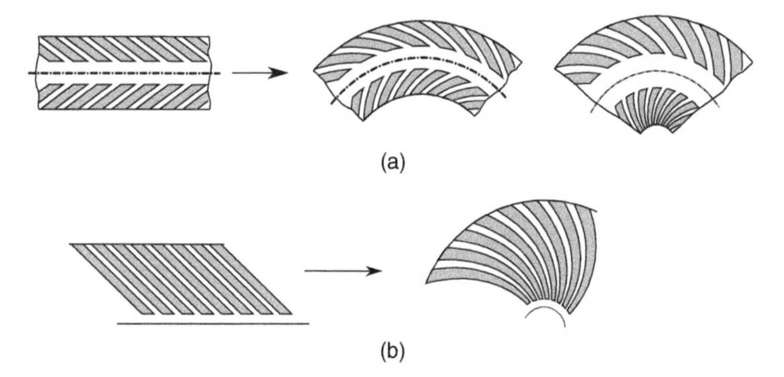

Figure 8.27 Conformal mapping of (a) rectangular herringbone bearing into annular or disc configuration, (b) a half-herringbone into a spiral groove.

Figure 8.28 Various embodiments of grooved bearings.

8.7 Rotary Bearings

All slider-bearing principles treated so far in this chapter, in addition to other possible variants, can be implemented as annular thrust bearings or as journal bearings or any other degree between them: conical, spherical, etc.

Thrust bearings can be treated as slider bearings with or without periodic boundary conditions in the sliding direction, see section 8.10, depending on the design. They are conformal surface bearings (as opposed to JB's which are not, see Chapter 9) and as such, generally not demanding from point of view of dynamic stability, except in the presence of misalignment, which introduces a cross-coupling in the stiffness coefficients, as in JB's.

8.7.1 Journal Bearings

JB's will be treated systematically in Chapter 9. In this section, we indicate briefly the aerodynamic, wedge principle of action in a plain JB. Thus, referring to Figure 8.29, we see that when the shaft is eccentric, by an amount e, there will arise: (i) a convergent gap yielding positive pressure build-up and (ii) a divergent gap having negative, or sub-ambient pressure. These will generally speaking strengthen one another so as to result in a net force in the vertical direction that will support the load W. When bearing number and eccentricity are small, we have the Sommerfeld bearing situation; i.e. the attitude angle ϕ approaches 90°, since the pressure distributions in both parts are almost identical. As the bearing number increases, the pressure in the convergent part increases, while in the divergent part, the pressure remains bounded by zero absolute pressure, i.e. absolute vacuum, so that the attitude angle decreases until it reaches zero at infinite sliding numbers.

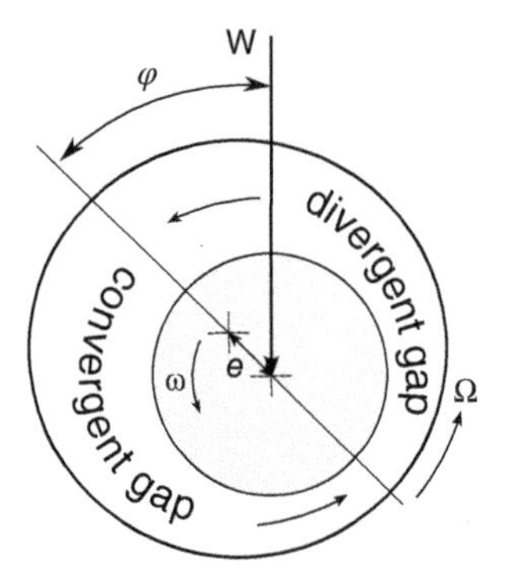

Figure 8.29 Section in journal bearing showing principle and indicating notation.

8.8 Dynamic Characteristics

Contrary to intuition, self-acting bearings are not exempt from dynamic instability, being also prone to negative damping in the same way EP bearings are. The general trend is also similar: if a slider bearing is designed to have high stiffness, then the damping reduces. In the extreme case, it can become negative. Since conventional slider types: tapered pad, Rayleigh step, etc. have moderate to low stiffness, owing to large side leakage, we should expect them generally to be positively damped.

Dynamic stability of spiral groove bearings has been treated e.g. in (Constantinescu and Galetuse 1987; Malanoski and Pan 1965; Wilcock 1969) and that of step or single slider in e.g. (Ono 1975; Tang 1971). Dynamic stability of tilting-pad bearings is amply discussed in Chapter 11.

We can determine the dynamic stiffness (and damping) $k + jvc$ by applying the method of perturbation explained in Chapter 7. We shall take as example the same two grooved-bearing cases examined in section 8.5.1 and whose static characteristics have been there presented.

The dynamic stiffness and damping of first case are plotted in Figure 8.30. Here, owing to the relatively small sliding speed and the large groove angle, the damping is initially negative only for the very small gap heights of

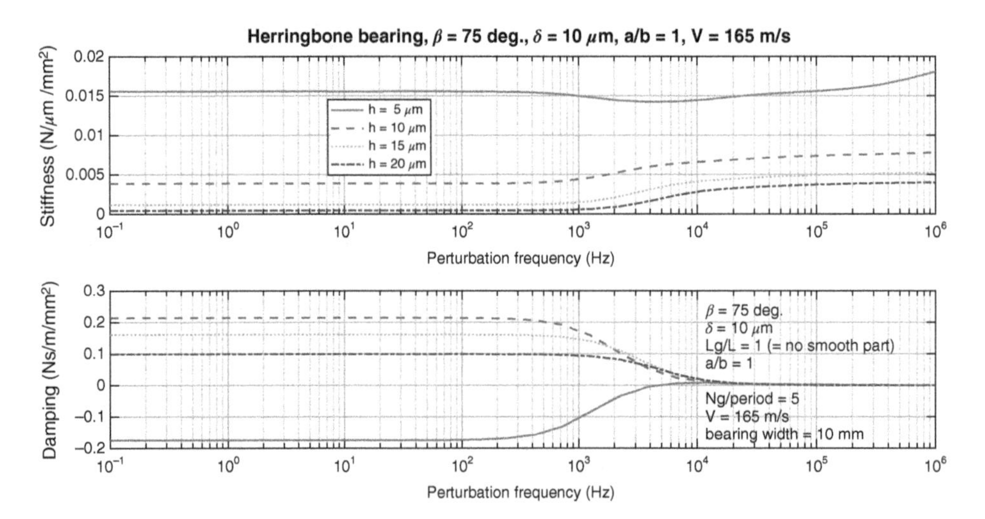

Figure 8.30 Evolution of the dynamic stiffness and damping as a function of the frequency for a herringbone bearing with $L = 10$ mm, $\beta = 75$ deg., $\delta = 10$ µm, $V = 165$ m s^{-1}, number of grooves per period $N_g = 5$, with periodic boundary conditions in the sliding direction, for various values of the nominal gap height h.

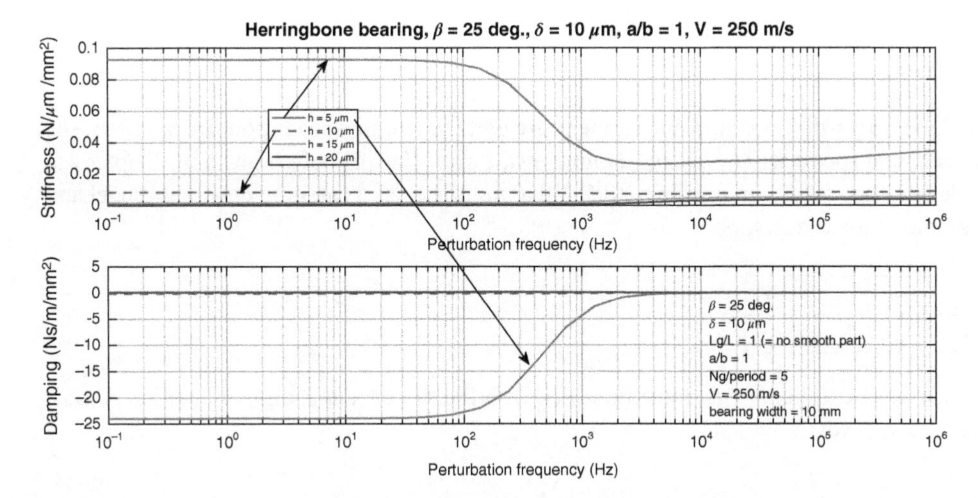

Figure 8.31 Evolution of the dynamic stiffness and damping as a function of the frequency for a herringbone bearing with $L = 10$ mm, $\beta = 25$ deg., $\delta = 10$ μm, $V = 165$ m s^{-1}, number of grooves per period $N_g = 5$, with periodic boundary conditions in the sliding direction, for various values of the nominal gap height h.

around 5 μm. This corresponds to the case of maximum static stiffness. As the gap height becomes larger the stiffness lowers and the damping becomes positive.

We see that the behaviour stereotyped in Chapter 7 holds true also for self-acting bearings:

- when k is positive and increasing in the frequency, then c is positive and decreasing.
- when k is positive and decreasing in the frequency, then c is initially negative and increasing.

In the second example, the sliding velocity is high and the groove angle is small. This combination yields a high load capacity and static stiffness, that are coupled with negative, or very small damping. Let us remark that for the case of 5 μm gap, the specific stiffness is considerable, being equal to just about one bar mean over-pressure per micrometer. (We saw in Figure 8.17 that the specific load capacity is also considerable at this gap height together with the friction power.)

8.9 Similarity and Scale Effects

We have seen that the static bearing dimensionless characteristics depend on three parameters:

$$\Lambda, \quad \frac{L}{B}, \quad H(X,Y),$$

where, $H(X,Y)$ denotes the dimensionless gap function. One can show that for two bearings to have the same gap function, only one parameter is involved, e.g. h_0/h_a since all the other parameters describing the function should have equal values in order to fulfill the geometrical similarity condition.

Since geometrical similarity is a prerequisite of full similarity, we are left only with the bearing number to determine behaviour of similar bearings. Thus, for two bearings to be similar, we must have the same value of Λ (with L/B and $H(X,Y)$ being identical). In that case, two similar bearings will share the same dimensionless load, static stiffness, static damping, and mass flow rate.

Thus, with the neglect of thermal similarity by assuming isothermal process, we can state that two bearings will be statically similar if they share the following dimensionless numbers:

$$L/B, \qquad H(X,Y), \qquad \Lambda$$

In that case, they will share also the following characteristics:

$$\bar{W}, \qquad K, \qquad C, \qquad \bar{\dot{m}}$$

If we now preclude the gas properties, the ambient temperature and the sliding speed, from the analysis, the only absolute parameters at our disposal are $(L,\ h_a,\ p_a)$. Following the same line of analysis given in Chapter 2 to ascertain the difference in behaviour between a scale-model and a prototype, let us assume now that model and prototype differ by the following α-scale factors:

model	L	h_a	p_a
prototype	$\alpha_L L$	$\alpha_{\rm h} h_a$	$\alpha_{\rm p} p_a$

In order that the two bearings be similar, $B,\ h_{\rm o}$ must be scaled with the same factors, and, in addition Λ must be the same. This last condition yields:

$$\frac{\alpha_L}{\alpha_{\rm p}\alpha_{\rm h}^2} = 1 \tag{8.35}$$

Also, the bearing characteristics will be:

model	W	$\dot{m}$	k	c
prototype	$\alpha_L^2\alpha_{\rm p}F$	$\alpha_{\rm h}^3\alpha_{\rm p}^2\dot{m}$	$(\alpha_L^2\alpha_{\rm p}/\alpha_{\rm h})k$	$(\alpha_L^4/\alpha_{\rm h}^3)c$

If we assume that the ambient pressure remains unchanged, and wish to examine the influence of the bearing size on the characteristics, then we set $\alpha_{\rm p} = 1$ to obtain:

$$\alpha_{\rm h}^2 = \alpha_L$$

Let us now assume that the prototype bearing has double the area of the model, i.e. $\alpha_L = \sqrt{2}$, then we have:

model	L	h_a	W	$\dot{m}$	k	c
prototype	$\sqrt{2}L$	$2^{1/4}h_a$	$2W$	$2^{3/4}\dot{m}$	$2^{3/4}k$	$2^{5/4}c$

In other words, doubling the bearing size places relatively more stringent tolerance requirements on the gap-to-length ratio; it doubles the load capacity, while less than doubling the air flow and the stiffness; and it more than doubles the damping.

Let us note here though that from fabrication point of view, relatively tighter tolerances could be achieved with larger parts, since tolerance scales approximately with the cube root of size, optimality ($\sim$ large $L/h_{\rm a}^2$) favours larger bearing sizes. Thus, relatively higher bearing loads and damping values favour larger sizes of bearings, as indeed we have also discovered in Chapter 6 with regard to EP centrally-fed bearings.

8.10 Hybrid Bearings

One accessible way to implement a thrust bearing is by means of an annular (i.e. ring type) bearing, as depicted in Figure 8.32, which might be made up following the aerostatic or aerdynamic design or both. Though somewhat misleading, the term "hybrid bearing" is commonly used to refer to either (i) an aerostatic bearing that is being operated at high speeds so that the aerodynamic effect might, often unintentionally or with adverse effects, become appreciable, as shown in Figure 8.33(a), (b); or (ii) less frequently, a hydrodynamic bearing that is equipped with external pressurisation provision (feed holes, etc.) so as to overcome the start-stop problem, the so-called "pressure

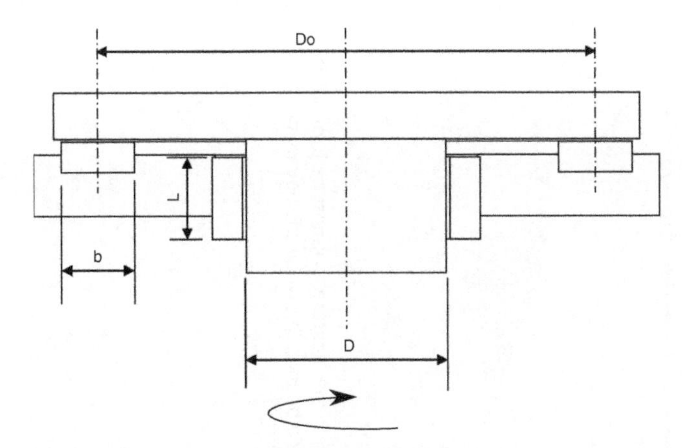

Figure 8.32 Rotary system incorporating an annular thrust bearing.

Figure 8.33 Annular aerostatic bearing or "hybrid" bearing, with either (a) continuous surface, (b) separated sectors or (c) a self-acting bearing (tapered pad as example) equipped with a feed hole. (Not to scale.)

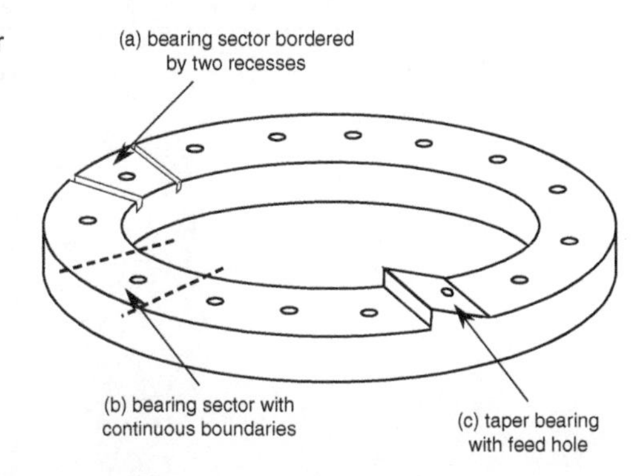

jacking", or to acquire higher operating load capacity and stiffness, Figure 8.33(c). In regard to the latter case, let us note that, except for spiral groove or herringbone groove bearings at high speeds, other self-acting bearings are inherently limited in pressure generation, so that design requirements might dictate an EP-bearing solution. With regard to the utilisation of hybrid bearings, Grassam and Powell (1964) have recommended the following guidelines:

- if the specific load is less than about 0.15 bar—use self-acting bearings;
- if the specific load is between about 0.15 bar and 0.9 bar—use self-acting bearings with provision of "pressure jacking", i.e. external pressure feed for starting and stopping purposes;
- if the specific load is greater than about 0.9 bar—use EP bearings, i.e. full hybrid bearings.

There could be numerous configurations of these bearings as a function of the applications envisaged. However, in most cases, there will be no particular advantage in combining the aerodynamic with the aerostatic effect. Thus, if one opts for implementing an EP-bearing solution, a nominally flat bearing, i.e. plain or with conicities around the feed holes, would be the best solution. Such a solution, moreover, is independent of the direction of rotation of the thrust disc.

Rather, we should turn our attention to the effect of sliding speed on the performance of the (otherwise aerostatic) bearing. For this purpose, let us compare the two types (a) and (b) of Figure 8.33; type (c) is discussed later as a special case.

Referring then to Figure 8.34(a) we see the same phenomenon described in Chapter 6 for circular flat bearings, namely that increasing sliding speed diminishes the pressure build-up in the gap so that both load capacity

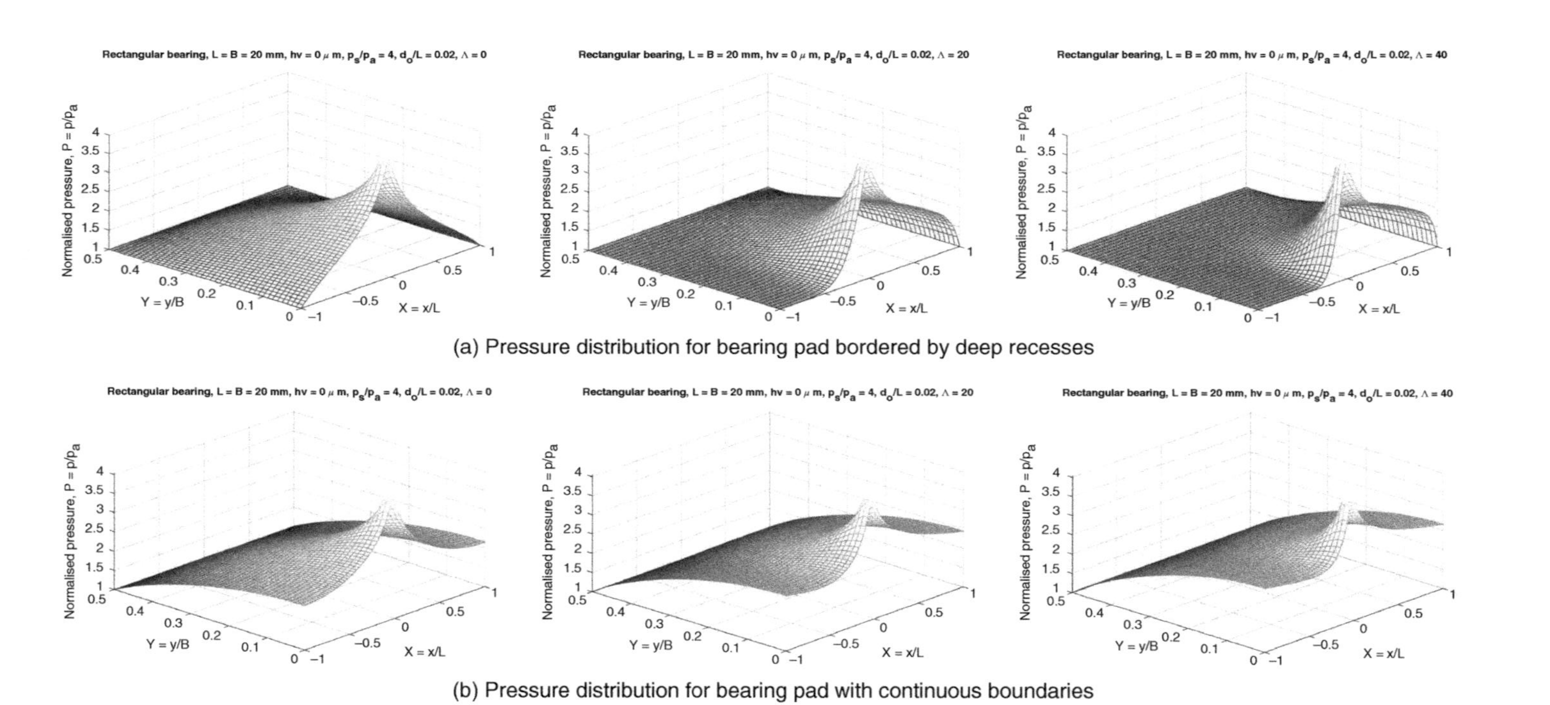

Figure 8.34 The pressure distributions in a rectangular EP, plain bearing pad for different sliding numbers Λ; (a) when the pressure is atmospheric on all four boundaries owing to the deep recesses before and after the pad, (b) when the pressure is atmospheric only on the side boundaries, while the boundary conditions on the other two are periodic, owing to the continuity of the bearing surface.

and stiffness may reduce drastically. In particular, the latter may become negative, as we shall see further. In Figure 8.34(b), on the other hand, owing to the continuous boundaries, the pressure is only slightly effected, resulting in increased load capacity and damping but in reduced stiffness.

It turns out that choosing the right bearing configuration is a trade-off between load capacity, stiffness and damping as a function of the sliding number Λ which is dictated by the maximum rotational speed of the system. To see this, we examine the values of these for four cases corresponding to bearing pad of type (a) and (b) (Figure 8.34) with either plain (i.e. flat) surface, or having small conicities (i.e. conically-shaped shallow pockets) around the feed holes, as delineated in Table 8.1. In the lase row of this table, the characteristics of case (c) is also given for comparison.

We make the following remarks:

- Type (a) pad, i.e. that bordered by atmospheric pressure from all sides, in its plain version, scores the worst; especially for higher sliding numbers when its stiffness becomes negative, which would lead to static instability (i.e. collapse) of the bearing system.
- Type (b) pad, i.e. that having continuous boundaries in the sliding direction, or periodic pressure boundary conditions in that direction, has superior force and damping characteristics. It has comparable stiffness to type (a) for $\Lambda = 0$. Most importantly, its stiffness increases with increasing Λ; i.e. it does not suffer from static instability.
- The stiffness, in both types can be appreciably increased by incorporating conicities. However, this will reduce the damping and should thus be moderate in depth.
- Type (c) pad, i.e. an originally self-acting pad that is now provided with a feed hole for external pressurisation, and which has continuous boundaries is the sliding direction, scores generally worse than type (b) that is provided with a conicity.
- the air consumption (not shown in the table) of type (a) is a bit less than the double of type (b).

Table 8.1 Evaluation of hybrid annular thrust bearing.

		Sliding number Λ		
Bearing-pad type	Characteristic	0	20	40
(a)	Force (N)	23.48	16.11	8.56
plain	Stiffness (N μm^{-1})	0.12	−0.24	−1.57
	Damping (10^3 Ns m^{-1})	3.23	5.34	3.06
(b)	Force (N)	30.92	32.98	34.99
plain	Stiffness (N μm^{-1})	0.13	1.10	1.56
	Damping (10^3 Ns m^{-1})	16.69	14.71	13.82
(a)	Force (N)	45.57	36.41	24.61
with 5 μm	Stiffness (N μm^{-1})	4.83	0.46	−1.54
conicity	Damping (10^3 Ns m^{-1})	2.60	4.13	3.31
(b)	Force (N)	63.65	66.66	71.49
with 5 μm	Stiffness (N μm^{-1})	5.23	7.14	9.28
conicity	Damping (10^3 Ns m^{-1})	5.72	4.76	4.03
(c)	Force (N)	16.41	26.56	27.74
with 5 μm	Stiffness (N μm^{-1})	0.94	3.50	2.55
taper height	Damping (10^3 Ns m^{-1})	2.50	1.81	2.22

See Figure 8.34 for explanation of types (a), (b) and (c). The feed hole is at the centre in cases (a) and (b). In case (c), it is situated one fourth of the pad width closer to the leading edge.

8.11 Conclusions

This chapter has firstly provided general considerations on the aerodynamic action showing its basic characteristics; advantages and limitations. One of the main shortcomings of using gases in self-acting bearings is their compressibility, which places a limitation on the maximum pressure generated; another is the low energy efficiency obtaining by the aerodynamic action.

Second, this chapter over-viewed a number of slider-bearing principles including herringbone-grooved bearings showing their basic static and dynamic characteristics.

Third, owing to its importance, this chapter provided a derivation of the narrow-groove theory (NGT) of grooved bearings, which treats the grooves as a continuum whose effect is described by an equivalent Reynolds equation. It presented also a discussion of the basic properties of that solution.

Finally, the dynamic characteristics of self-acting bearings have also been exposed showing that these bearings are generally not exempt of developing negative damping; especially in grooved bearings where the compressibility effect is at its highest.

References

Al-Bender, F., (2011). A generalized coefficient of friction for assessing energy efficiency of bearing systems *European Conference on Tribology*, Vienna, Austria.

Bonneau, D., Huitric, J. and Tournerie, B., (1993). Finite element analysis of grooved gas thrust bearings and grooved gas face seals. *Trans. ASME - Journal of Tribology* 115(3): 348–354.

Bootsma, J., (1975). *Liquid-lubricated spiral-groove bearings* PhD thesis Technische Hogeschool Delft.

Constantinescu, V.N., (1969). *Gas Lubrication*. ASME.

Constantinescu, V.N. and Galetuse, S., (1987). On the dynamic stability of the spiral-grooved gas-lubricated thrust bearing. *Transactions of the ASME - Journal of Tribology* 109(1), 183–188.

Dowson, D., (1979). *History of Tribology*. Longman, London.

Halling, J. (ed.) (1978). *Principles of Tribology*. McMillan.

Grassam, N.S. and Powell, J.W., (1964). *Gas Lubricated Bearings*. Butterworths, London.

Gross, W.A., (1962). *Gas Film Lubrication*. John Wiley & Sons.

Gross, W.A., (1980). *Fluid Film Lubrication* chapter Chapter Compliant Bearings, 483–549.

Gross, W.A., Matsch, L.A., Castelli, V., et al. (1980). *Fluid Film Lubrication*. John Wiley & Sons, New York.

Malanoski, S.B. and Pan, C.H.T., (1965). The static and dynamic characteristics of the spiral-grooved thrust bearing. *Journal of Basic Engineering Transactions of the ASME* 87(1): 547–558.

Muijderman, E. (1966). *Spiral Groove Bearings*. Philips Technical Library, Eindhoven, The Netherlands.

Muijderman, E.A., (1967). Analysis and design of spiral-groove bearings. *Transactions of the ASME - Journal of Lubrication Technology* 89(3): 291–306.

Muyderman, E., (1966). Constructions with spiral-groove bearings. *Wear* 9(2): 118–141.

Ono, K., (1975). Dynamic Characteristics of Air-Lubricated Slider Bearing for Noncontact Magnetic Recording. *Journal of Lubrication Technology* 97(2): 250–258.

Tang, T., (1971). Dynamics of Air-Lubricated Slider Bearings for Noncontact Magnetic Recording. *Journal of Lubrication Technology* 93(2): 272–278.

Vohr, J. and Chow, C., (1965). Characteristics of the herringbone-grooved, gas-lubricated journal bearings. *Transactions of the ASME, Journal of Basic Engineering*, 86(3): 568–578.

Whipple, R., (1958). *The inclined groove bearing*. Technical report, AERE Report T/R 622, (Revised). UKAEA, Research Group, Harwell, Berkshire, England.

Wilcock, D.F.E, (1969). *Design of Gas Bearings*. Mechanical Technology Inc. (MTI), Latham NY.

Wilcock, D.F.E., (1972). *MTI Gas Bearing design manual* Mechanical Technology Inc. New York.

Wildmann, M., Glaser, J., Gross, W.A., et al. (1965). Gas-lubricated stepped thrust bearing a comprehensive study. *Journal of Basic Engineering, Transactions of the ASME, Series D*, 87(1), 213–229.

Xue, Y. and Stolarski, T.A., (1997). Numerical prediction of the performance of gas-lubricated spiral groove thrust bearings. *Proceedings of the Institution of Mechanical Engineers, Part J: Journal of Engineering Tribology* 211(2): 117–128.

9

Journal Bearings

9.1 Introduction

The term journal bearings (JB) refers to a specific type of radial bearing wherein the bearing forms a sort of sleeve around the shaft, see Figure 9.1; which is sometimes referred to as *full JBs*. This is in contrast to bearings formed of separate pads, which are sometimes referred to as *partial JBs*.

The working principle of a JB may be self-acting, EP, or both; i.e. hybrid. Owing to simplicity of construction, which is often merely a shaft in a bush, a full JB is the most accessible way to support a shaft radially. Consequently, it is a bearing type that has been subject to much investigation; in particular the liquid-lubricated version, which has always been a vital engineering machine component.

In the majority of applications, the bearing surface is essentially fixed to the housing of the machine, being sometimes flexure-mounted to achieve self-alignment. In other applications, the shaft, or journal, might be fixed while the bearing revolves around it. Finally, although rare, both members might be required to rotate around one another simultaneously.

Self-acting bearings may be *plain*, in which both journal and bearing have a nominally cylindrical surface, or *grooved*, where surface features are machined either on the bearing or journal surfaces, or both, with the purpose of achieving additional behavioural properties.

The plain cylindrical journal bearing possesses the extreme qualities of both geometrical simplicity and vulnerability to self-excited, "whirl" instability, a type of orbital motion of the shaft, which is treated in detail in Chapter 10. Its characteristics are completely known, although it is not often used in practice, without taking additional measures to enhance dynamic stability, on account of the occurrence of self-excited whirling. Knowledge of the characteristics of self-acting plain journal bearings is very helpful in preliminary design studies; its static stiffness is often representative of the value of what can actually be achieved, while its instability-threshold speed is most likely a conservative estimate, but nevertheless indicative of typical values. Some of the classical works treating them include but are not restricted to (Ausman 1961; Castelli and Elrod 1965; Cheng and Pan 1965; Cheng and Trumpler 1963; Constantinescu 1969; Gross 1962; Pan and Sternlicht 1962; Raimondi 1961; Sternlicht 1959; Sternlicht and Winn 1963; Whitley et al. 1962; Wildmann 1956).

The *herringbone grooved*, or helical grooved journal bearing (HGJB) is the most well-known type of grooved bearing, which was largely developed in the 1950's and 1960's with the purpose of achieving higher performance than the plain type. It appeared to researchers to possess most of the virtues but less of the vices of the plain cylindrical journal bearing. In particular, it is claimed to enjoy a broader dynamic stability margin. Some of the classical works treating them include (Cunningham et al. 1969; Wilcock 1969; Muijderman 1967; Vohr and Chow 1965) more recent ones, containing numerical modelling and analysis, include (Bonneau and Absi 1994; Faria 2001; Schiffmann 2008; Tomioka et al. 2007) and many others.

Air Bearings: Theory, Design and Applications, First Edition. Farid Al-Bender.

Companion website: www.wiley.com/go/AlBender/AirBearings

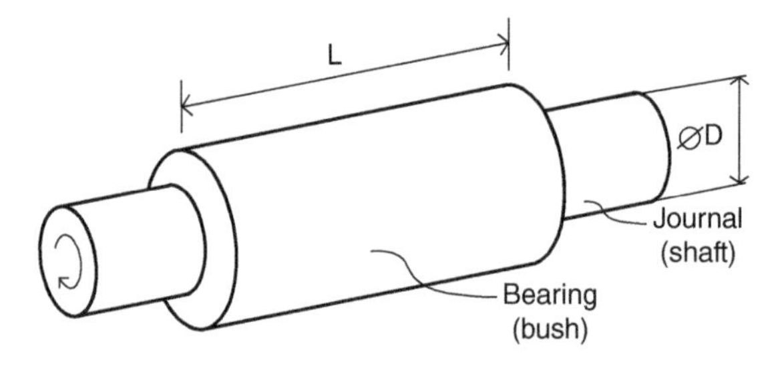

Figure 9.1 Typical journal bearing arrangement.

In this chapter, we shall treat both of these self-acting bearing types, in their generic forms, to gain a general idea about their behaviour, comparing, at the same time, their most salient vices and virtues.

In contrast to the self-acting type, *externally-pressurised* (EP) journal bearings rely for pressure build-up in the gap on external pressurised gas feed. In this way, they can operate at very low speeds with high stiffness and reasonable damping, typically for high-precision applications, where they are also referred to as *aerostatic* JBs. Their stiffness and damping characteristics depend on the number and size of feed holes and their restriction method, and on the presence or absence of surface features such as shallow pockets or conicities around the feed holes. As the rotation speed increases, an aerodynamic effect is generated, which could become as appreciable as, or stronger than the aerostatic effect, giving rise eventually to whirl instability problems. In that case, the EP bearing is commonly referred to as *hybrid* JB. The advantage of this type of bearing is that it ensures high stiffness and adequate damping at the lower speed range; loosing its dynamic-stability advantage at very high speeds. Typical applications are in high-speed, high-accuracy machining and positioning spindles, and similar systems. EP JBs, including the aerodynamic and whirl effects, have also been researched from the 1950's to the present, see e.g. (Franchek 1992; Kazimierski and Jarzecki 1979; Liu et al. 2005; Lund 1967; Mori et al. 1980a,b; Osborne and San Andrés 2006; Pink and Stout 1981; Robinson and Sterry 1958; Rohde and Ezzat 1967; San Andrés and Ryu 2008; San Andres 1991; Stout and Rowe 1974; Stout and Tawfik 1983; Su and Lie 2003, 2006; Tawfik and Stout 1982; Waumans 2009).

The literature contains a large volume of different solutions to JB problems, and it is not the purpose of this chapter to overview them or re-derive them, although some of them will be restated for the sake of completeness, comparison, or summarisation. Besides some analytical methods, journal-bearing design solutions are most conveniently obtained by numerical methods, in particular, the FD method. The main purpose of the present chapter is rather to formulate the problem, sketch and outline basic solutions and performance characteristics, and discuss relevant design issues.

Other types of JBs such as the foil, porous and tilting-pad will be treated in dedicated chapters.

9.1.1 Geometry and Notation

Referring to Figure 9.2, nominally the geometry of an idealised JB, i.e. assuming perfectly cylindrical bearing and shaft, is determined by:

- Shaft radius R (or diameter $D = 2R$).
- Bearing radial clearance c.
- Bearing length L.
- Eccentricity of shaft centre relative to bearing centre e.
- When the bearing is externally pressurised, the number, position, and size of the feed holes should additionally be specified.
- When the bearing or shaft is not plain, the geometry of the surfaces should be specified with respect to a nominal reference cylinder residing in either the journal or the bearing or both. This is particularly relevant to two types

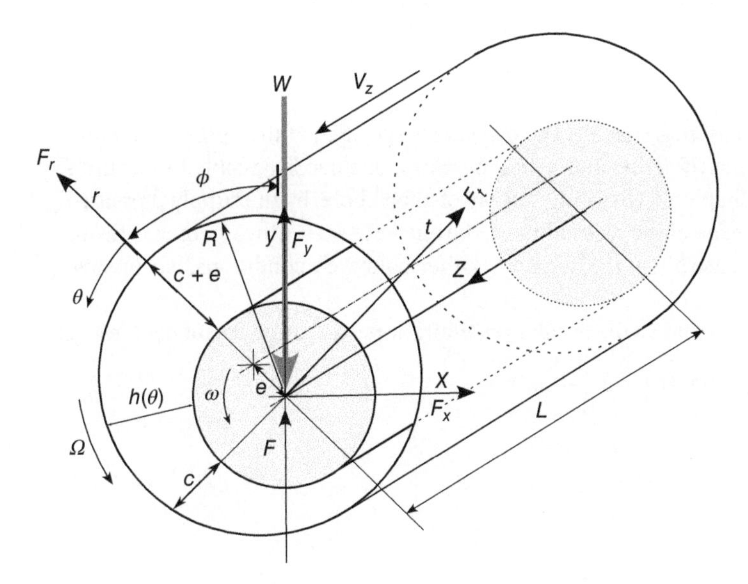

Figure 9.2 Journal bearing notation.

of bearings: (i) herringbone-grooved self-acting bearings (HGJB) and (ii) EP bearings with shallow pockets, or "conicities", surrounding the feed holes.

The JB is depicted with an eccentricity e in an arbitrary radial direction. The vector pointing in the opposite direction is conveniently designated as the "radial" r direction, while that orthogonal to it is the "tangential" t direction. Let us note that the (r, t) coordinates play the role of a canonical system that facilitates the definition and calculation of principal quantities. We have set θ, together with the journal and bearing rotational speeds ω and Ω respectively, in the counter-clockwise sense. In that way, the bearing force F shall lie somewhere in the positive the $r - t$ quadrant, subtending an angle of magnitude ϕ with the r axis, being referred to as the load angle or attitude angle. Furthermore, x and y axes are also be defined for the purpose of force analysis, being attached to a stationary reference frame, in any convenient orientation. We have set them here, as is most often done, so that the y-axis points in an opposite direction to the load line, as depicted on the figure. However, any other choice of convenient axes is also viable.

The Gap Function

The nominal gap between journal and bearing when their axes remain parallel but displaced by an eccentricity e may be shown to be equal to

$$h = c + e\cos\theta + O\left(\frac{c^2}{R}\right),$$

so that for small c/R, typically $\ll 10^{-2}$, the last term may be neglected to obtain:

$$h = c + e\cos\theta.$$

9.1.2 Basic Equation

The (dimensional) Reynolds equation for a journal bearing, as derived in Chapter 4, reads:

$$\frac{\partial}{\partial z}\left(\frac{h^3}{12\mu}\rho\frac{\partial p}{\partial z} - \rho\frac{V_z h}{2}\right) + \frac{1}{R}\frac{\partial}{\partial\theta}\left(\frac{h^3}{12\mu}\rho\frac{1}{R}\frac{\partial p}{\partial\theta} - \rho\frac{R(\Omega+\omega)h}{2}\right) = \frac{\partial}{\partial t}(\rho h) \tag{9.1}$$

with

$$-L/2 \leq z \leq L/2$$

As boundary conditions for solving this equation, (i) usually the pressure is specified at the outlet open boundaries as a constant p_a; (ii) at shear flow inlet, in self-acting and hybrid bearings, one needs possibly to account for inertia effects on the entrance pressure, see Chapter 3; (iii) at the supply of a feed hole, by the supply pressure p_s, for EP or hybrid bearings, so that, with knowledge of the restrictor law, the entrance pressure p_o, which is the ultimate boundary condition, may be determined as shown in Chapter 5. Other boundary conditions, e.g. involving the flow rate, are also possible.

Normalising with respect to R (or D), c, L, p_a μ and ν (the oscillation, whirl or perturbation frequency), we get

$$\frac{\partial}{\partial Z}\left(H^3\wp\frac{\partial P}{\partial Z} - \Lambda_z H\right) + \frac{\partial}{\partial \theta}\left(H^3\wp\frac{\partial P}{\partial \theta} - \wp\Lambda_r H\right) = \sigma\frac{\partial}{\partial \tau}(\wp H) \tag{9.2}$$

with

$$-L/D \leq Z = z/L \leq L/D$$

where,

$$H = h/c, \quad \wp = \rho/\rho_a, \quad P = p/p_a, \quad \tau = \nu t,$$

and

$$\Lambda_z = \frac{6\mu V_z}{p_a}\left(\frac{R}{c}\right)^2$$

is the axial sliding number, which is mostly equal or very close to zero, and will therefore be neglected in the rest of the treatment.

$$\Lambda_r = \frac{6\mu(\Omega + \omega)}{p_a}\left(\frac{R}{c}\right)^2$$

is the peripheral sliding number, being also called the "bearing number" or "compressibility number". In the remainder of this chapter, we shall treat only the case of

$$\Omega = 0$$

and employ the symbol Λ, i.e. without the subscript, in the rest of the treatment, as is done in most other publications. Thus,

$$\Lambda = \frac{6\mu\omega}{p_a}\left(\frac{R}{c}\right)^2.$$

Further,

$$\sigma = \frac{12\mu\nu}{p_a}\left(\frac{R}{c}\right)^2$$

is the squeeze number, where ν is characteristic angular frequency, such as the perturbation frequency in case of harmonic excitation.

The gap function, for very small c/R, in the absence of tilt, may be approximated as

$$H = \frac{h}{c} = 1 + \epsilon\cos\theta \tag{9.3}$$

with $\epsilon = \frac{e}{c}$, $0 \leq \epsilon < 1$ being the eccentricity ratio.

Inspecting Eq. 9.2, We see that, for a plain, self-acting bearing, the problem is characterised by the following dimensionless parameters:

$$\frac{L}{D}, \ \epsilon, \ \Lambda, \ \sigma,$$

which makes it relatively simple to analyse, as compared with bearings having more design parameters

Important Remark

For an axisymmetric JB, the shaft is in static equilibrium at its concentric position; that is, the net load is then zero. On the other hand, the radial stiffness, for an EP bearing at least, is at its maximum value in that position. It is therefore more logical and useful to consider the radial stiffness in the concentric position as a bearing characteristic, rather than the load, which can be obtained by integrating the stiffness w.r.t. the eccentricity.

If the radial stiffness k_r has its maximum in the concentric position then the load a JB can sustain is

$$W \leq k_r\, e$$

and

$$W_{max} \leq k_r.c,$$

which can be a useful design guideline for assessing the load capacity of a JB.

On the other hand, the bearing load close to concentric position is given by

$$W \approx k_r\, e.$$

9.2 Basic JB Characteristics

Film Forces

Refering to Figure 9.2, the film forces, in the principal directions (r, t), are given by

$$F_r = -\int_{-L/2}^{L/2}\int_0^{2\pi} (p - p_a)\cos\theta\; Rd\theta dz \tag{9.4a}$$

$$F_t = \int_{-L/2}^{L/2}\int_0^{2\pi} (p - p_a)\sin\theta\; Rd\theta dz \tag{9.4b}$$

$$W = F_r^2 + F_t^2 \tag{9.4c}$$

$$\phi = \arctan\frac{F_t}{F_r}. \tag{9.4d}$$

From these, we can calculate the forces in the (x, y) directions from

$$F_x = -F_r\sin\phi + F_t\cos\phi \tag{9.5}$$

$$F_y = \;F_r\cos\phi + F_t\sin\phi. \tag{9.6}$$

Let us note that with this choice of (x, y) coordinate system, we have for this static situation that:

$$F_x = 0 \tag{9.7}$$

$$F_y = -W. \tag{9.8}$$

Generally, however, if the eccentricity is altered, both the bearing load and the attitude angle will change. It might therefore be expedient to adopt the (r, t) system for calculating the principal forces. In that case, we formally choose the (x, y) to be identical with the (r, t) coordinates.

Stiffness and Damping

There are four stiffness components, defined as follows, for any suitable orthogonal x, y system of coordinates, cf. e.g. Figure 9.2:

$$k_{xx} = -\frac{\partial F_x}{\partial x} \tag{9.9a}$$

$$k_{xy} = -\frac{\partial F_x}{\partial y} \tag{9.9b}$$

$$k_{yx} = -\frac{\partial F_y}{\partial x} \tag{9.9c}$$

$$k_{yy} = -\frac{\partial F_y}{\partial y}. \tag{9.9d}$$

These are usually arranged in a 2×2 matrix. Similarly, there are four damping components, defined as the negative derivatives of the forces w.r.t. the velocities $\dot{x}$ and $\dot{y}$. Thus, for $\vec{e} = (\mathrm{d}x, \mathrm{d}y), \dot{\vec{e}} = (\dot{x}, \dot{y})$, the dynamic force $\vec{f} = (f_x, f_y)$ on the journal is given by:

$$\begin{Bmatrix} f_x \\ f_y \end{Bmatrix} = \begin{bmatrix} k_{xx} & k_{xy} \\ k_{yx} & k_{yy} \end{bmatrix} \cdot \begin{Bmatrix} x \\ y \end{Bmatrix} + \begin{bmatrix} c_{xx} & c_{xy} \\ c_{yx} & c_{yy} \end{bmatrix} \cdot \begin{Bmatrix} \dot{x} \\ \dot{y} \end{Bmatrix}, \tag{9.10}$$

in which k_{ij} represents the film reaction in the i-direction to a displacement in the j-direction. The damping coefficient c_{ij} is defined accordingly, while f_x and f_y are external dynamic forces acting on the rotor.

For convenience, we can, as we did in Chapter 7, write the total dynamic stiffness component as the complex number

$$k_{\mathrm{dyn(ij)}} = k_{ij} + j\nu c_{ij}, \tag{9.11}$$

where ν is the oscillation, or perturbation angular frequency.

The dynamic film characteristics are obtained by integration of the dynamic perturbation pressure ΔP over the bearing surface. This dynamic pressure may for this purpose be written as:

$$\Delta p = \Delta p_{\mathrm{k}} + j\nu \Delta p_{\mathrm{c}}, \quad j = \sqrt{-1}, \tag{9.12}$$

with Δp_{k} the in-phase component of the dynamic pressure yielding the stiffness, and Δp_{c} the out-of-phase component responsible for the damping coefficient, w.r.t. a real-valued perturbation of the gap $\Delta\vec{e} = (\Delta x, \Delta y)$.

The complete set of dynamic coefficients is then, for a given rotational speed and steady-state working eccentricity, obtained by a perturbation of the rotor position in both x-direction and y-direction, by small increments Δx and Δy respectively. The coefficients are in the former situation calculated as:

$$k_{xx} = \frac{-\int_{-L/2}^{L/2} \int_0^{2\pi} \Delta p_{\mathrm{k}} \cos\theta \; r\mathrm{d}\theta\mathrm{d}z}{\Delta x} \tag{9.13a}$$

$$k_{yx} = \frac{-\int_{-L/2}^{L/2} \int_0^{2\pi} \Delta p_{\mathrm{k}} \sin\theta \; r\mathrm{d}\theta\mathrm{d}z}{\Delta x} \tag{9.13b}$$

$$c_{xx} = \frac{-\int_{-L/2}^{L/2} \int_0^{2\pi} \Delta p_{\mathrm{c}} \cos\theta \; r\mathrm{d}\theta\mathrm{d}z}{\Delta x} \tag{9.13c}$$

$$c_{yx} = \frac{-\int_{-L/2}^{L/2} \int_0^{2\pi} \Delta p_{\mathrm{c}} \sin\theta \; r\mathrm{d}\theta\mathrm{d}z}{\Delta x} \tag{9.13d}$$

$$k_{yy} = \frac{-\int_{-L/2}^{L/2} \int_{0}^{2\pi} \Delta p_{\mathrm{k}} \sin\theta \; r d\theta dz}{\Delta y} \tag{9.13e}$$

...

Δx and Δy represent in these expressions a perturbation of the rotor centre in respectively the x- and y-direction leading to a corresponding perturbed height distribution $\Delta h(\theta)$.

Let us note that the minus sign in the previous equations means that we consider the force variation in the direction opposed to the motion.

For a symmetric system (i.e. steady-state working eccentricity $\epsilon = 0$) with small $\omega > 0$, and small excitation frequency ν, it can be shown that:

$$k_{xx} = k_{yy} > 0 \tag{9.14a}$$

$$c_{xx} = c_{yy} > 0 \tag{9.14b}$$

$$k_{xy} = -k_{yx} > 0 \tag{9.14c}$$

$$c_{xy} = -c_{yx} < 0. \tag{9.14d}$$

These relationships can help us better understand the development and evolution of whirl instability.

As mentioned above, these coefficients depend on the steady-state working conditions of the film, which include the eccentricity ratio ϵ, attitude angle ϕ and rotational speed ω, and, in particular for a compressible film, also on the perturbation frequency ν. Thus for a given plain, self-acting bearing,

$$k_{ij} = f(\epsilon, \phi, \omega, \nu) \tag{9.15a}$$

$$c_{ij} = f(\epsilon, \phi, \omega, \nu). \tag{9.15b}$$

When the static bearing load has only little effect on the working eccentricity ratio and direction, this relationship is reduced to: $k_{ij} = f(\omega, \nu)$ and $c_{ij} = f(\omega, \nu)$. Or, generally, in dimensionless form:

$$K_{ij} \ldots = f(\sigma, \Lambda).$$

Normalisation of Load and Stiffness

The load is rendered dimensionless by

$$\bar{W} = \frac{W}{LDpa}.$$

The stiffness and damping may be made dimensionless in various ways. We have adopted the following normalisation for this chapter:

$$K_{ij} = k_{ij}\frac{c}{p_{\mathrm{a}}LD},$$

$$C_{ij} = c_{ij}\frac{1}{3\mu L}\left(\frac{c}{D}\right)^3$$

We shall show examples of these for each bearing types below. For an EP JB, the dynamic stiffness and damping will depend, in addition to the previous, on the supply pressure, feed-hole diameters, conicity dimensions, etc.

The complete dynamic characterisation of a journal bearing consists therefore in the successive evaluation of the film properties at different combinations of rotational speed and perturbation frequency. The tabulated representation of the obtained results is referred to as a dynamic bearing map, which serves in most cases as input for the rotordynamic study of the complete gas bearing system. In the above formulations, the rotor axis is assumed to remain parallel with the bearing axis. In a completely analogous manner to the one outlined above, the dynamic tilt stiffness of the film may be defined as resulting from rotor misalignment, Ausman (1960). This will not be treated further, however.

Mass Flow Rate

For a self-acting bearing, the net flow into or out of the bearing is obviously zero. For an EP bearing on the other hand, the flow out of the bearing is equal to the flow through the feeding holes into the bearing. Consequently, the formulas derived in Chapter 2 can be utilised.

The dimensionless mass flow rate is defined as:

$$\bar{\dot{m}} = \frac{\dot{m}\Re T_a 12\mu}{c^3 p_a}.$$

Friction and Hydrodynamic Torque and Power

For simplicity of treatment, we assume in the following that only the shaft is rotating, at angular velocity ω, while the bearing is stationary, i.e. $\Omega = 0$.

We have seen from Chapter 8 that the shear traction on a bearing surface consists of two components: (i) that owing to viscous friction; and (ii) that due to aerodynamic action, occurring both in self-acting and EP bearings, regardless of the type, owing to the pressure build-up. Applying those formulas to a JB, we obtain that the elemental surface traction stress is given by:

$$\tau_t = \left(-\frac{\mu}{h}\omega R \mp \frac{h}{2R}\frac{\partial p}{\partial \theta} \right) \tag{9.16a}$$

$$= \frac{p_a c}{2R}\left(-\frac{\Lambda}{3H} \mp H\frac{\partial P}{\partial \theta} \right). \tag{9.16b}$$

($\mp$ corresponds to the shear stress being evaluated at the journal or at the bearing surface, respectively)

Note that we have used the sign convention whereby the surface traction stress is negative when it is in the opposite sense to that of the rotation. Thus, the first term is the dissipative friction term as postulated by Petrov's formula; the second is owing to the aerodynamic action and will be zero for a concentric smooth-surface bearing.

Let us note that the fact that the traction stresses are not equal on both surfaces does not violate Newton's third law, since the difference between them, being equal to

$$\frac{h}{R}\frac{\partial p}{\partial \theta},$$

is the stress required to accelerate the fluid around the bearing.

Integration of the traction stress τ_t over the entire bearing surface $A = \{0 \leq z \leq L, 0 \leq \theta < 2\pi\}$ multiplied by the radius, being equal to the arm of the couple, yields the total friction torque

$$T_f = R\int_A \tau_t \, dA, \tag{9.17}$$

The power is the product of this quantity with the angular velocity ω

$$P_f = T_f\, \omega. \tag{9.18}$$

Important Remarks

- The dissipative, Petrov's friction component of the traction torque will be present in all JBs whether they be of the self-acting or EP type. It is proportional to the Somerfeld variable, viz. the viscosity times the velocity divided by the film thickness. Thus, the friction power dissipation increases with the square of the speed. It is responsible for generating the major portion of heat in the bearing.
- For the case of a self-acting or hybrid JB, the aerodynamic power is that which is responsible for generating the load capacity. This is, strictly speaking, not dissipative in the same sense as the previous item, since some of it is lost by side leakage outside the bearing. The amount of energy expended owing to this depends on the eccentricity ratio, or pressure generation. However, this component is much smaller than the viscous friction one, as we also noticed in flat bearings, see Chapters 8 and 6.
- In an EP bearing, there is an additional power loss component, viz. that needed to pump the fluid through the bearing; the so-called "pumping power". It does not cause heat dissipation in the gap, as shown in Chapter 17.
- Thus, from an energy-cost point of view, the choice for a self-acting or an EP-hybrid bearing represents an optimisation problem that is not straightforward to solve owing in particular to the strong dependence on the speed. Within the category of self-acting bearings, optimisation should ensure minimum viscous-friction power loss as compared to aerodynamic power loss.

In the following, we shall consider the characteristics of the four generic JB types: plain, herringbone-grooved JB (HGJB), EP and hybrid bearings.

9.3 Plain Self-acting Journal Bearings

Also referred to as aerodynamic, the bearing characteristics depend upon:

$$(\epsilon, \Lambda, \frac{L}{D}, \sigma).$$

The general solution, i.e. for any range of the design parameters above, can be obtained only numerically, which is not so demanding when using e.g. the FD method. One of the oldest numerical solutions is that by Raimondi (1961). However, it is always useful and instructive to have some or other form of an analytic solution at ones disposal in order to evaluate general trends and obtain more insight into the design problem, in particular at its early stages.

We shall start by such a solution and complement its results by the numerical solution for appraisal and comparison.

9.3.1 Small-eccentricity Perturbation Static-pressure Solution

This problem was first solved by Ausman (1961) as reported in Wilcock (1969), which we retrace here below.

We assume no axial motion, $\Lambda_z = 0$, isothermal compression, $\wp = P$ and steady state, $\sigma = 0$, and solve for small perturbation ϵ around concentric position. Equation 9.2, becomes

$$\frac{\partial}{\partial Z}\left(H^3 P \frac{\partial P}{\partial Z}\right) + \frac{\partial}{\partial \theta}\left(H^3 P \frac{\partial P}{\partial \theta} - \Lambda P H\right) = 0. \tag{9.19}$$

We shall consider only radial motion of the shaft, leaving out the case of tilt, being of minor importance, which can be treated in a similar manner. In this case, the eccentricity ratio is a constant w.r.t. Z. Thus, putting

$$H = 1 + \epsilon e^{i\theta}, \qquad \epsilon \ll 1,$$

the pressure may be expressed as

$$P(\theta, Z) = 1 + \epsilon P_p + O(\epsilon^2) \approx 1 + \epsilon f(Z) e^{i\theta},$$

where P_p is the perturbed pressure w.r.t. the steady state film pressure, being equal to zero for the concentric bearing. We note that the perturbation pressure has the same form as the gap in θ.

Substituting these expressions into Eq. 9.2 and neglecting terms of $O(\epsilon^2)$ reduces the p.d.e. to the o.d.e:

$$\left(\frac{\mathrm{d}^2}{\mathrm{d}Z^2} - (1 + i\Lambda)\right) f(Z) - i\Lambda = 0, \tag{9.20}$$

with periodic boundary conditions in θ, while in the z-direction, we have:

$$P_\mathrm{p}(\pm L/D, \theta) = 0 \Rightarrow f(\pm L/D) = 0$$

Integrating this linear o.d.e. with the application of zero boundary conditions at the ends, yields:

$$f(Z) = -\frac{i\Lambda}{1 + i\Lambda}\left(1 - \frac{\cosh(\sqrt{1 + i\Lambda}\, Z)}{\cosh(\sqrt{1 + i\Lambda}\, L/D)}\right) \tag{9.21}$$

Considering only real solutions (in the steady state), we have:

$$P_\mathrm{p} = \mathrm{Re}(f(Z)e^{i\theta}) \tag{9.22}$$

The (r, t) components of the bearing force, are then given by

$$\begin{aligned}
F_\mathrm{r} &= p_\mathrm{a}R^2 \int_{-L/D}^{L/D} \mathrm{d}Z \int_0^{2\pi} -\epsilon P_\mathrm{p} \cos\theta \mathrm{d}\theta \\
&= \epsilon 2\pi p_\mathrm{a}R^2 \int_0^{L/D} \mathrm{d}Z\mathrm{Re}\left\{\left(\frac{i\Lambda}{1 + i\Lambda}\right)\left[1 - \frac{\cosh(\sqrt{1 + i\Lambda Z})}{\cosh(\sqrt{1 + i\Lambda L/D})}\right]\right\} \\
&= \epsilon 2\pi p_\mathrm{a}R^2\ \mathrm{Re}\left\{\left(\frac{i\Lambda}{1 + i\Lambda}\right)\left[(L/D) - \frac{\tanh(\sqrt{1 + i\Lambda L/D})}{\sqrt{1 + i\Lambda}}\right]\right\} \\
&= \epsilon p_\mathrm{a}LD\left(\frac{\pi}{2}\right)\mathrm{Re}\left\{\left(\frac{i\Lambda}{1 + i\Lambda}\right)\left[1 - \frac{\tanh(\sqrt{1 + i\Lambda L/D})}{(\sqrt{1 + i\Lambda L/D})}\right]\right\}
\end{aligned} \tag{9.23}$$

$$\begin{aligned}
F_\mathrm{t} &= p_\mathrm{a}R^2 \int_{-L/D}^{L/D} \mathrm{d}Z \int_0^{2\pi} \epsilon P_\mathrm{p} \sin\theta \mathrm{d}\theta \\
&= \epsilon p_\mathrm{a}LD\left(\frac{\pi}{2}\right)\mathrm{Im}\left\{\left(\frac{i\Lambda}{1 + i\Lambda}\right)\left[1 - \frac{\tanh(\sqrt{1 + i\Lambda L/D})}{(\sqrt{1 + i\Lambda L/D})}\right]\right\}
\end{aligned} \tag{9.24}$$

Or, rewriting the bearing force as a complex number, in a compact way:

$$\begin{aligned}
\bar{F}_r + i\bar{F}_t &= \frac{F_r + iF_t}{\epsilon p_\mathrm{a}LD} \\
&= \left(\frac{i\Lambda}{1 + i\Lambda}\right)\left\{1 - \frac{\tanh(\sqrt{1 + i\Lambda L/D})}{(\sqrt{1 + i\Lambda L/D})}\right\}\left(\frac{\pi}{2}\right)
\end{aligned} \tag{9.25}$$

Remarks

This exceedingly elegant, closed-form solution is very useful for estimating the static and the dynamic characteristics, as we shall see in an extension below, of plain aerodynamic journal bearings. It can serve as verification to numerical solution procedures. But most importantly, it can be used as a first design step, albeit for small eccentricities only, to estimate JB behaviour and to dimension it in a preliminary way for various values of the slenderness ratio.

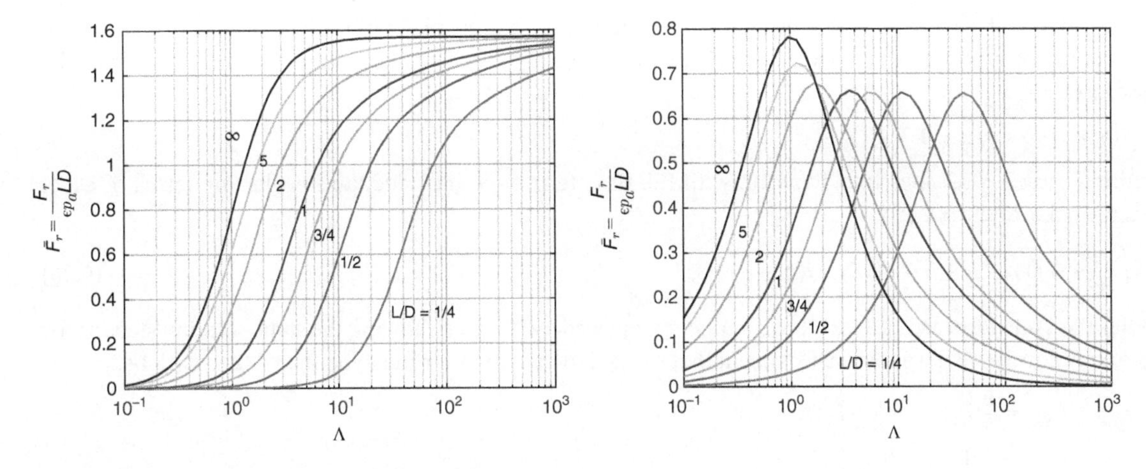

Figure 9.3 F_r and F_t plotted for for the full range of Λ and L/D.

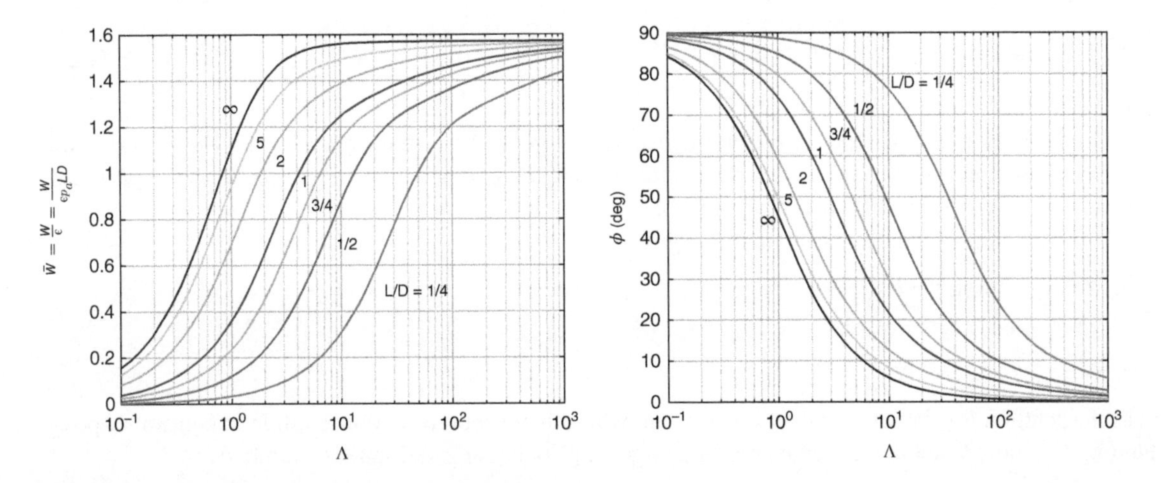

Figure 9.4 $\bar{W}$ and ϕ plotted for for the full range of Λ and L/D.

Graphs for $\bar{F}_r$ and $\bar{F}_t$ are provided in Figure 9.3.

Let us note that, when these functions are extended to negative values of Λ, F_r will appear to be even while F_t will appear to be odd. We need this knowledge to assess the dynamic stiffness of the bearing as will be given in the following section.

The radial and tangential forces can be converted into bearing load and load angle as given in Figure 9.4, which are even and odd functions in the same manner.

9.3.2 Dynamic Characteristics

The solution above can be easily extended to the dynamic case, when only small perturbations, in the radial direction around concentric position, are considered also in the time solution. In this case the film function is written:

$$H = 1 + \epsilon e^{i\theta} e^{\pm i\tau},$$

where we have used $\pm\tau$ for the time term since the time perturbation may be in phase or out of phase with film function, corresponding to forward and backward whirling respectively. The pressure may likewise be expressed as

$$P(\theta, Z) \approx 1 + \epsilon P_{\mathrm{p}} = 1 + \epsilon f(Z) e^{i\theta} e^{i\tau},$$

Substituting these expressions into Eq. 9.19, with the time term now included, and neglecting terms of $O(\epsilon^2)$ reduces the p.d.e. to the o.d.e:

$$\left(\frac{\mathrm{d}^2}{\mathrm{d}Z^2} - (1 + i(\Lambda \pm \sigma))\right) f(Z) - i(\Lambda \pm \sigma) = 0 \tag{9.26}$$

which has the same form as Eq. 9.20 but with Λ being replaced by $\Lambda \pm \sigma$. Following the same solution steps as for the static solution above, it can be easily shown that the general solution for the dynamic stiffness will be:

$$\begin{aligned} K_{RR} &\triangleq K_{XX} = \frac{1}{2}(\bar{F}_r(\Lambda - \sigma) + \bar{F}_r(\Lambda + \sigma)) \\ C_{RR} &\triangleq C_{XX} = -\frac{1}{2\sigma}(\bar{F}_t(\Lambda - \sigma) - \bar{F}_t(\Lambda + \sigma)) \\ K_{TR} &\triangleq K_{YX} = \frac{1}{2}(\bar{F}_t(\Lambda - \sigma) + \bar{F}_t(\Lambda + \sigma)) \\ C_{TR} &\triangleq C_{YX} = -\frac{1}{2\sigma}(\bar{F}_r(\Lambda - \sigma) - \bar{F}_r(\Lambda + \sigma)) \end{aligned} \tag{9.27}$$

where, $\bar{F}_{\mathrm{r}}$ and $\bar{F}_{\mathrm{t}}$ have the same forms as before.

N.B. In the above notation, we have, for convenience, made the (X, Y) system coincide with the (r, t) system. In whichever chosen orthogonal coordinate system, the XX and YY components are referred to as the principal components, while the XY and YX are the cross-coupled ones.

Note that the remaining dynamic coefficients are given by

$$K_{XX} = K_{YY}, \quad K_{XY} = -K_{YX}$$

and

$$C_{XX} = C_{YY}, \quad C_{XY} = -C_{YX}$$

That is, the solution is symmetric, which is consistent with it being a perturbation solution in concentric position. Figures. 9.5 and 9.6 plot these coefficients for a large range of bearing and squeeze numbers.

General Remarks

- For small Λ's the functions K_{XX} and C_{XX} behave as a phase-lead system (see Chapter 7), i.e. a positively damped system in the principal direction.
- For larger Λ, a lag–lead behaviour starts to develop, again similar in trend to that shown in Chapter 7, for flat aerostatic bearings with negative static damping, for which the region of negative damping increases as Λ increases.
- In this case, K_{XX} has a minimum value at a point $\Lambda = \sigma$, which will appear to be the point corresponding to half-speed whirl, or $\nu = \omega/2$.
- We can, in a similar way, conveniently fit the remaining two functions K_{XY} and C_{XY}. Such fits, when applied to other bearing types, may prove very handy devises to asses whirl stability in a systematic way; see the next item.
- Forestalling Chapter 10, let us remark that whirl instability, which is not triggered by negative principal damping as in flat aerostatic bearings, sets in, actually, when the following condition is violated:

$$|k_{xy}| < \sqrt{\frac{k_{xx}}{m}} c_{xx}.$$

(where, m is the rotating mass.)

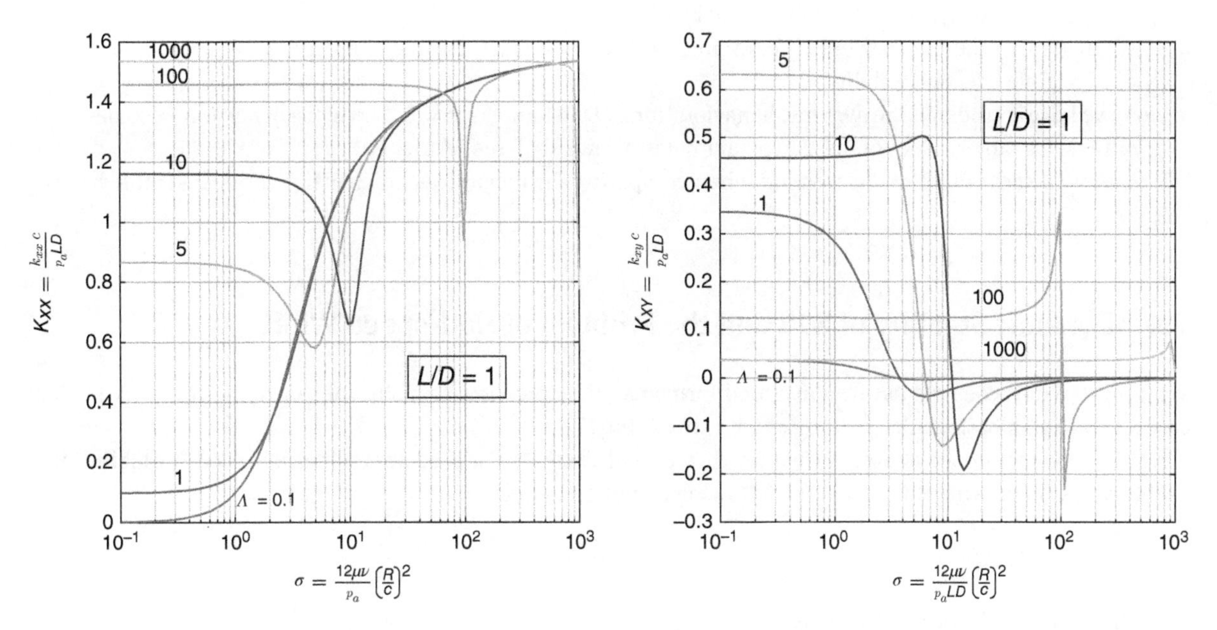

Figure 9.5 The dimensionless stiffness coefficients K_{XX} and K_{XY} as a function of the squeeze number σ for different values of the bearing number Λ. Self-acting journal bearing with $L/D = 1$ and small ϵ perturbation.

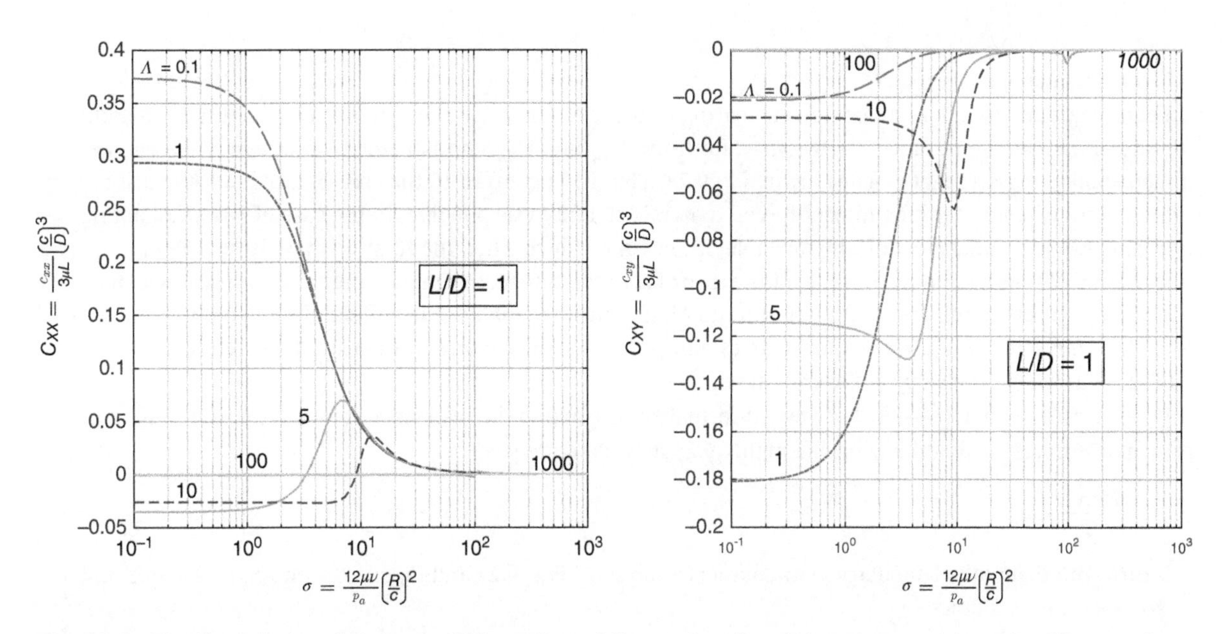

Figure 9.6 The dimensionless damping coefficients C_{XX} and C_{XY} as a function of the squeeze number σ for different values of the bearing number Λ. Self-acting journal bearing with $L/D = 1$ and small ϵ perturbation.

Thus, the smaller the magnitude of the "cross-coupled" stiffness coefficient k_{xy} is, the higher is the whirl-onset speed; i.e. the more dynamically stable is the bearing system.

- We shall see that the "cross-coupled" damping coefficient c_{xy} plays the role of an added mass and has, in most cases negligible effect on stability.
- Here we have considered the dynamic behaviour for $L/D = 1$ and for $\epsilon = 0$. The whirl behaviour becomes more favourable for larger values of these two parameters, as we shall see further.

It is useful to bear these points in mind when comparing the dynamic coefficients of various possible bearing solutions.

9.4 Dynamic Stability of a JB and the Problem of Half-speed Whirl

With the dynamic coefficients given in closed-form analytic expressions, a plain, concentric JB furnishes an ideal case-study to gain basic understanding of whirl instability.

As has been mentioned above, in Chapter 10, we shall see that, to a good approximation, whirl instability, in a 2D bearing system, will set in, when the following condition is violated:

$$|k_{xy}| < \sqrt{\frac{k_{xx}}{m}} c_{xx}.$$

In dimensionless form, this becomes:

$$|K_{XY}| < \sqrt{\frac{K_{XX}}{M}} C_{XX}, \tag{9.28}$$

where,

$$M = \frac{m p_{\mathrm{a}}}{9 L \mu^2} \left(\frac{c}{D}\right)^5$$

is the dimensionless mass.

Let us examine the situation for a given bearing number Λ, for which the dynamic coefficients are determined for any given squeeze number σ. Obviously, for finite K_{XX} and C_{XX} and non-zero K_{XY}, there will always exist a dimensionless mass M that violates inequality 9.28. That is, in contrast to the case of a single EP thrust bearing examined in Chapter 7, where unconditional dynamic stability was possible, in the case of JBs, the rotating system can always become dynamically unstable, i.e. begin to whirl, when the rotating mass is sufficiently high; i.e. unconditional stability is only possible for the case of zero rotating mass. On the other hand, one can see that, for a positive C_{XX}, there should be a certain minimum M for which inequality 9.28 will hold true. That value is referred to as the *critical mass* M_{crit}. We would like to establish the frequency σ or, equivalently, the speed Λ at which whirl will set in as well as the value of the critical mass.

Let us first note that $\sqrt{\frac{K_{XX}}{M}}$ in our present formalism is equal to the resonance frequency σ_{r} of the shaft on the principal stiffness, so that the whirl inequality 9.28 may be written as:

$$\frac{|K_{XY}|}{C_{XX}} < \sigma_{\mathrm{r}}.$$

It turns out that with elementary mathematical analysis of Eqs. 9.27 and Figure 9.3, one can solve this problem in an exact way as follows.

1. we have that

$$\frac{|K_{XY}|}{C_{XX}} = -\sigma \frac{|(\bar{F}_t(\Lambda - \sigma) + \bar{F}_t(\Lambda + \sigma))|}{(\bar{F}_t(\Lambda - \sigma) - \bar{F}_t(\Lambda + \sigma))},$$

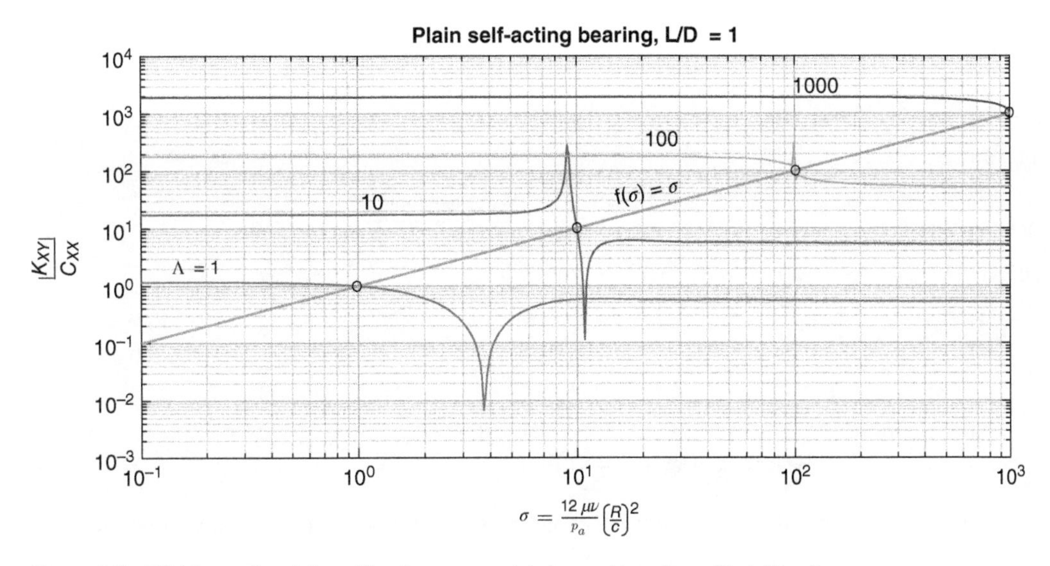

Figure 9.7 Whirl map for plain self-acting, concentric journal bearing with $L/D = 1$.

with $\bar{F}_t(0) = 0$, so that, for any given $\Lambda > 0$, we have

$$\sigma = \sigma_r = \Lambda \Rightarrow \frac{|K_{XY}|}{C_{XX}} = \sigma_r$$

(for $C_{XX} > 0$, which will be shown to be so later on).

Thus, whirl can occur when the dimensionless pulsation $\sigma_{\text{crit}} = \Lambda$, that is at a dimensional frequency equal to half of the rotation speed, since for $\nu = \omega/2$, we have that $\sigma = \Lambda$. This is also depicted in the whirl map of Figure 9.7 where we see that the line $f(\sigma) = \sigma$ intersects the curves $\frac{|K_{XY}|}{C_{XX}}$ at the points $\sigma = \Lambda$ as indicated by the small circles.

2. One can show easily that for $\sigma > \Lambda$ then $\frac{|K_{XY}|}{C_{xx}} < \sigma$, as is also evident in Figure 9.7. That is, if the system is stable at $\sigma_{\text{crit}} = \Lambda$ it will remain so at all higher frequencies.
3. The function $\bar{F}_t(\Lambda)$ is equal to zero at $\Lambda = 0$ and attains its maximum at some value of $\Lambda = \Lambda_m > 0$. Moreover, $\bar{F}_t(\Lambda)$ is an odd function in Λ. Thus, one can easily show that
   i) $\Lambda < \Lambda_m \Rightarrow C_{XX} > 0$ for all σ;
   ii) $\Lambda \geq \Lambda_m \Rightarrow C_{XX} \leq 0$ for $\sigma < \Lambda$ and $C_{XX} > 0$ for $\sigma \geq \Lambda$.
   This last property ensures that the principal damping is always positive for $\sigma \geq \sigma_{\text{crit}}$; a condition which we needed for the first step. This is also what we observe in the left panel of Figure 9.6.
4. For the sake of completeness, since $\bar{F}_r$ is an even, positive function, with a point of inflexion at some $\Lambda = \Lambda_i > 0$, having its minimum at the origin, we can easily show that, for a given $0 \leq \Lambda \leq \Lambda_i$, K_{XX} will have its minimum value at $\sigma = 0$ and for a given $\Lambda > \Lambda_i$, K_{XX} will have its minimum value at $\sigma = \Lambda$. This is what we observe in the left panel of Figure 9.5.
5. The value of M that will ensure the system is on the verge of instability will then be given by:

$$M_{\text{crit}} = \frac{K_{XX}(\sigma_{\text{crit}})}{(\sigma_{\text{crit}})^2} = \frac{K_{XX}(\Lambda)}{\Lambda^2} = \frac{1}{2}\frac{\bar{F}_r(2\Lambda)}{\Lambda^2}.$$

Figure 9.8 plots this critical mass as a function of Λ for different values of the slenderness ratio L/D. We note that (i) M_{crit} increases with increasing slenderness ratio and (ii) for very high Λ, the critical mass converges to an asymptotic value for all values of L/D.

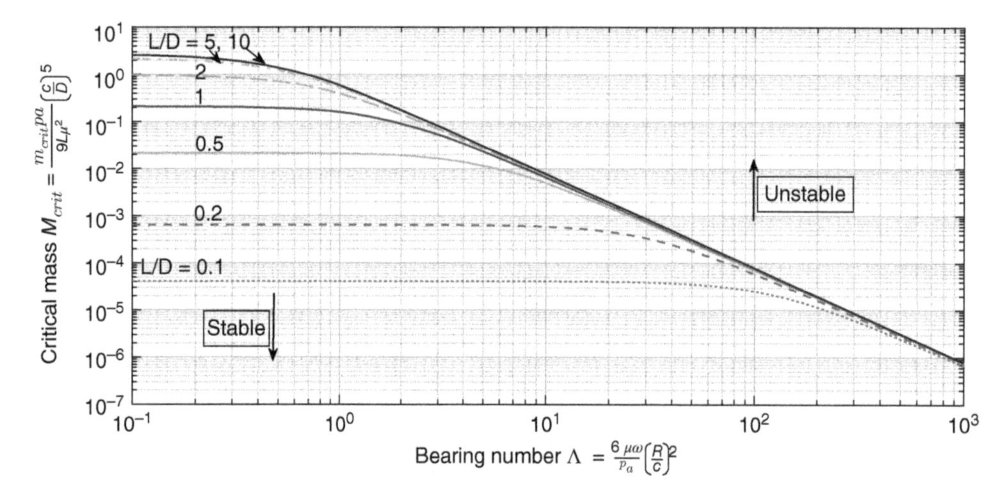

Figure 9.8 The critical mass as a function of Λ for different values of the slenderness ratio L/D for a plain, self-acting journal bearing.

Note on the Utility of Whirl Occurring at Half-speed

When neglecting the effect of the cross-coupled damping C_{XY}, which is justifiable at least for high squeeze numbers, we have seen for a plain concentric bearing that whirl occurs exactly at half-speed, i.e. at $\sigma_r = \Lambda_c$. In that case, the critical mass is entirely determined by the value of the principal stiffness at that point, by:

$$M_c = \frac{K_{XX}}{\sigma_r^2} = \frac{K_{XX}}{\Lambda_c^2}.$$

We have seen, furthermore, for this type of bearing, that the principal stiffness, for higher Λ values, happens to have a local minimum, namely the "dip" at that same point. In the traditional state of the art, the depth of this dip was taken as an indication of the proneness of a bearing type to self-excited whirl, or its lack of dynamic stability. This explains perhaps why plain JBs score worst with regard to whirl instability: they appear to have the deepest dip as compared to other bearing types.

If whirl should occur at half-speed for all bearings, then we need only compare the principal stiffness of those bearings at $\Lambda = \sigma$, i.e. the K_{XX}-dip value. Bearing systems that have higher principal stiffness at that point will be more stable, i.e. will have a higher critical mass. We shall see that this is certainly the case for EP bearings, but that its not so straightforward for HGJB, since whirl does not always occur at or near half-speed.

9.4.1 General Numerical Solution

The previously derived solutions are only valid for small eccentricity ratios, up to $\epsilon \approx 0.2$ as we shall see later. The behaviour for higher eccentricities may be obtained by a numerical solution, which is computationally not so demanding. Below, we shall present such solutions for the particular case of $L/D = 1$. Other cases can be similarly calculated or estimated with the aid of the perturbation solution.

We start by showing the pressure distributions depicted in Figure 9.9, where the bearing number is driven from very low to very high values from (a) to (c) producing pressure distributions in the bearing mid-plane that vary between "sinusoidal" and "co-sinusoidal" ones, respectively. It can be verified that the solution presented in (c) corresponds to $PH \rightarrow$ constant, i.e. the infinite bearing number solution, see Chapter 8.

Next, we consider the load capacity and load angle behaviour in the bearing number and the eccentricity. This is depicted in Figure 9.10. This type of numerical solution was first obtained by Raimondi (1961). In particular, as

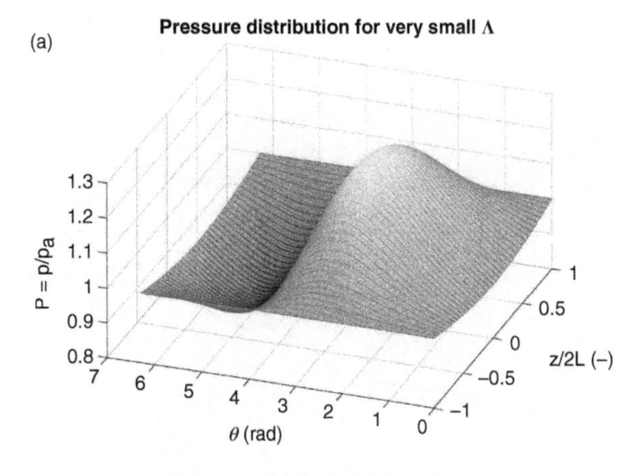

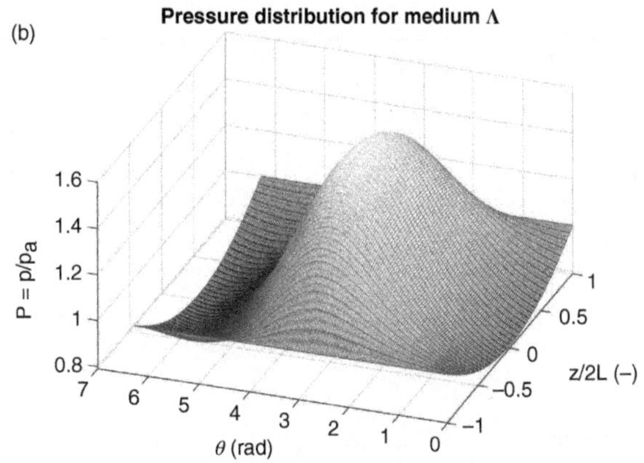

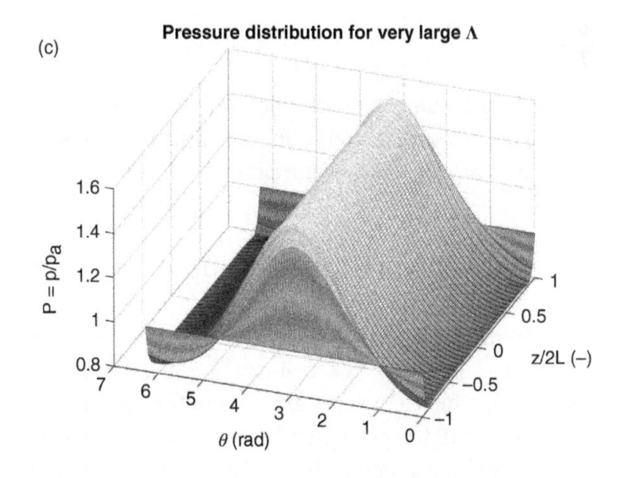

Figure 9.9 The dimensionless pressure distribution in a self-acting journal bearing for $L/D = 1$, $\epsilon = 0.3$ and (a) $\Lambda = 2$, (b) $\Lambda = 10$ and (c) $\Lambda = 1000$.

ϵ becomes higher, the eccentricity vector aligns more with the load vector, i.e. the attitude angle approaches zero more quickly.

For small eccentricity ratios, the solution is close to the perturbation solution as we can verify from Figure 9.11. In particular, for small bearing numbers, the two solutions are close to one another up to $\epsilon \approx 0.3$. As Λ gets higher, the solutions deviate from each other. One can notice, however, that the solutions remain close to one another up to $\epsilon = 0.2$ for all practical Λ values.

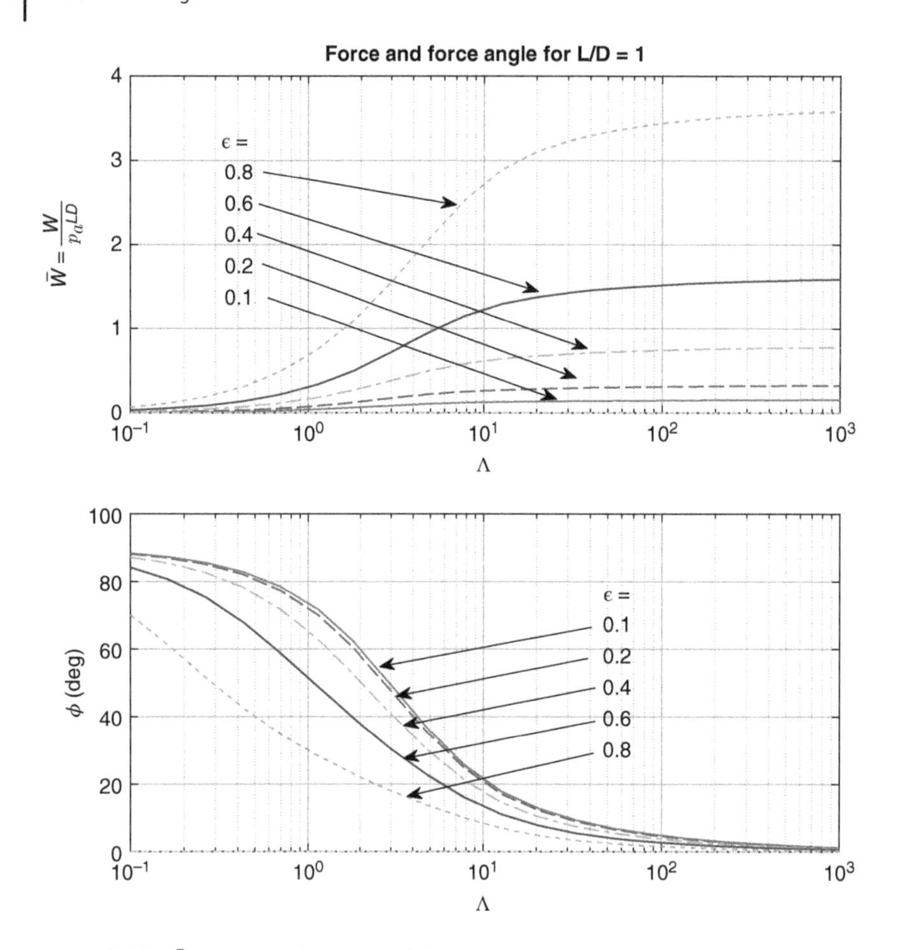

Figure 9.10 $\bar{W}$ and ϕ obtained by the FD numerical method plotted for for the full range of Λ and ϵ for $L/D = 1$.

Finally, the numerical solution can show us how the stiffness behaves for high eccentricities. Thus, from Figure 9.12, we see that the stiffness is almost constant up to $\epsilon \approx 0.3$, thereafter, it increases much faster, reaching high values at very high eccentricities.

9.5 Herringbone Grooved Journal Bearings

This type of bearing was apparently first presented in its thrust form by Whipple (1958) in connection with air lubrication. The first paper treating the journal version appears to be that of Vohr and Chow (1965). Other works followed, of which a number are cited in the introductory section.

This type of JB has been proposed as the one with a large whirl-stability range, while not suffering the construction and assembly complexity of a Tilting-Pad JB (TPJB), which is famed for its quasi-whirl-free operation.

The geometry and notation are depicted in Figure 9.13, which are essentially the same as those pertaining to the flat case, described in Chapter 8, the main difference being that the grooves are now a helix-form either on the shaft or the bush. When projected on a flat, the grooves pattern resembles the shape of herringbone, Figure 9.13 (c) left.

The general Reynolds equation has been derived for infinite number of grooves, i.e. the narrow-groove theory (NGT), in Chapter 8.

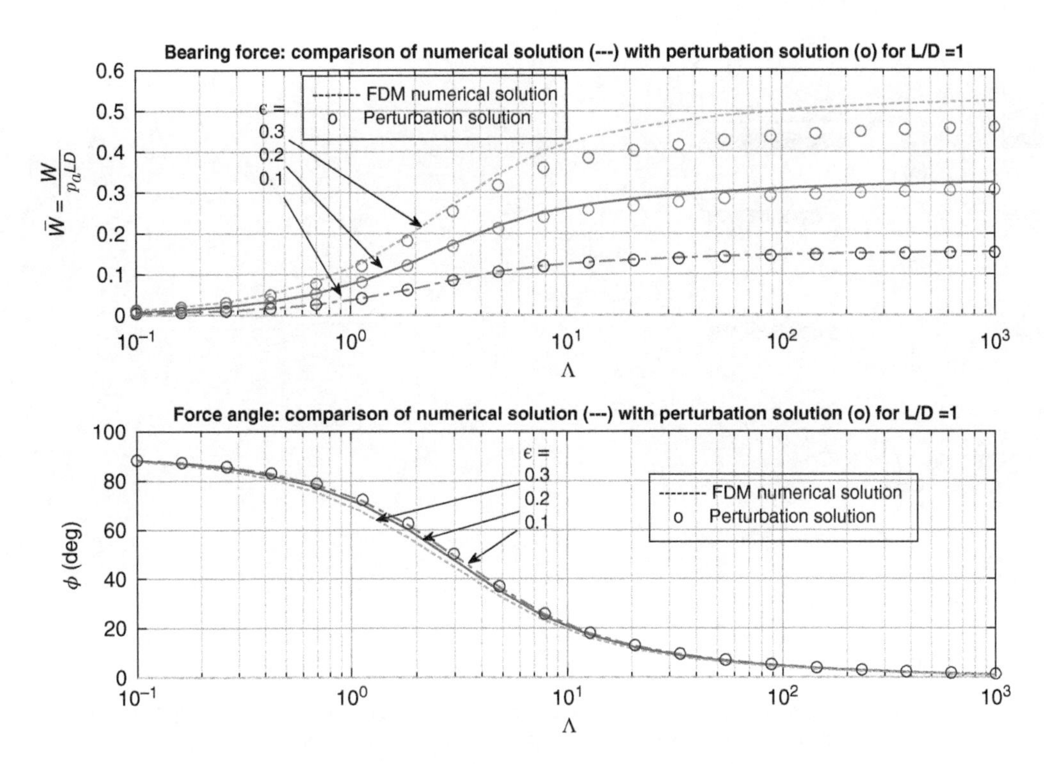

Figure 9.11 Comparison of the numerical solution (continuous lines) with the small-perturbation solution (circles) for small values of ϵ and for $L/D = 1$.

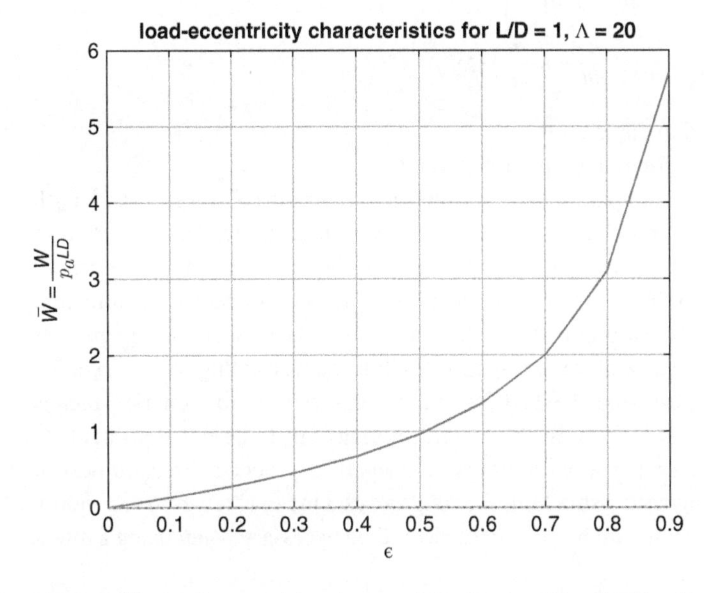

Figure 9.12 Load-eccentricity characteristics for $\Lambda = 20$ and $L/D = 1$.

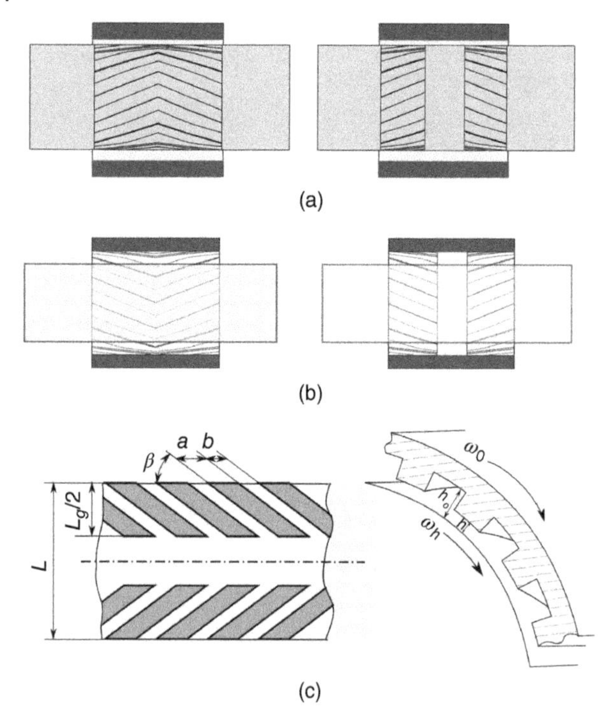

Figure 9.13 Herringbone grooved journal bearings with and without smooth central part (Not to scale). (a): grooves on shaft (full, left; partial, right). (b): grooves on bearing (bush) (full, left; partial, right). (c): notation.

Writing it in a dimensionless way for the grooved portion of the bearing, we have:

$$\begin{aligned}&\frac{\partial}{\partial\theta}\left(\wp\left(\bar{f}_\theta\frac{\partial P}{\partial\theta}+\bar{f}_c\frac{\partial P}{\partial Z}\right)\right)+\frac{\partial}{\partial Z}\left(\wp\left(\bar{f}_c\frac{\partial P}{\partial\theta}+\bar{f}_z\frac{\partial P}{\partial Z}\right)\right)\\&+\bar{c}_s\Lambda^*\left(\sin\beta\frac{\partial}{\partial\theta}-\cos\beta\frac{\partial}{\partial Z}\right)(\bar{f}_s\wp)=\left(\Lambda\frac{\partial}{\partial\theta}+\sigma\frac{\partial}{\partial\tau}\right)(\wp\bar{f}_v),\end{aligned}\tag{9.29}$$

where the variables and parameters are defined in Table 9.1.

For the smooth portion of the bearing, the usual Reynolds equation is valid.

The solution will then depend on these variables and parameters in addition to the ratios of the grooved length to total bearing length L_g/L and L/D. for the general characteristics of this equation, the reader is referred to Chapter 8. In particular, we mention here that (i) the term in Λ^* is responsible for the groove pumping action; (ii) this term is appreciably smaller in magnitude than the term in Λ; (iii) when the shaft is in concentric position and the grooves are oriented so that pumping is in the inward direction for all cases, then the case of rotating grooves will result in the same pressure field as that for the case of stationary grooves, but that is rotating with the shaft.

Let us note that the NG theory does not yield an easy-to-solve PDE, so that one has to resort to numerical means in order to achieve even the simple perturbation solutions. In this regard, it turns out to be not so difficult to pursue a numerical, e.g. FD solution right from the start; i.e. without resort to NG theory. Such a direct numerical solution may be useful, in particular when dealing with a small number of grooves. One may resort to NG theory for the case of very high number of grooves, as then too many grid points would be necessary when using a direct numerical solution.

The results that shall be presented below have been obtained by a direct FD solution using a total number of 24 grooves around the perimeter of the bearing.

Table 9.1 Definition of the variables and parameters of the HGJB Reynolds equation

Symbol	Definition	Symbol	Definition
σ	$\frac{12\mu v}{p_a}\left(\frac{R}{c}\right)^2$	$P, \wp$	$p/p_a, \rho/\rho_a$
Λ	$\frac{6\mu(\omega_h+\omega_0)}{p_a}\left(\frac{R}{c}\right)^2$	$\bar{g}_1$	$\bar{h}_o^3 \bar{h}^3$
Λ^*	$\frac{6\mu(\omega_h-\omega_0)}{p_a}\left(\frac{R}{c}\right)^2$	$\bar{g}_2$	$\bar{a}\bar{b}(\bar{h}_o^3 - \bar{h}^3)^2$
$\bar{f}_\theta$	$\frac{\bar{g}_1+\bar{g}_2\cos^2\beta}{\bar{g}_3}$	$\bar{g}_3$	$\bar{a}\bar{h}_o^3 + \bar{b}\bar{h}^3$
$\bar{f}_c$	$\frac{\bar{g}_2}{\bar{g}_3}\sin\beta\cos\beta$	$\bar{h}_o$	$\frac{h+\delta}{c}$
$\bar{f}_Z$	$\frac{\bar{g}_1+\bar{g}_2\sin^2\beta}{\bar{g}_3}$	$\bar{h}$	$\frac{h}{c}$
$\bar{c}_s$	$\bar{a}\bar{b}\bar{\delta}\sin\beta$	$\bar{\delta}$	$\frac{\delta}{c}$
$\bar{f}_s$	$\frac{\bar{h}_o^3-\bar{h}^3}{\bar{g}_3}$	$\bar{a}$	$\frac{a}{a+b}$
$\bar{f}_v$	$\bar{a}\bar{h}_o + \bar{b}\bar{h}$	$\bar{b}$	$\frac{b}{a+b}$
ω_h	speed of smooth member	ω_0	speed of grooved member

Correspondence between NG Theory and the Case of Finite Groove Number

In Wilcock (1969), results from Muijderman (1967) are utilised to show that when the number of grooves is finite, the situation involves additional leakage flow from the bearing on account of the "zig-zag" pressure distribution not conforming to uniform atmospheric pressure at the boundaries. This leads to an HGJB having finite-number of grooves to loose a percentage of its force (therefore also stiffness) as compared to the NG case. This loss may be approximated by the following formula:

$$\frac{\Delta F}{F} \approx 3.44(\frac{D}{L}\frac{1}{N_g \sin\beta})$$

Applied to the example that we shall use below: $L/D = 1$, $\beta = 30^o$, $N_g = 24$ grooves, we obtain:

$$\frac{\Delta F}{F} \approx 0.29,$$

which is appreciable and must be kept in mind when comparing results from different methods.

9.5.1 Static Characteristics

For the derivation and discussion of the general characteristics of the NGT Reynolds equation, the reader is referred to Chapter 8. Whereas for a thrust bearing with parallel nominal gap, the behaviour is independent of whether the grooved or the smooth member be rotating, for a journal bearing arrangement with a non-zero eccentricity, the behaviour is generally different when the grooves are stationary, i.e. on bearing, or when they are rotating, i.e. machined on shaft (which might be the easier option from fabrication point of view).

Figure 9.14 plots the static load characteristics of HGJB for the case of stationary grooves, in order to expose the main features of this type of bearing as compared to the plain (smooth) type. Comparing with Figure 9.10, we identify two fundamental differences. Firstly, whereas in the plain bearing case, the bearing force quickly saturates, around $\Lambda \sim 100$, in the HGJB bearing the force continues to increase, saturating at much higher values $\Lambda > 1000$. Thus, from about $\Lambda \approx 20$, the load capacity of an HGJB becomes much larger than that of a plain bearing. Below that value, however, the plain bearing might be more superior. Secondly, the attitude angle ϕ is almost insensitive

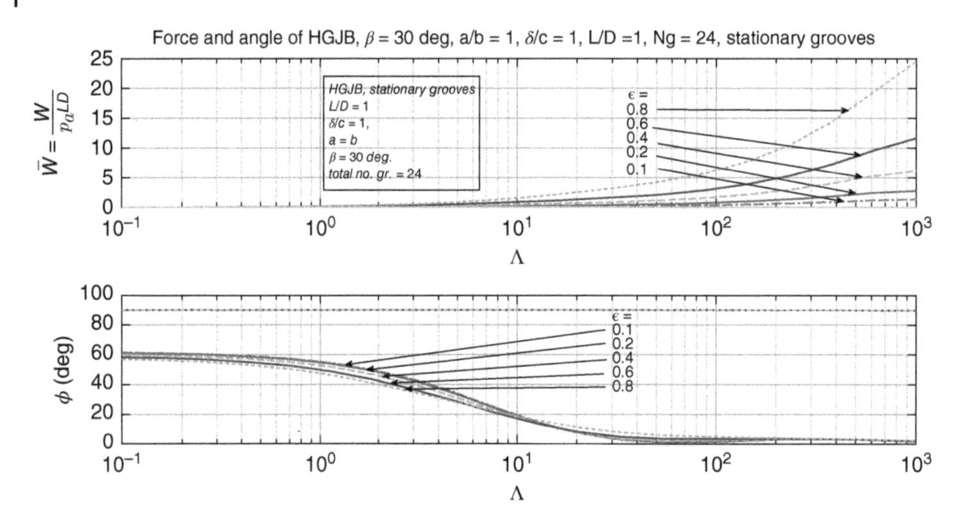

Figure 9.14 $\bar{W}$ and ϕ obtained by the FD numerical method plotted for for the full range of Λ and ϵ for $L/D = 1$.

to the eccentricity ratio ϵ but depends largely on Λ. Perhaps one can understand the behaviour of HGJB with the aid of the following two remarks: (i) for very small Λ, the behaviour is similar to a plain bearing having the same equivalent gap; (ii) for large Λ, the bearing behaves like a an EP bearing with $p_s = f(\Lambda)$, where f is an increasing function.

9.5.2 Dynamic Characteristics

We consider the stiffness and damping coefficients, as in the plain case, for a bearing in concentric position. In that case, these characteristics are independent of whether the grooves are rotating or stationary, although in the FD simulation we employed stationary grooves, i.e. on the bearing surface.

Figures 9.15 and 9.16 summarise the results.

We note the following.

- For small Λ, the principal stiffness K_{XX} is generally smaller than that for a plain bearing (Figure 9.38), which is consistent with the static results and may be directly attributed to the fact that at low sliding speeds, the pumping action is less effective as compared to increase in mean film thickness owing to the presence of grooves.
- For this same reason, viz. higher equivalent film thickness, the principal damping C_{XX} is likewise much smaller than in the plain case at low Λ.
- At higher Λ, the "dip" in K_{XX} is now much smaller in magnitude. That is why, in traditional theory, HGJB were, not quite justifiably, thought of as being dynamically more stable than their plain counterparts, see for example Vohr and Chow (1965). This is, however, generally not true, being in particular dependent on the whirl ratio, as we shall see further below and in Chapter 10.
- Generally then, as in the static case, the advantages in the dynamics of HGJB are situated in the high Λ range, where they are indeed superior to plain bearings in regard to principal stiffness, though generally with questionable whirl stability, since $|K_{XY}|/C_{XX}$ is not small.
- Thus, if we plot $|K_{XY}|/C_{XX}$ as a function of σ, for various values of Λ, as is depicted in the whirl maps of Figures 9.17 and 9.18, and we let them intersect with the line $f(\sigma) = \sigma$, then the critical pulsations are situated at the indicated intersection circles. However, we note also that for some cases, pertaining to $\Lambda = 50$ through 100, the curves almost touch the lines $f(\sigma) = \sigma$, which we have labelled "tangency". (One can verify that $C_{XX} > 0$ at the intersection points and beyond, but not necessarily for the tangency points). Moreover, between the tangency

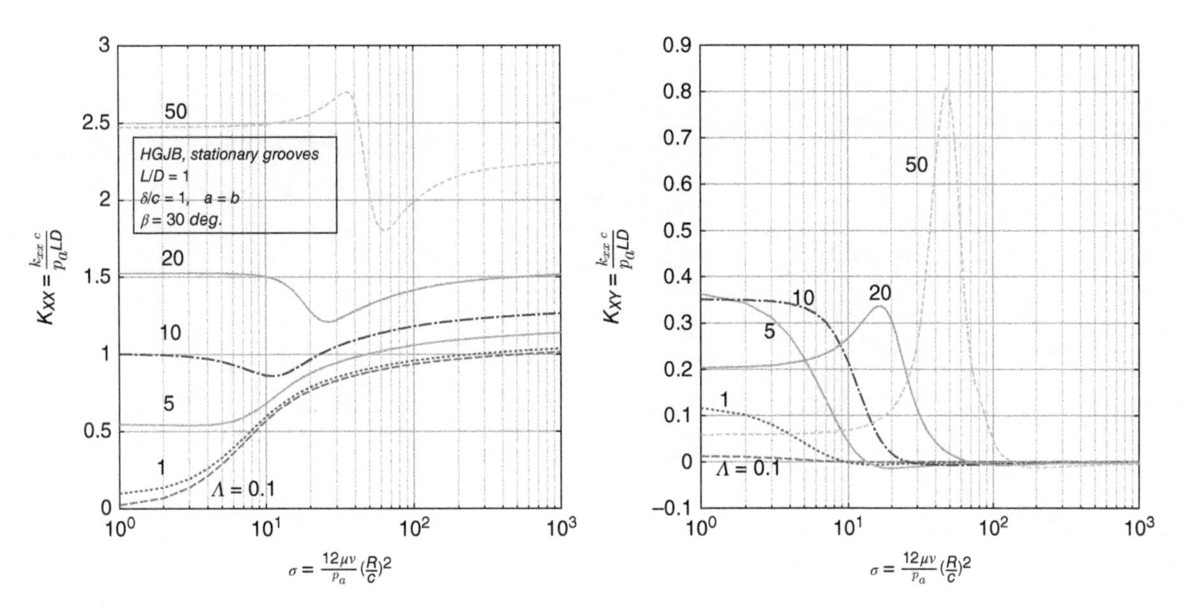

Figure 9.15 The dimensionless stiffness coefficients K_{XX} and K_{XY} as a function of the squeeze number σ for different values of the bearing number Λ. Self-acting HGJB journal bearing with $L/D = 1, H/h = 2, \beta = 30^o$, and small ϵ perturbation; 24 grooves on the stationary bearing (bush).

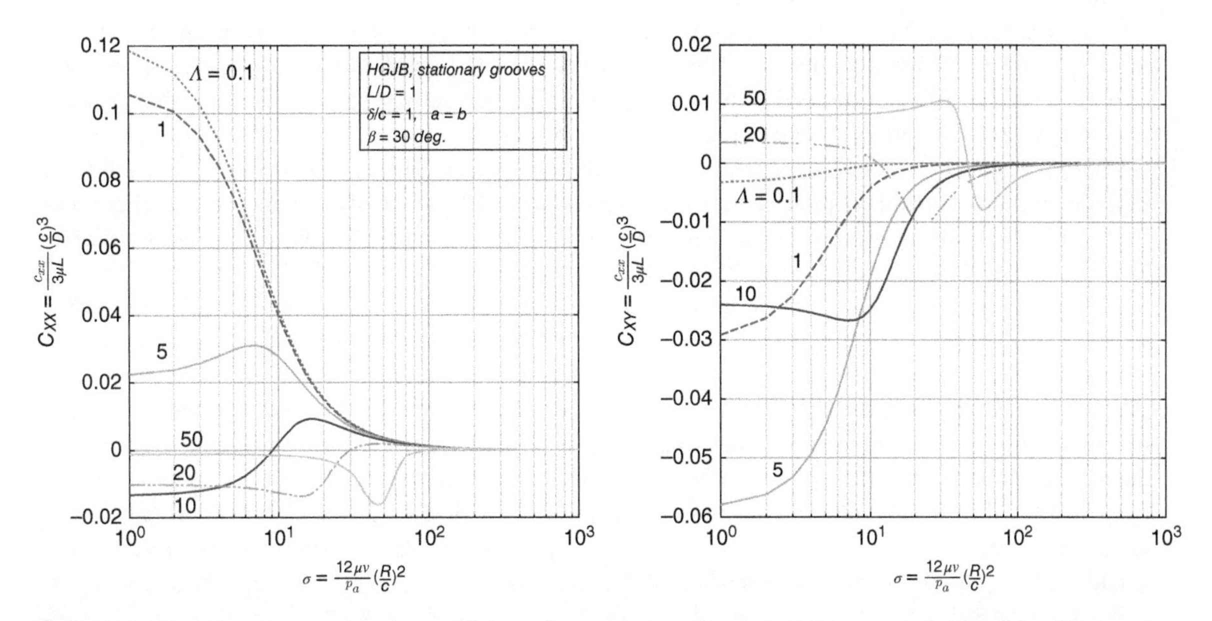

Figure 9.16 The dimensionless damping coefficients C_{XX} and C_{XY} as a function of the squeeze number σ for different values of the bearing number Λ. Self-acting HGJB journal bearing with $L/D = 1, H/h = 2, \beta = 30^o$, and small ϵ perturbation; 24 grooves on the stationary bearing (bush).

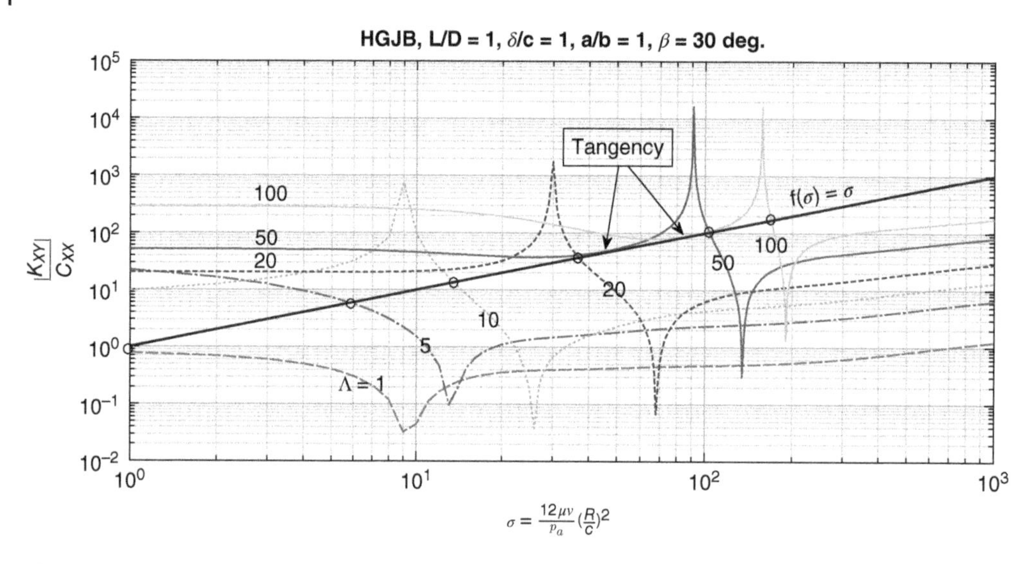

Figure 9.17 Whirl map for self-acting HGJB journal bearing with $L/D = 1, H/h = 2, \beta = 30^o$, and small ϵ perturbation; 24 grooves on the stationary bearing (bush).

and the intersection points, the $|K_{XY}|/C_{XX}$ curve lies above the $f(\sigma) = \sigma$ curve, which makes them invalid as half-speed whirl points. We further note the following:

- For very small Λ, the critical σ_r is approximately equal to Λ as in the plain-bearing case.
- As Λ increases, the intersection takes place at progressively higher values of Λ. In this case, one cannot then speak of "half-speed" whirl as such, but rather of "asynchronous". Indeed, in Figure 9.17, corresponding to a groove angle $\beta = 30^o$, the intersection, for $\Lambda = 50$, takes place beyond $\sigma = 100$, which means that the whirl frequency, for this case, is larger than the rotational frequency, i.e. "supersynchronous". Note however, that the tangency point lies around $\Lambda = 50$, pertaining to possible half-speed whirl. Unfortunately, there are only scant experimental data to confirm this behaviour. Cunningham et al. (1969) carried out a range of experiments on HGJB and pointed to some departure from half-speed whirl behaviour, showing quasi-synchronous behaviour in some cases, which might confirm our theoretical observation here.
- Repeating the same analysis for a groove angle $\beta = 60^o$, as depicted in Figure 9.18, we note that the intersection as well as the tangency points now occur at yet other places. It is therefore reasonable to assume that this behaviour will generally depend on the groove parameter values. Extending this argument slightly to the case of an imperfectly manufactured plain bearing, we might conclude that whirl, for a rotordynmically unstable system, will occur, generally speaking, not necessarily at half speed. Thus, the term "asynchronous" might be the more appropriate term to employ for all bearing types.
- If we calculate the corresponding critical mass values, we shall discover, as the readers might verify for themselves, that there are no dramatic differences there as compared to the plain-bearing case. In actual fact, one would need to optimise the gap geometry of an HGJB quite laboriously in order to come up with a dynamically stable, practicable design, see e.g. Schiffmann (2008).

To the general features of HGJB we must add that they are more energy efficient since they have less friction losses for the same load and stiffness characteristics than their plain counterparts, (see Chapter 8).

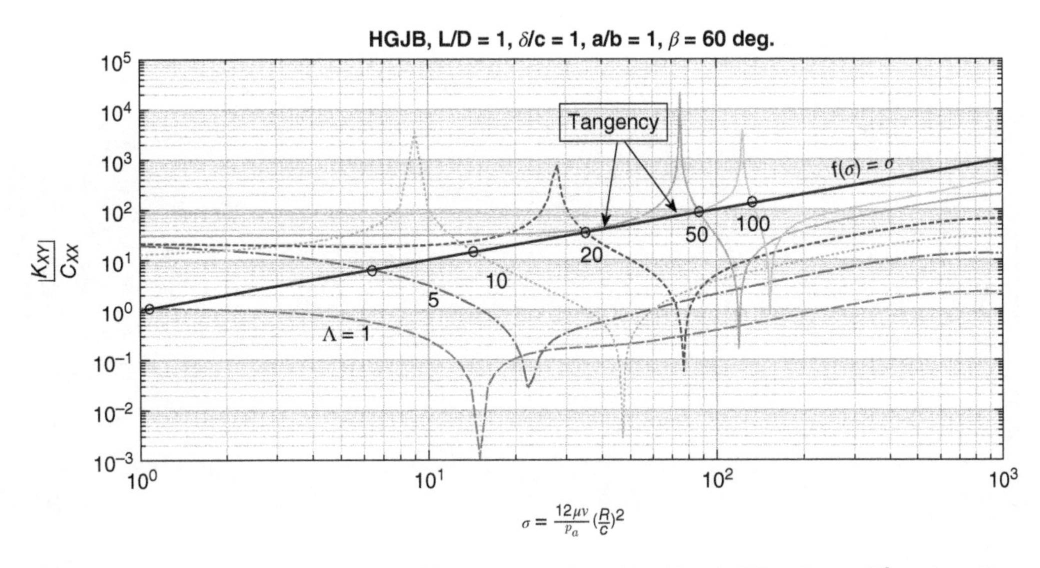

Figure 9.18 Whirl map for self-acting HGJB journal bearing with $L/D = 1, H/h = 2, \beta = 60^o$, and small ϵ perturbation; 24 grooves on the stationary bearing (bush).

9.6 EP Journal Bearings

EP journal bearings have the same basic structure as plain journal bearings with the exception that the air film is fed by pressurised fluid from an external source. There could be a single feed plane or multiple ones as depicted schematically in Figure 9.19. Each feed plane will contain several feed holes, usually more than three, but preferably more than four, for ensuring static stability at high eccentricity, that are provided with some type of restriction (see Chapter 5), the simplest of which being inherent, i.e. a plain feed hole, which will be the type that is adopted for further treatment in this chapter. To this type also belongs the case of slot-entry bearing, see e.g. (Stout and Tawfik 1983; Stout et al. 1978; Tawfik and Stout 1982), which represents, in actual fact, the case of quasi-infinite number of feed holes, or what is sometimes known also as *line feed*.

The purpose or advantage of opting for an EP bearing is threefold. Firstly, they are most suitable for high-precision applications, e.g. rotary table etc., at relatively small speeds, see e.g. Stout and Barrans (2000); secondly, in order to overcome start-stop friction and wear; thirdly, to provide additional dynamic stability in high-speed applications, as we shall see further below.

The governing Reynolds equation is the same as Eq. 9.1 or Eq. 9.2. However, the boundary conditions involve, in addition to the atmospheric pressure at the outer boundaries, the pressures p_0 just downstream of the feed holes. This latter is generally obtained by solving for the flow balance at each feed hole having a given supply pressure p_s and restriction law as given in the method of Chapter 5. With the restriction law being expressed simply as that

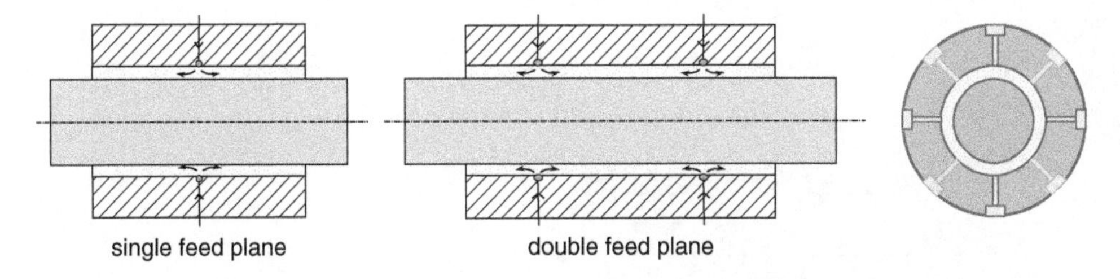

Figure 9.19 Schematic of EP journal bearings with single and double feed planes.

of a nozzle with a given coefficient of discharge, the flow balance is given by:

$$\dot{m}_o = C_d A_o \sqrt{\frac{2\kappa}{\kappa - 1}} \frac{p_s}{\sqrt{\Re T_s}} \Phi_e(\bar{p}_o), \qquad A_o = \text{orifice area} \tag{9.30}$$

where,

$$\Phi_e(\bar{p}_o) = \begin{cases} \sqrt{\bar{p}_o^{\,2/\kappa} - \bar{p}_o^{\,(\kappa+1)/\kappa}} & \text{for } \bar{p}_o \geq \bar{p}_c \\ \sqrt{\bar{p}_c^{\,2/\kappa} - \bar{p}_c^{\,(\kappa+1)/\kappa}} & \text{for } \bar{p}_o \leq \bar{p}_c \end{cases} \tag{9.31}$$

with,

$$\bar{p}_o = \frac{p_o}{p_s} \tag{9.32a}$$

$$\bar{p}_c = (\frac{2}{\kappa + 1})^{\kappa/(\kappa-1)} \tag{9.32b}$$

9.6.1 Single Feed Plane

The characteristics depend on the parameters

$$(\epsilon, \Lambda, \Lambda_f, \frac{L}{D}, N_f, \frac{d_o}{D}, P_s),$$

where,

$$\Lambda_f = 12 \frac{\mu R}{p_a c^2} \sqrt{\frac{2\kappa \Re T_s}{\kappa - 1}}$$

and N_f is the number of feed holes, which are assumed to be symmetrically disposed around the axis of the bearing, $P_s = p_s/p_a$, and d_o is the feed-hole diameter assumed to be identical for all feed holes; $\frac{d_o}{D} = r_o/R$ is feed-hole radius to bearing radius ratio, which is also termed R_f.

Figure 9.20 illustrates a possible embodiment of an EP JB with eight inherent restrictors.

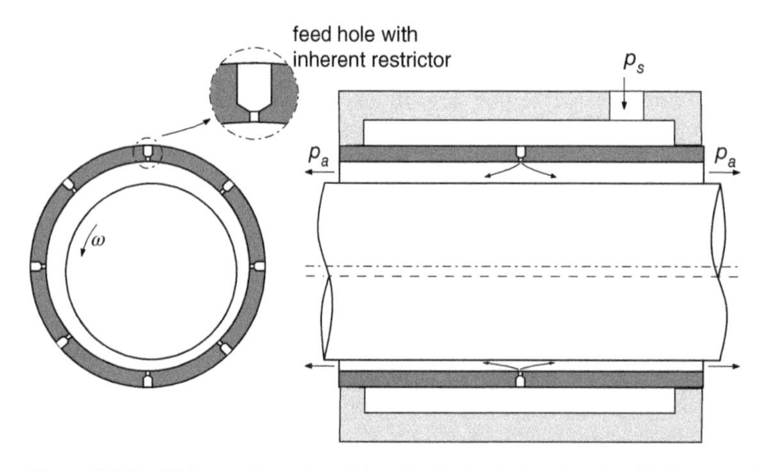

Figure 9.20 EP journal bearing with a single feed plane having 8 inherent orifices (not to scale).

Note on the Conformity Issue in an EP JB and the Problem of Static Stability

The working principle of an EP JB may be likened to a system of opposed aerostatic bearing pads. When the shaft, or journal, is concentric with the bearing, the net load on it is null. As the eccentricity increases, a net load is generated in an opposite direction to the eccentricity. Figure 9.21(a) and (b) shows a problem that is inherent to plane bearings, that is, upon increase of eccentricity, one side of the bearing will acquire a convergent gap, while the opposing one, a divergent gap. In other words, the force gain on the divergent-gap side will steadily decrease cf. the force loss on convergent side. Eventually, this will lead at some point to static instability, the so-called "shaft sticking" (see e.g. Neale (1973)), wherein the system will suddenly collapse. Obviously, for this to happen, a certain bearing-load threshold has to be exceeded, i.e. there will generally be no problem of lift-off when the bearing is lightly loaded.

This situation can be overcome by one or both of the following measures: (i) increase of the number of feed holes to six or more, (ii) the addition of shallow pockets, or "conicities" around each feed hole, while retaining a small number (four or even three) of them. This latter situation is illustrated in Figure 9.21(c) and (d), where we see that, with a proper choice of conicity depth, the opposed bearing gaps will remain convergent even at extreme eccentricities.

Finally, Figure 9.22 illustrates the problem of static stability in terms of positive and negative static stiffness, while Figure 9.23 plots the bearing load for an actual EP JB with four feed holes, showing that the stiffness becomes

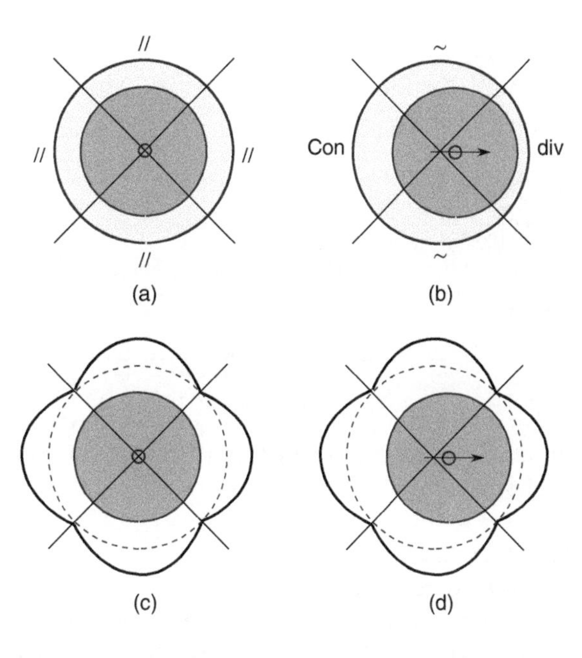

Figure 9.21 Illustration of the conformity problem in EP journal bearings. (a) zero-eccentricity journal in a plain bearing; the gap is nominally parallel in all four quadrants, (b) eccentric journal; the gap is now convergent in the left quadrant and divergent in the right quadrant, (c) zero-eccentricity journal in a bearing having shallow pockets; the gap is convergent in all four quadrants, (d) the gap remains convergent even after eccentricity of the journal.

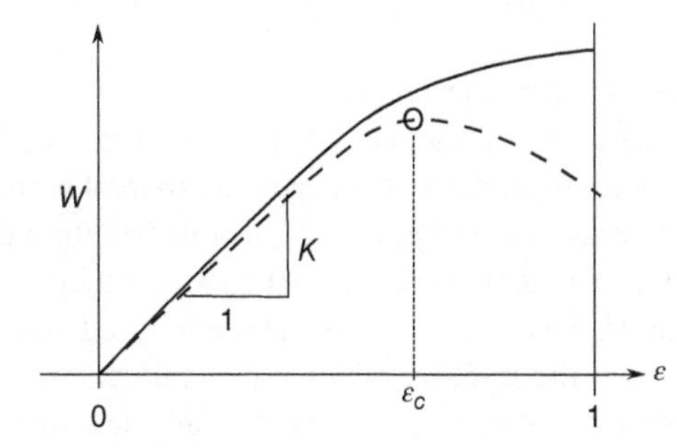

Figure 9.22 Qualitative behaviour of the bearing load as a function of the eccentricity ratio. The stiffness is positive for all eccentricities (solid line). The stiffness becomes negative beyond a critical eccentricity ratio ϵ_c leading to static instability of the bearing characterised by a sudden collapse of the shaft.

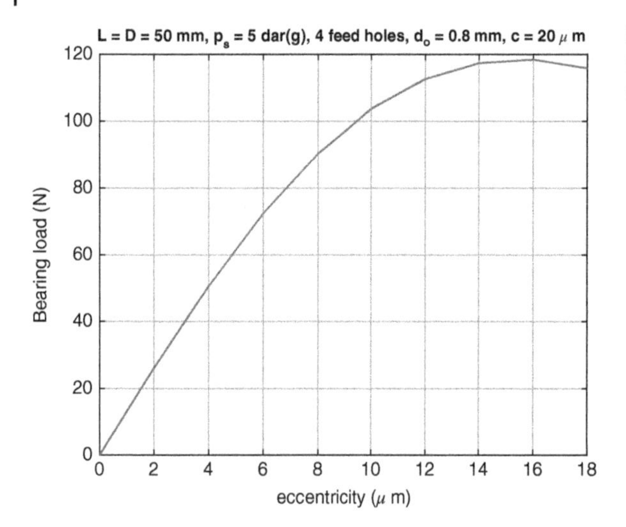

Figure 9.23 Bearing load as a function of the eccentricity for a bearing with four feed holes. Static stability becomes critical around an eccentricity of 15-16 μm.

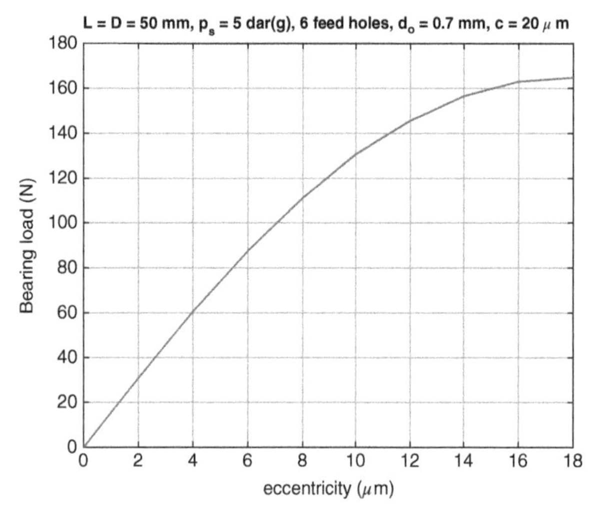

Figure 9.24 Bearing load as a function of the eccentricity for a bearing with six feed holes. The stiffness is positive throughout, though it reaches very small value at large eccentricity.

negative beyond an eccentricity of 15–16 μm. This can be overcome by increasing the number of feed holes to six as shown in the plot of Figure 9.24.

Basic Static Characteristics

(A) The Pressure Distribution

Let us first gain some idea about the pressure distribution in an EP JB for the types: (i) plain with single feed plane, (ii) plain with double feed planes, and (iii) bearing with shallow pockets or conicities having a single feed plain.

Single Plane, Plain Bearing

Figure 9.25 presents a 3D-plot of a typical pressure distribution in a single-feed-plane concentric plain EP JB, where we can note the symmetry thereof both in the axial and in the peripheral directions. When the eccentricity increases, as in Figure 9.26, this distribution departs from symmetry so that a net radial force will be generated. Let us note that despite the high eccentricity ratio of 0.8, the pressure gradients in the peripheral direction are not as high as those we have observed in self-acting bearings, see e.g. Figure 9.9. This behaviour, which limits not so much the stiffness, which is generally higher than that of a self-acting bearing, but the load capacity, is typical of passive EP bearings, which can only be remedied by using active restrictors or active compensation generally.

Double Feed Plane, Plain Bearing

Figures 9.27 and 9.28 depict the situation above for a double feed plane bearing. The behaviour is similar except for the presence of a quasi-uniform pressure ridge in the axial direction between each pair of feed holes.

Thus, the bearing characteristics will be qualitatively similar to those of a single feed plane bearing but with: (i) higher stiffness, e.g. in this example the maximum dimensionless stiffness $K_{XX} \approx 2.1$, cf. 1.6 for single plane feed; (ii) lower damping: e.g. dimensionless static damping $C_{XX} \approx 0.23$ cf. 0.26; (iii) higher air consumption, dimensionless flow rate $\bar{m} \approx 242$ (at maximum Λ_f) cf. 121 for single plane; i.e. a doubling.

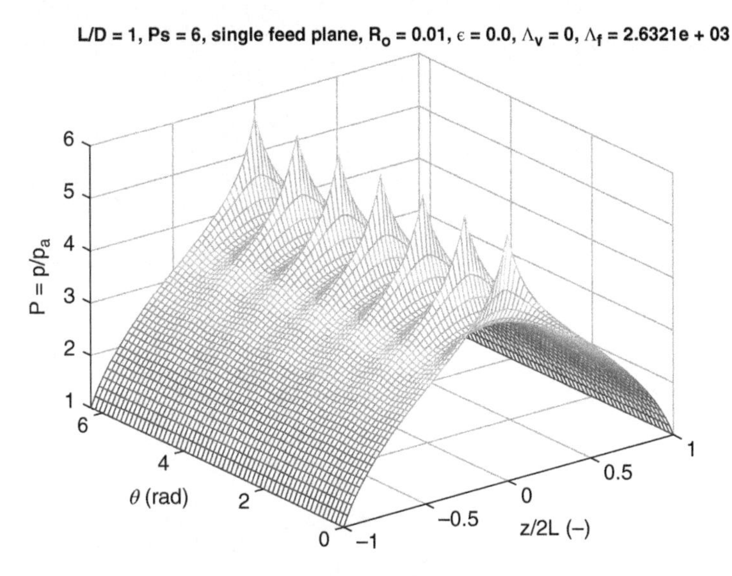

Figure 9.25 Typical pressure distribution in a concentric journal bearing with a single feed plane.

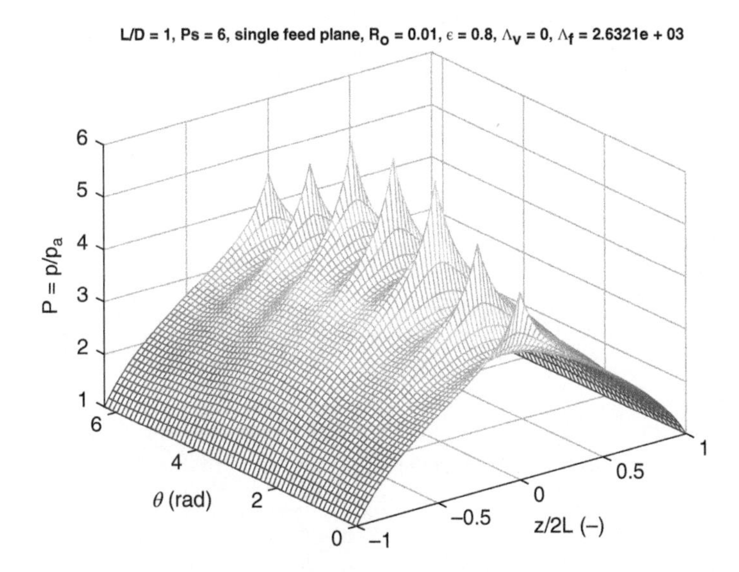

Figure 9.26 Typical pressure distribution in a eccentric journal bearing with a single feed plane.

Bearing Having Shallow Pockets or Conicities

Figure 9.29 shows the gap-height function having sinusoidal cavities, or so-called conicities around each feed hole. The corresponding pressure distributions for concentric and eccentric journals are depicted in Figs. 9.30 and 9.31 respectively.

Example

With the ratio of conicity depth h_v to clearance c, namely $H_v = h_v/c = 1$, we obtain:

- for $\epsilon = 0$: stiffness $K_{XX} = 2.04$, damping $C_{XX} = 0.085$ and flow rate $\bar{m} = 298$.
- for $\epsilon = 0.7$: $K_{XX} = 1.94$, $C_{XX} = 0.12$ and $\bar{m} = 326$.

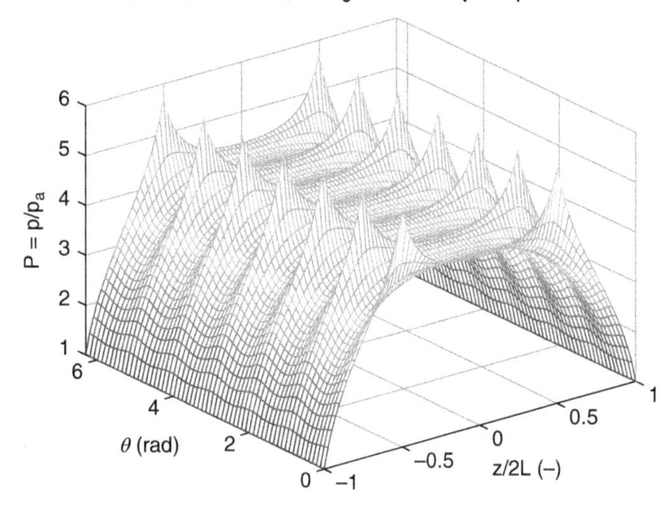

Figure 9.27 Typical pressure distribution in a concentric journal bearing with double feed planes.

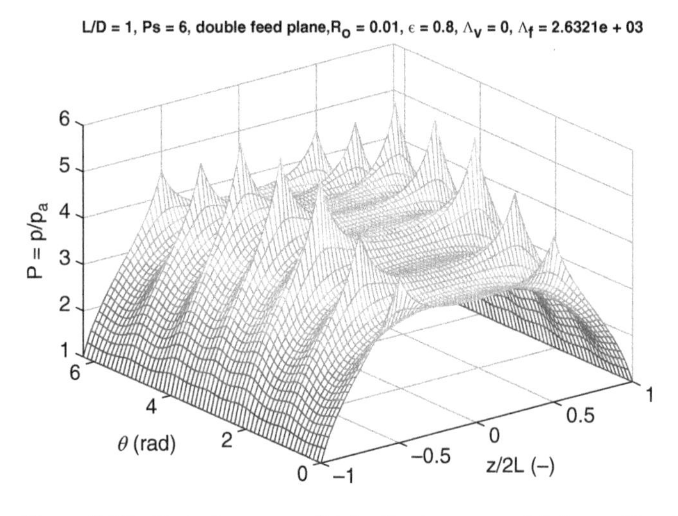

Figure 9.28 Typical pressure distribution in an eccentric journal bearing with double feed planes.

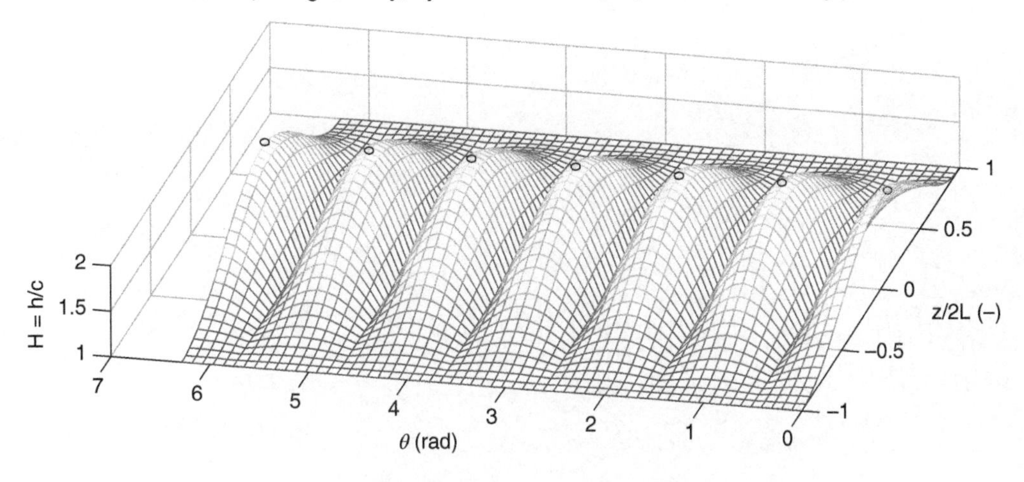

Figure 9.29 Bearing film profile showing sinusoidal conicities around each feed hole.

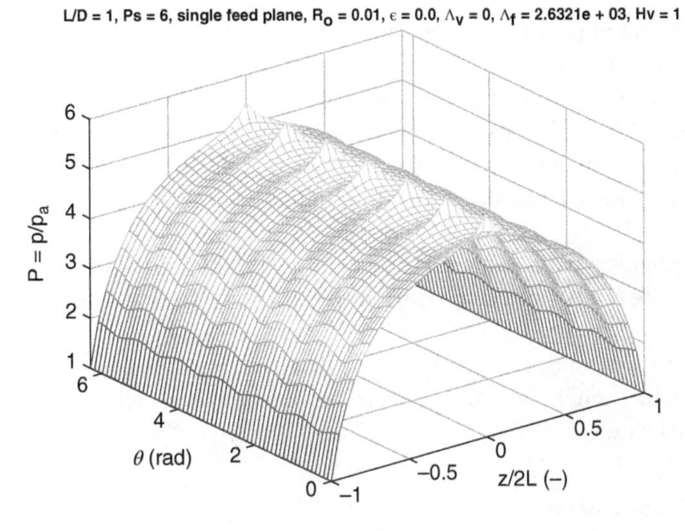

Figure 9.30 Pressure distribution for bearing with conicities. Concentric journal, $\epsilon = 0$

Thus, we see that (i) there is not a strong difference between the values as a function of the eccentricity, which may be viewed as an advantage, and (ii) that the stiffness and air consumption have increased and, as should be expected as general rule, the damping has decreased in comparison with the plain EP bearing case. This latter means a bearing with large conicity, while being stiffer, is more prone to pneumatic-hammer instability.

Note on the Dynamic Stability of EP Bearings

At relatively low rotational speeds, an EP bearing fits the description of "Aerostatic" bearing in that there is no appreciable aerodynamic action. In that case, the dynamic behaviour of an EP bearing system resembles qualitatively that of an aerostatic pad as described in Chapter 7. That is, the system will become unstable, in

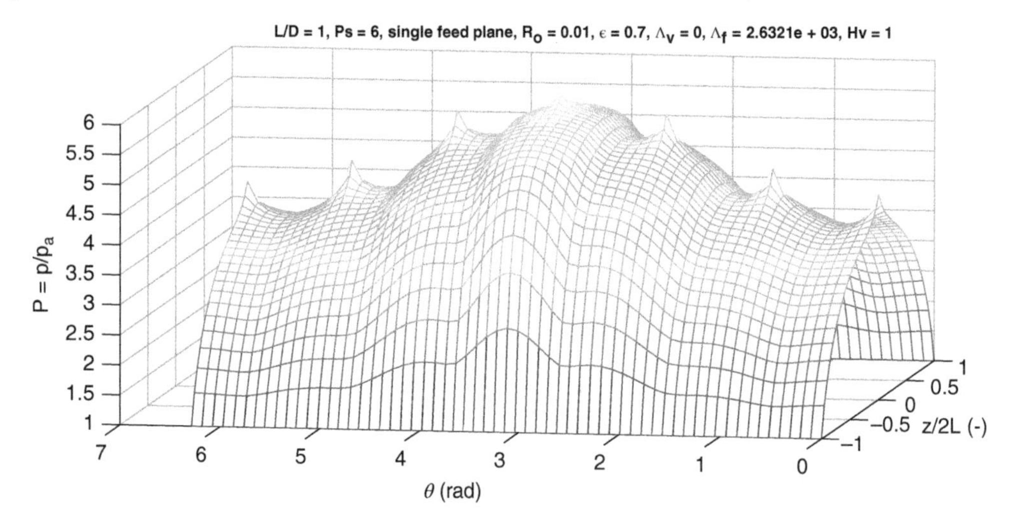

Figure 9.31 Pressure distribution for bearing with conicities. Eccentric journal, $\epsilon = 0.7$

a pneumatic-hammer sense, when the principal damping corresponding to the resonance frequency of the system becomes negative. Furthermore, just as in the case of an aerostatic pad, (i) when the bearing is plain, i.e. without conicities, the damping will always be positive and thus a bearing system will be unconditionally stable (independent of the rotor mass), (ii) increasing the stiffness, by larger supply pressure, or by smaller feed holes, reduces the damping to smaller values; when adding conicities around the feed holes, damping may become negative, (iii) the design of an appropriate, positively damped bearing consists thus in finding a trade-off between stiffness and damping.

As has been alluded to earlier, this instability mechanism is qualitatively different from the whirl instability phenomenon, which finds its root cause in the dominance of the cross-coupled stiffness when the principal damping is not sufficiently large.

As the rotational speed increases, appreciable aerodynamic action will arise. In that case, the otherwise aerostatic bearing will now be referred to as hybrid, which might be prone to whirl instability in the same way as self-acting (aerodynamic) bearings, as we shall see below.

Static Stiffness, Static Damping and Mass Flow Rate Characteristics

The static characteristics have been investigated and optimised by, among others, Pink and Stout (1978, 1981), Stout and Barrans (2000), Pink and Stout (1981). We consider first the dimensional characteristics in order to gain basic understanding about behaviour. Thereafter, we provide the dimensionless general characteristics. We shall confine the treatment to concentric bearings with $L/D = 1, p_s/p_a = 6$ and single feed plane, as an example. Results for the other cases can be obtained in a similar manner.

Figure 9.32 plots the principal static stiffness as a function of the radial clearance, for a range of feed-hole diameters. We note that (i) for each feed-hole diameter, there is a corresponding optimum clearance value that yields maximum stiffness, (ii) that this optimum stiffness increases with decreasing feed-hole diameter.

As regards the principal static damping, Figure 9.33 shows, for the same bearing, that the damping coefficients for all feed-hole diameters lie approximately on the same curve, varying approximately as $1/c^3$, with the damping being slightly higher for larger feed hole diameters. This is peculiar to plain bearings (i.e. without pockets/conicities) having inherent restrictors. Comparing to flat aerostatic bearings, see Chapter 7, we may expect the damping to reduce with the presence of conicities to the point of becoming negative for certain values of the bearing design parameters.

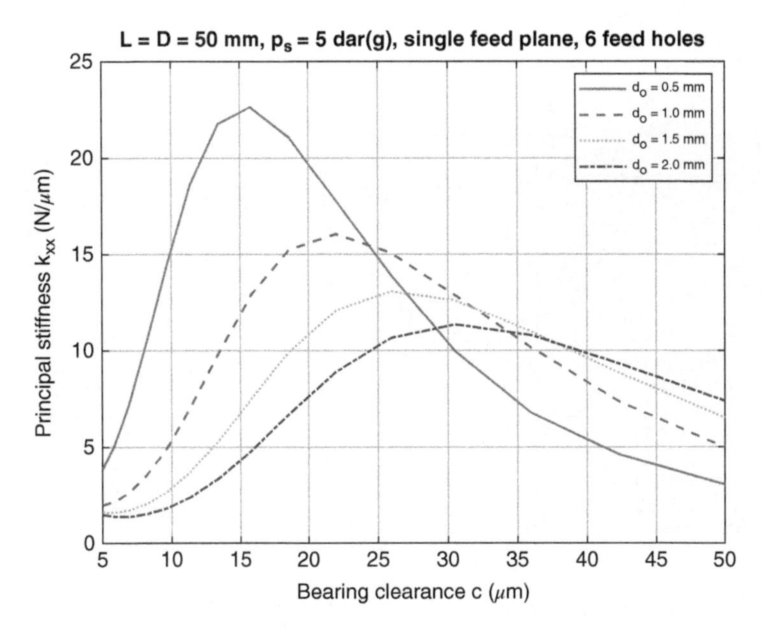

Figure 9.32 The principal stiffness k_{xx} as a function of the bearing clearance c for different values of the feed-hole diameters d_o. Single feed plane with 6 feed holes, $L = D = 50$ mm, $p_s = 5$ bar(g), $p_a = 1$ bar.

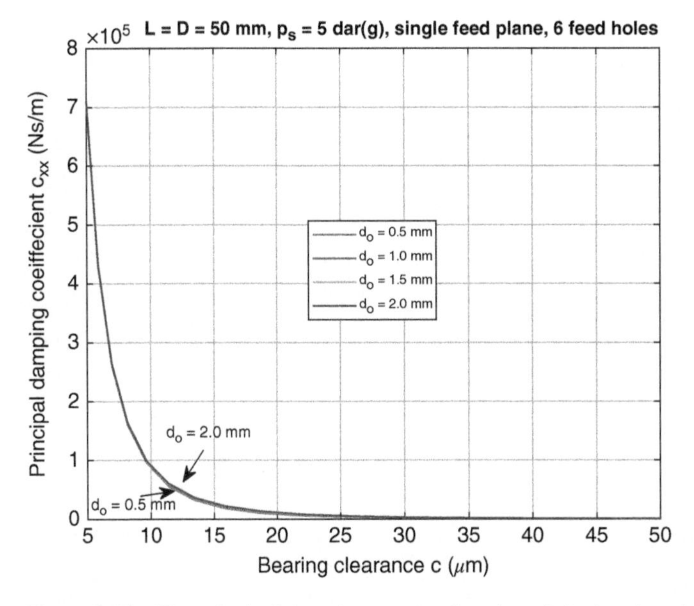

Figure 9.33 The principal damping c_{xx} as a function of the bearing clearance c for different values of the feed-hole diameters d_o. Single feed plane with 6 feed holes, $L = D = 50$ mm, $p_s = 5$ bar(g), $p_a = 1$ bar.

Finally, Figure 9.34 plots the air consumption or the mass flow rate through the gap. For very small clearance values, the flow is insensitive to the feed-hole diameter. This corresponds to the case in which the inlet pressure is very close to the supply pressure so that the flow varies purely as c^3. As the clearance increases, the values for the various feed-hole diameters begin to diverge from one another considerably and to deviate from $\sim c^3$ since the

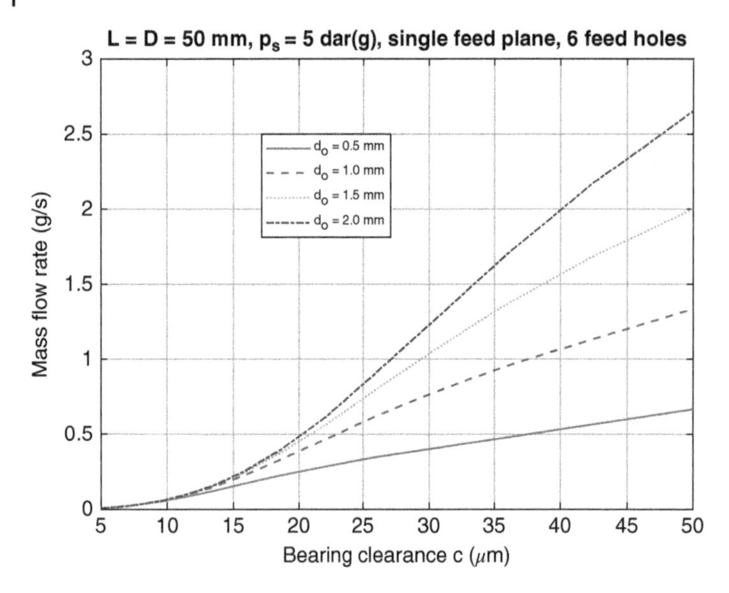

Figure 9.34 The mass flow rate $\dot{m}$ as a function of the bearing clearance c for different values of the feed-hole diameters d_o. Single feed plane with 6 feed holes, $L = D = 50$ mm, $p_s = 5$ bar(g), $p_a = 1$ bar.

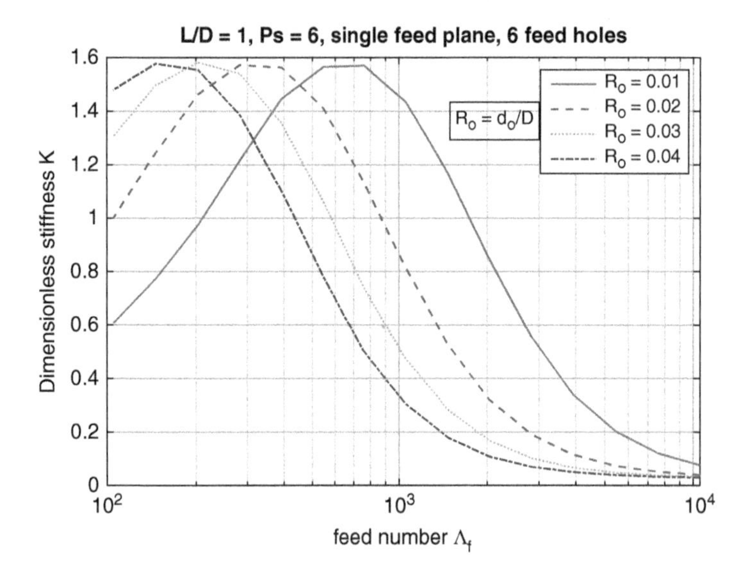

Figure 9.35 The dimensionless principal static stiffness K_{XX} as a function of the feed number Λ_f for various values of the feed-hole diameter ratio R_o.

orifice flow becomes more significant now. For extremely large clearances, the flow becomes almost proportional to the feed-hole diameter, which is consistent with the inherent-orifice flow formula.

Dimensionless Characteristics

It is more convenient, from design point of view to express the results above in dimensionless form. In that way, we can generalise them to other bearing sizes. Thus, we provide the characteristics above in dimensionless form in Figures 9.35, 9.36 and 9.37, which are self-explanatory. Note however the different shapes of the curves owing

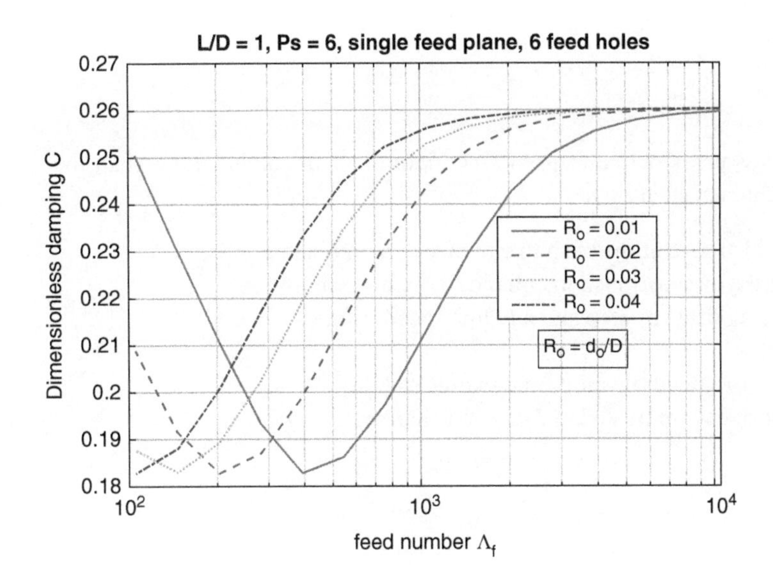

Figure 9.36 The dimensionless principal static damping C_{XX} as a function of the feed number Λ_f for various values of the feed-hole diameter ratio R_o.

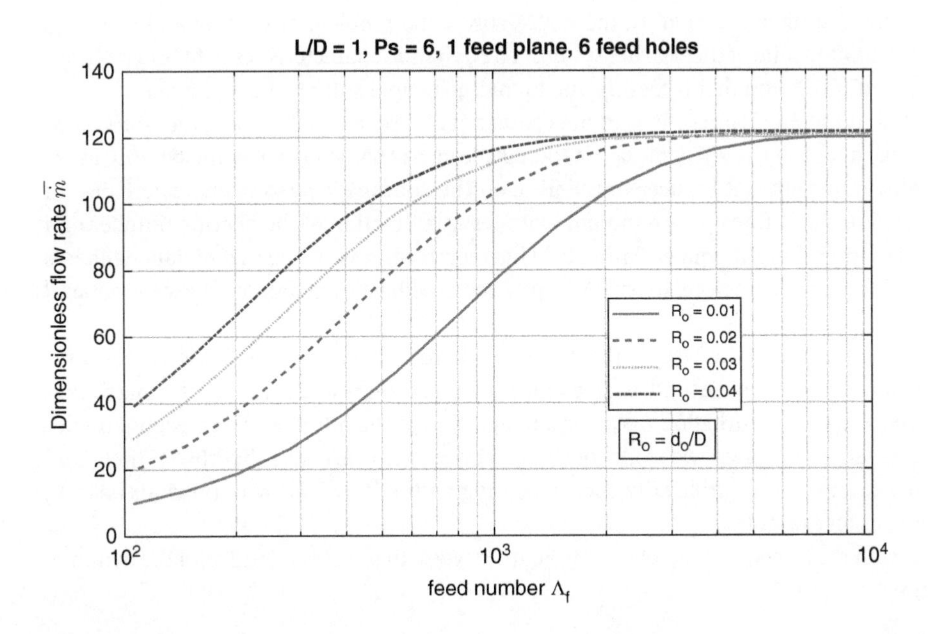

Figure 9.37 The dimensionless mass flow rate $\bar{\dot{m}}$ as a function of the feed number Λ_f for various values of the feed-hole diameter ratio R_o.

to the involvement of the clearance c in their definition. In particular, the damping curves appear now distinctly separated from one another.

9.6.2 Other Possible Combinations

Although not very common, it is nevertheless possible to affect other configurations or combinations than the generic ones that have been treated in this chapter, such as

- plain bearings equipped with a longitudinal groove in the bush to enhance dynamic stability;
- plain bearings having lemon-shaped or antisymmetric bush to enhance dynamic stability;
- HGJB bearings equipped with feed holes to assist start-stop and add low-speed stiffness;
- HGJB bearings with irregular grooves;
- EP bearings peripheral grooves in the bush to extend the range of dynamic stability;
- bearings with compliant or with porous surfaces are treated in dedicated chapters;
- …

9.7 Hybrid JBs

The somewhat misleading term, "hybrid bearing" is used in this context to denote externally pressurised JBs that are, however, used for high-speed applications (not necessarily always of high-precision requirements), e.g. for machining spindles. External pressurisation is utilised in the first instance so that the stiffness of the bearing is less dependent on the rotation speed, being considerably higher than the value for self-acting bearings, including HGJB, when the supply pressure is sufficiently high. It, moreover, solves the problem of starting and stopping, which might otherwise severely limit the life of the bearing system. The dynamic characteristics of hybrid bearings are somewhat better than those of HGJB and that is mainly due to their principal stiffness being higher.

The EP JB has been discussed in the previous section, in the context of centric-journal quasi-static, zero speed operation. In this section we take a look at the effect of the rotational speed, as characterised by the bearing number Λ and the additional aerodynamic effect it induces in the film, on the concentric bearing characteristics. In particular, we shall be interested in the stiffness and damping matrices as a function of the bearing number and the perturbation frequency ν comprised in the squeeze number. This represents a very large set of data, which is very difficult to present in a compact way. Therefore, we shall adopt some expediences in order to show the general trends.

Firstly, owing to the rotational symmetry of the system, the flow rate is hardly affected by the rotation of the journal, and we shall pay no further attention to it. Second, each element of the stiffness and the damping matrices is function of the rotational speed and the perturbation frequency, in addition to the other bearing design parameters. In order to facilitate presentation, however, we shall consider a concentric bearing so that the stiffness and damping matrices are symmetric, and take an arbitrarily chosen bearing with $L/D = 1$ (as with previous bearing types), $d_0/D = 0.02$ and $P_s = 6$, as an example.

Thus, to enable comparison with the foregoing two self-acting bearing types, Figs. 9.38 and 9.39 plot the dynamic coefficients of hybrid bearings.

We note the following.

- The principle stiffness is significantly increased cf. the self-acting types.
- The cross-coupled stiffness, which is the main culprit in whirl instability, has also increased, apparently in the same proportion.

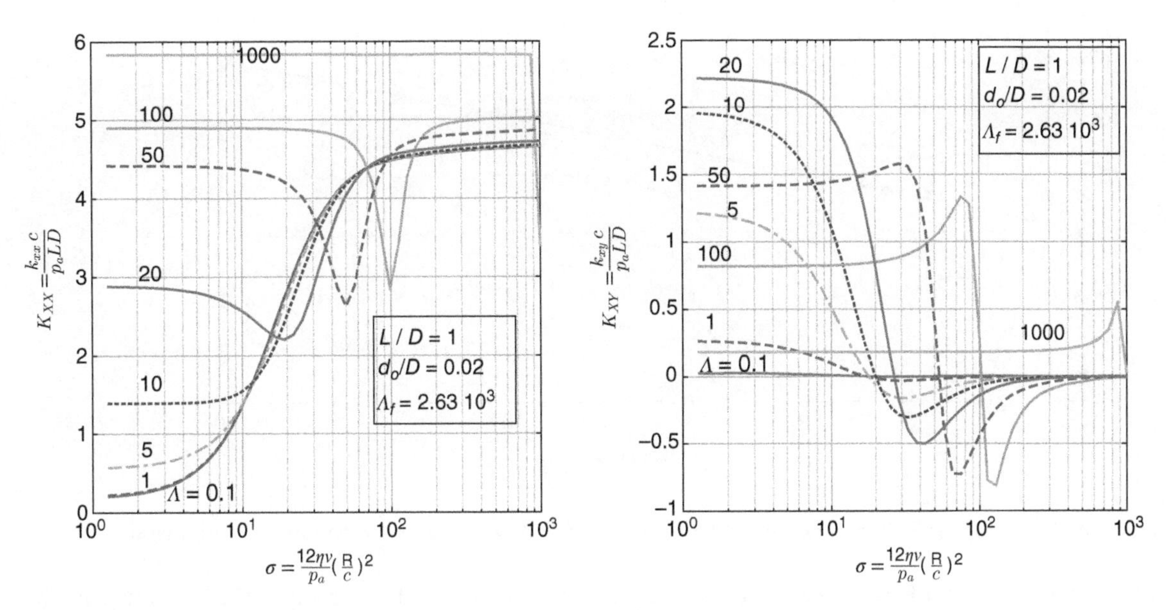

Figure 9.38 The dimensionless stiffness coefficients K_{XX} and K_{XY} as a function of the squeeze number σ for different values of the bearing number Λ. EP journal bearing with $L/D = 1$, $d_o/D = 0.02$ $\Lambda_f = 2.63\ 10^3$ and small ϵ perturbation.

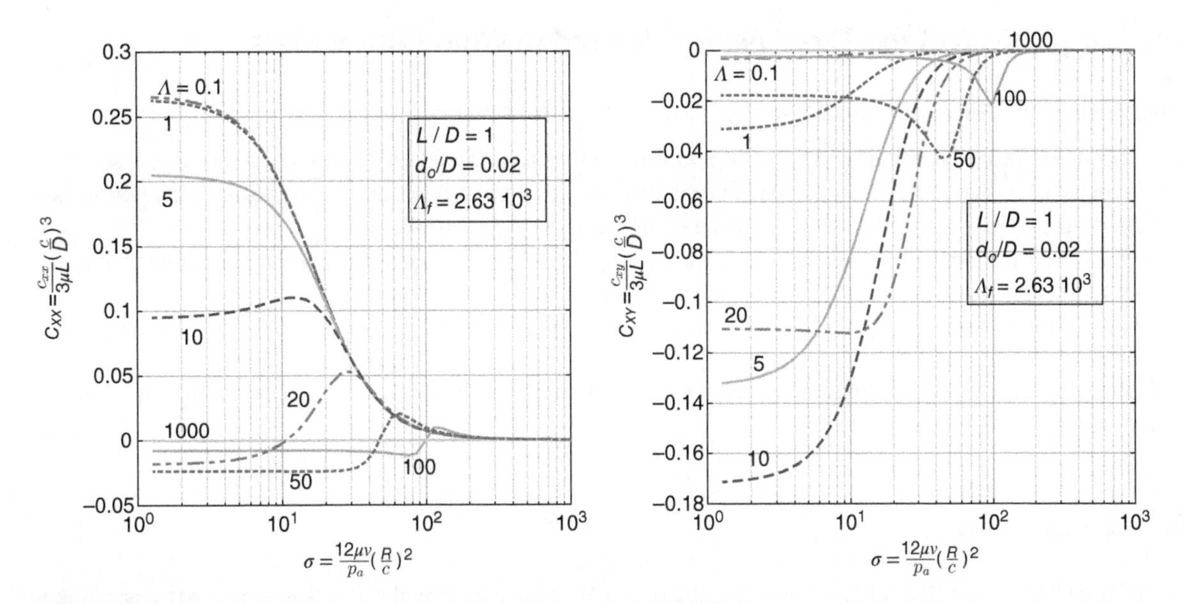

Figure 9.39 The dimensionless damping coefficients C_{XX} and C_{XY} as a function of the squeeze number σ for different values of the bearing number Λ. EP journal bearing with $L/D = 1$, $d_o/D = 0.02$ $\Lambda_f = 2.63\ 10^3$ and small ϵ perturbation.

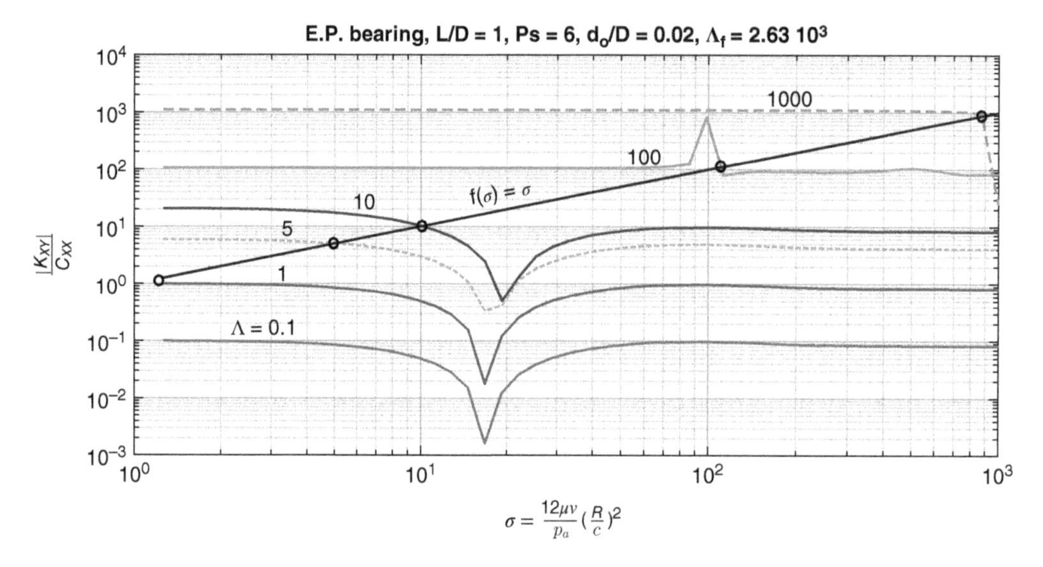

Figure 9.40 Whirl map for EP journal bearing with $L/D = 1$, $d_o/D = 0.02$ $\Lambda_f = 2.63\ 10^3$ and small ϵ perturbation.

- Notwithstanding this, EP hybrid have generally better stability than HGJB, especially in the low Λ region. The whirl map of Figure 9.40 shows that the resonance pulsations σ_r are at, or very close to the values of Λ, just as in the case of a plain bearing. This is due to the fact that k_{xy} arises almost purely from the aerodynamic effect while c_{xx} is almost independent of external pressurisation (as we saw earlier). However, the critical mass is much higher especially at lower speeds since the local minimum in k_{xx} (the dip) has increased cf. the other two types.

9.8 Comparison of the Three Types in Regard to Whirl Critical Mass

summarising the findings in regard to whirl of the three bearing examples, in concentric position, we see that:

- If whirl takes place, then that will be with a pulsation approaching half the rotation angular speed, so-called half-speed, i.e. a "whirl ratio" of 1/2; almost exactly, in case of plain self-acting bearings; approximately, in case of EP bearings; and at progressively increasing whirl ratio, in case of HGJB.
- Since EP bearings, in our example, have higher principal stiffness, their critical mass will be higher than the other two examples, which are not appreciably different from each other.
- At low Λ values, plain JBs will have higher critical mass (i.e. they are better damped) than the HGJB type, and vice-versa.
- The increase in critical mass difference becomes more significant at higher Λ values, but will remain within an order of magnitude.

9.9 Conclusions

This chapter has treated JBs and their basic characteristics. We have considered two categories of bearings; namely, self-acting and EP bearings. The first category is represented by two types: plain and spiral grooved, which significantly differ in their static and dynamic behaviour. The second one comprises various variants of EP bearings, which are essentially used for low-speed precision applications. As such, they are characterised by high stiffness and, depending on the design, also high damping. Such bearings are also often used for high-speed applications, where they are sometimes referred to as hybrid journal bearings. This type has certain important advantages over the self-acting types, particularly in regard to high stiffness and dynamic stability. When ultra-high speeds are envisaged, it is generally not possible, from this analysis, to select a particular bearing type: the process entails optimisation to avoid the occurrence of whirl instability. In this chapter, we have introduced the phenomenon of half-speed, or more generally, asynchronous whirl and the associated critical mass above which the system will become dynamically unstable. This problem is more systematically treated in Chapter 10.

References

Ausman, J.S., (1960). Torque produced by misalignment of hydrodynamic, gas lubricated journal bearings. *ASME - Journal of Basic Engineering* 335–341.

Ausman, J.S., (1961). An improved analytical solution for self-acting, gas-lubricated journal bearings of finite length. *Transactions of the ASME - Journal of Basic Engineering* 83(2): 188–194.

Bonneau, D. and Absi, J., (1994) Analysis of aerodynamic journal bearings with small number of herringbone grooves by finite element method. *Transactions of the ASME - Journal of Tribology* 116(4): 698–704.

Castelli, V. and Elrod, H.G., (1965) Solution of the stability problem of 360 degree self-acting gas-lubricated bearings. *Transactions of the ASME - Journal of Basic Engineering* 87(2): (199–212.

Cheng, H.S. and Pan, C.H.T., (1965) Stability analysis of gas-lubricated, self-acting, plain, cylindrical, journal bearings of finite length, using Galerkin's method. *Transactions of the ASME - Journal of Basic Engineering* 87(1): 185–192.

Cheng, H.S. and Trumpler, P.R., (1963) Stability of the high-speed journal bearing under steady load - 2: the compressible film. *Transactions of the ASME - Journal of Engineering for Industry* 85, 274–280.

Constantinescu, V.N, (1969). *Gas Lubrication*. ASME.

Cunningham, R.E., Fleming, D.P. and Anderson, W.J., (1969). Experimental stability studies of the herringbone-grooved gas-lubricated journal bearing. *Transactions of the ASME - Journal of Lubrication Technology* 91, 52–59.

Wilcock, D.F. (ed.) (1969). *Design of Gas Bearings*. Mechanical Technology Inc. (MTI), Latham NY.

Faria, M.T.C, (2001). Some performance characteristics of high speed gas lubricated herringbone groove journal bearings. *JSME International Journal - Series C* 44(3): 775–781.

Franchek, N.M., (1992). *Theory versus experimental results and comparisons for five orifice-compensated hybrid bearing configurations* Master's thesis Texas A&M University - Mechanical Engineering Department College Station, TX.

Gross, W.A., (1962). *Gas Film Lubrication*. John Wiley & Sons.

Kazimierski, Z. and Jarzecki, K., (1979). Stability threshold of flexibly supported hybrid gas journal bearings. *Transactions of the ASME - Journal of Lubrication Technology* 101, 451–457.

Liu, L.X., Teo, C.J., Epstein, A.H. and Spakovszky, Z.S., (2005). Hydrostatic gas journal bearings for micro-turbomachinery. *Journal of Vibration and Acoustics* 127(2): 157–164.

Lund, J.W., (1967). A theoretical analysis of whirl instability and pneumatic hammer for a rigid rotor in pressurized gas journal bearings. *Transactions of the ASME - Journal of Lubrication Technology* 89(2): 154–166.

Mori, A., Aoyama, K. and Mori, H., (1980a). Influence of the gas-film inertia forces on the dynamic characteristics of externally pressurized, gas lubricated journal bearings—part I: Proposal of governing equations. *Bulletin of the JSME* 23(178): 582–586.

Mori, A., Aoyama, K. and Mori, H., (1980b). Influence of the gas-film inertia forces on the dynamic characteristics of externally pressurized, gas lubricated journal bearings –part ii: Analysis of whirl instability and plane vibrations. *Bulletin of the JSME* 23(180): 953–960.

Muijderman, E.A., (1967). Analysis and design of spiral-groove bearings. *Transactions of the ASME - Journal of Lubrication Technology* 89(3): 291–306.

Neale, M.J., (1973). *Tribology Handbook*. Butterworths, London.

Osborne, D.A. and San Andrés, L., (2006). Experimental response of simple gas hybrid bearings for oil-free turbomachinery. *Transactions of the ASME - Journal of Engineering for Gas Turbines and Power* 128(3): 626–633.

Pan, C.H.T. and Sternlicht, B., (1962).On the translatory whirl motion of a vertical rotor in plain cylindrical gas-dynamic journal bearings. *Transactions of the ASME - Journal of Basic Engineering* 84(March): 152–158.

Pink, E. and Stout, K., (1978). Design procedures for orifice compensated gas journal bearings based on experimental data. *Tribology International* 11(1): 63–75.

Pink, E.G. and Stout, K.J., (1981). Characteristics of orifice compensated hybrid journal bearings *Proceedings of the 8th Gas Bearing Symposium*, 29–44 number 3 Leicester Polytechnic. BHRA Fluid Engineering, Cranfield, U.K.

Raimondi, A.A., (1961) A numerical solution for the gas lubricated full journal bearing of finite length. *ASLE Transactions* 4, 131–155.

Robinson, C.H. and Sterry, F., (1958). The static strength of pressure fed gas *journal bearings. AERE ED/R.*

Rohde, S.M. and Ezzat, H.A., (1967). On the dynamic behavior of hybrid journal bearings. *ASME - Journal of Lubrication Technology* 98, 90–94.

San Andrés, L. and Ryu, K., (2008). Hybrid gas bearings with controlled supply pressure to eliminate rotor vibrations while crossing system critical speeds. *Journal of Engineering for Gas Turbines and Power* 130(6): 062505 (10 pages).

San Andres, L.A., (1991). Effect of compressibility on the dynamic response of hydrostatic journal bearings. *Wear* 146, 296–283.

Schiffmann, J., (2008). *Integrated design, optimization and experimental investigation of a direct driven turbocompressor for domestic heat pumps* PhD thesis École Polytechnique Fédérale de Lausanne.

Sternlicht, B., (1959). Elastic and damping properties of cylindrical journal bearings. *Transactions of the ASME - Journal of Basis Engineering* 81, 101–108.

Sternlicht, B. and Winn, L.W., (1963). On the load capacity and stability of rotors in self-acting gas lubricated plain cylindrical journal bearings. *Transactions of the ASME - Journal of Basic Engineering* 83, 139–144.

Stout, K. and Barrans, S. (2000). The design of aerostatic bearings for application to nanometre resolution manufacturing machine systems. *Tribology International* 33(12): 803–809.

Stout, K. and Rowe, W., (1974). Externally pressurized bearings–design for manufacture part 1–journal bearing selection. *Tribology* 7(3): 98–106.

Stout, K. and Tawfik, M., (1983). Graphical design procedures for slot entry hybrid gas journal bearings. *Wear* 87(1): 51–68.

Stout, K., Pink, E. and Tawfik, M., (1978). Comparison of slot-entry and orifice-compensated gas journal bearings. *Wear* 51(1): 137–145.

Su, J.C. and Lie, K.N., (2003). Rotation effects on hybrid air journal bearings. *Tibol. Int.* 36, 717–726.

Su, J.C. and Lie, K.N., (2006). Rotor dynamic instability analysis on hybrid air journal bearings. *Tibology International* 39, 238–248.

Tawfik, M. and Stout, K., (1982). Optimisation of slot entry hybrid gas bearings. *Tribology International* 15(1): 31–36.

Tomioka, J., Miyanaga, N., Outa, E., et al. (2007). Development of herringbone grooved aerodynamic journal bearings for the support of ultra-high-speed rotors. *Transactions of the JSME - Machine Elements and Manufacturing* 73(730): 1840–1846.

Vohr, J. and Chow, C. (1965). Characteristics of the herringbone-grooved, gas-lubricated journal bearings. *Transactions of the ASME, Journal of Basic Engineering*, 86(3): 568–578.

Waumans, T., (2009). On the Design of High-Speed Miniature Air Bearings: Dynamic Stability, Optimisation and Experimental Validation PhD thesis K. U. Leuven, Department of Mechanical Engineering, (2009D16).

Whipple, R., (1958). The inclined groove bearing. Technical report, AERE Report T/R 622, (Revised). UKAEA, Research Group, Harwell, Berkshire, England.

Whitley, S., Bowhill, A.J. and McEwan, P., (1962). Half speed whirl and load capacity of hydrodynamic gas journal bearings. *Proceedings of the Institution of Mechanical Engineers* 176, 554.

Wildmann, M., (1956). Experiments on gas lubricated *journal bearings. ASME paper No. 56-LU B-8.*

10

Dynamic Whirling Behaviour and the Rotordynamic Stability Problem

10.1 Introduction

Any gas bearing design should start with the prediction of the static bearing characteristics. This first step guarantees that characteristics such as load carrying capacity, gas consumption and frictional losses match the specifications given by the application. It is therefore a necessary (but not a sufficient) condition for obtaining a successful gas bearing application.

In a second and equally essential design step, the dynamic behaviour of the total system should be studied. A literature survey on the dynamic stability of gas bearing systems clearly reveals the relevance of this topic. A basic understanding and a correct prediction of this aspect is a prerequisite to avoid dynamic instabilities which limit the operation range.

In general, two types of instability can be encountered when dealing with gas lubricated rotor-bearing systems: pneumatic hammering and asynchronous rotor whirl. For an externally pressurised gas bearing system, independent of whether they be rotating or stationary, the possible occurrence of pneumatic hammering exists. This self-excited vibration sets in when the direct, or principal damping in the gas film becomes negative. This loss of damping is caused by a time-lag effect due to the compressible nature of gases. Pneumatic hammering usually occurs when the design focuses on increasing the load carrying capacity and stiffness characteristics of a gas bearing by providing feeding pockets. This creates large volumes in which the compressible lubricating film can store energy and is able to release this energy in phase with the motion. This first type of instability has already been treated in Chapter 7.

We have seen that for a rotary system, such as a JB, there are besides the principal quantities; the stiffness and damping, also cross-coupled quantities. In what follows, we shall denote, as in Chapter 9:

- the principal quantities by the subscript *ii* (or *jj*),
- the cross-coupled quantities by the subscript *ij* (or *ji*),

and so avoid undue reference to the coordinate system adopted.

A second type of instability, which is of greater importance in rotating systems, is generally referred to as half-speed, or more generally sub- or non-synchronous whirling. It is comparable to pneumatic hammering in the sense that they are both self-excited phenomena. However, the underlying source of this type of instability involves rotation which causes a cross-coupling effect in the gas film, i.e. the dynamic stiffness coefficient $k_{ij} \neq 0$. It will be shown later, that this cross-coupling has the effect of reducing the overall damping ratio of the rotor-bearing system. This can lead to sub-synchronous shaft whirling being very destructive in nature. Postponing, i.e. increasing the onset speed of this whirling poses the greatest challenge in a high-speed gas bearing design.

This chapter starts with a short discussion on the nature and classification of whirl motion encountered in rotating machinery supported on gas bearings. Thereafter, the asynchronous whirling phenomenon is discussed extensively. To gain insight in the underlying mechanisms causing instability and to identify the key elements of

Air Bearings: Theory, Design and Applications, First Edition. Farid Al-Bender.

Companion website: www.wiley.com/go/AlBender/AirBearings

a stable bearing design, a subsequent section clarifies what a stable bearing means in terms of dynamic bearing coefficients.

The remaining part of the chapter gives an overview of different techniques which can improve the stability performance e.g. by optimising the gas film geometry itself or by taking measures external to the gas film.

10.2 The Nature and Classification of Whirl Motion

When dealing with rotating machinery on gas lubricated bearings, the gas film supporting the rotating shaft is rarely in a steady-state condition. Shaft whirling can always be observed. Whirl can be defined as the relative orbital motion between the geometrical axis of rotation of the shaft and the axis defined by its supporting bearing centres. Both cylindrical and conical whirl may exist. This whirling, however, does not automatically preclude the stability of the non-steady-state working condition. Therefore, it is important to distinguish between different types of whirling and their implications on stability.

In general, whirling is said to be stable if quasi-identical and sufficiently small whirl orbits are traced out with successive rotations; i.e. a *limit-cycle* behaviour. If the whirl orbit increases in size with successive rotations, without seeking a stable condition, the whirl is unstable. Unstable whirling of the rotor has led to many catastrophic failures in the past. Good rotor-bearing dynamic design is therefore characterised by small, stable rotor whirl orbits, and by the absence of instabilities from the operating range of the machine.

A more detailed classification of whirling motion and its stability is given by Czolczsynki (1999). He describes six stability states which can be encountered in rotating machinery supported on gas bearings. A first and main distinction is based on the stability of the static equilibrium position of the system (steady-state working point). Due to the non-linear gas film behaviour, systems with an unstable steady-state working point may possess a stable limit cycle. And vice versa, a stable static equilibrium position can be accompanied by an unstable limit cycle. Figure 10.1 depicts these six states of stability: when the rotational speed of the system is low, its static equilibrium position is stable (a); when the static equilibrium position loses stability, a stable limit cycle can occur. If the cycle's orbit is situated within the clearance circle (b), this instability is not dangerous and may be acceptable. But, the orbit pattern of this "stable" whirling may be rather unpredictable due to its dependence on many parameters, making it difficult to assure safe operation. A further increase in speed may result in a limit cycle exceeding the clearance as illustrated by situation (c). The stable static working condition may also be accompanied by an unstable limit cycle. This is not dangerous if the limit cycle lies outside the clearance circle (d) and leads to a situation that is comparable to (a). A totally different situation exists if this unstable limit cycle falls within the clearance (e). This is the most dangerous state as the stability analysis, which is mostly based on a linearised bearing model, cannot conclude on the existence of such an unstable limit cycle. However, the existence of this case has not been, to the knowledge of the present writer, observed in any experiment. The chaotic whirling (f) is also not acceptable since the whirling behaviour is unpredictable.

The following section deals with the whirling motion that is caused by residual imbalance in the system. It will be shown that if the imbalance is kept low, no dangerous or unstable situation can originate from this synchronous whirling. Hereafter, the phenomenon of self-excited whirling is described, which in contrast to synchronous whirling, is very destructive in nature and is caused by the loss of stability of the steady-state equilibrium position.

10.2.1 Synchronous Whirl

In practice, there exists no case where the geometrical centre line of a rotating shaft is identical to its mass line. A rotation around the geometrical centre line will impose an unbalanced force/torque to the shaft. This will introduce whirling which is synchronous with the rotational speed. A general study on this synchronous whirling behaviour due to rotor imbalance can be found in any reference work on rotordynamics (Childs 1993; Vance 1987).

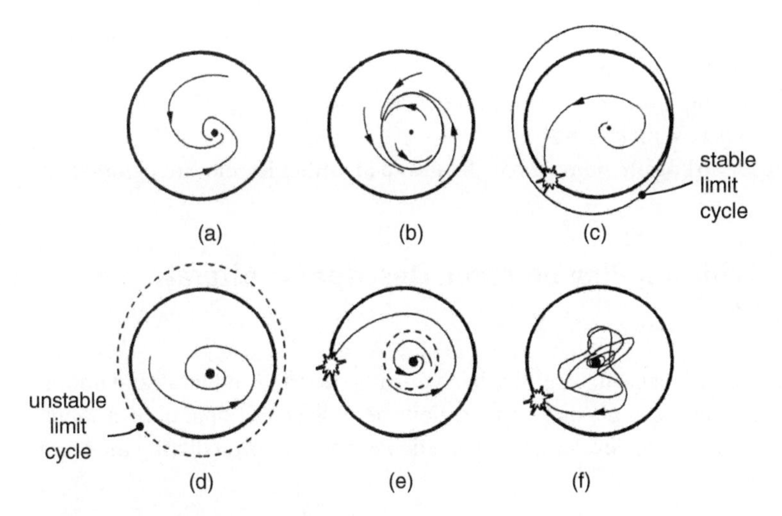

Figure 10.1 Classification of stability states according to Czolczsynki (1999). Source: Czolczsynki K 1999 Reproduced with permission from Springer Nature.

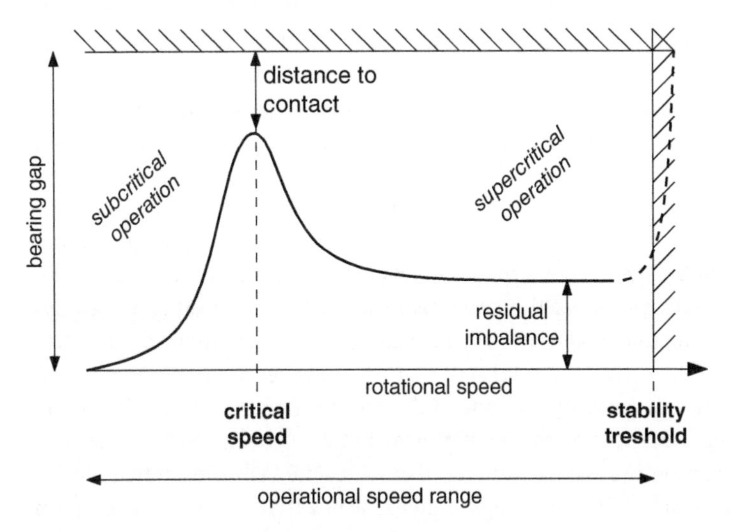

Figure 10.2 Typical (and qualitative) behaviour of the synchronous bearing response for a system with a supercritical operational speed range. Source: Liu L 2005.

There are however some specific and particular facts typical to the rotordynamic behaviour of an unbalanced rotor supported on gas bearings.

When compared to rolling element bearings or hydrodynamic/hydrostatic bearings of similar size, gas bearings usually possess lower stiffness values. This often makes that for a high-speed application, the operational speed range contains at least one synchronous resonance (which is referred to as a critical speed). Figure 10.2 shows the typical synchronous response which is observed in a system with a supercritical speed range.

10.2.2 Self-excited Whirl

The synchronous whirl described above is, in effect, the response of a passive linear rotor-bearing system to excitation induced by imbalance. This definition implies that, by limiting the amount of imbalance to within prescribed limits, no unstable behaviour can arise from this synchronous whirling. It should therefore be possible to accelerate such a system to any desired operational speed, if we ignore limitations such as centrifugal loading etc.

In practice, it has been found that an upper limit on the speed is set by the occurrence of a self-excited whirling phenomenon (as shown in Figure 10.2). This type of whirling is usually referred to as half-speed whirling, or more generally sub- or non-synchronous whirling. It can be very destructive in nature, setting in by only a minor increase of the rotational speed and often leads to shaft seizure.

The next section discusses this self-excited whirling phenomenon in detail and identifies its relevant parameters.

10.3 Study of the Self-excited Whirling Phenomenon: Description, Literature Overview and Relevant Parameters

In the early days of gas lubrication, bearing failures caused by self-excited whirling occurred suddenly and unpredictably. The successful utilisation of gas bearings for high-speed applications became conditional upon a sound understanding of the related phenomena and upon the development of a reliable method for predicting the onset speed of this type of whirling.

10.3.1 Description and Terminology

Grassam and Powell (1964, ch. 3) define *half-speed* or *half-frequency* whirling as: "Fairly suddenly whirling of either the shaft or the bearings in the direction of rotation at nearly half-shaft speed. The speed of transition between steady running to whirling is called the onset speed or threshold speed of half-speed whirl. If the speed is then further increased the amplitude of whirl increases rapidly until the shaft and the bearings touch and cause failure."

There are however some inadequacies in this formulation as it applies only to one of the many, though the most common forms of self-excited whirling. First of all, self-excited whirling has the tendency to set in at other frequencies than half the rotational speed. Powell (1970) reports the occurrence of one-third speed, one-quarter speed up to one-seventh speed whirl for externally pressurised bearing. He states that the onset speed always remains an exact multiple of the lower natural frequency and the whirl frequency is always equal to this lower natural frequency. This has led to the usage of the term *fractional speed* whirl. However, there exists no valid argument why the threshold speed should be an exact multiple of one of the natural frequencies. So, it is more correct to refer to it as *subsynchronous* whirling. Recent experiments performed at KULeuven PMA on a special type of externally pressurised journal bearing, reveal the possible occurrence of self-excited whirling at frequencies above the synchronous frequency, see (Waumans 2009). These findings point to the fact that the most general term is *non-synchronous* self-excited whirling.

Another remark on the aforementioned definition of self-excited whirling would be that, although whirling is mostly observed in the direction of rotation; i.e. *forward whirl*, it is also possible to appear opposite to the direction of rotation, being caused by the *backward whirling* mode which reaches the point of instability.

Grassam and Powell (1964) attribute the tendency of a shaft to whirl to the tangential component of the restoring force which is present after a perturbation. This is visually represented in Figure 10.3. This component is responsible for accelerating the shaft forward into whirl. According to them, the physical mechanism causing an instability to occur at half-speed can be given qualitatively as follows:

1. In an aerodynamic bearing, the average velocity of lubricant being pumped around in the bearing clearance and maintaining the pressure distribution is a half the rotational velocity and is in the direction of rotation.
2. If the shaft centre now orbits at half-shaft speed in the direction of rotation, the position of the converging-diverging gap system formed between the shaft and the bearing also moves forward at half-shaft speed.
3. Thus the lubricant fluid is pumped into the gap system at just the same average speed as the gap system moves forward so no pressure distribution can be set up and so the lubricant forces disappear.

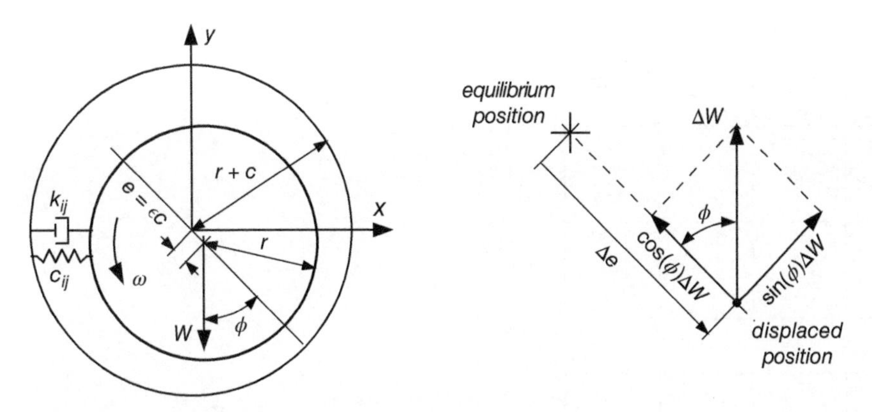

Figure 10.3 Steady-state working condition of a journal bearing and the restoring force ΔW at an angle ϕ (attitude angle) to the perturbation Δe. Note the tangential component $\sin(\phi)\Delta W$ causing the shaft to accelerate forward into whirl. Source: Adapted from Grassam NS and Powell JW 1964.

These physical arguments show that the load capacity of a journal bearing disappears, or at least is considerably reduced, if there is a relative half-speed whirl between bearing and shaft. Following this argument, a key role in the stability performance of a bearing is played by the attitude angle.

Some considerations have to be mentioned regarding the correctness of this qualitative and intuitive explanation. The basic idea originates from the research field of incompressible fluids, where there exists a similar self-excited instability known as "oil whip", as first reported by Newkirk (1924) and Newkirk and Taylor (1925). Frêne et al. (1997) show by applying the conservation of mass flow rate, the existence of half-speed whirling for an unloaded oil bearing. The effect of compressibility is therefore not considered in the explanation by Grassam and Powell (1964). Another point of comment regards the difference between the steady-state film behaviour and its dynamic properties. Although the value of the attitude angle may foreshadow the stability behaviour, it is *in se* a steady-state bearing characteristic. For a correct interpretation of the stability, one has to consider the full set of dynamic film coefficients, being the direct and cross-coupled stiffness and damping values.

It is generally accepted that the cross-coupled stiffness and (to a lesser degree) the cross-coupled damping effects in the gas film, are responsible for the phenomenon of the self-excited whirling instability (Frêne et al. 1997, p. 173). Rotation induced cross-coupling in the gas film gradually lessens the overall damping of the system until the self-excited whirl onset speed is reached. At this point, the system is in state of neutral stability (undamped mode). A further speed increase turns this undamped condition into a self-excited state. Violent whirling will set in, at a frequency equal to the eigenfrequency of the system. In most plain bearing geometries, the eigenfrequency happens to lie close to half the onset speed. An interesting reference with respect to this loss of damping due to rotation, is the experimental work performed by (Tully 1966). He measured the natural frequency and damping of a two-bearing system by observing the decaying oscillations of the rotor following a shock loading and repeated the measurements at several speeds up to the whirl onset speed. Some of his results are shown in Figure 10.4.

The figure clearly indicates the lack of overall damping at the onset speed of whirling. In a following section, this will be confirmed by the stability analysis on the simple case of a Jeffcott, i.e. two-dimensional[1], rotor-bearing configuration. The subsequent study will reveal the relative importance of all involved dynamic film properties.

Although it had been already clear at an early stage that the value of the attitude angle played a crucial role, the whirling instability was not directly attributed to the cross-coupling effect in the gas film. Even Marsh (1965) or Lund (1965) who both used cross-coupled dynamic properties in their theoretical stability theory, did not identify it as the main culprit of instability.

1 Though perhaps not quite accurate, we shall use the term "Jeffcott rotor" to indicate an idealised, two-dimensional rotor system, as is commonly also termed in the literature

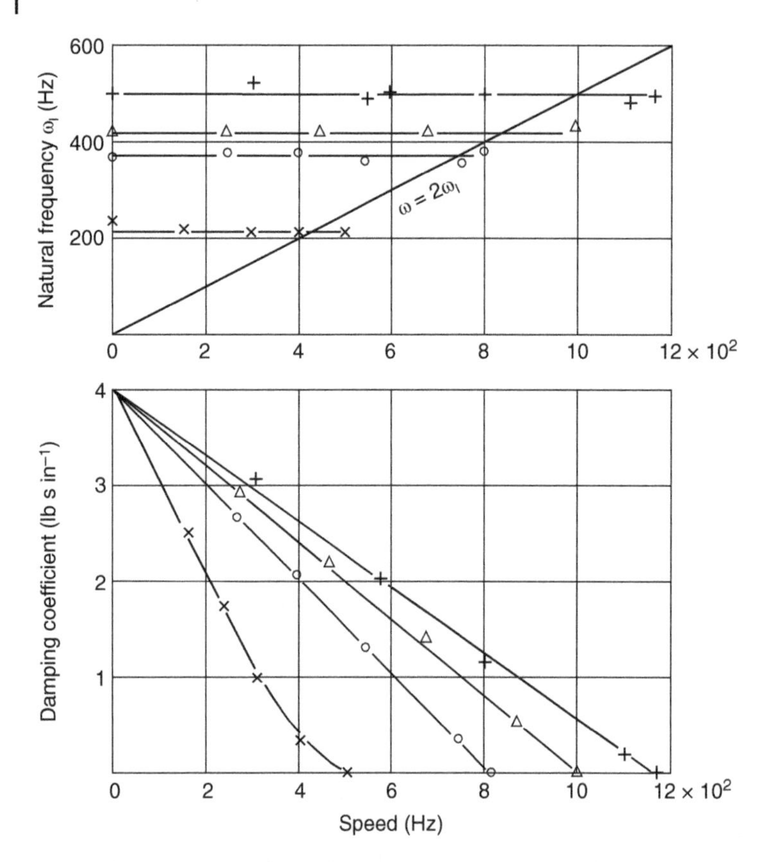

Figure 10.4 Variation of damping and natural frequency with speed for aerostatic bearings with a large clearance ratio (after Tully (1966)). $D = 0.75$ in, $L = 1.50$ in (= 38.1 mm), $c = 0.0011$ in. (= 28 μm) ×, $p_s - p_a = 10$ lbf in^{-2}; ∘, $p_s - p_a = 30$ lbf in^{-2}; △, $p_s - p_a = 50$ lbf in^{-2}; +, $p_s - p_a = 80$ lbf in^{-2} (10 lbf in^{-2} = 0.69 bar). Source: Tully, Neville (1966). The reference to the diagram in the original thesis is Fig 8.3.2 'Damping and Natural Frequency vs Speed'.

10.3.2 Half-speed Whirl in Literature

Due to its unpredictable occurrence and destructive nature, half-speed or half-frequency whirling, as it was originally labelled, received considerable research interest. Many researchers studied the issue both theoretically and experimentally to gain insight in the matter and to determine design guidelines.

Although similar whirl instabilities have been known to be associated with oil lubricated bearings, the majority of contributions regarding dynamic instabilities of rotor-bearing systems, is linked to gas bearing applications. The reason for this concentration of research towards gas film instabilities can be explained by the fact that gas lubricated bearings tend to operate at a higher peripheral speed than oil bearings, and because, in the absence of effective boundary lubrication, the consequence of an inadvertent high speed touchdown[2] is necessarily much more traumatic for gas lubricated bearings.

Experimentally Based Methods

In the 1950s, research was mainly experimental. A good example is the work of Whitley et al. (1962) who showed how the whirl onset speed was influenced by rotor mass, load, clearance, length-to-diameter ratio and gas viscosity.

2 A rather euphemistic description as given by Pan (1980), for what is generally referred to as a fatal bearing crash. This most often results in shaft seizure and it also implies the annihilation of a set of meticulously manufactured components.

To broaden the usage of experimentally obtained data he proposed a method in which a freely-supported bearing sleeve was supported on a shaft which was in turn supported in two other slave bearings. The shaft was slowly accelerated until the test bearing began to whirl and the onset speed was noted. It was then possible to predict onset speeds of whirl in any machine in which the tested bearing was employed. Whitley's method allowed machines to be successfully produced with self-acting gas bearings but in some cases the attainment of the desired operating speed was only achieved after a lengthy period of development.

Other interesting experimental work was performed by Reynolds and Gross (1962). Their investigation on self-acting bearings focused upon the effects of slenderness ratio and clearance ratio on the threshold speed and upon the character of the observed whirl. Their findings confirmed many previously published observations, but revealed some hitherto unreported results. Large-amplitude self-excited whirling was suppressed by centrifugally loading due to imbalance. Gross (1962) investigated the whirling behaviour in externally pressurised bearings in a similar way.

Theoretical Prediction of Onset Speed

Around 1960, there was a great deal of experimentally obtained design knowledge available on how to avoid or postpone self-excited whirling. Yet, there remained a need for a correct method of theoretically predicting the onset of half-speed whirl. For the case of aerodynamic bearings, three reliable methods of analysis were developed around 1965 to predict the onset speed of instability (Fuller 1969; Marsh 1980):

(i) Solution of the time-dependent equations of motion for the rotor.

To simulate the behaviour of dynamic systems of which gas bearings are part, one can rely on the simultaneous integration of both the time-dependent Reynolds equation and the equations of motion. The core of this class of methods is to update through a small time interval the distribution of pressure while the film geometry and its local rate of change are known. This can be done in an explicit, semi-implicit or implicit way while for the latter two an alternation direction technique (ADI) is often applied to simplify the solution in each time step. Castelli and Elrod (1965) made use of this technique to study the stability of a full self-acting gas bearing.

Depending on the simplifications made to obtain the pressure distribution at each time step, accurate predictions are possible because the inherent non-linearities in the fluid film are preserved. Among the disadvantages are the required computing time and the lack of insight into the underlying sources of instability. Recently, Czolczsynki (1999) used a slightly modified approach to this technique to study the rotordynamic behaviour of gas-lubricated systems.

(ii) Solution of the Reynolds equation using Galerkin's method.

The difficulty of obtaining an accurate solution for the time-dependent, non-linear Reynolds equation in the case of a self-acting bearing was dealt with by using the method of Galerkin to the *PH*-function. This approximation can be seen as a kind of series expansion of a certain function and has been applied extensively to many problems in lubrication engineering. Cheng and Trumpler (1963) were the first to apply this method to solve the stability problem encountered in self-acting bearings. Cheng and Pan (1965) extended this theory to bearings of finite length. Most results were in good agreement with existing theoretical and experimental data, but the proposed theory failed to predict a meaningful threshold speed for slender bearings working at high eccentricities and elevated bearing numbers.

(iii) Linearised theory for the dynamic behaviour of bearings.

The linearised theory is based on the full solution of the Reynolds equation to obtain the steady-state pressure distribution, and a local linearisation of the dynamic pressure field caused by a small harmonic distortion of the rotor. The linearisation is applied only to the dynamic pressure field and the essential non-linearity of the steady-state pressure distribution is preserved. This approach leads to the formulation of a dynamic stiffness and damping matrix containing both direct and cross-coupled terms which depend on the steady-state working conditions and perturbation frequency. This method is similar to the "eight coefficient" approach for oil bearings as introduced by Stodola (1925) and Hummel (1926).

Marsh (1965) published his theory on this topic nearly simultaneously with Lund (1965). Although the fundamental idea stated in these two publications is identical, strangely enough no cross-references exist between the work of both researchers. The method proposed by both authors can be seen as one of the first successful attempts in predicting the onset speed of self-excited whirling. Lund incorporated in his theory the effect of flexible and damped supports. Later on, he applied the same method to tilting pad bearings (Lund 1968) and to externally pressurised bearings (Lund 1967), but in the latter he made questionable assumptions regarding the pressure distribution in the vicinity of feeding sources.

By utilising this linearised theory (also referred to as the ϵ-perturbation theory), the total modelling process is split up into a gas film modelling step and a rotordynamic step. The dynamic film properties serve as input for the second step in the form of tabulated values. Moreover, this linearised method allows us to identify certain dynamic coefficients as main source of self-excited whirling, as is clarified by the sensitivity analysis of next section.

10.3.3 Sensitivity Analysis to Identify the Relevant Parameters

In order to gain insight in the underlying mechanism causing self-excited whirl instability and to identify its relevant parameters, the rotordynamic behaviour of a Jeffcott rotor-bearing configuration will be evaluated. The stability study on this simple case enables us to isolate the dynamic bearing coefficients responsible for inducing self-excited whirling. Further simplifications will lead to a general formula which is readily used to conclude on the stability of a system. The obtained results help in designing stable gas bearings and provide information on the optimal relative proportions of their dynamic properties.

The Jeffcott Rotor-bearing Configuration

The perturbation method used in Chapter 9 is applied to determine the dynamic film characteristics and employed to analyse the simple case of a Jeffcott rotor. This section has the purpose of mainly giving a physical interpretation to the underlying mechanisms which are responsible for the self-excited whirling phenomenon.

Figure 10.5 shows the dynamic model of a Jeffcott rotor with mass m supported on a single gas film which is represented by eight dynamic coefficients. The equations of motion are given by:

$$m\ddot{x} + c_{xx}\dot{x} + c_{xy}\dot{y} + k_{xx}x + k_{xy}y = 0 \tag{10.1a}$$

$$m\ddot{y} + c_{yy}\dot{y} + c_{yx}\dot{x} + k_{yy}y + k_{yx}x = 0, \tag{10.1b}$$

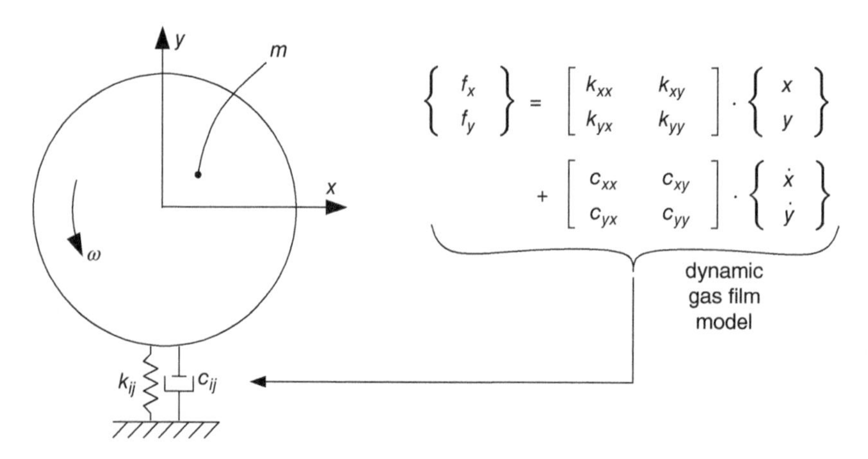

Figure 10.5 Dynamic model of a Jeffcott rotor with mass m. Source: Waumans T 2009.

where all dynamic coefficients k_{ij} and c_{ij} depend on the steady-state working condition (rotor speed ω, eccentricity ratio ϵ and attitude angle ϕ) and on the perturbation frequency ν. In the general case, all eight coefficients are distinct. When the steady-state eccentricity approaches zero (a condition which is true for unloaded vertical rotors, or by approximation also applicable to other situations), symmetry exists in the gas film geometry, which reduces the full set of eight coefficients to four by the fact that:

$$k_{ii} = k_{xx} = k_{yy} \tag{10.2a}$$

$$c_{ii} = c_{xx} = c_{yy} \tag{10.2b}$$

$$k_{ij} = k_{xy} = -k_{yx} \tag{10.2c}$$

$$c_{ij} = c_{xy} = -c_{yx}. \tag{10.2d}$$

By converting the system to the Laplace domain and by applying this symmetry, one obtains:

$$\begin{bmatrix} ms^2 + c_{ii}s + k_{ii} & c_{ij}s + k_{ij} \\ -c_{ij}s - k_{ij} & ms^2 + c_{ii}s + k_{ii} \end{bmatrix} \cdot \begin{Bmatrix} x \\ y \end{Bmatrix} = 0. \tag{10.3}$$

The homogeneous solution can be expressed as:

$$\lambda_{1,2} = \frac{-(c_{ii} + jc_{ij}) \pm \sqrt{(c_{ii} + jc_{ij})^2 - 4m(k_{ii} + jk_{ij})}}{2m}. \tag{10.4}$$

As will be shown later, the cross-coupled damping coefficient c_{ij} has, in most realistic cases, only a negligible effect on the solution. This justifies the simplification of the analytical solution to:

$$\lambda_{1,2} \approx \frac{-c_{ii} \pm \sqrt{c_{ii}^2 - 4m(k_{ii} + jk_{ij})}}{2m}, \tag{10.5}$$

where one solution represents the forward whirling mode and the other the backward whirling mode. As discussed earlier, an iterative solution strategy is required since all dynamic film coefficients not only depend on the rotational speed ω, but also on the perturbation frequency ν.

In general, the homogeneous solution of any linear system, $L(\mathbf{x}) = 0$, can be characterised by the eigenvalues (Den Hartog 1985):

$$\lambda = \eta + j\omega_d, \tag{10.6}$$

with ω_d representing the damped natural frequency and

$$\zeta = -\frac{\eta}{|\lambda|} = -\frac{\eta}{\omega_n} \tag{10.7}$$

being the damping ratio or relative damping of the solution. The latter quantity is a measure describing how oscillations in a system die down after a disturbance. For the purpose of assessing the stability performance of gas bearing systems, ζ is a better and more general measure of stability than the real part of the solution η since it represents a dimensionless property scaled with respect to the natural frequency ω_n of the system. The damping ratio ζ of a system is related to its logarithmic decrement δ through the following equation (Den Hartog 1985):

$$\delta = \frac{2\pi\zeta}{\sqrt{1 - \zeta^2}}. \tag{10.8}$$

Cross-coupling as Destabilising Factor

In Figure 10.6, the forward whirling solution of Eq. 10.5 has been evaluated for increasing values of the rotor speed ω. The calculations have been performed for an arbitrary aerostatic rotor-bearing configuration. An iterative solution strategy is used since all dynamic coefficients not only depend on the rotational speed, but also on the

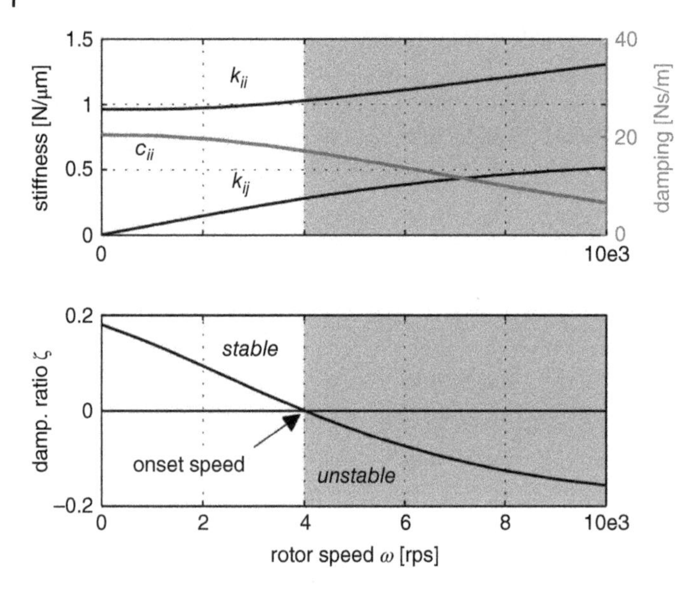

Figure 10.6 Typical behaviour of the dynamic coefficients k_{ii}, k_{ij} and c_{ii} with respect to rotor speed ω (top); and resulting damping ratio ζ (bottom). All quantitative values are obtained from calculations on a Jeffcott rotor supported on plain aerostatic bearings. Source: Waumans T 2009.

perturbation frequency. The top part of the figure shows the behaviour of the most important dynamic film coefficients with respect to the rotor speed. Due to compressibility and aerodynamic effects, the direct stiffness k_{ii} increases slightly from its stationary-rotor value. More significant is the rise in cross-coupling k_{ij} at higher speeds. The opposite can be said about the direct damping c_{ii} which gradually decreases, asymptotically reaching zero for an infinitely high perturbation frequency. In the bottom part of the figure, the damping ratio ζ starts at a maximal value and gradually decreases until it crosses the zero line. This point marks the onset speed of self-excited whirling and can be compared with an undamped eigenmode of the system.

The curves of this figure clarify in a visual way the origin of self-excited whirling. At low speeds, there exists nearly no cross-coupling in the gas film, while the damping capacity in the film is sufficient to keep the system stable. As we have seen in Chapter 9, an increase in rotational speed results in a reduction of the direct damping while the cross-coupling stiffness increases and acts as destabilising factor which reduces the overall damping of the system. In that sense (and by looking at Eq. 10.5) the cross-coupled coefficients act as negative damping to the system (Lund 1987). It can also be seen from this figure, that a further increase in speed only results in an ongoing deterioration of the system's stability as the gas film damping is insufficient to oppose the increasingly destabilising effect of the cross-coupling.

Influence of Other Parameters

It has been amply demonstrated that self-excited whirling is caused by the loss of damping due to rotation induced cross-coupling in the gas film. It would be interesting, not only from a theoretical point of view but particularly to provide some guidelines for practical bearing design, to know the influence of other parameters on the stability: more specifically, how they affect the threshold value of cross-coupling that leads to self-excited instabilities.

By evaluating Eq. 10.4 for different values of the cross-coupled stiffness k_{ij} while keeping all but one other parameter constant, it is possible to reveal the influence of that single parameter on the maximum allowable degree of cross-coupling. The results of this analysis are represented in Figure 10.7. In this figure, the damping ratio ζ of both the forward and the backward whirling mode have been plotted with respect to the amount of cross-coupling. It

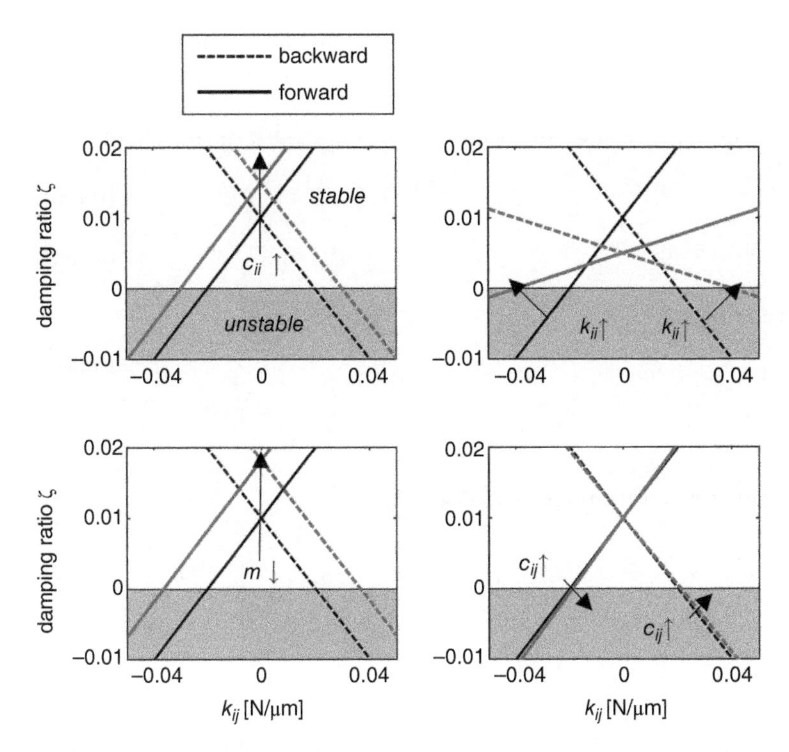

Figure 10.7 Qualitative stability study on a Jeffcott rotor-bearing system to reveal the influence of the dynamic bearing coefficients c_{ii}, k_{ii}, c_{ij} and rotor mass m on the maximum allowable amount of cross-coupling k_{ij}. The rotor-bearing is stable if the damping ratio ζ of both the forward and the backward mode is positive. Source: Waumans T 2009.

can be observed that an increase of the direct damping c_{ii} or a decrease in rotor mass m both improve the stability. The effect of a higher direct stiffness k_{ii} enlarges the span of allowable cross-coupling k_{ij} although with a higher minimal achievable value of the damping ratio. The contribution of the cross-coupled damping c_{ij} is minimal. Even in case of extreme values (an order of magnitude larger than the direct damping), only a small shift of both curves is noticeable. The asymmetry which is consequently created, will result in a quasi negligible difference between the positive and negative threshold value of k_{ij}. For this reason the aforementioned assumption, which leads to the simplified formula 10.5, is justifiable.

A Simple Stability Criterion

In order to derive a simple stability criterion, let us restate formula 10.5:

$$\lambda_{1,2} = \frac{-c_{ii} \pm \sqrt{c_{ii}^2 - 4m(k_{ii} + jk_{ij})}}{2m}.$$

It would be interesting to find an analytical expression for the value of k_{ij} which makes the real part of this solution zero. This would represent the threshold value of cross-coupled stiffness $k_{ij,\text{thres}}$ which, for a given combination of direct stiffness k_{ii}, direct damping c_{ii} and rotor mass m, results in an undamped eigenmode of the system as indicated by the zero-crossing in Figure 10.7.

This condition is fulfilled if:

$$\mathbf{R}\left(\sqrt{c_{ii}^2 - 4m(k_{ii} + jk_{ij})}\right) = c_{ii}. \tag{10.9}$$

N.B. We assume throughout this chapter that c_{ii} is positive. This is not always the case for hybrid bearings with pocketed orifices. In that case, however, the bearing might become unstable, in the pneumatic-hammering sense, owing to negative direct damping.

To find a closed-form expression for $k_{ij,\text{thres}}$ we represent a complex number by $x + jy$ and write the identity:

$$\mathbf{R}\left(\sqrt{(x+jy)(x+jy)}\right) = x \tag{10.10a}$$

$$\mathbf{R}\left(\sqrt{x^2 - y^2 + 2jxy}\right) = x. \tag{10.10b}$$

By comparison with Eq. 10.9, it is easily seen that:

$$x = c_{ii} \tag{10.11a}$$

$$y = \frac{2mk_{ij,\text{thres}}}{c_{ii}}. \tag{10.11b}$$

This leads to the following closed-form formulation for the threshold value of k_{ij}:

$$|k_{ij,\text{thres}}| = \sqrt{\frac{k_{ii}}{m}}\, c_{ii}. \tag{10.12}$$

The necessary condition for stability is therefore given by:

$$|k_{ij}| \leq \sqrt{\frac{k_{ii}}{m}}\, c_{ii}. \tag{10.13}$$

Let us note that Pan (1980) derived a similar condition for stability through the formulation and subsequent manipulation of the isotropic impedance matrix.

Now, defining the ratio $\kappa = k_{ij}/k_{ii}$, one obtains a formulation of the threshold value κ_{thres} and corresponding stability criterion that is very simple in form:

$$|\kappa_{\text{thres}}| = \left|\frac{k_{ij,\text{thres}}}{k_{ii}}\right| = 2\zeta_{\text{n}}, \quad |\kappa| \leq 2\zeta_{\text{n}}. \tag{10.14}$$

Note that ζ_{n} represents the damping ratio defined by $c_{ii}/(2m\omega_{\text{n}})$. It is different from the overall damping ratio ζ of the system, except for the condition of zero speed $\omega = 0$, in which then $\zeta = \zeta_{\text{n}}$).

This closed-form expression and corresponding stability criterion serves different purposes. On the one hand, the expression helps in understanding the stability issue encountered in high-speed gas bearing systems by providing a theoretical insight in the involved parameters. It clearly indicates how and to what extent all these parameters affect the stability performance. On the other hand, this simple stability criterion allows us to perform a quick and fairly accurate stability evaluation while designing or optimising a gas bearing application.

Although valuable both from theoretical and practical point of view, Eq. 10.13 has been derived for the case of a Jeffcott rotor-bearing configuration. It is however possible to derive similar closed-form expressions for other configurations as demonstrated by Figure 10.8. In case of a conical mode, k_{ii}/c_{ii} represents a tilt stiffness/damping value. The involved parameters and analysis of the latter case are completely analogous to the cylindrical case.

Whirl Ratio Ω_w

The previously derived criterion relates the values of the involved dynamic film coefficients (k_{ii}, c_{ii} and k_{ij}) at the onset of instability. It does not provide any information on the value of the onset speed ω_{thres} itself. To do so, the behaviour of the dynamic coefficients with respect to rotor speed and perturbation frequency should be known, as has been presented in Chapter 9, where it has been shown that there exists no general analytical expression which describes this relationship in a closed-form way, except for the case of concentric, plain, self-acting JB, obtained by the perturbation solution.

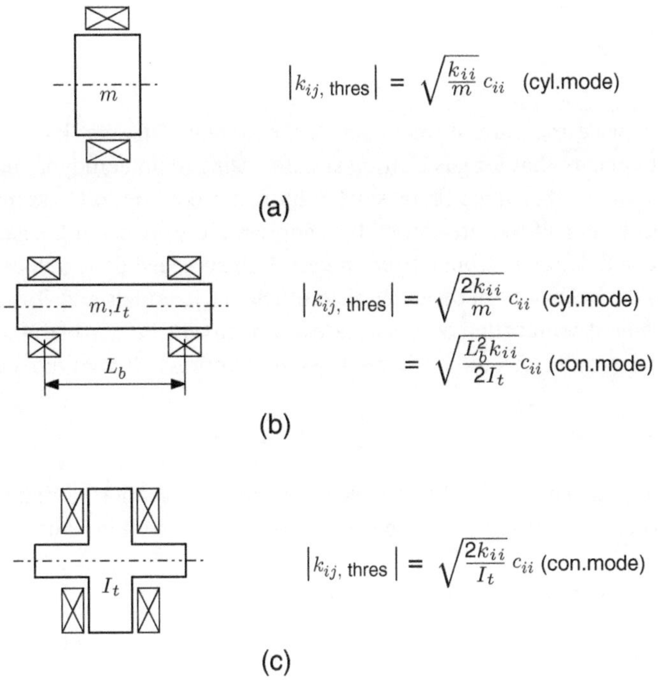

$$\left|k_{ij,\text{ thres}}\right| = \sqrt{\frac{k_{ii}}{m}}\, c_{ii} \quad \text{(cyl.mode)}$$

(a)

$$\left|k_{ij,\text{ thres}}\right| = \sqrt{\frac{2k_{ii}}{m}}\, c_{ii} \quad \text{(cyl.mode)}$$
$$= \sqrt{\frac{L_b^2 k_{ii}}{2I_t}}\, c_{ii} \quad \text{(con.mode)}$$

(b)

$$\left|k_{ij,\text{ thres}}\right| = \sqrt{\frac{2k_{ii}}{I_t}}\, c_{ii} \quad \text{(con.mode)}$$

(c)

Figure 10.8 Different rotor-bearing configurations for which a closed-form expression of the threshold cross-coupled stiffness $k_{ij,\text{thres}}$ can be derived: (a) Jeffcott – cylindrical mode; (b) rigid shaft on two journal bearings at a distance L_b apart – cylindrical and conical mode; (c) shaft with thrust disc – conical mode. In all cases m represents the rotor mass and I_t stands for the transversal moment of inertia. Source: Waumans T 2009.

When dealing with incompressible flows, one can assume that the cross-coupled stiffness k_{ij} grows linearly with the rotation speed while the direct coefficients k_{ii} and c_{ii} remain constant:

$$k_{ij}(\omega) = k_{ij}(0)\,\omega \tag{10.15a}$$

$$k_{ii}(\omega) = k_{ii} \tag{10.15b}$$

$$c_{ii}(\omega) = c_{ii}. \tag{10.15c}$$

Substitution of this relationship in Eq. 10.12 defines a ratio Ω_w between the onset speed ω_{thres} and the natural frequency ω_{n} of the system:

$$\Omega_w = \frac{\omega_{\text{thres}}}{\omega_{\text{n}}} = \frac{c_{ii}}{k_{ij}(0)}. \tag{10.16}$$

Due to the fact that whirling always occurs at the natural frequency of the system, this ratio Ω_w is also the ratio between the rotational frequency and whirling frequency. Although still a ratio and not an absolute formula for the maximum attainable speed, it forms a more tangible criterion for the onset speed of whirl.

Let us now look at the value of this whirl ratio for the simple case of a hydrodynamic (incompressible) plain journal bearing. The full Sommerfeld solution applies to this situation and the following formulas can be derived to quantify the stiffness and damping properties of the fluid film at zero eccentricity (Hamrock et al. 2004).

$$k_{xx} = \frac{6\pi\mu\omega r^3 L}{c^3} \triangleq k_{ij} \quad \text{(attitude angle } \phi = 90°\text{)} \tag{10.17a}$$

$$c_{xx} = \frac{12\pi\mu r^3 L}{c^3} \triangleq c_{ii} \tag{10.17b}$$

The whirl ratio Ω_w is then easily found to be:

$$\Omega_w = \frac{c_{ii}}{k_{ij}(0)} = 2, \tag{10.18}$$

which means that self-excited whirling sets in at twice the natural frequency of the system. This fact has led to the term half-speed or half-frequency whirl. It appears that for gas bearing systems using plain aerodynamic or aerostatic bearings, the whirl ratio is usually close to this value (between 1.7 and 2.2 according to Grassam and Powell (1964)). Deviations from this theoretical value of two are caused by compressibility effects in the gas film which cause the behaviour of the dynamic coefficients to deviate from the ideal linear speed dependence. Figure 10.6 shows this effect of compressibility on the behaviour of the dynamic coefficients. Reynolds and Gross (1962) attributed a whirl ratio below two to metal–metal contact that occurs in case of violent whirling. In extreme situations characterised by a large value of the bearing number Λ (also referred to as the compressibility number), the whirl ratio can differ significantly from two.

More Systematic and Exact Stability Analysis

The next part utilises another (and perhaps more elegant) method to investigate the stability aspect of rotors supported on gas bearings. The method merely confirms the previously derived conclusions and expressions, but it is nevertheless valuable to look at the analysis itself.

Let us therefore restate the basic equations of motion of the Jeffcott rotor-bearing system:

$$m\ddot{x} + c_{xx}\dot{x} + c_{xy}\dot{y} + k_{xx}x + k_{xy}y = 0 \tag{10.19a}$$

$$m\ddot{y} + c_{yy}\dot{y} + c_{yx}\dot{x} + k_{yy}y + k_{yx}x = 0, \tag{10.19b}$$

which in case of a steady-state working condition with zero eccentricity can be simplified to:

$$m\ddot{x} + c_{ii}\dot{x} + c_{ij}\dot{y} + k_{ii}x + k_{ij}y = 0 \tag{10.20a}$$

$$m\ddot{y} + c_{ii}\dot{y} - c_{ij}\dot{x} + k_{ii}y - k_{ij}x = 0. \tag{10.20b}$$

In the subsequent analysis this set of coupled equations will be converted to one single equation by introducing a whirling vector $z = x \pm iy$, where $i = \sqrt{-1}$. The dynamic behaviour described by this equation is governed by only two parameters.

To see this, first divide through the rotor mass m and by using $\kappa = k_{ij}/k_{ii}$ and $\gamma = c_{ij}/c_{ii}$ the equations become:

$$\ddot{x} + \frac{c_{ii}}{m}\dot{x} + \frac{c_{ij}}{m}\dot{y} + \omega_n^2 x + \kappa\omega_n^2 y = 0 \tag{10.21a}$$

$$\ddot{y} + \frac{c_{ii}}{m}\dot{y} - \frac{c_{ji}}{m}\dot{x} + \omega_n^2 y - \kappa\omega_n^2 x = 0. \tag{10.21b}$$

Then, by normalising the time t to $\tau = \omega_n t$, where $\omega_n^2 = k_{ii}/m$, and denoting a derivative, e.g. of x, with respect to τ by x', we obtain:

$$x'' + 2\zeta_n x' + 2\gamma\zeta_n y' + x + \kappa y = 0 \tag{10.22a}$$

$$y'' + 2\zeta_n y' - 2\gamma\zeta_n x' + y - \kappa x = 0. \tag{10.22b}$$

Finally, the whirl vector $z = x + iy$ is introduced. To this a physical interpretation can be attributed by assuming forward (or backward, in case a minus sign is used) circular whirling in the (complex) xy-plane. When dealing with a sinusoidal whirling response, the first and second derivative of z appear as:

$$z' = i\bar{v}z = \bar{v}(ix - y) \tag{10.23a}$$

$$z'' = i\bar{v}z' = \bar{v}(ix' - y'), \tag{10.23b}$$

where $\bar{v}$ stands for the perturbation-frequency ratio v/ω_n.

A summation of Eq. 10.22a and i times Eq. 10.22b then yields the single equation:

$$\left(1-\frac{2\gamma\zeta_n}{\bar{\nu}}\right)z''+\left(2\zeta_n-\frac{\kappa}{\bar{\nu}}\right)z'+z=0. \tag{10.24}$$

The two natural frequencies Ω_n of this system equal:

$$\Omega_n=\gamma\zeta_n\pm\sqrt{\gamma^2\zeta_n^2+1} \tag{10.25}$$

where the $\pm$ corresponds to forward and backward whirling, respectively.

We notice that the cross-coupled damping appears as a mass contribution, and therefore only changes the natural frequency of the system. In that sense it is comparable to the gyroscopic coupling effect acting on a long rotor. There is no direct effect on the stability characteristics of the system: the indirect effect would be that the gas film properties have to be evaluated at a slightly different value of the perturbation frequency. Taking into account that $\gamma\zeta_n \ll 1$, the effect is somewhat limited. By setting γ equal to zero, we obtain, for $\bar{\nu}=1$, a more simplified equation that corresponds to the previously derived results:

$$z''+(2\zeta_n-\kappa)z'+z=0. \tag{10.26}$$

This one equation characterised by only two parameters ζ_n and κ, describes the (linear) dynamic behaviour of a Jeffcott rotor-bearing configuration. Although simple in form, it embodies the basic features which distinguish self-excited whirling: (i) the cross-coupled stiffness acts as negative damping to the system; (ii) this compromises the overall damping and causes an instability to set in when $\kappa=2\zeta_n$ (as in Eq. 10.14); (iii) this instability will appear as self-excited whirling at the natural frequency ω_n of the rotor-bearing system.

By defining a backward whirling vector $z=x-iy$, or alternatively, substituting $-\bar{\nu}\rightarrow\bar{\nu}$ in Eq. 10.24, it is easily found that:

$$\left(1+\frac{2\gamma\zeta_n}{\bar{\nu}}\right)z''+\left(2\zeta_n+\frac{\kappa}{\bar{\nu}}\right)z'+z=0, \tag{10.27}$$

or when $\gamma\zeta_n=0$ and $\bar{\nu}=1$,

$$z''+(2\zeta_n+\kappa)z'+z=0. \tag{10.28}$$

These equations describe the backward whirling behaviour of the system. If κ is positive, no self-excited instability can arise from this mode because cross-coupling contributes to the overall damping in this case.

10.4 Techniques for Enhancing Stability

The previous sections have dealt with the basic description and characteristics of self-excited whirling. A parameter study has revealed the detrimental influence of cross-coupling on the overall damping of the gas bearing system. By introducing a simple stability criterion, the effect of all involved parameters has been studied and quantified. However, judging on the trends of the involved dynamic film coefficients for increasing rotor speed, an instability will inevitably set in. By proper bearing design, it should be possible to postpone the onset point in order to prevent self-excited whirling from occurring within the operational speed range.

This section discusses different techniques and bearing types which can be applied to enhance the stability of a high-speed gas bearing system and as a direct consequence increase its maximal attainable speed. A literature overview outlines the currently existing design guidelines and ways of stabilisation together with a comment on their potential added value to the domain of high-speed bearing design.

After this, in a dedicated section, the optimal combination of design parameters for the case of a plain aerostatic bearing will be sought for by means of a systematic dimensionless stability study. The results from this study, applied to a case study, seem to yield bearing geometries characterised by an extremely small clearance ratio and large length-to-diameter ratio. Owing to issues such as centrifugal or thermal rotor growth and manufacturing or alignment challenges, the practical use of those bearing geometries is very limited.

This fact obliges us to tackle the stability problem at its source by focusing on the cross-coupling effects in the gas film. Firstly, the issue of reducing the cross-coupled stiffness is treated where the various techniques and their limitations to achieve that are overviewed.

Second, as it will appear from all the above techniques that one remains limited in enhancing the inner damping in the bearing film, we turn our attention to the question of introducing damping from outside the film. To this end, the principle of external damping to stabilise the rotor motion is overviewed.

10.4.1 Literature Overview on Current Techniques

As the existence of self-excited whirling was already known in the early days of gas lubrication, many researchers have put a significant effort in gaining insight into the influence of the involved design parameters and in the development of bearing types with improved stability characteristics. Although a complete and thorough understanding of the phenomenon was not always present, research focused on the determination of design guidelines collected in many cases by conducting lengthy series of meticulous experiments.

In the 1960s, research mainly focused on the stability of plain aerodynamic bearings. This has led to numerous papers available in open literature, from which only the most important conclusions will be outlined in the first part of this section. Hereafter, the classic methods for improving the stability are discussed. These methods can be categorised into techniques which rely on film geometry modifications, such as stabilising surface features on the one hand, and bearings with a conformable gap geometry on the other hand. A good overview dating from 1980 on these stabilisation techniques can be found in (Pan 1980). This reference describes the status of gas bearing technology at that time and its applicability towards aero propulsion machinery.

The remaining part of this survey focuses on four categories of techniques (A–D) to enhance stability drawn also from the more recent advances in high-speed gas bearing technology relevant to the field of micro-turbomachinery, and more specifically for application in micro gas turbines. This utmost demanding application example requires state-of-the-art bearing technology. An overview and critical evaluation of the bearing solutions proposed for this purpose are given in A–C. Finally, in paragraph D, the effect of the gas properties on the stability is considered.

A. General Considerations and Design Rules on Stability

In the foregoing, the effect of all involved dynamic film coefficients on the stability has been amply illustrated. This theoretical study contributes to the basic understanding of the problem, but does not provide information that is directly suitable to practical bearing design. It would therefore be more interesting to determine the relationship between the geometrical design parameters and a criterion for stability. This criterion can take the form of the onset speed or maximum attainable speed, or the damping ratio of the rotor-bearing system at a certain working condition.

For the simple case of a plain aerodynamic bearing geometry, the number of design parameters may be limited to the clearance ratio c/r and slenderness ratio L/D. The poor stability characteristics however oblige us to investigate other and more advanced geometries such as externally pressurised bearings, spiral groove bearings, lobed geometries or flexibly supported bearing types. The large number of design parameters and their complex interaction make it well-nigh impossible to determine a direct correlation with the stability performance of a rotordynamic bearing system. This quickly leads to a complex optimisation problem that would require considerable computational effort since each stability evaluation involves an iterative solution procedure. Furthermore, there is no guarantee on the existence of an optimal combination of design parameters that is compatible with other requirements or limitations.

The following will therefore list well-known design rules that tend to improve the stability performance. Taking into account the higher stated considerations on the complex interaction between the design parameters and stability behaviour, we will make use of the dynamic coefficients to reason on their effect on the stability. Figure 10.9

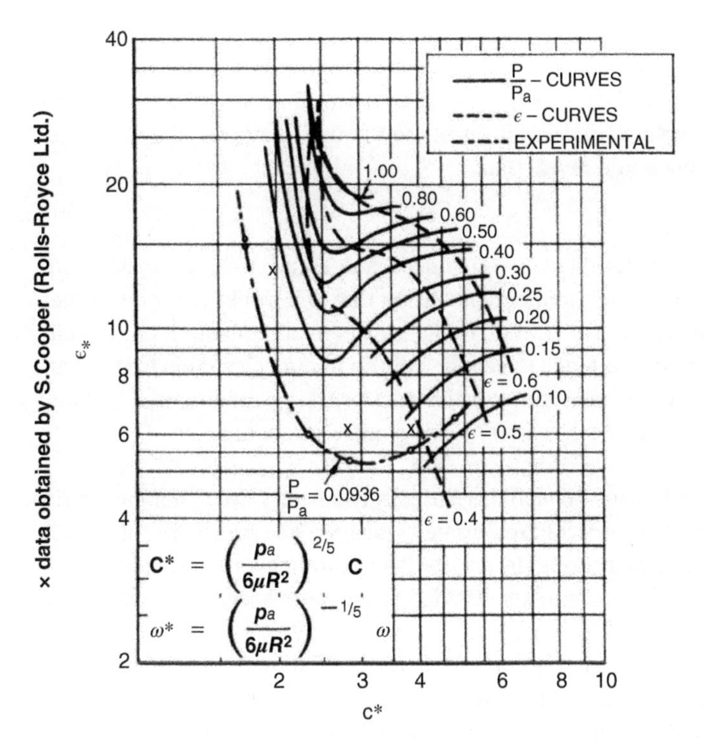

Figure 10.9 Stability curves C^* versus ω^* for $Mg/W = 1$ where M represents the rotor mass, g the gravitational constant and W the bearing load (in the figure the bearing radius is denoted by R, and the radial clearance by C). Stable operation requires values of (C^*, ω^*) under the curve corresponding to the actual steady-state working condition. The p/p_a-curves represent a dimensionless bearing load. Source: Sternlicht B and Winn LW 1963 by permission of ASME.

visually summarises most of these design rules and although derived for plain aerodynamic bearings, the observed trends still hold true for other bearing types[3].

- *Decrease of radial clearance*

  Rentzepis and Sternlicht (1962) analytically determined the regions of stability for plain aerodynamic bearings. Their results pointed to a very rapid ascend of the stability curves when lowering the radial clearance value. Supported by earlier experimental findings of Wildmann (1956), they concluded the existence of a clearance value under which one may never encounter instabilities.

  Let us now try to reason on this statement by observing the behaviour of the involved dynamic coefficients when the clearance value is reduced. The overall damping ratio ζ of a Jeffcott rotor system, when $\gamma\zeta_n = 0$ and $\bar{\nu} = 1$, is given by:

$$\zeta = 2\zeta_n - \kappa, \tag{10.29}$$

  where

$$\zeta_n = \frac{c_{ii}}{2m\sqrt{\frac{k_{ii}}{m}}} \tag{10.30a}$$

$$\kappa = \frac{k_{ij}}{k_{ii}}. \tag{10.30b}$$

3 The stability curves as presented by Rentzepis and Sternlicht (1962) and Sternlicht and Winn (1963) use a parameter set were the rotor speed ω appears only on one axis, though it was not non-dimensional. This greatly facilitates the interpretation of the charts.

Based on Eqs 10.17, both the direct damping c_{ii} and the cross-coupled stiffness k_{ij} scale with the inverse of the third power of the radial clearance c. In general, the maximal achievable direct stiffness value also increases significantly as a consequence of a clearance reduction[4]. Let us now assume that the direct stiffness k_{ii} is solely generated by an aerodynamic effect and therefore also behaves inversely to the third power of the clearance. This would make κ constant with respect to c and would reveal that:

$$\zeta \sim c^{-3/2} - c^{0}. \tag{10.31}$$

This partly confirms the statement of Rentzepis and Sternlicht (1962). Indeed, a clearance reduction leads to an improvement of the overall damping. But, and this relates to their conclusion that one would never encounter instabilities below a certain value of the clearance, there is no such thing as infinite and unconditional stability since the damping capability of a compressible film tends to zero at infinite frequencies. This trend of an improvement of the stability does however hold true when comparing identical working conditions. And thus, this implies that the theoretical optimal bearing geometry has an infinitely small clearance.

- *Increase of steady-state eccentricity*
  Theoretical predictions by Pan and Sternlicht (1962) have shown that a perfectly balanced vertical shaft in a plain aerodynamic journal bearing is always unstable with respect to self-excited whirling. The reason for this is that κ is at its highest in concentric position. By loading the bearing (by gravitation or through an external load), the steady-state working eccentricity is increased which results in an elevation of the onset speed as confirmed by the experimental results of Reynolds and Gross (1962).
  This eccentric operation yields a lower value of the attitude angle ϕ and therefore tends to reduces the stiffness ratio κ. The net effect on the film damping is positive since $c \sim 1/h^3$. As it will be shown further on, this approach is nearly impossible to apply in practice as a stability solution for miniature high-speed gas bearings.
  Sternlicht and Winn (1963) reported an increase of the onset speed with rotor mass. A distinction has to be made on the origin of the bearing load. If originating from an external source, it will always increase the onset speed. Gravitational loading due to the rotor mass, will however also increase the dynamic mass lowering the damping ratio ζ_n of the system.
- *Threshold eccentricity ratio*
  Multiple researchers have reported on the existence of a threshold eccentricity ratio. This represents the minimal steady-state eccentricity that should be maintained to guarantee stable operation. The threshold value depends on the geometrical parameters and on the rotational speed. Experiments by Reynolds and Gross (1962) indicate a higher threshold eccentricity ratio for slender geometries with a large clearance ratio.
- *Clearance value corresponding to a minimum onset speed*
  Figure 10.9 obtained by Sternlicht and Winn (1963) shows a clearance value which yields, for a given bearing load, a minimum onset speed.

B. Modifications to the Film Geometry: Stabilising Surface Features

The stability of self-acting gas bearings may be enhanced by surface features, brought either on the rotating or non-rotating bearing member. A large number of possibilities exists, from simple axial grooves to lobed geometries or more complex pumping groove patterns. Before discussing the different types of stabilising surface features found in literature, a general explanation for the improvement of the stability is given.

The principal consequence of nearly all types of surface features consists in the increase in direct stiffness due to the aerodynamic film action at wedge-shaped film sections. In contrast to the case of a plain self-acting bearing,

---

4 In case of an aerostatic bearing type, this would also require an adaptation of the feeding geometry. For an inherently restricted bearing, this can be obtained by decreasing the feed-hole diameter.

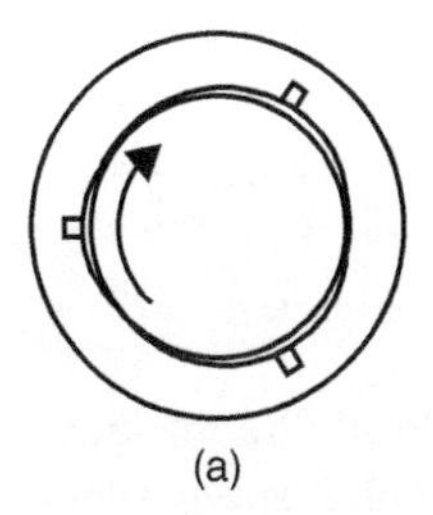

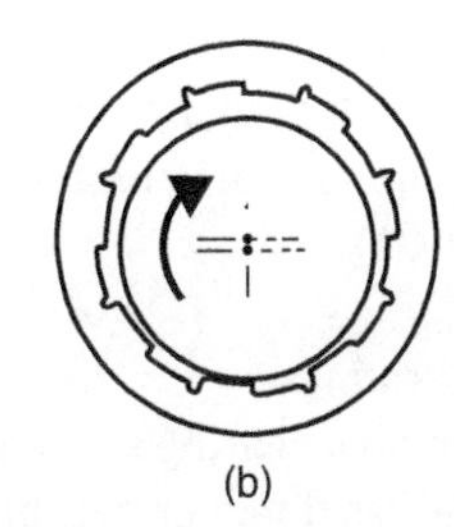

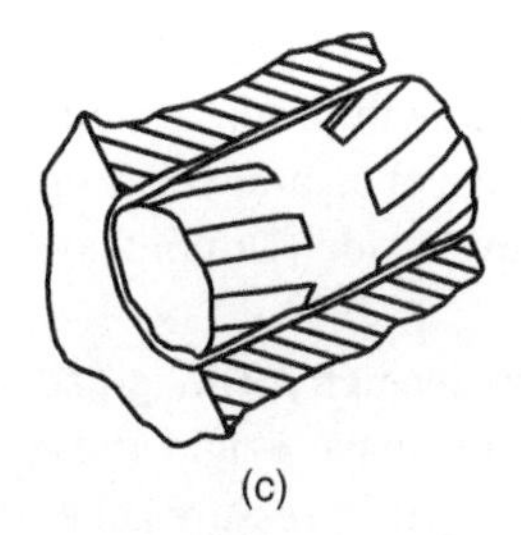

Figure 10.10 Journal bearing geometries with stabilising surface features: the three-lobe bearing (a), the Rayleigh step bearing (b), and the herringbone groove bearing (c). Source: Pan CHT 1980.

this stiffness generation allows for stable operation at low eccentricity values. Some types of surface features also attack the root cause of instability by reducing the cross-coupled stiffness generated by the gas film.

A wide variety of stabilising surface features is found in literature, as evidenced by (Fuller 1969; Grassam and Powell 1964; Gross et al. 1980; Pan 1980). The following brief overview will only address the most common types (Figure 10.10).

- *Axially grooved* journal bearing. A single axial groove machined in the stationary bearing member can affect the threshold speed to a certain extent despite the slight reduction in load-carrying capacity. Only when the groove opposes the direction of the load, the threshold speed of self-excited whirling is raised. The amelioration of the stability is in this case attributed to the increase in steady-state working eccentricity caused by the film interruption in the subambient pressure region. In addition, the groove can act as a physical depository for the collection of dirt and debris that might find its way into the bearing film (Fuller 1969).
- *Lobed or wave-shaped* journal bearing geometries (Figure 10.10(a)). Various non-circular bearing geometries with improved stability characteristics have been developed. Multi-lobe bearings are the most known and consist of two or more off-centred circular sections. Although two-lobe (or elliptical) bearings offer the advantage of ease of manufacturing through the assembly of two bearing halves, the three-lobe bearing is regarded as the most stable lobed bearing geometry (Pinkus 1959). Dimofte proposes a wave-shaped variant of this bearing type. The film geometry describes a sinusoidal profile in which the wave amplitude is usually a fraction of the radial clearance. A comparison of its characteristics with other bearing geometries is given in (Dimofte 1995a,b).
  In section 10.7, a modified wave-shaped bearing geometry is introduced with favourable stability properties. By removing the diverging areas of each wave section, the cross-coupling is reduced considerably.
- *Rayleigh step* journal bearing (Figure 10.10(b)). The configuration is an adaptation of the celebrated optimum plane slider geometry. This stepped profile is the most efficient circumferential pumping geometry, although it is a difficult bearing to fabricate. Design guidelines are provided in (Grassam and Powell 1964, pp. 100–103).
- *Spiral groove or herringbone groove* journal bearing (HGJB) (Figure 10.10(c)). This type of surface pattern is also known as herringbone or helical grooves. The spiral groove bearing comprises the most widespread viscous pumping geometry found in numerous gas or liquid film bearings, both journal and thrust type. As also mentioned in Chapter 9 concerning this type of bearings, Muijderman (1967) provides a good overview of the design methodology for optimum load-carrying capacity with incompressible films. The work of Bootsma (1975) deals with additional aspects such as frictional losses, stability and no-leakage conditions for various liquid-lubricated spiral groove bearings, including conical and spherical geometries.
  The effect of the design parameters (helix angle, groove depth, groove number etc.) on the stability of a gas-lubricated journal bearing has been validated experimentally by Cunningham et al. (1969) and Kaneko et al. (1974). Their results show that the stability is most affected by the radial clearance value. Spiral groove gas bearings have found application in high-speed rotating equipment such as cryogenic turbo expanders and

inertial guidance systems. The desired operating speed was reached by reverting to a small radial clearance value. Misalignment and rotor growth problems are circumvented by relying on alignment tolerant spherical bearing surfaces, and by the usage of ceramic bearing materials or conical configurations with a gap geometry that is invariant to centrifugal expansion (Dupont 2005; Risse 2001).

C. Bearings with a Conformable Gap Geometry: Tilting-pad and Foil Bearings

All bearing types discussed until now have a rigid and non-deformable bearing surface. The local film thickness and film geometry are not affected by the pressure generated in the gas film. For the support of high-speed rotors, bearings with a conformable gap geometry offer several advantages. In bearings with tilting pads or flexibly supported foils, the film geometry can change considerably due to the pressure developed in the bearing gap. The following overview will discuss the superior stability properties attributed to this class of bearings.

- *Tilting-pad* journal bearings (Figure 10.11(a)). This bearing type may be regarded as consisting of multiple pivoted sections. The wedge-shaped film geometry under each pad arises from the torque balance around the pivot or tilting point. In some cases, one pad has a radially compliant support which allows for centrifugal and thermal rotor growth. The superior stability behaviour of tilting-pad bearings can be attributed to: (i) the tilting action leads to an optimal pressure distribution with limited cross-coupling, and this for different working conditions; (ii) the configuration with one or more radially compliant pads makes it possible to reduce the nominal radial clearance value; and (iii) friction of the pivot may introduce some external damping to the rotor-bearing system as studied in (Dmochowski et al. 2009). Chapter 11 is dedicated to the treatment of this type of bearing.
  Various interesting realisations of the tilting-pad concept may be found in (Gross et al. 1980, section 4.15). Also worthwhile mentioning is the floating pad concept found in (Schmid 1974) for a cryogenic helium turbo expander with a journal diameter of 22 mm running at 220 000 rpm (= 4 884 000 DN = mean diameter(mm)× rpm). The pads are suspended by gas which is tapped from the interface between the rotor and inner pad surface. This thin gas film acts as a squeeze-film damper. The idea of introducing additional damping to the pad support is also adopted by Kim and Rimpel (2009) and Ertas (2009). The clearance behind the flexure pivot is in both cases fitted with a damping material. At a miniature scale, tilting pad designs have been proposed with flexure pivots manufactured by lithographical processes (Kozanecki et al. 2006; Sim and Kim 2009).
- *Foil bearings* (Figure 10.11(b, c)). This class of bearings has a compliant surface which is deformed under the pressure generated in the gas film. The research domain of foil bearing technology has received a lot of attention during the last 35 years, leading to a large number of foil bearing types. Chapter 12 is dedicated to the treatment of this type of bearing.

D. The Effect of Gas Properties

There exist several application domains of high-speed gas bearings were gases other than air are used as lubricant. In cryogenic turbo-expanders, the working fluid (helium, oxygen, nitrogen, argon, krypton…) is often also the lubricant gas for the bearings as this significantly simplifies their design and reduces the risk of leakage and contamination. Beams (1937) applied hydrogen fed bearings for his ultra-centrifuging experiments to minimise the frictional losses and to increase the sonic speed at the outlet of the driving nozzles. Although the motivation for using other gases originated from purely practical reasons, the achievement of the desired operational speed was never suspected to be owing to a possible enhancement of the stability behaviour. It would therefore be interesting to examine the effect of the gas properties on the stability performance.

The gas properties relevant to this investigation are the density and viscosity, which are listed for a number of gases in Table 10.1.

Let us first consider the case of an aerodynamic rotor-bearing configuration. For a given set of geometrical bearing parameters, only the dynamic viscosity μ of the gas has an effect on the dynamic film coefficients and hence on the stability.

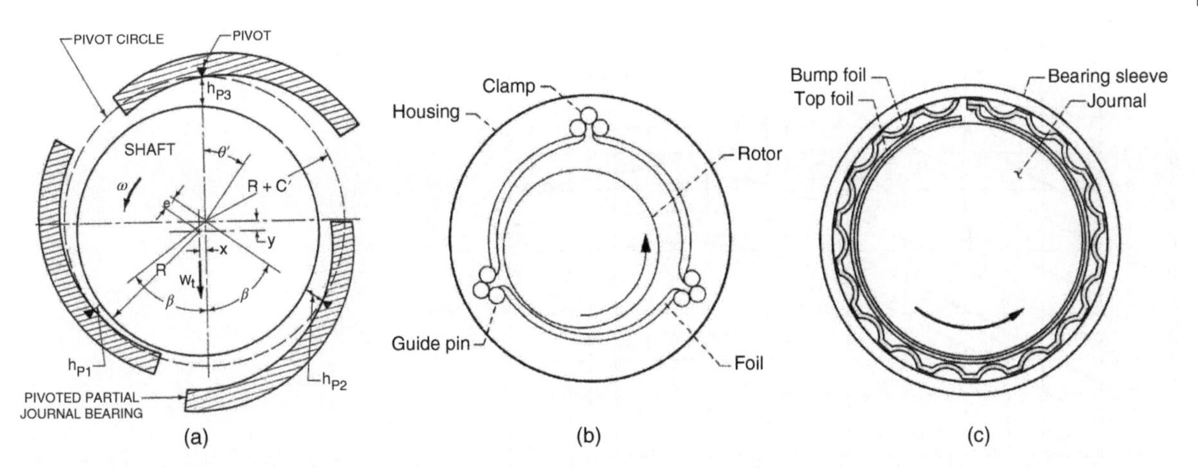

Figure 10.11 Journal bearings with a conformable gap geometry: tilting-pad bearing (a), tension-type foil bearing (b), and corrugated (or bump) foil bearing (c). Source: Waumans T 2009; Gross et al. (1980).

Table 10.1 Physical properties of different gases at $T = 273$ K in order of increasing viscosity.

	dyn. viscosity $\mu \times 10^{-6}$ **[Pa s]**	**density** ρ **[kg m^{-3}]**
hydrogen	8.4	0.090
nitrogen	17.0	1.25
air	17.1	1.29
helium	18.6	0.178
oxygen	18.9	1.43
argon	21.0	1.78
krypton	23.3	3.74

Source: Leijendeckers PHH, Fortuin JB, van Herwijnen F and Schwippert G 2002

The stability chart of Figure 10.12 depicts the threshold speed as a function of the bearing number Λ for a plain aerodynamic geometry. The bearing or compressibility number is a combination of geometrical parameters, gas properties and the rotational speed and thus also accounts for the dynamic viscosity of the lubricant gas. All curves are obtained by Cheng and Pan (1965) and confirm the affine relationship with respect to the gas viscosity up to moderate values of the bearing number (note the logarithmic scale for both abscissa and ordinate).

We note that the viscosity appears only in the bearing number Λ, which is proportional to the viscosity times the rotational speed. From this, one can conclude that, with everything else being the same, decreasing the viscosity, will increase the onset speed of whirl in the same proportion. For instance, using hydrogen instead of air can more than double the whirl onset speed.

The analysis should hold true in its main lines for the case of hybrid journal bearings. However, the principal stiffness in this case is effected by feed flow, which in its turn depends on the density and the viscosity. A more careful analysis is necessary in this respect.

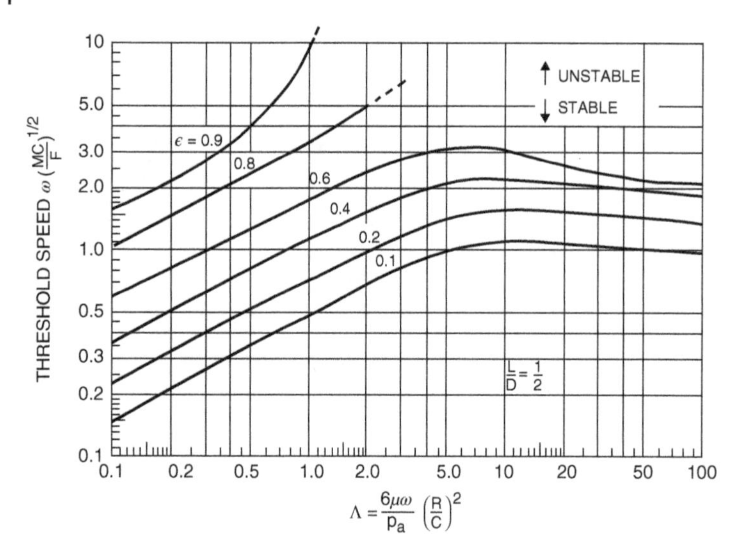

Figure 10.12 Stability chart for a plain aerodynamic bearing with a length-to-diameter ratio $L/D = 1/2$ as obtained by Cheng and Pan (1965). R represents the bearing radius, C the radial clearance value, M the rotor mass and F the bearing load ($F = Mg$, with g the gravitational constant). Source: Cheng HS and Pan CHT 1965 by permission of ASME.

10.5 Optimum Design of Externally Pressurised Journal Bearings for High-Speed Applications

Aerostatic bearings have always been an obvious choice for high-speed gas bearing applications since they suffer less from self-excited instabilities than their aerodynamic counterpart. Nevertheless, most aerostatic gas bearing applications require a sound and well-considered bearing design to prevent self-excited whirling from occurring within the operation range.

To this end, this section therefore derives guidelines for the optimal design of plain aerostatic bearings. Plain refers in this case to bearing surfaces that are free from any form of surface features. The entrance geometry in this study is limited to the inherently restricted type. The criteria of this design study solely regard the stability aspect of the bearing. This reflects in many situations also the maximal rotational speed that can be attained without encountering self-excited whirling. Other bearing requirements such as load-carrying capacity and stiffness are usually less stringent and thereby easily fulfilled in an aerostatic high-speed gas bearing application. More practical implications and considerations such as the frictional losses in the gas film, alignment requirements or manufacturing tolerances have to be taken into account in the total optimisation process.

The previous stability study has explained the influence of the involved dynamic film coefficients. The gained insight helps us to understand the underlying mechanism responsible for self-excited whirling and it allows us to reason on the qualitative effect of design parameter variations. Still, it seems not straightforward to translate this knowledge into practical design rules that lead to optimal bearing design. For this reason, the following part will attempt to formulate guidelines that directly relate design parameters and stability performance.

Literature contains numerous publications on the stability aspect of aerodynamic bearings. But, when regarding aerostatic journal bearings, only little valuable work has been done to investigate the effect of their design parameters on the stability behaviour. The work of Lund (1967) is one of the few examples on this topic available in open literature. This may be attributed to the fact that a suboptimal bearing design suffices to meet the requirements of most aerostatic gas bearing applications. In the 1960s, research initially oriented itself towards the phenomenon of pneumatic hammering and the study of self-excited whirling enjoyed therefore less attention. Although both being self-excited vibrations, their design methodology is considerably different. More recently, (Colombo et al. 2009) have investigated the effect of supply hole configurations on the bearing stability.

Instead of an entrance-flow model, the authors treat the downstream pressure level as a design variable, without loss of generality. The performance comparison is carried out for two different supply configurations, for a given set of other design parameters such as slenderness ratio and radial clearance.

Besides this lack of reference work, inadequacies, inaccuracies or limitations exist in the currently available design data: (i) the pressure profile is obtained by means of an analytical or semi-analytical method which makes the obtained results questionable at high values of the compressibility number; (ii) the entrance flow is modelled by empirical formulas which do not hold true over the entire range of working conditions; and (iii) the results are not presented in a dimensionless way or its representation has only a limited value for practical bearing design. By applying the entrance and film modelling process of Chapter 3 and by formulating a suitable set of dimensionless parameters together with a clear representation of the obtained results, it might be possible to contribute to the design methodology of high-speed aerostatic journal bearings.

The stability study of this section commences by defining a set of dimensionless design parameters. This makes the further derived stability charts more generally applicable. The main conclusion drawn from the stability study are: (i) for a given clearance value and slenderness ratio, there exists a feed-hole diameter that leads to optimal stability; (ii) stability always improves if the radial clearance is reduced or if the slenderness ratio is increased; and (iii) an increase of the bearing supply pressure has a positive effect on the stability performance.

Set of Dimensionless Design Parameters

This set contains both geometrical design parameters and fluid related properties. A correct (but otherwise arbitrarily chosen) combination of all related dimensional values results in a set of dimensionless parameters that fully defines a class of similar aerostatic journal bearings. When operated under similar conditions, these bearings yield identical dimensionless static and dynamic characteristics.

The formulation of a suitable set of dimensionless parameters is based on the normalisation of all involved quantities, or following the procedure of Buckingham Pi-theorem. Chapter 9 derives the normalisation procedure and the formulation of the parameter expressions. Let us restate the expressions introduced there for a journal bearing with radius r, radial clearance value c, length L and with N_f feeding sources of radius r_f. Its design is fully described by the subsequent dimensionless numbers:

$$L/D, \quad R_\mathrm{f}, \quad \Lambda_\mathrm{f}, \tag{10.32}$$

with Λ_f the feed number:

$$\Lambda_\mathrm{f} = 12\frac{\mu r}{p_\mathrm{a} c^2}\sqrt{\frac{2\kappa \Re T_\mathrm{s}}{\kappa - 1}}. \tag{10.33}$$

If all dimensionless working conditions are the same, i.e.:

$$\Lambda, \quad \sigma, \quad \epsilon, \quad p_\mathrm{s}/p_\mathrm{a} \tag{10.34}$$

then two geometrically similar bearings will also share the following dimensionless characteristics:

$$\bar{W}, \quad K_{ij}, \quad C_{ij}, \quad M_\mathrm{thres}. \tag{10.35}$$

The definition of the threshold rotor mass M_thres will be given further on.

Stability Charts

The stability charts hold information on the stability threshold of a rotor-bearing configuration. This is usually presented by plotting the value of the threshold rotor mass M_{thres} for different values of the bearing number Λ. The charts can be interpreted in different ways to obtain the maximum rotor mass for a given rotational speed or vice-versa as will be demonstrated below. It is important to mention that a stability chart is only valid for a certain rotor-bearing configuration and is calculated for one specific bearing geometry characterised by its set of dimensionless design parameters.

The stability study will be performed for the case of a Jeffcott rotor-bearing configuration. This, of course, confines the obtained results to this particular (simple) configuration. Nevertheless, it allows us to draw appropriate conclusions, either quantitatively or qualitatively, on the effect of the involved design parameters by comparing the stability performance of various bearing designs. The presented stability charts are furthermore easily transformed to hold information of equivalent rotor-bearing configurations.

To come to a dimensionless evaluation of the stability, we follow a nearly identical procedure as Lund (1967). The equations of motion for a Jeffcott rotor are given by:

$$m\ddot{x} + c_{xx}\dot{x} + c_{xy}\dot{y} + k_{xx}x + k_{xy}y = 0 \tag{10.36a}$$

$$m\ddot{y} + c_{yy}\dot{y} + c_{yx}\dot{x} + k_{yy}y + k_{yx}x = 0, \tag{10.36b}$$

where $k_{ij}(\omega, \nu)$ and $c_{ij}(\omega, \nu)$.

By normalising the time as $\tau = \nu t$ and by using the previously outlined definitions of K_{ij} and C_{ij}, this set can be converted into:

$$M\Omega_w^2 x'' + C_{xx}x' + C_{xy}y' + K_{xx}x + K_{xy}y = 0 \tag{10.37a}$$

$$M\Omega_w^2 y'' + C_{yy}y' + C_{yx}x' + K_{yy}y + K_{yx}x = 0, \tag{10.37b}$$

in which $K_{ij}(\Lambda, \sigma)$, $C_{ij}(\Lambda, \sigma)$ and the whirl ratio $\Omega_w = \sigma/(2\Lambda)$. The dimensionless mass M is hereby defined as (note that the expression for M contains ω rather than ν):

$$M = \frac{mc\omega^2}{p_a LD}. \tag{10.38}$$

The threshold of instability is found by transforming the system to the Laplace domain and by subsequently setting the real and imaginary part of the system's determinant to zero. This results in the following equations:

$$(-M\Omega_w^2 + K_{ii})(-M\Omega_w^2 + K_{jj}) - C_{ii}C_{jj} - K_{ij}K_{ji} + C_{ij}C_{ji} = 0 \tag{10.39a}$$

$$M = \frac{1}{\Omega_w^2}\frac{C_{ii}K_{jj} + C_{jj}K_{ii} - C_{ji}K_{ij} - C_{ij}K_{ji}}{C_{ii} + C_{jj}}. \tag{10.39b}$$

Since the coefficients K_{ij} and C_{ij} are functions of the whirl ratio Ω_w, a closed form solution for M_{thres} is not possible. Instead the solution is obtained numerically as follows: (a) choose a start value Ω_w^*; (b) calculate the coefficients K_{ij} and C_{ij}[5]; (c) determine herefrom the corresponding value for M through equation (10.39b); (d) substitute the obtained value into equation (10.39a) and evaluate its left-hand side, which then denotes the error. If the error is not zero, adapt Ω_w and proceed until the error becomes zero. The value $M_{\text{thres}}(\Lambda, \Omega_w)$ obtained in this way corresponds to the dimensionless rotor mass at the stability threshold. M_{thres} is also referred to as the *critical* rotor mass. A complete stability chart is obtained by repeating this procedure for different values of Λ. The total solution scheme is outlined in Figure 10.13.

The stability charts are most commonly presented by plotting the threshold value of the rotor mass M for different values of the bearing number Λ (as in the left of Figure 10.14). This way of representation has some

---

5 The required bearing coefficients $K_{ij}(\Lambda, \sigma)$ and $C_{ij}(\Lambda, \sigma)$ are usually obtained by interpolation of formerly tabulated values (look-up table). Experience indicates that in some situations, the bearing data features strong local variations. Because of this, interpolation yields an unsmooth curve and the stability chart is best derived by an in-situ calculation of the required dynamic coefficients.

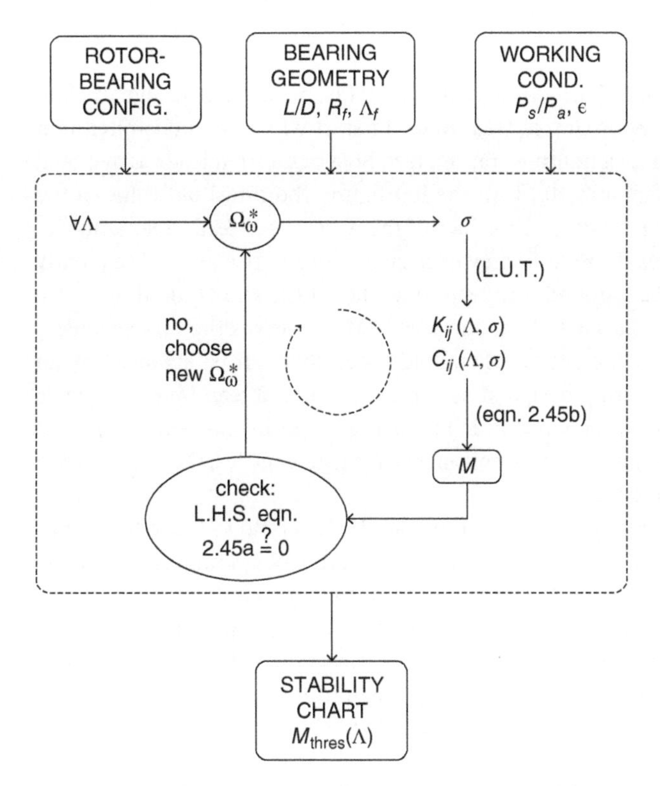

Figure 10.13 Solution scheme for calculating the stability charts. Source: Waumans T 2009.

disadvantages. First, by looking at equation 10.38 it appears that both the abscissa and the ordinate contain the rotational speed ω. This hinders an easy interpretation of the charts. By dividing M through the square of Λ, a more convenient and still dimensionless expression is derived:

$$\frac{72M}{\Lambda^2} = \frac{mp_a}{\mu^2 L}\left(\frac{c}{r}\right)^5. \tag{10.40}$$

By doing so, the stability charts can be interpreted to obtain the threshold value of the rotational speed for a given rotor mass or, vice-versa, to determine the maximum allowable rotor mass for a given operational speed.

A second and final remark on the way of presenting the stability curves, concerns the usage of the bearing number Λ for the abscissa. When comparing stability curves from bearings with a different clearance ratio c/r, the actual rotational speed ω matters rather than the bearing number Λ. The latter value describes the working condition as imposed to the gas film and is not a rotordynamically relevant representation of the system's rotational rate. The rotational speed ω is then again a dimensional number and does not take into account the size of system. Therefore, it would be interesting to look for a normalised value of the circumferential speed $\omega \times r$. This is easily found by dividing Λ through the feed number Λ_f to obtain:

$$\frac{\Lambda}{\Lambda_f} = \frac{1}{2}\omega r\sqrt{\frac{\kappa - 1}{2\kappa \Re T_s}}. \tag{10.41}$$

This dimensionless expression for the circumferential speed is a measure for the achieved rotational speed and is in that sense comparable to the well-known DN-number. The DN-number is defined as the product of the diameter D in mm and the rotational speed in rpm. Although not dimensionless, it is regarded as a size-independent bearing performance indicator.

Design Guidelines

Let us now apply this solution procedure to derive guidelines for the optimal, that is, most stable design of plain aerostatic journal bearings, fed by inherent restrictors or simple feed holes. First, it will be demonstrated that there exists, for a given value of the radial clearance and slenderness ratio, a feed hole radius that leads to maximal stability. This is illustrated by the stability charts of Figure 10.14. In the left figure, the threshold value of the rotor mass M is plotted against the bearing number Λ for different values of the normalised feed-hole radius R_{f}. By gradually increasing this value, the stability curves, from light grey over grey to black, rise and subsequently fall after the optimal value has been reached. The dash-dotted line represents the whirl ratio Ω_w and lies close to 0.5 for high values of Λ. The right figure displays the same data, but uses $72M/\Lambda^2$ versus the dimensionless circumferential speed $\Lambda/\Lambda_{\mathrm{f}}$. The arrows clearly mark the gain in threshold speed that can be achieved by an optimal selection of the feed hole radius. For most bearing design situations, this optimal feed hole also yields a nearly optimal, i.e. maximal film stiffness. Now, since the whirl ratio Ω_w, being the ratio between the natural frequency and onset speed of whirling, approaches 0.5 for most conventional bearing geometries, the optimisation towards stability also amounts to maximal stiffness and vice-versa.

A similar optimisation study has been performed for varying values of the slenderness ratio L/D and feed number Λ_{f}. Table 10.2 holds the optimal value of the feed-hole radius R_{f} for each combination of L/D and Λ_{f}. This table can be used as a design aid to derive the combination of L/D, Λ_{f} and R_{f} that leads to maximal stability.

The stability curves of Figure 10.15 summarise the influence of the feed number Λ_{f} (left) and slenderness ratio L/D (right). An increase of both design parameters can, if combined with a proper selection of the feed-hole radius, lead to a more stable bearing design. When translated to practical bearing design, this means that by reducing the radial clearance and/or by increasing the bearing length it is always possible to increase the threshold speed

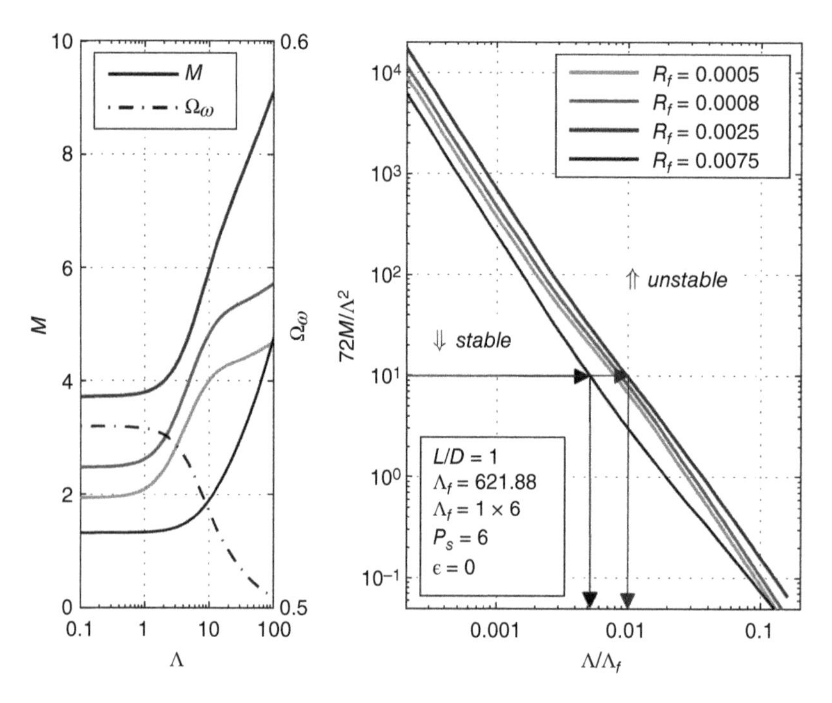

Figure 10.14 Stability chart for a Jeffcott rotor-bearing configuration and plain aerostatic bearing type. All information concerning the bearing design parameters and working conditions is provided in the figure. Both left and right plots contain the same information, but presented in a different way. It is clear from both figures that there exists a value of the normalised feed-hole radius R_{f} for which the stability is optimal. Source: Waumans T 2009.

Table 10.2 Optimal value of the feed-hole radius (tabulated as $R_f \times 10^3$) for different combinations of slenderness ratio L/D and feed number Λ_f. The values have been calculated for an aerostatic journal bearing with a supply pressure ratio $P_s = 6$ and one central row of six feed holes ($N_f = 1 \times 6$). The steady-state eccentricity $\epsilon = 0$, and the bearing number $\Lambda = 10$. (n.a. = optimal value does not exist).

		slenderness ratio L/D			
		.25	.5	1.	2.
feed number Λ_f	9.7	n.a.	n.a.	n.a.	62.5
	38.9	112.5	75.0	22.5	12.5
	155.5	10.0	6.25	4.375	3.125
	621.9	1.50	1.25	1.00	0.625
	1243.8	0.750	0.625	0.500	0.375

Source: Waumans T 2009

of self-excited whirling. These conclusions are in line with the previously defined design rules on aerodynamic bearings.

Instead of using the dimensionless rotor mass M or its more convenient expression $72M/\Lambda^2$ for the ordinate, these stability charts plot the dimensional rotor mass m versus the circumferential speed Λ/Λ_f. The appearance of both the clearance c and bearing length L in the expressions of M and $72M/\Lambda^2$ obliges to use m in this case. Although not leading to a dimensionless presentation of the stability charts, the derived conclusion and design guidelines remain generally applicable.

Effect of Supply Pressure

The onset speed of self-excited whirling can be postponed by increasing the supply pressure ratio. This has been experimentally observed by, among others, Osborne and San Andrés (2003) and Zhu and San Andrés (2004). Experiments conducted at KULeuven PMA and described in (Waumans 2009) also confirm these findings.

The postponement of self-excited whirling is mainly caused by the increased direct stiffness of the gas film. Since the cross-coupled stiffness k_{ij} arises from aerodynamic film effects, it stays largely unaffected by a change in supply pressure. This way, the destabilising cross-coupling ratio $\kappa = k_{ij}/k_{ii}$ reduces. The stiffening effect however also leads to a reduction of the damping ratio $\zeta_n = c_{ii}/(2m\omega_n)$ as experimentally verified by Osborne and San Andrés (2003)[6]. At the same time, Zhu and San Andrés (2004) report a decrease in direct damping c_{ii} in case of an elevated supply pressure which further decreases the damping ratio. Most situations show the former stiffening effect and the hereby reduced cross-coupling ratio, to be predominant, leading to a postponement in the onset speed of self-excited whirling. The stability performance as given by the overall damping ratio $\zeta = \zeta_n - \kappa$ and evaluated at the same rotational speed, can however deteriorate by an increase of supply pressure.

The aforementioned stability analysis has been performed for increasing values of the supply pressure ratio P_s. While the feed number Λ_f is always kept constant, the study is carried out both with and without an optimal selection of the feed-hole radius R_f. The former analysis represents a design situation that evaluates the possible gain in threshold speed due to an increase in supply pressure. The latter analysis is encountered in situations where the bearing has already been optimally designed (and manufactured) for a certain value of the supply pressure ratio. The stability analysis reveals in this case the additional improvement of the stability that can be expected by simply feeding the bearing with a higher supply pressure.

6 Powell (1963) reports an increase in damping ratio at a high supply pressure. These contradicting results can be attributed to the fact that he used a bearing with a rather large clearance ratio ($c/r = 0.0082$) for his experiment. Because of this, the test bearing operated far from the optimal design point and most probably close to the point at which pneumatic hammering sets in.

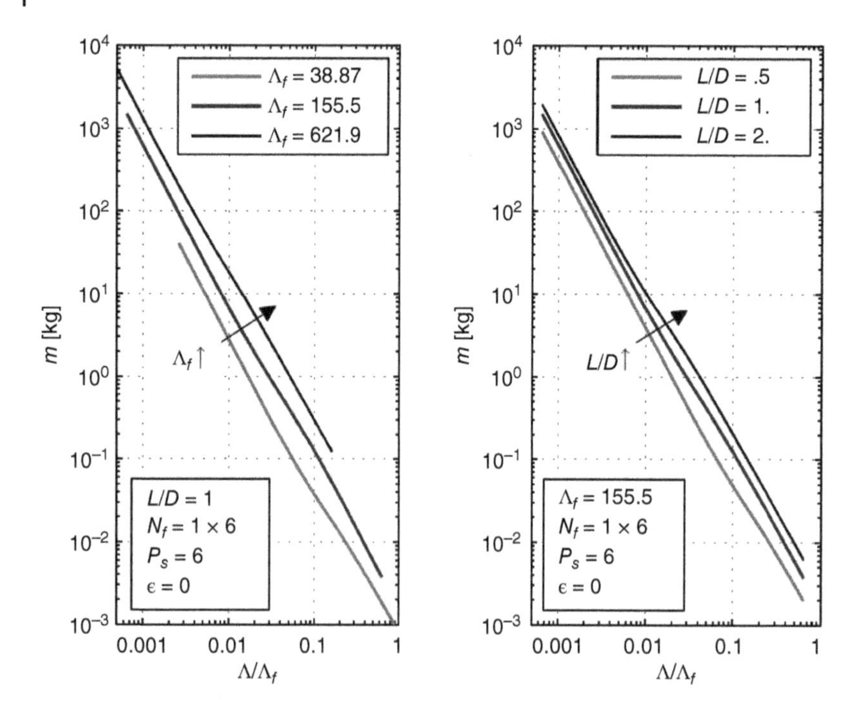

Figure 10.15 Stability charts showing the effect of the feed number Λ_f (left) and slenderness ratio L/D (right) on the threshold rotor mass m_{thres}. The stability curves reveal that: (i) an increase of the feed number (by for example a reduction of the clearance) leads for a given rotor mass to a gain in threshold speed; and (ii) a larger slenderness ratio also increases the threshold speed. For each of the curves, the feed-hole radius R_f has been chosen according to the optimal values of Table 10.2. Source: Waumans T 2009.

The left part of Figure 10.16 shows the stability curves calculated for various values of the supply pressure. The solid curves represent the situation where the feed-hole radius is chosen optimally each time, while all dash-dotted curves are evaluated for a fixed value of the feed-hole radius ($R_f = 0.0075$). It can be noticed that by increasing the supply pressure ratio, the curves shift upwards corresponding to an improvement of the stability performance. To further clarify this trend, the right bar plot evaluates the stability curves for an arbitrary value of the dimensionless rotor mass ($72M/\Lambda^2 = 1$). Since the stability curves run nearly parallel, the trends displayed in this plot will also hold true for other values of the rotor mass. The threshold value $(\Lambda/\Lambda_f)_{thres}$ rises with a higher supply pressure ratio P_s and this gain in stability is always larger when the feed-hole radius R_f is varied accordingly. The gain in threshold speed tends to become smaller for high values of P_s due to compressibility effects in the gas film.

Practical Bearing Design and Implications

When designing a gas bearing for a high-speed application, the rotor-bearing configuration and its overall dimension are predominantly dictated by other than bearing related specifications. This usually leads to a design environment where either the diameter and/or the bearing length is already prescribed. From the original set of design parameters composed of L/D, Λ_f and R_f, only two remain to be determined. In practice, the design therefore amounts for a given radius r to the choice of the radial clearance value c and the feed-hole radius r_f. Through

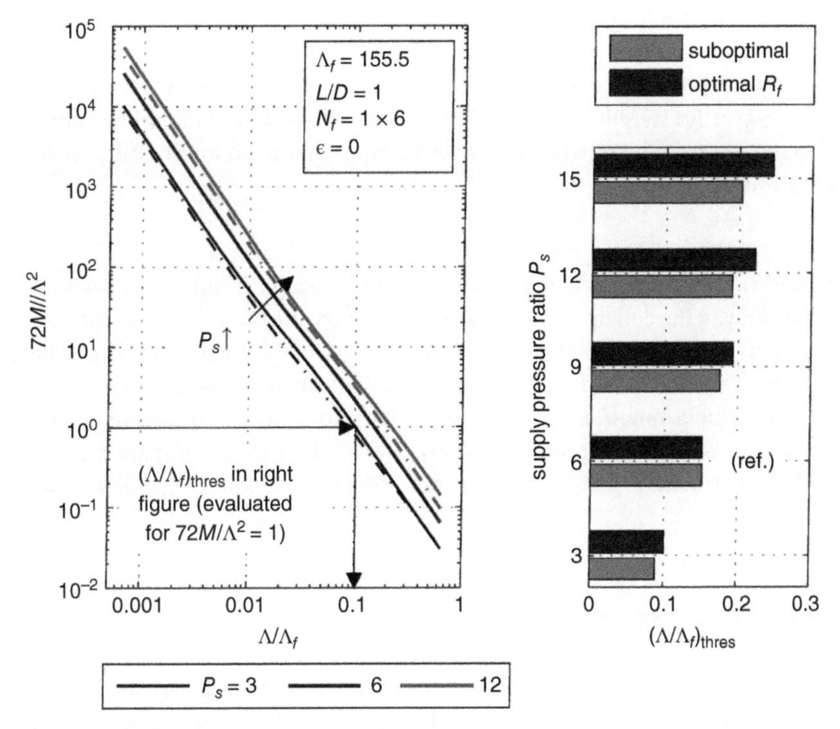

Figure 10.16 Stability curves calculated for various values of the supply pressure P_s. The solid curves represent the situation where the feed-hole radius is chosen optimally each time (for increasing P_s respectively $R_f = 0.0037, 0.0075$ and 0.02) while all dash-dotted curves are evaluated for a fixed value of the feed-hole radius ($R_f = 0.0075$) (left); for both situations the threshold value $(\Lambda/\Lambda_f)_{thres}$ has been evaluated when $72M/\Lambda^2 = 1$ (right). Source: Waumans T 2009.

consideration of the previously discussed relationship

$$R_f = f_1(L/D, \Lambda_f), \tag{10.42}$$

or its dimensional equivalent

$$r_f = f_2(L, c) \tag{10.43}$$

as provided by Table 10.2, the set of design parameters is further reduced to one parameter, namely the radial clearance value c.

It has been amply demonstrated that stability is unconditionally improved by reducing the bearing clearance. The achievement of the required operation speed comes then down to the selection of the radial clearance c that leads to a sufficiently high threshold speed of whirling. The corresponding feed-hole radius r_f is finally obtained through relationship 10.43 or Table 10.2.

This simplified design methodology does not take into account other bearing requirements and manufacturing related limitations. The attainment of the desired rotational speed through a reduction of the radial clearance value has several consequences. It not only increases the viscous frictional losses, which scale inversely proportional to the clearance, but it also complicates the manufacturing process as the tolerances become tighter. Besides this, the ability to deal with alignment errors, differential thermal expansion or centrifugal rotor growth also reduce significantly.

A final remark to conclude this section regards the usage of the stability threshold as a criterion for the design and optimisation of bearing geometries. This threshold point represents a neutral state of stability, i.e. an undamped system. To provide some robustness in the design and to cope with modelling errors and uncertainties, it is thus advisable to choose a somewhat more conservative stability criterion. This can be achieved by maintaining a minimal damping ratio over the entire operation range. The previously derived stability charts can for this purpose be recalculated to display curves of equal damping ratio.

Concluding Remark

The foregoing has served as a general study into the approach and the trends of optimising EP journal bearings for a Jeffcott, i.e. 2D rotor. A more elaborate study of the concrete design cases should take into account other aspects, in particular including 3D systems with their multiple whirling modes in the presence of the gyroscopic effect. The interested reader may consult (Waumans 2009), where a case study of an ultra-high speed EP bearing is worked out. This study, however, seems to yield bearing geometries characterised by an extremely small clearance ratio and large length-to-diameter ratio, which makes a bearing undesirable from the point of view of manufacture, assembly and robustness. Thus, it might be pertinent to examine two more techniques to enhance stability in the subsequent two sections.

10.6 Reducing or Eliminating the Cross-coupling

According to the preceding stability study on plain aerostatic, as well as aerodynamic, journal bearings, stability is always improved by a reduction of the radial clearance value. The attainment of the desired rotational speed should therefore be possible through the application of this basic design rule. The drawbacks involved have already been mentioned several times and they invoke the need to consider alternative design strategies.

Let us restate the expression for the overall damping ratio to determine the various approaches for improving bearing stability:

$$\zeta = 2\zeta_{\mathrm{n}} - \kappa, \tag{10.44}$$

in which

$$\zeta_{\mathrm{n}} = \frac{c_{ii}}{2m\sqrt{\frac{k_{ii}}{m}}} \tag{10.45a}$$

$$\kappa = \frac{k_{ij}}{k_{ii}}. \tag{10.45b}$$

The foregoing discussion on the optimal design of aerostatic bearings mainly aims at maximising the direct stiffness k_{ii} through a proper selection of the feed-hole radius. By observing the above expression, two other strategies remain to follow for improving the stability of a gas bearing (Reinhoudt 1972):

1. Unstable whirling can be postponed by a reduction of the cross-coupling ratio and this technique can, in that sense, be regarded as a *remedy* to the root cause of instability. This first method will try to eliminate or reduce the cross-coupling effects in the gas film by working on the film/entrance geometry itself. In most cases, this works by a counteraction of the Couette-induced film contribution.
2. Self-excited whirling can be "stabilised" by providing sufficient damping. This has effect by *compensating*, somewhere in the system, the destabilising forces induced by the gas film. In this second method, damping will be introduced to the rotor-bearing system. Since the film damping seems insufficient, external damping must be considered. This latter way of stabilisation is discussed in the following section.

In the following, a survey of open literature discusses implementations of the proposed stabilisation strategies for fluid film bearings. This is followed by newly conceived method, proposed by the author Waumans (2009), to demonstrate the feasibility of eliminating cross-coupling completely.

Existing Implementations

The most widespread implementation of the counteraction strategy was first suggested by Tondl (1967). By orienting the supply holes in a tangential way and in opposition to the journal rotation, the Couette induced cross-coupling effect is reduced. Liu and Chen (1999) have proposed a mathematical model to investigate the dynamic characteristics of a gas bearing with holes of tangential supply. Brown and Hart (1986) applied the same concept to a novel damper for turbomachinery applications. The jet damper consists of a number of tangential nozzles discharging against the rotor surface speed. In experiments on a viscous driven gas bearing, the so-called turbo-bearing, Bennett and Marsh (1974) a difference is clearly observed in whirling onset speed when the rotor is rotated along or against the orientation of the inclined supply holes. Angled injection has been applied by Doty et al. (1996) for the stabilisation of a nuclear magnetic resonance (NMR) sample spinner. This large clearance bearing was fed with different gases and recent findings show rotational speeds up to 30 kHz with a 3 mm rotor ($= 5.4 \times 10^6$ DN).

Since inertial forces play an important role in the stabilising strategy, the concept of tangential or angled injection seems more suited for application in liquid film bearings. This was performed by Franchek (1992) who tested a hybrid bearing in water with 45-degree injection against rotation. Later, San Andrés and Childs (1997) analysed the dynamic coefficients of a fluid film hybrid journal bearing with tangential injection at the centre of a recess. As an application example, the advantages of tangential orifice injection were demonstrated on a liquid oxygen hybrid bearing. An improvement of the stability was found without performance degradation on direct stiffness and damping coefficients. The benefits of near-tangential injection at the trailing edge of the recess were evaluated by Laurant and Childs (1999), while the pressure pattern encountered in the recess in case of angled injection has been modelled by Hélène et al. (2005).

Also interesting to mention in this context is the work done by Matsuda et al. (2007). They obtained an optimal fluid film clearance configuration through an optimisation procedure, whereby the clearance is represented by a Fourier series and the performance index is chosen as the sum of the whirl frequency ratios. An optimisation problem is then formulated to find the Fourier coefficients which minimise the performance index. The authors were hereby able to stabilise the hydrodynamic bearing up to elevated rotational speeds and over a wide range of eccentricity ratios. This unconditional stability, as claimed by the authors, can be attributed to a significant decrease in cross-coupled stiffness generated by the stability-optimised clearance geometry.

Another stabilising technique which might partly be classified under this category, comprises the tilting-pad bearings (Gross et al. 1980, ch. 4.15). In the most widespread implementation, the shaft is supported by three individual pivoted pads, which ideally are thus free to adjust their orientation without stiffness, friction or inertia hindrance. The "optimal" tilt angle of the pads follows from the torque balance around their pivot point. This results in an almost symmetrically static pressure distribution around the pivot point, i.e. a very small "attitude" angle of the shaft, which in turn accounts for limited cross-coupling between the dynamic film coefficients. The tilt angle adapts itself with bearing load and speed, leading to an optimal clearance geometry over the entire operating range. For a more elaborate treatment, see Chapter 11.

These implementations of the counteraction strategy all reduce the cross-coupling effect to a certain extent, but do not allow canceling or eliminating the destabilising force completely over a wide range of operating conditions. The bearing geometry introduced below has the unique feature of a controlled and complete counteraction of the cross-coupling, which makes it possible to eliminate the driving force of self-excited whirling up to high values of the rotational speed.

Proposed Bearing Geometry

While the basic idea of the strategy and its underlying motivation are similar, the implementation proposed in this section differs considerably from techniques found in literature. Instead of relying on fluid inertia forces to oppose cross-coupling, as is the case for angled injection strategies, the counteracting effect stems from a sectioned and strongly asymmetric film and entrance geometry. The bearing surface is hereby divided into separate sections by means of axial grooves. Each section contains a feeding recess which is placed asymmetrically and shifted against the shaft rotation. The feed hole itself is again shifted with respect to the feeding pocket. Figure 10.17 provides more insight in the geometrical details of the proposed bearing design.

A word of qualification is in order here: although this implementation may not be useful from practical point of view, it demonstrates clearly, however, that, for each rotation speed, one applies a supply pressure for which the cross-coupling is made null.

The improvement of the stability due to a decrease in cross-coupling, can be attributed to the following geometry specific effects:

- The pressure profile would normally show a crest towards the trailing edge of each section. By shifting the feeding pocket forward (i.e. against the rotational direction), this effect is opposed resulting in a somewhat flattened static pressure distribution. Although stability is actually governed by the dynamic bearing coefficients, the static pressure profile and hereto related attitude angle, can be regarded as a foreshadow of the degree of cross-coupling.
- The asymmetric position of the entrance geometry within each section creates a pressure-induced Poiseuille flow which counteracts the rotation induced Couette flow. This cancellation effect can considerably reduce the cross-coupled stiffness.

Based on the gained experience and throughout the evaluation of numerous bearing designs, it was believed that for each working condition, given by a combination of the rotational speed ω and perturbation frequency ν, there seems to exist a film geometry which yields minimal cross-coupling. This assumption indeed holds true for the proposed bearing configuration as will be demonstrated later, and also implies that this optimal film geometry is different for each working condition. A geometry can, therefore, be found which allows for stable operation at

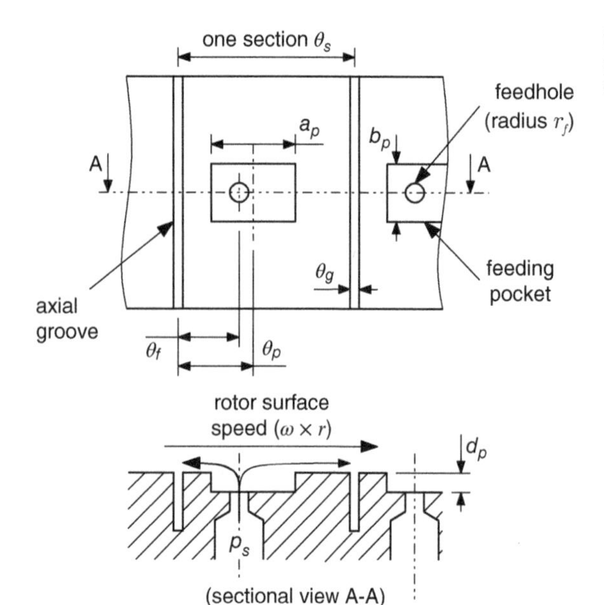

Figure 10.17 Proposed bearing configuration and its notation (partial view of the developed bearing surface). Source: Waumans T 2009.

Table 10.3 Design parameters of the evaluated bearing. The notation is described in Figure 10.17.

Design parameter	***Value***	***Design parameter***	***Value***
Bearing radius r	3 mm	Pocket depth d_p	25 μm
Slenderness ratio L/D	1	Section angle θ_s	60°
Radial clearance c	10 μm	Pocket centre angle θ_p	23.75°
Feed-hole radius r_f	0.15 mm	Feed-hole angle θ_f	20.15°
Pocket length a_p	1.44 mm	Groove angle θ_g	7.64°
Pocket width b_p	0.81 mm		

Source: Waumans T 2009

one single value of the rotational speed, being for example the target operational speed. The attainment of this desired speed remains however impossible since the optimised geometry will feature unstable behaviour in the intermediate speed range.

To overcome this fundamental drawback, two solutions are suggested: (i) the use of an actively or passively controlled variable bearing configuration would allow for a constantly adjustable film geometry (such as the tilting-pad bearing design discussed later); and (ii) by controlling the supply pressure, which is actually part of the bearings working condition, a fixed geometry suffices to keep the bearing stable over a wide range of operational speeds. This latter approach is, of course, only applicable in the case of an aerostatic or hybrid bearing.

The proposed bearing design relies on an adaptation of the bearing supply pressure to minimise the cross-coupled stiffness over a wide range of the rotational speed.

However, by increasing the rotational speed ω, the stable operating range not only shifts to a higher supply pressure, but also provides a limited damping ratio. This immediately leads us to the biggest shortcoming of this concept. The adaptation of the film and entrance geometry may lead to an elimination of the cross-coupled stiffness, but it also compromises the (already limited) damping capability of the gas film. This loss of damping is mainly caused by the axial sectioning grooves and shifted feeding recesses, both essential for the counteraction strategy.

Figure 10.18 clearly illustrates this reduction or elimination of cross-coupled stiffness k_{ij} at the cost of a loss in direct damping c_{ii}. To stress this fact and to provide a reference for comparison, the dynamic coefficients of a plain aerostatic journal bearing are plotted in the same figure. Both the plain (dash–dotted line) and newly proposed bearing (solid line) share the following values: $r = 3$ mm, $L/D = 1$ and $c = 10$ μm, but the feed-hole radius of the plain design is chosen different (meaning optimal with regard to stability), namely $r_f = 0.1$ mm. To keep the comparison fair, the plain bearing is supplied with $P_s = 6$, while the supply pressure ratio increases nearly linearly between $P_s = 2.37$ ($\omega = 0$) and $P_s = 6.36$ ($\omega = 500\,000$ rpm) for the proposed design. The plotted curves in the top part of the figure show a difference in direct damping of up to a factor 4, while the bottom part compares the damping ratio ζ of the forward cylindrical whirling mode. The damping ratio and with that the stability, deteriorates in a linear way for the plain design due to the rising cross-coupling effect. The proposed design features a positive, although limited, damping ratio over the full operating range and would hereby guarantee stable behaviour op to high values of the rotational speed.

Experimental validation confirms the basic working principle of the concept, and rotational speeds up to 438 600 rpm were attained for a miniature rotor of diameter 6 mm (= 2 631 600 DN). Nevertheless, high-speed operation points to the lack of film damping. More details on this method, its implementation and experimental verification can be found in Waumans (2009), Waumans et al. (2011b).

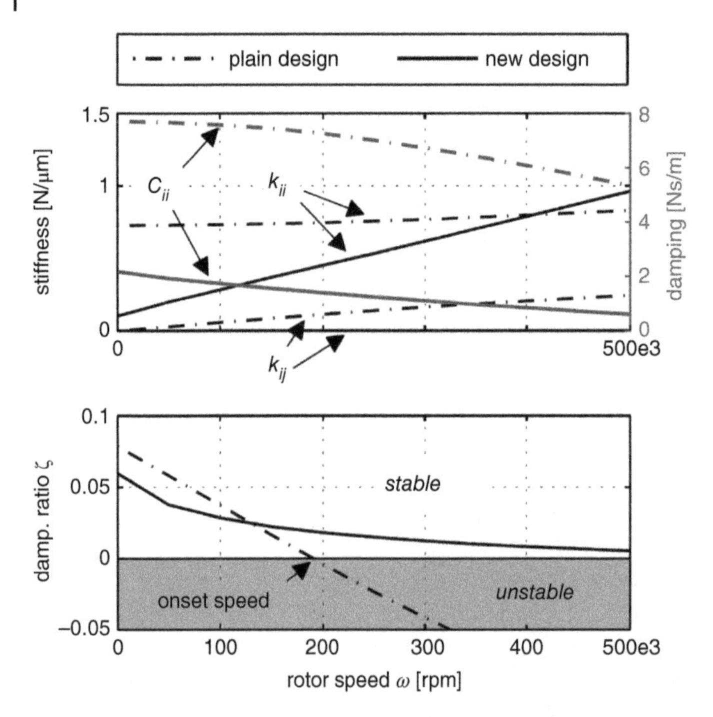

Figure 10.18 Comparison of the dynamic coefficients as a function of the rotational speed ω (top); and damping ratio ζ of the forward cylindrical whirling mode (bottom). The dynamic coefficients are evaluated at the forward cylindrical critical speed of the system. Source: Waumans T 2009.

10.7 Introducing External Damping

The insight gained through the numerical and experimental evaluation of various stabilisation techniques and the lessons learned from the foregoing sections conclusively reveal the necessity of (direct) damping to "solve" the stability problem of high-speed gas bearings. The intrinsic low viscosity of gas films obliges the bearing designer to revert to small and therefore impractical values of the nominal bearing clearance. The compressible nature of gases and its consequence in terms of a decline in damping versus frequency, further complicates the design situation. Since the gas film damping itself is insufficient to prevent self-excited whirling from occurring within the desired operational speed range, the potential of "external' damping has to be considered to stabilise the rotor-bearing system. This approach therefore works through a compensation, or suppression of the destabilising forces rather than by taking the effort to prevent their occurrence them. Although, it will appear that external damping has most effect if applied to a fairly intrinsically stable system, which indicates that a combined approach would most probably be the best option. The external damping may come forth from various sources within the system. Its practical implementation and realisation exists in many forms and configurations.

The section commences with a literature survey on the past treatment of this stabilisation technique and overviews at the same time the different approaches to realise its practical implementation. Thereafter, a linear study is proposed to establish guidelines for the optimal design of the external support parameters. The rotor-film-bush model is introduced and the resulting optimisation problem, to determine the support parameters, is indicated.

Literature Overview

It was reported by several researchers that self-excited whirling can be postponed or even run through, without damage occurring to the bearing surfaces, by mounting the bearings in soft rubber O-ring. This observation of running through self-excited whirling, is however inappropriate and stems from the faulty comparison to a typical resonance phenomenon. Nevertheless, this compliant support enables the attainment of the desired operational speed through the introduction of damping.

Montgomerey and Sterry (1956) were probably the first to appreciate the ability of rubber to suppress self-excited whirl when they developed an air bearing turbine for rotating the mirror of a high-speed camera. Both Grassam and Powell (1964, p. 177) and Powell and Tempest (1968) report the achievement of a dental drill running at respectively 650 000 rpm and 720 000 rpm through the usage of hybrid bearings supported by O-rings. Although it would seem an evident choice to provide damping by means of a compliant rubber support, the usage of rubber O-rings is also a very convenient choice for use with externally pressurised bearings as they provide a means of sealing the annular air supply reservoir surrounding the bearing.

The first valuable effort made to analyse the effect of a flexible, damped support, was done by Lund (1965). Although his contribution to the matter gives proof of a thorough and complete understanding of the stability problem, only qualitative design guidelines and conclusion are presented. The analysis method introduced by him can however be regarded as a milestone in the research field of fluid film bearings. Tondl (1971) investigated the effect of foundation mass tuning, but with very great simplifications indeed of the bearings. Mori and Mori (1969, 1971) examined the effect of a support by means of an aerostatic sub-bearing in a greatly simplified way.

The beneficial effect of a rubber O-ring support was studied in great detail by Kazimierski and Jarzecki (1979). First, they performed an experimental investigation of the dynamic properties of O-rings made of different kinds of rubber. In the test rig employed for this purpose, the O-rings are mounted in an identical way as in the final bearing system. The influence of the supply pressure on the dynamic properties is hereby taken into account. Hereafter, they adopted a time-based solution of the Reynolds equation, although combined with a questionable entrance flow model. The experimental validation on the flexibly supported bearing setup, yielded remarkably good agreement with the predicted stability thresholds. They conclude by describing a diagram that displays the improvement of stability as a function of both the support stiffness and damping. This diagram is however only applicable to the investigated setup bearings and does not seem to contain information on the optimal value of external damping.

The same modelling strategy was adopted by Czolczyński and Marynowski (1996) and Czolczyński et al. (1996) to study in detail the effect of the support stiffness and damping value. They describe the existence of always-stable loops, but since the support parameters were not related to the gas film properties, the value of the extensive amount of diagrams remains limited to an ad-hoc implementation bounded by their test conditions. A general applicable and clearly presented conclusion on the effect of all three support parameters, i.e. the support mass, stiffness and damping, was however not included in their paper. The authors also propose several alternative implementations of a flexibly damped support, such as: air rings (i.e. an aerostatic sub-bearing) or air rings with rubber seals or baffles to further increase the damping, as described in (Czolczsynki 1999, ch. 6).

More recent is the work of Tomioka et al. (2007) who compare the maximal attainable speed of a rigidly supported and an elastically damped herringbone bearing. With the latter configuration, they were able to reach a rotational speed of 509 000 rpm for a simple shaft of diameter 6 mm (3.05×10^6 DN), and without observing any subsynchronous whirling. Belforte et al. (2008) were able to operate a high-speed rotor of diameter 37 mm up to 75 000 rpm on elastically supported aerostatic bearings (2.78×10^6 DN). Their mainly experimental work compares the rotational response using rubber O-rings of three different materials and describes the influence of the supply pressure on the stability threshold, presenting Fourier spectra of the signals for rotor displacement.

Rimpel and Kim (2009) analyse the rotordynamic performance of a flexure pivot tilting pad bearing supported by an elastic bump foil structure. The latter support structure introduces damping due to Coulomb friction. The same idea is applied in (Kim and Rimpel 2009) where the flexure pivots are fitted with elastomeric dampers. Ertas (2009) makes use of an integral wire mesh damper to stabilise a hybrid tilting pad bearing and demonstrates stable operation up to 40 000 rpm for a 70 mm rotor (2.8×10^6 DN).

To conclude the literature survey, it would be interesting to refer to the stability properties of foil bearings. This bearing type with a fully compliant surface, is claimed to possess a virtually unlimited stability. As discussed earlier in this chapter, the claimed stability properties can be attributed to, among others, the introduction of external damping which arises from dry friction in the support structure.

Rotor-film-bush Model

Let us first look at the different implementations that exist for introducing external damping to the rotor-bearing system. The most widespread implementation is shown in Figure 10.19(a) and can best be described as a flexible supported floating bush/bearing. The support stiffness k_e and support damping c_e work, in actual fact, *in series* with the gas film through the mass of the floating bush m_b. Although fairly easy to realise in practice, a few drawbacks have to be considered: (i) it will be demonstrated that, in order to improve the stability considerably, the external support stiffness k_e must be smaller than the film stiffness by at least a factor two or three. This will then reduce the total stiffness as defined by the chain from rotor to foundation; and (ii) if the alignment of the (two) journal bearings relies on a form-fitted construction, the incorporation of a compliant element can compromise the overall alignment. This latter problem can be relieved by reverting to a single-bush implementation that comprises two journal bearing surfaces.

The former remark on the degradation of the system's total stiffness is of particular importance for high-speed machine tool applications. To prevent this loss of stiffness, an implementation as in Figure 10.19(b) is suggested. The damper acts in this case *in parallel* with the existing rigidly supported bearing. This second implementation comes at the cost of a somewhat more complex practical realisation.

Firstly, it would be interesting to look at the different ways to characterise the support parameters k_e and c_e. This dynamic characterisation depends of course on the chosen implementation and it is therefore difficult to propose a single dynamic support model. Still, it is useful to outline the different ways of characterisation. The by far simplest model, would be to consider a single and constant value for both the support stiffness k_e and damping c_e which is linear and the same for all directions of perturbation:

$$k_e = \text{const.} \tag{10.46a}$$

$$c_e = \text{const.} \tag{10.46b}$$

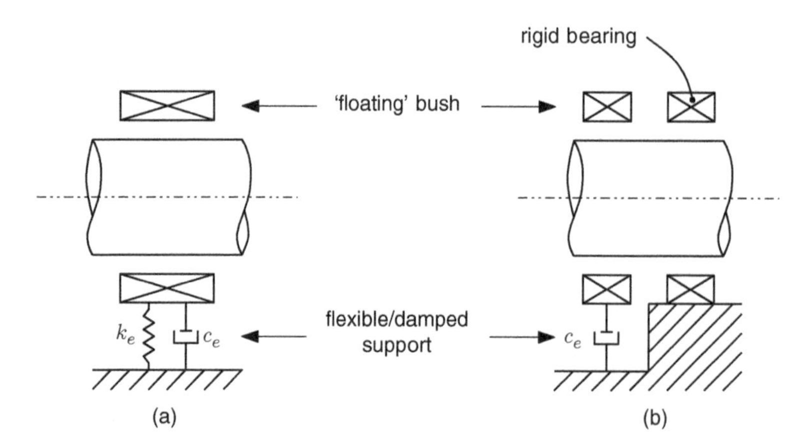

Figure 10.19 Schematic representation of two implementations for introducing external damping to a rotor-bearing system. Source: Waumans T 2009.

This representation does not allow however for an adequate characterisation of the most popular practical realisation of a flexible support, namely by means of rubber O-rings. In this case, the actual value of the external stiffness and damping would be in the ideal situation uniform, but strongly dependent on the perturbation frequency ν.

$$k_e = f(\nu) \tag{10.47a}$$

$$c_e = f(\nu) \tag{10.47b}$$

An even more accurate model would take into account the non-linearity and temperature effects of the rubber O-rings, see e.g. (Al-Bender et al. 2017).

The complexity of the dynamic support model is only bounded by the fact that it may not exceed the complexity of the dynamic gas film model, since that would require a whole new coupled solution procedure. The value of a complex support model is however always limited by the accuracy of the input parameters that describe its behaviour. And, as is certainly the case for rubber, the accurate determination of these parameters seems far from easy (Green and English 1992).

The previous study on the rotordynamic behaviour of a Jeffcott rotor-bearing system was based on the assumption of a rigidly supported bearing. Figure 10.20 extends this rotordynamic model to a Jeffcott rotor mounted on a flexibly supported bearing. The dynamic gas film model remains unchanged and is coupled in series to a dynamic support model through the mass of the floating bush m_b. For the following stability study we assume a constant support stiffness coefficient k_e and damping coefficient c_e, which acts in a uniform way for all directions of perturbation. The position of the floating ring is denoted by (x_b, y_b).

By assuming the gas film to be symmetric, which is the case for steady-state operation at a small eccentricity, the following relations exist: $k_{ii} = k_{xx} = k_{yy}$ and $k_{ij} = k_{xy} = -k_{yx}$ etc., and the set of equations of motion for this 4-dof system may be easily composed as:

$$m\ddot{x} + c_{ii}(\dot{x} - \dot{x}_b) + c_{ij}(\dot{y} - \dot{y}_b) + k_{ii}(x - x_b) + k_{ij}(y - y_b) = 0 \tag{10.48a}$$

$$m\ddot{y} + c_{ii}(\dot{y} - \dot{y}_b) - c_{ij}(\dot{x} - \dot{x}_b) + k_{ii}(y - y_b) - k_{ij}(x - x_b) = 0 \tag{10.48b}$$

$$\begin{aligned} &m_b\ddot{x}_b + c_e\dot{x}_b + k_e x_b \\ &\quad -c_{ii}(\dot{x} - \dot{x}_b) - c_{ij}(\dot{y} - \dot{y}_b) - k_{ii}(x - x_b) - k_{ij}(y - y_b) = 0 \end{aligned} \tag{10.48c}$$

$$\begin{aligned} &m_b\ddot{y}_b + c_e\dot{y}_b + k_e y_b \\ &\quad -c_{ii}(\dot{y} - \dot{y}_b) + c_{ij}(\dot{x} - \dot{x}_b) - k_{ii}(y - y_b) + k_{ij}(x - x_b) = 0. \end{aligned} \tag{10.48d}$$

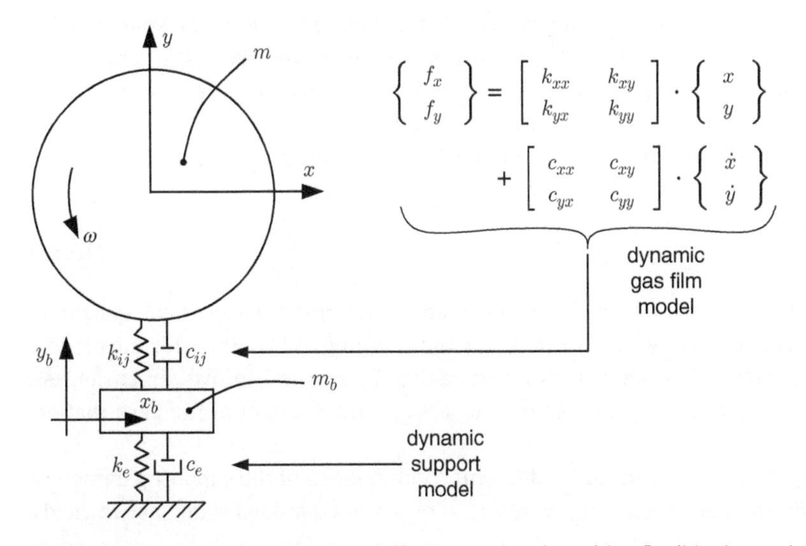

Figure 10.20 Dynamic model for a Jeffcott rotor-bearing with a flexible damped support. Source: Waumans T 2009.

This set of coupled equations can be solved by performing an eigenvalue analysis at various values of the rotational speed ω. This amounts to an iterative solution procedure since the dynamic film and/or support parameters depend on the perturbation frequency ν. An example of such an eigenvalue analysis and subsequent discussion of the thereby obtained results, will be given by means of a case study further on.

Optimal Design of the Support Parameters

The foregoing literature overview unveils the existence of numerous studies, all dealing with the influence of support compliance on the stability behaviour of a gas bearing system. Notwithstanding this vast amount of work, it would still be useful to perform a stability study ourselves, one that could further contribute in the following way:

- By combining the support model with an advanced dynamic gas film model, new and more accurate insights can be gained in the matter.
- The improvement of the stability performance is based on a proper selection of the support parameters. By relating the optimal values of the involved parameters to the rotor and gas film properties, a set of dimensionless guidelines will be formulated that is widely applicable.
- As has been mentioned several times, the usage of the damping ratio rather than the stability threshold, yields not only more information on the stability behaviour, but it might also lead to a more robust bearing design process.

Let us now try to formulate design guidelines for the optimal choice of m_b, k_e and c_e given the dynamic gas film properties and rotor mass. To minimise the number of variables and to arrive to conclusions which are widely applicable, a set of dimensionless parameters is searched for. The gas film and rotor properties are, with the neglect of the cross-coupled damping coefficient c_{ij}, best expressed as:

$$\omega_\mathrm{n} = \sqrt{\frac{k_{ii}}{m}} \tag{10.49a}$$

$$\zeta_\mathrm{n} = \frac{c_{ii}}{2m\omega_\mathrm{n}} \tag{10.49b}$$

$$\kappa = \frac{k_{ij}}{k_{ii}}, \tag{10.49c}$$

with respectively ω_n the natural frequency of the rotor-bearing system, ζ_n the damping ratio of the system and κ the ratio of cross-to-direct stiffness. The latter is a measure for the intrinsic quality of the bearing in terms of its tendency towards self-excited whirling. It is important to mention that these three parameters reflect the properties of the rotor-gas film only, without consideration of the external support structure.

The bush and support properties are, for this purpose, best related to the rotor and gas film properties in the following way:

$$M_b = \frac{m_b}{m}; \quad K_\mathrm{e} = \frac{k_\mathrm{e}}{k_{ii}}; \quad C_\mathrm{e} = \frac{c_\mathrm{e}}{c_{ii}}. \tag{10.50}$$

This amounts in total to six dimensionless parameters which fully characterise the rotor-bearing-bush system. It appears that by expressing the parameters in this way and by evaluating the damping ratio of the whirling modes, the natural frequency ω_n becomes redundant. The set of parameters is thereby reduced to five dimensionless parameters. Finally, suppose for the sake of simplicity that all parameters are independent of the perturbation frequency ν.

Under these conditions, the damping ratio ζ is evaluated for different combinations of the support parameters (K_e, C_e), while varying the other remaining parameters: M_b, κ and ζ_n. The set of equations stated above yields

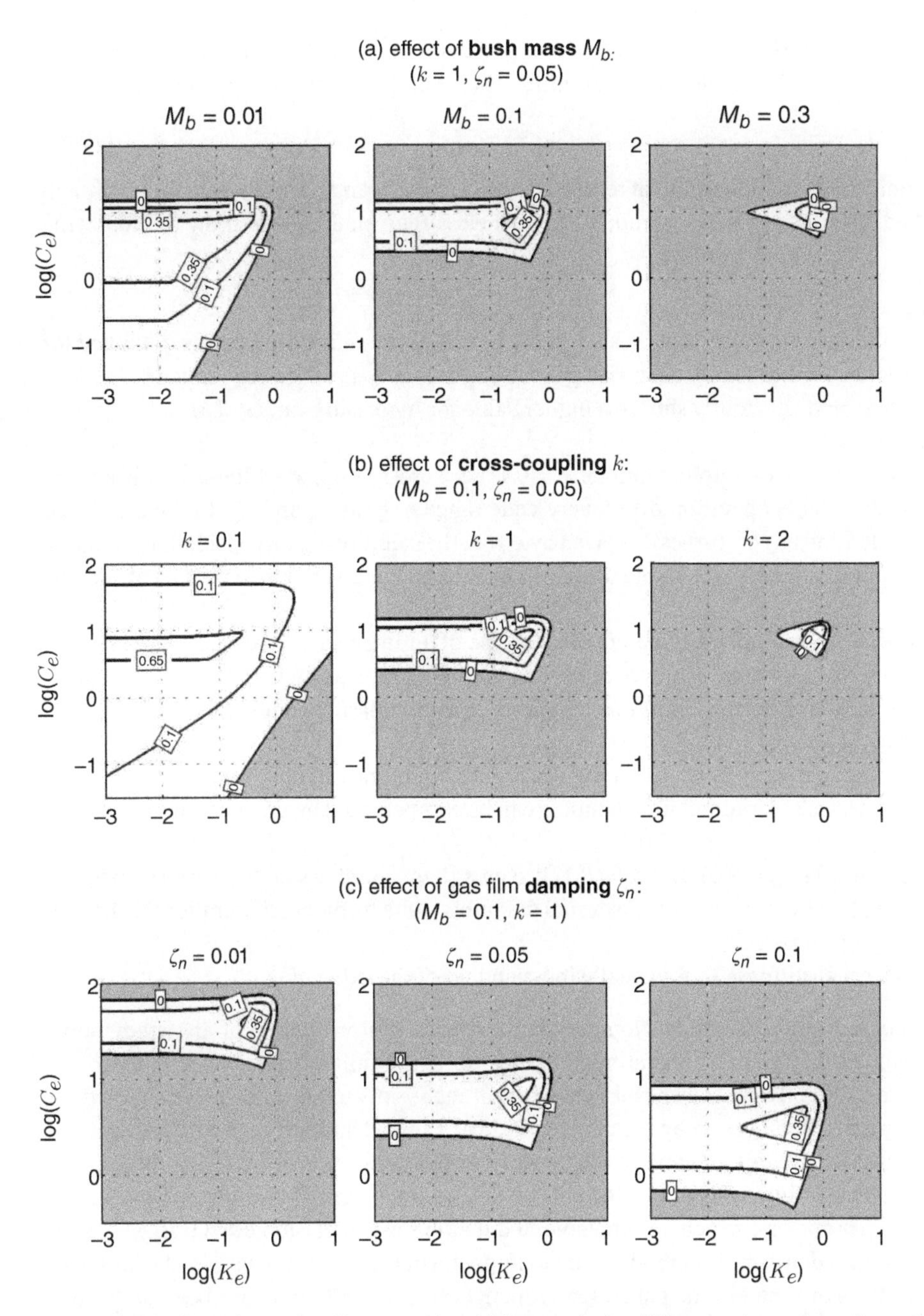

Figure 10.21 Damping ratio of least stable whirling mode for different values of the dimensionless support stiffness K_e and damping C_e. The grey-shaded areas reflect unstable whirling behaviour. (Source: Waumans T 2009).

four solutions, each representing an eigenmode or whirling mode of the system. The stability of the system is determined by the least stable, i.e. by our convention the mode with the lowest damping ratio. The hereby obtained result is summarised in Figure 10.21 by means of iso-damping lines (lines of equal damping ratio). This way of presenting the data not only provides information on the stability region itself, which is contained with the lines of $\zeta = 0$, but also reveals the degree of stability within this region. The top, middle and bottom row of figures clarify respectively the effect of the bush mass M_b, cross-coupling ratio κ and damping ratio ζ_n on the optimal choice of (K_e, C_e). The set of figures should, as a whole, allow for the optimal choice of the support parameters:

$$(K_e, C_e) = f(M_b, \kappa, \zeta_n) \tag{10.51}$$

The results are evaluated for only three discrete values of respectively M_b, κ and ζ_n and have been obtained through the assumption of a simplified gas film and support model. Nevertheless they enable us to draw the following conclusions:

- Effect of the *bush mass* M_b:
  - By increasing the mass of the floating bush, as indicated in the figure by the stability diagrams evaluated for $M_b = 0.01$, $M_b = 0.1$ and $M_b = 0.3$, the region comprising stable behaviour is significantly reduced.
  - The optimal value for both K_e and C_e slightly shift to a higher value for increasing values of M_b.
- Effect of the *cross-coupling* ratio κ:
  - A gas film characterised by a low cross-coupling ratio of $\kappa = 0.1$, i.e. a cross-coupled stiffness of only a tenth of the direct stiffness, features stable operation over a very wide range of both K_e and C_e. In the case of an equal amount of cross-coupled and direct stiffness, as given by $\kappa = 1$, the stability region considerably reduces, but still remains of practical value. For $\kappa = 2$, stable operation only exists for a very narrow range of support parameters.
  - An optimal choice of the external support parameters can, in case of an intrinsic stable bearing, result in a superior stability performance as shown by values of the overall damping ratio as high as $\zeta = 0.65$ and more.
  - The optimal value for both K_e and C_e slightly shift to a higher value for increasing values of κ.
- Effect of the gas film *damping* ζ_n:
  - The stability diagrams obtained for various values of the gas film damping ratio, going from $\zeta_n = 0.01$, $\zeta_n = 0.05$ up to $\zeta_n = 0.1$, prove that a lack of internal film damping can be compensated by providing more external damping.
  - When comparing the diagrams for $\zeta_n = 0.01$ and $\zeta_n = 0.1$, it seems that a similar stability performance can be reached by providing a tenfold higher amount of external damping to the former configuration. Although, with a somewhat reduced range of stability.
  - The optimal value of the external stiffness K_e is virtually independent of the value of ζ_n.

Waumans et al. (2011a) reports on the design, development and experimental validation of an aerodynamic journal bearing with a flexible, damped support operating at speeds up to 1.2 million rpm (= 7.2 million DN, journal diameter 6 mm). In terms of the DN-number, this achievement represents to our knowledge a record for an air bearing of the self-acting type. Stabilisation by means of a flexible, damped support therefore proves to be a promising solution to the dynamic stability problem of high-speed gas bearings. In order to select the support parameters in an optimal way, the stability analysis outlined above has been performed leading to the selection of an appropriate set of support parameters, which are implemented making use of elastomeric O-rings in combination with a tunable squeeze-film damper. A method for the manufacturing of miniature bearing bushes with a wave-shaped film height profile is outlined in the paper. Experiments up to 683 280 rpm have been performed with an air-driven turbine and up to 1203 000 rpm with a helium-driven turbine. Deceleration experiments have also been conducted in order to obtain an estimation of the frictional losses of the system.

10.8 Summary

The dynamic whirl behaviour of a gas lubricated rotor-bearing system has been discussed in this chapter. Particular interest is thereby given to the study of the self-excited whirl phenomenon and its remedies.

First, the various kinds of rotor whirl have been classified into synchronous whirl due to imbalance and self-excited whirl which sets in at high values of the rotational speed. Hereafter, the cross-coupled stiffness is identified as the root cause or the driving force of self-excited whirling. A simple but effective stability criterion is formulated that relates the maximum allowable cross-coupled stiffness to the other dynamic film parameters. Based on this stability criterion, some stabilising techniques from literature are assessed.

Three methods to overcome (or at least postpone the occurrence of) self-excited whirl have been proposed. In a first method, the design rules for achieving optimal stability with plain aerostatic bearings have been outlined in a dimensionless way. The second stabilising technique tries to eliminate the driving force of self-excited whirl, i.e. by reducing the cross-coupled stiffness. This approach, in addition to being limited in effect, is not easy to implement in practice. Finally, a third and last strategy works by compensating for the destabilising effects from without the gas film; that is by tuning the damped bearing support dynamic characteristics. This method is most popularly implemented in the form of O-ring supported bearing bush.

References

Al-Bender, F., Colombo, F., Reynaerts, D., et al. (2017). Dynamic characterization of rubber o-rings: Squeeze and size effects. *Advances in Tribology*.

Beams, H.W., (1937). The air turbine ultracentrifuge, together with some results upon ultracentrifuging the eggs of fucus serratus. *Journal of the Marine Biological Association of the United Kingdom* 21(2): 571–588.

Belforte, G., Colombo, F., Raparelli, T. and Viktorov, V., (2008). High-speed rotor with air bearings mounted on flexible supports: test bench and experimental results. *Transactions of the ASME - Journal of Tribology* 130(2): 021103 (7 pages).

Bennett, J. and Marsh, H., (1974). The steady state and dynamic behaviour of the turbo-bearing (paper C4) *Proceedings of the 6th International Gas Bearing Symposium*, Southampton, England.

Bootsma, J., (1975). *Liquid-lubricated spiral-groove bearings* PhD thesis Technische Hogeschool Delft.

Brown, R.D. and Hart, J.A., (1986). A novel form of damper for turbomachinery *Proceedings of the Workshop on Rotordynamic Instability Problems in High Performance Turbomachinery (NASA CP 2443)*, 325–348 Texas A&M University.

Castelli, V. and Elrod, H.G., (1965). Solution of the stability problem of 360 degree self-acting gas-lubricated bearings. *Transactions of the ASME - Journal of Basic Engineering* 87(2): 199–212.

Cheng, H.S. and Pan, C.H.T., (1965). Stability analysis of gas-lubricated, self-acting, plain, cylindrical, journal bearings of finite length, using Galerkin's method. *Transactions of the ASME - Journal of Basic Engineering* 87(1): 185–192.

Cheng, H.S. and Trumpler, P.R., (1963). Stability of the high-speed journal bearing under steady load - 2: the compressible film. *Transactions of the ASME - Journal of Engineering for Industry* 85, 274–280.

Childs, D., (1993). *Turbomachinery Rotordynamics - Phenomena, Modeling and Analysis.* John Wiley & Sons, New York.

Colombo, F., Raparelli, T. and Viktorov, V., (2009). Externally pressurized gas bearings: a comparison between two supply holes configurations. *Tribology International* 42(2): 303–310.

Cunningham, R.E., Fleming, D.P. and Anderson, W.J., (1969). Experimental stability studies of the herringbone-grooved gas-lubricated journal bearing. *Transactions of the ASME - Journal of Lubrication Technology* 91, 52–59.

Czolczsynki, K., (1999). *Rotordynamics of gas-lubricated journal bearing systems.* Springer-Verlag, New-York.

Czolczyński, K. and Marynowski, K., (1996). Stability of symmetrical rotor supported in flexibly mounted, self-acting gas journal bearings. *Wear* 194(1): 190–197.

Czolczyński, K., Kapitaniak, T. and Marynowski, K., (1996). Stability of rotors supported in gas bearings with bushes mounted in air rings. *Wear* 199(1): 100–112.

Den Hartog, J.P., (1985). *Mechanical Vibrations.* Dover Publications.

Dimofte, F., (1995a). Wave journal bearing with compressible lubricant - Part I: the wave bearing concept and a comparison to the plain circular bearing. *STLE Tribology Transactions* 38(1): 153–160.

Dimofte, F., (1995b). Wave journal bearing with compressible lubricant - Part II: a comparison of the wave bearing with a groove bearing and a lobe bearing. *STLE Tribology Transactions* 38(2): 364–372.

Dmochowski, W.M., Dadouche, A. and Conlon, M.J., (2009). Pivot friction effects on the dynamic properties of tilting pad *journal bearings (C1-122) Proceedings of the World Tribology Congress*, Kyoto, Japan.

Doty, F.D., Hacker, L.G. and Spitzmesser, J.B., (1996). Supersonic sample spinner. *US Patent*.

Dupont, R., (2005). *Isotrop und fliehkraftinvariant gestaltetes, gasgeschmiertes Spiralrillenlager in Kegelbauform für höchste Drehfrequenzen* PhD thesis Fachbereich Maschinenbau und Verfahrenstechnik der Technischen Universität Kaiserslautern.

Ertas, B.H., (2009). Compliant hybrid journal bearings using integral wire mesh dampers. *Journal of Engineering for Gas Turbines and Power* 131(2): 022503 (11 pages).

Franchek, N.M, (1992). *Theory versus experimental results and comparisons for five orifice-compensated hybrid bearing configurations* Master's thesis Texas A&M University - Mechanical Engineering Department College Station, TX.

Frêne, J., Nicolas, D., Degueurce, B., et al. (1997). *Hydrodynamic Lubrication - Bearings and Thrust Bearings*. Elsevier, Amsterdam.

Fuller, D.D., (1969). A review of the state-of-the-art for the design of self-acting gas-lubricated bearings. *Journal of Lubrication Technology* 91(1): 1–16.

Grassam, N.S. and Powell, J.W., (1964). *Gas Lubricated Bearings*. Butterworths, London.

Green, J. and English, C., (1992). Analysis of elastomeric O-rings seals in compression using the finite element method. *STLE Tribology Transactions*. 83–88.

Gross, W.A., (1962). Investigation of whirl in externally pressurized air-lubricated journal bearings. *Transactions of the ASME - Journal of Basic Engineering* 84, 132–138.

Gross, W.A., Matsch, L.A., Castelli, V., et al. (1980). *Fluid Film Lubrication*. John Wiley & Sons, New York.

Hamrock, B.J., Schmid, S.R. and Jacobson, B.O., (2004). *Fundamentals of fluid film lubrication*. Marcel Dekker, New York.

Hélène, M., Arghir, M. and Frêne, J., (2005). Combined Navier–Stokes and bulk-flow analysis of hybrid bearings: radial and angled injection. *Transactions of the ASME - Journal of Tribology* 127(3): 557–567.

Hummel, C., (1926). Kristische Drehzahlen als Folge der Nachgiebigkeit des Schmiermittels im Lager. *VDI-Forschungsheft*.

Kaneko, R., Mitsuya, Y. and Oguchi, S., (1974). High speed magnetic storage drums with grooved hydrodynamic gas barings (paper C2) *Proceedings of the 6th International Gas Bearing Symposium*, Southampton, England.

Kazimierski, Z. and Jarzecki, K., (1979). Stability threshold of flexibly supported hybrid gas journal bearings. *Transactions of the ASME - Journal of Lubrication Technology* 101, 451–457.

Kim, D. and Rimpel, A., (2009). Experimental and analytical studies on flexure pivot tilting pad gas bearings with dampers applied to radially compliant pads (GT2009-59285) *Proceedings of the ASME Turbo Expo (2009: Power for Land, Sea and Air*, Orlando, Florida, USA.

Kozanecki, Z., Dessornes, O., Kechana, F., et al. (2006). Tilting pad bearings for microturbine *Proceedings of the 6th International Workshop on Micro and Nanotechnology for Power Generation and Energy Conversion Applications*, 49–52, Berkeley, USA.

Laurant, F. and Childs, D.W., (1999). Rotordynamic evaluation of a near-tangential-injection hybrid bearing. *Transactions of the ASME - Journal of Tribology* 121(4): 886–891.

Leijendeckers, P.H.H., Fortuin, J.B., van Herwijnen, F. and Schwippert, G., (2002). *Poly-Technisch Zakboek*. Elsevier, Arnhem.

Liu, L., (2005). *Theory for hydrostatic gas journal bearings for micro-electro-mechanical systems* PhD thesis Massachusetts Institute of Technology - Department of Mechanical Engineering.

Liu, L.Q. and Chen, C.Z., (1999). Mathematical model for gas bearing with holes of tangential supply. *Transactions of the ASME - Journal of Tribology* 121(2): 301–305.

Lund, J.W., (1965). The stability of an elastic rotor in journal bearings with flexible, damped supports. *Transactions of the ASME - Journal of Applied Mechanics* 87(4): 911–920.

Lund, J.W., (1967). A theoretical analysis of whirl instability and pneumatic hammer for a rigid rotor in pressurized gas journal bearings. *Transactions of the ASME - Journal of Lubrication Technology* 89(2): 154–166.

Lund, J.W., (1968). Calculation of stiffness and damping properties of gas bearings. *Transactions of the ASME - Journal of Lubrication Technology* 90(4): 793–803.

Lund, J.W., (1987). Review of the concept of dynamic coefficients for fluid film journal bearings. *Transactions of the ASME - Journal of Tribology* 109(1): 37–41.

Marsh, H., (1965). *The stability of aerodynamic gas bearings.* Institution of Mechanical Engineers, London.

Marsh, H., (1980). Stability and rotordynamics for gas lubricated bearings. *Tribology International* 13(5): 219–221.

Matsuda, K., Kijimoto, S. and Kanemitsu, Y., (2007). Stability-optimized clearance configuration of fluid-film bearings. *Transactions of the ASME - Journal of Tribology* 129(1): 106–111.

Montgomerey, A.G. and Sterry, F., (1956). AERE Report No. E.D./R.1671.

Mori, H. and Mori, A., (1969). A stabilizing method of the externally pressurized gas journal bearings (vol. 2, paper no. 29) *Proceedings of the International Gas Bearing Symposium*, Southampton, England.

Mori, H. and Mori, A., (1971). A stabilizing method of the externally pressurized gas journal bearings, succeeding report: quantitative study (vol. 1, paper no. 4) *Proceedings of the International Gas Bearing Symposium*, Southampton, England.

Muijderman, E.A., (1967). Analysis and design of spiral-groove bearings. *Transactions of the ASME - Journal of Lubrication Technology* 89(3): 291–306.

Newkirk, B.L., (1924). Shaft whipping. *General Electric Review* 27, 169–178.

Newkirk, B.L. and Taylor, H.D., (1925). Shaft whipping due to oil action in journal bearings. *General Electric Review* 28, 559–568.

Osborne, D.A. and San Andrés, L., (2003). Experimental response of simple gas hybrid bearings for oil-free turbomachinery (GT2003-38833) *Proceedings of the ASME Turbo Expo (2003: Power for Land, Sea and Air*, Atlanta, GA, USA.

Pan, C.H.T., (1980). Rotor-bearing dynamics technology design guide. Part VI. Status of gas bearing technology applicable to aero propulsion machinery. Report ADA094167, SHAKER RESEARCH CORP, Ballston Lake NY.

Pan, C.H.T. and Sternlicht, B., (1962). On the translatory whirl motion of a vertical rotor in plain cylindrical gas-dynamic journal bearings. *Transactions of the ASME - Journal of Basic Engineering* 84(March): 152–158.

Pinkus, O., (1959). Analysis and characteristics of the three-lobe bearing. *Transactions of the ASME - Journal of Basic Engineering* 81, 49–55.

Powell, J.W., (1963). Unbalance whirl of rotors supported in gas journal bearings. *Engineer, London*

Powell, J.W., (1970). A review of progress in gas lubrication. *Review of Physics in Technology* 1(2): 96–129.

Powell, J.W. and Tempest, M., (1968). A study of high speed machines with rubber stabilized air bearings. *Transactions of the ASME - Journal of Lubrication Technology* 90, 701–708.

Reinhoudt, J.P., (1972). *On the stability of rotor-and-bearing systems and on the calculation of sliding bearings* PhD thesis Technische Hogeschool Eindhoven.

Rentzepis, G.M. and Sternlicht, B., (1962). On the stability of rotors in cylindrical journal bearings. *Transactions of the ASME - Journal of Basic Engineering* 84(3): 521–532.

Reynolds, D.B. and Gross, W.A., (1962). Experimental investigation of whirl in self-acting air-lubricated journal bearings. *ASLE Transactions* 5, 392–403.

Rimpel, A. and Kim, D., (2009). Rotordynamic performance of flexure pivot tilting pad gas bearings with vibration damper. *Transactions of the ASME - Journal of Tribology* 131(2): 021101 (12 pages).

Risse, S. (2001). *Ein Beitrag zur Entwicklung eines dopplessphärischen Luftlagers aus Glaskeramik* PhD thesis Fakultät für Maschinenbau der Technischen Universität Ilmenau.

San Andrés, L. and Childs, D., (1997). Angled injection - hydrostatic bearings analysis and comparison to test results. *Transactions of the ASME - Journal of Tribology* 119(1): 179–187.

Schmid, C. (1974). Gas bearing turboexpanders for cryogenic plants (paper B1) *Proceedings of the 6th International Gas Bearing Symposium*, Southampton, England.

Sim, K. and Kim, D., (2009). Design and manufacturing of mesoscale tilting pad gas bearings for 100-200 W class PowerMEMS applications. *Journal of Engineering for Gas Turbines and Power* 131(4): 042503 (11 pages).

Sternlicht, B. and Winn, L.W., (1963). On the load capacity and stability of rotors in self-acting gas lubricated plain cylindrical journal bearings. *Transactions of the ASME - Journal of Basic Engineering* 83, 139–144.

Stodola, A. (1925 Kritische Wellenstrung infolge der Nachgiebigkeit des Ölpolslers im Lager (critical shaft perturbations as a result of the elasticity of the oil cushion in the bearings). *Schweizerische Bauzeitung* 85(21): 265–266.

Tomioka, J., Miyanaga, N., Outa, E., et al. (2007). Development of herringbone grooved aerodynamic journal bearings for the support of ultra-high-speed rotors. *Transactions of the JSME - Machine Elements and Manufacturing* 73(730): 1840–1846.

Tondl, A., (1967). Bearings with a tangential gas supply (paper no. 4) *Proceedings of the International Gas Bearing Symposium*, Southampton, England.

Tondl, A., (1971). The effect of an elastically suspended foundation mass and its damping on the initiation of self-excited vibrations of a rotor mounted in air presurised bearings (vol. 1, paper no. 1) *Proceedings of the International Gas Bearing Symposium*, Southampton, England.

Tully, N., (1966). *Damping in externally pressurised gas bearing journals (Vibrational damping of externally pressurized orifice journal bearings undergoing free vibrations with and without shaft rotation)* PhD thesis Southampton University.

Vance, J.M., (1987). *Rotordynamics of Turbomachinery*. John Wiley & Sons, New York.

Waumans, T., (2009). *On the Design of High-Speed Miniature Air Bearings: Dynamic Stability, Optimisation and Experimental Validation* PhD thesis K.U. Leuven, Department of Mechanical Engineering, (2009D16).

Waumans, T., Peirs, J., Al-Bender, F. and Reynaerts, D., (2011a). Aerodynamic journal bearing with a flexible, damped support operating at 7.2 million DN. *Journal of Micromechanics and Microengineering* 21(10): 104014.

Waumans, T., Peirs, J., Reynaerts, D. and Al-Bender, F., (2011b). On the dynamic stability of high-speed gas bearings: stability study and experimental validation *Sustainable Construction & Design, Ghent University*, http://www.scad.ugent.be/journal/2011/SCAD_2011_2_2_342.pdf.

Whitley, S., Bowhill, A.J. and McEwan, P., (1962). Half speed whirl and load capacity of hydrodynamic gas journal bearings. *Proceedings of the Institution of Mechanical Engineers* 176, 554.

Wildmann, M., (1956). Experiments on gas lubricated journal bearings. *ASME paper No. 56-LU B-8*.

Zhu, X. and San Andrés, L., (2004). Rotordynamic performance of flexure pivot hydrostatic gas bearings for oil-free turbomachinery (GT2004-53621) *Proceedings of the ASME Turbo Expo (2004: Power for Land, Sea and Air*, Vienna, Austria.

11

Tilting Pad Air Bearings

11.1 Introduction

A pivoted pad, or tilting pad journal bearing consists of multiple curved-bearing-segments as shown in Figure 11.1. Each of these segments, often called pads, can perform a small rotation about a pivoting point, located at the pad back side. The herewith obtained rotational degree of freedom gives the individual pads the capability to align themselves to the rotating journal with minimum attitude angle as illustrated in Figure 11.2. The small attitude angle results in superior dynamic stability in comparison to a 360° plain journal bearing or any other types of rigid-wall bearing. In addition, the self-alignment of the pads eases the required manufacturing tolerances of the bearing surfaces. That is, a rigid bearing must be machined with high precision to obtain a gap-height of the order of 10 μm, while for a tilting pad bearing, the pads align themselves readily to a position for near optimal gap-geometry.

The benefits of tilting pad bearings were already recognised more than a century ago. Two of the first pioneers were Anthony G M Michell and Albert Kingsbury who developed, at the beginning of the 20th century, and independently of each other, a hydrodynamic axial thrust bearing of the tilting pad type (Dimond et al. 2011). These axial bearings where used in a range of applications such as steam turbines, centrifugal pumps, and hydroelectric generators. Michell also invented the oil lubricated tilting pad journal bearing around 1916 (Dimond et al. 2011; Simmons and Advani 1987). It was, however, not until the 1960s that the first tilting pad journal bearings started to appear with gas as a lubricant. One of the first successful implementations of gas lubricated tilting pad journal bearings was for a helium circulator, developed by the French company Société Rateau in the early 1960s (Grassam and Powell 1964; Powell 1970; Shaw 1983). The helium circulator rotated at 12.000 rpm while supported by two tilting pad gas bearings with an internal diameter of 160 mm and 140 mm respectively. These bearing were reported to perform in a robust manner and to have an outstanding start–stop behaviour, being at that time a good example of the benefits of aerodynamic bearings in large gas circulators as compared to to EP bearings.

Research on tilting pad gas bearings started to increase in the 1960s, which resulted in numerous theoretical models to predict their behaviour. One key contribution from this era is a paper published by Lund (1964). Lund presented an elegant analytical methodology to compute synchronous stiffness and damping coefficients for non-flexible tilting pad bearings. His method is now known as Lund's assembly method and has been implemented in many models since then (Nicholas 2003). Up until the early 1970s, studies on tilting pad journal bearings focused primarily on the steady-state properties such as load capacity and power losses. The few dynamic non-synchronous models, for whirl behaviour and stability prediction, were at that time restricted to bearings with an infinite length, or were mathematically incorrect due to the use of synchronous frequency in their stability calculations (Dimond et al. 2011). It would take almost a decade more till eventually Lund pointed out in 1978 that not the synchronous but the (damped) natural frequency should be used for stability calculations (Dimond et al. 2011; Nicholas 2003).

Air Bearings: Theory, Design and Applications, First Edition. Farid Al-Bender.

Companion website: www.wiley.com/go/AlBender/AirBearings

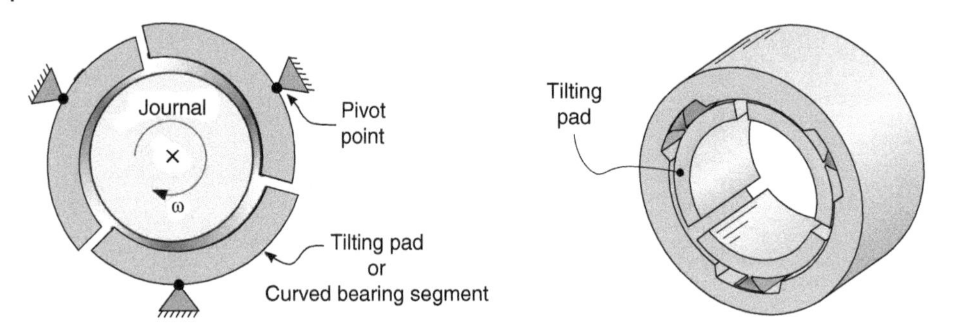

Figure 11.1 Schematic illustration of a tilting pad bearing with three pads. Source: Nabuurs M 2020.

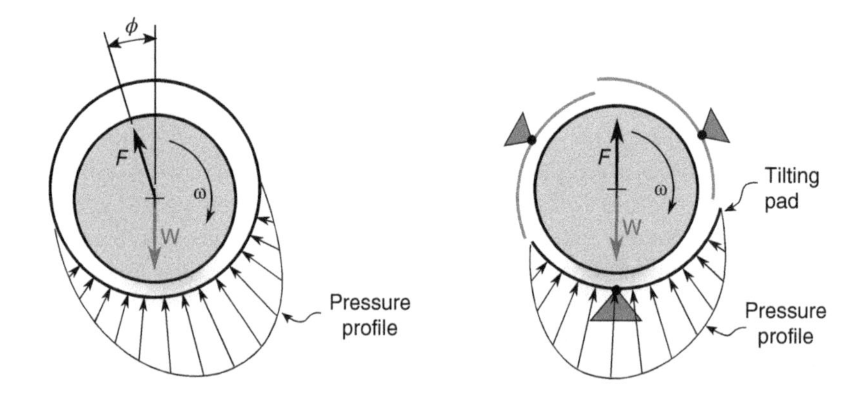

Figure 11.2 Attitude angle ϕ of a 360° plain bearing versus a titling pad bearing. Source: Nabuurs M 2020.

This statement started new research on the perturbation frequency dependency of gas journal bearings. This, in combination with the ever rising computing power, resulted in the calculation of non-synchronous coefficients with inclusion of various effects. Worthy of mention is the work of Czolczynski (1999), who published an orbit-based method to compute the frequency-dependent, non-linear coefficients of flexibly mounted plain journal bearings. It was however Lihua et al. (2007) who published for the first time the frequency-dependent coefficients for an aerodynamic tilting pad bearing. Unfortunately the initially published results were incorrect but were partly corrected and republished again in 2010 (Lihua et al. 2010). As far as is known, this work is the only publication providing both a theoretical model and the actual dynamic coefficients for a tilting pad gas bearing. Nevertheless, the model is only suitable for the most basic tilting pad journal bearing geometries, which are rarely found in practice. It does not include pivot point flexibility, film frictional forces, and pivot point offset. These aspects could, however, have a significant effect on the dynamic coefficients of the bearing and must therefore be taken into account.

As such, it is worthwhile to treat the full theory and the mathematical modeling of tilting pad bearings in the remainder of this chapter. For simplicity the principle behind a tilting pad bearing is first clarified starting with a fixed inclined sliding or plane slider bearing and its pressure profile. From there the model is expanded by including a pivoting point and the effect of a curved bearing surface. The model is then further expanded to characterise tilting pad journal bearings with expressions being given for the bearing load capacity, power losses, and dynamic stiffness. Next, the rotor/bearing dynamic stability is treated. Expressions are given for the system onset-speed of instability based on the bearing dynamic stiffness coefficients. At the end of this chapter, some remarks are given on the fabrication methods and aspects to produce the actual tilting pad bearing.

Table 11.1 Characteristics bearing quantities used in this chapter.

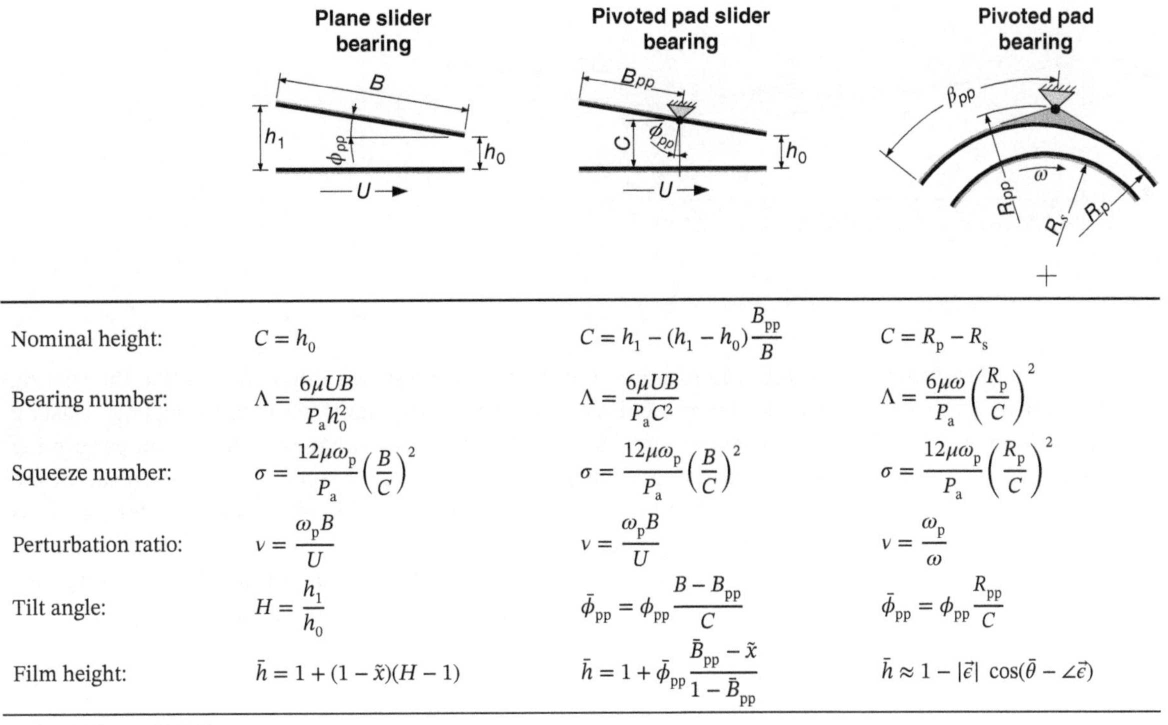

	Plane slider bearing	Pivoted pad slider bearing	Pivoted pad bearing
Nominal height:	$C = h_0$	$C = h_1 - (h_1 - h_0)\frac{B_{pp}}{B}$	$C = R_p - R_s$
Bearing number:	$\Lambda = \frac{6\mu UB}{P_a h_0^2}$	$\Lambda = \frac{6\mu UB}{P_a C^2}$	$\Lambda = \frac{6\mu\omega}{P_a}\left(\frac{R_p}{C}\right)^2$
Squeeze number:	$\sigma = \frac{12\mu\omega_p}{P_a}\left(\frac{B}{C}\right)^2$	$\sigma = \frac{12\mu\omega_p}{P_a}\left(\frac{B}{C}\right)^2$	$\sigma = \frac{12\mu\omega_p}{P_a}\left(\frac{R_p}{C}\right)^2$
Perturbation ratio:	$\nu = \frac{\omega_p B}{U}$	$\nu = \frac{\omega_p B}{U}$	$\nu = \frac{\omega_p}{\omega}$
Tilt angle:	$H = \frac{h_1}{h_0}$	$\bar{\phi}_{pp} = \phi_{pp}\frac{B - B_{pp}}{C}$	$\bar{\phi}_{pp} = \phi_{pp}\frac{R_{pp}}{C}$
Film height:	$\bar{h} = 1 + (1 - \tilde{x})(H - 1)$	$\bar{h} = 1 + \bar{\phi}_{pp}\frac{\bar{B}_{pp} - \tilde{x}}{1 - \bar{B}_{pp}}$	$\bar{h} \approx 1 - \lvert\vec{\epsilon}\rvert\cos(\bar{\theta} - \angle\vec{\epsilon})$

Source: Nabuurs M 2020.

Normalised quantities are used throughout the chapter to make the results as uniform as possible. Definitions of the most important normalised quantities are provided in Table 11.1 for a plane-slider, pivoted-slider, and pivoting-pad bearing.

11.2 Plane Slider Bearing

The most basic aerodynamic bearing is the plane-slider bearing with fixed taper angle and an infinite length L[1]. Its uncomplicated geometry can be defined with just a few parameters, which simplifies the analysis of its properties. Nevertheless, studying such simplified plane-slider bearing provides a qualitative understanding of the performance of the more complex tilting pad journal bearing. For this reason, the analysis of tilting pad journal bearings starts with the analysis of the plane-slider bearing depicted in Figure 11.3.

The gas film height at the leading and trailing edges is denoted by distances h_1 and h_0 respectively, while B is the width of the sliding bearing. Notice that the inclination angle ϕ of the depicted slider bearing is exaggerated. In reality the inclination, or tilt angle, will be small such that $\phi = (h_1 - h_0)/B$. For further analysis and without loss of generality, the top surface is stationary while the bottom surface slides with velocity U along the x-axis.

1 Note that the terms "width" and "length" have been interchanged in this chapter as compared to chapter 8. In this chapter the length is perpendicular to the sliding direction.

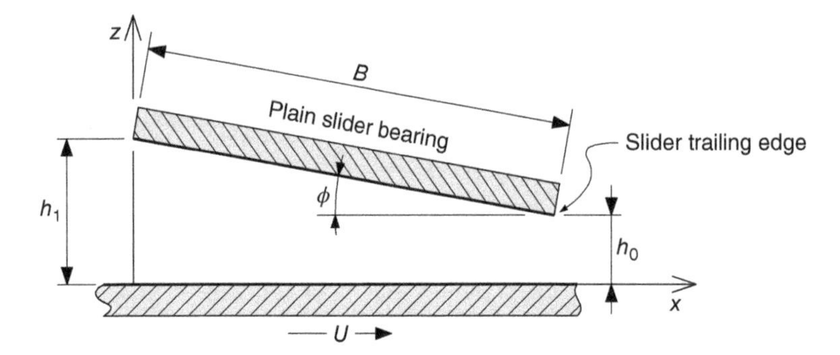

Figure 11.3 Cross-section of a plain-slider bearing. Source: Nabuurs M 2020.

It is common in the study of fluid-film lubrication to use dimensionless quantities[2] to describe the bearing geometry and its performance. The definition of these dimensionless quantities varies throughout existing fluid-lubrication literature. This inconsistency makes it difficult to directly compare results without performing the necessary transformation from one definition to another. Nevertheless, an attempt has been made in the remainder of this section to normalise the quantities in such a way that the number of required transformations is kept to a minimum. That is: the dimensionless inclination ratio $H = h_1/h_0$ is used together with the dimensionless position $\tilde{x} = x/B$, while the gas film height $h(x)$ is made dimensionless by dividing it through the trailing edge height h_0. With these definitions, the dimensionless gas film height beneath the slider plane becomes

$$\bar{h}(\tilde{x}) = \frac{h(x)}{h_0} = H - \tilde{x}(H - 1) \tag{11.1}$$

Pressure profile $\bar{p}(\tilde{x}) = p/P_a$ within the gas film can be derived from the Reynolds equation for steady, isothermal films of infinite length. Using the notation of (Gross 1980, p. 76), the Reynolds equation may be written as

$$\frac{\partial}{\partial \tilde{x}} \left(\bar{p}\bar{h}^3 \frac{\partial \bar{p}}{\partial \tilde{x}} \right) = \Lambda \frac{\partial \bar{p}\bar{h}}{\partial \tilde{x}} \tag{11.2}$$

with the dimensionless plane slider bearing number Λ as

$$\Lambda = \frac{6\mu UB}{P_a h_0^2} \tag{11.3}$$

Several pressure profiles for a range of bearing numbers are given in Figure 11.4 and Figure 11.5. The two figures are for a fixed inclination ratio H of 1.5 and 3 respectively. As seen from these pressure-curves, the location of the maxim pressure depends on the bearing number Λ and inclination ratio H. It approaches $\tilde{x} = H/(H + 1)$ for $\Lambda \rightarrow 0$ and $\tilde{x} = 1$ for $\Lambda \rightarrow \infty$ (Hamrock et al. 2004, p. 219). The location of maximum pressure will therefore shift towards the trailing edge for an increasing bearing number, see also Chapter 8.

Bearing load capacity

The load capacity of the sliding bearing can be obtained by integrating the pressure over the slider surface. For generality, the load capacity is made dimensionless by dividing it through the product of the bearing area and the ambient pressure P_a so that the load capacity $\bar{W}$ of the bearing becomes

$$\bar{W} = \frac{W}{P_a LB} = \int_0^1 (\bar{p}(\tilde{x}) - 1) d\tilde{x} \tag{11.4}$$

2 Dedicated symbols or symbols with an overbar accent are used throughout this chapter for dimensionless quantities.

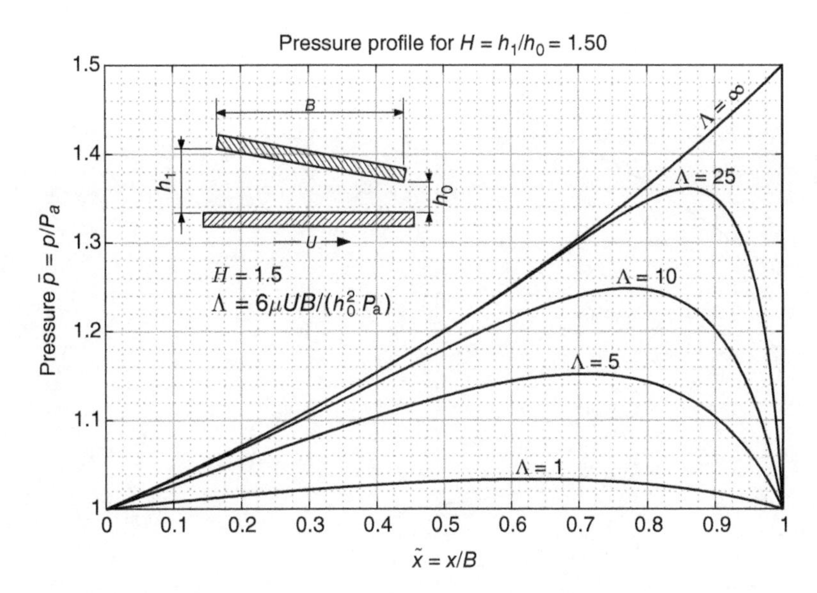

Figure 11.4 Pressure profile for a plane slider bearing with infinitely length and an inclination ratio of $h_1/h_0 = 1.5$. Source: Gross W 1980. Reproduced with permission from John Wiley & Sons.

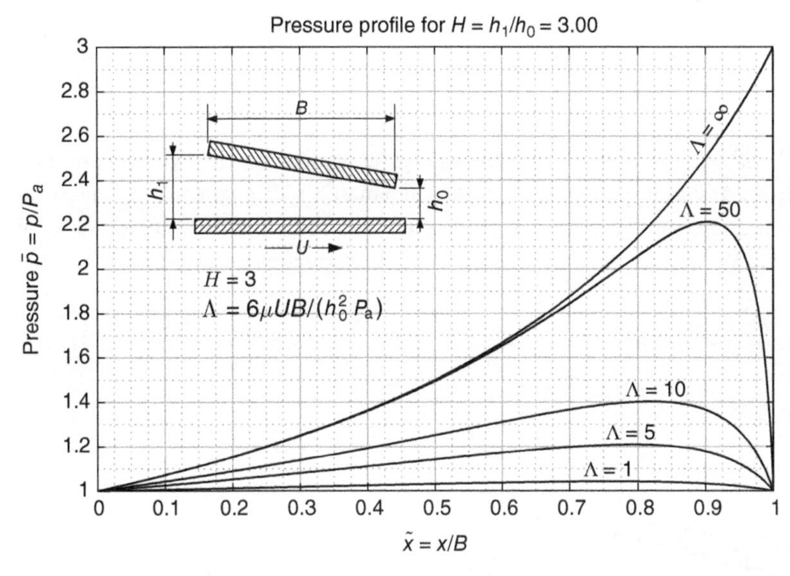

Figure 11.5 Pressure profile for a plane slider bearing with infinitely length and an inclination ratio of $h_1/h_0 = 3$. Source: Gross W 1980. Reproduced with permission from John Wiley & Sons.

The load capacity $\bar{W}$, or the force perpendicular to the bearing surface, will therefore depend on the aerodynamic pressure $\bar{p}$, which in its turn depends on the bearing number Λ and the inclination ratio H. Such dependency is shown in Figure 11.6 for a range of bearing numbers.

Centre of pressure

The centre of pressure $\tilde{x}_c$ is a point, along the film $\tilde{x}$-axis, around which the net-moment of pressure is zero. The location of $\tilde{x}_c$ is important in the study of pivoted sliding-pad bearings. That is, under the assumption that the

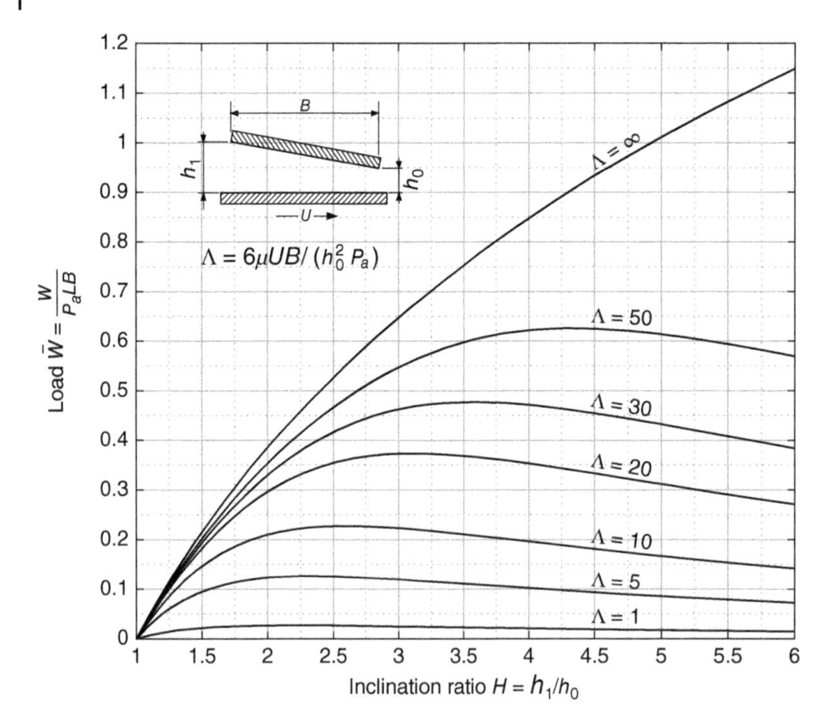

Figure 11.6 Load capacity of a plane slider bearing with infinite length. Source: Gross W 1980. Reproduced with permission from John Wiley & Sons.

pivoting point lies effectively on the slider face (such that shear forces can be neglected), then the steady-state centre of pressure must coincide with the pivoting point location. Any mismatch will result in a moment on the pad, changing the slider inclination angle until the centre of pressure and pivot point coincide. Such an equilibrium point might not always exist. Which results in malfunction or permanently damage due to contact

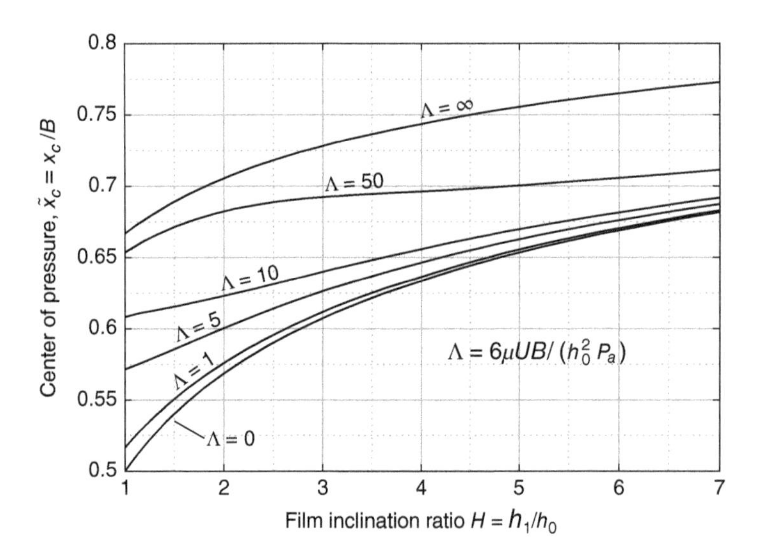

Figure 11.7 Centre of pressure for plane slider bearing with infinite length and compressible fluid. Source: Nabuurs M 2020.

of the surfaces. Hence, it is essential that the pivot point is positioned such that an equilibrium exists for all occurring sliding speeds.

For an incompressible fluid, the centre of pressure location $\tilde{x}_c$ depends only on the inclination ratio H and not on bearing number Λ. Any change in sliding speed will therefore not change $\tilde{x}_c$ so that the inclination angle (and thus the optimal pivot location) of a pivoted slider bearing remains constant. However, the centre of pressure for compressible films depends on both the inclination ratio H and the bearing number Λ. Generally, there exists no analytical solution for $\tilde{x}_c(H, \Lambda)$ in case of compressible films, except for the extreme cases when $\Lambda \to 0$ or $\Lambda \to \infty$. For these cases, a closed form solutions for $\tilde{x}_c(H, \Lambda)$ is derived (Gross 1980, p. 82), i.e.

$$\tilde{x}_c = \frac{2H(H+2)\ln(H) - 5H^2 + 4H + 1}{2(H-1)[(H+1)\ln(H) - 2(H-1)]} \quad \text{when} \quad \Lambda \to 0 \tag{11.5}$$

$$\tilde{x}_c = \frac{H}{H-1} - \frac{H-1}{2H(\ln(H)-1)+2} \quad \text{when} \quad \Lambda \to \infty \tag{11.6}$$

The lack of an analytical solution for intermediate bearing numbers complicates the study of pivoted-slider gas bearings significantly. For example, Figure 11.7 provides $\tilde{x}_c$ values for intermediate bearing numbers. The dependency of $\tilde{x}_c$ on Λ means that any pivoted-slider gas bearing will have a sliding-speed-dependent inclination angle ϕ, which cannot be analytically expressed. This will be further explained in the next section where the focus will be on the slider bearing with pivoting action.

11.3 Pivoted Pad Slider Bearing

The plane slider bearing theory, as discussed in the previous section, can be extended to the pivoted pad slider bearing. Here the plane bearing is supported by a single pivot point as illustrated in Figure 11.8. Under the assumption that the pad orientation is in equilibrium for a given set of operation conditions, i.e. bearing number Λ and load W, then any change in these conditions will alter the pressure distribution beneath the pad and thus momentarily shift the centre of pressure x_c away from the pivoting point position B_{pp}. The offset between B_{pp} and x_c will create a moment acting on the pad, which will in turn tilt the pad to a new equilibrium inclination angle. The pivoting pad

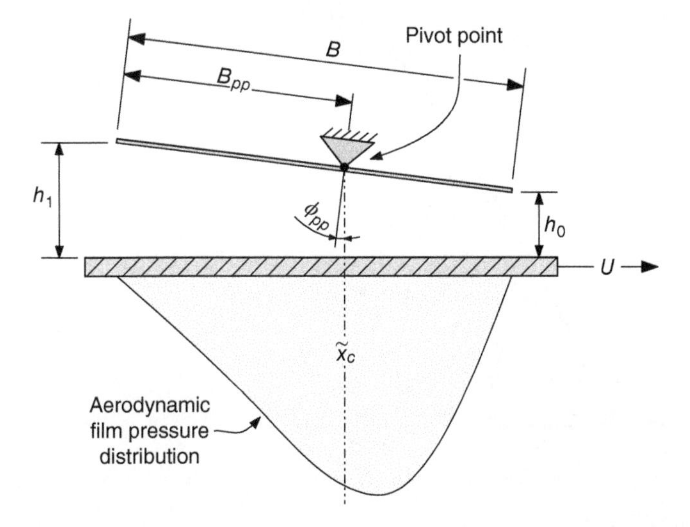

Figure 11.8 Side view of pivoted pad slider bearing. (Not to scale). Source: Nabuurs M 2020.

angle of inclination ϕ_{pp} is therefore variable, being a function of the bearing operation conditions. It is typically approximated by $\phi_{pp} \approx (h_1 - h_0)/B$.

Note that the inclination angle in Figure 11.8 is exaggerated. In reality the inclination angle will be on the order of milliradians. A normalised form of this inclination angle will be used in the remainder of this section and is defined as

$$\bar{\phi}_{pp} = \phi_{pp}\frac{B - B_{pp}}{C} \tag{11.7}$$

Here C is the gas-film height beneath the pivoting point, which is independent of the tilt angle $\bar{\phi}_{pp}$. Because of this independence, the distance C is used to normalise the gas film height h according to

$$\bar{h} = \frac{h}{C} = 1 + \bar{\phi}_{pp}\frac{\bar{B}_{pp} - \tilde{x}}{1 - \bar{B}_{pp}}. \tag{11.8}$$

where, $\bar{B}_{pp} = B_{pp}/B$.

Note that this is different from the fixed slider bearing gas-film case, where the distance h_0 was chosen to normalise the gas-film height. The trailing edge film height h_0 will not be constant under the operating conditions of a pivoted slider bearing and therefore not suitable to normalise the gas-film height. For the same reason, the distance C is also preferably used in the bearing number instead of h_0. By taking this into account, the pressure profile can be computed in accordance to (11.2) and the following definition of Λ

$$\Lambda = \frac{6\mu UB}{P_a C^2} \tag{11.9}$$

If the assumption is made that the pivoting plane can freely rotate without any friction or tilt stiffness, and we further assume the pivot point to be located at the bearing surface, then pad inclination angle $\bar{\phi}_{pp}$ depends only on the bearing number Λ and lateral pivot location B_{pp}. Or otherwise stated, for each particular sliding speed and pivot location B_{pp}, there exists a certain tilt angle ϕ_{pp} which results in a moment equilibrium around the pivoting point. This equilibrium tilt angle and the corresponding bearing load capacity $\bar{W}$ are plotted in Figure 11.9 for a range of pivot locations $\bar{B}_{pp}$ and bearing numbers.

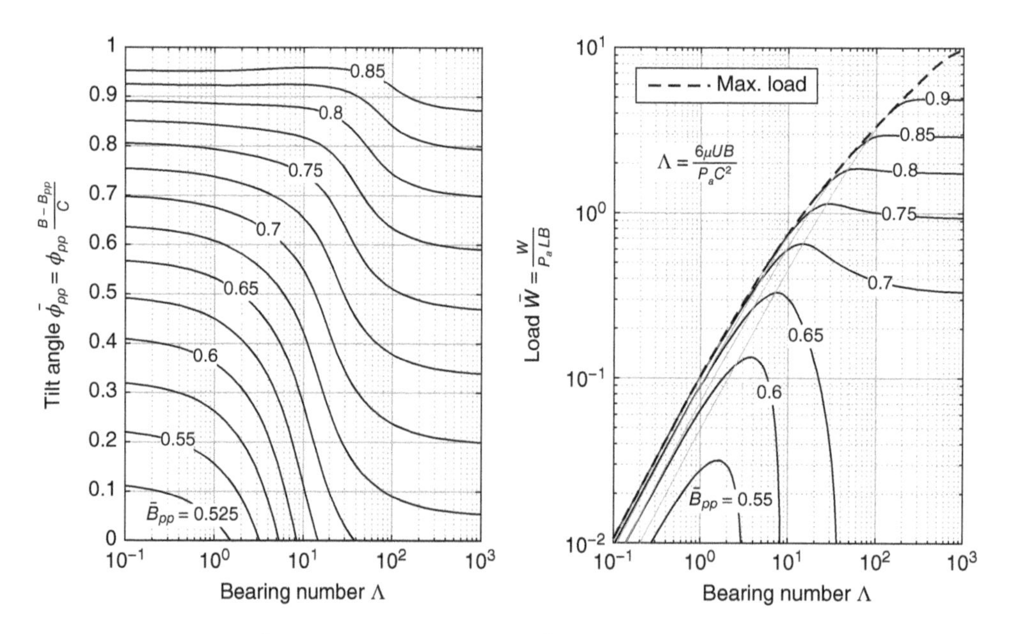

Figure 11.9 Tilt angle $\bar{\phi}_{pp}$ and bearing load capacity $\bar{W}$ of an infinitely long pivoted sliding bearing at equilibrium. Source: Nabuurs M 2020.

Note that the tilt angle is asymptotic to zero for the case $\bar{B}_{pp} < 2/3$ as $\Lambda \rightarrow \infty$, which results in parallel oriented sliding surfaces with no pressure or lift generation. Bearings with $\bar{B}_{pp}$ above $\frac{2}{3}$ will always have a certain load capacity. One can also notice from Figure 11.9 that there is no optimal value for $\bar{B}_{pp}$ that results in maximum load capacity for any bearing number. As such, there is no optimum pivot location as is the case for pivoted pad bearings with incompressible films. Incompressible films have their maximum load capacity (for any value of Λ) when $\bar{B}_{pp} = 0.714$ (Cameron 1971; Halling 1976).

For further reference, some compressible film equilibrium points (of the set Λ, $\bar{B}_{pp}$ and $\bar{\phi}_{pp}$, which result in maximum $\bar{W}$), are provided in Table 11.2 and Figure 11.10.

Dynamic gas-film stiffness

Besides load capacity, the bearing stiffness is another important aspect. Under the assumption that the bearing surfaces are rigid and will not deform due to a change in gas-film pressure, the bearing stiffness and damping are solely determined by the gas-film characteristics. One must be aware that gas is a compressible medium, resulting in frequency-dependent stiffness and damping coefficients. These dynamic coefficients are often determined by perturbation of the time-dependent Reynolds equation

$$\frac{\partial}{\partial \tilde{x}}\left(\bar{p}\bar{h}^{3}\frac{\partial \bar{p}}{\partial \tilde{x}}\right) = \Lambda\frac{\partial \bar{p}\bar{h}}{\partial \tilde{x}} + \sigma\frac{\partial \bar{p}\bar{h}}{\partial \tau^{o}} \tag{11.10}$$

Table 11.2 Pivoted pad configurations and orientation for maximum load-carrying capacity.

Λ	**0.01**	**0.1**	**1**	**10**	**100**
$\bar{B}_{pp}$	0.714	0.715	0.722	0.771	0.887
$\bar{\phi}_{pp}$	0.733	0.732	0.730	0.807	0.984
$\bar{W}$	1.07e-3	1.07e-2	1.02e-1	7.52e-1	3.34e + 0

Source: Nabuurs M 2020.

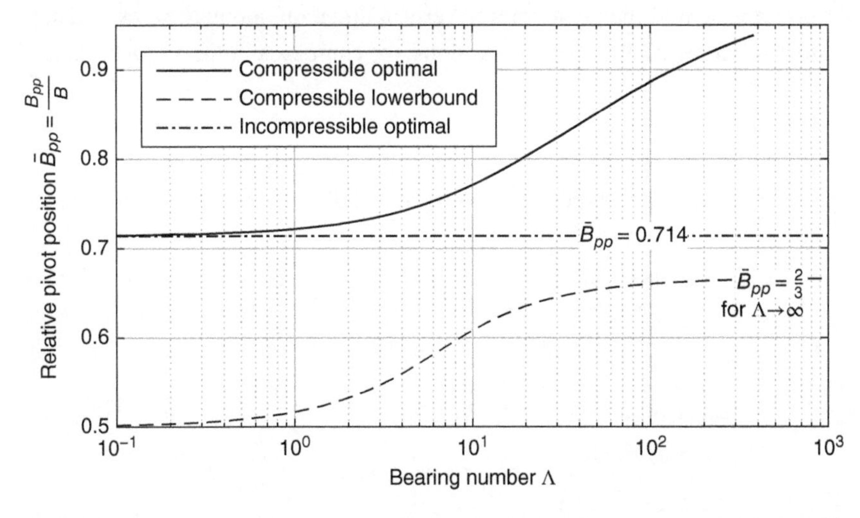

Figure 11.10 Optimal pivot position for maximum load-carrying capacity of a pivoted pad slider bearing with infinite length. Source: Nabuurs M 2020.

with the bearing number Λ according to (11.9) and the so-called squeeze number σ

$$\sigma = \frac{12\mu\omega_{\mathrm{p}}}{P_{\mathrm{a}}}\left(\frac{B}{C}\right)^2 \tag{11.11}$$

here τ^o represents the dimensionless time as $\tau^o = \omega_{\mathrm{p}} t$ with ω_{p} being the film perturbation frequency in rad s^{-1}. For an aerodynamic bearing, the squeeze number may also be expressed as $\sigma = 2\nu\Lambda$. The newly introduced symbol ν expresses the film perturbation ratio, which is the ratio between the film normal perturbation frequency and its lateral sliding frequency:

$$\nu = \frac{\omega_{\mathrm{p}}}{U/B} \tag{11.12}$$

The principle behind the perturbation method is to determine the change in film pressure due to a harmonically-changing film geometry. That is, the harmonic film-height variation is superposed on the static film geometry, thus

$$\bar{h} = \bar{h}_{(0)} + \Delta\bar{h}e^{\tau^o i}, \tag{11.13}$$

where $\bar{h}_{(0)}$ is the static gas film height according to (11.8) and $\Delta\bar{h}$ the complex amplitude of the harmonic film variation. The corresponding film pressure will then be

$$\bar{p} = \bar{p}_{(0)} + \Delta\bar{p}e^{\tau^o i}, \tag{11.14}$$

with $\bar{p}_{(0)}$ being the static gas film pressure and $\Delta\bar{p}$ the complex response amplitude. Substituting (11.13) and (11.14) into (11.10), subtracting the steady-state solution, discarding all non-linear (higher-order) terms, and equating the harmonic terms, yields a linear second-order partial differential equation of the form

$$\begin{aligned} &\frac{\partial}{\partial\tilde{x}}\left([3\bar{p}_{(0)}\bar{h}_{(0)}^2\Delta\bar{h} + \Delta\bar{p}\bar{h}_{(0)}^3]\frac{\partial\bar{p}_{(0)}}{\partial\tilde{x}} + \bar{p}_{(0)}\bar{h}_{(0)}^3\frac{\partial\Delta\bar{p}}{\partial\tilde{x}}\right) = \\ &\Lambda\left(\frac{\partial\bar{p}_{(0)}\Delta\bar{h}}{\partial\tilde{x}} + \frac{\partial\Delta\bar{p}\bar{h}_{(0)}}{\partial\tilde{x}}\right) + \sigma i(\bar{p}_{(0)}\Delta\bar{h} + \Delta\bar{p}\bar{h}_{(0)}). \end{aligned} \tag{11.15}$$

In order to obtain the full pivoted pad stiffness matrix, the gas film is perturbed in both the vertical (normal) and in the tilt direction, i.e. ϵ_{pp} and $\bar{\phi}_{\mathrm{pp}}$ respectively, as is further clarified in Figure 11.11.

Note that ϵ_{pp} is the dimensionless displacement of the pivot point in vertical direction (along the ϵ-axis). That is, $\epsilon_{\mathrm{pp}} = \frac{e_{\mathrm{pp}}}{C}$ with e_{pp} being the absolute displacement. With these definitions, the complex amplitude $\Delta\bar{h}$ becomes

$$\Delta\bar{h}e^{\tau^o i} = \Delta\bar{\phi}_{\mathrm{pp}}e^{\tau^o i}\frac{\bar{B}_{\mathrm{pp}} - \tilde{x}}{1 - \bar{B}_{\mathrm{pp}}} - \Delta\epsilon_{\mathrm{pp}}e^{\tau^o i}, \tag{11.16}$$

with $\Delta\epsilon_{\mathrm{pp}}$ and $\Delta\bar{\phi}_{\mathrm{pp}}$ being the complex perturbation amplitude in the vertical and tilt directions, respectively.

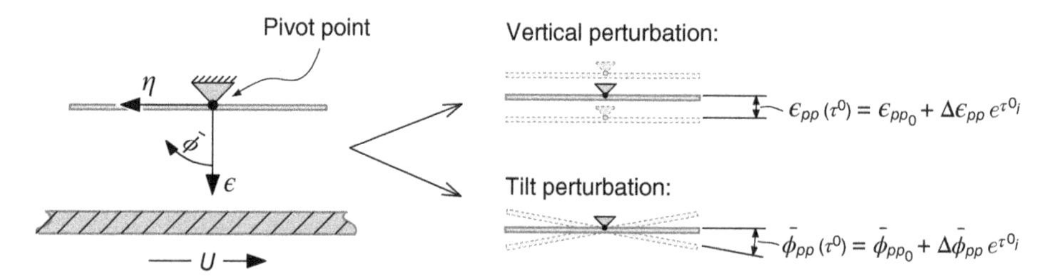

Figure 11.11 Film coordinate system and perturbation directions. Source: Nabuurs M 2020.

Substituting (11.16) into (11.15), such that $\Delta\bar{p} = f(\Delta\epsilon_{pp}, \Delta\bar{\phi}_{pp})$, gives us eventually the harmonic reaction force amplitude $\Delta\bar{F}_{gp}$ and moment amplitude $\Delta\bar{M}_{gp}$ around the pivot point, i.e

$$\Delta\bar{F}_{gp} = \int_0^1 (\Delta\bar{p}(\tilde{x}))\mathrm{d}\tilde{x} \tag{11.17}$$

$$\Delta\bar{M}_{gp} = \frac{1}{1-\bar{B}_{pp}} \int_0^1 (\Delta\bar{p}(\tilde{x}))(\tilde{x} - \bar{B}_{pp})\mathrm{d}\tilde{x} \tag{11.18}$$

The dimensionless dynamic stiffness matrix $\bar{\mathbb{Z}}_{gp}(\sigma) \in \mathbb{C}^{2\times2}$ can then be defined as

$$\bar{\mathbb{Z}}_{gp} = \begin{bmatrix} \frac{\Delta\bar{F}_{gp}}{\Delta\epsilon_{pp}} & \frac{\Delta\bar{F}_{gp}}{\Delta\epsilon_{pp}} \\ \frac{\Delta\bar{M}_{gp}}{\Delta\epsilon_{pp}} & \frac{\Delta\bar{M}_{gp}}{\Delta\epsilon_{pp}} \end{bmatrix} = \begin{bmatrix} \bar{z}_{\epsilon\epsilon} & \bar{z}_{\epsilon\bar{\phi}} \\ \bar{z}_{\bar{\phi}\epsilon} & \bar{z}_{\bar{\phi}\bar{\phi}} \end{bmatrix} = \underbrace{\begin{bmatrix} \bar{k}_{\epsilon\epsilon} & \bar{k}_{\epsilon\bar{\phi}} \\ \bar{k}_{\bar{\phi}\epsilon} & \bar{k}_{\bar{\phi}\bar{\phi}} \end{bmatrix}}_{\bar{\mathbb{K}}_{gp}} + \underbrace{\begin{bmatrix} \bar{d}_{\epsilon\epsilon} & \bar{d}_{\epsilon\bar{\phi}} \\ \bar{d}_{\bar{\phi}\epsilon} & \bar{d}_{\bar{\phi}\bar{\phi}} \end{bmatrix}}_{\bar{\mathbb{D}}_{gp}} i \tag{11.19}$$

Note that the dynamic stiffness matrix depends on the perturbation frequency or squeeze number σ. It is further a function of many other design parameters, such as $\bar{\mathbb{Z}}_{gp} = f(\Lambda, \sigma, \epsilon_{pp(0)}, \bar{\phi}_{pp(0),\bar{B}_{pp}})$. For reference purpose only, the values for $\bar{\mathbb{Z}}_{gp}(\sigma)$ are given for one particular set of variables in the graphs of Figure 11.12 and Figure 11.13. This set of variables is chosen from Table 11.2

11.3.1 Equivalent Bearing Stiffness

The overall dynamic stiffness of a pivoted sliding bearing is not only determined by the gas film but also by the pivot point mechanism and the mass distribution within the tilting pad. The assumption made earlier regarding

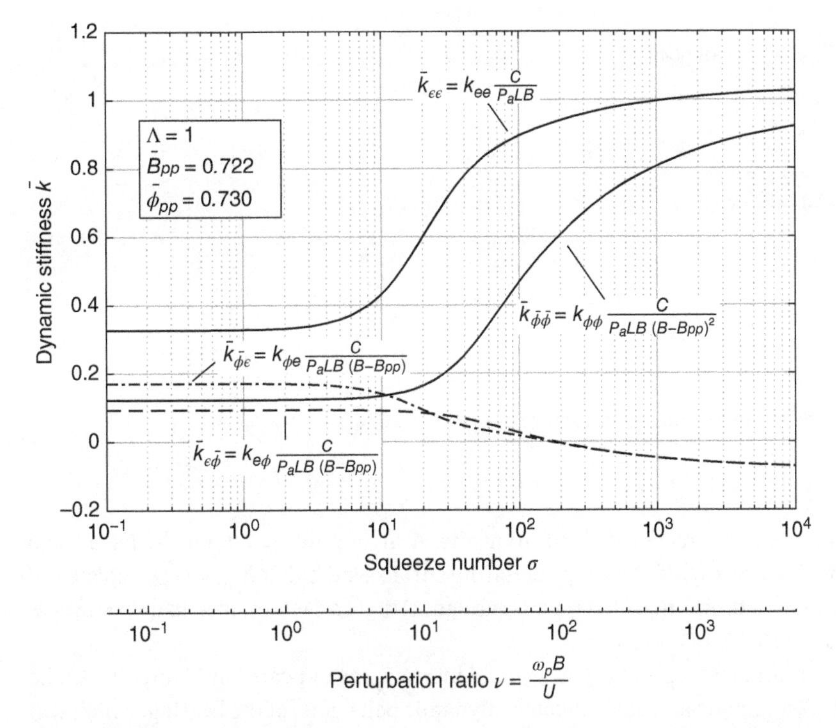

Figure 11.12 Dynamic stiffness of a pivoted sliding bearing with infinite length. Source: Nabuurs M 2020.

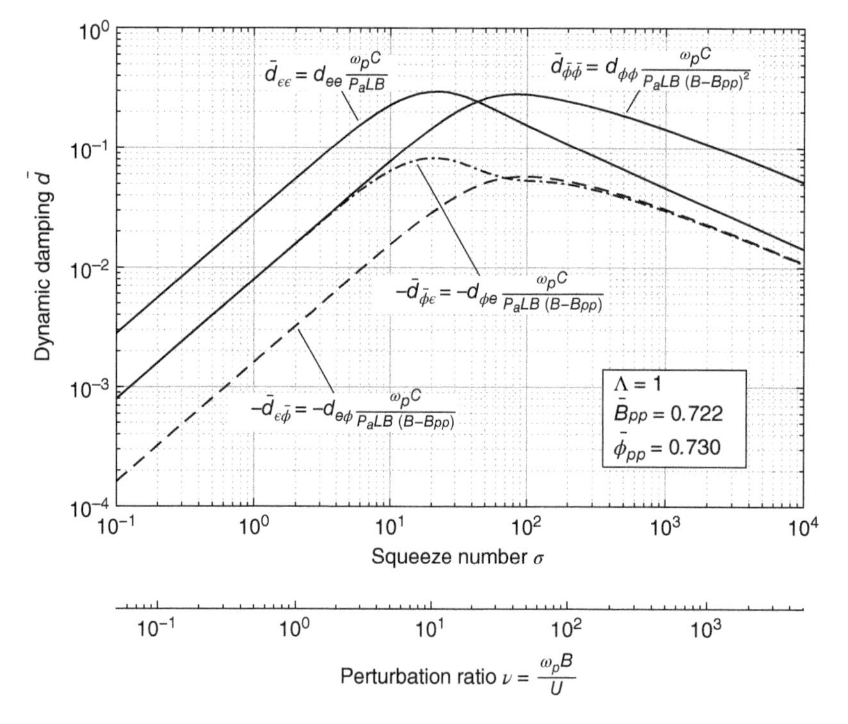

Figure 11.13 Dynamic damping of a pivoted sliding bearing with infinitely length. Source: Nabuurs M 2020.

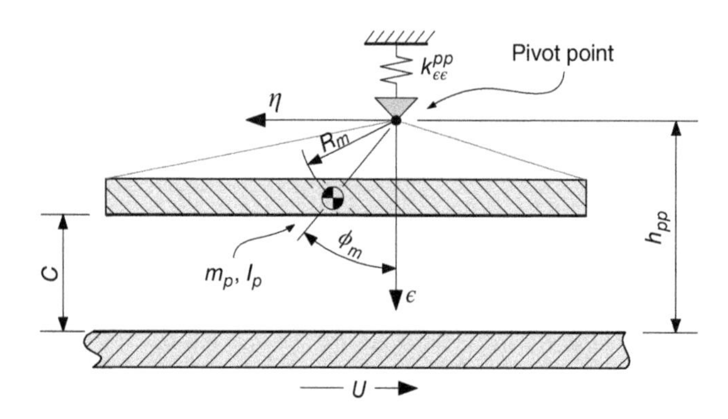

Figure 11.14 Pivoted slider bearing with pad mass and inertia. Source: Nabuurs M 2020.

an ideal pivot point with no tilt stiffness is in practice difficult to justify. A pivot point will typically have some form of tilt stiffness and is located with a certain offset from the bearing surface, such that $h_{pp} \neq C$ as depicted in Figure 11.14. The pivot point itself is sometimes mounted on a compliant structure, which has its own stiffness and damping as is further shown in Figure 11.14 and 11.15.

In practice, the pivoted pad has a certain mass m_p and tilt moment of inertia I_p, as indicated in Figure 11.14. The pivot offset, its compliance (/stiffness), and inertia, will all affect the dynamic behaviour of the bearing, which will become clear when deriving the equation of motion in the next paragraph.

Figure 11.15 Bearings equivalent mass-spring-damper system. Source: Nabuurs M 2020.

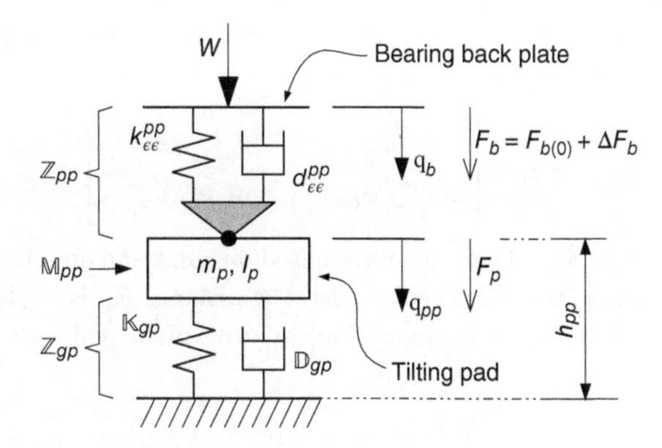

Equation of motion

For the remainder of this section, the pad position and orientation is expressed by the column vector $\mathbf{q}_{pp} = [e_{pp}, \phi_{pp}]$. For generality, a second column vector $\mathbf{q}_b = [e_b, 0]$ is introduced to express the position of the pivot point back-plane structure. This back-plane structure carries the bearing load W as is also illustrated in Figure 11.15. The equations of motion for the system may be written as

$$\begin{aligned}\vec{F}_b &= \mathbb{D}_{pp}(\dot{\mathbf{q}}_b - \dot{\mathbf{q}}_{pp}) + \mathbb{K}_{pp}(\mathbf{q}_b - \mathbf{q}_{pp}),\\ \vec{F}_p &= \mathbb{M}_{pp}\ddot{\mathbf{q}}_{pp} + \mathbb{D}_{pp}(\dot{\mathbf{q}}_{pp} - \dot{\mathbf{q}}_b) + \mathbb{K}_{pp}(\mathbf{q}_{pp} - \mathbf{q}_b) + \vec{F}_{gp}.\end{aligned} \tag{11.20}$$

Here $\mathbb{M}_{pp}$, $\mathbb{D}_{pp}$ and $\mathbb{K}_{pp}$ are the compliant pivot point mass, damping and stiffness matrices respectively. The definition of these matrices can be found in (11.26), (11.28) and (11.29). The column vector $\vec{F}_{gp}$ characterise the force and moment due to the aerodynamic gas-film pressure, i.e. $\vec{F}_{gp} = [F_{gp}, M_{gp}]$. These forces are found (in dimensionless form) by integrating the gas-film pressure $\bar{p}$ according to

$$\bar{F}_{gp} = \frac{F_{gp}}{P_a LB} = \int_0^1 (\bar{p}(\tilde{x}) - 1)d\tilde{x}, \tag{11.21}$$

$$\bar{M}_{gp} = \frac{M_{gp}}{P_a LB(B - B_{pp})} = \frac{1}{1 - \bar{B}_{pp}} \int_0^1 (\bar{p}(\tilde{x}) - 1)(\tilde{x} - \bar{B}_{pp})d\tilde{x}, \tag{11.22}$$

or, for small amplitudes, using the dynamic stiffness and damping matrix from (11.19), i.e.

$$\vec{\bar{F}}_{gp}(\tau^o) = \vec{\bar{F}}_{gp(0)} + \bar{\mathbb{D}}_{gp}\dot{\bar{\mathbf{q}}}_{pp} + \bar{\mathbb{K}}_{gp}(\bar{\mathbf{q}}_{pp} - \bar{\mathbf{q}}_{pp(0)}). \tag{11.23}$$

Here $\bar{\mathbf{q}}_{pp(0)}$ is the tilting pad steady state equilibrium position and the position where $\bar{\mathbb{K}}_{gp}$ and $\bar{\mathbb{D}}_{gp}$ are computed about. Note that $\bar{\mathbf{q}}_{pp}$ is dimensionless and, in this section, defined as

$$\bar{\mathbf{q}}_{pp} = \begin{bmatrix} \epsilon_{pp} \\ \bar{\phi}_{pp} \end{bmatrix} = \begin{bmatrix} \frac{e_{pp}}{C} \\ \frac{\phi_{pp}(B - B_{pp})}{C} \end{bmatrix} \tag{11.24}$$

Substituting (11.23) into the dimensionless form of (11.20) gives

$$\begin{aligned}\vec{\bar{F}}_b &= \bar{\mathbb{D}}_{pp}(\dot{\bar{\mathbf{q}}}_b - \dot{\bar{\mathbf{q}}}_{pp}) + \bar{\mathbb{K}}_{pp}(\bar{\mathbf{q}}_b - \bar{\mathbf{q}}_{pp})\\ \vec{\bar{F}}_p &= \bar{\mathbb{M}}_{pp}\ddot{\bar{\mathbf{q}}}_{pp} + \bar{\mathbb{D}}_{pp}(\dot{\bar{\mathbf{q}}}_{pp} - \dot{\bar{\mathbf{q}}}_b) + \bar{\mathbb{K}}_{pp}(\bar{\mathbf{q}}_{pp} - \bar{\mathbf{q}}_b)\\ &\quad + \vec{\bar{F}}_{gp(0)} + \bar{\mathbb{D}}_{gp}\dot{\bar{\mathbf{q}}}_{pp} + \bar{\mathbb{K}}_{gp}(\bar{\mathbf{q}}_{pp} - \bar{\mathbf{q}}_{pp(0)})\end{aligned} \tag{11.25}$$

In accordance to the previous normalisation rules of (11.21)–(11.22) and (11.24), the pivot point normalised mass matrix $\bar{\mathbb{M}}_{\text{pp}} \in \mathbb{R}^{2\times2}$ is defined as

$$\bar{\mathbb{M}}_{\text{pp}} = \begin{bmatrix} \bar{m}_{\text{p}} & -\bar{m}_{\text{p}}\bar{R}_{\text{m}} sin(\phi_{\text{m}}) \\ -\bar{m}_{\text{p}}\bar{R}_{\text{m}} sin(\phi_{\text{m}}) & \bar{m}_{\text{p}}\bar{R}_{\text{m}}^2 + \bar{I}_{\text{p}} \end{bmatrix} \tag{11.26}$$

with $\bar{R}_{\text{m}}$ and ϕ_{m} the normalised distance and angular direction from pivot-point to pad-centre-of-mass, respectively, see also Figure 11.14. The distance R_{m} is made dimensionless as $\bar{R}_{\text{m}} = R_{\text{m}}/(B - B_{\text{pp}})$. The mass $\bar{m}_{\text{p}}$ and inertia $\bar{I}_{\text{p}}$ are the dimensionless forms of the pad mass and tilt inertia (at centre of mass), which we defined as

$$\bar{m}_{\text{p}} = m_{\text{p}}\frac{\omega_{\text{p}}^2 C}{P_{\text{a}}LB} \qquad \bar{I}_{\text{p}} = I_{\text{p}}\frac{\omega_{\text{p}}^2 C}{P_{\text{a}}LB(B - B_{\text{pp}})^2} \tag{11.27}$$

The dimensionless pivot stiffness $\bar{\mathbb{K}}_{\text{pp}}$ and damping $\bar{\mathbb{D}}_{\text{pp}}$ matrix in (11.25) are defined in a similar way to (11.26), that is

$$\bar{\mathbb{K}}_{\text{pp}} = \begin{bmatrix} \bar{k}_{\epsilon\epsilon}^{\text{pp}} = k_{ee}^{\text{pp}}\frac{C}{P_{\text{a}}LB} & \bar{k}_{\epsilon\bar{\phi}}^{\text{pp}} = k_{e\phi}^{\text{pp}}\frac{C}{P_{\text{a}}LB(B-B_{\text{pp}})} \\ \bar{k}_{\bar{\phi}\epsilon}^{\text{pp}} = k_{\phi e}^{\text{pp}}\frac{C}{P_{\text{a}}LB(B-B_{\text{pp}})} & \bar{k}_{\bar{\phi}\bar{\phi}}^{\text{pp}} = k_{\phi\phi}^{\text{pp}}\frac{C}{P_{\text{a}}LB(B-B_{\text{pp}})^2} \end{bmatrix} \tag{11.28}$$

$$\bar{\mathbb{D}}_{\text{pp}} = \begin{bmatrix} \bar{d}_{\epsilon\epsilon}^{\text{pp}} = d_{ee}^{\text{pp}}\frac{\omega_{\text{p}}C}{P_{\text{a}}LB} & \bar{d}_{\epsilon\bar{\phi}}^{\text{pp}} = d_{e\phi}^{\text{pp}}\frac{\omega_{\text{p}}C}{P_{\text{a}}LB(B-B_{\text{pp}})} \\ \bar{d}_{\bar{\phi}\epsilon}^{\text{pp}} = d_{\phi e}^{\text{pp}}\frac{\omega_{\text{p}}C}{P_{\text{a}}LB(B-B_{\text{pp}})} & \bar{d}_{\bar{\phi}\bar{\phi}}^{\text{pp}} = d_{\phi\phi}^{\text{pp}}\frac{\omega_{\text{p}}C}{P_{\text{a}}LB(B-B_{\text{pp}})^2} \end{bmatrix} \tag{11.29}$$

Taking (11.25) and assuming small harmonic oscillations of $\bar{\mathbf{q}}_{\text{pp}}$ and $\bar{\mathbf{q}}_{\text{b}}$ (i.e. $\bar{\mathbf{q}}_{\text{pp}}(\tau^o) = \bar{\mathbf{q}}_{pp(0)} + \Delta\bar{\mathbf{q}}_{\text{pp}}e^{\tau^o i}$ and $\bar{\mathbf{q}}_{\text{b}}(\tau^o) = \bar{\mathbf{q}}_{b(0)} + \Delta\bar{\mathbf{q}}_{\text{b}}e^{\tau^o i}$) then the complex response amplitude can be derived as

$$\begin{bmatrix} \Delta\bar{F}_{\text{b}} \\ \Delta\bar{F}_{\text{p}} \end{bmatrix} = \begin{bmatrix} \bar{\mathbb{Z}}_{\text{pp}} + (\bar{\mathbb{Z}}_{\text{gp}} - \bar{\mathbb{M}}_{\text{pp}}) & -\bar{\mathbb{Z}}_{\text{pp}} \\ -\bar{\mathbb{Z}}_{\text{pp}} & \bar{\mathbb{Z}}_{\text{pp}} \end{bmatrix} \begin{bmatrix} \Delta\bar{\mathbf{q}}_{\text{pp}} \\ \Delta\bar{\mathbf{q}}_{\text{b}} \end{bmatrix}. \tag{11.30}$$

Here $\bar{\mathbb{Z}}_{\text{gp}}$ is the dynamic gas-film stiffness according to (11.19) and $\bar{\mathbb{Z}}_{\text{pp}}$ is the dynamic stiffness of the pivot point structure. That is,

$$\bar{\mathbb{Z}}_{\text{pp}} = \bar{\mathbb{K}}_{\text{pp}} + \bar{\mathbb{D}}_{\text{pp}}i \tag{11.31}$$

By setting $\Delta\bar{F}_{\text{p}}$ equal to 0, the system of equations defined in (11.30) can be solved for $\Delta\bar{F}_{\text{b}} = f(\Delta\bar{\mathbf{q}}_{\text{b}})$. That is

$$\Delta\bar{F}_{\text{b}} = \underbrace{[\bar{\mathbb{Z}}_{\text{pp}} - \bar{\mathbb{Z}}_{\text{pp}}(\bar{\mathbb{Z}}_{\text{pp}} + \bar{\mathbb{Z}}_{\text{gp}} - \bar{\mathbb{M}}_{\text{pp}})^{-1}\bar{\mathbb{Z}}_{\text{pp}}]}_{\bar{\mathbb{Z}}_{\text{eq}}}\Delta\bar{\mathbf{q}}_{\text{b}} \tag{11.32}$$

Such that the equivalent dynamic bearing stiffness becomes:

$$\bar{\mathbb{Z}}_{\text{eq}}(\omega_{\text{p}}) = \bar{\mathbb{Z}}_{\text{pp}} - \bar{\mathbb{Z}}_{\text{pp}}(\bar{\mathbb{Z}}_{\text{pp}} + \bar{\mathbb{Z}}_{\text{gp}} - \bar{\mathbb{M}}_{\text{pp}})^{-1}\bar{\mathbb{Z}}_{\text{pp}} \in \mathbb{C}^{2\times2} \tag{11.33}$$

An example of the frequency dependency of $\bar{\mathbb{Z}}_{\text{eq}}(\omega_{\text{p}})$ is given in Figure 11.16. The figure only shows the direct stiffness and damping terms, since the cross terms are zero due to no pivot stiffness and no cross terms in the mass matrix $\bar{\mathbb{M}}_{\text{pp}}$. Note that the eigenfrequency of the system plays an important role in the dynamic stiffness of the bearing, causing even negative dynamic stiffness values.

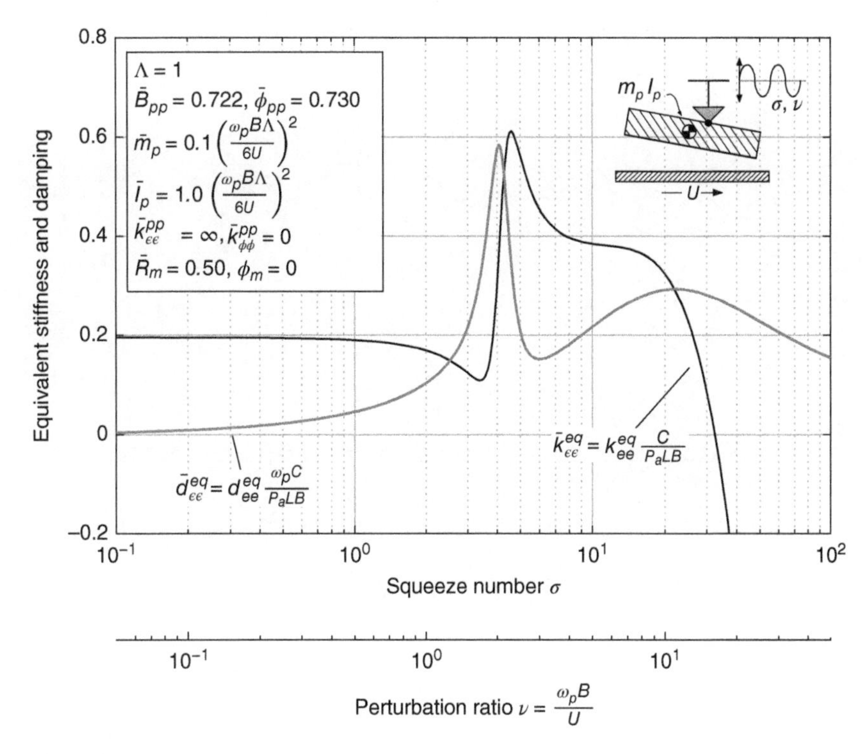

Figure 11.16 Equivalent and frequency depending stiffness and damping of a pivoting slider bearing. Source: Nabuurs M 2020.

11.4 Tilting Pad Journal Bearing

The focus in this section is mainly on the modeling of journal gas bearings of the tilting pad type. As for any other type of gas bearing, a mathematical model for load capacity, frictional losses and dynamic stiffness is essential for designing and implementing tilting pad gas bearings. However, the multiple degrees of freedom of a tilting pad bearing makes it more complicated in comparison with fixed geometry bearings: it has more degrees of freedom than a rigid plain or single pivoted-pad sliding bearing. That is, each individual pad can undergo a tilt movement around its pivot point. Such a pivot point may also be mounted on a compliant structure whereby it can move in the radial direction. The pad will have, in such a case, two degrees of freedom, i.e. one tilt and one translation. The gas film thickness within the bearing will therefore not only depend on the absolute position of the shaft, but also on the position and orientation of the individual tilting pads.

A tilting pad bearing comprising at least two pads will be considered in the remainder of this section. It is assumed that each pad has its own individual pivot point, about which the corresponding pad can tilt. The individual pivot points are either equally spaced around the circumference of the bearing, or at a certain angle ψ as depicted in Figure 11.17. Here a right-handed XY-coordinate system is used with a clockwise rotating shaft.

Similar to the pivoted-slider bearing, the pivot point itself can be mounted on a compliance structure as illustrated in Figure 11.18. This will give the pad, besides its pivot degree of freedom (DOF), also an additional radial

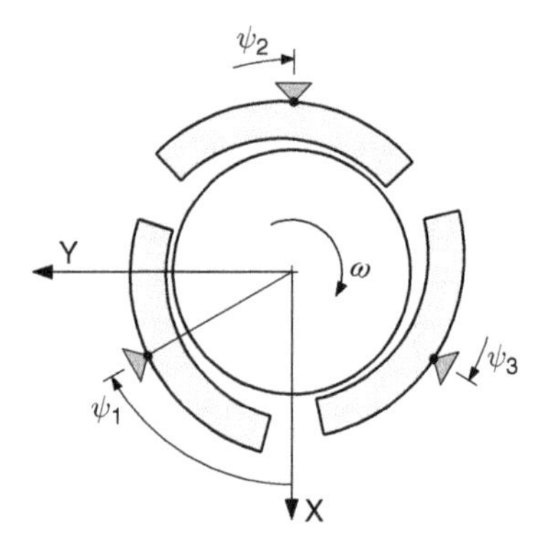

Figure 11.17 Global *XY*-coordinate system. Source: Nabuurs M 2020.

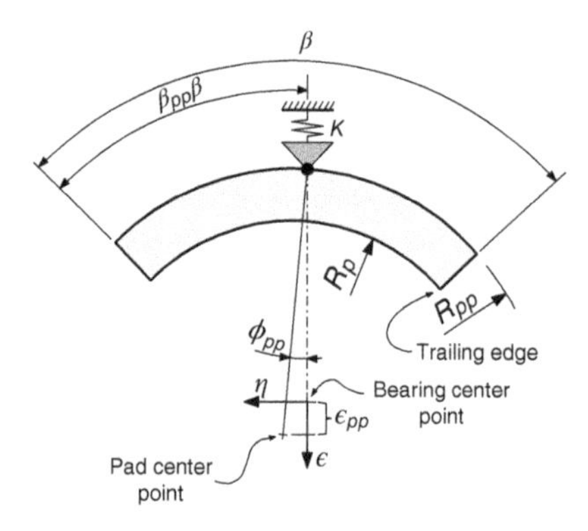

Figure 11.18 Local pad $\epsilon\eta$-coordinate system. Source: Nabuurs M 2020.

translation DOF. The parameters ϕ_{pp} and e_{pp} will be used to describe the pad tilt and radial position. The reference positions $\phi_{pp} = 0$ and $e_{pp} = 0$ stand for an orientation whereby the pad centre of curvature coincides with the bearing centre point. Note from Figure 11.18 that a positive value for ϕ_{pp} results in a displacement of the pad trailing edge towards the shaft and a positive value for e_{pp} results in a radial inwards displacement of the pad pivot. Similar to (11.24), the pad position parameters are made dimensionless and gathered in the generalised pad position column vector $\bar{\boldsymbol{q}}_{pp}$, according to

$$\bar{\boldsymbol{q}}_{pp} = \begin{bmatrix} \epsilon_{pp} \\ \bar{\phi}_{pp} \end{bmatrix} = \begin{bmatrix} \frac{e_{pp}}{C} \\ \frac{\phi_{pp} R_{pp}}{C} \end{bmatrix} \tag{11.34}$$

here R_p is the radius of the pad inner curvature and C the nominal clearance between pad and shaft curvature, i.e. $C = R_p - R_s$ (R_s being the shaft radius). The position of the shaft with respect to the global XY-coordinate system is describe with the generalised position vector $\boldsymbol{q}_s = [x_s, y_s]^T$. For convenience, $\boldsymbol{q}_s$ is made dimensionless by dividing it by the bearing clearance C to obtain

$$\bar{\boldsymbol{q}}_s = \begin{bmatrix} \bar{x}_s \\ \bar{y}_s \end{bmatrix} = \begin{bmatrix} \frac{x_s}{C} \\ \frac{y_s}{C} \end{bmatrix} \tag{11.35}$$

Gas-film geometry

The small gap between an individual tilting pad and the rotating journal determines the gas-film geometry of that particular bearing section. As such, any change in the position of either the journal or the adjacent pad will change the gas-film geometry and thus its aerodynamic behaviour. Within this work, the local gap or film-geometry beneath each individual tilting pad is defined as

$$\bar{h}(\bar{\theta}) = \frac{h(\bar{\theta})}{C} = 1 - |\vec{\epsilon}| \cos(\bar{\theta} - \angle\vec{\epsilon}) + \mathcal{O}(|\vec{\epsilon}|^2) \tag{11.36}$$

here $\bar{\theta}$ is the local tangential gas-film coordinate with the pad pivot point at $\bar{\theta} = 0$. The pad leading and trailing edges form the gas-film boundaries at respectively $\bar{\theta} = -\beta_{pp}\beta$ and $\bar{\theta} = (1 - \beta_{pp})\beta$.

Vector $\vec{\epsilon}$ express the gas-film eccentricity and is defined as the distance between pad center-of-curvature and journal center-of-curvature. Note that the position of an individual pad and the journal are mathematically expressed by the previously-defined position vector $\bar{\boldsymbol{q}}_{pp}$ and $\bar{\boldsymbol{q}}_s$ respectively. As such, for each pad $i \in \{1 \cdots N\}$ at angle ψ_i, the local gas-film eccentricity yields

$$\vec{\epsilon} = \bar{\boldsymbol{q}}_{pp} + \begin{bmatrix} \cos(\psi_i) & \sin(\psi_i) \\ -\sin(\psi_i) & \cos(\psi_i) \end{bmatrix} \bar{\boldsymbol{q}}_s \tag{11.37}$$

Note that angle ψ_i is the azimuth angle of the pivoting point of pad i, as further illustrated in Figure 11.17.

Pressure equation

The reduced form of the Reynolds equation is derived and discussed in Chapter 4, including the case of cylindrical coordinates. Here we show, the particular way it is made dimensionless for our purpose, which might be different from that expected by the reader.

Thus, in the remainder of this section, the following dimensionless form of the reduced Reynolds equation is used:

$$\frac{\partial}{\partial\bar{\theta}}\left(\bar{p}\bar{h}^3\frac{\partial\bar{p}}{\partial\bar{\theta}}\right)^2 + \frac{1}{\lambda^2}\frac{\partial}{\partial\bar{y}}\left(\bar{p}\bar{h}^3\frac{\partial\bar{p}}{\partial\bar{y}}\right) = \Lambda\frac{\partial\bar{p}\bar{h}\bar{u}}{\partial\bar{\theta}} + 2\Lambda\frac{\partial\bar{p}\bar{h}}{\partial\tau} \tag{11.38}$$

with the following dimensionless quantities

$$\Lambda = \frac{6\mu\omega}{P_a}\left(\frac{R_p}{C}\right)^2 \quad \bar{p} = \frac{p}{P_a} \quad \bar{h} = \frac{h}{C} \tag{11.39}$$

$$\lambda = \frac{L}{2R_p} \qquad \tau = \omega t \quad \bar{u} = \frac{\delta}{\delta + 1} \quad \bar{y} = \frac{z}{1/2L} \tag{11.40}$$

It is assumed that the gas-film is fully developed over the entire domain and has ambient pressure P_a at the boundaries, which results in the following Dirichlet boundary condition:

$$\bar{p}(\bar{\theta}, \bar{y}) = 1\,|_{\bar{y}=\{-1,1\}} \tag{11.41}$$

$$\bar{p}(\bar{\theta}, \bar{y}) = 1\,|_{\bar{\theta}=\{-\beta_{pp}\beta,(1-\beta_{pp})\beta\}} \tag{11.42}$$

Other boundary conditions which may apply are symmetry and continuity conditions. The former is frequently used to cut the domain (and thus also the computation time) in half, while the latter replaces (11.42) to obtain continuity of pressure $\bar{p}$ for domains without leading or trailing edges Waumans (2009).

$$\frac{\partial\bar{p}(\bar{\theta}, \bar{y})}{\partial\bar{y}} = 0\,|_{\bar{y}=0} \qquad \text{symmetry} \tag{11.43}$$

$$\bar{p}(\bar{\theta}, \bar{y})\,|_{\bar{\theta}=-\beta_{pp}\beta} = \bar{p}(\bar{\theta}, \bar{y})\,|_{\bar{\theta}=(1-\beta_{pp})\beta} \qquad \text{continuity} \tag{11.44}$$

All of the above boundary conditions can be implemented fairly easily, which is one of the few convenient aspects of gas-films in comparison to liquid-films, where effects as cavitation and two-phase flow also need to be considered (Gross 1980). Nevertheless, the complete boundary value problem is difficult to solve due to the non-linear nature of the Reynolds equation. It has become apparent that no analytical solutions exist for compressible gas-film

(Castelli and Pirvics 1968), so that solutions for $\bar{p}$ have to be found with suitable numerical methods as indicated in Chapter 2.

Gas film forces

In the study of tilting pad gas bearings it is required to compute the aerodynamic forces which act on the individual tilting pads. These forces must be in equilibrium with the remaining forces acting on the pads. If this is not the case, then the pads will change there tilt or radial position in order to restore force equilibrium. Thus, to further analyse the behaviour of a tilting pad bearing, we shall first derive a mathematical equation for the tilt moment M_{gp} and the radial force F_{gp} acting on an individual pivoting point within the local $\epsilon\eta$-frame. Here we shall use a dimensionless notation for the forces as:

$$\bar{F}_{\epsilon}^{gp} = \frac{F_{e}^{gp}}{P_{a}LR_{\mathrm{p}}} \qquad \bar{F}_{\eta}^{gp} = \frac{F_{n}^{gp}}{P_{a}LR_{\mathrm{p}}} \qquad \bar{M}_{\bar{\phi}}^{gp} = \frac{M_{\phi}^{gp}}{P_{a}LR_{\mathrm{p}}R_{\mathrm{pp}}} \tag{11.45}$$

It will be more convenient to first derive the forces with respect to the pad center-of-curvature and transform them afterwards to the pad pivot point. Thus, the force at the pad center-of-curvature $\vec{\bar{F}}_{\mathrm{pc}}$ can be found by integrating the friction $\bar{\tau}$ and pressure $\bar{p}$ profile according to

$$\vec{\bar{F}}_{\mathrm{pc}} = \begin{bmatrix} \bar{F}_{\epsilon}^{pc} \\ \bar{F}_{\eta}^{pc} \\ \bar{M}_{\bar{\phi}}^{pc} \end{bmatrix} = \frac{1}{2} \int_{\beta_{\mathrm{pp}}\beta}^{(1-\beta_{\mathrm{pp}})\beta} \int_{-1}^{1} \begin{bmatrix} -\cos(\bar{\theta}) & \sin(\bar{\theta}) \\ -\sin(\bar{\theta}) & -\cos(\bar{\theta}) \\ 0 & 1 \end{bmatrix} \begin{bmatrix} \bar{p}-1 \\ \bar{\tau} \end{bmatrix} d\bar{y}d\bar{\theta} \tag{11.46}$$

with $\bar{\tau}$ as the frictional term, defined as

$$\bar{\tau}(\bar{\theta},\bar{y}) = \frac{1}{\delta+1}\left(\frac{\bar{h}}{2}\frac{\delta\bar{p}}{\delta\bar{\theta}} + \frac{\Lambda}{6}\frac{\bar{u}}{\bar{h}}\right) \tag{11.47}$$

with the clearance ratio $\delta = Rp/C$. The frictional term $\bar{\tau}$ is often not mentioned in literature about tilting pad bearings. It is, however, incorrect to ignore frictional forces since they will contribute to the tilt moment around the pivot point. Especially at high rotational speed where the pressure term $\bar{p}$ will saturate to a certain level but the friction term $\bar{\tau}$ will keep increasing proportionally with bearing number Λ.

The aerodynamic forces at the pad centre point (11.46), may be transformed with an appropriate transformation matrix to obtain the forces on the pad pivot point $\vec{\bar{F}}_{\mathrm{gp}}$, that is

$$\vec{\bar{F}}_{\mathrm{gp}} = \begin{bmatrix} \bar{F}_{\epsilon}^{gp} \\ \bar{M}_{\bar{\phi}}^{gp} \end{bmatrix} = \begin{bmatrix} 1 & 0 & 0 \\ 0 & 1 & \frac{\delta+1}{\varsigma} \end{bmatrix} \vec{\bar{F}}_{\mathrm{pc}} \tag{11.48}$$

with $\varsigma = R_{\mathrm{pp}}/C$ being the dimensionless pivot point distance ratio and R_{pp} the absolute pivot point distance as shown in Figure 11.18.

Pad equation of motion

The equations of motion for each individual tilting pad may be written as:

$$\overbrace{\begin{bmatrix} F_{e}^{gp}(t) \\ M_{\phi}^{gp}(t) \end{bmatrix}}^{\text{Gas-film forces } \vec{F}_{\mathrm{gp}}} = \overbrace{\begin{bmatrix} m_{\mathrm{p}} & -m_{\mathrm{p}}R_{\mathrm{m}}sin(\phi) \\ -m_{\mathrm{p}}R_{\mathrm{m}}sin(\phi) & m_{\mathrm{p}}R_{\mathrm{m}}^{2} + I_{\mathrm{p}} \end{bmatrix}}^{\text{Mass matrix } \mathbb{M}_{\mathrm{pp}}} \begin{bmatrix} \ddot{\mathbf{q}}_{e}^{\mathrm{pp}} \\ \ddot{\mathbf{q}}_{\phi}^{\mathrm{pp}} \end{bmatrix} + \underbrace{\begin{bmatrix} d_{ee} & d_{e\phi} \\ d_{\phi e} & d_{\phi\phi} \end{bmatrix}}_{\substack{\text{Pad damping} \\ \text{matrix } \mathbb{D}_{\mathrm{pp}}}} \begin{bmatrix} \dot{\mathbf{q}}^{\mathrm{pp}}{}_{e} \\ \dot{\mathbf{q}}_{\phi}^{\mathrm{pp}} \end{bmatrix} + \underbrace{\begin{bmatrix} k_{ee} & k_{e\phi} \\ k_{\phi e} & k_{\phi\phi} \end{bmatrix}}_{\substack{\text{Pad stiffness} \\ \text{matrix } \mathbb{K}_{\mathrm{pp}}}} \begin{bmatrix} \mathbf{q}_{e}^{\mathrm{pp}} - \delta_{e}^{\mathrm{pp}} \\ \mathbf{q}_{\phi}^{\mathrm{pp}} - \delta_{\phi}^{\mathrm{pp}} \end{bmatrix}. \tag{11.49}$$

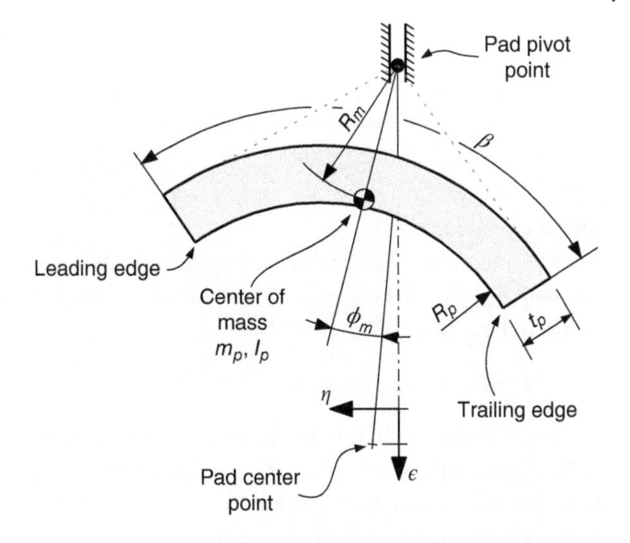

Figure 11.19 Pad centre of mass location. Source: Nabuurs M 2020.

The right-handed terms represent the pad dynamics with δ_e^{pp} and δ_ϕ^{pp} as the pivoting point geometrical preload. Note that the mass matrix $\mathbb{M}_{pp}$ includes the pad tilt inertia I_p. The inertia is computed at its centre of mass, which is assumed to be at an angle ϕ_m and distance R_m from the pivot point. (See also Figure 11.19 for clarification). The off-diagonal terms in the mass matrix are often ignored when studying tilting pad bearings. This is however only valid if the pad centre of mass is on the imaginary line between pad pivot point and the centre point of the pad, i.e. when $\phi_m = 0$. Which is, in practice, rarely the case so that the off-diagonal terms must be included in the mathematical model.

For simple pad geometries, such as the one shown in Figure 11.19, the pad mass m_p and its tilt inertia I_p may be analytically obtained from equations (11.50) and (11.51). That is, by using basic geometry, in combination with parallel axis theorem, and assuming a uniform pad thickness t_p and volumetric mass density ρ, the pad mass m_p and tilt inertia I_p are obtained by solving

$$m_p = 1/2\rho L\beta R_p^2(\alpha^2 - 1) \tag{11.50}$$

$$I_p = m_p\left[1/2R_p^2(\alpha^2 + 1) - \left(\frac{2R_p(\alpha^3 - 1)}{3\beta(\alpha^2 - 1)}\right)^2 (2 - 2\cos(\beta))\right] \tag{11.51}$$

where $\alpha = 1 + \frac{t_p}{R_p}$.

Similar to the above equations, the cross term $R_m \sin(\phi)$ within the mass matrix may be obtained by solving

$$R_m \sin(\phi) = \frac{2R_p(\alpha^3 - 1)}{3\beta(\alpha^2 - 1)}\sqrt{2 - 2\cos\beta}\,\sin\left(\beta_{pp} - \frac{1}{2}\beta\right). \tag{11.52}$$

The equation of motion (11.49) can be simplified and written in dimensionless form by performing a Fourier transformation $\mathscr{F}(\nu\omega)$ and substituting the following dimensionless quantities:

$$\bar{\mathbb{K}}_{pp} = \mathbb{K}_{pp}\circ\mathbb{U} \qquad \bar{\mathbb{D}}_{pp} = \mathbb{D}_{pp}\circ\nu\omega\mathbb{U} \qquad \bar{\mathbb{M}}_{pp} = \mathbb{M}_{pp}\circ\nu^2\omega^2\mathbb{U} \quad , \tag{11.53}$$

where the film perturbation ratio $\nu = \omega_p/\omega$ and the normalisation matrix $\mathbb{U}$ is

$$\mathbb{U} = \begin{bmatrix} \frac{C}{P_a L R_p} & \frac{C}{P_a L R_p R_{pp}} \\ \frac{C}{P_a L R_p R_{pp}} & \frac{C}{P_a L R_p R_{pp}^2} \end{bmatrix} \tag{11.54}$$

and $\circ$ represent the Hadamard product of two matrices.

Now, with (11.53), the equation of motion can be rewritten in dimensionless form as:

$$\vec{\bar{F}}_{\text{gp}}(\bar{\boldsymbol{q}}_{\text{pp}}, \bar{\boldsymbol{q}}_{\text{s}}) = \underbrace{(\bar{\mathbb{K}}_{\text{pp}} + \bar{\mathbb{D}}_{\text{pp}} i - \bar{\mathbb{M}}_{\text{pp}})}_{\bar{\mathbb{Z}}_{\text{pp}}(\nu)\in\mathbb{C}^{2\times2}}\bar{\boldsymbol{q}}_{\text{pp}} - \bar{\mathbb{K}}_{\text{pp}}\bar{\delta}_{\text{pp}} \tag{11.55}$$

Force equilibrium

In order to analyse the bearing performance at its operation point. It is first required to find the static position of the shaft $\bar{\boldsymbol{q}}_{\text{s}(0)}$ and the pad $\bar{\boldsymbol{q}}_{pp(0)}$ where the forces are in equilibrium. That is, to find the static value $\bar{\boldsymbol{q}}_{\text{s}(0)}$ and $\bar{\boldsymbol{q}}_{pp(0)}$, for $\bar{\boldsymbol{q}}_{\text{s}}$ and $\bar{\boldsymbol{q}}_{\text{pp}}$ respectively, which satisfy:

$$\vec{\bar{F}}_{\text{gp}}(\bar{\boldsymbol{q}}_{\text{pp}}, \bar{\boldsymbol{q}}_{\text{s}}) = \bar{\mathbb{K}}_{\text{pp}}(\bar{\boldsymbol{q}}_{\text{pp}} - \bar{\delta}_{\text{pp}}) \tag{11.56}$$

It is worthy of mention that there are more equilibrium points possible as is graphically shown with Figure 11.20.

As illustrated in this figure, there could exist two equilibrium positions for each pad. The left-most equilibrium position has a negative value for $\bar{\phi}$, meaning that the pad is tilted backwards and a diverging gap is formed. Although this is a mathematically valid solution, and could, in theory, even be stable, in practice it is assumed that the bearing operates with a converging gap such that all possible solutions with negative $\bar{\phi}_{\text{pp}}$ values are discarded. The diverging film geometry could also be mechanically avoided by using mechanical tilt stops or torsion springs Boyd and Raimondi (1962). The second equilibrium point is asymptotically stable and has a positive value for $\bar{\phi}_{\text{pp}}$. It is therefore typically selected to be the pad equilibrium point $\bar{\boldsymbol{q}}_{\text{pp}(0)}$.

11.4.1 Steady State Bearing Characteristics

Until now, only the individual gas-film segments and there associated tilting pads have been characterised. Namely, the pressure and exerted force of a individual gas-film segment is described by (11.38) and (11.46),

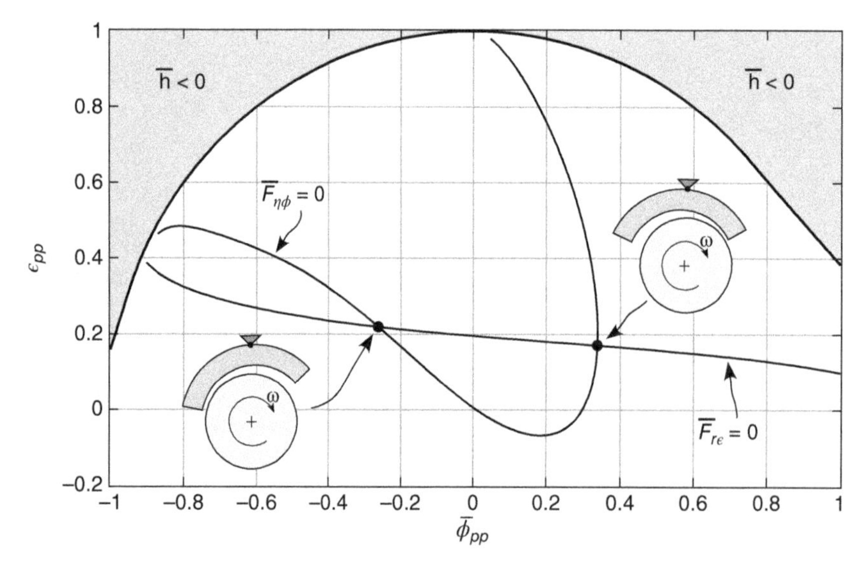

Figure 11.20 Locus of the equilibrium points. The curved lines represent the position of the the pad where the residue force $F_{r\epsilon} = 0$ and $\bar{F}_{\eta\phi} = 0$. The intersection point of the two curves, at $\bar{q}_{\text{pp}} = [0.219, -0.262]$ and $\bar{q}_{\text{pp}} = [0.170, 0.338]$, represent the position of the pad where there is force balance. This graph is made for a 3-pad bearing with $\beta = 120°$, $\beta_{\text{pp}} = 0.6$, $\lambda = 1$, $\delta = 200$, $\varsigma = 300$, $\delta_\epsilon = 0.2$, $\bar{k}_{\text{pp}}^{\phi\phi} = 4$, $\bar{\boldsymbol{q}}_{\text{s}} = \boldsymbol{0}$ and $\Lambda = 1$. Source: Nabuurs M 2020.

while the dynamics of the a single pad is described by (11.55). However, none of these equations describe the characteristics of the complete tilting pad bearing which has in general multiple pads. Hence, the purpose of this section is to characterise the complete tilting pad bearing.

That is, the bearing overall load capacity and power losses are first expressed with use of Lund's assembly method, Nicholas (2003). Next, the dynamic gas-film stiffness is derived and combined with the corresponding pad dynamics to obtain the equivalent dynamic pad stiffness $\mathbb{Z}_{eq}$. Subsequent summation of the individual equivalent stiffness matrices provides the global bearing stiffness $\mathbb{K}_b$ and damping $\mathbb{D}_b$ matrices. The coefficients of these matrices depend on the perturbation ratio ν and bearing number Λ. This dependency is significantly more complex for tilting pad bearings than for plain or lobe-shaped bearings, as will be further clarified at the end of this section.

Bearing load capacity

In a tilting pad bearing with number of pads N, the total gas-film force on the rotating journal is the force summation of the N individual gas-film sections. For example, the tilting-pad bearing illustrated in Figure 11.21 has multiple pads and thus multiple gas-film sections.

Each of these gas-film sections can be modeled separately with the previously proposed model to obtain the aerodynamic force $\vec{\bar{F}}_{pc}$ according to (11.46). Notice that this force vector is with respect to the pad local $\epsilon\eta$-coordinate system. As such, an additional coordinate transformation needs to be performed, prior to summation, to obtain the overall bearing force $\vec{\bar{F}}_b$, that is

$$\vec{\bar{F}}_b = \begin{bmatrix} \bar{F}_x^b \\ \bar{F}_y^b \\ \bar{M}_{\bar{\phi}}^b \end{bmatrix} = \sum_{i=1}^{N} \begin{bmatrix} -\cos(\psi_i) & +\sin(\psi_i) & 0 \\ -\sin(\psi_i) & -\cos(\psi_i) & 0 \\ 0 & 0 & 1 \end{bmatrix} \left\{ \vec{\bar{F}}_{pc} \right\}_i \tag{11.57}$$

The resulting bearing forces vector $\vec{\bar{F}}b = \{\bar{F}_x^b, \bar{F}_y^b, \bar{M}_{\bar{\psi}}^b\}$ is with respect to the global XY-coordinate system and normalised according to

$$\bar{F}_x^b = \frac{F_x^b}{P_a L R_p} \qquad \bar{F}_y^b = \frac{F_y^b}{P_a L R_p} \qquad \bar{M}_{\bar{\phi}}^b = \frac{M_b}{P_a L R_p^2} \tag{11.58}$$

The normalised bearing load-carrying capacity $\bar{W}$ is typically defined as the resultant of the radial forces $\bar{F}_x^b$ and $\bar{F}_y^b$, which is given quite simply as

$$\bar{W} = \frac{W}{P_a L R_p} = \sqrt{(\bar{F}_x^b)^2 + (\bar{F}_y^b)^2} \tag{11.59}$$

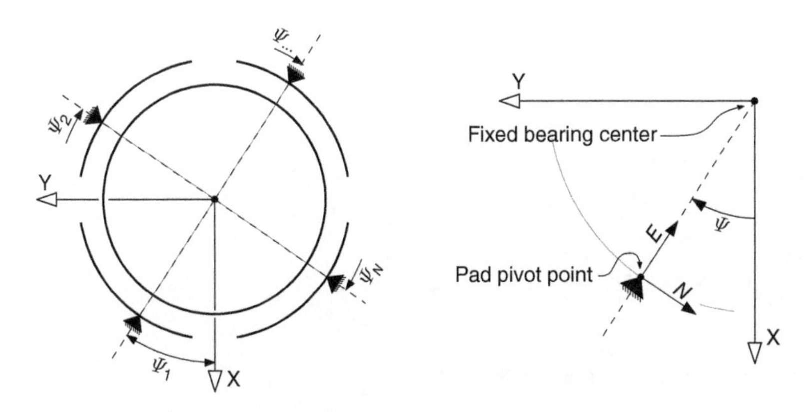

Figure 11.21 Pad assembly angles. Source: Nabuurs M 2020.

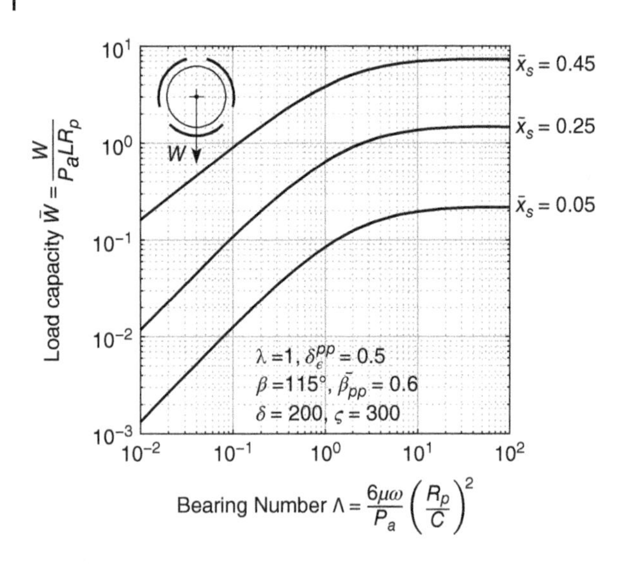

Figure 11.22 Bearing load capacity, load on pad. Source: Nabuurs M 2020.

The tilting pad bearing load capacity depends (besides the bearing geometry) mainly on the journal eccentricity and rotational speed. This dependency is shown in Figure 11.22 and 11.23 for an arbitrary tilting pad bearing with 3-pads.

Figure 11.22 shows the bearing load capacity when the load is placed directly at one of the pads, in line with one of the pad pivot points. The three lines represent the load curve for a journal eccentricity of respectively 5%, 25%, and 45% of the nominal film clearance C. A slightly different situation occurs when the load is placed between two adjacent pads. This situation, as shown in Figure 11.23, provides a similar load capacity for small eccentricity ratios. However, if eccentricity ratios become large, the bearing load capacity becomes significant lower in comparison to the load on pad situation. Meaning that the load capacity of a tilting pad (and thus also its stiffness) is position and direction depending.

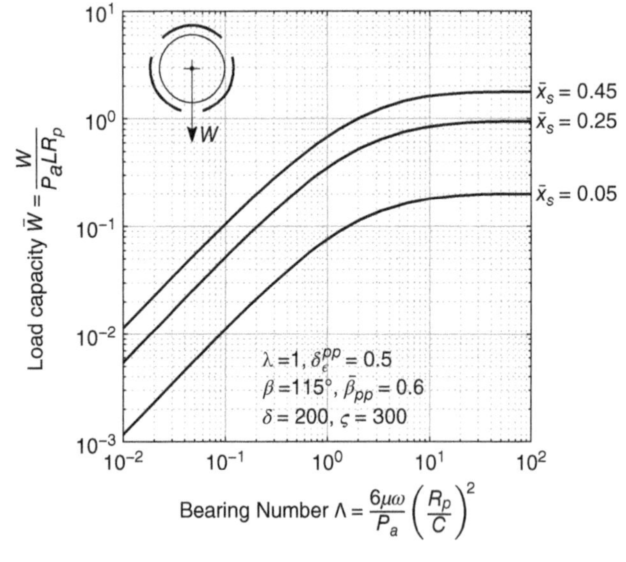

Figure 11.23 Bearing load capacity, load between pads. Source: Nabuurs M 2020.

Power loss

The bearing total power loss P is simple the summation of film losses of each individual pad, or in dimensionless form

$$\bar{P} = \sum_{i=1}^{N} \bar{M}_{\phi_i}^{pc} = \frac{1}{2}\sum_{i=1}^{N} \int_{\beta_{pp}\beta}^{(1-\beta_{pp})\beta} \int_{-1}^{1} \bar{\tau}_i \, d\bar{y} d\bar{\theta}, \tag{11.60}$$

where $\bar{M}_{\phi}^{pc}$ can be obtained from (11.46) or $\bar{\tau}$ directly from (11.47).

Note that the power loss $\bar{P}$ is normalised in this section according to

$$\bar{P} = \frac{P}{P_a L R_s^2 \omega} \tag{11.61}$$

11.4.2 Dynamic Stiffness of a Tilting Pad Bearing

The dynamic stiffness of a tilting pad gas bearing may be characterised by first deriving the individual pad equivalent stiffness as previously performed for the tilled sliding bearing. That is, the dynamic gas-film stiffness underneath each pad is first characterised by a dynamic stiffness matrix $\mathbb{Z}_{pc} \in \mathbb{C}^{3\times 2}$, which could be derived by assuming a harmonic excitation of the gas-film eccentricity vector $\vec{\epsilon}$

$$\vec{\epsilon}(t) = \vec{\epsilon_{(0)}} + \mathrm{Real}(\Delta\vec{\epsilon}e^{i\nu\tau}), \tag{11.62}$$

where $\Delta\vec{\epsilon}$ is a complex amplitude.

The resulting forces will have the form:

$$\bar{F}_{pc}(t) = \bar{F}_{pc(0)} + \mathrm{Real}(\Delta\bar{F}_{pc}e^{i\nu\tau}) \tag{11.63}$$

so that the dynamic stiffness matrix is defined as:

$$\Delta\bar{F}_{pc} = \bar{\mathbb{Z}}_{pc}\Delta\vec{\epsilon}. \tag{11.64}$$

In other words, $\bar{\mathbb{Z}}_{pc}$ is the Jacobian matrix of $\bar{F}_{pc}(t)$ with respect to $\Delta\vec{\epsilon}(t)$.

Equivalent pad stiffness

Similar to the pivoted slider bearing equivalent stiffness (11.33), the equivalent tilting pad journal bearing stiffness can also be derived as

$$\bar{\mathbb{Z}}_{eq} = -\bar{\mathbb{Z}}_{gs}(\bar{\mathbb{Z}}_{pp} + \bar{\mathbb{Z}}_{gp})^{-1}\bar{\mathbb{Z}}_{gp} + \bar{\mathbb{Z}}_{gs} \in \mathbb{C}^{2\times 2} \tag{11.65}$$

where the dynamic gas film stiffness contribution to the pad is given as:

$$\bar{\mathbb{Z}}_{gp} = \begin{bmatrix} 1 & 0 & 0 \\ 0 & 1 & \frac{\delta+1}{\varsigma} \end{bmatrix} \bar{\mathbb{Z}}_{pc} \tag{11.66}$$

and the dynamic gas film stiffness contribution to the shaft is given as:

$$\bar{\mathbb{Z}}_{gs} = \begin{bmatrix} 1 & 0 & 0 \\ 0 & 1 & 0 \end{bmatrix} \bar{\mathbb{Z}}_{pc} \tag{11.67}$$

and $\bar{\mathbb{Z}}_{pp}$ is according to equation (11.55)

Note that $\bar{\mathbb{Z}}_{pp}$ is a function of the perturbation frequency ratio ν, dimensionless bearing number Λ, shaft position $\bar{\boldsymbol{q}}_s$, in addition to the (many) other bearing design parameters.

Overall dynamic bearing stiffness

As for the overall static bearing load and power-loss characteristics, we can also derive the overall dynamic bearing stiffness $\bar{\mathbb{Z}}_b(\nu, \Lambda)$. Thus, by utilising the well-established Lund's assembly method Lund (1968), Nicholas (2003), the overall dynamic stiffness is expressed as

$$\bar{\mathbb{Z}}_b(\nu) = \sum_{i=1}^{N} \{[\mathbb{T}_{\psi_i}(\psi_i)]^T \bar{\mathbb{Z}}_{eq_i}(\nu)\mathbb{T}_{\psi_i}(\psi_i)\} \text{ with } \mathbb{T}_{\psi_i} = \begin{bmatrix} \cos\psi_i & +\sin\psi_i \\ -\sin\psi_i & \cos\psi_i \end{bmatrix} \tag{11.68}$$

The linearised dynamic force amplitude, on a harmonically perturbed journal, is then expressed by

$$\Delta\vec{\bar{F}}_b(\nu) = \bar{\mathbb{Z}}_b(\nu)\Delta\bar{q}_s. \tag{11.69}$$

Notice that $\bar{\mathbb{Z}}_b(\nu)$ is a complex 2×2 matrix. Separating it into its real and complex parts provides the frequency depending stiffness $\bar{\mathbb{K}}_b(\nu)$ and damping $\bar{\mathbb{D}}_b(\nu)$ coefficients. For small harmonic journal motions, the bearing exerted force may now be expressed with the linear model

$$\vec{\bar{F}}_b(\tau) = \vec{\bar{F}}_{b(0)}(\bar{q}_{s(0)}) + \bar{\mathbb{K}}_b(\nu, \bar{q}_{s(0)})(\bar{q}_s(\tau) - \bar{q}_{s(0)}) + \bar{\mathbb{D}}_b(\nu, \bar{q}_{s(0)})\dot{\bar{q}}_s(\tau), \tag{11.70}$$

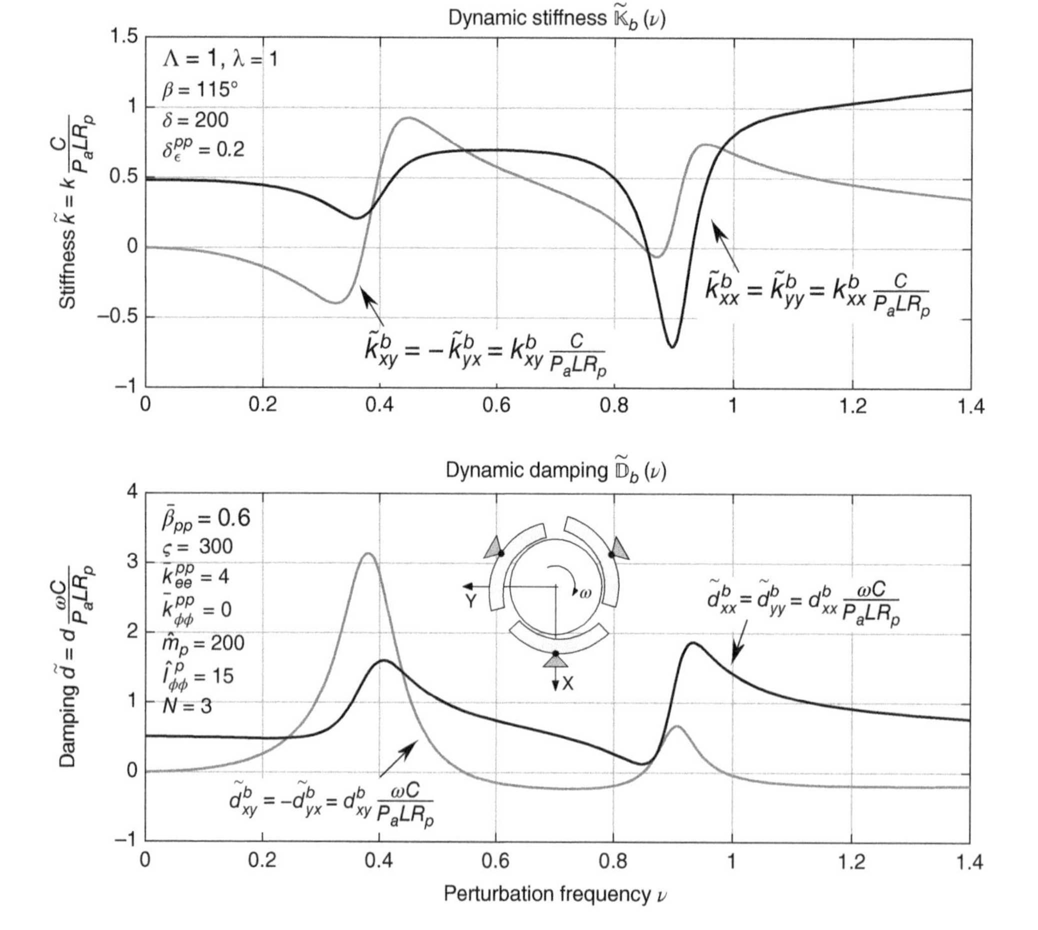

Figure 11.24 Dynamic stiffness and damping coefficient values for an axisymmetric three-pad bearing with arbitrary geometry as a function of the perturbation ratios. Coefficients are obtained with the proposed mathematical model. Source: Nabuurs M 2020.

with $q_{s(0)}$ being the journal static position. The stiffness $\bar{\mathbb{K}}_b$ and damping $\bar{\mathbb{D}}_b$ matrices occurring in (11.70) are actually a function of many parameters. Namely:

$$\{\bar{\mathbb{Z}}_b, \bar{\mathbb{K}}_b, \bar{\mathbb{D}}_b\} = f(\Lambda, \lambda, \nu, \delta, \beta, \bar{q}_{s(0)}, \beta_{pp}, \varsigma, \bar{\mathbb{M}}_{pp}, \bar{\mathbb{D}}_{pp}, \bar{\mathbb{K}}_{pp}, \bar{\delta}_{pp}, \Psi_n, N) \tag{11.71}$$

Most of these parameters are even multidimensional and differ for each individual pad. The bearing coefficients, for a three-pad bearing, can therefore easily depend on more than a hundred individual variables. Thus, it is practically impossible to present the coefficients for all generic or practical bearing configurations. Nevertheless, for illustration purpose, Figure 11.24 presents the stiffness and damping coefficients for a three-pad bearing as obtained with the model presented in this section.

As shown in the figure, the bearing coefficients are strongly dependent on the perturbation frequency ratio ν due to the pad's natural frequencies. The first dynamics, at $\nu \approx 0.4$ is in this particular case caused by the pad rotational mode while the dynamics at $\nu \approx 0.7$ is caused by the pad translational mode. Notice that due to the pad dynamics the bearing stiffness may even become negative for some perturbation frequencies while the direct damping coefficients may vary an order of magnitude. This coefficient fluctuation is typical for titling pad bearings, especially when they operate at high rotational frequencies and thus within the dynamics of the individual tilting pads. The rotor dynamic stability analysis is relatively complicated for such bearings with fluctuation and frequency depending coefficients. The next section is therefore dedicated to the stability analysis and treats the derivation of the system onset-speed of instability.

11.5 Dynamic Stability

Analysing the rotor dynamic stability is an important aspect during the development phase of gas bearings. Half-speed, or more generally asynchronous, whirl instability and other destructive whirl motions must be avoided in order to achieve a successful bearing design. The most critical type of instability, which is also the topic of this section, is generally referred to as half-speed whirling, see Chapters 9 and 10. It is a self-excited phenomenon with destructive consequence to the bearing and rotor. The underlying source of self-excited whirl instabilities are the cross-coupling effects within the gas bearing. These destabilising effects become more severe when the rotational speed increases, eventually exceeding the damping stabilising effect so as to result in a violent and unstable rotor motion, which easily consumes the available bearing gap. Contact between journal and bearing surface is then unavoidable, resulting in permanent damage to the bearing.

Destructive situations like this must therefore be avoided by predicting the half-speed whirling onset-speed in advance and designing bearings with an onset-speed beyond the system nominal operation speed. Thus, the first part of this section treats the theoretical prediction of onset-speed instability, hereby mainly focusing on tilting pad gas bearings. The second part considers some practical design strategies and rules to increase the onset-speed beyond a safe level.

Onset-speed of instability

A full rotor-bearing model may be utilised to study and predict the onset-speed of instability. The aim of such a model is typically to describes the actual rotordynamics as closely as possible. That is, rotor flexibility, bearing locations, and gyroscopic effects are all taken into account. A large body of literature on rotordynamics exists which describes in detail how to model these effects, either in an analytical or a numerical way (Childs 1993; De Kraker 2009; Ishida and Yamamoto 2013; Muszynska 2005). However, these full rotor-models are somewhat complex and may initially be too comprehensive for a first estimation of whirling behaviour and dynamic stability analysis. Simplified models are therefore proposed in the past, like the Laval–Jeffcott rotor model De Kraker (2009) or the single mass model (Pan 1964). It has been shown that the latter is sufficiently accurate for a first characterisation of dynamic rotor response and its stability (Vleugels 2009; Waumans 2009). For this reason, and due to its simplicity

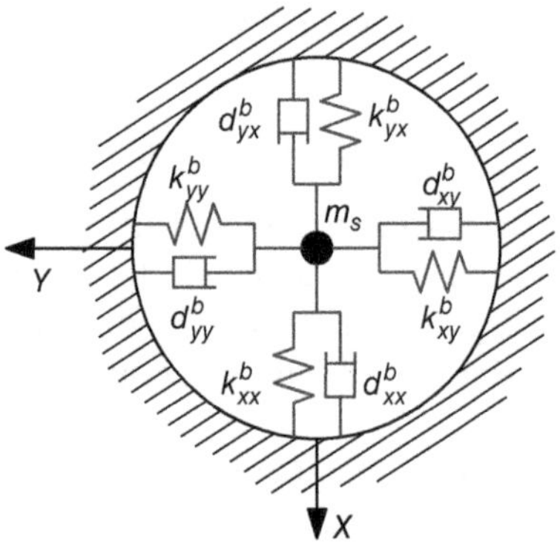

$$\mathbb{K}_b = \begin{bmatrix} k^b_{xx} & k^b_{xy} \\ k^b_{yx} & k^b_{yy} \end{bmatrix}$$

$$\mathbb{D}_b = \begin{bmatrix} d^b_{xx} & d^b_{xy} \\ d^b_{yx} & d^b_{yy} \end{bmatrix}$$

$$q_s = \begin{bmatrix} x_s \\ y_s \end{bmatrix}$$

Figure 11.25 Rotor model with concentrated rotor mass m_s. Source: Nabuurs M 2020.

in comparison with other models, the single-mass model is utilised in this section. That is, the rotor is represented by an equivalent point mass, placed within a 2D representation of a single bearing, as shown in Figure 11.25. The bearing itself is represented by four springs and dampers, which are characterised by the 2×2-coefficient matrices $\mathbb{K}_b$ and $\mathbb{D}_b$ as defined in the previous section.

The equivalent point mass will have two degrees of freedom, i.e. along the X and Y-axes respectively. By assuming small harmonic motions and applying the previously introduced normalisation rules, the equation of motion for the equivalent point mass is expressed as

$$\vec{\bar{F}}_s(t) = \bar{\mathbb{K}}_b \bar{q}_s(t) + \bar{\mathbb{D}}_b \dot{\bar{q}}_s(t) + \bar{\mathbb{M}}_b \ddot{\bar{q}}_s(t), \tag{11.72}$$

with $\bar{q}_s$ being the generalised coordinate column vector and $\vec{\bar{F}}_s$ the forces exerted on the equivalent point mass. Both are with respect to the global and fixed XY-coordinate frame as shown in Figs. 11.17 and 11.25. The normalised mass matrix is defined as

$$\bar{\mathbb{M}}_b = \begin{bmatrix} \bar{m}_s & 0 \\ 0 & \bar{m}_s \end{bmatrix} \quad \text{with } \bar{m}_s = m_s \frac{v^2 \omega^2 C}{P_a L R_p}, \tag{11.73}$$

where the equivalent rotor mass m_s depends on the whirl motion under investigation and the locations of the supporting bearings. For example, in Figure 11.26 the rotor shown may undergo cylindrical or conical whirl motions within its two supporting bearings. Both whirl motions may be studied by the single-mass model.

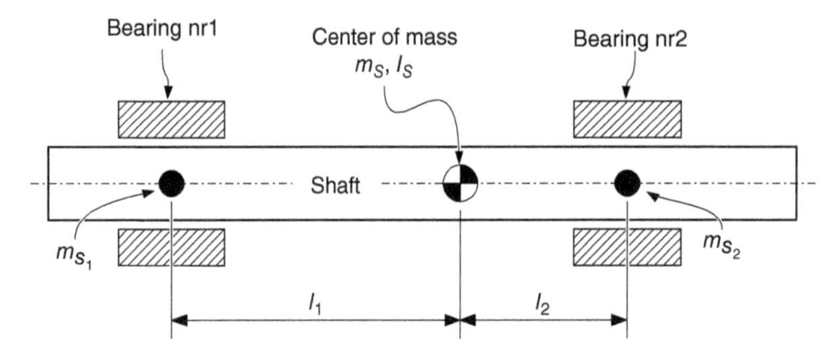

Figure 11.26 Equivalent rotor mass of a rigid unsymmetrical rotor supported by two bearings. Source: Nabuurs M 2020.

That is, the overall rotor mass m_S and inertia I_S is redistributed to two single point masses at the location of the two bearings. For investigating cylindrical whirl, the sum of these two point masses must be equal to the total mass m_S while not changing the original centre of mass location. Taking this into account, the equivalent mass m_s for both bearings may be found by solving the system:

$$\begin{bmatrix} 1 & 1 \\ l_1 & -l_2 \end{bmatrix} \begin{bmatrix} m_{s_1} \\ m_{s_2} \end{bmatrix} = \begin{bmatrix} m_S \\ 0 \end{bmatrix}. \tag{11.74}$$

In a similar manner, the equivalent masses for the conical whirl must be such that the overall tilt inertia of both masses is equal to I_S without changing the centre of mass. Hence, the equivalent masses for conical whirl studies is found by solving the system:

$$\begin{bmatrix} l_1^2 & l_2^2 \\ l_1 & -l_2 \end{bmatrix} \begin{bmatrix} m_{s_1} \\ m_{s_2} \end{bmatrix} = \begin{bmatrix} I_S \\ 0 \end{bmatrix}. \tag{11.75}$$

Substituting either the cylindrical (11.74) or conical (11.75) equivalent point mass into (11.73) yields the mass matrix $\bar{\mathbb{M}}_b$.

Notice that the proposed single mass model and its equation of motion (11.72) is a linear model, therefore not describing any possible non-linear or gyroscopic effects. In addition, the model assumes a rigid rotor such that effects arising from rotor flexibility are neglected. Despite these limitations, the single mass model is still suitable for the majority of rotor dynamic analyses, including half-speed whirl analysis (Adams and Padovan 1981).

Two methods to determine the system onset-speed of instability are discussed in this section. The first method is based on an eigenvalue analysis and may be utilised to compute rotordynamic stability. Computing the eigenvalues of a system with gas bearings is, however, time consuming and inefficient when the goal is to obtain stability charts. A second method is, therefore, discussed at the end of this section. This second method is suitable to compute stability charts in a more direct way with less computational effort.

System stability by eigenvalue analysis

The system onset-speed of instability may be studied by computing the system eigenvalues and evaluating the real parts. Such real eigenvalue analysis is, however, not straightforward to perform due to the frequency dependency of the bearing stiffness matrix $\mathbb{K}_b$ and damping matrix $\mathbb{D}_b$. Coefficients of both matrices depend on the system perturbation frequency ω_p and, thus, the system eigenvalues $\lambda(\omega_p)$, (not to be confused with the slenderness ratio). The perturbation frequency ω_p is generally irrelevant for a stable system. However, for a system with an unstable whirl motion, the whirl frequency ω_v will become the dominant system perturbation frequency. That is, $\omega_p = \omega_v$ in case of an unstable system (Nicholas 2003). This implies that the only feasible solution to the eigenvalue problem, at the threshold of instability, is a solution for λ with an imaginary part equal to ω_p. Taking this into account, a stability criterion for frequency-dependent gas bearings may be expressed as

$$\mathrm{Re}(\lambda(\omega_p)) < 0 \bigcup |\mathrm{Im}(\lambda(\omega_p))| \neq \omega_p \qquad \forall \quad \omega_p \in (0, \omega] \tag{11.76}$$

Note that, beside the perturbation frequency, the bearing coefficients are also a function of the system rotational frequency ω. The stability criterion (11.76) must therefore be evaluated for all possible rotational frequencies which may occur for the system under investigation. In other words, the rotating system may be classified as stable if the stability criterion (11.76) is satisfied for all rotational frequencies. Computational time may be reduced by evaluating the normalised form of (11.76) and for a range of dimensionless equivalent point masses $\bar{m}_s$. Combinations of Λ and $\bar{m}_s$, which results in a stable system, are typically compiled in the form of a stability chart.

For example the stability chart of Figure11.27 is for an three-pad, tilting-pad bearing with a geometry according to Table 11.3. The journal static position is at the bearing center, i.e. $\bar{q}_{s(0)} = \{0, 0\}$ which is equivalent to a vertically orientated rotor with no radial load. Combination of equivalent point mass m_s and rotational speeds Λ, which violate the proposed stability criteria, are mark by the grey area. The hatched lines indicate the border between

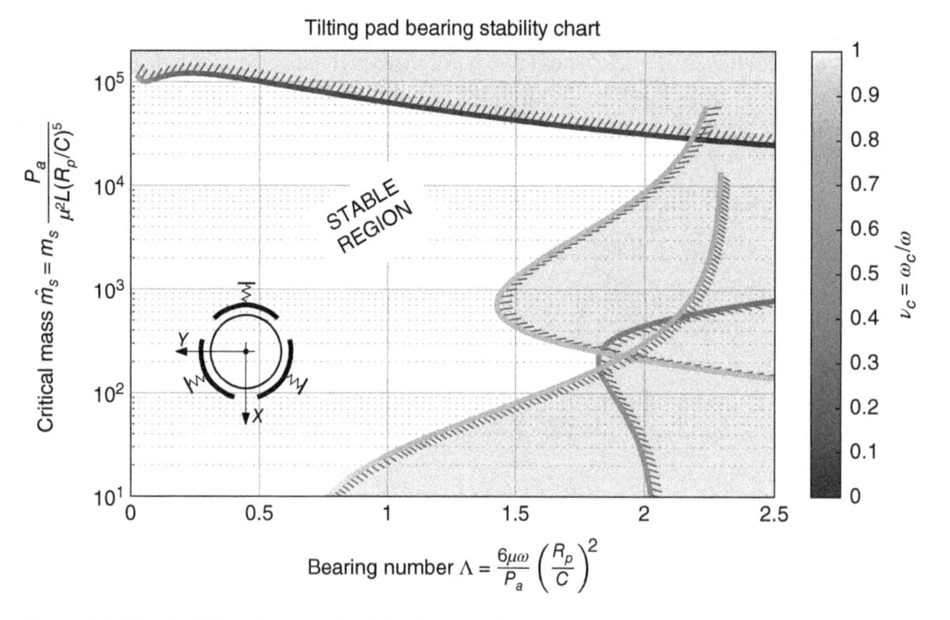

Figure 11.27 Stability chart for flexible tilting pad bearings with a geometry according to Table 11.3. Source: Nabuurs M 2020.

Table 11.3 Normalised geometry of a 3-pad tilting pad bearing with flexible pivot.

Description	Value	Definition
Clearance ratio, δ	200	$R_s/C \approx R_p/C$
Pivot point distance ratio, ς	300	R_{pp}/C
Length to diameter ratio, λ	1	$L(2R_p)^{-1}$
Pad arc length, β	120	[deg]
Pivot tangential position, $\bar{\beta}_{pp}$	0.7	β_{pp}/β
Pad geometrical preload, δ_ϵ^{pp}	0.75	e/C
Pad radial stiffness, $\bar{k}_{\epsilon\epsilon}^{pp}$	8	$k_{ee}^{pp} C(P_a L R_p)^{-1}$
Pad tilt stiffness, $\bar{k}_{\phi\phi}^{pp}$	0.05	$k_{\phi\phi}^{pp} C(P_a L R_p R_{pp}^2)^{-1}$
Pad mass, $\hat{m}_p$	600	$m_p \mu^{-2} L^{-1} (R_p/C)^{-5} P_a$
Pad tilt inertia, $\hat{I}_{\phi\phi}^p$	50	$I_{\phi\phi}^p \mu^{-2} L^{-1} R_{pp}^{-2} (R_p/C)^{-5} P_a$

Source: Nabuurs M 2020.

stable and unstable operation with the line grey-scale representing the normalised frequency ratio of the unstable whirl motion.

It is relatively straightforward to derive the bearing onset-speed of instability once a bearing stability chart is constructed. The first step is to compute the equivalent rotor mass according to (11.74) or (11.75). By subsequently normalising the equivalent rotor mass $m_s \rightarrow \hat{m}_s$, and utilise the stability chart to find the lowest associated bearing number Λ which destabilise the system. As a final step, the system onset-speed of instability is obtained by converting the lowest Λ value back to a rotational speed ω with the help of (11.39).

For example, assume a symmetrical rotor of 350 gram, with a diameter of 16 mm, and supported by two flexible tilting pad bearings according to Table 11.3. For such rotor, the dimensionless and speed-independent cylindrical rotor mass is: $\hat{m}_s = m_s \mu^{-2} L^{-1} (R_p/C)^{-5} P_a = 1e4$. If in the chart of Figure 11.27 a horizontal line is drawn from left to right at $\hat{m}_s = 1e4$, it will intersect the first critical curve at approximately $\Lambda = 2$ and entering the unstable region to the right of the curve. This rotor-bearing system will therefore have its first onset-speed at $\Lambda = 2$ or 430 krpm. At this particular speed, the rotor goes from a stable to an unstable cylindrical whirl mode with a whirl frequency of roughly 0.45× the rotational speed as is indicated by the grey-scale of the critical curve at the intersection point (Nabuurs 2018). Note that the onset-speed of instability of different rotor-bearing configurations (but with the same normalised bearing geometry) can be obtained with the same stability chart.

Direct computation of onset-speed

Stability charts are extremely valuable during the design and development phase of tilting pad bearings. They provide a quick way to obtain the onset-speed of instability. Nevertheless, computing the stability chart itself can be a comprehensive task, especially if the system eigenvalues (11.76) have to be evaluated for a large set of equivalent rotor masses $\hat{m}_s$ and bearing numbers Λ. Another way to derive the stability charts is to directly compute the critical rotor mass, or bearing number, which put the system on the threshold of stability. This method, as originally proposed by Lund (1965, 1967), assumed that the system is on its threshold of stability when at least one of the eigenvalues are of the form $\lambda = 0 \pm \omega_p$, thus $\mathrm{Re}(\lambda) = 0$. By taking this into account, the characteristic equation of (11.72) may be written as

$$\det\left(-\bar{m}_c \frac{\nu_c^2}{\nu^2}\mathbb{I} + \bar{\mathbb{D}}_b(\nu)\frac{\nu_c}{\nu} i + \bar{\mathbb{K}}_b(\nu)\right) = 0 \pm 0i \tag{11.77}$$

where $\bar{m}_c$ and ν_c are the critical rotor mass and whirl frequency ratio which put the system on the threshold of instability Lund (1967). Both may be directly computed by substituting $\omega_p = \omega_c$ into (11.77) and solving the equation for $\bar{m}_c$ and ν_c. Stability charts obtained in this way are published for aerostatic and aerodynamic plain journal bearings by (Czolczynski 1999; Czołczyński and Marynowski 1996; Lund 1965; Waumans 2009) and many others, see also Chapter 10. Stability charts for the more complicated foil bearings are published by Vleugels (2009, p. 78) (see also Chapter 12) and Schiffmann and Spakovszky (2013), while charts for tilting pad bearings are published by (Sim 2007, p. 66) and Lihua et al. (2011). It is important to note that equation (11.77) is non-linear, having thus, in general, more than one solution for $\bar{m}_c$ and ν_c. In addition, the above method is incapable of identifying all the possible bifurcation points or combination of $\bar{m}_c$ and ν_c, and as such, providing incomplete stability chart and an overestimation of system stability Nabuurs (2020). Special care must therefore be taken when computing stability charts with the direct method (11.77) instead of the complete, but more computational expensive, stability criterion (11.76).

Increasing the onset-speed of instability

The bearing onset-speed of instability, and thus the bearing maximum operation speed, depends on the capability of the individual pads to eliminate their attitude angle. Tilting pad bearings with zero attitude angle have in theory no onset-speed of instability and thus an infinitely high operation speed. In practice there will, however, be a certain attitude angle due to practical aspects like:

- tilt stiffness $k_{\phi\phi}^{pp}$ of the pivot points
- frictional shear stress $\bar{\tau}$ acting on the pad
- pivot damping $d_{\phi\phi}^{pp}$
- pad mass m_p and tilt inertia I_p
- cross-terms within the mass-matrix $\mathbb{M}_{pp}$.

It is, therefore, good practice to reduce these effect as much as possible to obtain a bearing with high onset-speed of instability. Tilt stiffness must, for example, be made at least an order of magnitude below the gas-film stiffness

$k_{\phi\phi}$ for an acceptable onset-speed of instability Nabuurs (2020). Tilt or pivot damping delays the pad reaction time to a changing shaft position and therefore causes an (dynamic) attitude angle. Damping the tilt movement of the pads is, therefore, not a way to increase the onset-speed of instability. The mass and inertia of the pads is practically limited by the minimum mass required to construct the metal or ceramic pads from. It must have sufficient thickness or strength to prevent appreciable deformation due to aerodynamic pressure. Nevertheless, cross-terms with the mass matrix can be completely eliminated by designing the pad geometry in such a way that $\phi_{\mathrm{m}} = 0$. Thus a geometry where pad centre of mass is on the imaginary line between the pivot point and pad centre of curvature. Taking the above points into account is imperative in order to obtain a successful tilting pad bearing design with a sufficient high onset-speed of instability.

11.6 Construction and Fabrication Aspects

Pivot point design

Different pivot-point construction systems can be found in the existing literature, most of them can be divided into one of the following five main design principles depicted in Figure 11.28. In principles (a) through (c), the pivot-pad and bearing housing are separate parts. This has the advantage that the pads can be fabricated and machined individually and, at a later stage, assembled in the bearing housing. This could be interesting as modular, large-series production method. Possible drawbacks of this principle is that the pads and housing have to be produced to high tolerances, or else one should have an assembly and adjustment procedure to ensure correct working. This might not be so attractive or evident. Another important issue is related to the wear of the hinge, e.g. knife-edge contact, with continued use, which can undermine the correct functioning of this system. This aspect is overcome in the principles (d) and (e), which are of the monolithic style, where bearing pads and housing are carved out a single, interconnected piece. Such a design is beneficial since it overcomes the problems of assembly,

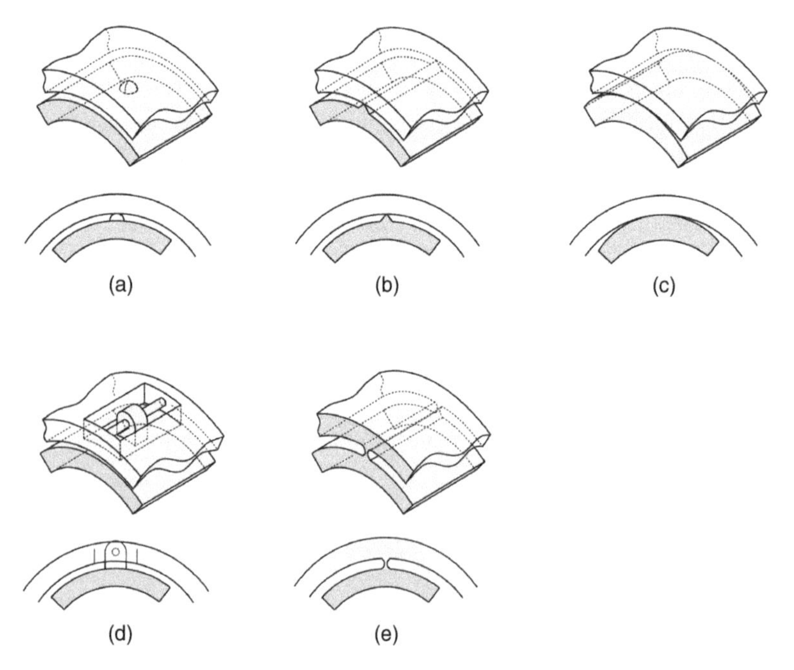

Figure 11.28 Pivot point designs, figure still need some sub-figure numbering in order to refer to the individual designs. i.e. (a) Spherical-pivot (or ball-in-socket pivot), (b) Hinge-pivot, (c) Rocker-pivot (Cylindrical-pivot), (d) Torsion-flexure-pivot, (e) Bending-flexure-pivot. Source: Nabuurs M 2020.

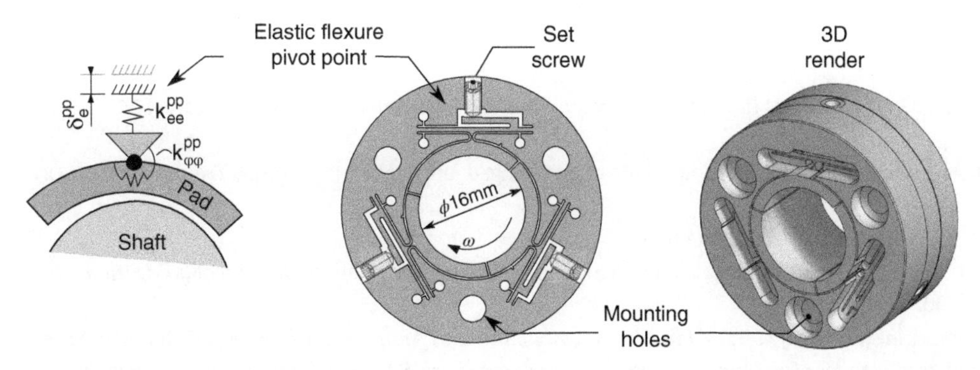

Figure 11.29 Design schematic and actual 2D/3D design of a flexible tilting pad bearing capable of operating beyond 300.000 rpm. Source: Nabuurs M 2020.

adjustment and integration of the individual pads with the bearing housing as is the case for the non-monolithic designs. Moreover, since the monolithic designs depend on elastic deformation for the pivoting action, no wear of the pivot point will occur, assuming no plastic deformation or metal fatigue. The downside of this principle, however, is the much higher production cost involved.

Finally, in each of the two categories, the different variants have their own advantages and disadvantages so that each solution should eventually be evaluated separately for the application at hand. It is imperative, however, that the hinge/pivot principle should accommodate the required kinematics. For example, we show in Figure 11.29 how the theoretical requirements on the pad and its hinge support are translated to a monolithic design that can be fabricated by Wire EDM.

Thermal aspects

In most bearing pad or bush types that are supported by thin constructions such as flexures, o-rings, etc., a thermal issue might arise. The heat generated in the gap by viscous friction can become quite considerable, while it is generally not possible or convenient to evict it via the rotating member or shaft, see Chapter 17. Thus, most of the heat has to be conducted through the (stationary) bearing pad or bush to the housing, which can be taken to be a heat sink. However, if the pad is connected to the housing through only a thin web having a very small cross section, this might not suffice as an effective passage to the large thermal flux. As a result, the temperature of the pad might increase leading possibly to thermal distortions, which, in their turn, might lead to failure of the bearing system. Thus, it might be prudent to take this aspect into account during the pad-design stage.

11.7 Summary

In this chapter, we considered the theory and practice of tilting-pad air bearings, in particular of the journal type. This bearing type is famed for its better dynamic, i.e. whirl stability as compared to other-rotary bearing principles. In order to build up the theory, we first looked at the plane, infinitely long inclined-pad sliding bearing case in order to establish its basic load, stiffness and damping characteristics. In a TPB, the inclined pad is supported by a pivot so that it can freely adjust its attitude. This aspect introduces additional dynamical aspects in the working of the pad, which need to be carefully formulated and dealt with in order to develop an effective design theory and practice. The greatest part of this chapter is then devoted to the case of journal TPB, where first the static characteristics are addressed followed by the whirl dynamics. Optimisation of the design of TPBs is discussed with the main aim of pushing the whirl-onset speed as far as possible beyond the working speed range of the system. Finally, a the fabrication aspects of these bearings is briefly discussed, since these might not be as evident as for other bearing types.

References

Adams, M., and Padovan, J., (1981). Insights into linearized rotor dynamics. *Journal of Sound and Vibration* 76(1): 129–142.

Boyd, J. and Raimondi, A., (1962). Clearance considerations in pivoted pad journal bearings. *ASLE Transactioons* 5(2): 418–426.

Cameron, A., (1971). *Basic Lubrication Theory*. Longman.

Castelli, V. and Pirvics, J., (1968). Review of numerical methods in gas bearing film analysis. *Journal of lubrication technology* 90(4): 777–790.

Childs, D., (1993). *Turbomachinery Rotordynamics: Phenomena, Modeling, and Analysis*. John Wiley & Sons, New York.

Czolczynski, K., (1999). *Rotordynamics of gas-lubricated journal bearing systems*. Springer Science & Business Media.

Czołczyński, K. and Marynowski, K., (1996). Stability of symmetrical rotor supported in flexibly mounted, self-acting gas journal bearings. *Wear* 194(1–2): 190–197.

De Kraker, B., (2009). *Rotordynamics*. Shaker Publishing.

Dimond, T., Younan, A. and Allaire, P., (2011). A review of tilting pad bearing theory. *International Journal of Rotating Machinery*.

Grassam, N.S. and Powell, J.W., (1964). *Gas lubricated bearings*. Butterworths, London.

Gross, W., (1980). *Fluid film lubrication*. Wiley.

Halling, J., (1976). *Introduction to Tribology Applied Mathematical Sciences*. Wykeham Publications.

Hamrock, B., Schmid, S. and Jacobson, B., (2004). *Fundamentals of Fluid Film Lubrication*. Taylor & Francis.

Ishida, Y. and Yamamoto, T., (2013). *Linear and Nonlinear Rotordynamics: A Modern Treatment with Applications* EngineeringPro collection. Wiley.

Lihua, Y., Huiguang, L. and Lie, Y., (2007). Dynamic stiffness and damping coefficients of aerodynamic tilting-pad journal bearings. *Tribology International* 40(9): 1399–1410.

Lihua, Y., Huiguang, L. and Lie, Y., (2010). Corrigendum to “dynamic stiffness and damping coefficients of aerodynamic tilting-pad journal bearings”. *Tribology International* 43(1): 518–521.

Lihua, Y., Shemiao, Q., Geng, H. and Lie, Y., (2011). Stability analysis on rotor systems supported by self-acting tilting-pad gas bearings with frequency effects. *Chinese Journal of Mechanical Engineering* 24(3): 380–385.

Lund, J., (1964). Spring and damping coefficients for the tilting-pad journal bearing. *ASLE transactions* 7(4): 342–352.

Lund, J., (1965). The stability of an elastic rotor in journal bearings with flexible, damped supports. *ASME Journal of Applied Mechanics* 32, 911–920.

Lund, J., (1967). A theoretical analysis of whirl instability and pneumatic hammer for a rigid rotor in pressurized gas journal bearings. *Journal of Lubrication Technology* 89(2): 154–165.

Lund, J., (1968). Calculation of stiffness and damping properties of gas bearings. *Journal of Lubrication Technology* 90(4): 793–803.

Muszynska, A., (2005). *Rotordynamics Mechanical Engineering*. CRC Press.

Nabuurs, M., Al-Bender, F. and Reynaerts, D., (2018). *On the dynamic stability of flexible tilting pad gas bearings* Technische Akademie Esslingen.

Nabuurs, M., (2020). *Tilting pad gas bearings for high speed applications: Analysis, Design and Validation* PhD thesis. KU Leuven.

Nicholas, J.C., (2003). Lund’s tilting pad journal bearing pad assembly method. *Journal of vibration and acoustics* 125(4): 448–454.

Pan, C., (1964). Spectral analysis of gas bearing systems for stability studies. Technical report, Mechanical Technology incorporated, latham NY.

Powell, J., (1970). A review of progress in gas lubrication. *Review of Physics in Technology* 1(2): 96.

Schiffmann, J. and Spakovszky, Z., (2013). Foil bearing design guidelines for improved stability. *Journal of Tribology* 135(1): 011103.

Shaw, E., (1983). *Europe's Nuclear Power Experiment*. Pergamon.

Sim, K.H., (2007). *Rotordynamic and thermal analyses of compliant flexure pivot tilting pad gas bearings* PhD thesis.

Simmons, J. and Advani, S., (1987). Michell and the development of tilting pad bearing *Fluid Film Lubrication–Osborne Reynolds Centenary: Proceedings of the 13th Leeds-Lyon Symposium on Tribology*, pp. 49–56 Elsevier Science. *On the dynamic stability of flexible tilting pad gas bearings*

Vleugels, P., (2009). *Development of Aerodynamic Foil Bearings for Micro Turbomachinery* PhD thesis.

Waumans, T. (2009). *On the design of high-speed miniature air bearings: dynamic stability, optimisation and experimental validation* PhD thesis. Blockeel, Hendrik (supervisor).

12

Foil Bearings

12.1 Introduction

Foil bearings are a special case of fluid-film lubrication in which the (compliant) bearing surface may be deformed by the pressure prevailing in the air gap, which results in a situation similar to iso-viscous Elasto-Hydrodynamic Lubrication (EHL) (or more appropriately perhaps, EAL: Elasto-Aerodynamic Lubrication). This situation might offer certain interesting advantages.

Foil bearings can be classified into two major types: (1) bearings with (at least one) bearing surface under tension (further called tension-type foil bearings) and (2) bearings with (at least one) bearing surfaces having a low modulus of elasticity material (Gross 1980) (from now called compliant surface foil bearings), see Figure 12.1. In the latter class, the low modulus of elasticity material must be interpreted very broadly, it can range from an elastic material, e.g. silicone rubber, up to a complex notch-type hinge mechanism.

The development of the tension type foil bearings started with the publication of Blok and Van Rossum (1953). Most of the tension-type foil bearing developments occurred in the early 1970s with the publications of Licht (1969), Eshel and Wildmann (1968) and Barlow (1967) for both the aerodynamic and aerostatic types. The major application of this foil bearing type is in tape recorders where the tape moves over a shaft. The aerostatic version is further developed by Al-Bender (Al-Bender and Smets 2004) who uses externally pressurised tension-type foil bearings as a linear sliding bearing on a non-rotating shaft. The tension-type foil bearings are no longer used as journal bearings for high speed applications.

In the late 1970s mainly two different types of flexible surface foil bearings were under development, see Figure 12.2. The cantilevered design (a), often also called a multi-leaf foil bearing, was developed by companies like AiResearch (Air Force Aero Propulsion Laboratory). These days, the cantilevered design is no longer in use. Probably this is because of the rather bad/strange height distribution. Each leaf gives a jump in height distribution of the foil thickness; compared to the gap height these jumps are huge. As a result the performance is rather bad.

The development of the first generation bump-type, also called the Hydresil design, was initially the work of Mechanical Technology Incorporated (MTI) (Walowit and Anno 1975) and later Mohawk Innovative Technologies (MiTi). The history of this of kind foil bearing is also nicely illustrated in the publication of Agrawal (1997). In the beginning the development was (almost) solely driven by companies and not by research institutes. This clarifies the lack of accurate modelling and knowledge in the literature. Most knowledge is still inside these companies and the applications are limited very specifically to where it is affordable to invest in the development of a specific design of foil bearing. The focus was clearly on the achieved results and advantages and not on the physical understanding of the bearing (at least not published in literature). Meanwhile, several generations of foil bearings exist, each yielding a higher load capacity. In the second generation foil bearings (see Figure 12.2c) a variable stiffness and top-foil stiffener is used. In the third generation, the variable stiffness is even more pronounced (see Figure 12.2d).

Air Bearings: Theory, Design and Applications, First Edition. Farid Al-Bender.

Companion website: www.wiley.com/go/AlBender/AirBearings

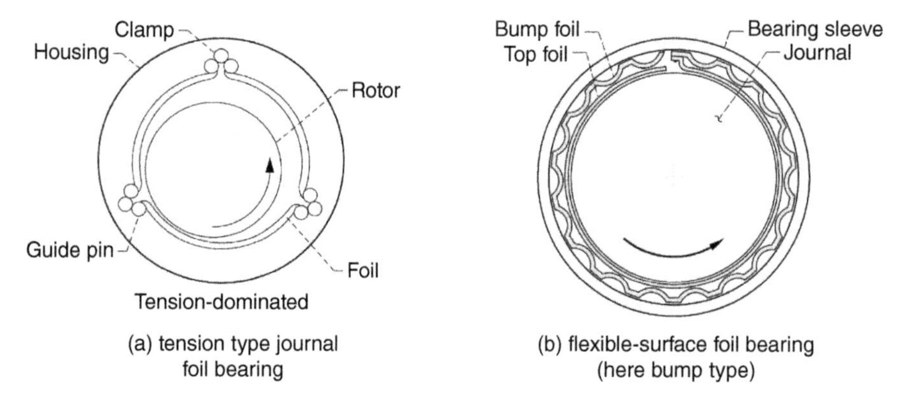

Figure 12.1 Overview foil bearing types: tension type and flexible surface. Source: From Dellacorte C, Radil K, Bruckner R and Howard S, 2008. Reprinted with permission from Taylor & Francis.

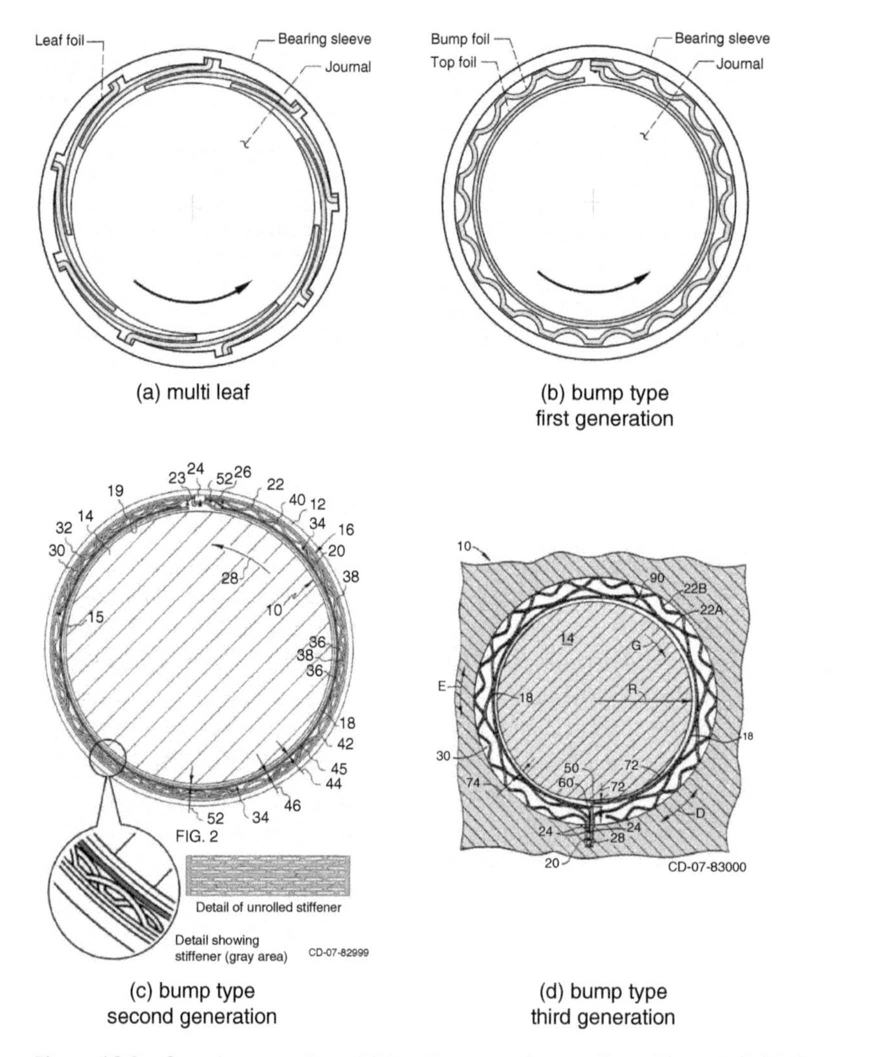

Figure 12.2 Overview compliant foil bearing types. Source: From Vleugels P, 2009.

In the (late) 1990s foil bearings became very popular, even hyped, in research, often driven by the need for oil-free turbomachinery and the high speed requirements. Most publications focus therefore on macro scale applications or shaft diameters of more than 20 mm and a speed limited to 150 000 r.p.m. Most of the knowledge of foil bearings is based on empirically obtained knowledge. However, recently more and more papers are appearing on the theoretical modelling of foil bearings in general and very small (shaft diameter < 20 mm), high speed applications ($> 150\ 000$ r.p.m.) in particular.

12.2 Compliant Material Foil Bearings: State-of-the-Art

12.2.1 Early Foil Bearing Developments

The earliest applications of foil bearings are in the Air Cycle Machine (ACM) which is at the heart of the Environmental Control System (ECS) in an aircraft. This system is responsible for cooling, heating and pressurising the aeroplane. Agrawal (1997) claims that in 1997 almost all ACM units in military and civil aeroplanes use foil bearings. He also shows, unfortunately without any bearing detail, results from several ACM's, like the F-18 which rotates up to 95 000 r.p.m. and for Cessna-550 rotating up to 105.000 r.p.m.

12.2.2 Recent Advances in Macro Scale Foil Bearings

MiTi Fuel Cell Compressor and High Temperature Turbo-machinery

Mohawk Innovative Technology Inc, or MiTi, is the world's largest supplier of foil bearings. Application of foil bearings is nowadays limited to specific applications in turbo-machinery for aeroplanes. In their latest publications they try to replace roller bearings in turbojets with foil bearings. These applications are far more difficult as speed and temperature are much higher. In 2005 Heshmat and Walton (2005) briefly reported on a foil bearing test rig up to $650°C$. For that specific turbojet, the cost reduction would be 20% and the weight could be reduced by 30%, due to the elimination of the complete lubrication system. In 2006 Heshmat (Heshmat and Tomaszewski 2006) reported about a redesigned 134 N thrust class, 120 000 r.p.m. turbojet, see Figure 12.3, with a high-temperature foil bearing and compliantly mounted ball bearing near the compressor. The conclusions are comparable with the ones from the previous publication.

Foil bearings can also be an interesting bearing solution for fuel cells. MiTi (Walton et al. 2006) reported on a preliminary system design study for the turbocharger in a Proton Exchange Membrane fuel cell. In this design they aim at a speed of 250 000 r.p.m. with a shaft in the range of 20 mm. An important point in this design is the thrust bearing. Several different thrust bearings were tested for load up to 98 N and a speed up to 250 000 r.p.m.

NASA Oil Free Turbomachinery Laboratory & Case Western Reserve University

NASA (Howard et al. 2001b, Radil et al. 2002) did a lot of experimental work in the past. Their current work focuses on a heat resistant low friction coating named PS304. This coating is applied on the rotor shaft and avoids excessive wear of the rotor and reduces friction. It is a mixture of a NiCr binder, chrome oxide hardener and silver. They tested this coating on a commercially available Capstone C30 microturbine at a speed of 100 000 r.p.m. or a DN-number of ± 3.1 MDN.

NASA performed a large number of experiments on the behaviour of foil bearings at high temperature. The static stiffness decreased by a factor two at $500°C$ compared to the stiffness at room temperature. Unfortunately they did not give a good explanation of the cause of this decrease. For these experiments they used a foil bearing of diameter 35 mm at a speed of 30 000 r.p.m., which is not so fast.

Recently, research started on the influence of cooling air on foil bearings. Due to the viscous losses a temperature gradient can be expected which can be large enough to have an influence on the bearing behaviour. Dykas (Dykas

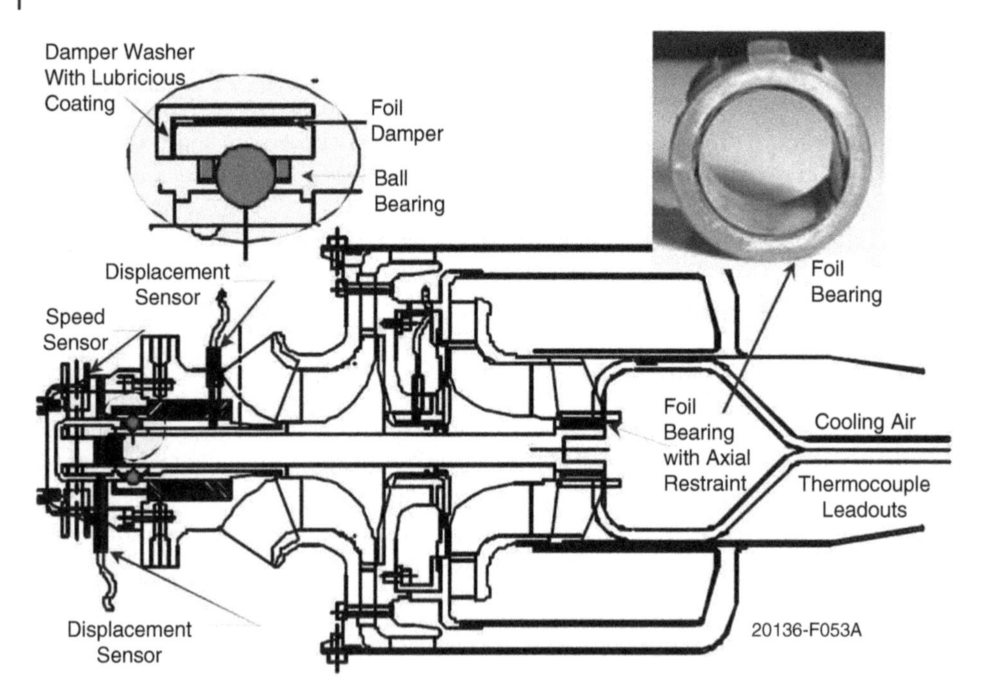

Figure 12.3 Oil-free turbojet with a foil bearing as high-temperature foil bearing and a compliantly mounted ball bearing. Source: From Heshmat H and Tomaszewski M, 2006, by permission of ASME.

et al. 2006) did some experiments on the influence of cooling air on a thrust bearing. They concluded that cooling air has a significant influence on the load capacity of the thrust bearing.

12.2.3 Recent Advances in Mesoscopic Foil Bearings

Mohawk Innovative Technologies mesoscopic foil bearings

Heshmat and his company Mohawk Innovative Technology developed a lot of foil bearings. Their research is mainly focused on large bearings (diameter > 20 mm). But in recent work they show some results of a 6 mm diameter foil bearing intended for a mesoscopic turbojet (Salehi et al. 2004). With their demonstrator (shown in Figure 12.4) they can reach up to 700 000 r.p.m. It is not clear what determines the upper limit. The limitation is certainly not the half-speed whirl which is often the limit in aerodynamic bearings. They clearly mention that there is no subsynchronous vibration. In their publications (and this paper is no exception) they do not give many details on the design of the foil bearings.

Elastic Material Foil Bearings at Xi'an Jiaotong University

Using roller bearings in cryogenic turbo-expanders is difficult because of the very low temperature. Therefore, the Chinese Xi'an Jiaotong University developed several concepts for foil bearings in cryogenic turbo-expanders. Initially, Hayashi(Hayashi 1994) published some concepts and afterwards Hou (Hou et al. 2004a) (Hou et al. 2004c) and Xiong (Xiong et al. 1997) did some experiments on those 12 mm diameter journal bearings. With those bearings they can reach 220 000 r.p.m. with only a vibration amplitude of a few micrometers. Their bearings have a quite simple geometry. They use a spirally wound foil with small wires as a spacer or a foil with an elastic material as support (see Figure 12.5).

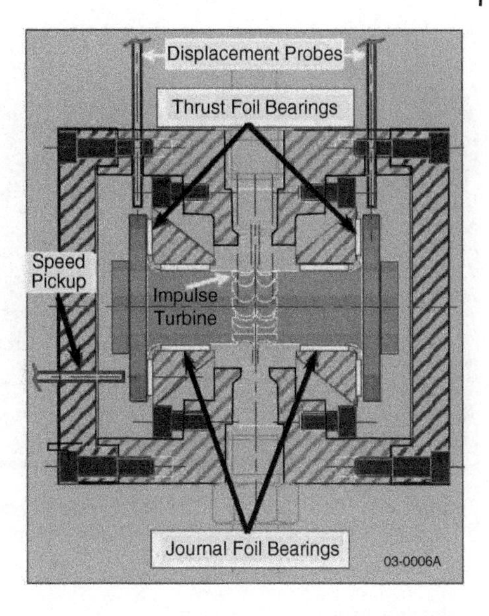

Figure 12.4 Cross section of a mesoscopic simulator with foil bearings capable of rotating at 700 000 r.p.m. Source: Salehi M, Heshmat H, Walton J and Tomaszewski M 2004 by permission of ASME.

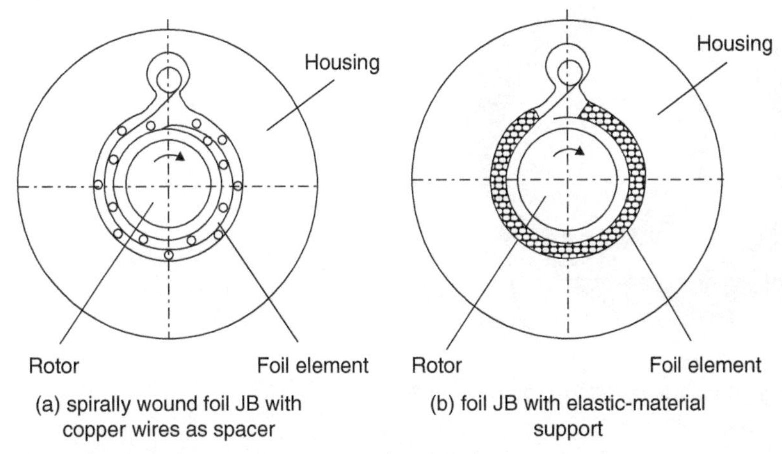

Figure 12.5 High-speed cryogenic foil bearings. Source: From Hou Y, Zhu Z and C.Z. C 2004c © 2004 Elsevier.

Characteristics of Micro Foil Bearings at KIST

The Korea Institute of Science and Technology (KIST) (Yong-Bok et al. 2006) reported about a turbo generator experiment, see Figure 12.6 with 7 mm diameter foil bearings and a speed up to 300 000 r.p.m.. The rotor vibration was mainly synchronous with an amplitude of about 5–20 μm. This large vibration is still smaller than the initial clearance of 30 μm, such that there is no contact between the rotor and journal. The foil bearing consists of top foil and a bump foil of 75 μm with 17 bumps.

Virtual Design of LIGA Fabricated Mesoscale Foil Bearing at TAMU

The flexible element in foil bearing is very often a corrugated bump foil as it incorporates "easy" production with a hysteretic friction source. However, by making the bearing small it becomes very difficult to machine the bumps. Texas A&M University (TAMU) plans to use the LIGA production technique for the production of the bump foil (Kim 2006). LIGA (the German acronym for lithography, electroplating and moulding) is a production technique

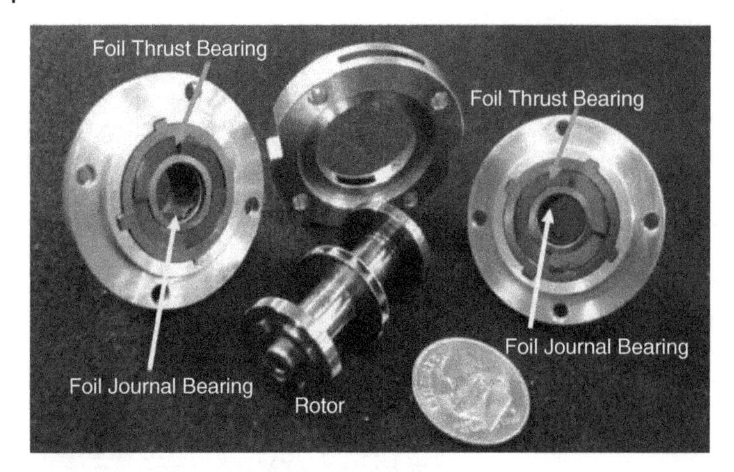

Figure 12.6 Components of the micro power system from KIST. Source: Yong-Bok L, Dong-Jin P, Chang-Ho K and Keun R 2006. Reproduced with permission from IOP Publishing, Ltd.

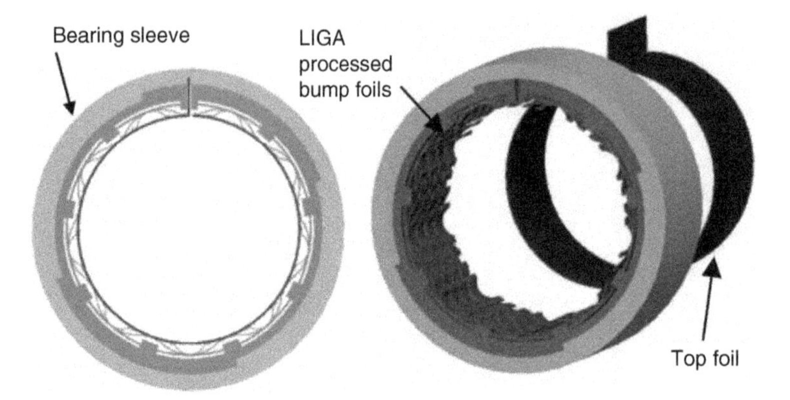

Figure 12.7 Proposed mesoscale foil bearing using LIGA by TAMU. (Source: (Kim 2006)).

typical of micro scale applications. Figure 12.7 shows a scheme of the foil bearing. Not much is known about their future plans with these LIGA based foil bearings.

12.3 Self-Acting Tension Foil Bearing

With the establishment of elastohydrodynamic lubrication as a fundamental branch of lubrication theory, the flexibility or elasticity of the bearing or constraining surface acquired significance as a design variable. Thus, developments in bearing design took the form of elastomeric and rubberised liners inside the bearing surface to permit a closer conformity between mating surfaces in the minimum film-thickness region. The extreme case of bearing surface flexibility is the foil bearing, shown schematically in Figure 12.8. This consists of a stationary, flexible band or foil wrapped partially around a rotating shaft as shown in Figure 12.8(a) with an interfacial lubricant between the foil and the shaft. In some cases the shaft is held stationary and the foil moves past it with velocity U, as we see in Figure 12.8(b), whereas in other circumstances both the shaft and the foil rotate.

The following analysis is based to a large extent on (Eshel and Elrod 1965; Gross 1980). Let the tension per unit width of foil be T on each side of the central contact region and let R denote the radius of the journal. The curvature

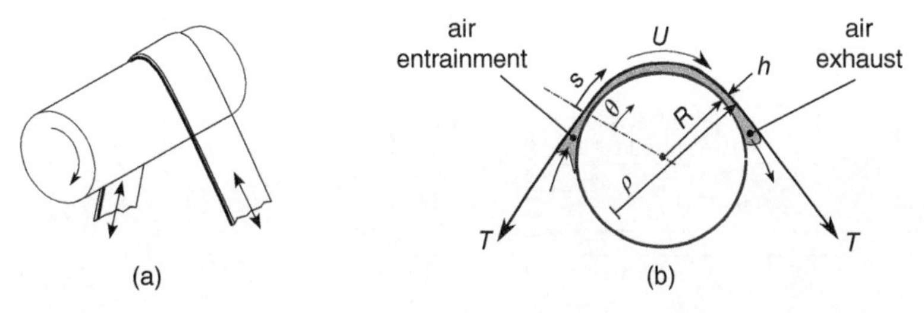

Figure 12.8 The foil bearing showing (a) general view, and (b) infinitely wide foil.

of the foil *with respect to* the journal may be expressed by the relationship

$$\frac{1}{r'} = \frac{\mathrm{d}^2 h}{\mathrm{d}s^2} \bigg/ \left[1 + \left(\frac{\mathrm{d}h}{\mathrm{d}s}\right)^2\right]^{3/2}, \tag{12.1}$$

where the coordinate s is measured in the circumferential or θ-direction. The relative curvature $1/r'$ in Eq. 12.1 can also be expressed as

$$1/r' = 1/R - 1/\rho, \tag{12.2}$$

where ρ is the radius of curvature of the foil. We can usually neglect the term $(\mathrm{d}h/\mathrm{d}s)^2$ in the denominator of Eq. 12.1, so that from the last two equations

$$\frac{1}{\rho} = \frac{1}{R} - \frac{\mathrm{d}^2 h}{\mathrm{d}s^2}. \tag{12.3}$$

Now, the Reynolds equation for incompressible, one-dimensional flow in the s-direction becomes

$$\frac{\mathrm{d}}{\mathrm{d}s}\left(h^3 \frac{\mathrm{d}p}{\mathrm{d}s}\right) = 6\mu U \frac{\mathrm{d}h}{\mathrm{d}s} \tag{12.4}$$

with the usual notation. Furthermore, the equilibrium of an elemental length of foil $\rho \mathrm{d}\theta$ within the angle of wrap shows clearly that the pressure force $p\rho \mathrm{d}\theta$ exerted on the element by the shaft is exactly matched by the tension component $T\mathrm{d}\theta$. Thus

$$p = T/\rho. \tag{12.5}$$

By substituting for R from Eq. 12.3 into Eq. 12.5, and then for p into Eq. 12.4, we obtain the foil-bearing equation

$$\frac{\mathrm{d}}{\mathrm{d}s}\left(h^3 \frac{\mathrm{d}^3 h}{\mathrm{d}s^3}\right) = -\frac{6\mu U}{T}\left(\frac{\mathrm{d}h}{\mathrm{d}s}\right). \tag{12.6}$$

It is convenient to write this in non-dimensional form by introducing the following parameters:

$$\epsilon = \frac{6\mu U}{T}; \quad \xi = \frac{s\epsilon^{1/3}}{h_0}; \quad H = h/h_0,$$

where h_0 is the constant film thickness at the centre of contact [see Figure 12.8(b)]. With these definitions and after integrating once, Eq. 12.6 becomes

$$\frac{\mathrm{d}^3 H}{\mathrm{d}\xi^3} = \frac{1 - H}{H^3}. \tag{12.7}$$

Relevant auxiliary conditions for this differential equation are: $H = 1$ and $H' = 0$ at some point $\xi = 0$ in the mid-region of the gap, and that $H'' = c_0$, some constant, at $\xi \to \mp\infty$ where the foil that is in tension is tangentially disposed to the shaft or cylinder.

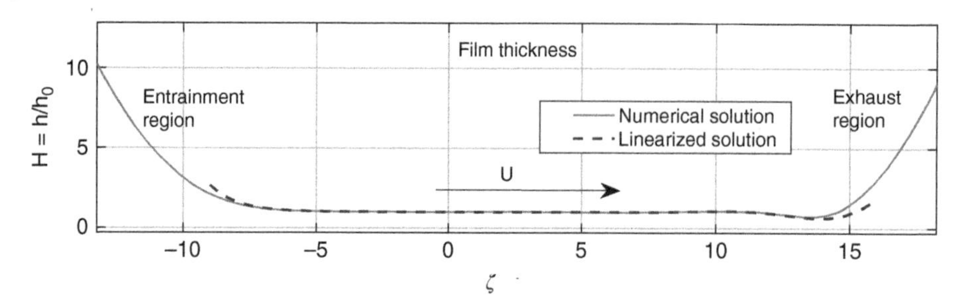

Figure 12.9 The dimensionless foil film profile showing the characteristic EHL ripple at film exhaust.

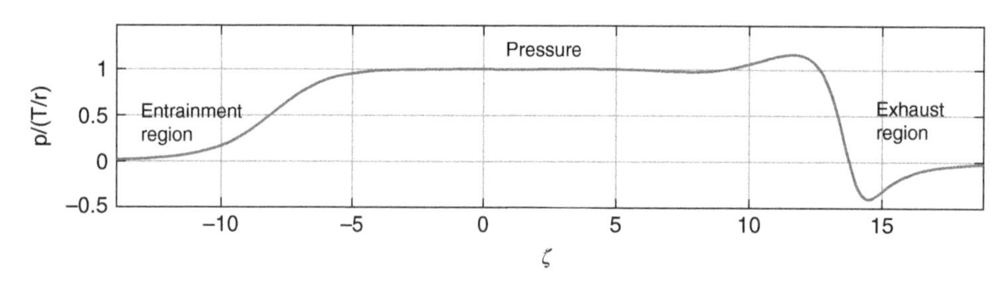

Figure 12.10 The dimensionless pressure distribution in the film.

This equation is non-linear in H but we can solve it numerically as an initial-value problem (e.g. by Euler's integration method) by assuming certain values for $H''(0)$ and evaluating $H''(\infty) = H''(-\infty)$. This scheme shows that there is a unique solution to Eq. 12.7, for which $H''(\infty) \approx 0.643$ (corresponding to $H''(0) = 0.6183\ 10^{-3}$). The film profile and the pressure distribution pertaining to this solution are depicted in Figures 12.9 and 12.10, respectively.

In the middle portion of the foil, we can obtain a linearised solution, if we assume that the central gap has only very small variations ΔH from the condition $H = 1$, i.e. $\Delta H \ll 1$. Thus, by putting $H = 1 - \Delta H$ in Eq. 12.7, and neglecting second-order terms, we obtain:

$$\frac{d^3}{d\xi^3}(\Delta H) + \Delta H = 0, \tag{12.8}$$

for which the general solution is given by

$$\Delta H = Ae^{-\xi} + Be^{\xi/2}\cos\frac{\sqrt{3}\xi}{2} + Ce^{\xi/2}\sin\frac{\sqrt{3}\xi}{2}. \tag{12.9}$$

If we examine this solution carefully, we observe firstly that the constants A, B, and C must be exceedingly small, so that ΔH is small in the central region. Recall that the parameter ξ is proportional to the distance s measured circumferentially around the bearing from an arbitrary datum point in the central region. Thus at relatively large values of ξ corresponding to the exhaust region in Figure 12.9 (dashed line), the sinusoidal terms in Eq. 12.9 dominate, and the film thickness exhibits a characteristic EHL "ripple" as shown. At relatively small ξ-values corresponding to the entrainment region in Figure 12.9, the first term in Eq. 12.9 dominates, so that the entrance region is exponential in form. For large values of $|\xi|$, corresponding to the regions outside the film region (dashed curve), the solution diverges, owing to the exponential terms.

The hydrodynamic pressure expressed in the dimensionless form pR/T varies in an approximately inverse manner to film thickness, and exhibits a constant value over the central region followed by a characteristic peak towards the exit/exhaust, as seen in Figure 12.10.

Using the value $H''(\infty) \approx 0.643$, it can be shown that the magnitude of the gap h_0 is given by

$$h_0 = 0.643R\left(\frac{6\mu U}{T}\right)^{2/3}. \tag{12.10}$$

This simple formula, which can be used for design purposes, is important because it applies not only to the infinitely wide, perfectly flexible foil bearing considered above, but also it will remain essentially the same when the effects of foil stiffness, finite width, and lubricant compressibility are taken into account.

Foil bearings are used in all cases where moving webs are supported, such as in the paper manufacturing and textile industries and in steel mills. They also find wide application in information storage devices such as tape recorders, computer reels, and flexible strip memories. As an example, consider a video tape moving against a reading drum of radius $R = 20$ mm. A typical entrainment speed (owing mainly to the rotation of the drum) is $U = 40$ m s^{-1}. If the required gap between tape and drum is $h_0 = 10$ μm, and the viscosity of air is 1.8 10^{-3} Pa s, then the required tension in the tape is given by

$$T = 6\mu U\left(\frac{0.643R}{h_0}\right)^{3/2} \approx 200 \text{ N m}^{-1}\text{-tape width.} \tag{12.11}$$

12.3.1 Effect of Foil Stiffness

The previous analysis was based on infinitely flexible foil, which will follow the contour of a shaft when wrapped around it even with the slightest tension. In practice, this will not always be the case and a portion of the wrap angle will be lost owing to stiffness. The previous theory will be more or less still applicable, if the bending stiffness of the foil satisfies a certain condition and it will do so only on the part of the foil that actually wraps the shaft (in the absence of an air film). The loss of wrap angle is depicted in Figure 12.3, where α is the actual wrap angle and $\Delta\alpha$ is the loss on each side of the shaft: $\alpha + 2\Delta\alpha$ is the theoretical wrap angle obtained when foil is infinitely flexible or when T is infinitely large. The wrap loss angle is given by Eshel and Licht (1971):

$$\Delta\alpha = \sqrt{\frac{D}{T\,R}} \tag{12.12}$$

where,

$$D = \frac{Et^3}{12(1-\nu^2)},$$

is the bending stiffness of the foil (E, t, ν are Young's modulus, foil thickness and Poisson's ratio respectively.)

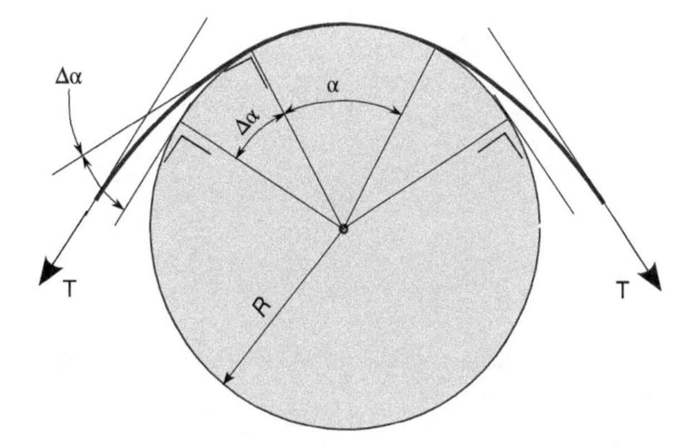

Figure 12.11 Loss of wrap. (Not to scale).

The (infinitely flexible foil) theory will be applicable when the stiffness parameter (Eshel and Licht 1971)

$$S = \frac{D}{T\,R^2\left(\frac{6\mu U}{T}\right)^{2/3}} < 0.8.$$

12.4 Externally-Pressurised Tension Foil Bearing

In this type of foil bearing, where insufficient or zero entrainment speed exists to build an aerodynamic film, external pressurisation is utilised to maintain that film. Here, we shall present a brief overview of design theory, fabrication and experimental validation. The bearing consists of a thin metal foil, equipped with feeding hole(s), which partially enfolds the shaft, and forms an effective, low-cost alternative for (rigid) cylindrical bearings, whose application might be hampered by technical as well as economical limitations. The basic design problem appears to be that of making the right trade-off between foil stiffness, load capacity, and total stiffness of the system. The latter is the resultant of the foil and air-gap stiffness, in series. Several prototype bearings have been built and tested for their load capacity, air consumption and stiffness characteristics (Al-Bender and Smets 2004; Gysegom and Smets 2002). Some of them proved to be far superior to conventional air bearings (owing to the foil being able to conform to the shaft very closely); while some were comparable to rigid bearings of the same dimensions. Measurements on a prototype bearing, (50 mm, diameter, 40 mm wide, 6 bar(g) supply pressure), shows a load capacity of more than 800 N, a radial stiffness of 8.5 N/μ, and air consumption of around 2 Normal-litre/min.

Thus, in contrast to the self-acting type, the foil bearing type described in this section is an externally pressurised one belonging to the tension-foil category. It combines the advantage of compliant surface with the high stiffness obtainable from external pressurisation. Possible applications of the new foil bearings are linear slideway systems, e.g. for high speed pick-and-place machines, where they could form a viable alternative to conventional precision bearing types, but also to a variety of radial and axial bearing applications where the loads are high, stiffness requirements medium and friction low. In the following, we shall begin with theoretical considerations, then discuss design aspects and end with experimental validation.

Owing to its complexity the design theory is only sketched, (see Gysegom and Smets (2002) for more detail).

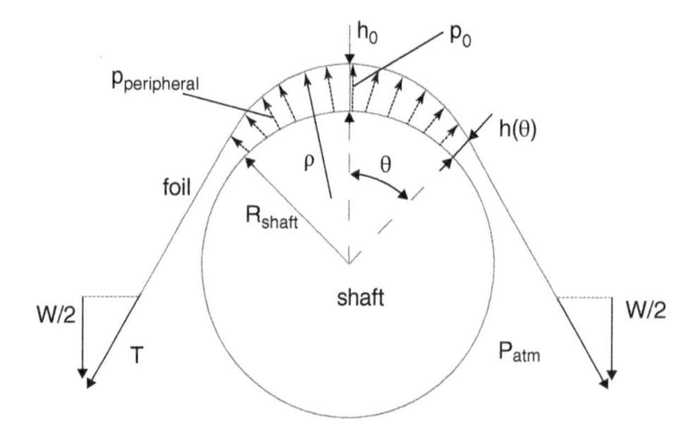

Figure 12.12 Schematic view of an EP tension foil bearing arrangement and notation. Source: Gysegom W and Smets K 2002.

12.4.1 Theoretical Analysis

A comprehensive theoretical model describing the static behaviour of the foil bearing may be quite complex (see Wildmann (1969)). Here we shall sketch this elasto-hydrodynamic problem in a simplified way. Figure 12.12 shows a 2D (infinitely wide) version of the bearing arrangement. If we consider the flow to be incompressible and laminar, and the foil to be perfectly flexible, then we have:

$$\frac{\mathrm{d}}{\partial\theta}\left(h^3\cdot\frac{\mathrm{d}p}{\mathrm{d}\theta}\right)\equiv 0 \qquad \text{(Reynolds' Equation)}$$

$$p=\frac{T}{\rho}\approx\frac{T}{R}\left(1-\frac{1}{R}\frac{\mathrm{d}^2h}{\mathrm{d}\theta^2}\right) \qquad \text{(Membrane equation)}$$

(where ρ is the local radius of curvature of the foil and R is the radius of the rigid shaft). Combining these two equations yields the flexible foil bearing equation:

$$\frac{\partial}{\partial\theta}\left(h^3\cdot\frac{\partial^3h}{\partial\theta^3}\right)=0$$

which is the same equation as in the previous section except that the velocity term has been set to zero and that we put $\theta\sim s$.

This equation has to be solved, to determine the pressure distribution and the gap geometry, subject to appropriate boundary conditions on the pressure and the foil. We have developed a semi-analytical method, which requires only one generic set of dimensionless solutions that can subsequently be used to generate any desired particular solution, by scaling and interpolation. Finally, correction of the solution to account for finite width effects is achieved using simplified theory. In this manner, the bearing load, gap geometry and flow rate are determined for any given set of design parameters, (see Gysegom and Smets (2002) for more detail).

12.4.2 Practical Design of a Prototype

The foil bearing consists of a thin metal foil, equipped with feeding hole(s), which enfolds the shaft (see Figure 12.13). A flexible hose is attached to each feeding hole to supply pressurised air, in such a way as not to affect the foil shape. The foil is fixed to the rigid frame (body of the bearing) at both ends, passing over alignment cylinders (kinematic support system with 2×3 d.o.f.'s). In this way, the foil can wrap itself around the shaft

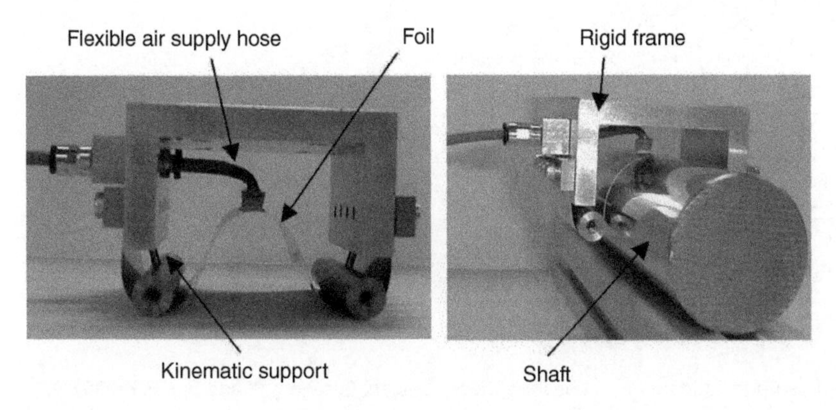

Figure 12.13 Practical embodiment of an EP foil bearing. (Source: (Gysegom and Smets 2002)).

correctly at all times. By virtue of its flexibility, the foil continuously adapts its geometry to different working conditions generating an optimal air gap. Furthermore such a bearing can accommodate a relatively wide range of shaft sizes without any alignment problem and is suitable for rotary as well as linear sliding applications.

The design parameters to be selected are: the supply air pressure, the foil geometry and material, the wrap angle, and the restrictor diameter(s) and location(s). Results show that very high load factors could be achieved; stiffness, however, is a major concern. The total stiffness of the bearing system can be represented as a serial connection of the air-gap stiffness and the foil (tensile) stiffness. A "thick" foil results in a higher foil stiffness, but the load capacity may deteriorate owing mainly to loss of wrap angle, and the air-gap stiffness may decrease owing to loss of flexibility of the foil. One measure to increase foil stiffness, without jeopardising load and film stiffness, is to minimise the total length of the foil. The final design problem is an optimisation whose main aspect is to make the right trade-off between foil stiffness, load capacity, and total stiffness of the system.

12.4.3 Experimental Validation

In order to verify the theoretical model, a test rig was built as depicted in Figure 12.14 to investigate the basic behaviour of aerostatic foils. On this simple set-up, the air is fed through the shaft and the foil extension as well as the gap geometry could be measured. Having refined and validated the design theory, a number of prototype bearings were built and tested for their load capacity, air consumption and stiffness characteristics. The first two turn out to be far superior to conventional bearings. The last is comparable to rigid bearings of the same dimensions. As an example, a foil bearing prototype (Figure 12.13) with a 50 mm wide and 0.08 mm thick, steel foil on a 40 mm diameter shaft, (wrap angle 180 degrees) was supplied with a 6 bar air pressure. The measurements show a load capacity of more than 800 N, a radial stiffness of 8.5 N/μm, and air consumption of around 2 Nl/min. Figure 12.15 illustrates the good agreement between theoretical prediction and experiment showing how the total stiffness is the resultant of the foil and air-gap stiffness, in series. Obviously, this bearing is not well optimised since the latter stiffness values are not well matched (the air gap is much stiffer than the foil). With better optimisation, a resultant stiffness of around 25 N/μm may be achievable.

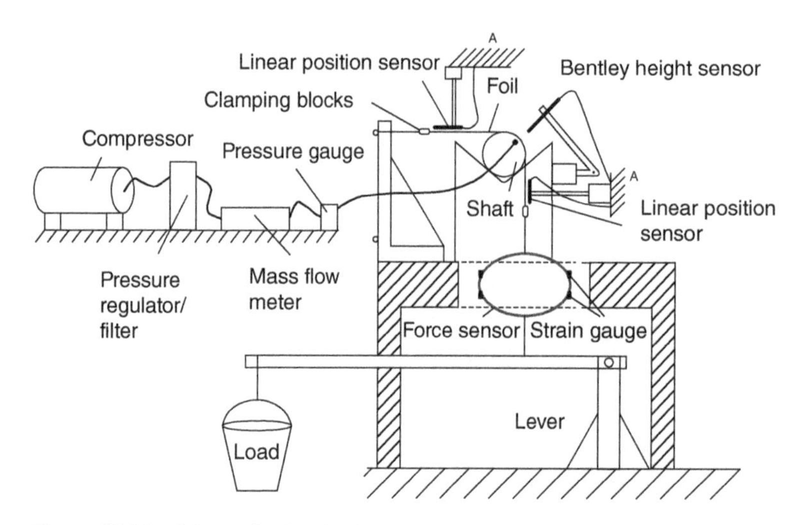

Figure 12.14 Schematic sketch of a test setup to validate EP foil bearing theory. Source: Gysegom W and Smets K 2002.

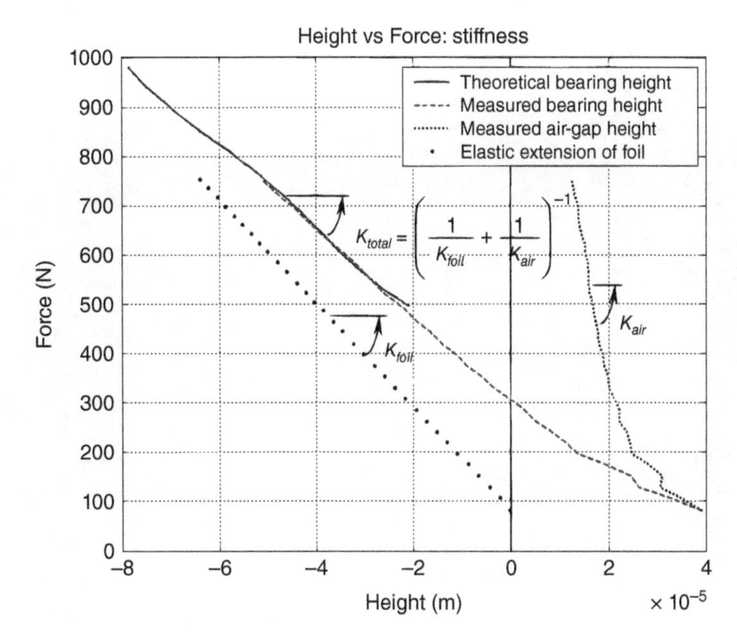

Figure 12.15 Experimental results of a prototype foil showing the bearing load as a function the bearing height (lift). The symbol k refers to stiffness. Source: Gysegom W and Smets K 2002.

12.5 Bump Foil Bearing

The rest of this chapter is based on (Vleugels 2009), parts of which are also to be found in (Vleugels et al. 2006).

12.5.1 Modeling of a Foil Bearing with an Idealised Mechanical Structure

Journal Foil Bearings

Although foil bearings (see Figure 12.16) are complex systems consisting of an air film, a top-foil, a flexible structure and in some cases a damping source (e.g. hysteretic friction or special coatings) it is possible to simulate some aspects of their behaviour relatively easily. A very popular model was first published by (Heshmat et al. 1983b) and is still in use. In this model the top-foil is modelled as an ideal top-foil: stiff enough to have no "sag" between the bumps or springs (see Figure 12.17) but without membrane or bending stiffness. This means that the deflection at each point only depends on the forces acting on that particular point.

The assumption of no membrane or bending stiffness is disputable because the top-foil is bent into a shell which gives it stiffness in the axial direction (Carpino et al. 1994). Nevertheless, this stiffening effect must not be overestimated. The flexibility of the top-foil in the axial direction is used in the second and third generation bump foil bearings to improve the performance by varying, in axial direction, the stiffness of the flexible element, i.e. the bumps. A better argument for not using the perfectly flexible top-foil assumption is the limited amount of flexible elements in the axial direction. In case of a "stiff" top-foil or only one flexible element in the axial direction, the gap height, which is the result of the eccentricity and the deflection on the flexible element, will be constant along the axial direction. The pressure is integrated along the axial direction to calculate this deflection.

The ideal top-foil assumption will overestimate the deflection in the middle section of the bearing, as the pressure is the highest in that region, and underestimate the gap height near the edges as the pressure is low there.

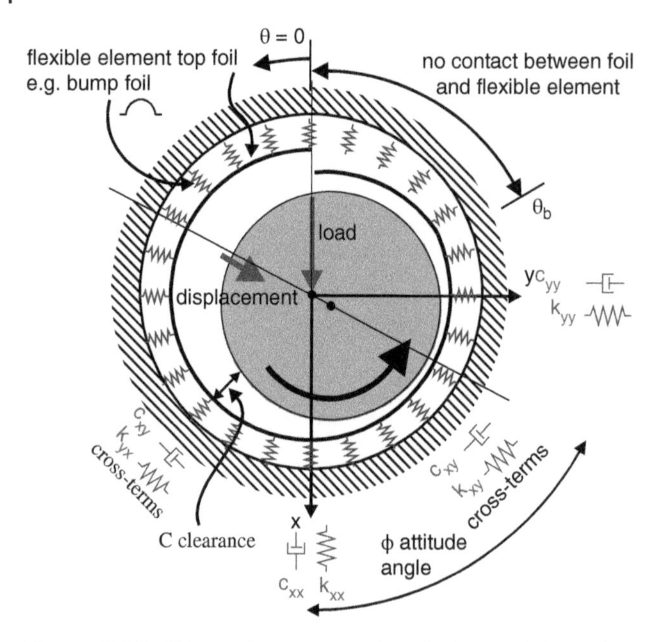

Figure 12.16 Schematic representation of a journal foil bearing. The flexible structure is represented by spring elements. Source: From Vleugels P, Waumans T, Peirs J, Al-Bender F and Reynaerts D 2006. Reproduced with permission from IOP Publishing Ltd.

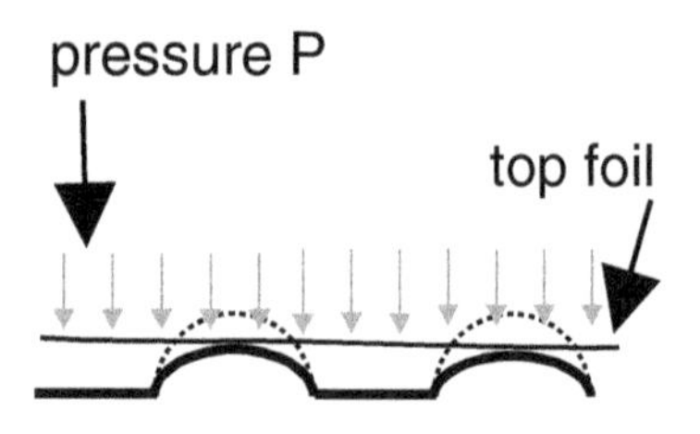

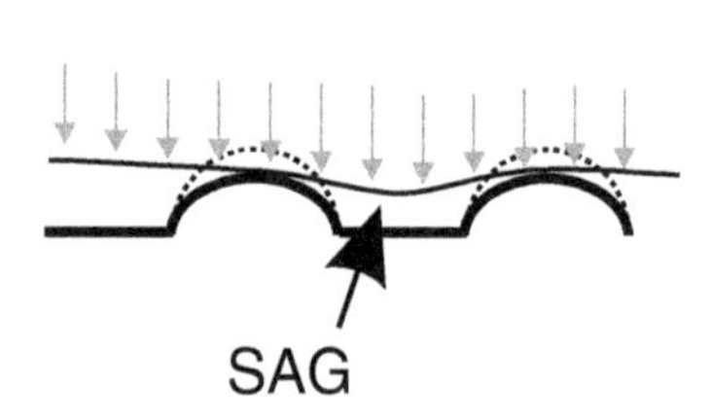

Figure 12.17 Bump-type flexible elements with a top-foil (undeformed bumps in dashed lines). The bottom figure shows the effect of sag between two bumps, caused by a top-foil with too low stiffness. Source: From Vleugels P, 2009.

Both effects will partially cancel out the errors made by the ideal top-foil assumption. The overall result will be that the gap height of a non-ideal top-foil is something between the maximum and minimum deflection of the ideal top-foil. However, the assumption of an ideal top-foil is still widely used and therefore it is also adopted here for better comparison with former simulations (Peng and Carpino 1993).

Some papers mention the foil bearing feature of operating with an eccentricity ratio $\epsilon > 1$. Theoretically this is not possible with an ideal top-foil. At the borders of the bearing the pressure is equal to the ambient pressure or the deflection of the flexible element is zero which results in a contact between top-foil and the journal for $\epsilon \geq 1$.

The flexible structure can be implemented by means of different types of elements such as corrugated foils or bump foils (Heshmat et al. 1983b), advanced elastic materials (Hou et al. 2004c) or spaced spirally wound foils

(Kitazawa et al. 2003). Nevertheless the modelling remains the same. The flexible structure is modelled as a set of springs, with or without dampers in parallel. In this research damping in the flexible element is neglected (for more details see section 12.9.4). The flexible element enters the normalised Reynolds equation via the pressure-dependent gap height $H + \alpha(P - 1)$:

$$\frac{\partial}{\partial \theta}\left(P(H + \alpha(P-1))^3 \frac{\partial P}{\partial \theta}\right) + \frac{\partial}{\partial \zeta}\left(P(H + \alpha(P-1))^3 \frac{\partial P}{\partial \zeta}\right) = \Lambda \frac{\partial}{\partial \theta}(P(H + \alpha(P-1))) + \sigma \frac{\partial}{\partial T}(P(H + \alpha(P-1))) \tag{12.13}$$

with

$$\alpha = \frac{p_a}{kC} \tag{12.14}$$

The bearing compliance parameter α is inversely proportional to the surface stiffness k (in N m^{-2} m^{-1}) and the clearance C (in m), and proportional to the ambient pressure p_a (in N m^{-2}). A foil bearing with a low α is thus almost comparable with a rigid-surface aerodynamic bearing.

Thrust Foil Bearings

The modelling of thrust foil bearings is comparable with journal bearings. The major difference between thrust and journal bearings is the region of the converging wedge. In journal bearings, this region is created by the eccentric position of the shaft. For thrust bearings however, an eccentric shaft position has no effect on the gap. Therefore, the converging gap is built in the flexible structure itself. Typically the complaint structure has multiple identical segments where each segment results in a converging wedge by gradually increasing the height of the structure. Of course, the production of these segments with gradually increasing height, which is only a few micrometers, is very challenging. The number of papers on foil thrust bearings is numerous, to name some: (Heshmat et al. 1983a; Iordanoff 1999; Kitazawa et al. 2003).

12.6 Numerical Analysis Methods for the (Compliant) Reynolds Equation

Despite the recent advances in computer aided simulation, the solution of the Reynolds equation is still not straightforward. The main difficulty is the non-linear character caused by the compressibility of the fluid which results in a non-constant density ρ. In case isothermal compression is assumed, the density ρ is represented by the pressure p. An excellent review of numerical methods for gas bearing simulation is given by Castelli and Pirvics (1968) and by Gross (1980).

Basically two approaches can be followed: time-dependent, or time-stepping, and time-independent or steady-state simulation. The first approach directly simulates the bearing behaviour, giving information on both steady-state and dynamic stiffness and damping properties. Major drawbacks are the computational cost and the complexity. However, a combination of a correct modelling of the bearing and the surrounding system (in this case the rotor) can give useful information on the dynamic behaviour of the system (in this case the maximum achievable speed). The steady-state or time-independent simulation does not provide information about the dynamic properties but it gives with a "limited" computational cost and complexity a lot of information about the bearing. The dynamic properties are found by applying e.g. the perturbation method (see section 12.9).

In this work, the approach of time independent simulation is followed. The oldest methods used for the time independent simulation are based on the combined analytic-numerical approach where the classical Reynolds equation is linearised before solving it. A first method is called the linearised *ph* method (Ausman 1960) where the Reynolds equation is written in terms of the dependent variable $\psi = ph$. Another method, also developed by Ausman (1959), uses the perturbation method to obtain an approximate solution. A difficulty in both methods are the systematic errors which appear in the solution, which are difficult to determine in magnitude. Therefore, nowadays only direct numerical solution methods are used.

For the direct numerical solution of the Reynolds equation several methods are in use: the finite difference method (FDM), finite volume method (FVM) or the finite element method (FEM). The latter is sporadically used (e.g. by Peng and Carpino (1993)) but most of the time the FDM is used as it is the easiest to implement without special tools.

12.7 Steady-State Simulation with FDM and Newton–Raphson

The steady-state solution is found by removing the time dependent σ term in equation 12.13. The resulting equation is solved by a numerical scheme as presented by Heshmat et al. (1983b). This method consists of solving the equation by finite differences using the Newton–Raphson method. However, care should be taken with the boundary conditions of the foil bearing problem. At the edges of the foil it is clear that the pressure is fixed or

$$P = 1 \text{ at } \zeta = 0 \text{ and } 1 \tag{12.15}$$

As the top-foil is in general not fixed to the flexible element, it could lose contact with the flexible element or "lift off" when a sub-ambient pressure is reached. This problem is analogous to cavitation, which takes place in liquid hydrodynamic journal bearings using oil/water. At the angle θ_b where the pressure becomes sub-ambient, this gives the following set of boundary conditions:

$$\frac{\partial P}{\partial \theta} = 0 \text{ and } P = 1 \text{ at } \theta = \theta_b \tag{12.16}$$

Boundary conditions 12.16 are generally known as the Reynolds or JFO (Floberg and Jakobsson 1957; Olsson 1965) boundary conditions. In fact, the mentioned boundary conditions are the film rupture conditions of the so-called JFO boundary conditions that also contain a set for film re-formation (making them different from the Reynolds boundary conditions, which only handle film rupture).

From angle θ_b on, the gap height H remains constant as the top-foil comes loose of the compliant structure. Numerical problems arise because this angle θ_b (and the corresponding height) is not known beforehand. A correct implementation of these JFO boundary conditions is not evident. In section 12.7.1 an overview about the different possible approaches is given.

A foil bearing contains a non-continuous top-foil making the bearing not circular symmetric and giving a pressure boundary condition at the groove of the top-foil:

$$P = 1 \text{ at } \theta = 0 \tag{12.17}$$

As a result of this, the bearing orientation with respect to the load vector is also a parameter. Unless otherwise stated the simulations are done with the groove at $\theta = 0$. An additional iteration step is thus needed in order to let the resulting attitude angle coincide with the applied load in the vertical direction. This is done by changing the angle with the minimum gap height.

12.7.1 Different Algorithms to Implement the JFO Boundary Conditions in Foil Bearings

As mentioned in the previous section, the loss of contact between top-foil and supporting flexible elements leads to additional boundary conditions, see equation 12.16, on an *a priori* unknown angle θ_b. This section discusses three different ways of treating these boundary conditions.

In the first method, the pressure in the region with sub-ambient pressure or the no-contact zone is simply neglected without taking the boundary conditions into account. Nevertheless most authors state the correct boundary conditions (Heshmat et al. 1983b). This is clearly not a correct implementation as the pressure gradient at the border is not zero and mass conservation is not fulfilled. The advantage of this method is, however, its simplicity.

In the second method, the pressure gradient becomes zero at the border by keeping the height constant in the no-contact zone and resetting the sub-ambient pressures. At first sight, this seems a correct implementation as the pressure is atmospheric and pressure derivative is zero. However, the constant height in the no-contact region is not sufficient to prevent pressure generation in the no-contact zone. The no-contact zone acts like two parallel surfaces with atmospheric pressure boundary conditions on three borders, while, on one remaining side a parabolic-like pressure distribution, which causes a pressure gradient although the height is constant. In order to prevent any pressure generation and to fulfill mass conservation it is necessary to modify the differential equation in the no-contact region by eliminating the diffusion term (see equation 12.13).

To conclude, although the second method gives good results, it has the problem that in the no-contact region the differential equation still allows pressure generation. In order to solve this fundamental problem with the previous methods, one needs to implement a slightly modified Elrod's cavitation algorithm which is often used to take cavitation into account in liquid film bearings (Elrod 1981). Elrod proposed to add a switching function to the differential equation which eliminates the diffusion term in the equation if the point of interest lies in the cavitated zone or in this case the no-contact zone. Furthermore he changed the variable P to a new variable θ representing the fractional film content in the cavitated zone and the density ratio in the full-film zone. After simulation, the density ratio is transformed back to the pressure.

In order to make use of Elrod's algorithm for the simulation of foil bearings, the algorithm is modified in this work. The basic principle of adding a switching function in order to eliminate the diffusion term from the Reynolds equation in the no-contact region remains. But instead of replacing the pressure variable with the fill-fraction variable, one has to keep the pressure as the variable (voids cannot exist in air). Furthermore the height is kept constant in the no-contact region, which represents its actual behaviour. It is important to mention the upwind difference scheme (UDS) for the convection term as also proposed by Elrod. For the diffusion term a central difference scheme can be used. An advantage of this method is its robustness. After only a few iteration steps, the mass errors are extremely small, meaning that full convergence for the complete bearing surface is obtained. Inherent to numerical simulations, the no-contact border is only known up to one grid point. This results in a border switching between 2 grid points. Therefore the algorithm blocks the border if it is stabilized. One or two additional iteration steps then lead to full convergence of the solution.

12.7.2 Simulation Procedure

The compliant non-linear Reynolds equation 12.13 with its boundary conditions 12.15, 12.16 and 12.17 is solved with the finite differences approach. The procedure here used is also employed by other authors, e.g. (Heshmat et al. 1983b), and will thus be only briefly explained. This elliptic equation is discretised with a central difference scheme for the diffusion terms and upwind scheme (UDS) for the convection term (see also section 12.7.1). Also a switch function is included to eliminate the diffusion term in the no-contact zone. The simulation scheme for a foil bearings is iterative because of the non-linear character, caused by the pressure-dependent density, represented by the pressure in case of isothermal compression.

There is also another difficulty in the simulation strategy of fluid film bearings. In practice one is interested in the resulting eccentricity, given the load and speed. However, the load capacity of the bearing is the output of the simulation. So, in practice, one has to perform multiple simulations in order to obtain the eccentricity, given the load and speed. For non-rotational symmetric bearings like foil bearings (the top-foil is a non-continuous foil), the procedure is a bit more complicated as the angle between the load vector and groove in the top-foil is constant, often even zero. As a result of this, the shaft position, and thus also the angle with minimum eccentricity, called $\theta_{min\ height}$, varies when the bearing number Λ changes (e.g. speed increases). In practice, the bearing housing orientation with respect to the load vector does not change during operation. In general, the groove will be on top of the bearing. This means, that the angle of minimum height needs to be equal to the attitude angle or $\theta_{min\ height} = \phi$.

Based on the aforementioned issues, the simulation procedure can be basically split up in two separate loops. Firstly,

(i) a loop for searching the angle of minimum gap height

1. initialise matrices
2. air film loop
   (a) calculate mass flow errors $\dot{m}$ and partial derivatives
   (b) calculate ΔP and update $P_i = P_{i-1} + \Delta P$
   (c) if $\Delta P > P_{\text{max air film error}}$ iterate again
3. if $P < 1$ and θ_b not yet constant: update switch function and keep H
4. if $\Delta P > P_{\text{max error}}$ start from air film loop
5. calculate steady-state properties W, ϕ.

This loop is very basic in structure. In the first simulation the minimum groove is placed at $\theta_{\text{min groove}} = 180°$. After simulation the value is corrected with the resulting attitude angle. In order to speed up the process, the first simulations are performed with a relatively large grid size, afterwards the grid is refined in order to predict the attitude angle ϕ more accurately (the angle predictions are relatively sensitive to the grid size). To update the groove angle, a Newton–Raphson method could be used but this results sometimes in bad predictions, therefore it is not used here. In general, after a few iterations the minimum gap angle is predicted up to 1°.

Secondly,

(ii) the solution of the compliant Reynolds equation:

1. search for minimum height angle $\theta_{\text{min height}}$
   (a) start (rough) calculation with $\theta_{\text{min height}} = \pi$
   (b) $\theta_{\text{min height}} = \phi + \pi$
   (c) recalculate
   (d) if $\phi \neq 0$ then $\theta_{\text{min height i+1}} = \phi + \theta_{\text{min height i}}$ and recalculate
   (e) recalculate with finer mesh
   (f) if $\phi \neq 0$ then $\theta_{\text{min height i+1}} = \phi + \theta_{\text{min height i}}$
2. calculate steady-state properties W, ϕ.

This loop for the simulation of the foil bearing itself in fact consists of two loops. After the initialisation of the matrices, the air bearing loop is started. In this loop the pressure distribution is calculated and updated. A major concern here is the prediction of the density by the pressure. In the first iterations, the density, represented by the pressure P, is badly predicted, making the simulation results less/not reliable. The loop is repeated until the pressure distribution becomes sufficiently accurate. If the air film geometry is predicted, the switching function and the height distribution matrix are updated and the air film loop is started again. As a general criterion for convergence, the maximum pressure change, from iteration to another is set to $1\ 10^{-6}$. This is typically reached after 10 to 20 iterations. The convergence speed highly depends on the input conditions; high pressures (thus high bearing numbers Λ or high eccentricities ϵ) tend to converge more slowly.

12.7.3 Steady-State Simulation Results and Discussion

The resulting pressure and height distribution at the middle plane of a foil bearing with $\alpha = 0$ and $\alpha = 1$ are shown in Figure 12.18. The case of $\alpha = 0$ is comparable with a usual, fixed-geometry aerodynamic bearing with a pressure

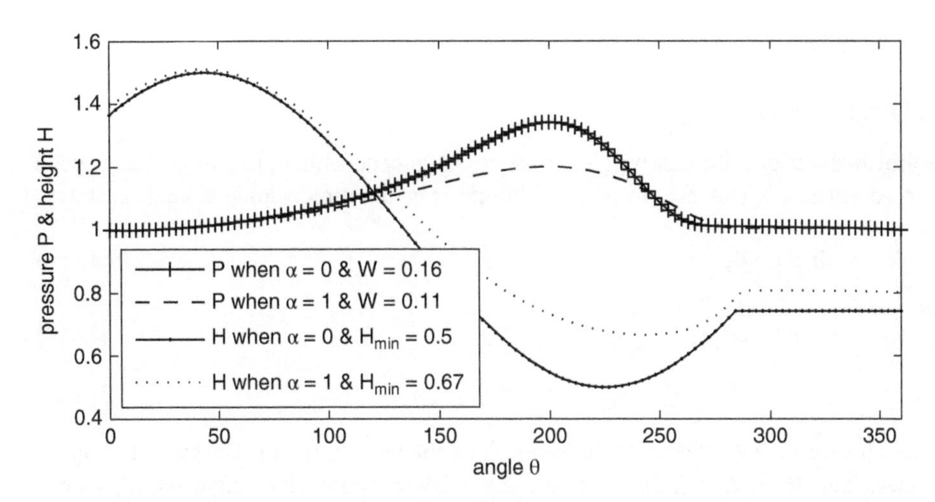

Figure 12.18 Pressure profiles for $\alpha = 0$ and $\alpha = 1$ with $\epsilon = 0.5$ and $\Lambda = 1$. Source: From Vleugels P, Waumans T, Peirs J, Al-Bender F and Reynaerts D 2006. Reproduced with permission from IOP Publishing Ltd.

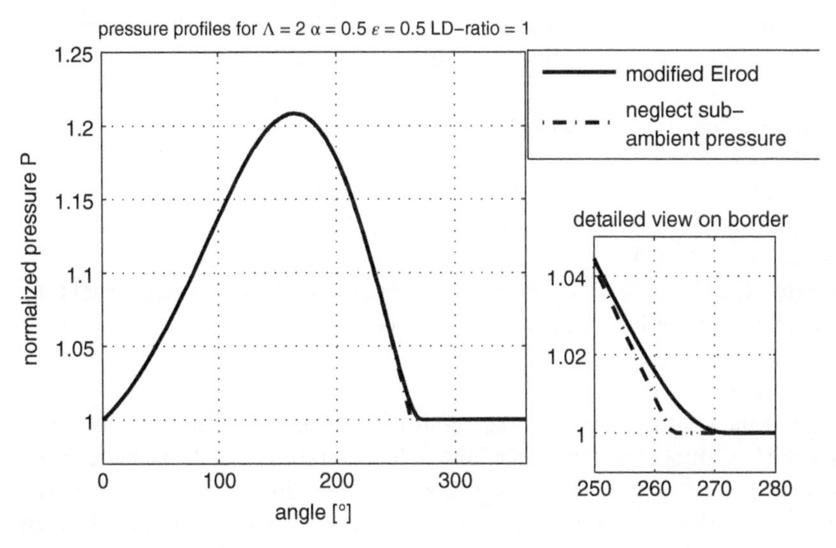

Figure 12.19 Pressure profiles for $\Lambda = 2$, $\alpha = 0.5$, $\epsilon = 0.5$ and $LD - ratio = 1$. Source: From Vleugels P, 2009.

groove on top and without the sub-ambient region. As can be seen on the figure, the minimum height increases from $(1 - \epsilon)C = 0.5C$ to $0.67C$. In this way, it is much more feasible to choose a high eccentricity value as working point. As the bearing contains a flexible element it is more "forgiving" to a hard contact between the rotor and bearing surface (e.g. due to a suddenly increased load or excessive rotor vibrations due to imbalance or instabilities). Top-foil coatings further improve this property.

Figure 12.19 shows the resulting pressure distribution for two different implementations of the JFO boundary conditions: only neglecting the sub-ambient pressure and the proposed modified Elrod's algorithm. Although the different algorithms handle the no-contact region differently, only the zone close to the border between contact and the no-contact is influenced. The load capacity does not change significantly. The advantage of the proposed modified Elrod's algorithm is its robustness and the more correct representation of the actual behaviour.

12.8 Steady-State Properties

12.8.1 Load Capacity and Attitude Angle

As already explained in the previous sections, the bearing performance is highly dependent to the geometry, speed and operating conditions (i.e. eccentricity). The most important parameter is the (dimensionless) load capacity:

$$W_x = \frac{F}{p_a BD} = \int_0^{L/D} \int_0^{2\pi} P \sin\theta \, d\theta \, d\zeta \tag{12.18}$$

$$W_y = \frac{F}{p_a BD} = \int_0^{L/D} \int_0^{2\pi} P \cos\theta \, rmd\theta \, d\zeta \tag{12.19}$$

$$W = \sqrt{W_x^2 + W_y^2} \tag{12.20}$$

with F the load, p_a the ambient pressure, $D = 2R$ bearing diameter and B the bearing width. The attitude angle ϕ is less important but it indicates where the region of the minimum gap height is located, it is calculated as follows:

$$\Phi = \arctan \frac{W_x}{W_y} \tag{12.21}$$

Figure 12.20 plots the load capacity for foil bearings with α equal to 0, 1 and 5 as a function of the bearing number Λ and eccentricity ϵ for L/D-ratio 1. Important to notice is the totally different magnitude scale of the load for each α. The load capacity for the same eccentricity is greatly reduced when the bearing compliance α increases. As mentioned before the $\alpha = 0$ case can be compared with a rigid bearing without sub-ambient region.

When the bearing number is small (i.e. low speed) the load capacity W is proportional with Λ and ϵ. Compressibility effects are negligible or the bearing behaves like a liquid (thus incompressible) film bearing. For large Λ, the compressibility becomes significant and the load capacity saturates. Further increasing the Λ or speed has no more influence on the load capacity (see also Chapter 8).

Also shown, on the right figures, is the attitude angle which, for all cases, tends to 90° for low Λ values (not fully visible in this graph) and drops to zero for high bearing numbers Λ and eccentricities ϵ.

Load Capacity Dependence on the Compliance α

As already shown in the previous graphs the steady-state properties highly depend on the bearing compliance α. In Figure 12.21 the load capacity is plotted as a function of the compliance α for several cases (note the logarithmic scale in the ordinate). In the left figure the bearing number Λ is kept constant at 1 while in the right figure the eccentricity is fixed to 0.5 and the bearing number Λ is varied. When Λ and ϵ are low, the dependency on α is much less pronounced. However, when the generated pressure is high, thus Λ and/or ϵ high, load capacity decreases very rapidly when also the compliance α increases. This is of course not surprising as a generated high pressure results in a larger top-foil deflection, which in turn decreases the generated pressure. Thus, eventually, the load capacity reduces. This figure once again stresses the dependency of the steady-state properties on the compliance α or, in practice, the need to know/control the clearance value C as it highly influences the properties, (note that the bearing stiffness k is much easier to control). Probably this is one of the major reasons why simulations and measurements do not coincide well from time to time in literature. Furthermore, it illustrates that the compliance value α should not be chosen too high as the required eccentricity ϵ will be high and thus the system becomes more sensitive to crashes.

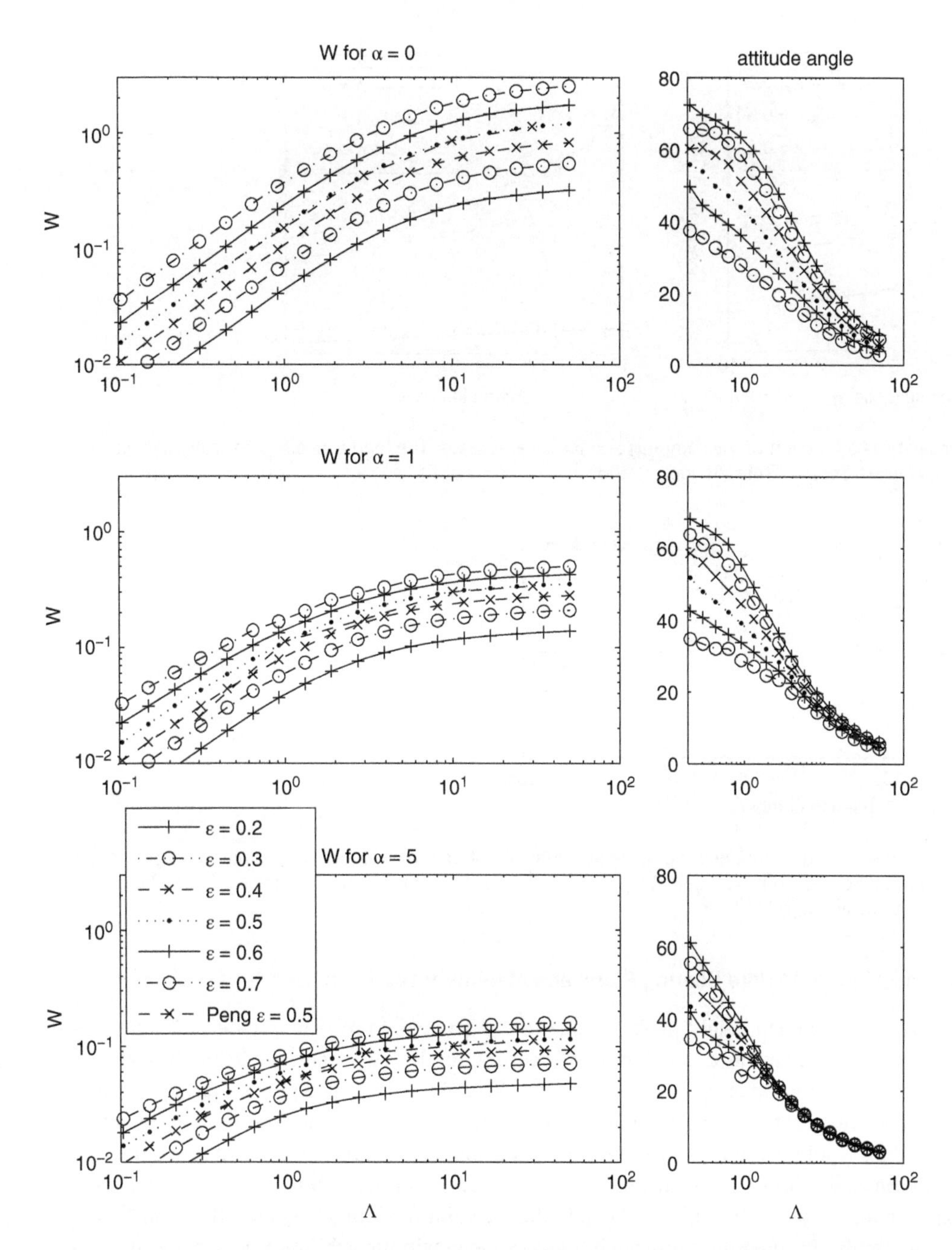

Figure 12.20 Load capacity and attitude angle for $\alpha = 0, 1$ and 5. Sources: Source: From Vleugels P, Waumans T, Peirs J, Al-Bender F and Reynaerts D 2006. Reproduced with permission from IOP Publishing Ltd. and From Vleugels P, 2009.

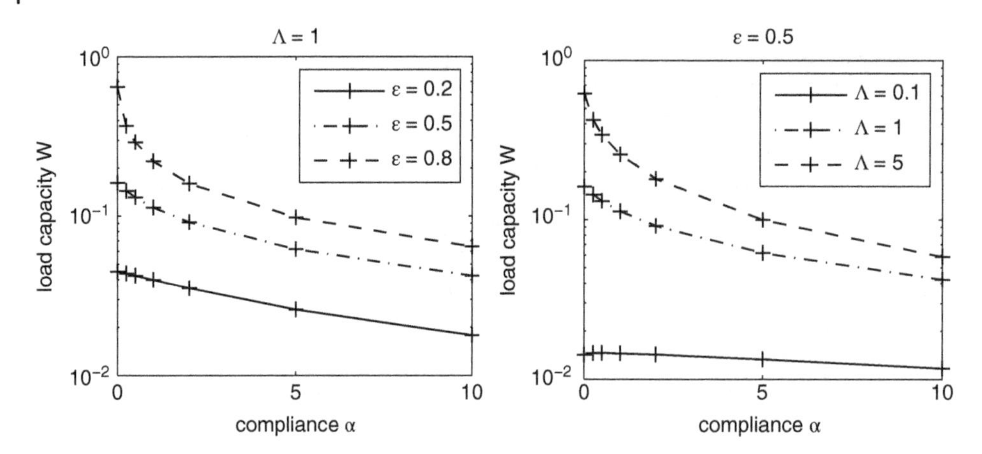

Figure 12.21 Load capacity as a function of the compliance α for several cases. The load capacity drastically decreases if the eccentricity ϵ or Λ increases. Source: From Vleugels P, 2009.

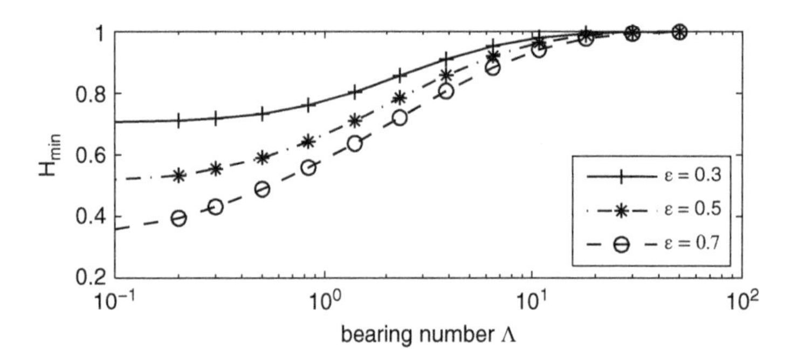

Figure 12.22 Minimum gap height in middle bearing plane as a function of bearing number Λ. The minimum gap height tends to 1 for large Λ. Sources: Source: From Vleugels P, Waumans T, Peirs J, Al-Bender F and Reynaerts D 2006. Reproduced with permission from IOP Publishing Ltd.

12.8.2 Minimum Gap Height in Middle Bearing Plane and Maximum Load Capacity

As explained before, the minimum gap height increases due to the generated pressure. Figure 12.22 shows the minimal gap height in the middle plane as a function of the bearing number Λ, eccentricity ϵ and bearing compliance α. For small Λ the minimum gap is $H_{\text{min}} = (1 - \epsilon)$ or the effect of the generated pressure is negligible. For large Λ, H_{min} tends to 1, i.e. the generated pressure makes the gap increase with ϵ. The minimum gap cannot be larger than 1 as larger values would result in the lack of a diverging channel. Increasing Λ further will not result in higher pressures and the load capacity (being the integrated pressure) saturates. Moreover, fixed geometry gas bearings have a maximum load capacity but this is related to the gas compressibility (Chapter 8). In foil bearings this occurs too, but additionally also the minimum gap influences the maximum achievable load capacity.

Some authors use a different definition of the eccentricity. They take into account the increased gap due to the generated pressure. This reduces the eccentricity value ϵ and the load capacity seems to increase. Based on this modified eccentricity they claim the improved load capacity compared to, for instance, rigid surface bearings. This is of course not a fair way to compare bearings. What they do not mention either is the fact that near the border, where there is much less pressure generation, the minimum gap has not (or has much less) increased. If of course, the top-foil is thick enough such that the deflection in the meridional plane is uniform, it can be interesting to look also at this modified eccentricity as it is a good indicator for the actual minimum gap.

12.8.3 Thermal Phenomena in Foil Bearings and Cooling Air

In the work presented here, temperature influences are neglected. However, some authors (Bruckner and Prahl 2005; Dykas et al. 2006) mention that the bearing performance depends on the amount of cooling air supplied to the system. The origin of this can be found in the viscous losses in the air gap (see Chapter 17). Foil bearings often operate with a very small film thickness, causing somewhat high viscous losses and thus quite a large heat generation, which can introduce a significant temperature elevation. The effect of this elevated temperature is multiple. In contrast to liquids, the gas viscosity increases with the temperature, which (theoretically) will lead to increased load capacity, when that lies below the saturation limit. However, the increased temperature also influences the gas density and the film geometry and thus complicates the simulation as also energy balances.ep needs to be incorporated. It is thus not possible to draw direct conclusions about the effect of temperature on the bearing performance without proper thermal simulation of the total system.

The generated heat needs to be evacuated. In rigid surface bearings this is no problem as most materials used have a good thermal conductivity and there is sufficient material near the air gap (the bearing is often made out of rigid block). However, in foil bearings the amount of material (e.g. in a bump foil bearing it is only a thin foil) is limited resulting in a bad path to evacuate the heat. Several authors reported about the heating of the (bump) foil. However, none of them really addressed the problem completely. It is of course extremely difficult as also the compliant structure itself deforms under the influence of temperature. If the air gap decreases as a result of this thermal deformation, even more heat will be generated, resulting in thermal instability.

It is thus very difficult to formulate proper conclusions about the influence of thermal aspects on foil bearings. Besides the computational difficulties, also the physical modelling of the thermal behaviour of the compliant structure is very important. As a general conclusion, it is always a good idea to supply some cooling air to the foil bearing in order to limit/stabilise temperature effects.

12.8.4 Variable Flexible Element Stiffness and Bilinear Springs

In the simulations shown, the bearing compliance α is kept constant over the total bearing surface. However, it can be interesting to vary the compliance in order to influence certain properties. The most popular way is by changing the stiffness of the flexible elements. This is done in what is called the "second generation foil bearings" from MiTi and NASA. The stiffness is varied axially such that the stiffness in the middle section is higher. This improves the load capacity significantly, which is not so surprising of course. Also along the circumference the stiffness can be altered. This more or less mimics a tilting pad bearing.

In a way, bump-type foil bearings with a constant pitch have also a variable compliance (Roger Ku and Heshmat 1992) (Le Lez et al. 2007). The corrugated foil is typically fixed on one side while the other end can move freely. The bump closest to the fixed end is the stiffest and the bump stiffness decreases in the direction of the free end. The cause of this is found in the friction force between housing and the flat part of the corrugated foil. In order to deflect, the stiffest bump has to overcome the friction force of all remaining bumps before it can move. This friction force depends thus on the amount of bumps in the direction of the free end, in other words, the bump stiffness decreases in the direction of the free end.

Another possibility is to use bi-linear springs (Heshmat et al. 1982); for small deflections only the first spring is used. From a certain deflection on, the second spring makes contact too which stiffens the flexible element locally. This is a very elegant way to improve the load capacity for (highly) compliant bearings.

It is clear from the simulations that the bearing compliance α has a significant influence on the performance (e.g. load capacity). It is thus no surprise that more stiff bearings improve the load capacity. Increasing the stiffness evidently results thus in better performing bearings. Eventually, however, the foil bearing will act as a rigid surface bearing without the advantages of a foil bearing.

12.8.5 Geometrical Preloading

As will be explained later, the onset speed of half speed whirl increases when the bearing operates at large eccentricities. So it is useful to design the bearing such that it operates at a large eccentricity; but not too large in order to prevent a (rubbing) contact between journal and housing. This can of course be achieved by increasing the rotor mass but this has a negative impact on the stability. Better is to increase the load without increasing the mass. In bump-type foil bearings this is (relatively) easily achieved by locally putting small shims between the housing and the corrugated foil. This local deformation of the top-foil will act as an additional load force on the bearing. During start-up this results in a rubbing contact thus a higher start-up torque and lift-off speed. If the shim dimensions are chosen well, from a certain moment on an air film will lift the top-foil such that there will be no rubbing contact anymore.

At zero or low speed there is thus possibly no air gap. This complicates the simulation procedure as this is theoretically not feasible. A method to solve this problem is to solve the problem first without the preload. If the solution is (almost) found, the preload is gradually applied.

12.9 Dynamic Properties

A major problem with aerodynamic bearings and fluid film bearings in general is their dynamic (in)stability. The steady-state solution does not provide information about the dynamic stability of a given working point. It is known from the literature that especially aerodynamic bearings are prone to half-speed whirl, a self-excited vibration which results in undamped whirling at approximately half the rotational frequency. In order to evaluate stability, the dynamic stiffness and damping properties should be known.

In conventional bearing systems, the determination of the stiffness properties is relatively easy, the (numerical) differentiation of the steady-state load characteristics is often sufficient. In fluid film bearings in general, and gas film bearings specifically, the calculation is much less straightforward. For specific cases (infinitely long or infinitely short bearings) and incompressible fluids, analytic formulas exist, which depend on the bearing geometry, viscosity, speed and load (Hamrock et al. 2004). However, for gas bearings the situation is much more complicated as analytic formulas cannot be derived. The dynamic properties become even frequency dependent.

12.9.1 Dynamic Properties Calculation with the Perturbation Method

In the past several methods were in use for the determination of the dynamic properties. A review can be found in Chapters 7 and 10. The most applicable method for the determination of the dynamic stiffness and damping properties is the perturbation method (Lund 1987, 1968; Peng and Carpino 1993). In this method the pressure P and height H functions are replaced by the steady-state solution $P_{(0)}$ and $H_{(0)}$ plus a small harmonic variation:

$$P = P_{(0)} + \kappa\, e^{i(\tau+\varphi)} \tag{12.22}$$

$$H = H_{(0)} + \eta\, e^{i\tau} \tag{12.23}$$

The dimensionless stiffness and damping pressures are then given by:

$$K = \frac{\kappa}{\eta} \cos\varphi \tag{12.24}$$

$$B = \frac{\kappa}{\eta} \sin\varphi \tag{12.25}$$

Combining formulas 12.13, 12.22, 12.23, 12.24 and 12.25, and neglecting the second order terms, the following set of linear equations is obtained (the steady-state solution $P(0)$ and $H(0)$ are assumed to be known):

$$\nabla[(3P_{(0)}G_{(0)}^{2} - \Lambda P_{(0)})\,\eta + (3P_{(0)}G_{(0)}^{2}\alpha\nabla P_{(0)} + G_{(0)}^{3}\nabla P_{(0)} - \Lambda(P_{(0)}\alpha + G_{(0)}))\,K + P_{(0)}G_{(0)}^{3}\nabla K]$$
$$= -\sigma(G_{(0)} + P_{(0)}\alpha)B$$
$$\nabla[(3P_{(0)}G_{(0)}^{2}\alpha\nabla P_{(0)} + G_{(0)}^{3}\nabla P_{(0)} - \Lambda(P_{(0)}\alpha + G_{(0)}))\,B + P_{(0)}G_{(0)}^{3}\nabla B]$$
$$= \sigma(P_{(0)}\eta + (G_{(0)} + P_{(0)}\alpha)K$$
$$G_{(0)} = H_{(0)} + \alpha(P_{(0)} - 1) \tag{12.26}$$

If the gap height is perturbed in the X direction, the pressure distributions of K and B can be calculated. After integration, they can be resolved into the stiffness and damping coefficients, K_{xx}, K_{yx}, B_{xx} and B_{yx}. Similarly, a perturbation in the Y-direction will yield the K_{yy}, K_{xy}, B_{yy} and B_{xy} coefficients.

Transformation of the dimensionless stiffness and damping coefficients back to dimensional properties is done via (Wilcock 1969):

$$K_{ij} = \frac{k_{ij}C}{p_a LD} \tag{12.27}$$

$$B_{ij} = \frac{b_{ij}C\omega}{p_a LD} \tag{12.28}$$

Let us note here that, in the notation adopted in this work, B_{ij} is the dimensionless damping **force** while b_{ij} is the dimensional damping coefficient.

12.9.2 Stiffness and Damping Coefficients

Figure 12.23 (notice the varying range of the ordinate!) shows the four synchronous stiffness coefficients for three different α values: 0, 1 and 5. The value has a large impact on the range and influences also the general trend of the values. The figures show that the flexible element becomes predominant from a certain bearing number Λ on (in this case mainly the speed) which is not surprising if one bears in mind that the calculated stiffness consists of the air film stiffness and the flexible element stiffness in series. For low bearing numbers the stiffness of the air gap is much lower than the stiffness contributed by the flexible structure, making the air gap stiffness predominant (for K_{xx} and $\alpha = 1$: $\Lambda = 0.1 \ldots 6$). However, as the bearing number Λ increases, the air gap stiffens and from a certain point on the stiffness of the flexible element becomes more important. The point where this occurs shifts to the left when the α value is increased (e.g. K_{xx} and $\alpha = 5$ for $\Lambda > 1$ instead of > 6 when $\alpha = 1$). This is not so surprising because high α values indicate a more compliant structure. This behaviour was also observed in the other direct stiffness term, K_{yy} and to a lesser extent also in the cross-coupling terms K_{xy} and K_{yx}. The $\alpha = 0$ case in Figure 12.23 shows the stiffness coefficients for a foil bearing with infinite surface stiffness. In this case, the air gap constitutes the weakest element. For completeness, also the damping coefficients are shown in Figure 12.24. The obtained values of the synchronous stiffness and damping coefficients agree very well with the values published by Peng and Carpino (1993).

Notice also that the direct stiffness terms K_{xx} and K_{yy} are proportional to the bearing number Λ. The sign of the cross term K_{yx} changes when Λ increases. The damping coefficients show a less predictable behaviour, some damping coefficients show an optimal value combined with low values for low and high bearing numbers Λ. All these effects can be explained when the frequency dependency of the dynamic coefficients is taken into account.

12.9.3 Influence of Compliant Structure Dynamics on Bearing Characteristics

In the discussion until now, the dynamics of the compliant structure itself are not taken into account. Depending on the physical implementation, this may be justified, e.g. when a thin almost massless corrugated foil as compliant structure is used. For some foil bearing types however the dynamics of the structure itself will have an influence

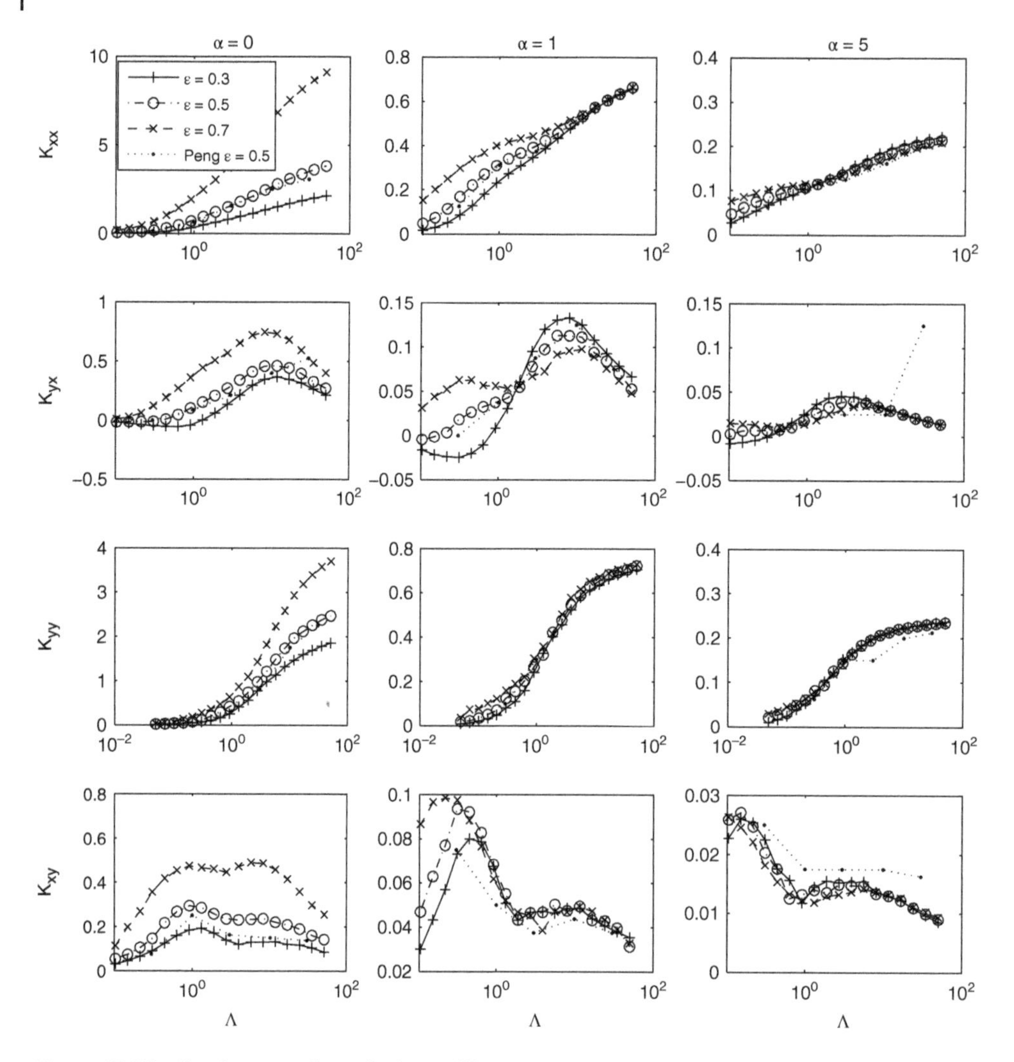

Figure 12.23 Synchronous dimensionless stiffness (note the varying ordinate scale). Sources: Source: From Vleugels P, Waumans T, Peirs J, Al-Bender F and Reynaerts D 2006. Reproduced with permission from IOP Publishing Ltd. and From Vleugels P, 2009.

on the bearing dynamics. In the literature, this is almost always neglected. Only (Howard et al. 2001a) reported on the influence of the compliant structure dynamics on the global bearing dynamics. Unfortunately, he discussed the effect only very briefly and mainly focused on the damping effects. He also did not take into account the compressibility, and thus the frequency dependency, of the bearing.

The major parameter to decide whether the dynamics should be taken into account is the compliant structure eigenfrequency with respect to the frequency range of interested. When this eigenfrequency is part of the frequency range of interest, the dynamics need to be taken into account. The effect on the overall performance depends on the frequency. For low frequencies, the influence of compliant structure dynamics will be negligible. For high frequencies however (thus far above the structural eigenfrequencies of the compliant structure), the compliant structure will act as a much stiffer spring. In the end, the global stiffness will thus mainly be determined by air gap stiffness or the foil bearing will resemble a rigid surface aerodynamic bearing.

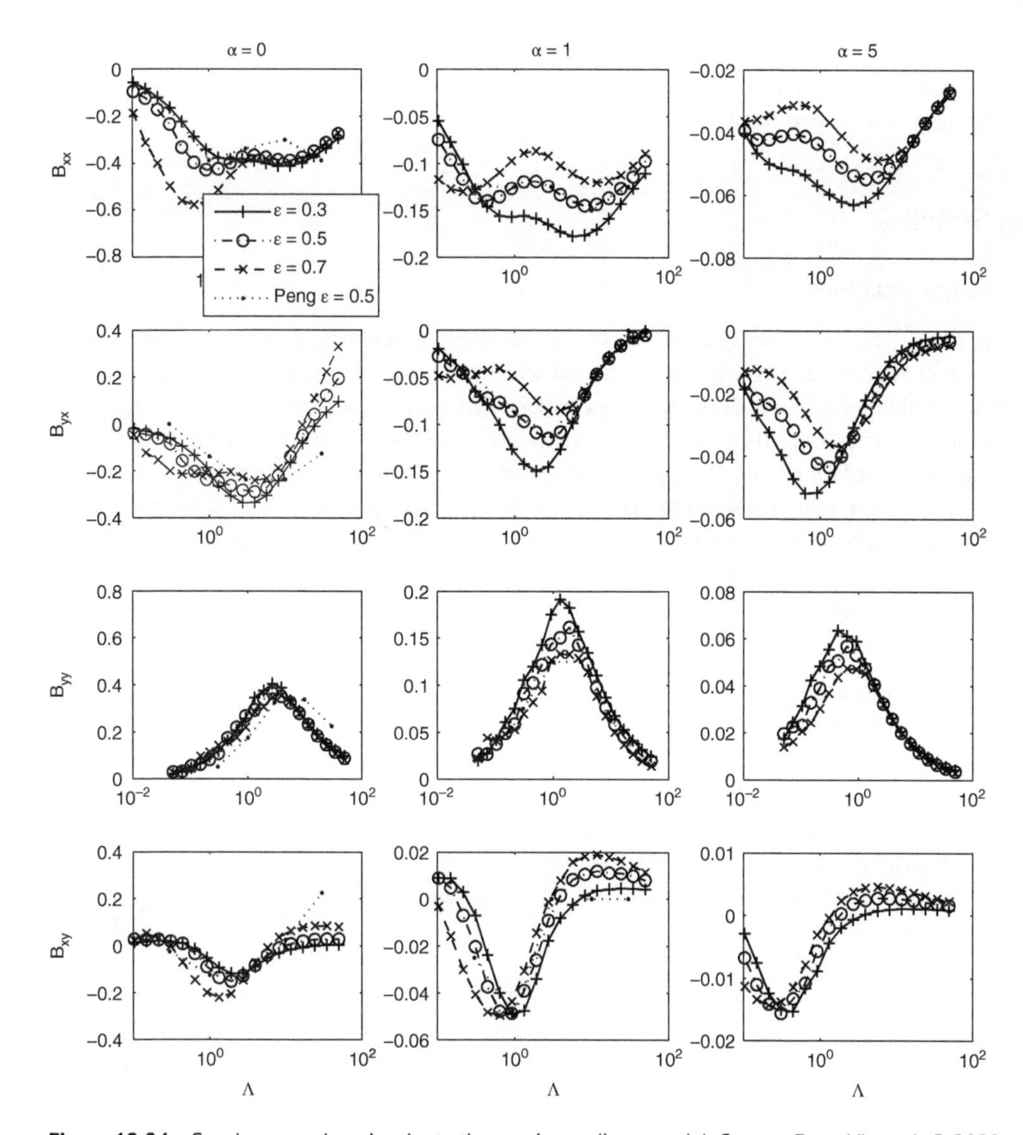

Figure 12.24 Synchronous damping (note the varying ordinate scale). Source: From Vleugels P, 2009.

12.9.4 Structural Damping in Real Foil Bearings

One of the most interesting advantages of foil bearings is the damping source which is often an integral part of the design. In bump-type foil bearings, for instance, the (dry) friction between the corrugated foil and housing is considered as the (major) damping source. This damping will have a positive effect on the damping, comparable with the flexible (and damped) supported air bearings (Guo and Kirk 2003a) (Guo and Kirk 2003b) (Czolczynski et al. 1996), see also Chapter 10. The modelling of damping effects in foil bearings is still not fully understood and rarely discussed in literature.

In this work, structural damping has been neglected as it is very difficult to quantify, even though, it will surely enhance the stability (see also 12.10). Taking damping into account during the simulations is not complicated. But it is very difficult to translate the required damping to a bearing design specification and vice versa. Through this, it is better to design the bearing for optimal performance without compliant structure damping. The damping in the structure will further improve the performance.

12.10 Bearing Stability

12.10.1 Bearing Stability Equations

Using the previously mentioned perturbation method it is possible to calculate the stiffness and damping properties of the bearings. However, it does not give any information at all about the bearing stability or more correctly about the system stability. Evaluation of the stability is only possible when the complete system, thus bearing and rotor, is analysed. This clearly complicates the usage of hydrodynamic bearings in general because the designer cannot simply select a bearing based on simple rules/graphs which provides the desired performance.

Nevertheless, it is very interesting to perform a stability analysis for a simple case like a symmetric rigid rotor supported by two identical journal bearings. The shown methodology is also applicable for other (non-)symmetric systems, of course the governing equations differ slightly.

The analysis is based on a symmetric system with a rigid rotor supported on two identical bearings. For simplicity, only half the system is modelled and the resulting system matrix is:

$$m\begin{bmatrix}\ddot{x}\\ \ddot{y}\end{bmatrix}+\begin{bmatrix}b_{xx} & b_{xy}\\ b_{yx} & b_{yy}\end{bmatrix}\begin{bmatrix}\dot{x}\\ \dot{y}\end{bmatrix}+\begin{bmatrix}k_{xx} & k_{xy}\\ k_{yx} & k_{yy}\end{bmatrix}\begin{bmatrix}x\\ y\end{bmatrix}=0 \tag{12.29}$$

The stability of this system can be evaluated by supposing that the shaft is whirling and then looking for the whirl frequency $\gamma\omega$. Finally the Routh–Hurwitz stability criterion can be applied to evaluate stability (Frêne et al. 1997). Nevertheless, it is more useful to normalise the equation as initially proposed by Lund (Lund 1967) because a very important parameter, the critical mass M_c, is herewith introduced:

$$M_c=\frac{m\ p_a}{\mu^2 L\left(\frac{R}{C}\right)^5}=\frac{K_{eq}}{72\Lambda^2} \tag{12.30}$$

with

$$K_{eq}=\frac{K_{xx}B_{yy}+K_{yy}B_{xx}-K_{xy}B_{yx}-K_{yx}B_{xy}}{B_{xx}+B_{yy}} \tag{12.31}$$

Ths stability criterion is provided by the equation:

$$(K_{eq}-K_{xx})(K_{eq}-K_{yy})-K_{xy}K_{yx}-B_{xx}B_{yy}+B_{xy}B_{yx}=0 \tag{12.32}$$

As can be found in (Lund 1967) a positive value of the L.H.S. of the stability criterion (Eq. 12.32) results in an unstable operation, and a negative value in stable behaviour. It is well known, from vibration theory, that stiffness cross-coupling terms are a source of instability (Frêne et al. 1997), see also section 12.10.1 and Chapter 10.

A major problem in solving these equations is the dependency of the stiffness and damping coefficients on the whirl frequency $\gamma\omega$, making the solution method for the whirl frequency ratio γ iterative. The parameter whirl frequency ratio γ is taken into account during the calculation of the stiffness and damping coefficients via the σ-parameter. Important to notice is the 5th power in the M_c parameter, which not only indicates that the R/C ratio is very important but also that a slightly different clearance value C, e.g. due to manufacturing errors or thermal expansion, can have a serious impact on the bearing stability. Prudence is called for in the evaluation during whirling. The whirling itself may become stable due to non-linearities which can induce sufficient additional damping to stabilise the whirling.

It is of course also possible to evaluate the stability with other methods, the result stays the same but the advantage of the method shown is that it gives a better insight in what the stability determines.

Role of the Cross-coupling Coefficients in the Stability

The aforementioned equations are generally applicable but are quickly complicated to give the reader a better insight in the stability problem. Frene (Frêne et al. 1997) discusses the bearing stability topic extensively. For better understanding, he simplifies the discussion to a system without damping (thus system 12.29 but without the b matrix). The stability conditions for a conservative system ($K_{xy} = K_{yx}$) can be simplified to:

$$\begin{aligned} &K_{xx} > 0 \\ &K_{yy} > 0 \\ &K_{xx}K_{yy} - K_{xy}K_{yx} > 0 \end{aligned} \tag{12.33}$$

Thus if the cross-coupling terms are zero, the system is always stable (assuming that the direct stiffness terms are positive). When there are cross coupling terms, stability is no longer guaranteed. When also the damping terms are considered, the analysis becomes more complex and also the system mass becomes a parameter for the stability. See also in this respect Chapter 10.

12.10.2 Foil Bearing Stability Maps

Using the aforementioned technique, proposed by Lund (Lund 1967), it is possible to calculate stability maps as shown in Figure 12.25. These maps relate M_c, Λ and load W. For example if the bearing geometry and rotor mass are fixed, W and M_c can be calculated and the intersection of both graphs determines the maximum bearing number Λ or the maximum speed. Given a bearing geometry (in case of foil bearings a L/D-ratio and α value), a high onset speed can be reached when M_c is as low as possible and W as high as possible. The lines of constant eccentricity ϵ are also shown as they are used for the calculation of the stability map.

Figure 12.25 shows the stability maps for foil bearings with α values 0, 1 and 5. There is only little difference between the maps with $\alpha = 0$ (almost comparable with a fixed geometry journal bearing with pressure groove) and the foil bearings with $\alpha = 1$ and 5. A comparison with the map for a normal plain journal bearing (Wilcock 1969) reveals that the maximum speed is only increased by 50% to 100%, not fully in agreement with literature where almost "unlimited" speed is promised.

However, foil bearings can safely operate at higher eccentricities compared to normal fixed geometry journal bearings. This makes it possible e.g. to decrease the clearance value C, lowering the M_c value drastically and consequently increasing the maximum achievable speed.

In order to make the stable region as large as possible, the critical rotor mass M_c should be as small as possible. However, the load W (often just the weight) should be as large as possible, often conflicting with the requirement of low M_c. This immediately clarifies the benefits of side pressure loading like used proposed by MIT (Piekos et al. 1997) and geometrical pre-loading, used in foil bearings (see 12.8.5). The determining factor for the critical rotor mass is the C/R-ratio which should be small. Typically it should be on the order of 10^{-3}. For shafts with a diameter of 6 mm this results in a clearance of 3 μm. Due to limitations in manufacturing capabilities (and thermal expansion or centrifugal rotor growth), this is very hard to realise for mesoscopic or microscopic bearings. However, for a clearance of 10 μm (thus a C/R-ratio of 3.33 10^{-3}) M_c becomes a factor of 400 worse. Also additional damping sources (e.g. a layer of rubber or bump friction dissipation) will improve the stability.

Also not taken into account in this discussion are the compliant structure dynamics (see also section 12.9.3), the compliant structure is thus considered massless. Depending on the mechanical implementation of the compliant structure (i.e. high or low eigenfrequency) it will have a significant influence on the overall bearing stability. Generally spoken, operating above the structural eigenfrequency results in much a stiffer global bearing stiffness or in a reduced (virtual) bearing compliance α (virtual because for low frequencies, i.e. thermal growth, the α remains unaffected).

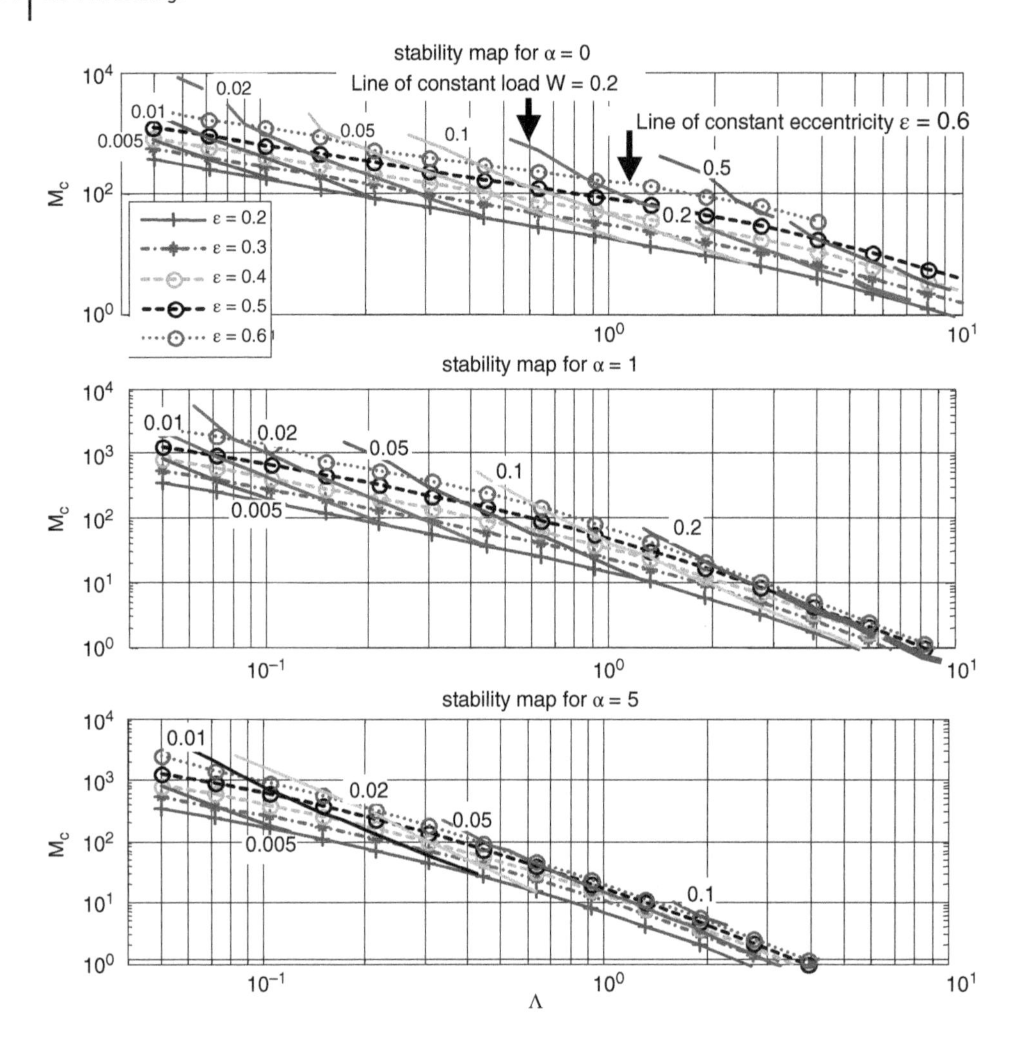

Figure 12.25 Stability maps for foil bearings. Sources: Source: From Vleugels P, Waumans T, Peirs J, Al-Bender F and Reynaerts D 2006. Reproduced with permission from IOP Publishing Ltd.

12.10.3 Fabrication Technology

The most popular design of a foil bearing consists of an underlying corrugated foil (called bump foil) as a flexible element and a top foil; see Figure 12.26(a). Another design consists of a spirally wound foil with spacers in between; see Figure 12.26(b). A crucial item in all foil bearing designs is the stiffness of the flexible element. For semi-cylindrical bumps with a negligible friction, the surface stiffness k can be calculated with (Walowit and Anno 1975) (see Figure 12.26 a,b, for notation)

$$k_{bump} = \frac{Ef^3}{2s(1-\nu^2)l_0^3}. \tag{12.34}$$

In the case of a spirally wound foil, the foil in the middle acts as the flexible element. The surface stiffness can be estimated with

$$k_{\text{spirally-wound}} = \frac{192Ef^3}{12s^4}, \tag{12.35}$$

in which the flexible element is modelled as a double-clamped beam; see Figure 12.26.

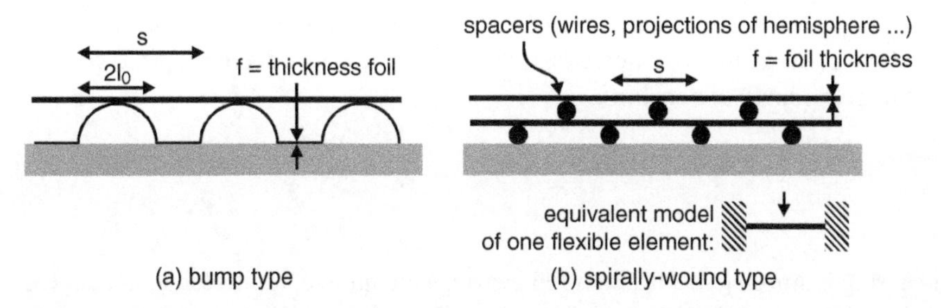

Figure 12.26 Design and embodiment alternative of a bump foil bearing. Sources: Source: From Vleugels P, Waumans T, Peirs J, Al-Bender F and Reynaerts D 2006. Reproduced with permission from IOP Publishing Ltd.

Table 12.1 Foil bearing properties.

Foil bearing type		**Bump type**			**Spirally wound foil**	
Reference	Reference 1 Rubio[a]	Reference 2 Heshmat[b]	Proposed design 1	Reference 3 Kitazawa[c]	Reference 4 Hou[d]	Proposed design 2
Bearing diameter D (mm)	38.17	35	8	30	12.02	8
Bearing width (mm)	38.10	31	8	30	14	8
Number of flexible elements	25		20	15	9	20
Clearance C (μm)	35	33–47	15	20	10	20
(Corrugated) foil thickness f (mm)	0.102	2 × 0.076	0.015	0.1[e]	0.060[e]	0.012
Pitch (mm)	4.572		1.257	6.17		
Steady stiffness of the flexible structure k_x (N μm^{-1})	1, … , 3[f]	3.5, … , 35[f]				
Surface stiffness k (Nmm^{-3})[g]	*0.8,… ,2.6*	*4.1,… ,41*	2.4	*1.1*	*1.1*	2.3
Estimated bearing compliance α	*3.5,… ,1*	*0.7,… ,0.07*	2.7	*4.5*	*8*	2.8

Values in italic are calculated.
a) Rubio and San Andres (2004).
b) Heshmat (1994).
c) S Kitazawa et al. (2003).
d) Hou et al. (2004b).
e) Phosphor bronze foil.
f) A range is given as the stiffness is highly non-linear.
g) Estimated with $k = \frac{4k_x}{\pi DL}$ (assuming that there is contact over 180°).
Source: From Vleugels P, Waumans T, Peirs J, Al-Bender F and Reynaerts D 2006. Reproduced with permission from IOP Publishing Ltd.

Table 12.1 shows the relevant properties of different foil bearing designs as they are found in the literature. In addition, two designs are proposed for an 8 mm shaft diameter: a bump and a spirally wound foil bearing. For the two (larger scale) bump-type foil bearings, the steady stiffness of the flexible structure can be found in the references. For all designs the surface stiffness k and bearing compliance α are derived. The range of α in the table is quite large (from 0.7 to 8). However, as a general trend it can be concluded that α is in the range of 0.5 to 3 for large diameters and a bit higher for smaller diameters.

The dimensionless compliance parameter α is independent of the bearing diameter. In general, the clearance C scales down with the diameter, resulting in a higher surface stiffness k if α is kept constant. This is advantageous as the available space (πD) for the integrated flexible elements decreases. On the other hand, due to the small dimensions it can be beneficial to increase α as the load decreases. This can be done by decreasing the surface stiffness k, a conflicting constraint with the small dimensions.

12.11 Conclusions

In this chapter, an idealised foil bearing is simulated. Although foil bearings are quite complicated systems, it is possible to simulate certain aspects with the modified Reynolds equation where the gap height not only depends on the eccentricity but also on the local pressure via the dimensionless compliance parameter α. Also studied in this chapter is the simulation of this modified Reynolds equation and more specifically the unconventional top-foil behaviour which loses contact with the supporting structure because there cannot be an area with sub-ambient pressure as in rigid surface bearings. Furthermore, steady-state properties such as the load capacity are simulated, which not only depend on the bearing number Λ and eccentricity ϵ but also on the bearing compliance α. The obtained results for foil bearings are inline with the results obtained for a rigid surface bearing (almost comparable with $\alpha = 0$) but are highly influenced by the compliance α. For instance, the more compliant the structure, the less the load capacity. Also, the dynamic properties are highly influenced by the compliance as the compliant structure can be considered as spring in series with the air gap stiffness. Furthermore, a rotor stability analysis is carried out and some methods to improve the stability are discussed.

References

Agrawal GL (1997 Foil air/gas bearing technology (97-GT-347) *Proc. of the International Gas Turbine & Aeroengine Congress & Exhibition*, Orlando, Florida, USA.

Al-Bender, F. and Smets, K., (2004). Development of externally pressurised foil bearings *Proceedings of the 4th Euspen International Conference*, 434–435.

Ausman, J., (1960). An improved analytical solution of self-acting gas lubricated journal bearings of finite length. *Journal of basic engineering* 83(2): 188–194.

Ausman, J.S., (1959). Theory and design of self-acting gas-lubricated journal barings including misalignment effects, 161–192.

Barlow, E., (1967). Derivation of governing equations for self-acting foil bearings. *Journal of lubrication technology* 7, 334–339.

Blok, H. and Van Rossum, J., (1953). The foil bearing - a new departure in hydrodynamic lubrication.

Bruckner, R. and Prahl, J., (2005). Analytic modeling of the hydrodynamic, thermal, and structural behavior of foil thrust bearings.

Carpino, M., Lynn, A. and Peng, J., (1994). Effects of membrane stresses in the prediction of foil beraing performance. *Tribology Transactions* 37, 43–50.

Castelli, V. and Pirvics, J., (1968). Review of numerical methods in gas bearing film analysis. *Trans. ASME - Journal of Lubrication Technology* 90(4): 777–792.

Czolczynski, K., Kapitaniak, T. and Marynowski, K., (1996). Stability of rotors supported in gas bearings with bushes mounted in air rings. *Wear* 199, 100–112.

Dellacorte, C., Radil, K., Bruckner, R. and Howard, S., (2008). Design, fabrication, and performance of open source generation i and ii compliant hydrodynamic gas foil bearings. *Tribology Transactions* 51(3): 254–264.

Dykas, B., Prahl, J., Dellacorte, C. and Bruckner, R., (2006). Thermal management phenomena in foil gas thrust bearings.

Wilcock, D.F. (ed) (1969). *Design of Gas Bearings*. Mechanical Technology Inc. (MTI): Latham NY.

Elrod, H., (1981). A cavitation algorithm. *Journal of lubrication technology* 103, 350–354.

Eshel, A. and Elrod, H.G.J., (1965). The theory of the infinitely wide, perfectly flexible, self-acting foil bearing. *J. of Basic Eng., Trans. ASME* 87(4): 831–836.

Eshel, A. and Licht, L., (1971). *FOIL BEARING DESIGN MANUAL* Ampex Corporation Research and Advanced Technology Division 401 Broadway zb. su, -Redwood City. California.

Eshel, A. and Wildmann, M., (1968). Dynamic behavior of a foil in the presence of a lubricating film. *Journal of applied mechanics* 35(2): 242–247.

Floberg, L. and Jakobsson, B., (1957). The finite journal bearing, considering vaporization. Transactions of Chalmers University of Technology (Goteborg) p. 190.

Frêne, J., Nicolas, D., Degueurce, B., et al. (1997). *Hydrodynamic Lubrication – Bearings and Thrust Bearings*. Elsevier, Amsterdam.

Gross, W.A., (1980). *Fluid Film Lubrication* chapter Chapter Compliant Bearings, 483–549.

Guo, Z. and Kirk, R., (2003a). Instability boundary for rotor-hydrodynamic bearing systems, part 2: rotor with external flexible damped support. *Journal of Vibration and Acoustics* 125(10): 423–426.

Guo, Z. and Kirk, R., (2003b). Instability boundary for rotor-hydrodynamic bearing systems, part1: jeffcott rotor with external damping. *Journal of Vibration and Acoustics* 125(10): 417–422.

Gysegom, W. and Smets, K., (2002). *Experimental study, design and realisation of aerostatic foil bearings (in dutch)*: Master's thesis Department of Mechanical Engineering, Kuleuven, 02EP08. 02EP08.

Hamrock, B.J., Schmid, S.R. and Jacobson, B.O., (2004). *Fundamentals of fluid film lubrication*. Marcel Dekker, New York.

Hayashi, K. and Hirasata, K., (1995). Developments of aerodynamic foil bearing for small high speed rotor. 30, 291–299.

Heshmat, H., (1994). Advancements in the performance of aerodynamic foil journal bearings: high speed and load capability. *Journal of tribology* 116, 287–295.

Heshmat, H. and Tomaszewski, M., (2006). Small gas turbine engine operating with high-temperature foil bearings *Paper GT2006-90791, presented at the International Gas Turbine Institute Turbo Expo Conference, Barcelona, Spain, May 2006*.

Heshmat, H. and Walton, J., (2005). Turbojet engine demonstration with a high temperature air foil bearing.

Heshmat, H., Shapiro, W. and Gray, S., (1982). Development of foil journal bearings for high load capacity and high speed whirl stability. *Journal of lubrication technology* 104, 149–156.

Heshmat, H., Walowit, J. and Pinkus, O., (1983a). Analysis of gas lubricated compliant thrust bearings. *Journal of lubrication technology* 105, 638–646.

Heshmat, H., Walowit, J. and Pinkus, O., (1983b). Analysis of gas-lubricated foil journal bearings. *Journal of lubrication technology* 105, 647–655.

Hou, Y., Xiong, L. and Chen, C., (2004a). Experimental study of a new compliant foil air bearing with elastic support. *Tribology Transactions* 47, 308–311.

Hou, Y., Zhu, Z. and Chen, C., (2004b). Comparative test on two kinds of new compliant foil bearing for small cryogenic turbo-expander. *Cryogenics* 44, 69–72.

Hou, Y., Zhu, Z. and Chen, C.Z., (2004c) Comparative test on two kinds of new compliant foil bearing for small cryogenic turbo-expander. *Cryogenics* 44, 69–72.

Howard, S., Dellacorte, C., Valco, M. et al., (2001a). Steady-state stiffness of foil air journal bearings at elevated temperatures. *Tribology Transactions* 44(3), 489–493.

Howard, S., Dellacorte, C., Valco, M.,et al. (2001b). Dynamic stiffness and damping characteristics of a high-temperature air foil journal bearing. *Tribology Transactions* 44(4): 657–663.

Iordanoff, I. (1999). Analysis of an aerodyamic compliant foil thrust bearing: method for a rapid design. *Journal of Tribology* 121, 816–822.

Kim, D., (2006). Virtual design of liga fabricated meso scale gas bearings for microturbomachinery *Proceedings Power MEMS, 2006*.

Kitazawa, S., Kaneko, S. and Watanabe, T., (2003). Prototyping of radial and thrust air bearing for micro gas tubine. 1–6.

Le Lez, S., Arghir, M. and Frêne, J., (2007). A new bump-type foil bearing structure analytical model. *ASME Journal of Engineering for Gas Turbines and Power* 129(4): 1047–1057.

Licht, L., (1969). An experimental study of high-speed rotors supported by air-lubricated foil bearings—Part 1: Rotation in pressurized and self-acting foil bearings. *ASME, Journal of lubrication technology.*

Lund, J., (1967). A theoretical analysis of whirl instability and pneumatic hammer for a rigid rotor in pressurized gas journal bearings. *Journal of lubrication technology* April, 154–166.

Lund, J., (1968). Calculation of stiffness and damping properties of gas bearings. *Journal of lubrication technology* (10): 793–803.

Lund, J., (1987). Review of the concept of dynamic coefficients for fluid film journal bearings. *Journal of tribology* 109, 37–41.

Olsson, K., (1965). Cavitation in dynamically loaded bearings. *Trans Chalmers University of Technology Goteborg* p. 308.

Peng, J. and Carpino, M., (1993). Calculation of stiffness and damping coefficients for elastically supported gas foil bearings. *Journal of tribology* 115, 20–27.

Piekos, E., Orr, D.J., Jacobson, S., et al. (1997). Design and analysis of microfabricated high speed gas journal bearings.

Radil, K., Howard, S. and Dykas, B., (2002). The role of radial clearance on the performance of foil air bearings. *Tribology Transactions* 45(4): 485–490.

Roger Ku, C. and Heshmat, H., (1992). Compliant foil bearing structural stiffness analysis: Part I—Theoretical model including strip and variable bump foil geomety. *Journal of tribology* 114 (2): 394–400.

Rubio, D. and San Andres, L., (2004). Bump-type foil bearing structural stiffness: experiments and predictions Proc. ASME Turbo Expo 2004 *T2004-53611.*

Kitazawa, S., Kaneko, S. and Watanabe, T., (2003). Prototyping of radial and thrust air bearing for micro gas turbine *Proceedings of the International Gas Turbine Congress TS-019*, 1–6.

Salehi, M., Heshmat, H., Walton, J. and Tomaszewski, M., (2004). Operation of a mesoscopic gas turbine simulator at speeds in excess of 700.000rpm on foil bearings *ASME Turbo Expo (2004: Power for Land, Sea, and Air*, vol. GT2004-53870.

Vleugels, P., (2009). *Development of aerodynamic foil bearings for micro turbomachinery* PhD thesis Katholieke Universiteit Leuven – Department of Mechanical Engineering.

Vleugels, P., Waumans, T., Peirs, J., et al. (2006). High-speed bearings for micro gas turbines: Stability analysis of foil bearings. *Journal of Micromechanics and Microengineering* 16, S282.

Walowit, J. and Anno, J., (1975). *Modern Developments in Lubrication Mechanics*. John Wiley and Sons, New York.

Walton, J., Tomaszewski, M., Heshmat, C. and Heshmat, H., (2006). On the development of an oil-free electric turbocharger for fuel cells. GT2006-90796, 395–400.

Wildmann, M., (1969). Foil bearings. *Journal of lubrication technology* 37–44.

Xiong, L., Wu, G., Hou, Y., et al. (1997). Development of aerodynamic foil journal bearings for a high speed cryogenic turboexpander. *Cryogenics* 37, 221–230.

Yong-Bok, L., Dong-Jin, P., Chang-Ho, K. and Keun, R., (2006). Rotordynamic characteristics of a micro turbo generator suported by air foil bearings. *Journal of Micromechanics and Microengineering* 17, 297–303.

13

Porous Bearings

13.1 Introduction

Porous bearings offer another possibility of introducing the (gas) lubricant into the bearing film; namely through an extremely large number of feeding holes that are of exceedingly small diameters; i.e. the pores. In this way, one achieves a distributed flow all over the bearing surface, which has certain important advantages, but also some drawbacks.

The use of porous air bearings dates back to the 1950s, where they were successfully used in the construction of machines and high-speed applications (Montgomery and Sterry 1955; Robinson and Sterry 1958; Sheinberg and Shuster 1960). The first publication on porous gas bearings was probably that of Montgomery and Sterry (1955). They demonstrated the practicability of a porous gas journal bearing by rotating a shaft at 250.000 rpm in a pair of porous sleeves mounted on rubber O-rings. From that point forward, numerous papers have become available in the open literature on the successful use of porous gas bearings in all sorts of applications. In 1970, Booser (1970) even asserted that porous gas bearings were produced at a rate of about 20 million a day.

In recent years, the use of porous surfaces as well as porous-media restrictors in aerostatic bearings has gained a great deal of popularity. Because of their higher load capacity, stiffness, inherent design simplicity and low initial cost, porous bearings might potentially offer the best alternative for orifice-compensated bearings (Kwan and Corbett 1998a; Singh et al. 1984; Sneck 1968). An additional aspect is the filtering effect of the porous media, which may be viewed as an advantage or a disadvantage. On the one hand, it prevents small particles passing through the narrow bearing gap, and so increasing the service life of the surfaces. On the other hand, the pores can get clogged over time resulting in a lower permeability, which is disadvantageous.

Because of the many applications, not only in gas bearings but also in more general applications such as industrial filters, a great deal of work has concentrated on the development of porous materials with improved characteristics. Typical materials for porous gas bearings are: sintered metals, ceramics (i.e. SiC) and graphite (porous carbon). The permeability of these materials must, for optimal static and dynamic characteristics, be between 1×10^{-16} and 1×10^{-14} m^2 according to Fourka and Bonis (1997). Most examples of porous aerostatic bearings published so far are based on metallic or graphite materials, because of their lower cost and ease of machining (Kwan 1996). Furthermore, porous bearings out of graphite are also very crash resistant. In 2013, Cracaoanu and Bremer (2013) demonstrated that a porous bearing out of graphite was still operational after 50 crashes, while a conventional gas bearing with orifices already failed after 8 to 16 crashes. This being said, one of the biggest problems of ductile materials, such as graphite, is the problem of pore smearing which leads to poor permeability control. Porous ceramics eliminate this problem (Kwan 1996).

Despite the many advantages of porous gas bearings, their widespread use is still hampered by a inhomogeneous and variable permeability. This leads to problems of repeatability and reproducibility of bearing characteristics such as load capacity, stiffness and damping (Fourka and Bonis 1997). At Cranfield University, longstanding research has been performed on the production of uniform single and two-layered porous ceramic structures for

Air Bearings: Theory, Design and Applications, First Edition. Farid Al-Bender.

Companion website: www.wiley.com/go/AlBender/AirBearings

aerostatic bearings. Kwan (1996) and Kwan and Corbett (1998b) showed that hot isostatic pressing is a suitable technique to produce uniform ceramics. Another interesting finding in that study was that a two-layered porous bearing, with a densified layer at the bearing interface, is more stable over a broad range of supply pressures and bearing gaps while a single-layered bearing suffered from instabilities. Another study at Cranfield University Durazo-Cardenas (2003); Durazo-Cardenas et al. (2010) showed that a porous-ceramic hydrostatic bearing can also be manufactured by a starch consolidation (SC) technique. They showed that the SC porous-ceramic bearing has an improved performance over more conventional hydrostatic bearings, namely the static and rotational stiffness were 95 % and 150 % higher, respectively. In addition, flow rate and pumping power decreased by 64 % while the heat generation in the porous ceramic bearing showed 50 % lower temperature rise.

Chien et al. (2012) also developed a dual restrictive porous bearing. However, instead of using two porous materials with a different permeability and pore size, they sealed the substrate layer at the bearing interface with a solid epoxy resin of 40 μm thickness. Then, a close pitch array of 50 μm holes was lasered in the epoxy for the air flow. Experimental results showed that the stiffness of the dual restrictive layer bearing is 3.6 times higher than a traditional porous bearing. Unfortunately, no data was provided regarding the damping or stability.

Porous materials, often of sintered metals, may also be used for constructing "laminar" restrictors for externally-pressurised bearings, which have certain advantages as compared to other restrictor types, see Chapter 5 and (Belforte et al. 2005; Fourka and Bonis 1997; Silveira et al. 2010). The latter studied the pressure distribution and mass flow rate of a circular bearing pad with a metal woven wire cloth feeding system and compared this with a feeding pocket and porous restrictor. Experiments showed that the bearing characteristics of the woven wire cloth were more repeatable.

Porous gas bearings are just as vulnerable to dynamic instabilities as conventional gas bearings (Rao and Majumdar 1979). For this purpose, some authors argued that porous inserts are preferable to a porous medium (Kwan and Corbett 1998a). However, these argumentations are not in line with the observations of Majumdar (1983). He showed that the damping of a journal bearing with porous inserts is comparable with that of a full porous bearing. Su and Lie (2003) demonstrated that an orifice-compensated bearing with five rows of feed-holes approximates the bearing performances of a porous bearing. In addition, Su and Lie (2006) analysed the dynamic stability of an aerostatic journal bearing with respect to the number of restrictor rows and compared this with a porous bearing. They found that the latter is more stable at low rotational speeds (i.e. $\Lambda < 0.1$) and high rotational speeds (i.e. $\Lambda > 1$).

With the emergence of laser beam machining and micro drilling, interest has grown in aerostatic bearings with very small feed-holes (viz. diameters from 30–50 μm), as a compromise between orifice-fed and porous bearings. For this purpose, Miyatake and Yoshimoto (2010) investigated the static and dynamic characteristics of gas lubricated bearings with such small feed-holes. From their numerical analysis it became clear that a higher maximum stiffness can be obtained, as well as a higher damping coefficient.

In the following we shall overview the static and dynamic modelling of externally-pressurised porous bearings with a focus on the journal type. Thrust bearings are also touched upon in regard to the static characteristics.

13.2 Modelling of Porous Bearing

Figure 13.1 shows a schematic presentation of a plain (or generic) porous gas bearing. The restriction of the gas occurs in a porous material (e.g. graphite, sintered metal or ceramic) instead of in a restrictor. The permeability of the porous material is typically between 1×10^{-14} and 1×10^{-16} m^2 for optimal static and dynamic bearing characteristics (Fourka and Bonis 1997), while the porosity is usually around 20 %. As illustrated in Figure 13.1, pressurised air at constant supply pressure p_s is fed from a plenum through the porous material of width B and thickness h_p. The air pressure reduces to $p'(x, y, z)$ in the porous media and drops further to $p(x, y)$ in the bearing

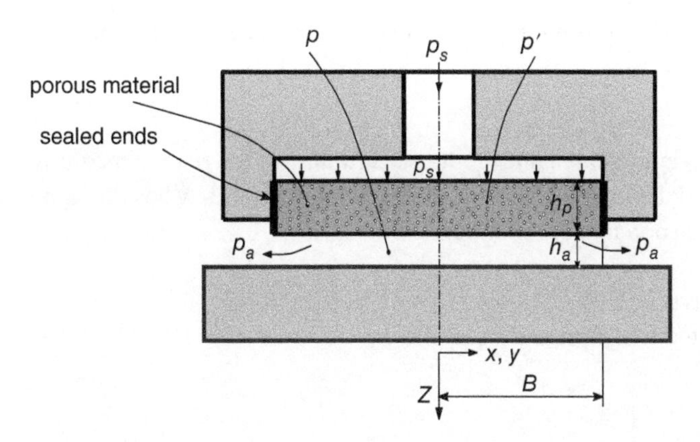

Figure 13.1 Configuration and notation of a porous aerostatic plain bearing (not to scale).

gap, of height h_a. Finally, the air exhausts to the atmosphere p_a at the bearing ends. The ends of the porous block are sealed to prevent end leakage.

This section describes the implemented analytical and numerical models for a porous gas bearing. Many theories have been published to model flow in a porous bearing (Majumdar 1975; Mori et al. 1968, 1969; Sneck and Elwell 1965; Sneck and Yen 1964, 1967; Sun 1975a). In general, the problem is divided into two flow sections which are jointly considered. Namely the flow in the porous medium (feed flow), where the pressure is given the symbol p' to distinguish it from the film pressure, and the flow in the lubricant film, (film flow over a porous surface), where the pressure is given the usual symbol p. The mathematics used to model the flow in either regions are derived in a unifying approach as for an orifice-compensated bearing. A detailed review pertaining to the theoretical modelling and applications of porous gas bearings may be found in (Kwan and Corbett 1998a; Majumdar 1976a; Sneck 1968).

13.2.1 Feed Flow: Darcy's Law

The literature on the flow in a porous restrictor is quite extensive, not merely for gas bearings, but also for the flow of oil, water and gas through all kind of porous media such as sand, rocks and industrial filters. A basic reference on this subject may be found in (Collins 1961; Dagan 1989; Scheidegger 1960). Hitherto, very complex equations have been derived describing this phenomena in hydrogeology. In gas-bearing analysis, the flow through the porous media is mostly described by Darcy's law. This law was experimentally defined by Darcy in the nineteenth century when he studied the (1D) flow of water through beds of sand. Darcy found that the discharge Q through a porous medium normal to the flow is proportional to the pressure drop $\partial p'/\partial z$ across the porous material. Later on, Robinson and Sterry (1958) demonstrated experimentally that this law is only valid for a laminar (viscous) flow. To overcome this limitation, several attempts have been made to account for the effect of inertial forces (Kwan 1996; Kwan and Corbett 1998b). The most widely accepted model to describe the effect of inertia flow in a porous media is the Forchheimer equation (Withaker 1996). This is, in essence, Darcy's law extended by an additional inertia term. However, several authors produced numerical and experimental evidence that there is no significant difference when inertial forces are neglected (Schroter and Heinzl 1994). Moreover, Taylor and Lewis (1975) and Cieslicki (1994) showed the existence of a regime at lower flow rate where the Forchheimer's equation appeared to be invalid and that Darcy's law would provide a much better fit to the experimental data. Taylor and Lewis (1974) demonstrated that this low-flow regime normally applies to aerostatic bearings. This was confirmed by Greenberg and Weger (1960).

As a result, most theoretical analyses of porous gas bearings are based on Darcy's law (Kwan and Corbett 1998a). In fact, this is an expression of conservation of momentum that can be derived from the Navier–Stokes equations via homogenisation. For a 1D incompressible flow and in the absence of body forces, Darcy's law may be written as:

$$Q = -\frac{k_p A}{\mu}\frac{\partial p'}{\partial z} \tag{13.1}$$

in which Q is the volumetric flow rate, A the cross-sectional area and k_p the *permeability* of the porous medium. The latter depends on the nature of the porous material and on its porosity. Let us note that this law has the same form as that pertaining to viscous flow through a pipe or a narrow channel.

Equivalence with flow through a system of capillary orifices

The volumetric flow through a capillary tube is related to the pressure gradient by

$$Q = \frac{\pi d^4}{128\mu}\frac{\partial p}{\partial z},$$

where d is the orifice diameter. If we have N orifices per unit area, then the resulting effect is, (Constantinescu 1969):

$$k_p = \frac{\pi d^4 N}{128},$$

which is an interesting equivalence, especially when considering unidirectional flow through porous media.

As was mentioned above, the permeability of a porous material can be highly non-uniform depending on the production process. Yet, by taking certain precautions and selecting the right material it is possible to obtain a permeability which is sufficiently uniform with deviations below 5 %, so that it can be regarded as constant (Constantinescu 1969). The minus sign in equation 13.1 indicates that the flow direction is from high to low pressure, so that Q is positive in the direction of the flow.

Dividing equation 13.1 by the cross-sectional area A, Darcy's law is transformed into its most general form, namely:

$$q = -\frac{k_p}{\mu}\frac{\partial p'}{\partial z} \tag{13.2}$$

in which q is the Darcy flux, which may be interpreted as the mean fluid velocity in the porous medium for unit porosity.

That is, the velocity in a porous restrictor is dependent on the effective porosity ϕ of the porous material, also called open porosity. This dimensionless parameter is defined as the ratio of the volume of the void spaces through which the fluid can flow to the total volume. This notion allows us to extend the 1D law to the more general 3D law.

Thus denoting the fluid velocity vector in the porous medium by $\mathbf{v}' = (u', v', w')$, then this velocity is determined by the ratio of the flux $\vec{q}$ to the effective porosity ϕ:

$$\mathbf{v}' = \frac{\mathbf{q}}{\phi} \tag{13.3}$$

Thus, the fluid velocity can be obtained as:

$$\mathbf{v}' = \frac{-k_p}{\mu\phi}\nabla p' \tag{13.4}$$

Substituting for $\mathbf{v}'$ in the continuity equation:

$$\frac{\partial \rho}{\partial t} + \nabla \cdot (\rho \mathbf{v}) = 0 \tag{13.5}$$

and assuming the permeability coefficient k_p of the porous material to be constant, yields:

$$\frac{\partial \rho}{\partial t} + \nabla \cdot \left(\rho \frac{-k_p}{\mu \phi} \nabla p' \right) = 0 \tag{13.6}$$

Further, supposing a uniform and isothermal process, obeying the ideal gas law, the density ρ can be replaced by the pressure p', thus:

$$\frac{\partial p'}{\partial t} + \nabla \cdot \left(p' \frac{-k_p}{\mu \phi} \nabla p' \right) = 0 \tag{13.7}$$

Finally, the pressure distribution p' inside the porous body, for an isothermal process should thus obey the law:

$$\frac{\partial p'}{\partial t} - \frac{k_p}{2\mu\phi} \nabla^2 p'^2 = 0 \tag{13.8}$$

Note that when the time term is set equal to zero (e.g. steady flow), then Eq. 13.8 states that p'^2 obeys Laplace's equation.

The variables may be normalised by defining:

$$(X, Y, Z, P', \tau) = (x/B, y/L_p, z/h_p, p'/p_a, \nu t) \tag{13.9}$$

whereupon, equation 13.8 is rewritten as:

$$\frac{\partial^2 P'^2}{\partial X^2} + \left(\frac{B}{L_p} \right)^2 \frac{\partial^2 P'^2}{\partial Y^2} + \left(\frac{B}{h_p} \right)^2 \frac{\partial^2 P'^2}{\partial Z^2} - 2\sigma\gamma \frac{\partial P'}{\partial \tau} = 0 \tag{13.10}$$

where, B, L_p, h_p are characteristic dimensions (e.g. width, length and height) of the porous block, and the two dimensionless parameters σ and γ are defined, via an expedient characteristic bearing film height h_a as:

$$\sigma = \frac{12\mu\nu B^2}{h_a^2 p_a} \tag{13.11}$$

is the squeeze number and,

$$\gamma = \frac{\phi h_a^2}{12 k_p} \tag{13.12}$$

the dimensionless porosity parameter.

13.2.2 Film Flow: Modified Reynolds Equation

The fluid flow in the bearing gap of a porous bearing can also be described by the Navier–Stokes equations, neglecting the effect of the inertial forces (under the conditions given above). Finally, this results in the well-known viscous flow equations:

$$\frac{\partial^2 u}{\partial z^2} = \frac{1}{\mu} \frac{\partial p}{\partial x} \tag{13.13}$$

$$\frac{\partial^2 v}{\partial z^2} = \frac{1}{\mu} \frac{\partial p}{\partial y} \tag{13.14}$$

$$\frac{\partial p}{\partial z} = 0 \tag{13.15}$$

Since the pressure and the normal component of the fluid velocity must be continuous across the film-bearing interface, it follows, according to the bearing configuration in Figure 13.1, where $z = 0$ at the porous surface, that:

$$\left.\begin{aligned} p(x,y,0) &= p'(x,y,0) \\ w(x,y,0) &= w'(x,y,0) \end{aligned}\right\} \text{ on the bearing surface} \tag{13.16}$$

$$\left.\frac{\partial u}{\partial z}\right|_{z=0} = \frac{\alpha}{\sqrt{k_{\mathrm{p}}}}[u(x,y,0) - u'(x,y,0)] \tag{13.17}$$

$$\left.\frac{\partial v}{\partial z}\right|_{z=0} = \frac{\alpha}{\sqrt{k_{\mathrm{p}}}}[v(x,y,0) - v'(x,y,0)] \tag{13.18}$$

where α is a dimensionless slip coefficient that depends on the material characteristics of the permeable material and not on the physical properties of the fluid (Beavers et al. 1970). Fair to good agreement was found between their model and experimental results for different porous materials.

Although, the above relationship was originally determined for incompressible fluids, namely water, Beavers et al. (1974) showed that their slip-flow boundary condition also applies to compressible fluids such as air. Based on these results, the Beavers-Joseph slip boundary condition has been widely used by many researchers in the field studying the flow in a porous gas bearing.

At the impermeable surface of the bearing counter surface, e.g. the shaft, a no-slip condition is imposed. Thus, if u_{s} represents the surface speed of the bearing platen, then, the following boundary conditions can be specified:

$$\begin{aligned} u(x,y,h_{\mathrm{a}}) &= u_{\mathrm{s}} \\ v(x,y,h_{\mathrm{a}}) &= 0 \\ w(x,y,h_{\mathrm{a}}) &= 0 \end{aligned} \tag{13.19}$$

Finally, we need to set boundary conditions at the bearing ends. Firstly, as the porous restrictor is sealed at it ends to prevent end leakage, the fluid velocity is zero in the y-direction:

$$\left.\begin{aligned} v'(x,0,z) &= 0 \\ v'(x,L_{\mathrm{p}},z) &= 0 \end{aligned}\right\} \text{ on } -B \le x \le B\,, -h_{\mathrm{p}} \le z \le 0 \tag{13.20}$$

Second, at gap exit ambient conditions are present:

$$\left.\begin{aligned} p(x,0,0) &= p_{\mathrm{a}} \\ p(x,L_{\mathrm{p}},0) &= p_{\mathrm{a}} \end{aligned}\right\} \text{ on } -B \le x \le B \tag{13.21}$$

Integrating equations 13.13 and 13.14, according to the boundary conditions 13.16 - 13.21, and substituting into the continuity equation while integrating over the film thickness, we obtain the modified Reynolds equation for a porous gas bearing:

$$\nabla \cdot \left[\frac{ph^3}{12\mu}(1+\Phi)\,\nabla p - \frac{u_{\mathrm{s}}}{2}(1+\Psi)ph\right] = \left.\frac{k_{\mathrm{p}}}{2\mu}\frac{\partial p'^2}{\partial z}\right|_{z=0} + \frac{\partial}{\partial t}(ph) \tag{13.22}$$

in which,

$$\Phi = \frac{3\left(\sqrt{k_{\mathrm{p}}}h + 2\alpha k_{\mathrm{p}}\right)}{h\left(\sqrt{k_{\mathrm{p}}} + \alpha h\right)} \tag{13.23}$$

$$\Psi = \frac{\sqrt{k_{\mathrm{p}}}}{\sqrt{k_{\mathrm{p}}} + \alpha h} \tag{13.24}$$

are two dimensionless parameters that account for the effect of slip flow at the film-bearing interface. Further, by comparing equation 13.22 with the Reynolds equation for a solid-walled bearing, it is apparent that an additional term is added to the right-hand side. This term describes the flow in the porous medium at the film-bearing interface, coupling the flow in the lubricant film with the flow in the porous restrictor.

To maintain dimensionless notation, the modified Reynolds equation 13.22 is normalised according to

$$(X, Y, Z, H, P, \tau) = (x/B, y/B, z/h_p, h/h_a, p/p_a, vt) \tag{13.25}$$

Substitution in equation 13.22 and rearranging, yields (cf. sec. 13.6.2):

$$\nabla \cdot [(1+\Phi)PH^3 \, \nabla P - (1+\Psi)\Lambda PH] = \beta \left.\frac{\partial P'^2}{\partial Z}\right|_{Z=0} + \sigma \frac{\partial}{\partial \tau}(PH) \tag{13.26}$$

in which,

$$\beta = \frac{6k_p B^2}{h_a^3 h_p} \tag{13.27}$$

is the dimensionless porous feeding parameter.

Introducing the dimensionless slip parameter S, defined by Goldstein and Braun (1971):

$$S = \frac{\sqrt{k_p}}{\alpha h_a} \tag{13.28}$$

and substituting in equations 13.23 and 13.24, we may then write:

$$\Phi = \frac{3(SH + 2\alpha^2 S^2)}{H(S+H)} \tag{13.29}$$

$$\Psi = \frac{S}{S+H} \tag{13.30}$$

Thus, setting the slip parameter S to zero reduces the modified Reynolds equation to the no-slip Reynolds equation for porous gas bearings. Generally, except for the feeding term in β, the difference between the Reynolds equation for porous media and the usual Reynolds equation depends on S and α. The most crucial factor in this might be the permeability. To see this, let us consider the following case: $h_a = 10\ \mu\text{m}$, $h_p = 5\text{mm}$, $B = 0.1\ \text{m}$, and $\alpha = 0.1$, then we have:

- $k_p = 10^{-14}$ yields $S \sim 0.1$
- $k_p = 10^{-16}$ yields $S \sim 0.001$

Thus, we see for the second case of very low k_p, that the we obtain the no-slip Reynolds equation for porous bearing.

Furthermore, we find that for both cases, we have

$$\Phi \approx 3\Psi,$$

so that we can define an equivalent film thickness $H^* = (1+\Psi)H > H$, in which $(1+\Psi)$ plays the role of the correction factor, and write:

$$\nabla \cdot [PH^{*3} \, \nabla P - \Lambda PH^*] = \beta \left.\frac{\partial P'^2}{\partial Z}\right|_{Z=0} + \frac{\sigma}{1+\Psi}\frac{\partial}{\partial \tau}(PH^*) \tag{13.31}$$

Let us remark here that, as mentioned earlier, many authors use the no-slip form of this equation without incurring appreciable error.

A further simplification is afforded if we assume that the flow through the porous media is unidirectional (cf. the analogy with capillary orifices). This situation obtains when the thickness of the porous block h_p is much smaller than its width, which is mostly the case; thus, we can write

$$\left.\frac{\partial P'^2}{\partial Z}\right|_{Z=0} = -({P_s}^2 - P^2),$$

so that

$$\nabla \cdot [PH^{*3} \nabla P - \Lambda PH^*] = -\beta({P_s}^2 - P^2) + \frac{\sigma}{1+\Psi}\frac{\partial}{\partial \tau}(PH^*) \tag{13.32}$$

In steady state ($\sigma \to 0$), no sliding and constant gap height case, this equation reduces to Poisson-type equation in $({P_s}^2 - P^2)$, which has closed form solutions for certain bearing configurations (Constantinesu 1969).

Thus, putting

$$\Pi = ({P_s}^2 - P^2)$$

yields

$$\nabla \cdot (H^{*3} \nabla \Pi) = 2\beta\Pi \tag{13.33}$$

Let us consider the following two special cases.

2D flat thrust bearing with constant gap height

Here, $H^* = 1$ so that

$$\frac{d^2\Pi}{dX^2} = 2\beta\Pi$$

With boundary conditions $\Pi(1) = P_s^2 - 1$ and $d\Pi(0)/dX = 0$ (symmetry), the solution is:

$$\Pi = ({P_s}^2 - P^2) = \frac{\cosh(\sqrt{2\beta}X)}{\cosh\sqrt{2\beta}}({P_s}^2 - 1) \tag{13.34}$$

or

$$\frac{(P^2 - {P_s}^2)}{(1 - {P_s}^2)} = \frac{\cosh(\sqrt{2\beta}X)}{\cosh\sqrt{2\beta}},$$

which finally yields

$$P = P_s\sqrt{1 + \frac{1 - P_s^2}{P_s^2}\frac{\cosh(\sqrt{2\beta}X)}{\cosh\sqrt{2\beta}}}.$$

This equation sums up all pressure distributions in flat, infinitely long thrust porous bearings. We see that the behaviour depends on the supply pressure ratio P_s and the dimensionless feeding parameter β. This is plotted in Figure 13.2, for $P_s = 5$, where we have normalised the dimensionless pressure in the gap further by $P_s - 1$ so that the behaviour will be similar for other supply pressure values. Let us remark here that this set of pressure distributions resemble, at least qualitatively, those obtaining in a centrally-fed bearing having a convergent gap (or conicity) as has been shown in Chapter 6.

Integrating the pressure w.r.t. X yields the load

$$W = \int_0^1 P\mathrm{d}X - 1.$$

This is plotted in Figure 13.3, for $P_s = 2 - 6$, being also further normalised by $P_s - 1$. We see that all load curves lie pretty close two one another, with a difference of less than 10% from the mean value for $\beta > 1$, so that we can take the mean value as a general design guideline for all values of P_s in this range.

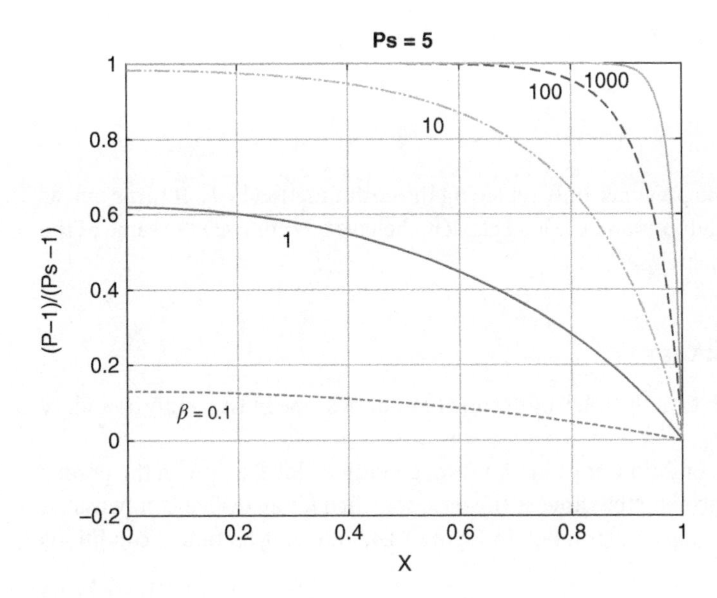

Figure 13.2 Pressure distribution in a 2D porous bearing as a function of β.

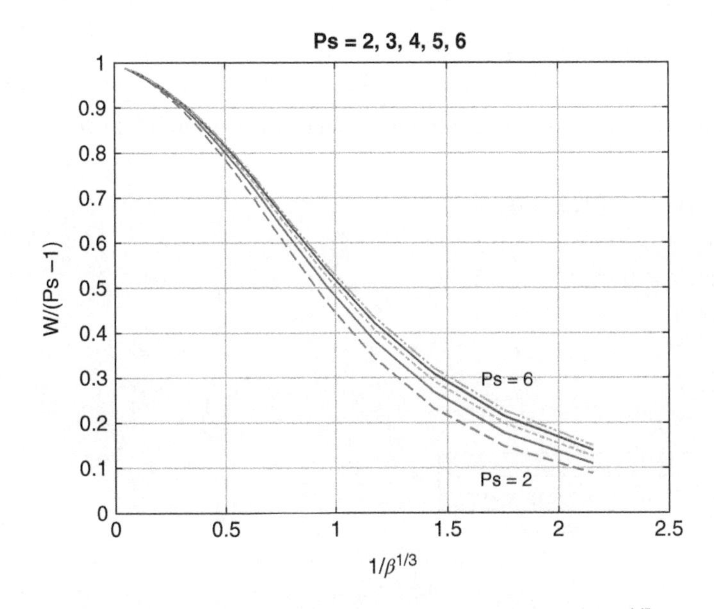

Figure 13.3 Load of a 2D porous bearing as a function of $1/\beta^{1/3}$ (being proportional to gap).

Flat radial thrust bearing with constant gap

This is similar to the previous case but with X replaced by $R = r/r_a$ ($B \mapsto r_a$) and the Reynolds equation reads:

$$\frac{1}{R}\frac{\partial}{\partial R}\left(R\frac{\partial \Pi}{\partial R}\right) = 2\beta\Pi, \qquad \left(\beta = \frac{6k_p r_a{}^2}{h_a^3 h_p}\right)$$

With boundary conditions $\Pi(1) = P_s^2 - 1$ and $\partial\Pi(0)/\partial X = 0$ (symmetry), the solution is (Constantinescu 1969):

$$\Pi = (P_s{}^2 - P^2) = \frac{I_0(\sqrt{2\beta}X)}{I_0(\sqrt{2\beta})}(P_s{}^2 - 1) \tag{13.35}$$

where I_0 is the zero-order modified Bessel function of the first kind. Rearrangement yields

$$P = P_s \sqrt{1 + \frac{1 - P_s^2}{P_s^2} \frac{I_0(\sqrt{2\beta}X)}{I_0(\sqrt{2\beta})}}.$$

We see that the solution has the same form as the previous, only replacing the cosh function by I_0. It turns out, as a matter of fact, that not only the pressure distributions and the load capacity behave qualitatively the same as the previous case, but they are quantitatively also very close.

13.2.3 Boundary Conditions for the General Case

Again, to match the conditions of the flow variables at the inlet, outlet and boundaries, one must specify boundary conditions on the pressure and its derivatives.

These comprise, in addition to the boundary conditions for the film flow, conditions for the flow in the porous restrictor. The conditions for the film flow are virtually the same as the ones specified for an orifice-compensated bearing, except at the inlet. According to the bearing configuration in Figure 13.4, following boundary conditions are set:

(i) $P = 1$ (ambient) $\quad 0 \le \theta \le 2\pi,\ Y = 0$ and $L/r,\ Z = 0$
(ii) $P_{\theta=0} = \lim_{\theta \to 2\pi} P_\theta$ (continuity) $\quad 0 \le Y \le L/r,\ Z = 0$
(iii) $\partial P/\partial Y = 0$ (symmetry) $\quad 0 \le \theta \le 2\pi,\ Y = L/(2r),\ Z = 0$

In order to determine the pressure distribution P' in the porous medium, following boundary conditions must be set:

(i) $P' = P$ (continuity) $\quad 0 \le \theta \le 2\pi,\ 0 \le Y \le 1,\ Z = 0$
(ii) $P'_{\theta=0} = \lim_{\theta \to 2\pi} P'_\theta$ (continuity) $\quad 0 \le Y \le 1,\ -1 \le Z \le 0$
(iii) $\partial P'/\partial Y = 0$ (sealed ends) $\quad 0 \le \theta \le 2\pi,\ Y = 0$ and $1,\ -1 \le Z \le 0$
(iv) $\partial P'/\partial Y = 0$ (symmetry) $\quad 0 \le \theta \le 2\pi,\ Y = 0.5,\ -1 \le Z \le 0$

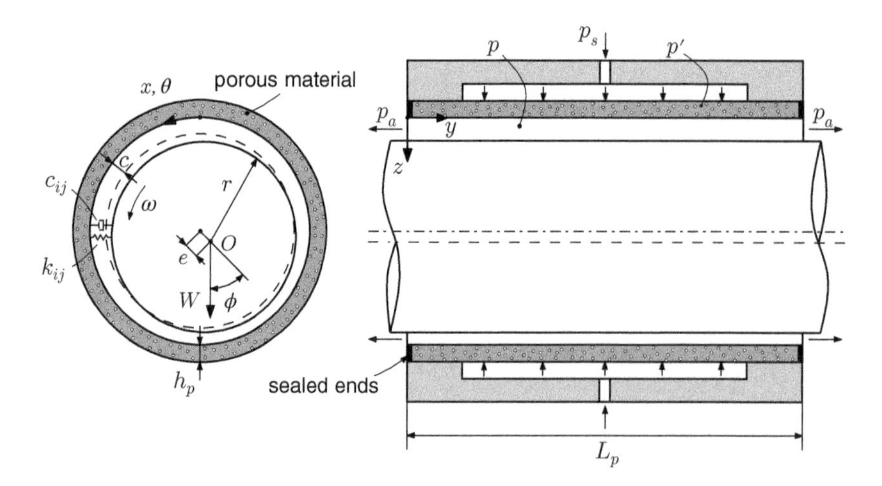

Figure 13.4 Configuration and notation of a porous aerostatic journal bearing (not to scale). Source: From Cappa S 2014.

As shown in Figure 13.4, the pressurised air in the plenum is not supplied over the entire length of the porous bushing because of assembly reasons. The supply boundary condition in the plenum is therefore given by,

(v) $P' = P_s$ (supply) $\quad 0 \le \theta \le 2\pi, Y_{\text{plenum}}, Z = -1$

while at the solid walls a Neumann boundary conditions is specified:

(vi) $\partial P'/\partial Z = 0$ (solid wall) $\quad 0 \le \theta \le 2\pi, Y_{\text{wall}}, Z = -1$

13.2.4 Solution Procedure

We further take the bearing configuration in Figure 13.4 as an example. Determination of the static pressure in the porous medium P', necessary for the subsequent calculation of the pressure distribution in the bearing gap P, is a difficult task. This is so because it involves the solution of equation 13.10 with the unknown boundary condition $P' = P$ at the film-bearing interface. This can only be done as the flow in the porous medium and the flow in the lubricant film are solved for simultaneously.

In order to avoid this difficulty, it is assumed, as a first approximation, that the flow in the porous media is only in the radial direction. This assumption is correct when the thickness of the porous restrictor is small compared to the radius of the bearing, i.e. $h_p/r \ll 0.1$ (Majumdar 1977; Rao 1979). The steady-state Laplace equation 13.10 can then be simplified to:

$$\frac{\partial^2 P'^2}{\partial Z^2} = 0 \tag{13.36}$$

Solving this simple PDE problem, together with the boundary conditions,

(i) $P' = P$ (continuity) $\quad 0 \le \theta \le 2\pi, 0 \le Y \le 1, Z = 0$
(ii) $P' = P_s$ (supply) $\quad 0 \le \theta \le 2\pi, Y_{\text{plenum}}, Z = -1$
(iii) $\partial P'/\partial Z = 0$ (solid wall) $\quad 0 \le \theta \le 2\pi, Y_{\text{wall}}, Z = -1$

results in the following expression for P' at the solid walls of the stator,

$$P'^2(Z) = P^2 \tag{13.37}$$

thus,

$$\left.\frac{\partial P'^2}{\partial Z}\right|_{Z=0} = 0 \tag{13.38}$$

while at the plenum,

$$P'^2(Z) = (P^2 - P_s^2)Z + P^2 \tag{13.39}$$

thus, as for the plane case (p. 410),

$$\left.\frac{\partial P'^2}{\partial Z}\right|_{Z=0} = P^2 - P_s^2 \tag{13.40}$$

Substituting equation 13.38 and 13.40 into the modified Reynolds equation 13.26, the resulting equations can be solved numerically for the static pressure distribution P over the calculation domain shown in Figure 13.5. Subsequently, the pressure P is used for the calculation of the static pressure distribution P' in the porous medium according to equation 13.37 and 13.39. Using P and P' of the 1D flow, the 3D fluid flow can be calculated by successively solving equation 13.26 and equation 13.10. The flow in the bearing gap is iteratively found with the Newton–Raphson method as was the case for an orifice-compensated bearing. The flow in the porous bushing, on the other hand, is solved iteratively by means of a successive over-relaxation (SOR) scheme as this offers great

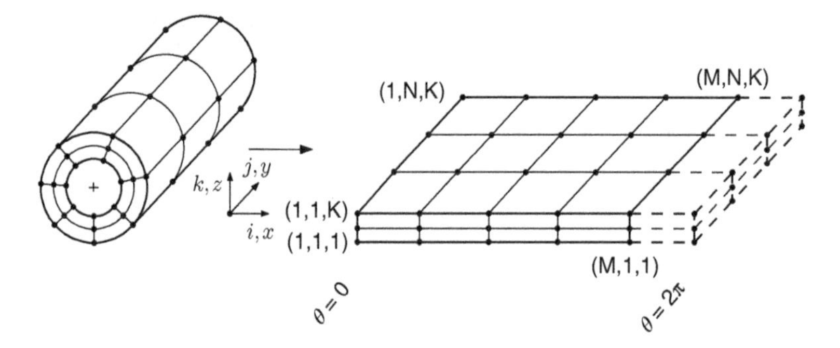

Figure 13.5 Discretised reference grid of the finite-difference model for a porous bearing (the number of grid points are only illustrative). Source: From Cappa S 2014.

speed of convergence (provided that the relaxation parameter is well chosen). The solution procedure, shown in Figure 13.6, is terminated when the following convergence criteria is met:

$$\frac{|P^n - P^{n-1}|}{|P^n|} < 0.001 \tag{13.41}$$

in which n is the number of iterations.

This numerical model and solution procedure, can also be used for modelling the fluid flow in a partially porous bearing, e.g. a porous ring bearing. To do so, one must reduce the modified Reynolds equation to its original form at the non-permeable wall by setting the permeability coefficient k_{p} to zero at the corresponding grid points in equation 13.26.

Remark

The curvature effect that occurs by transforming the physical grid of the porous bushing to a rectangular reference grid, as shown in Figure 13.5, can be neglected because h_{p}/r is sufficiently small (Rouleau and Steiner 1974).

13.3 Static Bearing Characteristics

Once the pressure distribution is determined, the static bearing characteristics can be calculated in the same way that was adopted in the previous chapters for orifice-fed and self-acting bearings.The same notation for the load capacity, stiffness, etc. will be adopted.

13.4 Dynamic Bearing Characteristics

Like other bearing types, porous bearings are also susceptible to the development of negative damping, which can lead to dynamic stability problems. We need, therefore, to be able to evaluate the dynamic characteristics of those bearings.

Many articles in the literature have dealt with this problem (Lund 1967; Majumdar 1976b; Majumder and Majumdar 1988; Rao 1977; Rao and Majumdar 1979; Sun 1975b).

In the following, a brief discussion on the dynamic gas film modelling of a porous gas bearing is given. The method of small perturbation, as discussed in Chapter 7 will be adopted. That is, a small amplitude harmonic perturbation of the height distribution is superposed on the steady-state height profile H_0,

$$H = H_0 + \Delta H e^{j\tau} \tag{13.42}$$

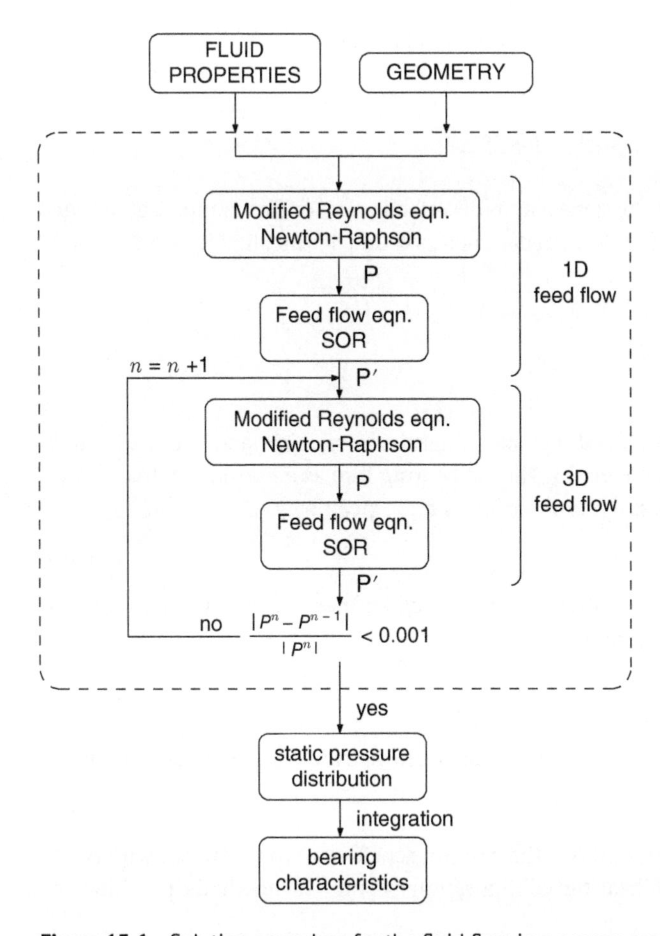

Figure 13.6 Solution procedure for the fluid flow in a porous gas bearing. Source: From Cappa S 2014.

where ΔH is complex-valued, in general, and one order lower than H_0. The corresponding film pressure will thus be of the form:

$$P = P_0 + \Delta P e^{j\tau} \tag{13.43}$$

with P_0 the static pressure distribution in the bearing gap.

The film dynamics of a porous bearing are found in a similar way by applying the perturbation method to both the feed flow equation 13.10 and the modified Reynolds equation 13.26. At first, the perturbed pressure in the porous medium $\Delta P'$ is calculated by substituting equation 13.43 in equation 13.10, cancelling the static terms out and neglecting all higher-order terms. In this way, we obtain:

$$\begin{aligned} P'_0 \frac{\partial^2 \Delta P'}{\partial \theta^2} + \Delta P' \frac{\partial^2 P'_0}{\partial \theta^2} + \left(\frac{r}{L_\text{p}}\right)^2 \left(P'_0 \frac{\partial^2 \Delta P'}{\partial Y^2} + \Delta P' \frac{\partial^2 P'_0}{\partial Y^2}\right) \\ + \left(\frac{r}{h_\text{p}}\right)^2 \left(P'_0 \frac{\partial^2 \Delta P'}{\partial Z^2} + \Delta P' \frac{\partial^2 P'_0}{\partial Z^2}\right) = j\sigma\gamma \Delta P' \end{aligned} \tag{13.44}$$

Solving this linear system for $\Delta P'$ and substituting into the perturbed Reynolds equation,

$$\begin{aligned}&\nabla \cdot \left[((3P_0H_0^2\Delta H + H_0^3\Delta P)\nabla P_0 + P_0H_0^3\nabla\Delta P)(1+\Phi)\right.\\&\left.-\Lambda(P_0\Delta H + H_0\Delta P)(1+\Psi)\right] = 2\beta\left.\frac{\partial(P_0'\Delta P')}{\partial Z}\right|_{Z=0} + j\sigma(P_0\Delta H + H_0\Delta P)\end{aligned} \tag{13.45}$$

finally leads to the dynamic pressure distribution ΔP in the same way as described for an orifice-compensated bearings in Chapter 7. Note that the perturbation of the dimensionless parameters Φ and Ψ is negligible as $\Delta\Phi/\Phi \ll 1$ and $\Delta\Psi/\Psi \ll 1$.

13.5 Dynamic Film Coefficients

The dynamic pressure distribution ΔP in itself gives no direct meaningful information about the dynamic behaviour of the gas film. This can be obtained by characterising the air-bearing film as a non-linear, frequency-dependent spring-damper element with a dimensionless dynamic stiffness K_{dyn}, given by:

$$K_{\text{dyn}}(v) = K_k(v) + j\sigma C_c(v) \tag{13.46}$$

This is a lumped-parameter formulation composed of a dimensionless spring stiffness K_k and a dimensionless damping force σC_c. The dynamic stiffness K_{dyn}, in turn, can be calculated as:

$$K_{\text{dyn}} = -\frac{\partial F}{\partial H} = \int_{\bar{A}} \left(\frac{\Delta P}{\Delta H}\right) \mathrm{d}\bar{A} \tag{13.47}$$

Now, if ΔH is taken to be real, say unity for simplicity, then the dynamic pressure distribution ΔP can be written as:

$$-\Delta P = \Delta P_k + j\sigma\Delta P_c \tag{13.48}$$

wherein ΔP_k yields the film stiffness (in phase with ΔH) and ΔP_c the damping coefficient (out of phase with ΔH).

These coefficients can be calculated for different bearing configurations using the methods provided in Chapters 7 and 9.

As has been discussed in Chapter 7, whenever possible, it is advisable to design an air bearing such that the static damping is positive, i.e. $c(0) > 0$. This is mostly so for a well-designed orifice-compensated bearing with a parallel bearing gap, but not always for a porous bearing (Rao 1977). A static negative damping in a porous bearing possibly occurs because the air can flow back through the permeable wall in phase with the dynamic motion of the rotor. The range of σ over which a negative damping occurs in a porous bearing increases with the supply pressure p_s, the porosity parameter γ and the porous feeding parameter β (Rao 1977).

The above trends are of great importance at the design stage of a gas bearing. This is because the static value of the film stiffness and film damping tell us something about the dynamic behaviour at higher perturbation frequencies. Nevertheless, a quantitative evaluation of the dynamic behaviour can only be made when the film properties are actually calculated at multiple frequencies. This can be time consuming if one, for example, needs to study the rotor dynamic behaviour of a rotor-bearing system.

Roblee (1985), Roblee and Mote (1990), see also Chapter 7, proposed a simple lumped parameter approximation, valid for a positively damped bearing, in which the film behaviour is represented as a phase-lead system when the static damping is positive, namely:

$$k_{\text{dyn}}(s) = k(0)\frac{1+\tau_1 s}{1+\tau_2 s} \tag{13.49}$$

in which $s = jv$ is the Laplace variable and where τ_1 determines the zero of the system,

$$\tau_1 = \tau_2\frac{k(\infty)}{k(0)} \tag{13.50}$$

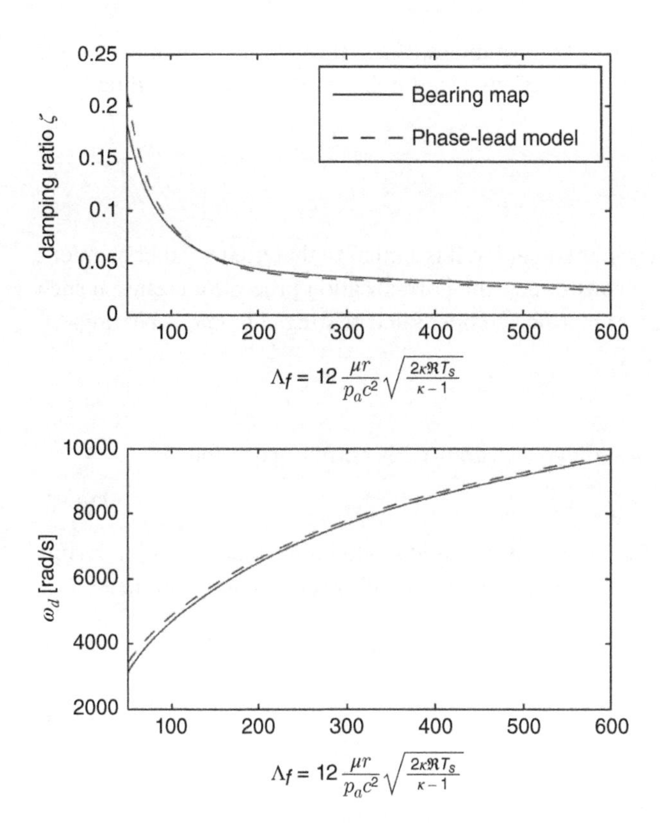

Figure 13.7 Damping ratio ζ and the damped natural frequency ω_d of an aerostatic journal bearing with inherent restrictors as a function of the feed number Λ_f. The solution from the phase-lead model (dashed line) is compared with the exact solution obtained from a dynamic bearing map (solid line). Bearing geometry: $R_o = 0.005$, $N_f = 8$, $L/D = 1$, $\Lambda = 0$ and $\epsilon = 0$ at $P_s = 6$. Source: From Cappa S 2014.

and τ_2 the pole of the system,

$$\tau_2 = \frac{c(0)}{k(\infty) - k(0)} \tag{13.51}$$

This simplified model requires only two calculations of k and c for a complete dynamic characterisation, namely for $\nu = 0$ and $\nu = \infty$. It is clear that this results in a large reduction of calculation time when the stability of the bearing system is analysed. Normally, this is done by the evaluation of the film properties at different perturbation frequencies and rotational speeds, as both have an influence on the stiffness and damping of the gas film. The obtained results are then tabulated in the form of a bearing map which serves as an input for the rotor dynamic analysis (Waumans 2009). This process can take up to several hours depending on the speed range of interest.

Besides the time advantage, it is of course of primary importance that a simplified model is still accurate, independently of the bearing parameters. To check this, we have calculated the damping ratio ζ and the damped natural frequency ω_d as a function of the feed number Λ_f in Figure 13.7. Once by means of the phase-lead model and once with the classical method (bearing map). The agreement between both models is good. Particularly if one takes into account that for one set of design parameters, i.e. one Λ_f value, the calculation time is almost 60 times faster! The small discrepancy between the results can possibly be explained by the fact that the dynamic stiffness k_{dyn} is a high order system rather than a simple one-pole/one-zero system (Licht and Elrod 1960). If necessary, the accuracy of the phase-lead model can be improved by fitting the dynamic stiffness through more than two points or by using the value of ω_d, determined with the phase-lead model, as a start point for the classical method.

13.6 Normalisation

This section defines the normalisation procedure that is in this chapter: it is similar to that used in other chapters, though with some particularities. In order to maintain consistency, the normalisation procedure is chosen such that the normalised bearing parameters are the same for an orifice-compensated bearing and porous bearing.

13.6.1 Aerostatic Porous Journal Bearing

An externally pressurised journal bearing with radius r and radial clearance c is normalised as follows:

$$(\theta, Y, Z, H, P, P', \tau) = (x/r, y/r, z/h_p, h/c, p/p_a, p'/p_a, \nu t) \tag{13.52}$$

Substitution of these normalised parameters into the Reynolds equation and modified Reynolds equation results in an expression for the dimensionless bearing number Λ, squeeze number σ and porous feeding parameter β:

$$\Lambda = \frac{6\mu\omega}{p_a}\left(\frac{r}{c}\right)^2 \tag{13.53}$$

$$\sigma = \frac{12\mu\nu}{p_a}\left(\frac{r}{c}\right)^2 \tag{13.54}$$

$$\beta = \frac{6k_p r^2}{c^3 h_p} \tag{13.55}$$

For the flow in porous medium a distinction is made for the normalisation of the y-coordinate, namely:

$$(Y) = (y/L_p) \tag{13.56}$$

Applied to the feed flow 13.8 results in the squeeze number σ given by equation 13.54 and a porosity parameter γ, defined as:

$$\gamma = \frac{\phi c^2}{12k_p} \tag{13.57}$$

Subsequently, the dimensionless load capacity $\bar{W}$, film stiffness K and damping C, are normalised as:

$$\bar{W} = \frac{W}{p_a LD} \tag{13.58}$$

$$K_{ij} = \frac{k_{ij}c}{p_a LD} \tag{13.59}$$

$$C_{ij} = \frac{c_{ij}}{24\mu L}\left(\frac{c}{r}\right)^3 \tag{13.60}$$

A journal bearing is fully characterised by three dimensionless bearing parameters[1]: the slenderness ratio L/D, the dimensionless feed-hole radius $R_o = r_o/r$ and the feed number Λ_f (Waumans 2009). The latter is defined as,

$$\Lambda_f = 12\frac{\mu r}{p_a c^2}\sqrt{\frac{2\kappa \Re T_s}{\kappa - 1}} \tag{13.61}$$

1 With the exception of the feeding configuration such as the number of feed-holes N_f and their arrangement in case of an orifice-compensated bearing.

for an orifice-compensated bearing and as,

$$\Lambda_f = \frac{3k_p A}{L_p c^3} \tag{13.62}$$

for a porous bearing. The dimensionless feed number Λ_f is introduced to normalise the entrance mass flow rate $\dot{m}_o$:

$$\bar{\dot{m}}_o = \frac{\dot{m}_o}{\pi p_a \rho_a c^3/(6\mu)} \tag{13.63}$$

$$= \Lambda_f R_o H P_s \Phi_e(\bar{p}_o) \tag{13.64}$$

13.6.2 Aerostatic Porous Thrust Bearing

The bearing parameters of a circular thrust bearing are also normalised according to the quantities given in equation 13.52. Only the radius r and bearing gap h are normalised with respect to the exit conditions:

$$(R, H) = (r/r_a, h/h_a) \tag{13.65}$$

Substitution into the feed flow and film flow equations leads to the following expression for the bearing number Λ, squeeze number σ, porous feeding parameter β and porosity parameter γ:

$$\Lambda = \frac{6\mu \mathbf{V}_t r_a}{p_a h_a^2} \tag{13.66}$$

$$\sigma = \frac{12\mu\nu}{p_a}\left(\frac{r_a}{h_a}\right)^2 \tag{13.67}$$

$$\beta = \frac{6k_p r_a^2}{h_a^3 h_p} \tag{13.68}$$

$$\gamma = \frac{\phi h_a^2}{12k_p} \tag{13.69}$$

Lastly, the load capacity W, axial stiffness k_{zz} and axial damping c_{zz} are normalised with respect to the bearing surface $A = \pi r^2$ and atmospheric pressure p_a:

$$\bar{W} = \frac{W}{\pi r_a^2 p_a} \tag{13.70}$$

$$K_{zz} = \frac{k_{zz} h_a}{\pi r_a^2 p_a} \tag{13.71}$$

$$C_{zz} = \frac{c_{zz}}{12\mu\pi r_a}\left(\frac{h_a}{r_a}\right)^3 \tag{13.72}$$

The normalisation of the entrance mass flow rate $\dot{m}_o$ is similar to that outlined above for a journal bearing.

13.7 Validation of the Numerical Models

Before closing this chapter, we will validate the porous gas film model with reference data found in the open literature. The validation process is focused on the newly developed porous model.

Firstly, the correctness of the static solution is verified by comparing the dimensionless load capacity $\bar{W}$ with experimental and theoretical data under different conditions. When going through the literature, it appears difficult to find reliable experimental data for porous gas bearings as most foregoing studies focused on the theoretical

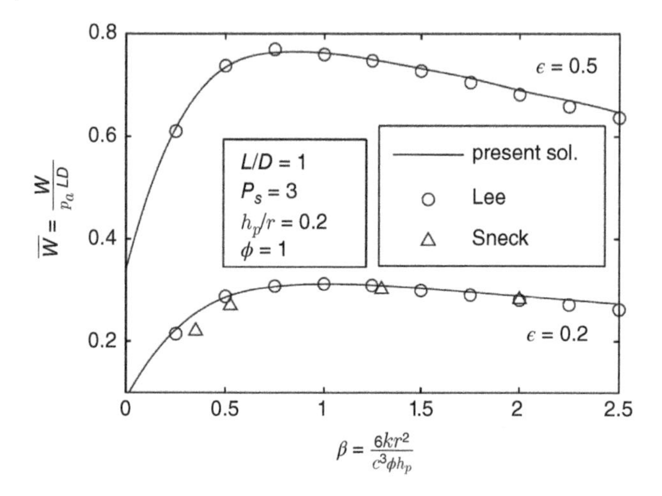

Figure 13.8 Comparison of the normalised load capacity for a non-rotating porous journal bearing at different eccentricities ϵ. Experimental data from Sneck and Elwell (1965) and theoretical data from Lee and You (2009). Source: From Cappa S 2014.

part. One of the few works that published experimental data is that of Sneck and Elwell (1965) and has therefore become the standard. Figure 13.8 compares the present solution (solid line) for a non-rotating porous gas journal bearing with the experimental data of Sneck and Elwell (1965) (triangular markers) and the theoretical data of Lee and You (2009) (circular markers). A no-slip condition is assumed in the analysis, i.e. $S = 0$, because of a lack of data in both references. This assumption is not entirely correct as slip flow has certainly occurred during the tests of Sneck and Elwell (1965). Nevertheless, very good to excellent agreement is observed for varying values of the porous feeding parameter β, even at high eccentricities $\epsilon = e/c$. For small values of β, a slight difference may be observed between our model and the experimental data of Sneck and Elwell (1965). It is difficult to explain this discrepancy, but it might be due to the action of slip flow which was not taken into account in these analyses. This view is supported by Singh et al. (1984) who showed that slip flow reduces the load-carrying capacity for low values of β.

A second comparison takes the effect of slip flow into account. The dimensionless load capacity $\bar{W}$ in Figure 13.9 is compared against data from Singh et al. (1984) (circular marker). Again, good to excellent agreement is found for different values of the slip coefficient α over the entire range of β. From this data it is seen that the load capacity increases with the slip coefficient α in the lower range of β, whereas, the opposite is observed for higher values of β. This finding is consistent with that of Rao (1982) who used a more simplified Beavers–Joseph slip velocity condition that was initially proposed by Saffmann (1971) and later inherited by Sparrow et al. (1972) and Wu and Castelli (1976, 1977).

Lastly, Figure 13.10 compares the dynamic solution of the porous model. This is done against theoretical data from Rao (1977) as no reliable reference data was found in the literature. As can be seen, our model (solid line) does not correspond with the data points of Rao (square markers), certainly not for lower perturbation frequencies.

A possible explanation for the discrepancy might be that the WKBJ-approximation[2] used by Rao for the solution of the dynamic pressure distribution in the porous restrictor is not accurate or inappropriate. Further, in order to apply the WKBJ-approximation, Rao assumed a purely radial flow in the porous medium instead of a three-dimensional flow. These possible explanations, along with the fact that the computer power was limited at the time Rao published his results, may explain the discrepancy with our model.

2 The Wentzel–Kramers – Brillouin–Jeffreys approximation is a method for approximating the solution of a linear partial differential equation.

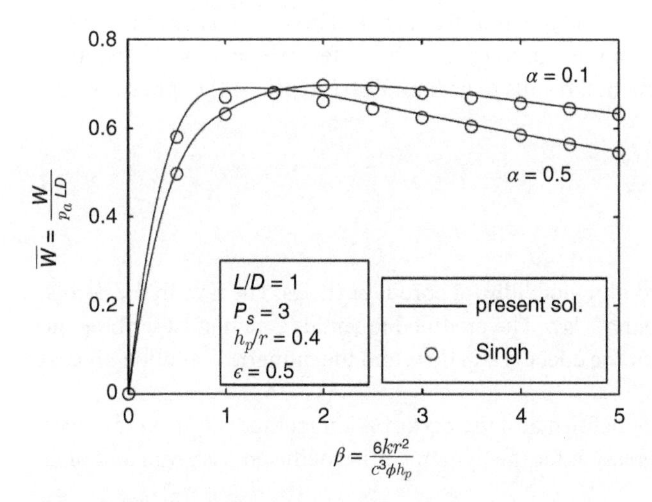

Figure 13.9 Comparison of the normalised load capacity for a non-rotating porous journal bearing at different values of the slip parameter α. Theoretical data from Singh et al. (1984). Source: From Cappa S 2014.

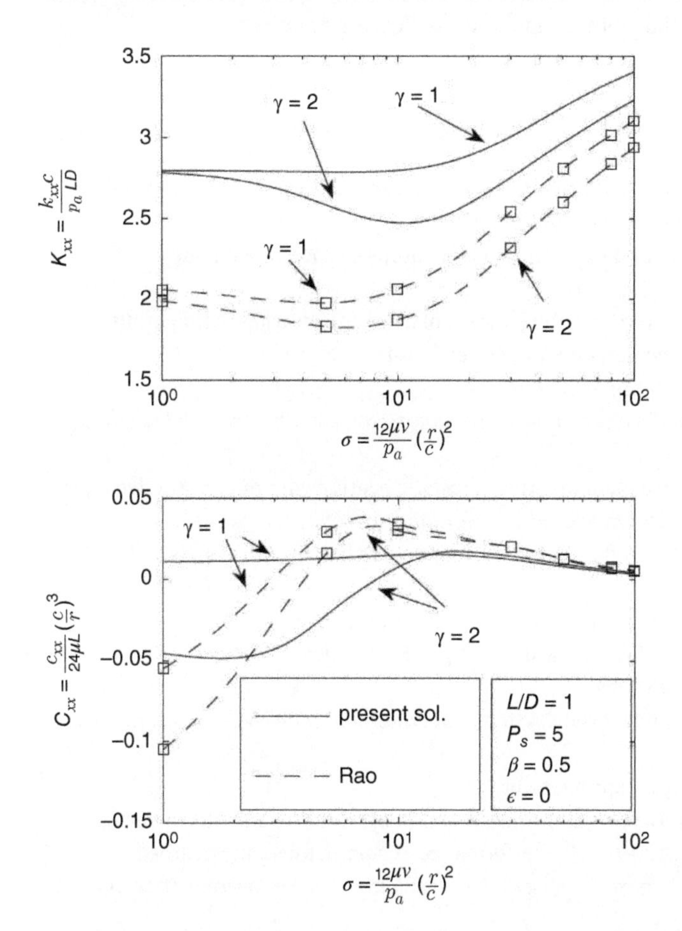

Figure 13.10 Comparison of the dimensionless stiffness and damping for a non-rotating porous journal bearing as a function of the squeeze number for different values of the porosity parameter. Theoretical data from Rao (1977). Source: From Cappa S 2014.

In conclusion, this comparative study with experimental and theoretical data found in the open literature proves the validity of the implemented porous model. Although it is not so easy to find reference data for the dynamic behaviour of porous gas bearings, the excellent experimental results to be found in (Cappa 2014) might serve as a more conclusive evidence.

13.8 Conclusions

This chapter has dealt with the static and dynamic fluid film modelling of porous bearings. The flow in the porous medium is assumed to be viscous and is governed by Darcy's law. The modified Reynolds equation, describing the flow of the lubricant film, is derived, taking into account the effect of slip flow, and the numerical solution process is explained.

In a second part, the static bearing characteristics are defined and the perturbation method has been reviewed and applied to the said bearing configurations. An expression for the dynamic film coefficients is given and their typical behaviour has been addressed.

The porous model has been validated with experimental and theoretical data found in the open literature. Although, no reliable experimental data was found for the validation of the dynamic bearing coefficients, good to excellent agreement was found for the static solution over a wide range of bearing parameters.

References

Beavers, G.S., Sparrow, E.M. and Magnuson, R.A., (1970). Experiments on couple parallel flows in a channel and a bounding porous medium. *ASME Journal of Basic Engineering* 92(4): 843–848.

Beavers, G.S., Sparrow, E.M. and Masha, B.A., (1974). Boundary condition at a porous surface which bounds a fluid flow. *AIChE Journal* 20(3): 596–597.

Belforte, G., Raparelli, T., Viktorov, V. and Trivella, A., (2005). Feeding system of aerostatic bearings with porous media. *Proceedings of the 6th JFPS International Symposium on Fluid Power* 126–131.

Booser, E.R., (1970). Plain-bearing materials. *Machine Design* 42, 14–20.

Cappa, S, (2014). *Reducing the error motion of an aerostatic rotary table to the nanometre level* PhD thesis KU Leuven, Department of Mechanical Engineering.

Chien, K., Tsai, C., Hsu, F., et al. (2012). Development of a planar porous aerostatic bearing with dual restrictive layer. *Proceedings of the 9th International Conference on Multi-Material Micro Manufacture (4M 2012)* 175–179.

Cieslicki, K., (1994). Investigations of the effect of inertia on flow of air through porous bearing sleeves. *Wear* 172, 73–78.

Collins, R.E., (1961). *Flow of Fluids Through Porous Materials.* Reinhold Publishing Corp., New York.

Constantinescu, V.N., (1969). *Gas Lubrication* First edn. The American Society of Mechanical Engineers.

Constantinesu, V.N., (1969). *Gas Lubrication.* ASME, New York.

Cracaoanu, I. and Bremer, F., (2013). Air bearings in high precision systems. *International conference on sustainable construction and design (SCAD).*

Dagan, G., (1989). *Flow and transport in porous formations.* Springer-Verlag.

Durazo-Cardenas, I.S., (2003). *Development of porous-ceramic hydrostatic bearings* PhD thesis Cranfield University.

Durazo-Cardenas, I.S., Corbett, J. and Stephenson, D.J., (2010). The performance of a porous ceramic hydrostatic journal bearing. *Proceedings of the Institution of Mechanical Engineering, Part J: Journal of Engineering Tribology* 224, 81–89.

Fourka, M. and Bonis, M., (1997). Comparison between externally pressurized gas thrust bearings with different orifice and porous feeding systems. *Wear* 210, 311–317.

Goldstein, M.E. and Braun, W.H., (1971). Effect of velocity slip at a porous boundary on the performance of an incompressible porous bearing. *NASA Technical Note.*

Greenberg, D.B. and Weger, E., (1960). An investigation of the viscous and inertial coefficients for the flow of gases through porous sintered metals with high pressure gradients. *Chemical Engineering Science* 12(1): 8–19.

Kwan, Y. (1996). *Processing and fluid flow characteristics of hot iso-statically pressed porous alumina for aerostatic bearing applications* PhD thesis Cranfield University.

Kwan, Y. and Corbett, J. (1998a). Porous aerostatic bearings – an updated review. *Wear* 222, 69–73.

Kwan, Y. and Corbett, J. (1998b). A simplified method for the correction of velocity slip and inertia effects in porous aerostatic thrust bearings. *Tibol. International* 31(12): 779–786.

Lee, C. and You, H., (2009). Characteristics of externally pressurized porous gas bearings considering structure permeability. *Tribology Transactions* 52, 768–776.

Licht, L. and Elrod, H., (1960). A study of the stability of externally pressurized gas bearings. *ASME Journal of Applied Mechanics*, 250–258.

Lund, J.W. (1967). A theoretical analysis of whirl instability and pneumatic hammer for a rigid rotor in pressurized gas journal bearings. *ASME Journal of Lubrication Technology*, 154–165.

Majumdar, B.C., (1975). Analysis of externally pressurized porous gas journal bearings - i. *Wear* 33, 25–35.

Majumdar, B.C., (1976a). Gas lubricated porous bearings: A bibliography. *Wear* 36, 269–273.

Majumdar, B.C., (1976b). Whirl instability of externally pressurized gas-lubricated porous journal bearings. *Wear* 141–153.

Majumdar, B.C., (1977). Porous gas journal bearings: a semi-analytical solution. *Journal of Lubrication Technology*, 487–489.

Majumdar, B.C., (1983). Stiffness and damping of externally pressurized gas journal bearings with porous inserts. *Proceedings of the Institution of Mechanical Engineering* 197, 25–29.

Majumder, M.C. and Majumdar, B.C., (1988). Theoretical analysis of pneumatic instability of externally pressurized porous gas journal bearings considering velocity slip. *ASME Journal of Tribology* 110, 730–733.

Massey, B., (1968). *Mechanics of fluids*. D. Van Nostrand Company Ltd., London.

Miyatake, M. and Yoshimoto, S., (2010). Numerical investigation of static and dynamic characteristics of aerostatic thrust bearings with small feed holes. *Tibology International* 1353–1359.

Montgomery, A.G. and Sterry, F., (1955). A simple air bearing rotor for very high rotational speeds. *AERE ED/R.*

Mori, H., Yabe, H. and Yamakage, H., (1968). Theoretical analysis of externallu pressurized porous gas journal bearings (1ste report). *Bulletin of JSME* 11, 527–532.

Mori, H., Yabe, H. and Yamakage, H., (1969). Theoretical analysis of externallu pressurized porous gas journal bearings (2nd report). *Bulletin of JSME* 12, 1512–1518.

Rao, N.S., (1977). Analysis of the stiffness and damping characteristics of an externally pressurized porous gas journal bearing. *ASME Journal of Lubrication Technology* 295–301.

Rao, N.S., (1979). Design of externally pressurized porous gas bearings with journal rotation. *Wear* 52, 1–11.

Rao, N.S., (1982). Analysis of aerostatic porous journal bearings using the slip velocity boundary conditions. *Wear* 76, 35–47.

Rao, N.S. and Majumdar, B.C., (1979). Analysis of pneumatic instability of externally pressurized porous gas journal bearings. *ASME Journal of Lubrication Technology* 101, 48–53.

Robinson, C.H. and Sterry, F, (1958). *The static strength of pressure fed gas journal bearings. AERE ED/R.*

Roblee, J., (1985). *Design of Externally Pressurized Gas Bearings for Dynamic Applications* PhD thesis University of California Berkely, CA.

Roblee, J.W. and Mote, C.D., (1990). Design of externally pressurized gas bearings for stiffness and damping. *Tribology International* 23(5): 333–345.

Rouleau, W.T. and Steiner, L.I., (1974). Hydrodynamic porous journal bearings—Part i: Finite full bearings. *ASME Journal of Lubrication Technology* 96(3): 346–353.

Saffmann, P.G., (1971). On the boundary condition at the surface of a porous medium. *Studies in Applied Mathematics* 50(2): 93–101.

Scheidegger, A.E., (1960). *Physics of Flow Through Porous Media*. MacMillan.

Schroter, A. and Heinzl, J., (1994). Air-bearings with areal disposed micro-orifices. *Proceedings 3rd International Conference on Ultraprecision in Manufacturing Engineering* 253–256.

Sheinberg, S.A. and Shuster, V.G., (1960). A porous thrust bearing which is stable under vibrations. *Stanki i instrument* 31, 23–27.

Silveira, Z.C., Nicoletti, R., Fortulan, C.A. and Purquerio, B.M., (2010). Ceramic matrices applied to aerostatic porous journal bearings: material characterization and bearing modeling. *Ceramica* 56, (201–211.

Singh, K.C., Rao, N.S. and Majumdar, B.C., (1984). Effect of slip flow on the steady-state performance of aerostatic porous journal bearings. *ASME Journal of Tribology* 106, 156–162.

Sneck, H.J., (1968). A survey of gas-lubricated porous bearings. *ASME Journal of Lubrication Technology* 804–809.

Sneck, H.J. and Elwell, R.C., (1965). The externally pressurized, porous wall, gas-lubricated journal bearing – ii. *ASLE Transactions* 8(4): 339–345.

Sneck, H.J. and Yen, K.T., (1964). The externally pressurized, porous wall, gas-lubricated journal bearing – i. *ASLE Transactions* 7(3): 288–298.

Sneck, H.J. and Yen, K.T., (1967). The externally pressurized, porous wall, gas-lubricated journal bearing – iii. *ASLE Transactions* 10, 339–347.

Sparrow, E.M., Beavers, G.S. and Hwang, I., (1972). Effect of velocity slip on porous walled squeeze films. *Journal of Lubrication Technology* 94(3): 260–265.

Su, J.C. and Lie, K.N., (2003). Rotation effects on hybrid air journal bearings. *Tibology International* 36, 717–726.

Su, J.C. and Lie, K.N., (2006). Rotor dynamic instability analysis on hybrid air journal bearings. *Tibol. International* 39, 238–248.

Sun, D.C., (1975a). Analysis of the steady state characteristics of gas lubricated, porous journal bearings. *ASME Journal of Lubrication Technology* 97, 44–51.

Sun, D.C., (1975b). Stability of gas-lubricated, externally pressurized porous journal bearings. *ASME Journal of Lubrication Technology* 494–505.

Taylor, R. and Lewis, G.K., (1974). *Proceedings of the 6th International Gas Bearing Symposium*. British Hydromechanics Research Association Fluid Engineering.

Taylor, R. and Lewis, G.K., (1975) Experience relating to the steady performance of aerostatic porous thrust bearings. *Proceedings of the Institute of Mechanical Engineers* 189, 383–390.

Waumans, T., (2009). *On the Design of High-Speed Miniature Air Bearings: Dynamic Stability, Optimisation and Experimental Validation* PhD thesis K.U. Leuven, Department of Mechanical Engineering, (2009D16).

Withaker, S., (1996). The forchheimer equation: A theoretical development. *Springer link: Transport in Porous Media* 25, 27–61.

Wu, E.R. and Castelli, V., (1976). Gas-lubricated porous bearings – infinitely long journal bearings, steady-state solution. *ASME Journal of Lubrication Technology* 98(3): 453–462.

Wu, E.R. and Castelli, V., (1977). Gas-lubricated porous bearings - short journal bearings, steady-state solution. *ASME Journal of Lubrication Technology* 99(3): 331–338.

14

Hanging Air Bearings and the Over-expansion Method

14.1 Introduction

The concept of a "hanging" bearing might provide the answer to an important engineering need, viz. the manipulation, in a plane, of objects suspended from that plain. This may, eventually, make the *ceiling* as important as the *floor*, thus allowing for a more efficient use of the work space, and opening up new production and instrumentation possibilities. It is reported (Anon 1989), e.g., that such an application could result in quadrupling throughput of an electronic system assembly robot cell. One might, likewise, contemplate the design of a 3D measurement machine based on this principle. Furthermore, once such bearing components are sufficiently well developed, understood, and commercially producible, many more applications will undoubtedly suggest themselves.

The hanging action, with a lubricating air film, may be realised in various ways, (Al-Bender and Van Brussel 1994); principally by (i) a positive-pressure thrust bearing combined with a magnet that attracts it to the ceiling, (ii) a vacuum bearing equipped with a suitable flow controller, and (iii) an "over-expansion" bearing, in which the over-expansion generates sub-ambient pressure in the gap. Since their basic theory is largely known, the first two types will be outlined with the purpose of determining their basic properties and characteristics. Thereafter, the over-expansion bearing, whose underlying theory is less known, is selected as the subject for further treatment: a substantial part of this chapter is therefore devoted to establishing basic theory prior to determining its essential characteristics. Since the flow in the gap of this type of bearing is predominantly turbulent, this treatment provides for a good case study in dealing with turbulent flow both on modelling and validation levels. The chapter commences with an outline in which the hanging-bearing problem is stated and the different solutions presented, evaluated and compared. Next, the problem of the over-expansion bearing is formulated, and the objectives of a development study are stated. A theoretical model is then proposed, which is used to solve for the pressure distribution in function of the bearing design parameters. This is followed by a description of an experimental programme developed and carried out to validate the theoretical model. Finally, other aspects, details and accessories are also discussed. Some general conclusions are then drawn.

14.2 Outline

Here, the hanging bearing problem will be stated, then different solutions are discussed.

14.2.1 Problem Statement

Referring to Figure 14.1, it is required to support a load W, attached to bearing part (a), by suspension from the part (b), such that:

Air Bearings: Theory, Design and Applications, First Edition. Farid Al-Bender.

Companion website: www.wiley.com/go/AlBender/AirBearings

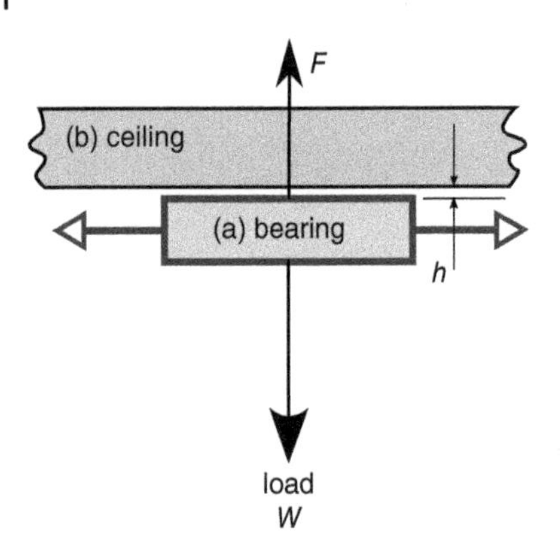

Figure 14.1 Hanging bearing arrangement.

1. The two parts (a) and (b) are not contacting, (usually then separated by a thin fluid film).
2. The resulting system should be statically stable, i.e. an increase in the gap width h must result in an increase in the bearing force F, or

$$\partial F/\partial h > 0\,,$$

over the design load range. In other words, the bearing must have *stiffness*, (usually, the higher the better).
3. The system should also be dynamically stable, i.e. no self-excited vibrations (of large amplitude) should arise.

In this way, the parts (a) and (b) are free to move relative to one another in the horizontal plane, with very small frictional resistance. We shall, for the present, consider that the lubricating fluid is a gas, or, more specifically, air, and that the bearing parts are nominally flat and parallel.

14.2.2 Possible Solutions

We consider three different, realistic, solutions, (although, other ones may exist). These are sketched in Figure 14.2: (1) magnet combined with thrust bearing; (2) vacuum pump or suction bearing; and (3) over-expansion bearing, respectively.

Magnet-Thrust Bearing

Here, the magnetic force F_m pulls the bearing toward the upper plate, while the thrust bearing force F_t pushes it away (in order to maintain an air film between the two parts). The resultant bearing force is then $F = F_m - F_t$. However, both F_m and F_t generally decrease with h, in roughly the same way, as shown in Figure 14.3. Therefore, in order to satisfy the stiffness requirement, (i.e. $F > 0$ and increasing with h), one is obliged to simultaneously

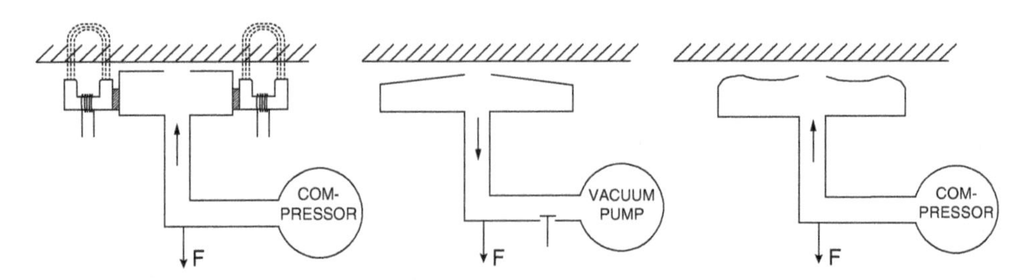

Figure 14.2 The various hanging bearing solutions. Source: From Al-Bender F and Van Brussel H 1994.

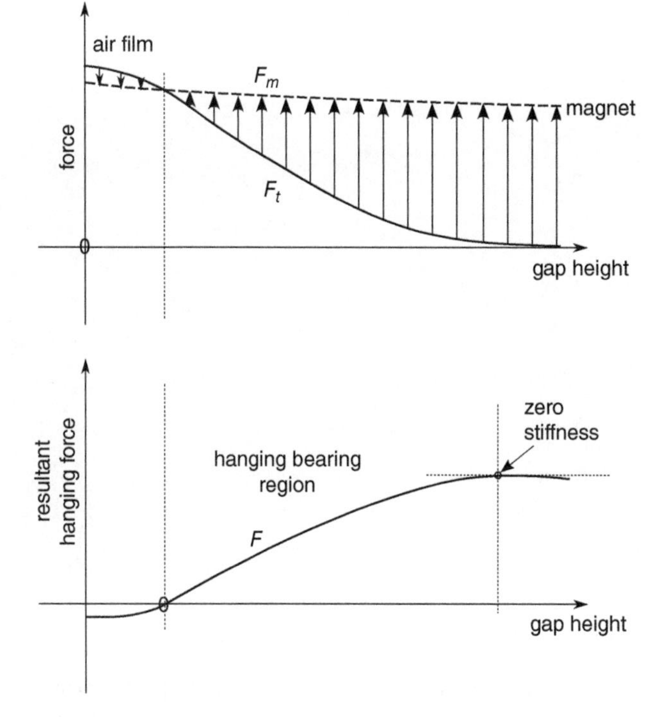

Figure 14.3 Force characteristics of magnet bearing.

take a large magnet gap, a high magnet force, and a small air bearing gap, as is also depicted in Figure 14.3, and confirmed by the remark in (Anon 1989) that the ratio of the magnetic force to the resultant force is of the order of 10^2. Considering the engineering implication of these results, we may be able to deduce the following evaluation:

Advantages

- relatively simple arrangement; component parts easily available,
- the magnet may be also used for the driving system,
- safety is ensured against air supply failure, (but not magnetic failure!).

Disadvantages

- high magnetic power needed and/or a heavy magnet,
- upper plate must not only be flat and smooth but also of a magnetic material,
- small air gaps necessary,
- assembly/adjustment may be difficult,

Example

Consider a thrust bearing with diameter of 60 mm, feed-hole diameter of 1.5 mm, conicity of 14 μm, supply pressure 5 bar (gauge) and ambient pressure of 1 bar. Calculation shows that this bearing is positively damped for any gap height.

The load and flow characteristics of this bearing are shown in Figure 14.4

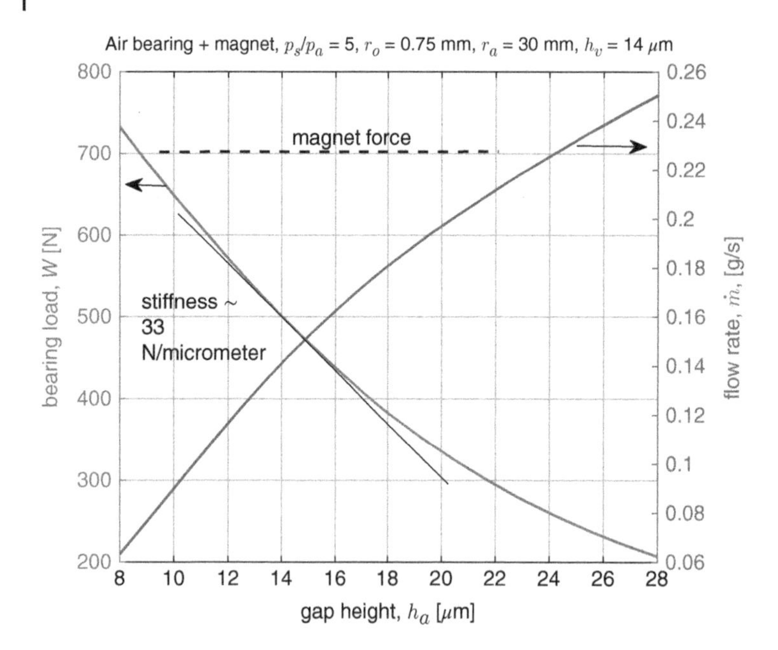

Figure 14.4 Load and flow rate as a function of the gap-height for a convergent-gap, thrust-magnet bearing.

If we set the minimum gap of the thrust bearing to approximately 8 μm, then we need a magnet force of approximately 700 N, at e.g. a magnet gap of around 1 mm. This can be easily achieved using a rare-earth permanent magnet of a bout 0.5 kg in mass.

Vacuum (or Suction) Air Bearing

This works by the same principle as the usual thrust aerostatic bearing; with the difference that it uses the ambient pressure p_a as supply pressure, and discharges to a sub-ambient (vacuum) reservoir, with pressure p_v. The topology of the problem demands then that flow should take place from the outer diameter to the inner one,[1] resulting in an *inverted* thrust bearing. Consequently, the pressure distribution characteristic, (in regard to the bearing force generation), is qualitatively the same in both cases.

A vacuum pump is needed to maintain the sub-ambient discharge pressure, together, possibly, with a check-valve system to provide for flow compensation in case a sufficiently large vacuum reservoir is not available. The evaluation of this solution may be summarised thus:

Advantages

- the bearing may be light in weight,
- upper plate does not have to be magnetic.

Disadvantages

- vacuum system may be costly and noisy,
- energy consumption may be inefficient (because of the need for flow compensation),
- pressure distribution is not favourable: the effective pressure decreases with the radius, i.e. with the area, resulting in limited mean bearing pressure,
- no safety against air failure.

1 Since it is very difficult and impracticable to maintain sub-ambient pressure at the outer diameter of the bearing but not in the whole work space.

Example

It turns out that, for small film thickness (and thus Reynolds numbers, so that the flow can be assumed to be viscous), one can apply the same equations and calculation procedures as for a usual thrust bearing (see Chapter 6) to determine the characteristics of a suction bearing. Considering a bearing with $p_v/p_a = 0.2$, $p_a = 1$ bar, $r_o = 5$ mm, $r_a = 30$ mm, and conicity $h_v = -20$ μm (the negative sign means that the gap diverges from inner to outer radius), yields the following characteristics.

Firstly, Figure 14.5 depicts the pressure distribution in the bearing film. When the gap height is small, the pressure approaches the atmospheric value (except within the suction hole). As the gap increases, the pressure decreases to that corresponding to a centrally-fed, flat bearing. However, owing to pressure loss at entrance to the gap, the pressure can decrease further, at the cost of excessive air consumption, as shown in table 14.1 for high gaps.

The load and air-flow characteristics are depicted in Figure 14.6. The behaviour is again similar to a thrust bearing except for the absolute values, i.e. qualitatively.

Over-Expansion Bearing

Figure 14.7 depicts the change of character of an EP bearing with increasing film thickness, see (Gross 1962). In the small gap range, one has the usual thrust bearing behaviour. as the gap height increases, the bearing force gradually diminishes to zero, after which a "hanging" or "pull" region ensues, characterised by a negative bearing force. This force is limited for a liquid bearing, being due to the Bernoulli effect, but can become appreciable in a compressible fluid bearing, owing to super-sonic over expansion. When the gap height becomes too large, this mode breaks down and goes over to positive (thrust) force caused by the jet issuing from the feed hole. In the following, we discuss the bearing action pertaining to the second (pull/hanging) region.

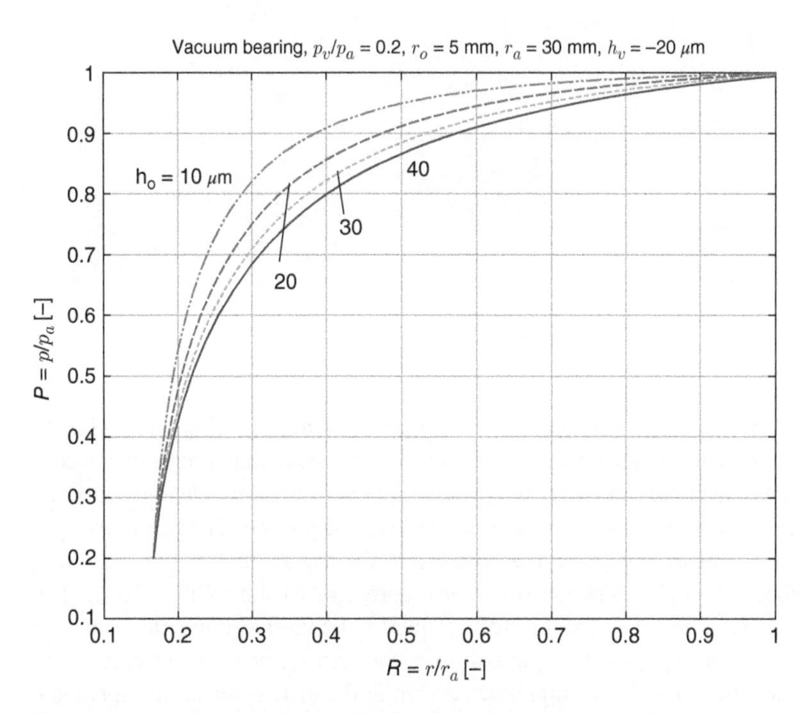

Figure 14.5 Typical pressure distribution in a divergent-gap, suction bearing.

Table 14.1 Load and flow rate in at large gap-heights for a divergent-gap, suction bearing, with $p_v/p_a = 0.2$, $p_a = 1$ bar, $r_o = 5$ mm, $r_a = 30$ mm, and conicity $h_v = -20$ μm.

Gap height μm	Load (N)	Mass flow (g s^{-1})
120	−51.2	−1.0
220	−98.7	−4.5
320	−140.8	−8.2
420	−167.5	−11.3
520	−183.5	−14.4

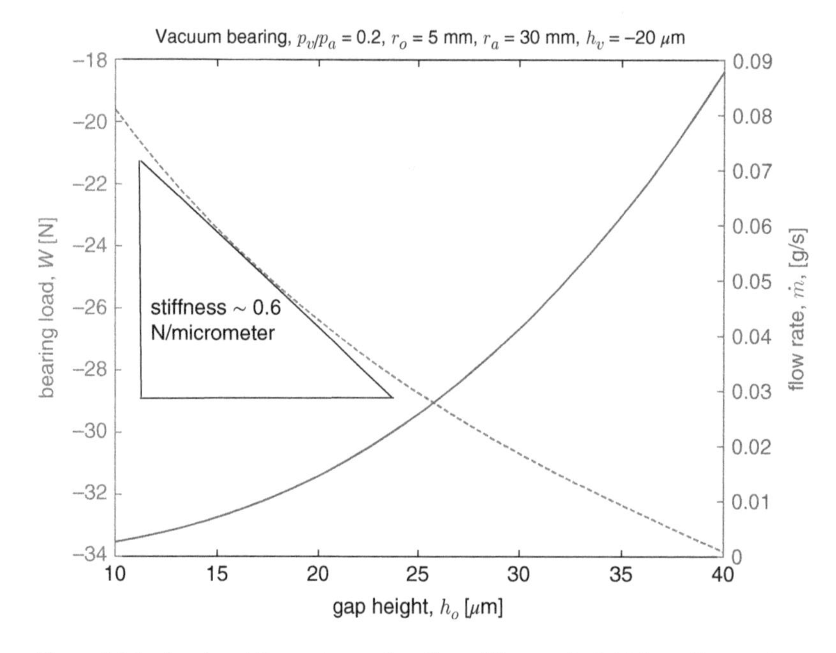

Figure 14.6 Load and flow rate as a function of the gap-height for a divergent-gap, suction bearing.

Principle of Operation

Pressurised air is fed through an orifice in the middle of the bearing pad into the bearing gap where it flows until the ambient outside diameter is reached. Considering the flow cross-sectional area, we see that it resembles a converging diverging nozzle, with the throat being at the gap entrance. Thus, when the flow is inviscid and subsonic, it will be an accelerating decelerating one, and the Bernoulli principle shows that the pressure is minimum at gap entrance, building up to ambient pressure at bearing exit, i.e. the pressure in the gap is sub-ambient, resulting in an attracting force between the surfaces. This phenomenon was first reported in 1828 by Willis (1830). The force generated in this case is somewhat small, however, for this arrangement to be used as a practical hanging bearing. When the flow is fast enough to attain sonic speed at gap entrance, the arrangement will resemble that of a de Laval nozzle. In this case the flow will further accelerate downstream of the orifice, reaching supersonic conditions, until, at some point, a shock wave will restore it back to subsonic condition whence it proceeds to exit.

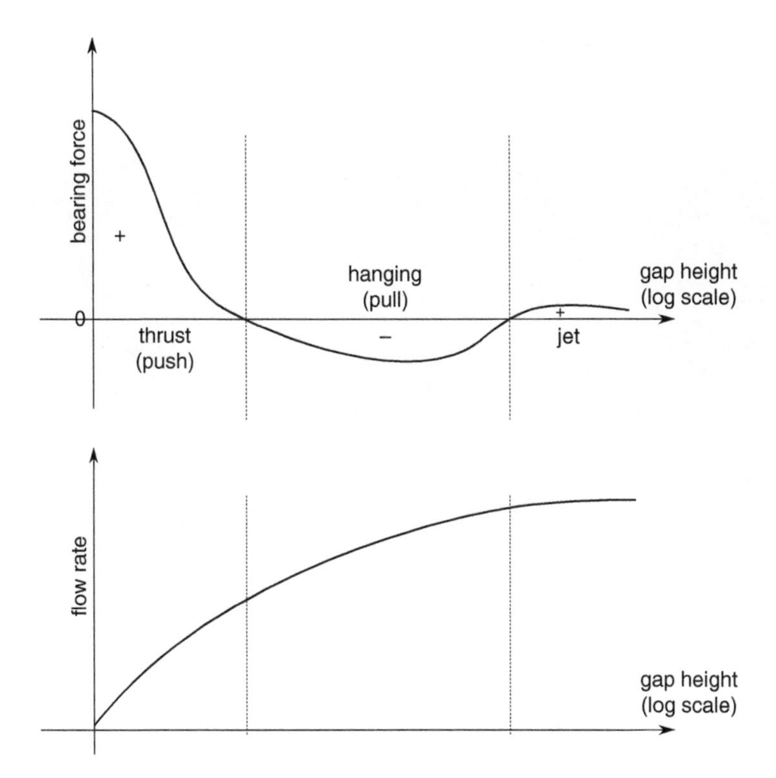

Figure 14.7 The various film regions of an E.P. bearing following Gross (1962).

The pressure distribution associated with this flow (see later) is particularly favourable since, in the supersonic region, the pressure decreases with increasing radius, and thus bearing area.

This type of bearing is expected to combine the advantages, and overcome the main drawbacks, of both of the two previous solutions. The arrangement is particularly simple: a circular disc supplied with pressurised air in the middle; i.e. the same as the usual thrust bearing. The principle of operation is, however, not the same. Instead of the slow viscous (subsonic) flow pertaining to thrust bearings, the flow in this type of bearing is required to be supersonic, (also turbulent then), over the greater part of the bearing area. In this way the bearing gap may be likened to a *de Laval nozzle*: if the flow is choked at gap entrance, (i.e. the air velocity is sonic there), then, due to the increasing flow area in the direction of the flow, the latter will be further accelerated, eventually resulting in sub-ambient pressures, until a shock wave discontinuity restores the flow to subsonic conditions whence it continues until the exit is reached, where ambient conditions prevail. (For a complete and systematic treatment of this problem, see (Massey 1968)).[2]

The bearing arrangement is shown in Figure 14.8 together with the expected pressure distribution. To ensure stiffness, the shock must lie some distance upstream of the exit, such that it can shift toward the exit when the film thickness is increased, thus providing the compensation action.

Summarising as before, we have:

2 The name of Bernoulli has become associated with the phenomenon of obtaining a lift force by inducing (inviscid) flow along the upper side of a body immersed in a fluid, (this being due to the reduction of the static pressure on the upper side with the increase of the velocity). Strictly speaking, this does not hold true in our application, since the suction action would yield only a very small force in radial flow using an incompressible fluid, even when no viscous friction is present. Rather, the crucial aspect in our case is compressibility and supersonic flow. Thus, it may be equally (if not more) appropriate to attach the name of Mach or de Laval to this bearing.

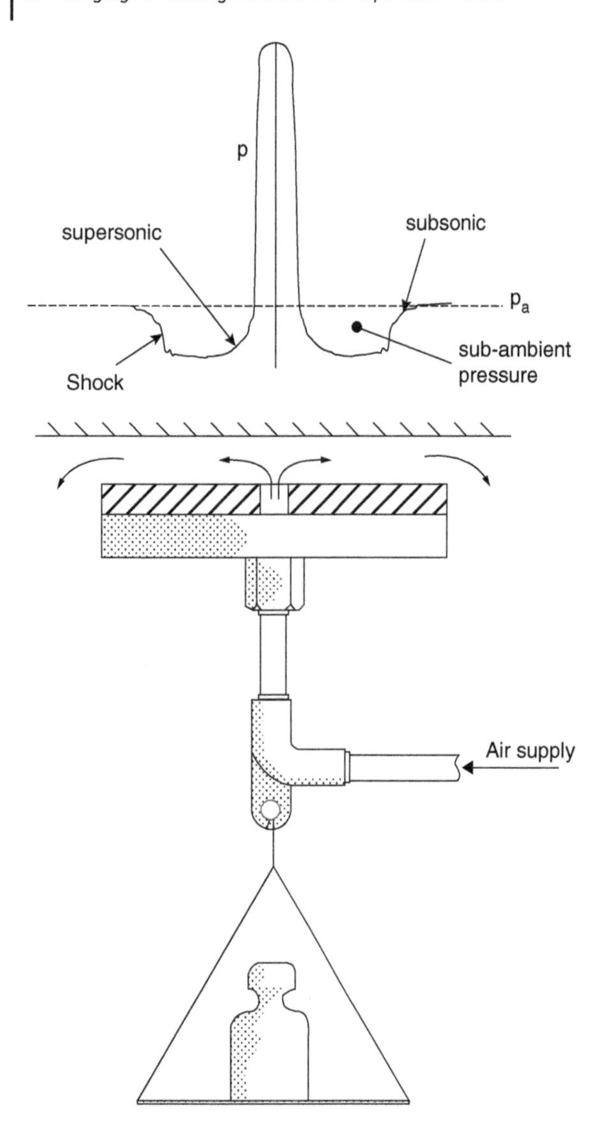

Figure 14.8 The over-expansion/de Laval bearing. Source: Dries J and Stevens B 1992.

Advantages

- the bearing may be light in weight,
- upper plate does not have to be magnetic,
- large film thickness possible,
- more favourable effective pressure distribution, (increasing with the radius),
- high tilt stiffness.

Disadvantages

- relatively high air consumption (cf. magnet bearing), and associated noise,
- no safety against air failure.

Nothing has been said so far about the dynamic stability of these different bearing systems. In the first one, since the magnet provides a quasi-constant force and hardly any damping, stability will be almost exclusively determined by the behaviour of the thrust air bearing (cf. Chapter 7). The vacuum bearing is expected, on account of the small pressure ratio across it, to be free from self-excited vibrations. It is difficult to say offhand, however, whether or when the third one would be unstable; we might expect that there will be little chance of instability in the useful design range, i.e. when the bearing force is large, by analogy with the previous case.

14.2.3 Choice of a Solution

It may be rewarding to carry out a quantitative and more detailed comparison between the three different solutions outlined in the previous section. However, even when the first two solutions are relatively easy to model, (as is indeed the case), the third one has still to be worked out. It will not be ungainly, therefore, to investigate the over-expansions bearing, especially when its potential superiority is kept in mind.

The remainder of this chapter is therefore devoted to the over-expansion bearing.

14.3 Problem Formulation

Although the basic theory had been formulated and a rough prototype of this bearing type had been constructed (Al-Bender 1991), the systematic development of such a bearing as an efficient and reliable component took some additional steps:

1. the construction of a good theoretical model,
2. experimental verification,
3. Design optimisation,

When these steps are satisfactorily addressed, attention may then be directed to the study of tilt stiffness, dynamic stability, and the effect of high relative speed between the bearing parts. Fabrication issues and safety aspects may also be relevant.

14.4 Theoretical Analysis

Referring to Figure 14.9, air flows axisymmetrically from a plenum at pressure p_s, through the feed-hole curtain at radius r_0 and average pressure p_0, into the bearing gap, whose height h is, in general a function of the radius r, and discharges into atmosphere at radius r_a and pressure p_a.

Several authors (Constantinescu 1969; Moller 1966; Mori 1961) have considered the case of supersonic flow in the bearing, (for the case of uniform gap only however). We find the third of these, viz. the investigation carried out by Moller (1966), to be the most realistic, complete, and verified by experiment. Modifying his theory to include the effect of the variable gap does not present any difficulty. Therefore, we will initially adopt his theoretical model, since it is also not very complicated, leaving the possibility of further modification until and if the experimental comparison should deem it necessary.

14.4.1 Basic Assumptions

The flow is assumed to be two-dimensional, (axisymmetric), for which the boundary-layer (B-L) simplification is applicable, viz. the static pressure p is constant across the gap, and only the (B-L) momentum equation in the direction of the flow is significant. Referring to Figure14.9(b), the flow may be divided into three regions:

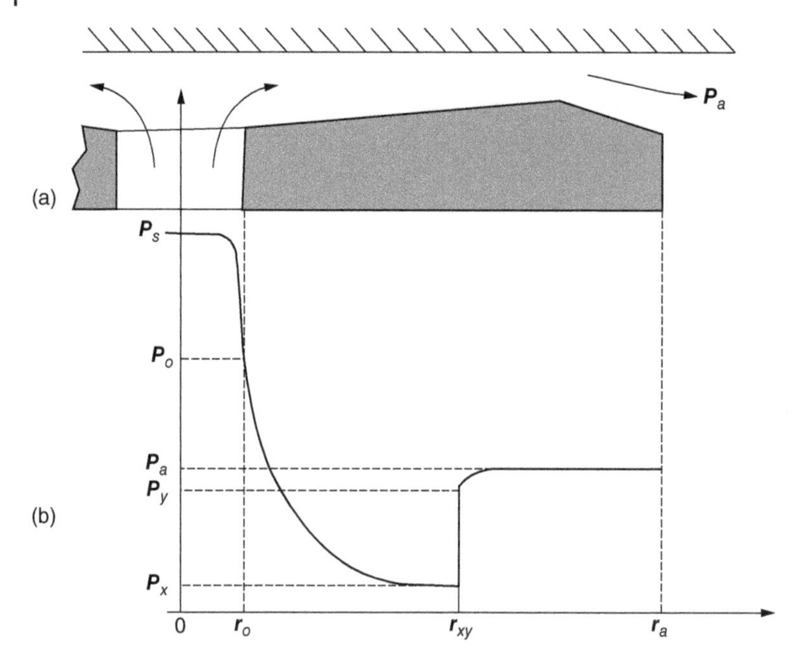

Figure 14.9 Schematic pressure-flow development in an over-expansion bearing. Source: Dries J and Stevens B 1992.

1. the feed region, $p_s \rightarrow p_o$, in which the flow is assumed isentropic, i.e. inviscid and adiabatic;
2. the supersonic region, $r_o \rightarrow r_{xy}$, $(p_o \rightarrow p_x)$, where the flow is assumed turbulent and adiabatic;
3. the subsonic region, $r_{xy} \rightarrow r_a$, $(p_y \rightarrow p_a)$, where the flow is assumed turbulent and isothermal.
   (A normal shock, at r_{xy}, is assumed to separate the two last regions.)

Furthermore, Moller (1966) shows that no significant errors are incurred if the flow is assumed to be one-dimensional, i.e. the fluid velocity u is uniform across the gap, (except in the immediate vicinity of the walls where a friction factor is deduced for the shear stress there). This simplifies the analysis significantly. Choking is assumed to occur approximately at the inlet curtain area. Further assumptions may be made in the course of the analysis.

14.4.2 Basic Equations and Definitions

The Integral Momentum Equation

Referring to Figure 14.10 for notation, since the radial velocity u is assumed uniform across the film except in the close proximity of the walls, the normal velocity v may be neglected, and the momentum equation is written:

$$\rho u \partial u/\partial r = -\mathrm{d}p/\mathrm{d}r + \mu \partial^2 u/\partial z^2 \,.$$

Making use of the assumed symmetry of the flow about the centre line ($z = h/2$), we can integrate this equation across the film to get:

$$\frac{\rho}{2}\frac{\mathrm{d}}{\mathrm{d}r}\left(\int_0^{h/2} u^2 \mathrm{d}z\right) = -\frac{\mathrm{d}p}{\mathrm{d}r}\frac{h}{2} + \mu\left(\frac{\partial u}{\partial z}\right|_{z=0}^{z=h/2}$$

or,

$$\frac{\rho}{2}\frac{\mathrm{d}}{\mathrm{d}r}\left(\frac{h}{2}\bar{u}^2\right) = -\frac{\mathrm{d}p}{\mathrm{d}r}\frac{h}{2} - \mu\frac{\partial}{\partial z}u(r,0)$$

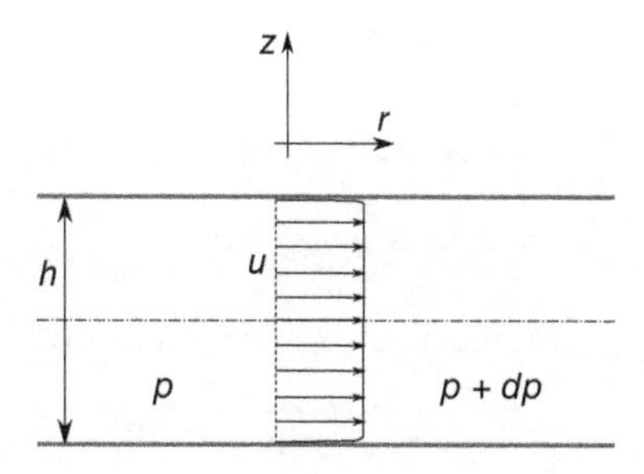

Figure 14.10 Notation.

where, $\bar{u}(r)$ is the average velocity, and

$$\mu \frac{\partial}{\partial z} u(r, 0) = \tau_w$$

is the viscous shear stress at the wall which will be explained later. Finally, dropping the bars for simplicity, and using a prime (′) to denote r-derivatives, the integral momentum equation is written:

$$p' + 2\tau_w/h + (\rho/2h)(hu^2)' = 0 \tag{14.1}$$

This is the turbulent-flow integral momentum equation that forms, together with continuity equation (here below) the basis of turbulent Reynolds equation explained in Chapter 4.

The Integral Continuity Equation

In view of the previous assumptions, we may write this equation directly:

$$2\pi rh\rho u = \dot{m} = \text{const.} \tag{14.2}$$

where, $\dot{m}$ is the mass flow rate.

The Energy Equation

Consistent with the assumption of uniform velocity across the film, the temperature of the fluid is likewise assumed uniform. We then have, for adiabatic expansion, (see (Massey 1968)):

$$c_p T + \frac{1}{2}u^2 = c_p T_s = \text{const., (adiabatic)} \tag{14.3}$$

where, c_p is the specific heat at constant pressure of a perfect gas, and T_s is the stagnation temperature, (in our case, equal to the supply temperature).

For isothermal expansion, we have:

$$T = T_w = \text{const., (isothermal)} \tag{14.4}$$

where, T_w is conveniently taken to be the wall temperature. Note that from practical considerations, we usually set

$$T_s = T_w = T_a$$

The Equation of State

For a perfect gas, this equation is written (Chapter 2):

$$p = \Re \rho T \tag{14.5}$$

where, $\Re$ is the universal gas constant, ($\Re = c_p - c_v$), c_v being the specific heat capacity at constant volume for a perfect gas).

Other Relationships and Definitions

The Mach number is the ratio of the gas velocity to the local sonic velocity, thus

$$\mathrm{M} = u/a,$$

where a is the sonic velocity given by:

$$a = \sqrt{\kappa p/\rho} = \sqrt{\kappa \Re T}$$

in which, $\kappa = c_p/c_v$.

The Reynolds number (based on the gap width) is the ratio of the inertial force to the viscous force:

$$\mathrm{Re} = \rho u 2h/\mu = \dot{m}/\pi r \mu, \qquad \text{(due to continuity).}$$

The wall shear stress τ_w is a key parameter in this system and is usually expressed as a fraction of the dynamic pressure $\rho u^2/2$, through multiplication by a *friction factor*, f_r, (that is, however, not constant). Thus, one writes:

$$\tau_\mathrm{w} = f_\mathrm{r} \frac{1}{2} \rho_\mathrm{r} u^2$$

where, the subscript $()_\mathrm{r}$ denotes a reference value for the variables to be assigned later, together with a formula for the friction factor, which is generally a power-law function of the Reynolds number, see further below.

14.4.3 Derivation of the Pressure Equations

Due to to the different flow characteristics in each of the three regions described previously, the corresponding pressure equations will be different. These will then be derived separately for each region; to be matched afterwards for the complete solution, (i.e. from supply to exit).

The Inlet Region

Here Euler's equation, (which is equation 14.1 with the friction term dropped), may be applied together with the equations of state and energy to give after integration:

$$\frac{p_s}{p_o} = (1 + (\kappa - 1)\mathrm{M}_o^2/2)^{\kappa/(\kappa-1)}, \qquad \mathrm{M}_o \leq 1 \tag{14.6}$$

When the flow is choked, as in our case of interest, then $\mathrm{M}_o = 1$, and p_o/p_s reaches the critical value

$$p_o/p_s = (2/(\kappa + 1))^{\kappa/(\kappa-1)} \approx .528 \qquad \text{(for air, } \kappa = 7/5\text{).}$$

Note, however, that, in practice, the flow is neither frictionless nor one-dimensional, such that we will have to introduce an empirical correction at the inlet in order to match the flow rates accurately. This may be done by assigning a coefficient of discharge C_d to the inlet flow, or equivalently assuming a total pressure loss coefficient K_o. Adopting the latter, for convenience, Eq. 14.6, at choked conditions ($\mathrm{M} = 1$) is rewritten:

$$\frac{p_s}{p_o} = (1 + (1 + K_o)(\kappa - 1)/2)^{\kappa/(\kappa-1)}, \tag{14.7}$$

where, typically, $K_o \approx 0.3$ (Moller 1966), which corresponds to $C_\mathrm{d} \approx 0.84$. (Let us note that this value agrees with that deduced for choked turbulent flow in Chapter 5).

The Supersonic Region

In this region, it is more convenient to derive an equation for the Mach number together with a relationship connecting it to the pressure.

We shall need, first, differentiated forms of the basic equations and relationships. Thus, the differential continuity equation is written:

$$\rho'/\rho + u'/u + 1/r + h'/h = 0 \tag{14.8}$$

where, the accents (′) indicate ordinary derivatives w.r.t. r. The Mach number relationship gives:

$$(\mathrm{M}^2)'/\mathrm{M}^2 = (u^2)'/u^2 - T'/T \tag{14.9}$$

The equation of state gives:

$$p'/p = \rho'/\rho + T'/T,$$

which, with equations 14.8 and 14.9 gives:

$$p'/p = \frac{1}{2}(T'/T - (\mathrm{M}^2)'/\mathrm{M}^2) - 1/r - h'/h \tag{14.10}$$

The energy equation, (for adiabatic expansion), gives, after manipulation:

$$T_s/T = 1 + \frac{1}{2}(\kappa - 1)\mathrm{M}^2 \tag{14.11}$$

which in differential form is:

$$\frac{T'}{T} = -\frac{\frac{1}{2}(\kappa - 1)(\mathrm{M}^2)'}{1 + \frac{1}{2}(\kappa - 1)\mathrm{M}^2} \tag{14.12}$$

Also, the wall shear stress is written:

$$\tau_w = f_r^* \rho u^2/2 \tag{14.13}$$

where,

$$f_r^* = f_r \rho_r/\rho$$

Substituting Eqs. 14.9 through 14.13 in Eq. 14.1 and simplifying, yields the Mach number differential equation:

$$(\mathrm{M}^2)' = 2\frac{\mathrm{M}^2\left(1 + \frac{1}{2}(\kappa - 1)\mathrm{M}^2\right)}{(\mathrm{M}^2 - 1)}\left(\frac{1}{r} + (1 - \kappa\mathrm{M}^2/2)\frac{h'}{h} - \kappa\mathrm{M}^2\frac{f_r^*}{h}\right) \tag{14.14}$$

This equation characterises the flow in the supersonic region. After it has been solved, the pressure may be obtained from the following relationship:

$$\frac{p}{p_o} = \frac{r_o}{r}\frac{h_o}{h}\frac{\mathrm{M}_o}{\mathrm{M}}\left(\frac{1 + \frac{1}{2}(\kappa - 1)\mathrm{M}_o^2}{1 + \frac{1}{2}(\kappa - 1)\mathrm{M}^2}\right)^{1/2} \tag{14.15}$$

which is obtained from the equations of continuity, state and energy.

The Subsonic Region

The only difference between this and the previous region is that the flow is assumed isothermal, i.e. equation 14.4 applies instead of equation 14.3. Although no significant errors might be incurred if the adiabatic assumption is upheld, (and thus the same equations are used), it will nevertheless be more convenient to write the flow equation in terms of the pressure rather than the Mach number. This is so because we have a boundary condition on the former, (viz. $p(r_a) = p_a$), but none on the latter.

Substituting Eqs. 14.2, 14.5 (also with $\rho'/\rho = p'/p$, $T = T_\mathrm{a}$), 14.8, and 14.13 in equation 14.1 and simplifying, yields the pressure equation:

$$(p^2)' = \frac{C\,p^2(2/r + h'/h - 2f_\mathrm{r}^*/h)}{r^2h^2p^2 - C} \tag{14.16}$$

where, we have put,

$$C = \left(\frac{\dot{m}}{2\pi}\right)^2 \Re T_\mathrm{a}$$

Equation 14.16 may be regarded as the subsonic, turbulent Reynolds equation.

Conditions Across the Normal Shock

Assuming the shock to be normal and adiabatic, we have (Massey 1968; Moller 1966):

$$\frac{p_y}{p_x} = \frac{2\kappa}{\kappa+1}\mathrm{M}_x^2 - \frac{\kappa-1}{\kappa+1} \tag{14.17}$$

where the subscripts $()_x$, $()_y$, denote conditions just upstream and just downstream of the shock, respectively, (the latter being assumed of negligible thickness).

The Friction Factor

In the literature, (see Chapter 4), a two-parameter, power-law relationship is often adopted for the friction factor, i.e.

$$f_\mathrm{r} = m\mathrm{Re}_\mathrm{r}^n,$$

where, m, n are empirically determined constants.

Moller (1966) shows empirically that $m \approx 0.079$ and $n \approx -1/4$ so the friction factor may be expressed in the form:

$$f_\mathrm{r} = 0.079/\mathrm{Re}_\mathrm{r}^{1/4}$$

where,

$$\mathrm{Re}_\mathrm{r} = u\rho_\mathrm{r}2h/\mu_\mathrm{r}$$

which, eventually, leads to:

$$f_\mathrm{r}^* = \frac{0.079}{(\mathrm{Re}_\mathrm{a})^{1/4}}\left(\frac{r}{r_a}\right)^{1/4}\left(\frac{T_\mathrm{r}}{T_\mathrm{w}}\right)^{1/4}\left(\frac{T}{T_\mathrm{r}}\right)^{3/4} \tag{14.18}$$

where $\mathrm{Re}_\mathrm{a} = \dot{m}/\pi r_a\mu_\mathrm{a}$ is the exit Reynolds number.

Taking, for simplicity, the reference temperature T_r to be the arithmetic mean of the fluid and the wall temperatures, one obtains:

$$f_\mathrm{r}^* = \frac{0.079}{(\mathrm{Re}_\mathrm{a})^{1/4}}\left(\frac{r}{r_a}\right)^{1/4} F(\mathrm{M}^2) \tag{14.19}$$

where, in the supersonic region, F is easily found from equation 14.11; and in the subsonic region, $F \equiv 1$.

14.4.4 Normalisation of the Final Equations

In order to facilitate understanding of the model's equations, and to enable a systematic solution of the problem, eventually through computer programming, it is found best that the variables are first rendered dimensionless. In this way, the number of design parameters is reduced and their relative influence better gauged.

Normalisation may be carried out, (equivalently), with respect to different sets of reference dimensional parameters. One may choose, for better consistency, the exit parameters as reference. Thus, we define:

$$R = r/r_a, \qquad H = h/h_a, \qquad P = p/p_a.$$

This will, however, be left out, for now, since it is lengthy and does not add new information.

14.4.5 Solution Procedure

Equations 14.7, 14.14–14.17 and 14.19, (with appropriate auxiliary conditions), are sufficient for the solution of the problem. This is carried out in the following sequence.

First, the inlet equation is solved to obtain the mass flow rate. Second, the Mach number equation is integrated (numerically) w.r.t. r, starting at the inlet radius onwards. Third, the pressure equation for the subsonic region, (with the mass flow rate now known), is integrated from the outer radius backwards. Finally, the position of the shock is determined as the point where the last two solutions satisfy the shock relationship.

After the pressure distribution has been determined, the load capacity may be calculated by integrating it over the area. Other bearing characteristics, such as axial and tilt stiffness are likewise calculated.

14.4.6 Matching the Solution With Experiment: Empirical Parameter Values

The model, as presented above, proved to be very close to experimentally obtained results, as will be seen further below. However, some adjustments and choice of particular parameter values were identified in order to make the agreement between model and experiment as good as possible. These are as follows:

- **Correction on the feed-hole diameter** Owing to the large gaps associated with this type of bearing coupled with the very high flow rates, it was deemed necessary to apply some correction factor on the feed-hole diameter in order to affect a better fit with experimental data. On the measured pressure profiles, it was actually observed that the radius at which the supply pressure sharply dropped to the inlet pressure value was slightly larger than the actual value of the feed-hole radius. Furthermore, this increase had the same proportion for feed-holes of different sizes. This factor was:

  $$\text{corrected } r_o = 1.02 \times \text{actual } r_o.$$

- **Entrance loss coefficient** In viscous flow entrance theory, see Chapter 3, we derived a coefficient of discharge coefficient to account for the entrance flow losses. This device is also applicable, *mutatis mutandis*, to the present flow situation. However, here we have followed the method of Moller (1966) in prescribing a pressure-correction coefficient K_o as defined in Eq. 14.7. A value of $K_o \approx 0.35$ proved to match the flow best.
- **Weighting factor for the reference temperature** This is needed for estimating the friction factor, for which Moller (1966) takes the arithmetical mean of the wall temperature T_w and the mean temperature of the fluid $T(r)$. We have generalised the approach of Moller to a weighted mean:

  $$T_{\text{ref}} = \alpha_T T + (1 - \alpha_T) T_w.$$

  This has little influence on the subsonic flow portion, but can improve the supersonic portion appreciably. It turned out that a good value for α_T is very close to zero. In other words, a good value for the reference temperature is the wall temperature itself.

- **Empirical coefficient for the friction factor** Moller (1966) uses 0.079 for m in Eq. 14.18 (and just before). We have found that this coefficient should optimally take different values for the supersonic and the subsonic flow regimes. This is motivated by the fluid-temperature differences in those two regions and their effect on the apparent viscosity. For the supersonic region, the value $m = 0.0625$, and for the subsonic region $m \approx 0.1$ appeared to yield best performance.

14.5 Experimental Verification

14.5.1 Test Apparatus

For accurate verification of the theoretical model, we should be able to measure the pressure distribution in the bearing gap. For this purpose, a test apparatus such as the one depicted in Figure 14.11 has been constructed. The arrangement comprises a sliding bearing table equipped with a pressure probe-hole connected to a pressure transducer, and a horizontally set LVDT to record the radial position. The test bearing pad is suspended by means of a loaded chord passing over two frictionless pulleys. The gap height is measured by non-contacting displacement transducers (of the Eddy-current type, from Bently, Nevada). The parallelism of the air gap is obtained by adjusting the three balancing nuts that are axisymmetrically disposed around the bearing pad. Dry and filtered air is supplied through a flow-meter as shown. The measurement details are shown in Figure 14.12.

14.5.2 Range of Tests

Without loss of generality, all tests were confined to bearing pads having an outer diameter of 60 mm. The design parameters that were varied are (i) the feed-hole diameter, (ii) the bearing-surface profile, and (iii) the supply pressure.

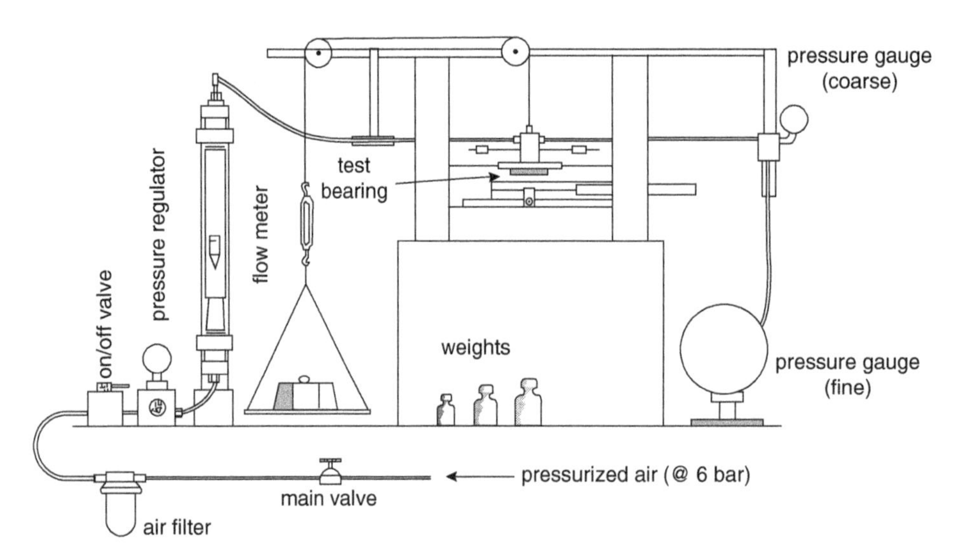

Figure 14.11 Experimental apparatus for the pressure distribution. Source: Dries J and Stevens B 1992.

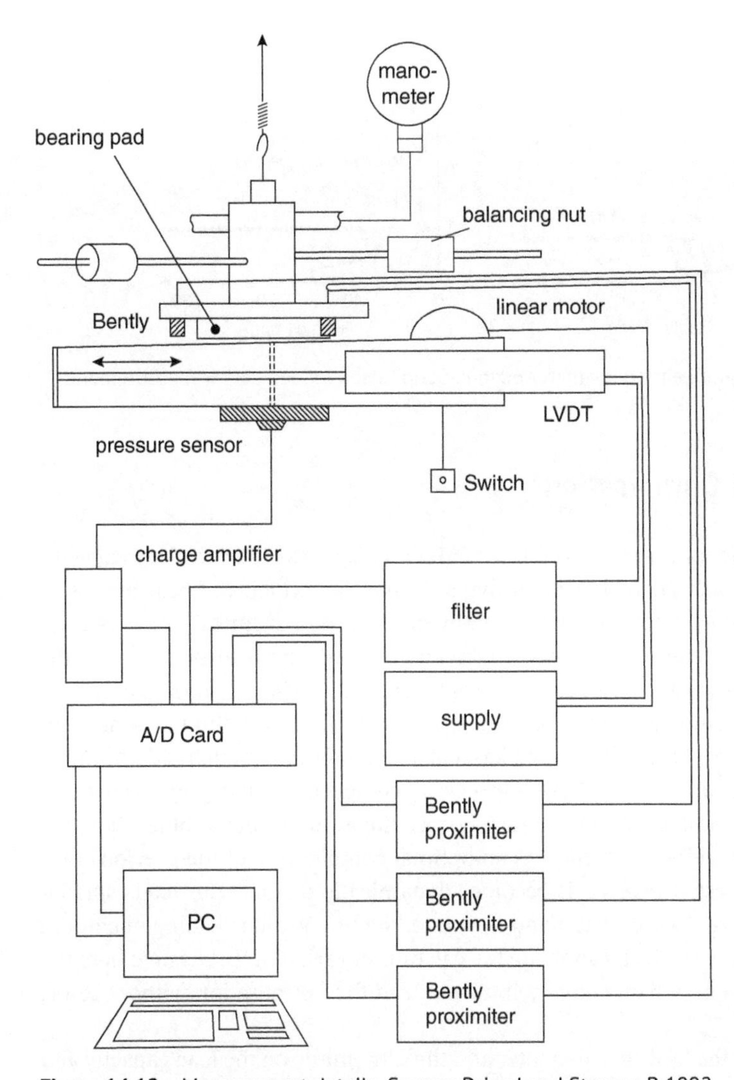

Figure 14.12 Measurement details. Source: Dries J and Stevens B 1992.

The performance characteristics that were measured are (i) the pressure distribution, (ii) the bearing load, and (iii) the air consumption; all of the foregoing for a range of gap heights up to about 0.4 mm.

Initially, uniform gap bearings were tested, in order to determine basic behaviour at different feed-hole diameters ratios and at various supply pressures.

The developed model showed very good agreement with the numerous measurements that were carried out. An example of these is given in figure 14.13 where the effect of varying, two values of (a) the feed pressure, (b) the film thickness, and (c) the feed-hole diameter are shown, (solid line = model, dots = measurement).

The model also showed that the flow solution is not always unique. In that case dynamic instabilities are likely to take place. Details of the model development, the experiment set-up, tests and the results are to be found in (Dries and Stevens 1992).

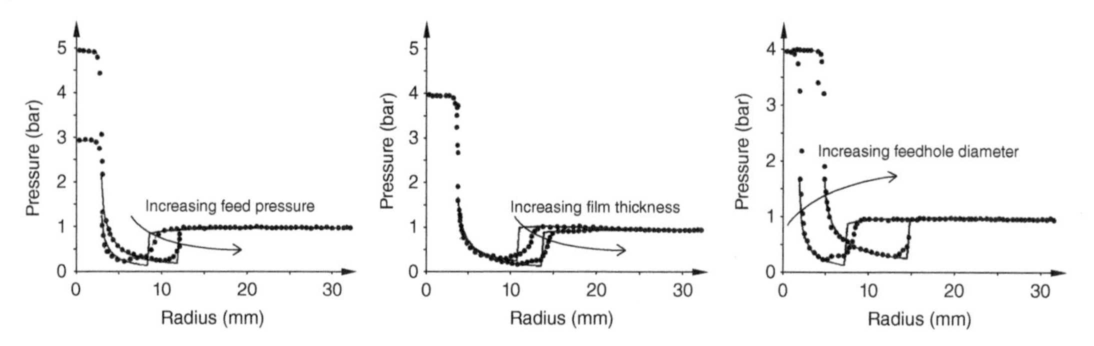

Figure 14.13 Pressure distribution in an EP hanging bearing, diameter 60 mm. Solid line = model, dots = measurement. Source: From Al-Bender F and Van Brussel H 1994.

14.6 Bearing Characteristics and Optimisation

The basic bearing characteristics are (Al-Bender and Van Brussel 1994; Dries and Stevens 1992): the load capacity, the stiffness and the air consumption. A fourth characteristic is dynamic stability which the bearing has to fulfill, and which imposes an extra constraint on the maximum attainable load and stiffness. For a circular centrally fed bearing, the design parameters determining these characteristics are: the feed pressure P_s, the feed-hole diameter $2R_0$, and the film geometry (profile $h(x)$ and nominal height h_{min}). The film profile may have an important influence on the bearing characteristics. However, it is not easy to find an optimum profile for general application. Many different profiles have been tried on the basis of a generalised parametric description. Figure 14.14 shows the pressure distribution in a sample of these bearings, all being fed by 3 bar (gauge) pressure. Profile 1 gives the highest load capacity of about 180 N, but requires exceedingly high flow. Profile 2 has the highest load-to-flow ratio, but possesses little stiffness. Profile 3 is an optimal combination of the previous two, combining high load-to-flow with high stiffness. However, it becomes dynamically unstable for loads smaller than 100 N. In any case, for a given bearing size, feed pressure and flow rate, the film profile yielding maximum load can, in theory, be determined from the flow model by applying the calculus of variation. Let us note here the close agreement between model and experiment, which could help us perform the optimisation without resort to experiment.

Finally, Figure 14.15 shows the influence of the feed-hole diameter and the film profile on the load capacity and the stiffness.

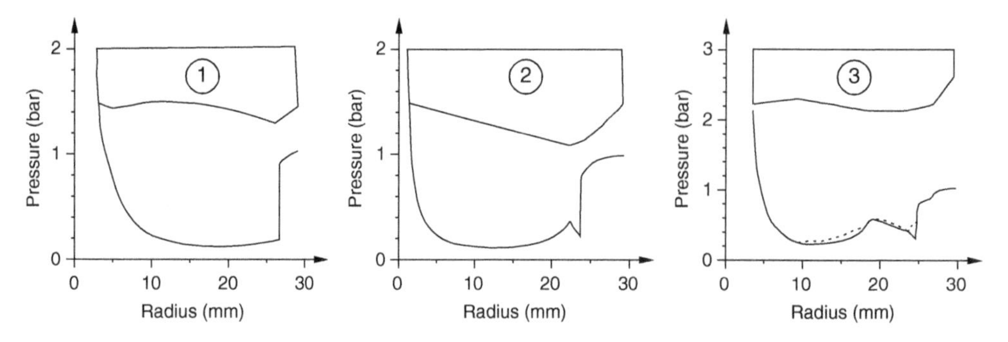

Figure 14.14 Influence of profile on pressure distribution. Source: From Al-Bender F and Van Brussel H 1994.

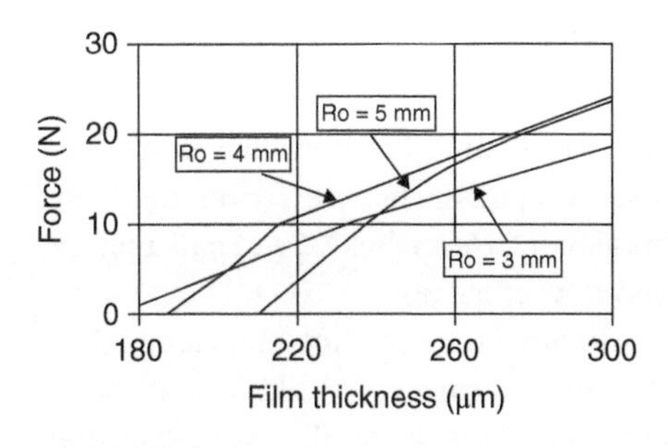

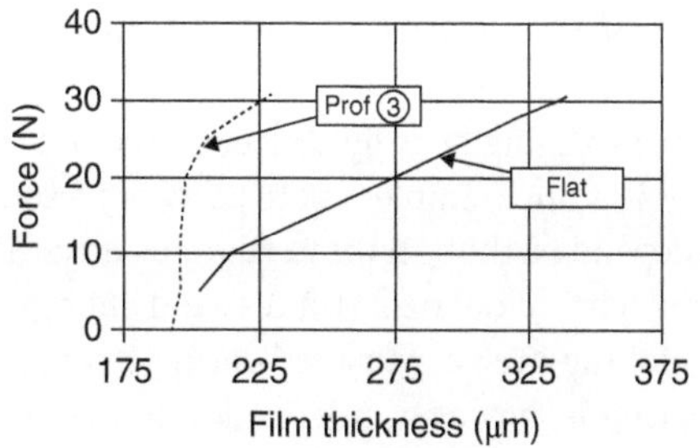

Figure 14.15 Load characteristics of hanging bearings, dia. 60 mm, feed press. 3 bar(g). Source: From Al-Bender F and Van Brussel H 1994.

The existence of an optimum R_0, at any given gap range, is evident. In regard to the profile, we may note that the stiffness obtained from profile 3 ($\sim$ 2 N/μm), in a certain gap range, can be considerably bigger than that of a flat bearing ($\sim$ 0.6 N/μm), but not so as compared with the thrust bearing-magnet solution. Also, as mentioned previously, the stiffness increases with increasing feed pressure.

The air consumption is of the order of normal litres per second for this bearing size and feed pressure. While this is reckoned to be more efficient than a vacuum bearing, nevertheless, the noise level caused by such flow cannot be ignored. This is perhaps one of the main drawbacks of this system that has to be treated in the future.

14.7 Design Methodology

The bearing may be optimised with respect to the load-to-flow ratio = Efficiency Factor (EF), which, for a fixed supply pressure, is proportional to the power consumption, and the stiffness/load ratio depending on the application requirement, i.e. a trade-off has to be sought.

If we ignore the profile and consider a flat bearing, we find that EF increases with the feed pressure (and so does the stiffness), and when the film thickness is specified, it becomes a simple matter to determine an optimal feed-hole diameter.

Finally, optimisation will mean a trade-off between the various bearing parameters and characteristics: load capacity, stiffness, power consumption, gap width, simplicity of gap profile fabrication, and dynamic stability; and as such should require a separate study.

14.8 Other Details

These will be stated in order of importance. First, a safety arrangement to guard against air supply failure may have to be included, since when that happens the bearing will fall off resulting in damage. This may be realisable in various ways: simply as an elbowed arm attached to and following the bearing; (it may simultaneously carry the air supply), or some arrangement based on sensing the pressure and triggering and emergency attachment device.

Secondly, possible methods for driving the bearing around may be suggested. The handiest of these may be an electromagnetic method, (e.g., a linear or a Sawyer motor). Wheel drive, (or even jet drive), can also be considered. External manipulation, through a lever arm, is of course always possible.

Thirdly, air supply facility to the mobile bearing may be considered. These, and other possible details, are, however, out of scope of the present text.

14.9 Brief Comparison of the Three Hanging-Bearing Solutions

We have discussed three types of hanging bearing concepts that are qualitatively different from one another in almost all details. Consequently, it is not a simple task to carry out a sensible comparison between them; the more so, since their application will depend on the specific needs and constraints at hand. Nevertheless, we shall attempt to make some general comparison based on load, stiffness and energy load-flow efficiency.

Firstly, the suction bearing and the over-expansion bearing have approximately the same load efficiency. The stiffness of the the suction bearing is, however, inferior to the over-expansion bearing. If we add to that that a vacuum pump is needed for the suction bearing, we can conclude that, at least for large loads, the over-expansion bearing should be preferred.

Second, comparing to the thrust-magnet bearing, we see that it is superior to the previous two in regard to load, stiffness and air consumption. The principal drawback of this solution is the requirement for the ceiling (or bearing counter surface) to be of a magnetic material, preferably soft iron. This might prove problematic in large-stroke general applications.

14.10 Aerodynamic Hanging Bearings

The author is not aware of any publications dealing with this topic or any suggested applications. However, in order to make the topic of hanging bearings more complete, it might be pertinent to consider also this bearing possibility. Since it is intended basically to be demonstrative, however, the analysis will be restricted to the inclined bearing case that is operating at sufficiently large bearing numbers.

14.10.1 Inclined and Tilting Pad Case

Let us consider the feasibility and basic characteristics of this type of bearing in a very simplified way. We assume therefore:

- **Subsonic entrance** i.e. that the mean flow velocity at entrance remains below the local speed of sound $c = \sqrt{\kappa p_a/\rho_a} = \sqrt{\kappa \Re T_a}$. In this way, owing to the divergent gap, the flow will remain subsonic throughout.
- **No slip** which could arise at the places in which the pressure becomes very low. However, in those regions, the gap-height will be large, so that the Knudsen number would remain small enough (refer to Chapter 4 for discussion on this issue).
- **High sliding number** Λ that is, we consider basically high-speed behaviour. In that case, the solution is enormously simplified to ph = const. on the largest portion of the bearing, see Chapter 8.
- **2D isothermal case** i.e. infinitely long bearing with well conducting walls.

We consider dimensionless equations and characteristics. Thus, referring to Figure 14.16 for notation, writing $P = p/p_a, H = h/h_a, X = x/L$, we have:

$$PH = 1, \qquad P(0) = H(0) = 1,$$

$$H = 1 + \nu X, \qquad \nu = \frac{h_o - h_a}{h_a} = H_v.$$

Substituting the second equation into the first, we obtain the sub-ambient (or vacuum) pressure

$$P_v = P - 1 = -\frac{\nu X}{1 + \nu X},$$

Figure 14.16 Hanging tapered pad notation. (Not to scale.)

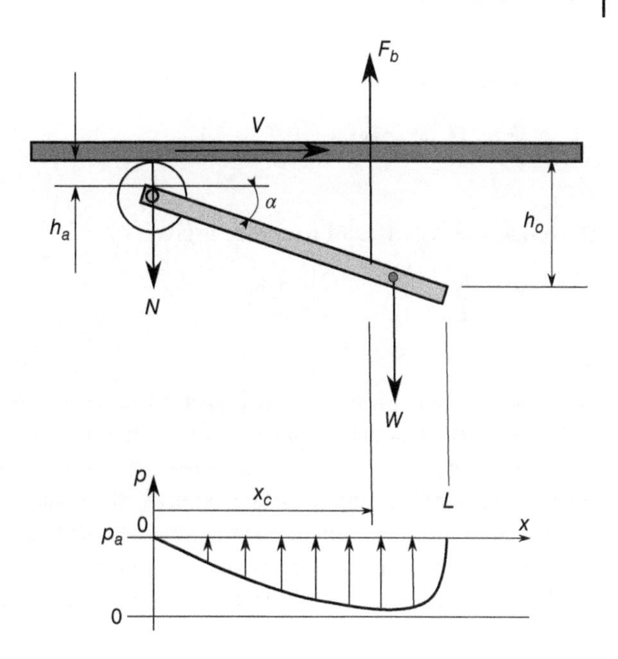

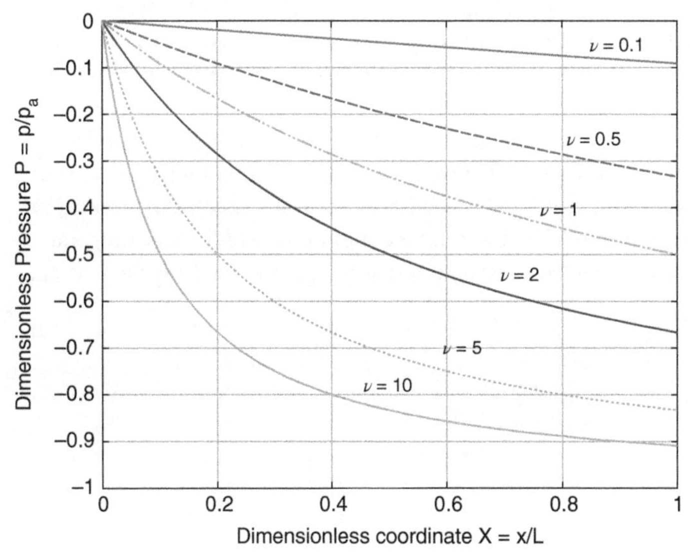

Figure 14.17 Pressure distribution in divergent self-acting taper bearing.

which is plotted in Figure 14.17 for different values of ν. Note that this pressure distribution is valid on the largest portion of the bearing except near exit where the pressure should return to atmospheric, as shown in Figure 14.16 (but not in Figure 14.17).

We wish to determine two basic characteristics of this bearing, namely (i) the load capacity and (ii) the position of the centre of pressure, which we need to establish static stability and optimum position of the hanging load or tilting hinge.

Dimensionless load is given by

$$W = \int_0^1 P_\nu \mathrm{d}X = 1/\nu \log(1+\nu) - 1.$$

Thus, $-1 > W \geq 0$, which corresponds to $\infty > \nu \geq 0$.

Dimensionless moment of pressure is given by

$$M_p = \int_0^1 P_\nu X \mathrm{d}X = -\left(\frac{W}{\nu} + \frac{1}{2}\right).$$

Then,

Dimensionless centre of pressure is given by

$$X_\mathrm{c} = \frac{x_\mathrm{c}}{L} = \frac{M_p}{W} = -\frac{2W + \nu}{2\nu W}.$$

Discussion

Firstly, we consider the case of fixed taper. Here, we note that the magnitude of the dimensionless load W increases with the dimensionless taper angle ν as shown in Figure 14.18. Therefore, for a fixed taper angle α, if h_a increases, e.g. as a consequence of increased load on the bearing, ν will decrease. Consequently, unlike in the case of a positive pressure bearing, an arrangement with constant taper angle will be statically unstable, i.e. will have a negative static stiffness, much like the force of a magnet, which increases with proximity. Thus, we might use a bearing with fixed taper as a pre-load mechanism. However, to be able to use it as a bearing, we should look for another configuration to achieve a bearing system with positive axial stiffness.

Secondly, we consider the effect of tilt. For this, let us note that, when $\nu \to 0$, $X_\mathrm{c} \to 2/3$ and when $\nu \to \infty$, $X_\mathrm{c} \to 1/2$. In other words, the centre of pressure shifts to the left with increased tilt (or taper) angle, as a consequence, e.g. of increased load. Thus, if we fix the position of loading point somewhere between those two extremes, allow the bearing to tilt, and we let the load vary, then this will lead to negative tilt stiffness. That is, the pad will keep on tilting until it makes contact with the ceiling.

Thus, in the absence of positive axial and tilt stiffness, it is impossible to achieve a statically stable application of a tilting or fixed pad hanging in a manner similar to the thrust bearing type, which is inherently stable statically. Therefore, we need to develop a way to attain static stability, if we wish to achieve a successful application of hanging self-acting bearings. To show the direction in which this can be done, we consider the arrangement depicted in Figure 14.19, where we have incorporated wheeled rotation point at the inlet to the bearing. One can

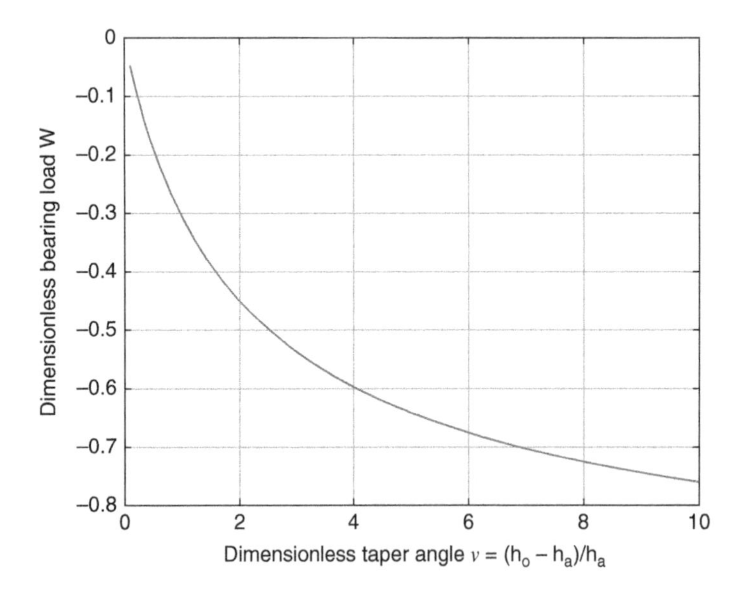

Figure 14.18 Dimensionless load as a function of the normalised taper angle.

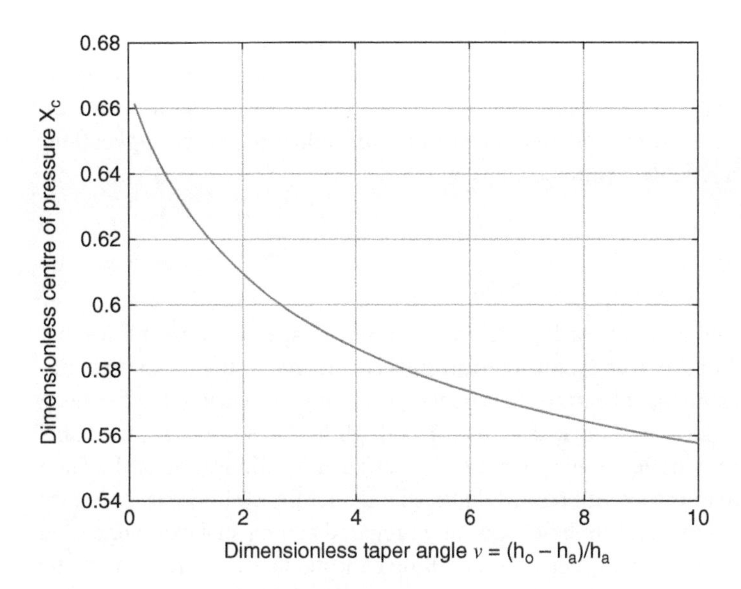

Figure 14.19 Dimensionless centre of pressure.

show that this leads to a statically stable arrangement when the point of load application is situated at or beyond the largest value of the centre of force, namely $x_w = 2/3\ L$, so that the reaction force N of the wheel is positive (i.e. in the same direction as the load). In that way, we have:

$$F_b = N + W,$$

$$F_b x_b = W x_w.$$

Since it might not be practical to implement a wheeled rotation point in the pad (except when these wheels are used to drive the bearing along the ceiling), we can replace this arrangement by an air-film device as shown in Figure 14.20. Here, we have equipped the tilting pad with convergent-gap part at the inlet, whose force, being equivalent to the reaction force of the wheel, increases with decreasing minimum gap value.

Figure 14.20 Statically stable hanging tapered pad with the wheel being replaced by a positive-pressure air gap.

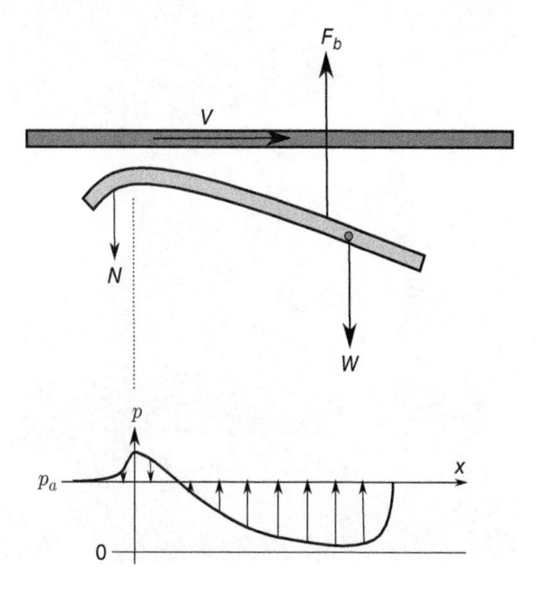

The equations describing this system are straightforward and we leave them to the interested reader to derive.

A more pertinent question concerns the possible applications of such a system. One obvious application is the use of this system as a means to pre-load a thrust bearing in order to obtain a large net axial stiffness, for instance. Flying-heads for data storage could be another. However, as mentioned in the Introduction, other applications could present themselves, once this becomes a reliable technology.

14.11 Conclusions

This chapter has roughly situated the problem of hanging bearings, showing the advantages and possibilities that they can offer. It has also shown the potential of the over-expansion bearing as compared to the other possibilities, namely, the vacuum bearing and the thrust-magnet bearing. This new type of hanging bearing that has been developed at KU Leuven is very simple and effective, though it suffers from high air consumption. A reliable working model has been derived, based on turbulent flow theory, for use as a design and optimisation tool, which at the same time furnishes a useful case study in turbulent pressure-induced channel flow. This bearing is easy to implement, requiring relatively large gap heights and thus not specially prepared ceiling surfaces. Suggested applications are: for hanging robot arms in micro assembly cells; for handling delicate objects such as magnetic discs, chocolate etc.; flat-wall climbing devices such as window-cleaning robots; or simply as elements to pre-load other thrust bearings, when light weight is required. A miniature model car that moves on a wooden-board ceiling (2 m × 3 m) has been constructed at KU Leuven/PMA for demonstration, Figure 14.21. Finally, some considerations pertaining to self-acting hanging bearings have been provided, which qualify and quantify basic behaviour and design issues. Moreover, basic considerations have been provided for designing self-acting hanging-bearings, which might be interesting for use as pre-loading systems.

Figure 14.21 Miniature car equipped with hanging bearing travelling on the ceiling (during the open days at KU Leuven, PMA 1992). Source: Photo Archive KULeuven, PMA

References

Al-Bender, F., (1991). On the development of Bernoulli-"hanging" aerostatic bearing. Internal report, K.U. Leuven, Faculty of Applied Sciences, Dept. Mech. Eng., Div. PMA.

Al-Bender, F. and Van Brussel, H., (1994). Development of hanging externally pressurized air bearings. In: *Proceedings of the 3rd International Conference on Ultraprecision in Manufacturing Engineering. ISBN: 3-926832-11-8*. (ed. Weck M and Kunzmann H): 183–186. Verlag Franz Rhiem Duisburg.

Anon (1989). Speed and precision from novel assembly robot. *Assembly Automation* 9(2): 85–87.

Constantinescu, V.N., (1969). *Gas Lubrication*. ASME.

Dries, J. and Stevens, B., (1992). Ontwerp en bouw van hangende luchtlagers, (design and construction of hanging air bearings) Master's thesis KU Leuven, Mechanical Engineering Department. (In Dutch).

Gross, W.A., (1962). *Gas Film Lubrication*. John Wiley & Sons.

Massey, B., (1968). *Mechanics of fluids*. D. Van Nostrand Company Ltd.,London.

Moller, P.S., (1966). Radial flow without swirl between parallel disks having both supersonic and subsonic regions. *ASME – Journal of Basic Engineering* 147–154.

Mori, H., (1961). A theoretical investigation of pressure depression in externally pressurized gas-lubricated circular thrust bearings. *ASME Journal of Basic Engineering* 201–208.

Willis, R., (1830). On the pressure produced on a flat plate when opposed to a stream of air issuing from an orifice in a plane surface. *Transactions of the Cambridge Philosophical Society* 3(1): 129–141.

15

Actively Compensated Gas Bearings

15.1 Introduction

We have seen that air bearings offer important advantages over conventional bearings, which emanate from their low friction and high speed capabilities combined with their high accuracy and long life. That is why aerostatic bearings are widely used in precision systems. Those advantages of air bearings cannot however always be realised to the full owing to some limitations, in particular, their low specific stiffness and their liability to develop negative damping, which may lead to pneumatic hammering in certain working conditions. The question then arises as to how to overcome these limitations, or even turn them into advantages.

The necessary first step in bearing design is the prediction of the static characteristics: this provides the general domain of feasibility of a bearing application, especially in regard to load carrying capacity, or mean bearing pressure. Such a domain is bounded by the minimum allowable film thickness and the maximum allowable pressure. Knowledge of the dynamic behaviour of the air film and its effective exploitation in modifying the dynamic bearing force may lead to extending the domain of feasibility, viz. by means of active bearing compensation.

We have also seen that the air film characteristics depend on such design parameters as feed (and exhaust) pressure and film geometry, which, although usually taken to be fixed, can be dynamically varied by applying suitable actuators. In this way, the dynamic characteristics can be modified by means of *active compensation* or servo control. This possibility, which is encouraged by recent developments in actuator/sensor/controls technology and the advances made in the mechatronics methodology, should lead to an extension of the range of application of air bearings into the realm of *smart devices*.

Another category pertaining to the exploitation of dynamic behaviour to generate load-carrying capacity is the use of a fluid film as a "squeeze" film bearing, damper, or both. Here, the fluid flow in the film is caused solely by the reciprocating motion of the bearing surfaces so as to generate a net load carrying force when the oscillation amplitude is sufficiently large (owing to the non-linear, compressible, behaviour of the air film).

The key to utilising all these aspects and possibilities is a good fundamental understanding of the dynamic behaviour of air bearing films, which has been treated in Chapters 7 and 10. The present chapter will utilise that knowledge in order to examine the various possibilities to achieve active compensation and subsequently to demonstrate the feasibility of this by designing and testing prototypes of a smart air bearing.

This chapter will thus present the general method of active dynamic compensation as a means of achieving new and higher levels of bearing performance.

In the following sections, we shall first formulate and discuss the essentials of active bearing film compensation. Thereafter, we will apply that knowledge to studying two prototype applications belonging to thrust and journal bearings, respectively. This topic is complemented by the case of active control of the traction force in the plane of the bearing, which finds important application e.g. in the field of wafer manipulation. Finally, we devote a section

Air Bearings: Theory, Design and Applications, First Edition. Farid Al-Bender.

Companion website: www.wiley.com/go/AlBender/AirBearings

to the treatment of squeeze-film levitation, which is yet another interesting application of active air bearings. A detailed case study of a precision slider on active thrust bearings is presented separately in Chapter 16.

15.2 Essentials of Active Bearing Film Compensation

The basic idea behind active compensation is to use external actuators in order to affect, generally enhance, the (dynamic) bearing forces. Active compensation belongs to the field of *mechatronics* that is opening new possibilities in system design (Al-Bender 2009). The past two decades have witnessed increased activity in active control of air bearings, in particular for precise positioning applications (Al-Bender and Van Brussel 1994, 1997, 1998; Horikawa et al. 1991; Lee and Gweon 2000; Sato and Harada 1988; Shimokohbe et al. 1986; van Rij et al. 2009). Recently, Raperelli et al. (2016) presented a review of active compensation of aerostatic thrust bearings.

Figure 15.1 defines the notation of a circular bearing with variable film and pressure parameters. Figure 15.2 depicts the general mechatronics framework of an active air bearing system. Figure 15.2(a) shows the different possibilities of "activating" an aerostatic bearing. The aerostatic bearing supports a bearing mass (platen) that constitutes part of a mechanical system with an assumed known dynamic behaviour. For the purpose of active compensation, the aerostatic bearing system comprises one or a combination of active elements, namely: *support* and *conicity* actuators, and *inlet* and *outlet* pressure/flow controller. These are responsible for inducing an active bearing force. Another possibility not taken up in this scheme, owing to its rarity, is to control the fluid rheological property, namely the viscosity, such as the possibility of lubrication by a conducting gas under the influence of a magnetic field, as given in Constantinescu (1967).

Figure 15.2(b) is a schematic of the active compensation system. The error signal in the bearing platen height variation, or an equivalent quantity, is fed into a controller that produces a signal, which is in turn amplified

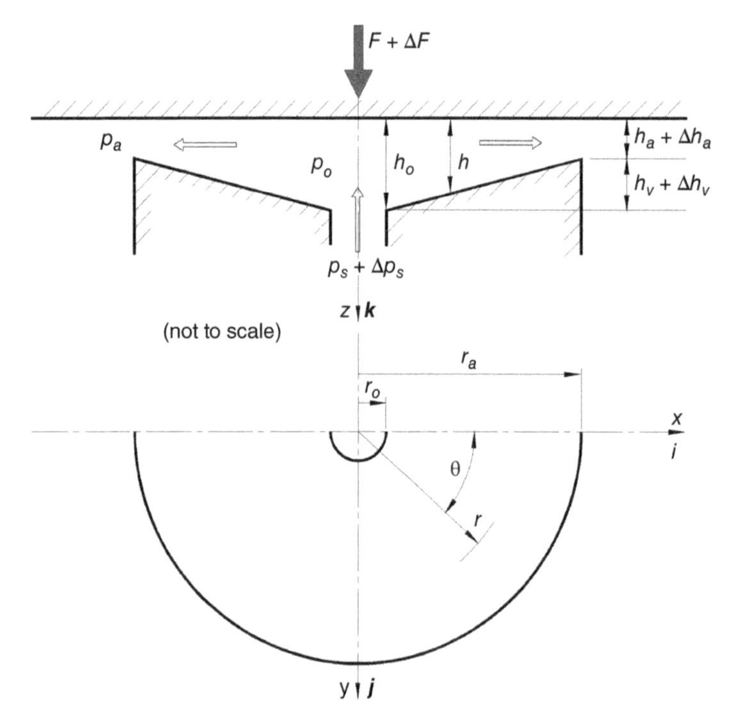

Figure 15.1 Schematic of active air bearing system. Source: Republished with permission of ELSEVIER, from Al-Bender F 2009; permission conveyed through Copyright Clearance Center, Inc.

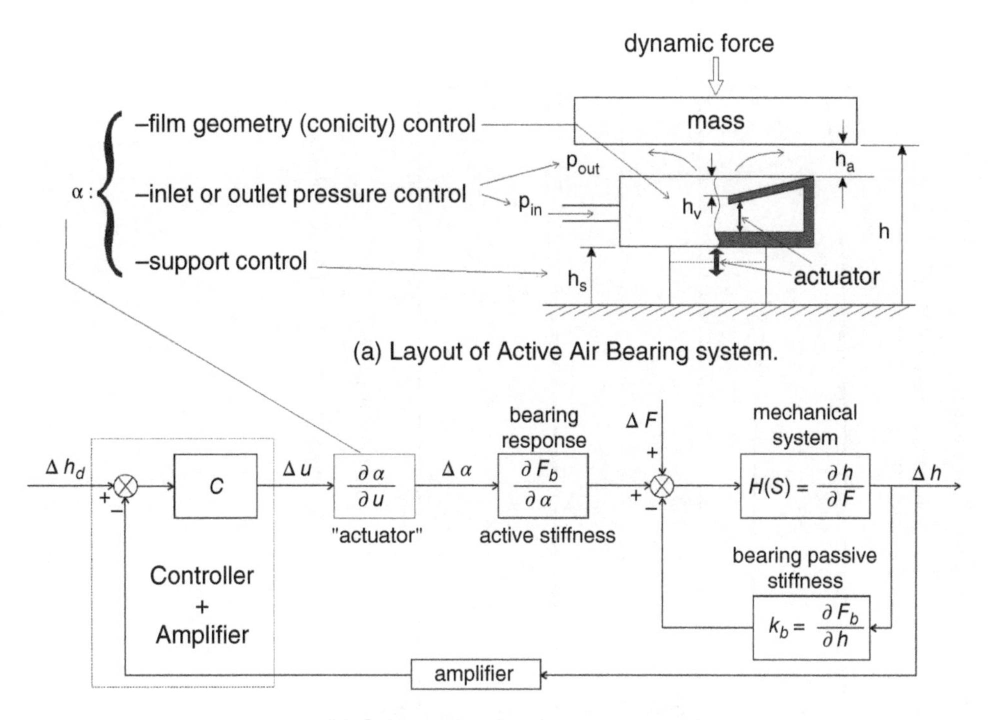

Figure 15.2 Schematic of active air bearing system. Source: Republished with permission of ELSEVIER, from Al-Bender F 2009; permission conveyed through Copyright Clearance Center, Inc.

and fed into the "actuator" to produce a variation in the respective bearing parameter. Depending on the bearing response to that parameter, a (dynamic) bearing force is created and fed into the aerostatic bearing system where it is also assumed that an external disturbance may act on the system. It will be evident below that the control problem is influenced by the choice of the actuation method, owing to the different bearing response characteristics. Details of the controller design, which belong to the field of system and control theory, are outside the scope of this treatment. For this matter, the reader may find numerous textbooks, e.g. (Goodwin et al. 2001). Generally, if the disturbance to be compensated is small in magnitude, which is generally the case pertaining to precision machines, the bearing system may be linearised around the nominal working point, and a conventional PID controller will suffice for control. Such a controller consists in a linear combination of three terms operating on the error signal, namely, (i) a proportional gain term P, (ii) an integral term I, and (iii) a derivative term D. Rules to tune this type of controller, i.e. to determine the values of the three terms may be found in many classical text books such as (Goodwin et al. 2001). On the other hand, when the amplitudes are large or when the working point is variable, other more complex, adaptive and sometimes non-linear control methods may be used (Isidori 1999; Macmillan 1962).

The different actuation methods and their properties have been investigated by various authors (Al-Bender 2009), and are discussed in more detail in (Al-Bender and Van Brussel 1994; Raperelli et al. 2016; Aguirre 2010). Obviously, the response of the bearing film force to the actuated parameter plays a crucial role in determining the performance of the complete system.

Film Geometry Actuation

On the basis of theoretical models outlined in Chapter 7, Figure 15.3 compares the simulated characteristics of that response for the generic cases depicted in Figure 15.4, i.e. (1) gap height variation, (2) conicity variation and

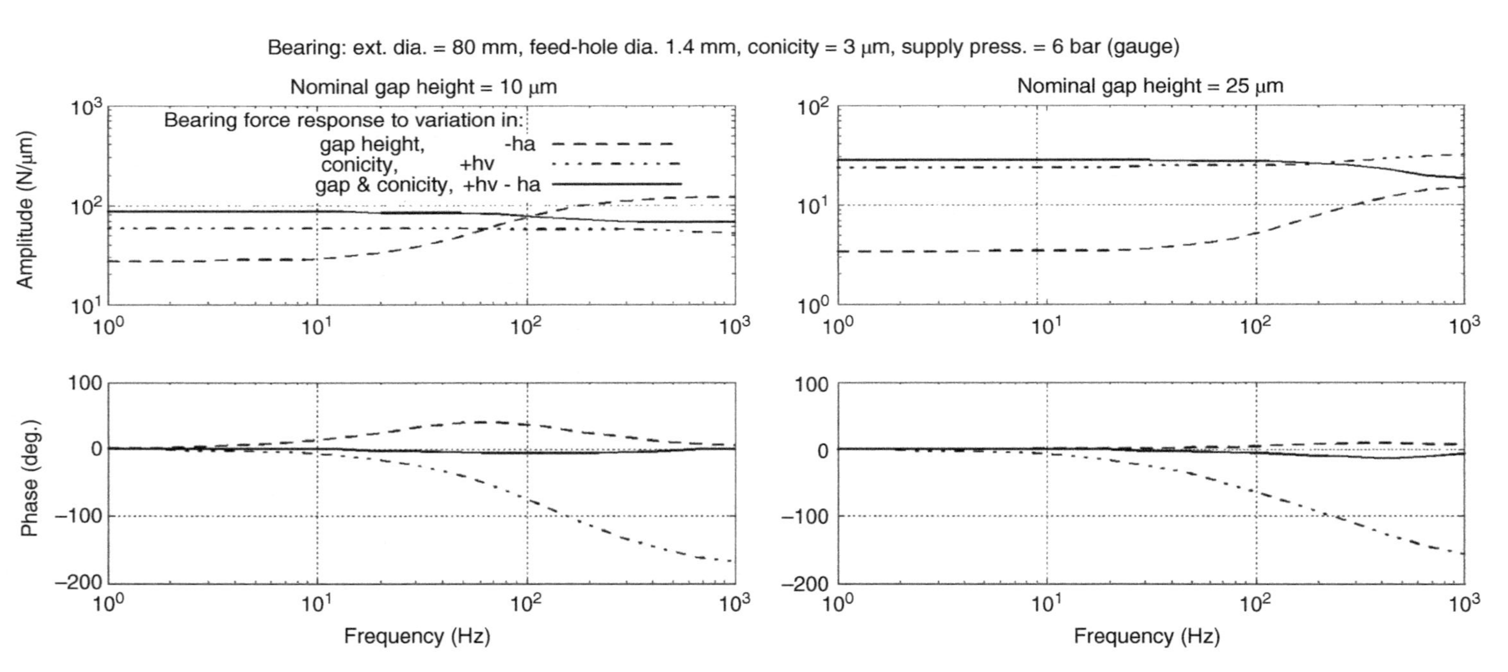

Figure 15.3 Bearing force response to variation in gap height, conicity, and their combination. Source: Republished with permission of ELSEVIER, from Al-Bender F 2009; permission conveyed through Copyright Clearance Center, Inc.

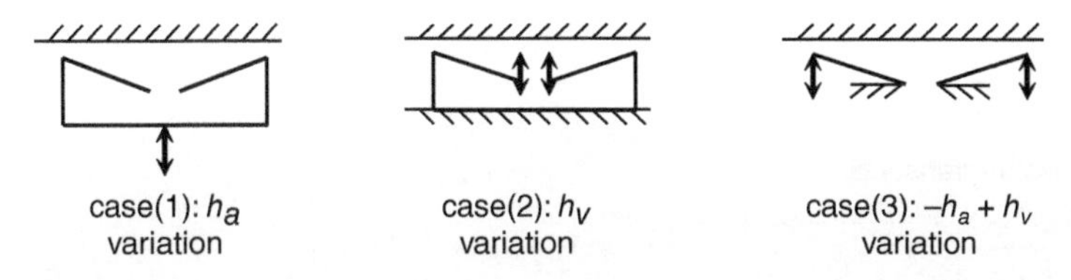

Figure 15.4 Gap variation types. Source: Republished with permission of ELSEVIER, from Al-Bender F 2009; permission conveyed through Copyright Clearance Center, Inc.

(3) an "offset" combination of the previous two which will be further treated in the following. (Other generic cases, pertaining to tilt etc. may be treated similarly.)

It is evident that the latter method (3) is superior to the other two on account of the high (practically constant) gain and the negligible phase shift that it offers over a large bandwidth. Another important aspect for the control problem is the actuated mass. Here, the last two methods are better than the first since only the bearing surface (a thin plate) has to be actuated.

Pressure Actuation

In regard to inlet/outlet pressure control, although the actuated mass can be made negligibly small (e.g. a small valve or diaphragm), the bearing force response has a very limited bandwidth (see Figure 15.5, and (Al-Bender and Van Brussel 1994)), unless (i) the gap is large, in which case the passive stiffness will be low, and (ii) the inlet pressure actuator is not succeeded by a large volume of air, which is difficult to realise in practice. Thus, this type of bearing control is mainly used in quasi-static situations (Al-Bender and Van Brussel 1994).

In the remainder of this chapter, various active control cases will be considered with the purpose of discovering and evaluating the various possibilities and potentials.

15.3 An Active Bearing Prototype with Centrally Clamped Plate Surface

Referring to Figure 15.6, (see also Al-Bender and Van Brussel (1997)), this active air bearing design is based on an already existing mechanically self-compensating bearing (of Dr. Blondeel) as described in Snoeys and Al-Bender (1987). It consists, in the present prototype, of a thin plate surface (80 mm diameter) clamped at the centre and free to move at the outer edge where it is reinforced by a thick collar that is pre-loaded on three, axisymmetrically disposed, piezo-electric actuators. An equal displacement of these actuators causes rigid-body motion of the collar and, consequently, a change of the conicity of the bearing (as indicated by the dashed line). Note that, for a fixed platen position, an increase in the conicity is accompanied by a decrease in the nominal (e.g. the external) film thickness. This design corresponds thus to case (3) of Figure 7.2(b). The first eigenfrequency of the plate is around 1.4 kHz. For the purpose of compactness, a capacitive displacement sensor (8 mm external diameter) is placed in the middle of the bearing (as an integral part) with a thin gap around it forming a feed annulus. This sensor has a sensitivity of 400 mV μm^{-1} and the analogue signal can be resolved to better than 1 mV. The control part of the system consists in a simple analogue PI controller.

15.3.1 Simulation Model of Active Air Bearing System with Conicity Control

Based on the knowledge gained in regard to the dynamic behaviour of the air film in passive and active conditions, it is possible to construct an equivalent linear representation of the active bearing-platen system as depicted in Figure 15.7. The input to the system is the actuator displacement, X_{in}, the output is the platen displacement, X_o. The active bearing stiffness is k_v, which is estimated to be 16 and 41 N μm^{-1} (of membrane displacement, X_1), for

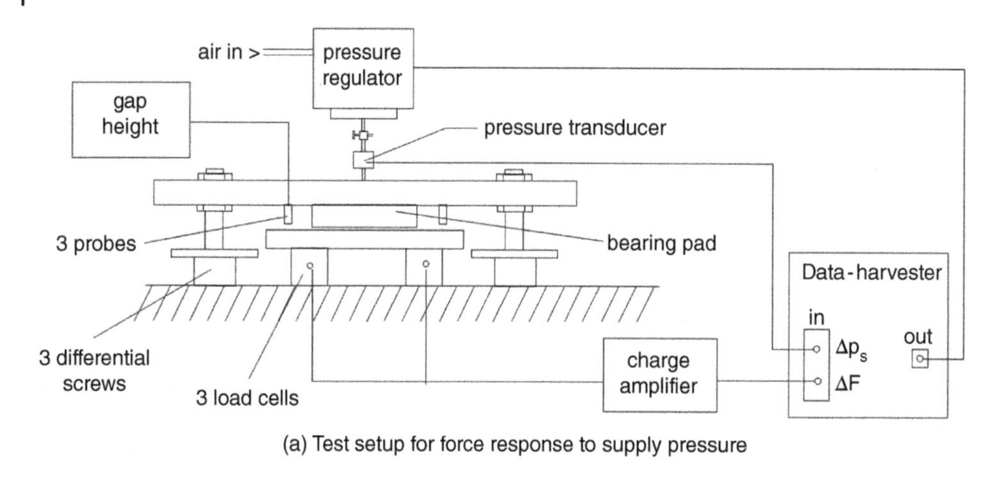

(a) Test setup for force response to supply pressure

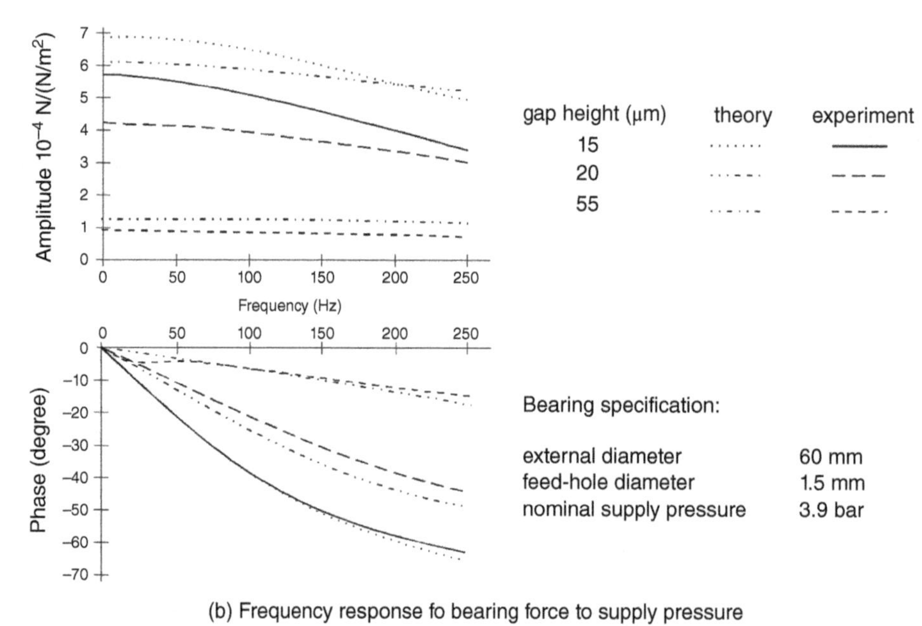

(b) Frequency response fo bearing force to supply pressure

Figure 15.5 (a) Test set-up for measuring force response to supply pressure, (b) frequency response of bearing force to supply pressure. Source: Republished with permission of ELSEVIER, from Al-Bender F 2009; permission conveyed through Copyright Clearance Center, Inc.

gap heights 25 and 10 μm, respectively. This model is then used to derive appropriate values for the proportional and integrative gains of the controller.

15.3.2 Tests, Results and Discussion of the Active Air Bearing System

Three types of test are carried out:

- Test 1: Stiffness at a given nominal gap of open-loop and closed-loop system, $F_{\text{disturb}}/X_{\text{out}}$, see Figure 15.8.
- Test 2: Transfer Function of the closed-loop system, $X_{\text{out}}/X_{\text{in}}$, see Figure 15.9.
- Micro trajectory tracking, see Figure 15.10.

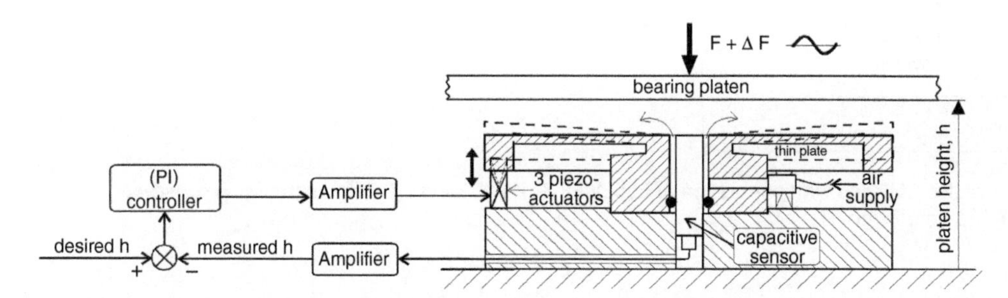

Figure 15.6 Schematic of test set-up comprising an active air bearing with centrally-clamped plate surface. Source: Republished with permission of ELSEVIER, from Al-Bender F 2009; permission conveyed through Copyright Clearance Center, Inc.

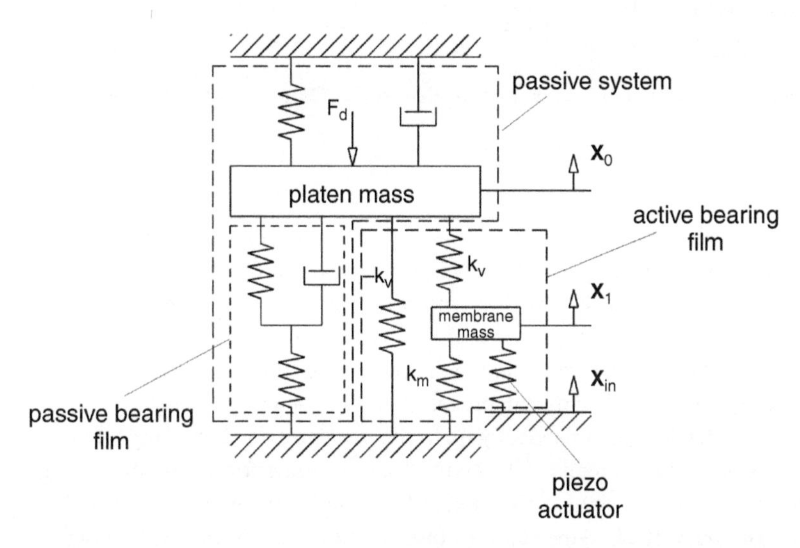

Figure 15.7 Schematic representation of the platen-active bearing system. Source: Republished with permission of ELSEVIER, from Al-Bender F 2009; permission conveyed through Copyright Clearance Center, Inc.

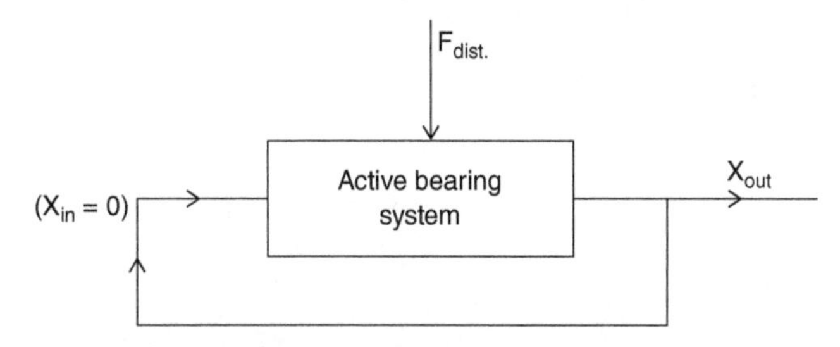

Figure 15.8 Test 1. Source: Republished with permission of ELSEVIER, from Al-Bender F 2009; permission conveyed through Copyright Clearance Center, Inc.

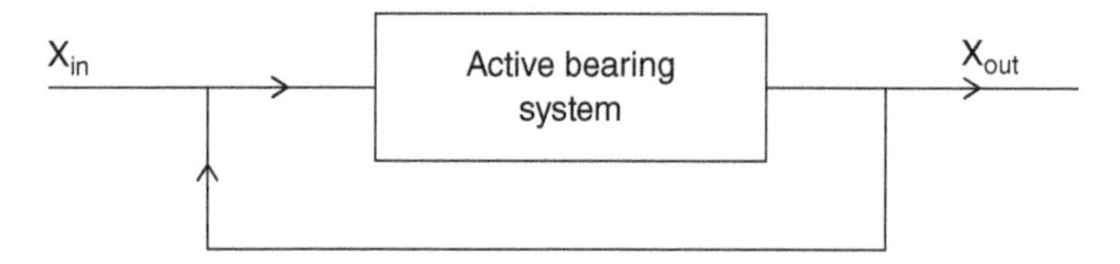

Figure 15.9 Test 2. Source: Republished with permission of ELSEVIER, from Al-Bender F 2009; permission conveyed through Copyright Clearance Center, Inc.

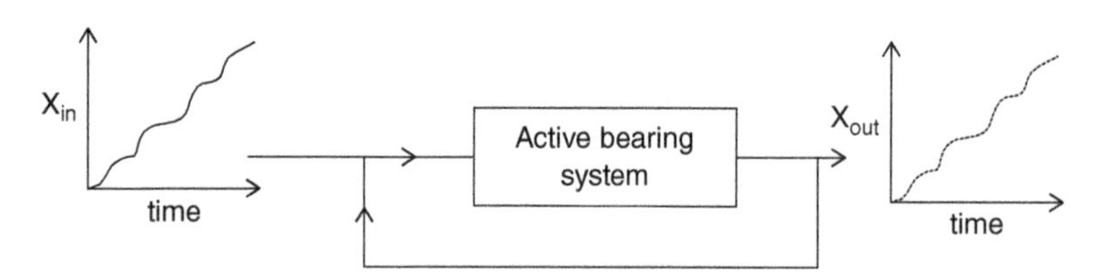

Figure 15.10 Test 3. Source: Republished with permission of ELSEVIER, from Al-Bender F 2009; permission conveyed through Copyright Clearance Center, Inc.

Test (1)

Before designing the PI controller the open-loop characteristics of the system were determined for two test gap heights viz. 10 and 25 μm. The latter values were chosen so that there was a significant difference in the passive stiffness of the bearing and, consequently, the mechanical system as a whole. It was found that, as expected, the system dynamics were governed solely by the first natural frequency of the platen system on the passive air film. Consequently, the closed-loop performance of the mechatronic system is governed (or limited) by these. The PI controller was tuned to obtain optimum compensation. (The tuning procedure is outside the scope of this work.) A test was carried out to examine the system behaviour, at a fixed air gap, in the presence of a disturbing force on the platen; in other words, the dynamic stiffness of the compensated system. The test was performed on the test rig described in Chapter 7, where the measurement accuracy is also discussed. The results are shown in Figure 15.11. By virtue of the use of integration action (I) in the controller, the static stiffness of the system is infinite for both of

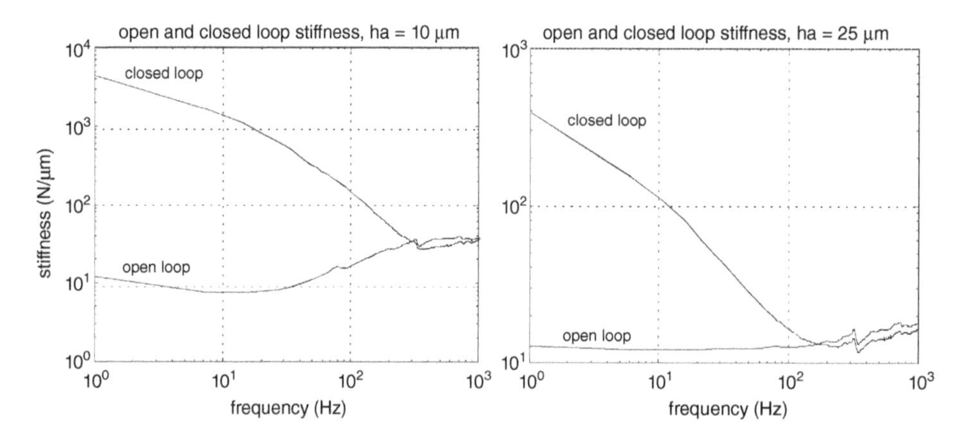

Figure 15.11 Test (1): Comparison between passive(open loop) and active (closed loop) stiffness characteristics. Source: Republished with permission of ELSEVIER, from Al-Bender F 2009; permission conveyed through Copyright Clearance Center, Inc.

the test cases. Beyond the cross-over frequencies, (300 and 200 Hz respectively,) the controller has no effect. It may be worthy of mention here that the infinite static stiffness may be obtained only on a limited bearing force range and may certainly not be confused with infinite load capacity (or bearing force). Infinite static stiffness around the working point, or gap height, means that a small change in the bearing load will leave the gap unaltered at zero frequency, i.e. that the gap will re-adapt its height at a rate that is faster than zero Hz. The range, or magnitude of bearing force variation that can be compensated depends firstly on the saturation levels of the actuators and their amplifiers, and secondly, on the ultimate force the bearing pad can sustain, which is limited by the bearing area times the supply (gauge) pressure.

Further examination of the case of 10 μm film thickness (which is the more realistic one in practice) reveals that even at a frequency of 100 Hz, there is a ten-fold increase in the stiffness in comparison with the passive value.

Test (2)

In order to confirm these characteristics, the transfer function of the closed loop system was measured. The results are depicted in Figure 15.12 and show that the closed loop system behaves like a pure, infinitely stiff spring up to almost 100 Hz (for the case of 10 μm).

Test (3)

This is the test of trajectory tracking relevant for nano-positioning systems. Here, the bearing platen was required to follow a prescribed reference path in the form of discrete steps of about 0.1 μm (each in 50 ms) over a range of 1.2 μm. (Note that on this range of measurement, the absolute accuracy of the capacitive sensor becomes irrelevant for the experiment; only the resolution matters.) Typical results are shown in Figure 15.13. The tracking error is less than 10 nm in the worst case, i.e. close to the resolution limit of the sensor, and the step response is completely free from overshoot. This same principle has been employed for the design and construction of active journal bearings for use in high speed, precision, electro-spindles for machine tool applications Al-Bender and Van Brussel (1998), see following section.

15.3.3 Conclusions

One can conclude from the above that there are various possibilities of applying active compensation methods to air bearings and that these can result in considerably enhancing their dynamic performance and diversifying their application. In particular, an active system that is based on the control of the film geometry (the conicity) has the advantages of highest gain per actuator displacement and very large bandwidth of the active air film. Moreover, infinite static stiffness is easily achieved using integrative action in the controller. This active air bearing system is particularly suited for high performance ultra-precision positioning applications. A systematic treatment of this type of bearing as well as its successful utilisation in an ultra-precision slide system has been recently presented in (Aguirre et al. 2010; Aguirre et al. 2008a, b, c, 2009; Aguirre 2010), which are summarised in Chapter 16.

15.4 Active Milling Electro-Spindle

This section describes the development of two high-frequency (HF) electro-spindles. One is equipped with passive air bearings and the other with an active front (radial) bearing to assess the effect and relevance of active compensation as an alternative to other technologies, in particular active magnetic bearings. It proposes the use of air bearings as an effective and competitive alternative to ball bearings for these applications. Air bearings are shown to lead to a spindle design that meets the required performance specifications as set by machine tool builders. In particular, owing to their relatively higher damping compared to ball bearings, another design strategy is possible

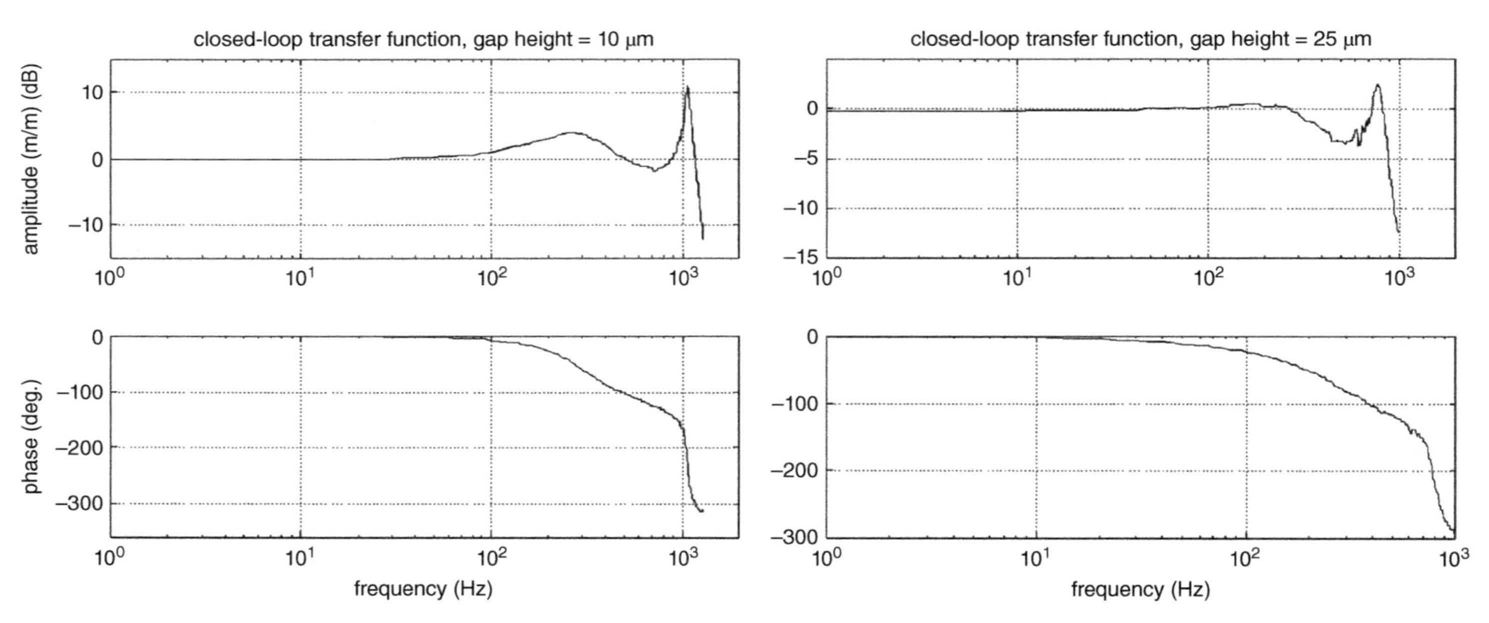

Figure 15.12 Test (2): Transfer function of the closed loop system. Source: Republished with permission of ELSEVIER, from Al-Bender F 2009; permission conveyed through Copyright Clearance Center, Inc.

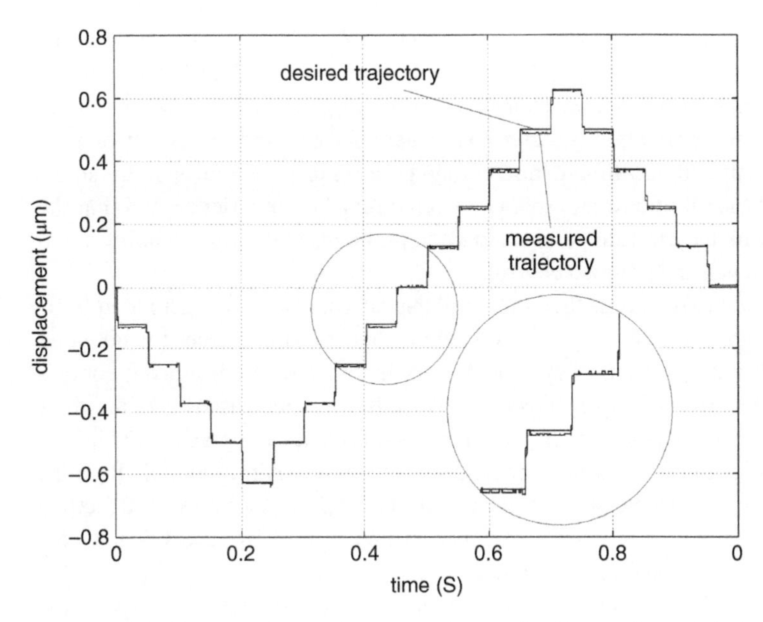

Figure 15.13 Test (3): Micro-staircase tracking results. Source: Republished with permission of ELSEVIER, from Al-Bender F 2009; permission conveyed through Copyright Clearance Center, Inc.

in which resonance of the lowest mode is allowed (with advantage) to take place within the working frequency range. Relevant to the topic of this chapter is that active bearing compensation principles, as described in the previous sections, can also be applied to the (radial) spindle air bearings in order to overcome their modest stiffness and, more importantly, to achieve high levels of precision. In that case, an infinite static stiffness is also obtained together with the possibility of enhancing the system's dynamics by feedback control. The two prototypes have been designed and built successfully yielding good performance results that are in agreement with the design values. The active front bearing of the second spindle shows infinite static stiffness and active control bandwidth of 700 Hz. It is further shown that nanometre-level axis-of-rotation error can be achieved by using repetitive-control techniques.

15.4.1 Context Sketch

Higher speed and better accuracy are becoming key words in the machine tool sector of industry. They can lead to a significant reduction in production times on more than one count: shortening machining time; reducing the number of machining and finishing stages (e.g. polishing); and, coupled with reliability, eliminating the need for quality control. High frequency (HF) electro-spindles play a central role in the drive towards this goal. In fact, a spindle represents the element of highest performance in the machine tool and at the same time the weakest link in the force chain. Apart from morphology classification (motor-spindle/electro-spindle), spindles are often classified according to the type of bearings they use since the rest is more or less common to all types. The most popular type of bearings used are rolling element bearings, more specifically angular contact ball bearings that can meet the demands of high speed/high accuracy. In discussing recent advances in ball bearing research and technology, Weck and Koch (1993) provide a table of comparison between the various bearing types. Although air bearings actually score the best in that comparison, they have given as the main reason for the use of ball bearings the fact that they have high performance density and enjoy high degree of standardisation. The speed performance of rolling element bearings is measured by the revolution factor Ndm (N = maximum speed in r.p.m., dm = bearing pitch-circle diameter in mm.) Nakamura (1996) has reviewed the evolution in rolling element bearings for high speed applications over the past 30 years showing that the Ndm factor has enjoyed a steady increase during the decade preceding that publication, reaching values higher than 3×10^6 r.p.m.×mm recently. This is owing to the intensive research and advances in bearing materials (ceramic elements and coatings) as well as new lubrication techniques such as oil-jet, oil-air and under-race lubrication. Apart from the increased cost and complexity of these new bearing systems, they appear to be reaching their ultimate limit of high performance. Moreover, in his study, Nakamura (1996) shows that nearly 20% of the input power is spent on overcoming bearing friction and that the heat generated in this process must be channelled carefully in order to avoid problems of thermal stability, not to mention questions of intricate assembly and bearing pre-load control.

With this picture in mind, it may be justifiable to examine the other available alternatives; in particular air bearings. In contrast to ball bearings, air bearings are limited by no Ndm factor: their maximum speed is bounded only by the shaft centrifugal growth and bursting speed. They provide very high accuracy for a given component/assembly accuracy owing to the air film error averaging effect. Thanks to the very low viscosity of air, their friction losses are among the lowest of all bearing types. Last, the use of air as lubricant can prove crucial in certain "clean" applications (e.g. machining of graphite) as well as being generally more environmentally friendly. Consequently, while ball bearings become too sophisticated and reach the limit of their performance, and (active) magnetic bearings remain complex and expensive, it would be worthwhile considering the comparatively low-cost and effective technology of air bearings for high speed electro-spindles.

In this context, we shall in the following, first discuss the development of two different high-speed (or high-frequency, so-called HF) spindles equipped with air bearings. The first of these was intended for wood-working which sets no high demands on accuracy, stiffness or load capacity: the basic requirement is high speed and simple (therefore low-cost) construction. The second spindle is intended for milling of moulds and dies requiring higher speed, accuracy and stiffness. Since air bearing stiffness is comparatively low, it was decided to apply the method of active compensation to (at least one of) the bearings of the latter spindle so as to enhance its dynamic stiffness whereby the static stiffness can be made infinite as is the case in active magnetic bearings. The advantages (over magnetic bearings) are (i) higher load capacity (approximately double that of magnetic bearings), (ii) simpler construction and control of the active air bearing, and (iii) that the spindle may be used with passive bearings having active control as an extra option.

We start by specifying typical required performance for each spindle, then examine the feasibility of using air bearings for these applications, outline the design procedure, and sketch the performance results obtained with each of the two prototypes.

15.4.2 Specifications of the Spindles

The force, stiffness, accuracy and power requirements for the two spindles are different owing to the different machining requirement of the two applications. The general specifications, (which have been supplied by machine tool builders), are summarised in Table 15.1 and Table 15.2 below.

Forces and Displacements

These show the typical values required for the respective applications based on experience with conventional spindles. However, as the spindle speed increases, the forces required become smaller for a given feed rate, so that we might expect these specified forces to be somewhat high for the maximum rotation speed required and the feed rates that can be realised on an advanced machine tool of today.

Speed, Power, Size and Mass

The most notable requirement here is that the spindle mass be limited to 30 kg so that the total mass of the z-axis of the machine tool is kept small enabling higher dynamics. Since the minimum size of the spindle is determined by the shaft size and motor power, a light construction material was necessary to meet this

Table 15.1 Specification of forces and displacements.

	Wood-Working spindle		Milling spindle		
	Radial	Axial	Radial	Axial	Remarks
Max. working force (N)	200	300	300	300	Force at accidents 1000 N
Distance from nose (mm)	50		100		can vary depending on tool
Max. displacement (μm)	50	50	30	10	
resulting stiffness (N μm^{-1})	4	6	10	30	

Table 15.2 Specification of speeds, power, size and mass.

	Wood-working spindle	Milling spindle
Operating speeds	0–36 000 rpm	0–45 000 rpm
Power	8 kW @ 12 000–36 000 rpm	12 kW @ 12 000–45 000 rpm
Max. tool mass	2 kg up to 9000 rpm	2 kg up to 9000 rpm
Min. tool mass	0.5 kg @ 36 000 rpm	0.5 kg @ 36 000 rpm
Max. spindle mass	30 kg	30 kg
Max. spindle length	500 mm	525 mm
Spindle diameter	165 mm (customised shape)	140 mm h6 (cartridge type)

requirement. Consequently, the spindle housing was made in aluminium. While it proved possible to meet the rigidity requirements using aluminium, it remains, owing to its softness, not suitable for building prototypes that have to be assembled and dissembled frequently.

Stiffnesses

Air bearing calculations show that the following bearing (passive) stiffnesses are feasible (using a supply pressure of 6 bar):

- Front bearing (dia. = length = 60 mm): 40–70 N μm^{-1}, depending on speed and eccentricity.
- Rear bearing (dia. = length = 45 mm): 30–50 N μm^{-1}, depending on speed and eccentricity.
- Axial bearing (area $\approx 6000 - 9000$ mm^2): 70–100 N μm^{-1}, depending on displacement and size.

When shaft bending (of steel, average dia. $\approx$ 50 mm) is included, the required stiffness (of 10 N μm^{-1}) at tool tip is easily obtained. The axial stiffness is well above the requirement.

When the front radial bearing is designed as an active element (see further below), its static stiffness becomes infinite thereby increasing the stiffness at the nose substantially. Ideally, the rear (radial) bearing as well as the axial bearing may be made active, as is the case in active magnetic bearing spindles. However, since the contribution of the rear bearing to the nose stiffness is limited, it is sufficient, in the first instance to use a passive bearing. Likewise, the stiffness and damping of the passive axial bearing are high enough for the desired performance; however, this can also be made active if infinite static axial stiffness is desired. Thus, it can be seen that the use of air bearings, in their passive or active forms, allows a high degree of flexibility in design.

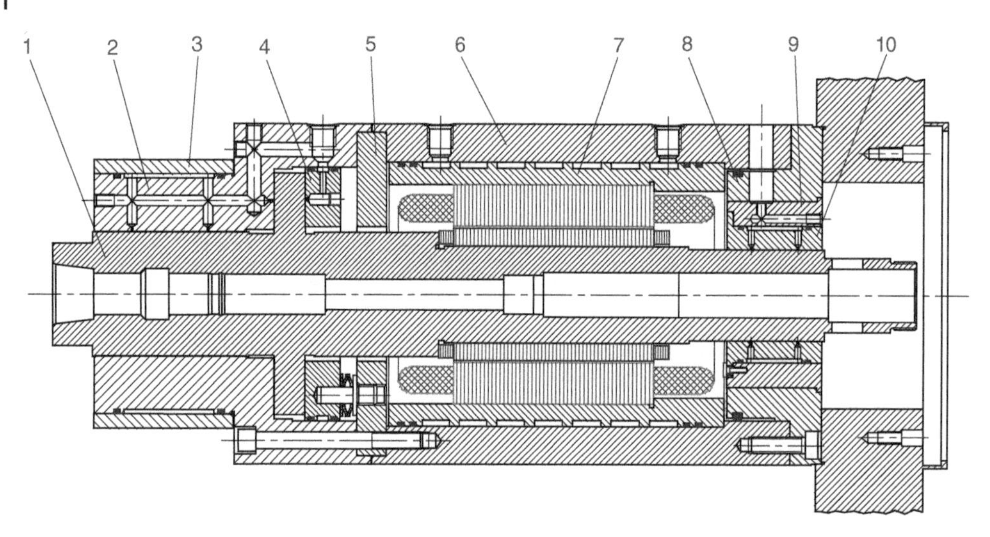

Figure 15.14 Spindle with passive air bearings for wood-working. Source: Al-Bender F and Van Brussel H 1998.

Discussion: Where Active Compensation Can Make the Difference

Regarding static characteristics, one can show that the use of air bearings can result in an optimal and proportionate spindle design. Let us note that the (very common) use of the stiffness at the nose as an indication of spindle stiffness (and hence performance) is somewhat misleading since the resulting stiffness at the tool tip is much less than nose stiffness: long slender tools together with tool holders (e.g. HSK) are commonly used in mould and die milling and their stiffnesses are connected in series. We may, therefore, pose the following two questions (and suggest possible answers to them):

1) Why is the use of air bearings not often considered for milling electrospindles? The answer lies probably in their low load capacity (rather than their low stiffness). However, for high-speed milling, especially in finishing operations, the required feed forces are relatively low, for a given cutting power. Therefore, the only remaining requirement would be to make the air bearings more robust against accidents.
2) What is the significance of high bearing stiffness despite its limited contribution to tool-tip stiffness? Here, the prime motivation appears to be the enhancement of spindle shaft dynamics, in particular, to shift the frequency of the first shaft modes (which are predominantly rigid body modes) outside, viz. above, the working frequency range: ball bearings are very poorly damped so that resonance amplitudes may attain very high values. In contrast, air bearings can show significantly higher damping values (e.g. low-frequency damping ratios of 15% and higher are typical) so that resonance within the working frequency range can be tolerated, when carefully designed to avoid whirl instability. In that case one can argue that a better spindle design can be obtained by increasing the shaft inertia and reducing the bearing stiffness. In so doing, a higher dynamic stiffness is obtained (by virtue of the inertia line) as depicted in Figure 15.15. Moreover, the frequencies of shaft bending (higher) modes are shifted further upwards. The only drawback of this approach would be the low static stiffness. This is exactly where active compensation can come on the scene: using actively controlled bearings, the static stiffness can be made infinite and the resonances actively damped, see dashed line in Figure 15.15, which conveniently bridges the low-frequency stiffness region to the high-frequency inertia one so that the compliance is acceptably low everywhere.

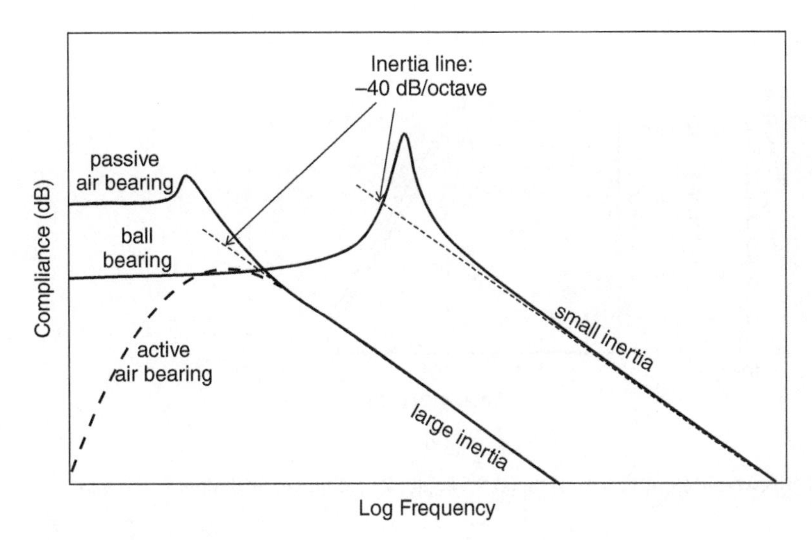

Figure 15.15 Comparison of different design possibilities employing passive and active bearings. Source: Al-Bender F and Van Brussel H 1998.

Passive Dynamic Characteristics

Radial Direction

Figure 15.16 shows the frequency response behaviour of a rigid shaft supported on radial air bearings for two different tool masses (0.5 and 2 kg). The front and rear bearing dimensions are:

- front bearing, dia. = 60 mm, length = 70 mm
- rear bearing, dia. =45 mm, length = 50 mm.

The first, and most prominent, natural frequency, which corresponds to a tilting mode, is in the range 400 to 450 Hz depending on the tool mass. With low tool mass, the resonance is better damped than with high tool mass. The damping ratio is about 20%. The static stiffness at the tool tip is about 45 N μm^{-1}. The actual stiffness will be about 14 N μm^{-1} when shaft and tool flexibility are taken into account. Therefore, a resulting stiffness of more than the specified value is expected. At resonance, which is sufficiently well damped, the rigid shaft stiffness will only be reduced by about a half in comparison with the static stiffness (in the case of the smaller tool mass), which would result in about 10 N μm^{-1} at the tool tip, i.e. still within the requirements.

The provision of active compensation of the front bearing can lead to infinite static stiffness of that bearing. However, since shaft and tool flexibility act in series, only a doubling of the stiffness at tool tip may be expected. The utility of active bearing compensation would therefore be more in the direction of: (i) fine tool positioning, for precision applications, and (ii) improving the dynamics of the system, e.g. adding more damping to the principal resonance mode (cf. Figure 15.15 above).

Axial Direction

Here, only the rigid shaft mode is relevant. The stiffness of the axial (thrust) bearing ranges between 70 and 100 N μm^{-1}, depending on the axial load. With a shaft mass of about 12 kg, the resonance frequency is in the range of approximately 385–460 Hz. The damping, which is both frequency and load dependent, is sufficient

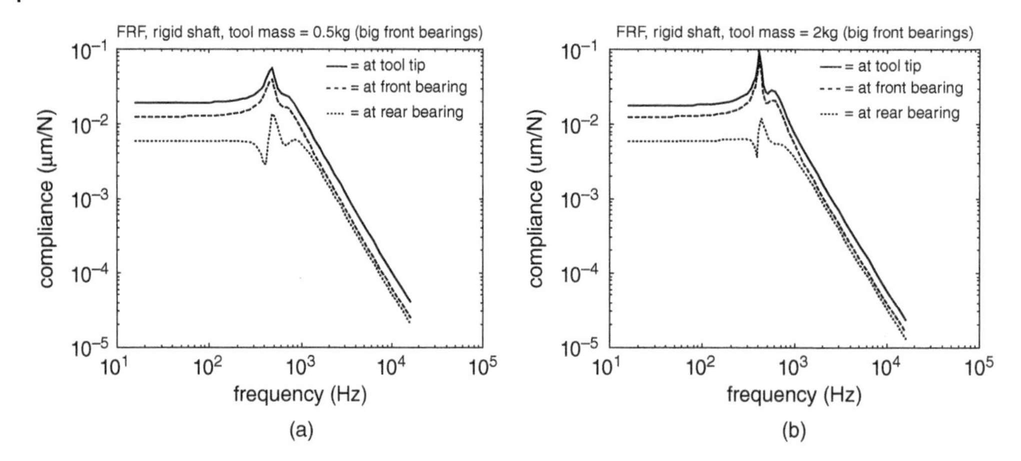

Figure 15.16 FRF of rigid shaft system, with two different tool masses, (a) 0.5 kg, (b) 2 kg. Source: Al-Bender F and Van Brussel H 1998.

to round the resonance peak off: a damping ratio of more than 20% can be obtained. Note that the damping (ratio) can be increased by reducing the stiffness, which seems to be a universal rule, especially with air bearings. Since the stiffness is more than the required value, it may be permissible to reduce it in order to obtain higher damping.

In the following sections the two prototype spindles are described together with their performance characteristics.

15.4.3 Spindle With Passive Air Bearings

A cross section of the spindle is shown in Figure 15.14 comprising a hollow shaft with an HSK 50F taper (1), front and rear journal bearings (2) and (10), and a thrust bearing (4). The journal bearings are of the double row inherently compensated type which have usually lower stiffness than orifice compensated bearings but much higher damping. The thrust bearing is preloaded on one side with low stiffness (constant force) springs for easy assembly and self-adjustment. This bearing is placed on the front side of the spindle so that thermal expansion of the shaft will have little effect on the tool position. The spindle is driven by an 8 kW asynchronous motor that is water-cooled at the stator. The maximum design speed of the motor is 36,000 r.p.m. The radial stiffness (measured at the nose of the spindle) and the axial stiffness are around 38 N μm^{-1} and 100 N μm^{-1}. Both radial and axial run-outs are below 1 μm.

During idle running tests, two resonance frequencies were identified: 394 Hz, and 467 Hz corresponding to tilt and axial modes of the shaft; both were well damped.

15.4.4 Active Spindle

This spindle differs from the one described in the previous section by having an active front radial bearing; though the passive characteristics of both spindles are comparable. This active bearing comprises a compliant bearing surface (10), see Figure 15.17, that is supported on four rows of two piezo-actuators (11). When these are powered, the bearing surface is deformed in a controlled manner to induce a radial force in the air film acting on the shaft (see below).

15.4.5 Repetitive Controller Design and Results

Including a repetitive controller in the feedback loop constitutes a common way to improve the attenuation of periodic disturbances, which are the most relevant when considering axis-of-rotation errors.

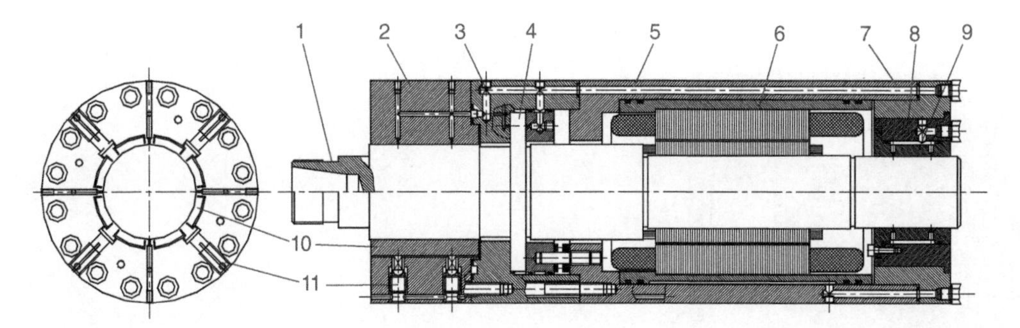

Figure 15.17 Spindle with active front air bearings intended for mould and die milling. Source: Al-Bender F and Van Brussel H 1998.

As periodic disturbances are also characteristic of spindle applications, repetitive control is not new in the field of active bearing control. In other active air bearing applications, repetitive control has been applied to overcome the low stiffness and damping of the air film (Horikawa and Shimokohbe 1990; Horikawa et al. 1992).

However, most repetitive controllers enforce perfect suppression of the periodic disturbances without considering the consequences of the performance trade-off on the overall performance.

In repetitive control, the Bode sensitivity integral dictates a trade-off between improved suppression of periodic disturbances and degraded performance for non-periodic inputs. Pipeleers et al. (2009) have experimentally demonstrated the implications of this trade-off by applying a recently developed repetitive controller design approach to reduce the error motion of the axis of rotation of our active spindle. This design methodology translates the performance trade-off into trade-off curves between a non-periodic and periodic performance index, of which the practical relevance is illustrated by the obtained experimental results.

These results show a suppression of the axis-of-rotation error from around 80–100 nm r.m.s., without control, down to about 10 nm r.m.s., with the well-tuned controller turned on, and this during run-up and run-down in the speed range 900–1200 r.p.m.

Let us thus conclude that, as in the flat-bearing case, active control can result in extremely fine performance also in the journal-bearing case.

15.5 Active Manipulation of Substrates in the Plane of the Film

The previous applications have been concerned with motion control in the direction normal to the plane of the air film. In this section, we examine the case of active motion control in the plane of the bearing. We have seen in Chapter 8 that the viscous hydrodynamic/aerodynamic action in a fluid film can be interpreted as a viscous pump in that the tangential motion of one bearing surface relative to the other results in pumping the fluid along the moving surface: this leads to a pressure build-up when the gap is convergent (or tapered), which forms the basis of hydrodynamic action. We saw that the reverse situation also occurs when one imposes a pressure gradient along the gap, (e.g. by feeding on one side of the bearing and exhausting on the other); i.e. this will induce a shear stress on the bearing surfaces causing a "viscous motor" action.

This principle has been employed in the past for constructing rotary viscous drives, or so-called "laminar motors", for use in precision lathe spindles so as not to induce any disturbance forces, see (Chen and DeBra 1987; Delhaes et al. 2009; Wegener et al. 2017; Wu et al. 2002).

Recently, this same principle has been applied, with success, to the planar, non-contacting manipulation of thin, delicate objects or substrates such as semiconductor wafers (van Rij et al. 2009). The principle is depicted in Figure 15.18. In this concept, the substrate is levitated on a thin air film between it and a system of square

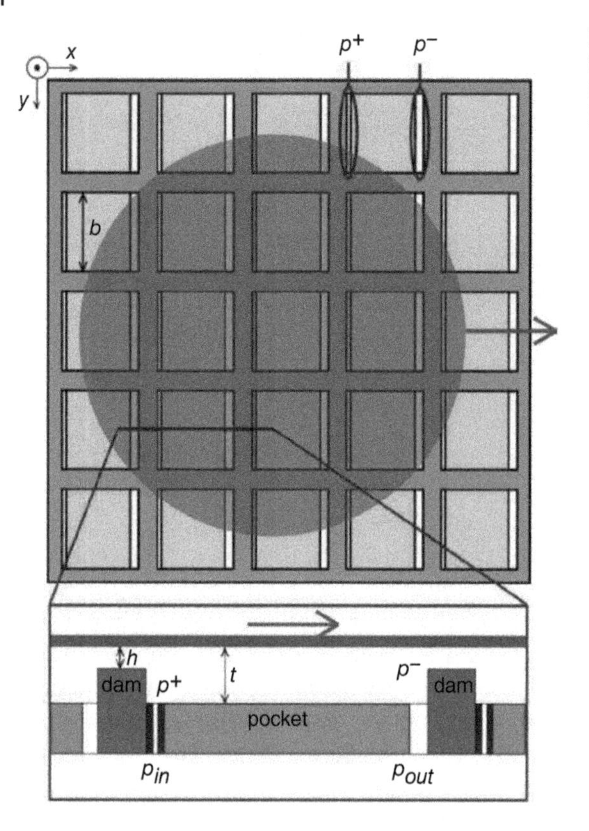

Figure 15.18 An array of air actuators for contactless wafer handling and positioning. Source: van Rij J, Wesselingh J, van Ostayen R, Spronck J, Schmidt RM and van Eijk J 2009 © 2009 Elsevier.

air-bearing pads, or cells. In each of these pads, the air is fed on one side, with a pressure p^+, and exhausted on another, with a pressure p^- in a controlled way. Assuming incompressible flow (which is valid for small working gauge pressures), and uniform conditions along the width of the pad, the shear traction force on one cell is given by:

$$F = \frac{b(t-h)}{2}(p^+ - p^-),$$

where, b is the width of the pad, t is the gap height in the recess of the pad and h is the gap height above the lands, or "dams" (we are in Holland!). This traction force acts on the substrate to accelerate it (in the absence of other external forces).

On the other hand, the load capacity is proportional to the mean pad pressure $(p^+ + p^-)/2$. Thus, by changing the pressure difference while maintaining the mean value constant, one can vary the traction force on the substrate, while keeping it afloat at a constant height. Furthermore, by incorporating cells with orientations in the four cardinal directions, one is able to apply the traction forces in any desired direction in the plane of the substrate. A laboratory prototype of this system, in which the pressures to the cells were controlled by servo valves, yielded the following characteristics:

- Maximum acceleration of 600 mm s^{-2}, in a positioning range of ±5mm.
- Positioning bandwidth of 50 Hz.
- 6 nm servo error for planar degrees of freedom.

Currently, in order to control the wafer in 6 degrees of freedom, the actuators are divided into four groups that each provides uni-directional actuation and part of the vertical support. The p_{in} and p_{out} of these groups are controlled separately by eight piezo valves. The size of valves must be increased when the number of actuators in

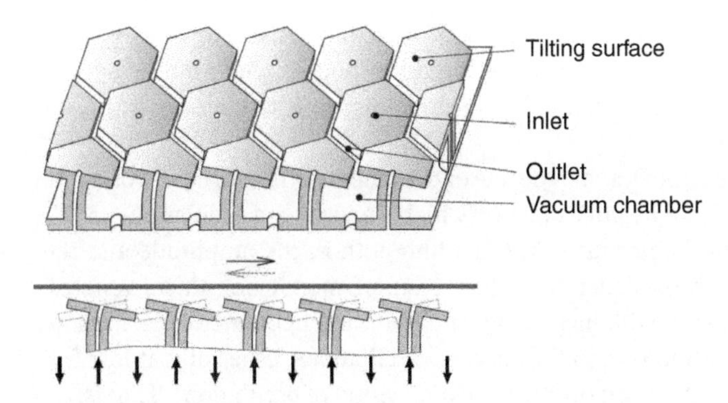

Figure 15.19 The flowerbed planar contactless actuator arrangement. Source: From Vuong PH, 2016.

these groups increases. This leads to a reduction of the valve bandwidth. Furthermore, because these valves are placed externally, relatively far upstream of the actuator, they might cause some delay in the system response, see Section 15.2. Hence, it is difficult to increase the control bandwidth using these external valves. Thus, a new concept was proposed that does not suffer this disadvantage, by depending on gap-geometry control rather than pressure control. In analogy with the results obtained in Section 15.2, we shall see here too that control of the film geometry (rather than the feed or exhaust pressures) offers the fastest possible response.

The New Concept

Examining the more general equation for the traction force acting on the bearing surfaces (see Chapter 8), reveals that this force depends also on the geometry of the air film. In particular, if the gap is tilted in one direction, the force will follow the same direction and increase with the amount of tilt angle.

After developing an analytical model, and comparing the performance and the structural complexity of various possible designs, one of them was chosen for further research and development. This variant uses a tilting action of surface parts to control the direction and the magnitude of flow, and thus the traction force, in the air film. This principle, called the "flowerbed" is depicted in Figure 15.19. The actuator is composed of hexagonal air-bearing cells, resembling flowers with long stems. Acting on these stems with differential shear plates enables tilting them in any desired direction. The cells can thus be collectively tilted using only a limited number of (piezo-electric) actuators. This system yields very high performance in terms of traction force range (which translates to acceleration), bandwidth and accuracy. Further details may be found in (Krijnen et al. 2017; Vuong 2016).

15.6 Squeeze-Film Bearings

Squeeze-film (SF) bearings are a special type of active air bearing in which load carrying capacity is induced by transverse oscillation of one or both bearing surfaces relative to one another, i.e. motion normal to the plane of the film.

This type of bearing requires then no external pressurisation or aerodynamic shear, but relies entirely on the squeeze action; that is, motion in the normal direction to the gas film. It thus has some advantages over the previously mentioned bearing types, namely the absence of a pressurised gas source or high sliding speed; but also some obvious disadvantages arising from its more complex structure, limited load capacity and costly components.

This class of gas bearings is not new, dating back to the 1960s (Wilcock 1969; Gross 1963; Langlois 1962) when elementary theory and applications were developed. In recent years, interest in this type of bearing has been renewed in the wake of the advent of mechatronics. The emphasis has now shifted mainly towards application of various bearing types and configurations (Shou et al. 2013; Stolarski 2010; Stolarski and Chai 2006a, b, 2008;

Yoshimoto et al. 2007). Let us also note here that the term "acoustic levitation" has (somewhat erroneously) been used to describe these bearings (Stolarski 2014; Stolarski et al. 2016). We should emphasise that SF bearings are not to be confused with ultrasonic (or acoustic) levitation, which relies on standing acoustic waves to trap light objects or particles in space, see e.g. (Whymark 1975).

The energy required to generate film pressure comes from forcing the one or the other bearing surface to oscillate with large amplitude (as compared with the nominal gap height) and a very high frequency of oscillation.

The principle of operation relies on the non-linear behaviour of the gas film both in the amplitude and the frequency of transverse (squeeze) oscillations, akin to the so-called "d.c.-shift" phenomenon in non-linear systems; viz. that the point of static equilibrium will gradually shift with increasing amplitude and frequency of oscillation.

Consider a uniform film of air at atmospheric conditions. Small transverse oscillations, especially at low frequency, will result in zero net pressure change (when averaged over time of one period of oscillation). That is:

$$\Delta p(t) \propto \Delta h(t+\phi) \leftrightarrow \int_0^T \Delta p(t)\mathrm{d}t = 0.$$

However, as the amplitude and frequency increase, the behaviour of the mean pressure will increasingly depart from being symmetric with the direction of the film change, resulting in a temporal average that exceeds ambient pressure. That is, the film will have a net (i.e. d.c.) load capacity to support an external static force.

Note that this behaviour is also valid for incompressible fluid bearings when cavitation (at sub-ambient pressures) is allowed for.

In the following, we shall try to gain a basic understanding of the squeeze-film action by considering the case of a plane, infinitely wide (i.e. 2D) film, and obtain an idea about the orders of magnitude of the various quantities concerned.

The principle can be (and has been) applied to other bearing configurations such as journal, conical and spherical bearings (see previous references).

Questions related to construction and electro-mechanical-control (or mechatronics) aspects of this type of bearing in practice are not dealt with, being considered outside the scope of this treatise. The interested reader can seek them in cited and other literature.

Principle of Operation

Consider an infinitely long flat plate (the bearing platen), of half-width L, separated by an air gap h_0 from a reference bottom rigid flat surface, oscillating with an amplitude $dh < h_0$ and angular frequency ω, as depicted in Figure 15.20.

We assume for the time being that the 2D Reynolds equation (see Chapter 4) is valid for this film situation and that there is, as yet, no tangential motion between the surfaces:

$$\frac{\partial}{\partial x}\left(\frac{h^3}{12\mu}\rho\frac{\partial p}{\partial x}\right) = \frac{\partial(\rho h)}{\partial t}, \qquad (-L \leq x \leq L). \tag{15.1}$$

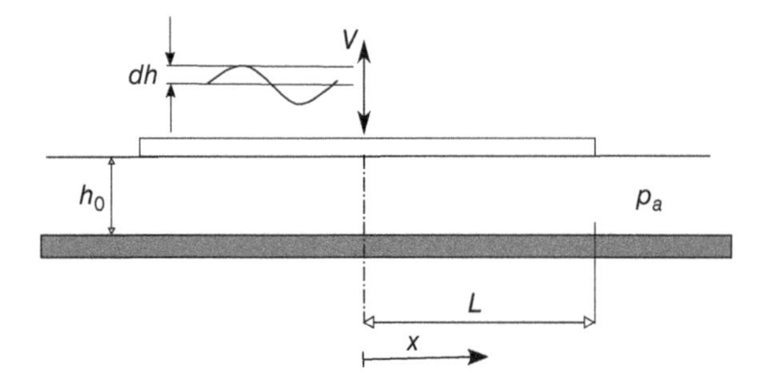

Figure 15.20 Schematic of squeeze film bearing configuration.

with b.c.:

$$p(-L) = p(L) = p_a; \qquad \rho(-L) = \rho(L) = \rho_a.$$

Normalising by $L, h_o, p_a, \rho_a, \omega$, this equation becomes:

$$\frac{\partial}{\partial X}\left(\frac{H^3}{12\mu}\wp\frac{\partial P}{\partial X}\right) = \sigma\frac{\partial \wp H}{\partial \tau}. \tag{15.2}$$

where

$$\sigma = \frac{12\mu\omega}{p_a}\left(\frac{L}{h_o}\right)^2$$

is the squeeze number.

Using the polytropic relationship between pressure and density:

$$P = \wp^{\chi} \qquad \text{or} \qquad \wp = P^{1/\chi} \qquad 1 \leq \chi \leq \kappa \tag{15.3}$$

results in

$$\frac{\partial}{\partial X}\left(\frac{H^3}{12\mu}\frac{\partial P^{\left(\frac{\chi+1}{\chi}\right)}}{\partial X}\right) = \sigma\left(\frac{\chi+1}{\chi}\right)\frac{\partial P^{\left(\frac{1}{\chi}\right)}H}{\partial \tau}. \tag{15.4}$$

with b.c.

$$P(-L) = P(L) = 1$$

or, when solving a symmetric half-bearing case,

$$P'_X(0) = 0, \qquad P(L) = 1$$

The Case $\sigma \to \infty$

This asymptotic case, which is however of great interest, has been discussed briefly in Chapter 7, for the small amplitude case. It results in a singular-perturbation problem yielding a uniform pressure distribution in the film, except the immediate vicinity of the outer edge (the mathematically so-called "boundary layer") where the pressure has to conform to the ambient value. The solution is thus (Wilcock 1969):

$$P^{\frac{1}{\chi}} = \frac{C}{H(\tau)}; \qquad C \text{ is some constant} \tag{15.5}$$

If we put

$$H(\tau) = 1 + \epsilon \sin \tau, \qquad \epsilon < 1$$

then

$$P^{\frac{1}{\chi}} = \frac{C}{1 + \epsilon \sin \tau} \tag{15.6}$$

Let us consider, without loss of generality, the case $\chi = 1$. Pan and Malanoski in (Wilcock 1969) show that the solution for this case is:

$$P = \frac{\sqrt{1 + \frac{3}{2}\epsilon^2}}{1 + \epsilon \sin \tau}; \qquad (\chi = 1). \tag{15.7}$$

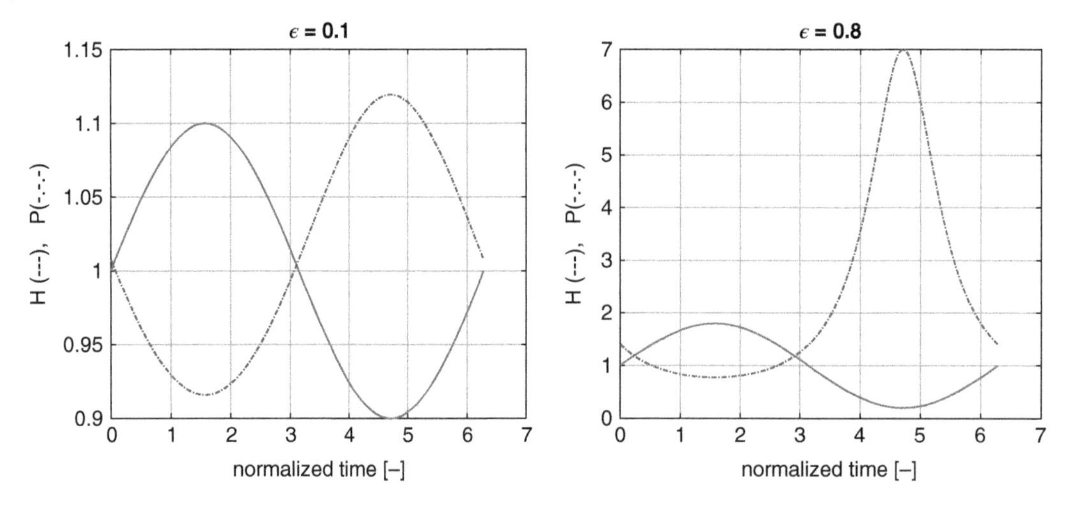

Figure 15.21 The mean pressure evolution with the gap height during one squeeze-film cycle for $\sigma \to \infty$. Left panel small ϵ; right panel large ϵ.

Finally, integrating the pressure over one period, we obtain the mean, temporally averaged, net pressure, i.e. the normalized load

$$\bar{P} = \frac{1}{2\pi}\int_0^{2\pi}(P(\tau)-1)d\tau = \frac{\sqrt{1+\frac{3}{2}\epsilon^2}}{\sqrt{1-\epsilon^2}} - 1 = \frac{W}{Ap_{\mathrm{a}}} \tag{15.8}$$

We see that for $\epsilon \ll 1$ we have

$$\bar{P} \approx 0$$

i.e. there is no net (d.c.) force or load capacity. (In actual fact, there is only stiffness and damping as we saw in Chapter 7, see left panel of Figure 15.21.)

However, for larger values of ϵ significant net pressure begins to develop, as in right panel of Figure 15.21.

The static stiffness may be obtained by derivation of the load capacity w.r.t. the nominal gap height. thus,

$$k = -\frac{\partial W}{\partial h_{\mathrm{o}}} = -Ap_{\mathrm{a}}\frac{\partial \bar{P}}{\partial \epsilon}\frac{d\epsilon}{dh_{\mathrm{o}}} = \frac{Ap_{\mathrm{a}}}{h_{\mathrm{o}}}\frac{5\epsilon^2}{2(1-\epsilon^2)\sqrt{1+\frac{3}{2}\epsilon^2}} \tag{15.9}$$

Validity and Relevance of the Asymptotic Solution

Most SF bearings are operated in the ultrasonic frequency range so as to avoid noise discomfort. The gaps used should be as small as manufacturing tolerances allow so that a small actuator/transducer stroke will suffice to generate sufficient pressure. If for example we consider an SF bearing with:

f	50	kHz
h_{o}	10	μm
L	20	mm
p_{a}	10^5	Pa
μ	210^{-5}	Pa s

then σ is well above 10^3.

As we shall see in the next section, in this range of σ the asymptotic solution is valid with good accuracy. We can thus, use the simple formulas in most cases to design flat SF bearings of various contours.

Numerical Solution for All σ Range

A feasible way to solve Eq. 15.4 with its b.c. might be by the method of finite differences (as has been performed by many authors including the present one). Applying this scheme on the grid

$$0 \leq X \leq 1 \qquad 0 \leq \tau < 2\pi$$

and using appropriate central difference formulation, the solution can be found iteratively with the aid of Newton–Raphson method. Other bearing configurations, possibly with non-uniform gaps, can be dealt with in a similar fashion.

An overview of typical results is presented in the following.

Pressure Distribution

Figure 15.22 depicts the pressure distributions along the film throughout one cycle of sinusoidal gap motion $0 \leqslant \tau \leq 2\pi$ for various values of σ and ϵ. We see that for small σ and ϵ, the pressure distribution along the film is quasi-"parabolic" in form with its amplitude (i.e. P at the centre of the bearing) being sinusoidal in the time. As σ increases, the distribution along the gap becomes more uniform (except at the outer edge where the pressure has to converge to atmospheric, often via further rising, falling and rising again), while its amplitude remains approximately sinusoidal. For high ϵ values on the other hand, this amplitude departs appreciably from sinusoidal (see also Figure 15.21), while the shape of the pressure distribution along the film changes from quasi-"parabolic" at small σ to almost uniform at large σs values of σ.

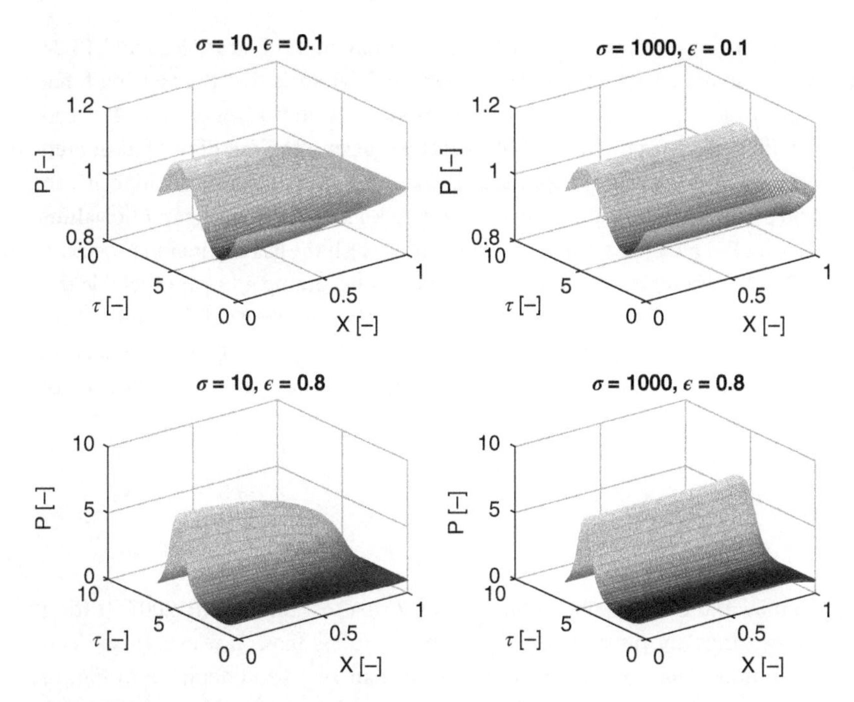

Figure 15.22 The pressure distribution over the width of the pad during one squeeze-film cycle for different σ and ϵ combinations.

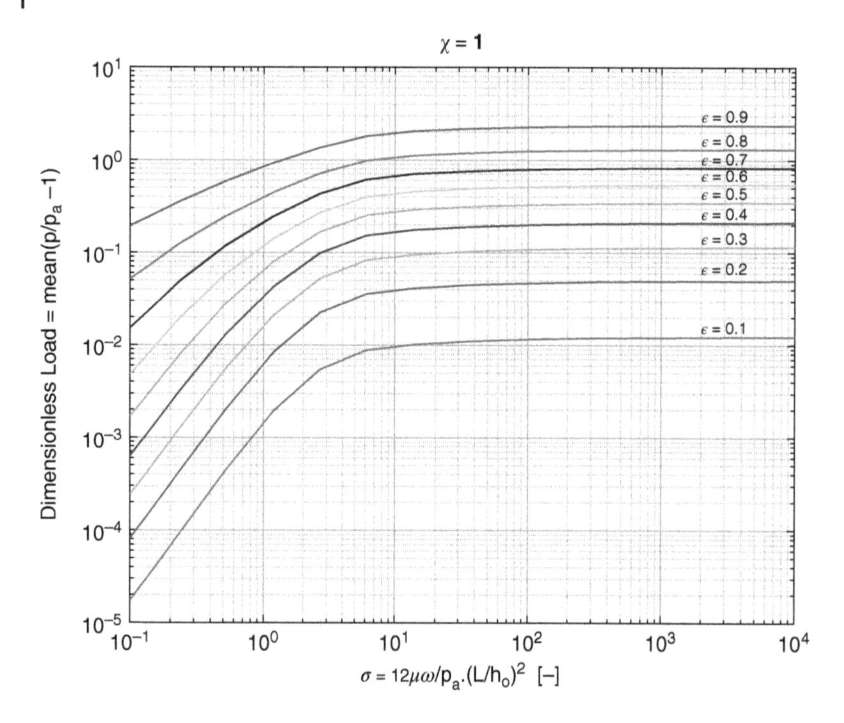

Figure 15.23 The dimensionless load as a function of σ and ϵ for isothermal process.

Global Results: Load-σ-ϵ

The global behaviour of a plain SF bearing can be summarised in the charts presented in Figures 15.23 and 15.24 assuming isothermal and isentropic gas expansion, respectively. The general behaviour is that the bearing force increases as a power law with σ for small values of σ, saturating to an asymptotic value from about $\sigma = 100$ and further. The bearing load increases also with the gap excursion amplitude ratio ϵ: at small values of this parameter, the load capacity is negligibly small; from $\epsilon = 0.5$ upwards, appreciable load capacity appears (for high σ), in particular for values of ϵ above 0.75, a mean pressure above 1 atmosphere (gauge) is built. The asymptotic values can be easily calculated from Eq. 15.8. All the $P - \sigma$ curves shift appreciably upwards if the gas expansion process is considered to be isentropic (Figure 15.24), the difference with the isothermal case becoming more appreciable the higher the ϵ value becomes (see Eq. 15.6 for explanation). In reality the gas expansion process will lie somewhere between the two cases, varying along the gap and through the film excursion cycle, depending on many factors including the thermal conductivity and thermal inertia/capacity of the bearing surfaces, and the frequency of excursion, (see e.g. Chapter 17).

15.6.1 Other Configurations

In (Shou et al. 2013; Stolarski 2010; Stolarski and Chai 2006b; Stolarski et al. 2016; Yoshimoto et al. 2007) (tilting type) plane slider bearings and journal bearings are presented. Other configurations have, however, also been known, such as spherical bearings for rotational systems such as reported in (Pan 1990) and depicted in Figure 15.25. The advantage of this design is that it is not sensitive to the alignment errors of the two end bearings. It goes without saying that many other useful configurations can be conceived, designed and constructed.

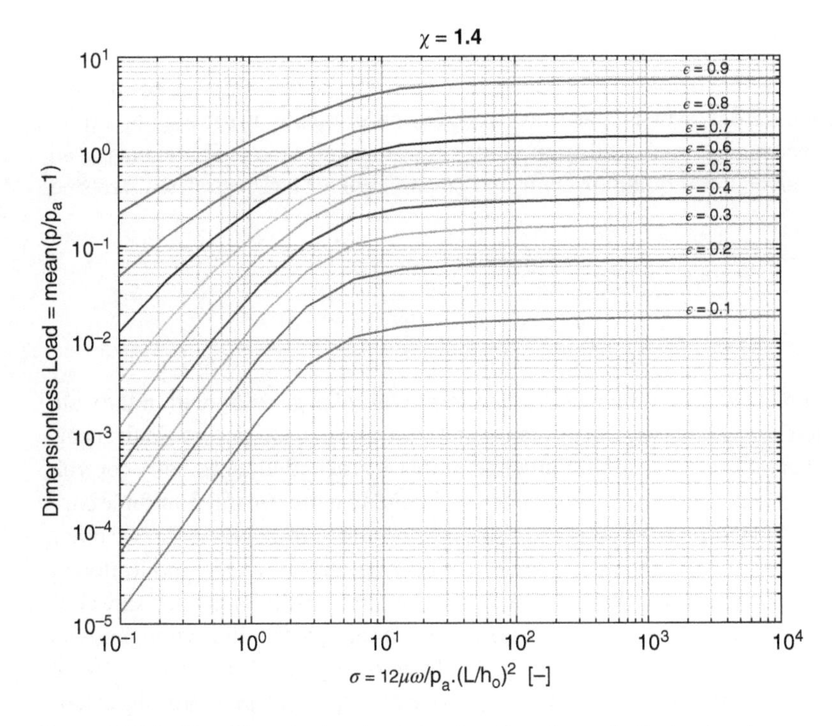

Figure 15.24 The dimensionless load as a function of σ and ϵ for isentropic process.

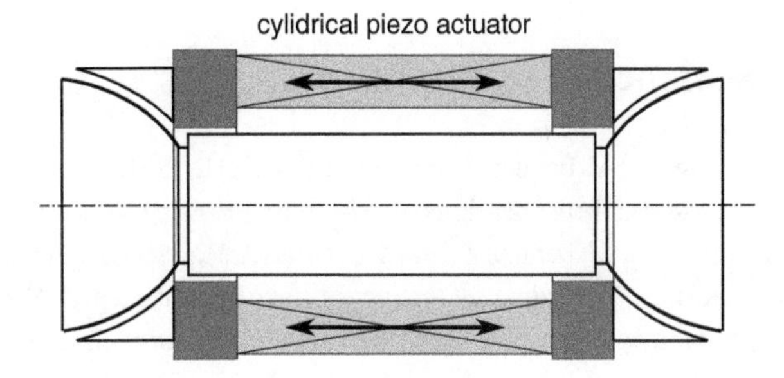

Figure 15.25 Prototype of spherical SF bearing (not to scale).

15.6.2 Assessment of Possible Inertia Effects

The question arises whether inertia forces in the gas flow might become appreciable at such high amplitudes of motion or velocity (high squeeze numbers) and whether we can assess that for a general situation with given system parameters (gap height, ambient pressure, etc.). This problem has been treated as an application example in Chapter 4. The main result reads that the effect of the inertia forces is determined by the squeeze Reynolds number:

$$\mathrm{Re}_{\mathrm{sq}} \triangleq \frac{\rho_{\mathrm{a}} \omega {h_{\mathrm{o}}}^{2}}{12\mu} \tag{15.10}$$

If $\mathrm{Re}_{\mathrm{sq}} \ll 1$, one would be justified in ignoring inertia effects.

As an example, if $\rho_{\mathrm{a}} = 1\ \mathrm{kg\ m^{-3}}$, $\omega = 10^{5}\ \mathrm{rad\ s^{-1}}$ ($\approx 15\,\mathrm{kHz}$), $h_{\mathrm{o}} = 10^{-5}$ m, $\mu = 2\ 10^{-5}$ Pa s, then $\mathrm{Re}_{\mathrm{sq}} = 0.05$ and thus inertia effects are negligible. However, if ω is increased by one order of magnitude, the inertia terms could become significant (but then only in the vicinity of the border of the bearing where the flow is non-zero.)

15.6.3 Ultrasonic Levitation and Acoustic Bearings

Ultrasonic levitation does not, strictly speaking, belong to the realm of air bearings since the mechanism does not involve fluid-film lubrication theory. Rather, acoustic bearings work by the principle of entrapping a thin, light object in the node of an acoustic wave generated, for instance, by a loudspeaker. Moreover, the envisaged applications for this type of so-called bearing differ qualitatively from conventional air bearings. The interested reader may refer to Whymark (1975).

15.7 Conclusions

In this chapter, we firstly situated the problem of active compensation or servo control of gas films within the state of the art of air bearings. We saw that there are various ways to influence and control the dynamic force in the bearing film, which can be narrowed down to two basic ways: inlet/outlet pressure control and film geometry control. The latter having generally a broader frequency range than the former. We formulated the general film-force control framework and discussed it in connection with examples pertaining to thrust and journal bearings. Two other special cases have been presented. Firstly, control of the traction force on the bearing surfaces and its application in wafer in-plane positioning has been presented together with two solutions belonging to inlet/outlet pressure control and to film-geometry control respectively. Secondly, load generation by means of squeeze-film action presents a way of constructing bearings that do not require a pressure supply. The basic theory pertaining to this type of active bearing has been overviewed. From all of this, we see that active gas bearings represent an interesting extension to conventional, passive bearings, all the more so with the increasing utility of the mechatronics engineering methodology.

References

Aguirre, G., Al-Bender, F. and Van Brussel, H., (2010 A multiphysics model for optimizing the design of active aerostatic thrust bearings. *Precision Engineering* 34 (3): 507–515.

Aguirre, G., Al-Bender, F. and Van Brussel, H., (2008a). dynamic stiffness compensation with active aerostatic thurst bearings *Proceedings of the International Conference on Noise and Vibration Engineering. International Conference on Noise and Vibration Engineering. Leuven, Belgium, September 15–17, 2008 (105–117).*

Aguirre, G., Al-Bender, F. and Van Brussel, H., (2008b). A multiphysics coupled model for active aerostatic thrust bearings *Proceedings of the IEEE/ASME International Conference on Advanced Intelligent Mechatronics. Xian, China, July 2–5*, 2008 (pp. 710–715).

Aguirre, G., Al-Bender, F. and Van Brussel, H., (2008c). Optimized design of active aerostatic thrust bearings for high precision positioning and disturbance rejection *Proceedings of the 10th Anniversary International Conference Euspen. 10th Anniversary International Conference Euspen. Zurich, Switserland*, May (19-22, 2008 (pp. (193–197).

Aguirre, G., Al-Bender, F. and Van Brussel, H., (2009). Development of an active aerostatic slide with sub-micrometer precision in six degrees of freedom *EUSPEN Conference Proceedings: Vol. 1. Euspen International Conference. San Sebastian, Spain*, 2–5 June 2009 (pp. 316–319).

Aguirre, G.A., (2010). Optimized Design of Active Air Bearings for Ultra-Precision Positioning Slides PhD thesis KU Leuven – Dept. Mechanical Engineering.

Al-Bender, F. (2009). On the modelling of the dynamic characteristics of aerostatic bearing films: From stability analysis to active compensation. *Precision Engineering* 33(2): 117–126.

Al-Bender, F. and Van Brussel, H., (1994). Active dynamic compensation of aerostatic bearings In *Tools for Noise and Vibration Analysis, Vol. 1, Proceedings ISMA19, Leuven, September (1994* (ed. Sas P): pp. 187–197.

Al-Bender F and Van Brussel H (1997). Active aerostatic bearing through control of film geometry *Proceedings of 9th IPES – 4th International Conference UME, Braunschweig, Germany, 1997*, Vol. 2, 389–392.

Al-Bender, F. and Van Brussel, H., (1998). Development of high frequency (hf) electrospindles with passive and active air bearings *Proceedings of International Seminar on Improving Machine Tool Performance, San Sebastian, 1998, ISBN84-8373-050-2, Vol. I, 175-184.*

Chen, C.J. and DeBra, D.B., (1987). A laminar flow motor for precision machining. *CIRP Annals* 36(1): 385–390.

Constantinescu, V.N., (1967). On the possibilities of magnetogasodynamic lubrication. *ASME – Journal of Lubrication Technology* 314–322.

Delhaes, G.M., van Beek, A., van Ostayen, R.A. and Schmidt, R.H.M., (2009). The viscous driven aerostatic supported high-speed spindle. *Tribology International* 42(11): 1550–1557. Special Issue: 35th Leeds-Lyon Symposium.

Wilcock, D.F. (ed) (1969). *Design of Gas Bearings*. Mechanical Technology Inc. (MTI): Latham NY.

Goodwin, G., Graebe, S. and Salgado, M., (2001). *Control System Design*. Prentice Hall, New Jersey.

Gross, W.A., (1963). Gas bearings: A survey. *Wear* 6, 423–443.

Horikawa, O. and Shimokohbe, A., (1990). An active air bearing. *JSME International Journal, Series 3* 33(1): 55–60.

Horikawa, O., Sato, K. and Shimokohbe, A., (1992). An active air journal bearing. *Nanotechnology* 3, 84–90.

Horikawa, O., Yasuhara, K., Osada, H. and Shimokohbe, A., (1991). Dynamic stiffness control of active air bearing. *International Journal of the Japanese Society of Precision Engineering* 25(1): 45–50.

Isidori, A., (1999). *Nonlinear Control Systems II*. Springer, London.

Krijnen, M., van Ostayen, R. and HosseinNia, H., (2017). The application of fractional order control for an air-based contactless actuation system. *ISA transactions* 82, 172–183.

Langlois, W., (1962). Isothermal squeeze films. *Quarterly of Applied Mathematics* 20, 131–150.

Lee, S.Q. and Gweon, D.G., (2000). A new 3-dof z-tilts micropositioning system using electromagnetic actuators and air bearings. *Precision Engineering* 24(1): 24–31.

Macmillan, R.H., (1962). *Non-Linear Control System Analysis*. Pergamon Press, Oxford.

Nakamura, S., (1996). High-speed spindles for machine tools. *International Journal of the Japanese Society of Precision Engineering* 30(4): 291–294.

Pan, C.H.T., (1990). Gas lubrication (1915–1990) In *Achievements in Tribology* (ed. L.B. Sibley and F.E. Kennedy) 31–55, Toronto, Ontario, Canada.

Pipeleers, G., Demeulenaere, B., Al-Bender, F., et al. (2009). Optimal performance trade-offs in repetitive control: experimental validation on an active air bearing setup. *IEEE Transactions on Control Systems Technology* 17(4): 970–979.

Raperelli, T., Viktorov, V., Colombo, F. and Lentini, L., (2016). Aerostatic thrust bearings active compensation: Critical review. *Precision Engineering* 44, 1–12.

Sato, Y. Murata, K. and Harada, M., (1988). Dynamic characteristics of hydrostatic thrust air bearing with actively controlled restrictor. *ASME – Journal of Tribology* 110, 156–161.

Shimokohbe, A., Aoyama, H. and Watanabe, I., (1986). A high precision straight-motion system. *Precision Engineering* 8(3): 151–156.

Shou, T., Yoshimoto, S. and Stolarski, T., (2013). Running performance of an aerodynamic journal bearing with squeeze film effect. *International Journal of Mechanical Sciences* 77, 184–193.

Snoeys, R. and Al-Bender, F., (1987). Development of improved externally pressurized gas bearings. *KSME Journal* 1(1): 81–88.

Stolarski, T., (2010). Numerical modeling and experimental verification of compressible squeeze film pressure. *Tribology International* 43(1): 356–360.

Stolarski, T., (2014). Acoustic levitation – a novel alternative to traditional lubrication of contacting surfaces. *Tribology Online* 9(4), 164–174.

Stolarski, T. and Chai, W., (2006a). Load-carrying capacity generation in squeeze film action. *International Journal of Mechanical Sciences* 48(7), 736–741.

Stolarski, T. and Chai, W., (2006b). Self-levitating sliding air contact. *International Journal of Mechanical Sciences* 48(6): 601–620.

Stolarski, T. and Chai, W., (2008). Inertia effect in squeeze film air contact. *Tribology International* 41(8): 716–723.

Stolarski, T., Gawarkiewicz, R. and Tesch, K., (2016). Acoustic journal bearing – performance under various load and speed conditions. *Tribology International* 102, 297–304.

van Rij, J., Wesselingh, J., van Ostayen, R., Set al. (2009). Planar wafer transport and positioning on an air film using a viscous traction principle. *Tribology International* 42(11): 1542–1549. Special Issue: 35th Leeds-Lyon Symposium.

Vuong, P.H., (2016). Air-based contactless actuation system for thin substrates, the concept of using a controlled deformable surface PhD thesis Delft University of Technology.

Weck, M. and Koch, A., (1993). Spindle-bearing systems for high-speed applications in machine tools. *Annals of the CIRP* 42(1): 445–448.

Wegener, K., Mayr, J., Merklein, M., et al. (2017). Fluid elements in machine tools. *CIRP Annals* 66(2): 611–634.

Whymark, R.R., (1975). Acoustic field positioning for containerless processing. *Ultrasonics* 13(6): 251–261.

Wu, H.C., Peirce, S. and DeBra, D., (2002). Blending quiet hydraulics, optics, precision mechanical design and digital control for precision machine tools. *IFAC Proceedings Volumes* 35(2): 657–662. 2nd IFAC Conference on Mechatronic Systems, Berkeley, CA, USA, 9-11 December.

Yoshimoto, S., Kobayashi, H. and Miyatake, M., (2007). Float characteristics of a squeeze-film air bearing for a linear motion guide using ultrasonic vibration. *Tribology International* 40(3): 503–511.

16

Design of an Active Aerostatic Slide

16.1 Introduction

Linear slides in high-precision machinery need to combine long-stroke motion with accurate position control. Traditionally, research effort has been focused on improving the positioning accuracy and resolution in the direction of motion. The use of non-contacting solutions (hydrostatic, aerostatic, magnetic), which avoid problems related to friction, together with iron-less motors and high resolution encoders make sub-micrometer accuracy in linear motion a state-of-the-art technique.

This level of precision is generally not achieved in the other five degrees of freedom (dof) of the motion of the slide. Three error sources that add uncertainty to the positioning of any point on the top surface of the carriage are distinguished here (see Figure 16.1), pertaining to the manufacturing and assembly limitations of the guiding surfaces, to the flexibility of the non-contacting support and to the structural flexibility of the carriage, (Aguirre 2010).

The first error source (a) is the flatness error of the guiding surfaces (typically 10 μm/m for granite blocks), which is transmitted to the motion of the carriage. This error becomes all the more relevant the longer the stroke of linear motion is.

The second error source (b) is owing to the flexibility of the bearing film, being relatively high for non-contacting bearings, which produces undesired motion in the carriage, on the order of 5 nm per Newton disturbance for aerostatic bearings, when external forces are applied to the carriage (e.g. in a machining process) or inertial forces arise due to acceleration.

When external forces are applied, the structural flexibility of the carriage also leads to local deformations (c), which can be of the same order of magnitude as in case (b).

The minimisation of these errors is constrained by manufacturing limitations and the intrinsic flexibility of non-contacting bearing films, so that the accuracy in the six dof cannot be brought in general to sub-micrometer level by further optimisation of current technologies. Even the traditional design approach of increasing the bearing stiffness is not sufficient in this respect, since the flatness error of the surfaces (error (a)) would be transmitted entirely to the carriage by an infinitely stiff bearing.

An unconventional approach for precision design is required, which combines the principles of classical design (high bearing stiffness is still desirable) with the benefits of active control and calibration for error compensation. The performance of a linear slide can be greatly improved by embedding extra sensors and actuators, which detect the deviations from the expected performance and act on the system to reduce the measured error.

Active control of the six dof of the slide can be a solution for achieving sub-micrometer accuracy in the six dof, and as a side advantage, for adding short stroke motion control capabilities. Magnetic levitation slides are intrinsically active slides with multi-dof control, since magnetic bearings need active control for stable operation, and

Air Bearings: Theory, Design and Applications, First Edition. Farid Al-Bender.

Companion website: www.wiley.com/go/AlBender/AirBearings

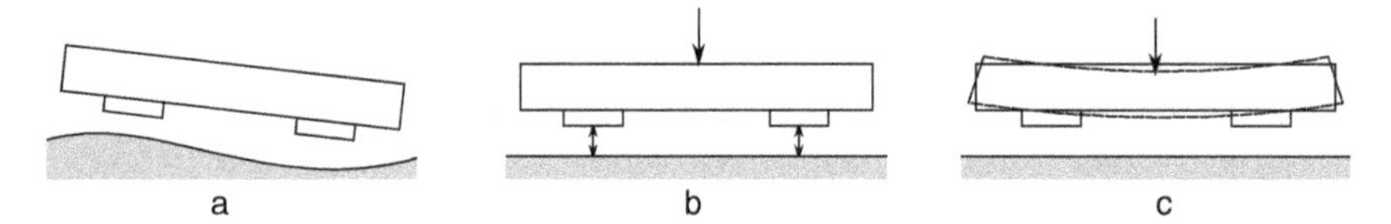

Figure 16.1 Positioning error sources in a linear slide. Source: From Aguirre, G. 2010.

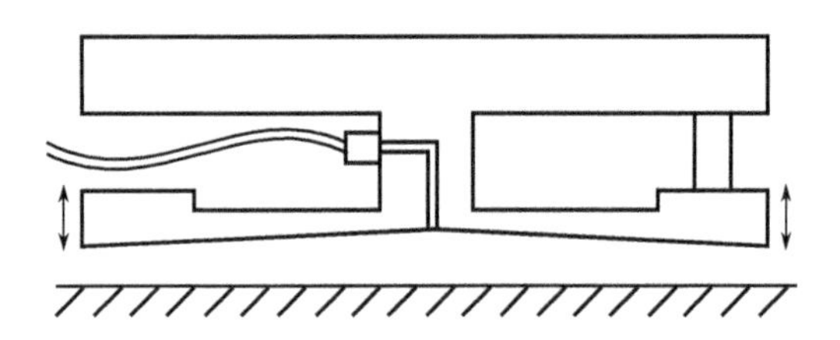

Figure 16.2 Principle of an active aerostatic bearing. Source: Republished with permission of Elsevier, from Aguirre G, Al-Bender F and Van Brussel H, 2010; permission conveyed through Copyright Clearance Center, Inc.

sub-micrometer accuracy applications have been developed (Peijnenburg et al. 2006). In aerostatic and hydrostatic slides, on the contrary, the bearing function is passive and thus actuators and sensors need to be added to implement active control. An example of a hydrostatic slide with five dof compensation for high precision applications is presented in (Park et al. 2006).

The design of an active aerostatic slide based on active air bearings (Aguirre 2010), such as the ones presented in Chapter 15, is presented here. The main design challenges for achieving optimal performance are discussed in detail in this chapter.

The design of the active air bearings is addressed first (see Figure 16.2), with the development of a multiphysics simulation model that considers its main design variables, and detailed experimental tests performed in order the verify their performance and the validity of the model. Finally, the design of an active aerostatic slide which integrates such bearings is discussed, and a prototype with its experimental results is presented.

16.2 A Multiphysics Active Bearing Model

We have seen that active air bearings show very promising results (see Chapter 15), but their performance depends on the coupled interaction between aspects such as the fluid dynamics in the bearing gap, the flexibility of the bearing surface, the piezoelectric actuators and the control. All these effects are relevant, and influence each other, in such a way that none of them can be optimised separately. Therefore, the fluid dynamics models discussed in Chapters 7 and 15 are not entirely sufficient for predicting the behaviour of active air bearings.

The interaction of an air bearing with the flexibility of its air film has been considered for the analysis of the head-disk interface in hard disk drives (HDD). In Jayson et al. (2003) the static effect of the aerostatic support is included in the structural finite element model of the HDD by means of lumped springs constraining vertical, roll and pitch degrees of freedom. In (Peng and Hardie 1995), (Tang 1971) and (Ono 1975) the opposite approach is followed, by solving the fluid dynamics model considering the effect of the flexible support with lumped mass and spring. None of these approaches is valid for modeling active air bearings based on gap shape control, as not only the flexibility of the support needs to be considered, but also the flexibility of the bearing surface.

A finite element multiphysics model that combines the model based on Reynolds equation presented in Chapter 15 and (Al-Bender 2009) with models for the structural flexibility, piezoelectricity and control, with strongly coupled formulation, is proposed here (Aguirre et al. 2010). The main goal of the model is to help in optimising the design of active air bearings. Therefore, the focus is not on representing in detail all the phenomena present in the bearing, but to analyse the effects that are relevant for the global performance of the bearing, with low calculation effort to allow its use in optimisation procedures.

16.2.1 General Formulation of the Model

The model is developed using finite element (FE) formulation, and it can be used to simulate static and dynamic operation of the active bearing. It has been implemented on a freely available open source finite element software for multiphysics problems (Elmer CSC - IT Center for Science (n.d.)).

As will be justified later in this section, a 2D finite element mesh of the bearing surface, plus an extra node to represent the motion of the rigid counter surface are sufficient to have good representation of the behaviour of the bearing model, avoiding the use of a more computationally demanding 3D mesh (see Figure 16.3). The mesh has to be adapted to have a higher node density around the restrictor or the feed hole area, where large pressure gradients are apt to take place, see Chapter 5, and four nodes define the contact surface between each piezoelectric actuator and the bearing surface.

The physical equations of the model are solved by the finite element method (Zienkiewicz et al. 2005), leading to a set of N linear equations for the N unknowns of the problem:

- deformation of the bearing plate δh_s
- change in the pressure distribution δp
- voltage applied to the actuators δv
- displacement of the counter surface δh_{cs}.

For a systematic treatment, these equations can be rewritten in matrix from, as follows:

$$[M] \cdot \begin{Bmatrix} \delta\ddot{p} \\ \delta\ddot{h}_s \\ \delta\ddot{h}_{cs} \\ \delta\ddot{v} \end{Bmatrix} + [C] \cdot \begin{Bmatrix} \delta\dot{p} \\ \delta\dot{h}_s \\ \delta\dot{h}_{cs} \\ \delta\dot{v} \end{Bmatrix} + [K] \cdot \begin{Bmatrix} \delta p \\ \delta h_s \\ \delta h_{cs} \\ \delta v \end{Bmatrix} = \{f\} \tag{16.1}$$

The formulation of only the vertical displacement (perpendicular to the bearing surface) is presented here for a simplified representation. The names of the system matrices (M, C and K) and vector (f) are related to their role as inertia, damping and stiffness matrices and force vector in structural FE models. In this multiphysics model, not all the components of the vector and matrices have those physical roles, but the name is kept out of convention. The contributions to M, C, K and f are discussed next.

16.2.2 Structural Flexibility

The analysis of the structural flexibility in the bearing mainly concerns the deformation of the bearing plate, as the support of the piezo-actuators is assumed to be relatively much stiffer than the bearing plate. Therefore, a 2D

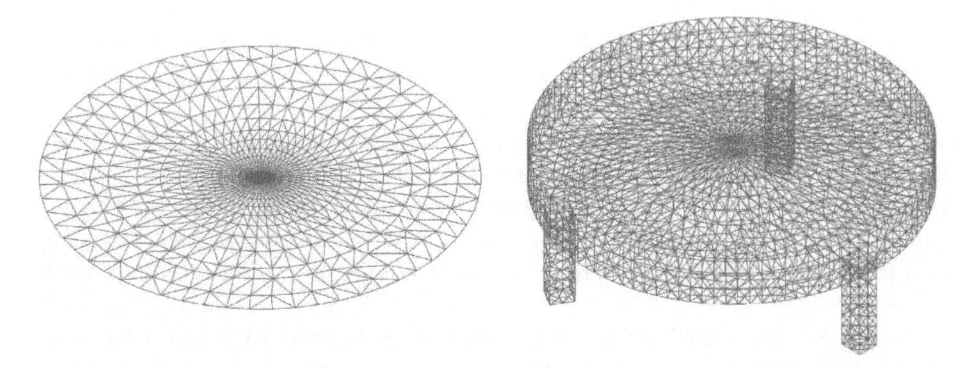

Figure 16.3 2D and 3D mesh of an active bearing. Source: Republished with permission of Elsevier, from Aguirre G, Al-Bender F and Van Brussel H, 2010; permission conveyed through Copyright Clearance Center, Inc.

mesh with triangular shell elements (available in most structural finite element software) is sufficient to represent this phenomenon, with the advantage of much lower calculation effort than a 3D grid. The shell elements assign translational and rotational degrees of freedom to the nodes

$$[M_s] \cdot \{\delta \ddot{h}_s\} + [C_s] \cdot \{\delta \dot{h}_s\} + [K_s] \cdot \{\delta h_s\} = 0 \tag{16.2}$$

where the contributions to the structural degree-of-freedom-equations in (16.1) are M_s, C_s and K_s, the inertia, stiffness and damping matrices, respectively, of the structural flexibility problem. No external force contribution is considered, as forces coming from the air pressure and from the piezo-actuators will appear as internal coupling forces.

16.2.3 Fluid Dynamics

The model of the pressure distribution is based on the one for dynamic analysis of passive (rigid surface) air bearings presented in Chapters 7 and 15. Some minor modifications are needed in the solution procedure, mainly due to the lack of axial symmetry in the pressure distribution and the need for a 2D representation. The time-dependent, planar Reynolds equation is used:

$$\nabla_{xy} \cdot \left(\frac{\rho h^3}{12\mu} \nabla p - \frac{\rho h}{2} V \right) = \frac{\partial}{\partial t} \rho h \tag{16.3}$$

where the nabla operator acts only on the bearing plane co-ordinates, ρ is the density of the air, p is the pressure, h is the air gap height, μ is the dynamic viscosity of the air, V is the relative velocity between the bearing surfaces and t is the time.

The Reynolds equation assumes the air flow to be constrained to the two dimensions of the bearing plane, with the boundaries of the problem as only flow sources or sinks. This is valid for representing the behaviour of the air film layer in between the non-permeable bearing surfaces, but not in the area under and immediately around the restrictor, where there is an air flow coming in perpendicular to the film. The effect of the feeding flow is included in the model as a flow source in the area under the restrictor, by extending the Reynolds equation: adding a scalar field representing the feeding mass flow rate *per unit area* $\dot{m}_A$.

$$\nabla_{xy} \cdot \left(\frac{\rho h^3}{12\mu} \nabla p - \frac{\rho h}{2} V \right) + \dot{m}_A = \frac{\partial}{\partial t} \rho h \tag{16.4}$$

$\dot{m}_A$ takes the value zero for all the bearing surface except for the area under the air entrance, where, for an inherent restrictor, it can be calculated as in Chapter 2 and (Al-Bender 1992).

$$\dot{m}_A = \frac{2\pi r_r h_r}{\pi r_r^2} C_d \sqrt{\frac{2\kappa}{\kappa - 1}} \frac{p_s}{\sqrt{RT_a}} \Phi_e \tag{16.5}$$

where r_r is the restrictor radius, h_r is the air gap height at the restrictor, C_d is the discharge coefficient, κ is the ratio of specific heats (=1.4 for air), p_s is the feeding pressure, Φ_e is the nozzle function (Al-Bender 1992) (Chapter 5), R is the specific gas constant and T_a the ambient temperature.

The main requirement for the extension of the model is to estimate properly the pressure drop produced in the restrictor area, so that the pressure in the rest of the bearing surface can be estimated properly by the Reynolds equation. The error in the estimation of the pressure inside the restrictor area (due to assuming viscous flow and applying Reynolds equation) is not relevant since the air entrance region is relatively very small (1 mm restrictor diameter for 80 mm diameter pad in the bearings used in study) and it has thus almost no influence in the total force.

For the required model accuracy, the flow in the bearing can be considered isothermal (Al-Bender 1992), and the air is compressible, so that from the ideal gas equation, the density can be replaced by the pressure through:

$$p = \rho R T_a \tag{16.6}$$

obtaining the following expression

$$\nabla_{xy} \cdot \left(\frac{ph^3}{12\mu} \nabla p - \frac{ph}{2} V \right) + RT_{\mathrm{a}} \dot{m}_{\mathrm{A}} = \frac{\partial}{\partial t} ph \tag{16.7}$$

The fluid-dynamics model is represented by a second order non-linear partial differential equation (16.7), with constant ambient pressure outside the bearing as boundary condition. A step preceding to the solution by finite elements is the linearisation of the model with respect to the two unknowns, the pressure and the air gap height distribution. The linearised model is used in the static analysis for an iterative approach to the solution of the non-linear equation, while in the harmonic case the linearised model is assumed to be valid for small disturbances

$$\begin{aligned} p &= p_0 + \delta p \\ h &= h_0 + \delta h \end{aligned} \tag{16.8}$$

Substituting (16.8) in (16.7), δh and δp become the new unknowns of the problem, with p_0 and h_0 the current guess in the iterative procedure for the static analysis, or the static solution for the harmonic analysis. Applying a standard finite element procedure to discretise the model (Zienkiewicz et al. 2005), the following system is obtained.

$$[C_{\mathrm{p}}]\{\delta \dot{p}\} + [K_{\mathrm{p}}]\{\delta p\} = \{f_{\mathrm{p}}\} + c_{\mathrm{ps}}\{\delta \dot{h}\} + k_{\mathrm{ps}}\{\delta h\} \tag{16.9}$$

where C_{p} and K_{p} represent the contributions to the C and K matrices of the full system, f_{p} is the vector containing the components not depending on δh or δp, and c_{ps} and k_{ps} are the air gap height to pressure coupling coefficients.

Deformation-To-Pressure Coupling

It must be noted that δh does not coincide with any of the degrees of freedom defined in (16.1). However, δh can be redefined as the relative displacement between the bearing and counter surfaces, leading to the following expression.

$$\delta h = \delta h_{\mathrm{cs}} - \delta h_{\mathrm{s}} \tag{16.10}$$

This change of variable leads to a representation of the fluid dynamics model in terms of the degrees of freedom of the global model

$$[C_{\mathrm{p}}]\{\delta \dot{p}\} - [C_{\mathrm{ps}}]\left\{ \begin{array}{c} \delta \dot{h}_{\mathrm{s}} \\ \delta \dot{h}_{\mathrm{cs}} \end{array} \right\} + [K_{\mathrm{p}}]\{\delta p\} - [K_{\mathrm{ps}}]\left\{ \begin{array}{c} \delta h_{\mathrm{s}} \\ \delta h_{\mathrm{cs}} \end{array} \right\} = \{f_{\mathrm{p}}\} \tag{16.11}$$

where the contributions to the pressure degree of freedom equations in (16.1) are C_{p} and C_{ps} to C, K_{p} and K_{ps} to K, and f_{p} to f.

The matrices C_{ps} and K_{ps} represent the coupling between the deformation of the bearing plate or the displacement of the counter surface and the pressure distribution. Any change in δh_{s} or δh_{cs} affects δp, and vice-versa. Therefore, the effect of the pressure on the deformation needs also to be considered in order to obtain the strongly coupled formulation.

Pressure-To-Deformation Coupling

The pressure in the bearing gap applies a force on both bearing surfaces that leads to deformations and displacements. Pressure can be converted to force on each node using an equivalent area A_{eq} as a conversion factor. This contribution could be considered as an external force in the structural flexibility equations, but as it depends on δp it can be treated as an internal force, defining a coupled contribution K_{sp}, which multiplies the pressure degrees of freedom with their equivalent area

$$[K_{\mathrm{sp}}] = \{A_{\mathrm{eq}}\}[I] \tag{16.12}$$

where I is the identity matrix and the sum of all the equivalent areas A_{eq} is equal to the bearing area.

In case of static analysis, an iteration is performed to find the equilibrium between deformation and pressure distribution. In this case, the force that is applied to the bearing plate is the sum of the pressure distribution in the previous step p_0 and the change in pressure obtained in the current one δp. Therefore, a contribution to the external force vector must be added to consider the force due to p_0

$$\{f_{\text{sp}}\} = \{A_{\text{eq}}\}[I]\{p_0 - p_{\text{a}}\} \tag{16.13}$$

where p_{a} is the ambient pressure.

16.2.4 Dynamics of the Moving Elements

The dynamics of the aerostatic film and the bearing plate are already included in the models described above, but the dynamic behaviour of other moving elements in the system is not considered yet. The bearing pad is assumed to be fixed, while inertia m_{cs}, stiffness k_{cs}, damping c_{cs} and external force f_{cs} contributions can be added to the node representing the counter surface to the bearing (see Figure 16.4). This node can also be used to apply external forces to the bearing. This modeling approach is also valid when the moving element is the bearing pad and the counter surface (or platen) is fixed

$$m_{\text{cs}} \cdot \delta\ddot{h}_{\text{cs}} + c_{\text{cs}} \cdot \delta\dot{h}_{\text{cs}} + k_{\text{cs}} \cdot \delta h_{\text{cs}} = f_{\text{cs}} \tag{16.14}$$

16.2.5 Piezoelectric Actuators

Piezoelectric stack actuators are used to deform the bearing plate in order to control the bearing force. These actuators consist of a stack of ceramic disks separated by thin metallic electrodes, providing large strokes with low operating voltages, compared to monolithic actuators. They can withstand high axial compressive forces, but they are quite brittle in the other directions. They must be thus mounted so that only compressive loads are applied to them.

There exist finite element models that represent the coupling between the electric and mechanical fields in piezoelectric material based on its geometry, thus using 3D elements (Piefort 2001). These models are specific to monolithic actuators, but they can be adapted to multilayer actuators to give an equivalent axial behaviour.

However, since the contribution of the piezo-actuator is expected to be constrained to the axial direction, it is more meaningful to represent the stack with a single degree of freedom linear model. Such a model can be found in (Preumont 2006)

$$\begin{aligned} D &= \epsilon^T E + d_{33} T \\ S &= d_{33} E + s^E T \end{aligned} \tag{16.15}$$

where D is the electric displacement (charge per unit area, expressed in Cm^{-2}), E the electric field (V m^{-1}), T the stress (Nm^{-2}) and S the strain, ϵ^T is the dielectric constant under constant stress, s^E is the compliance under

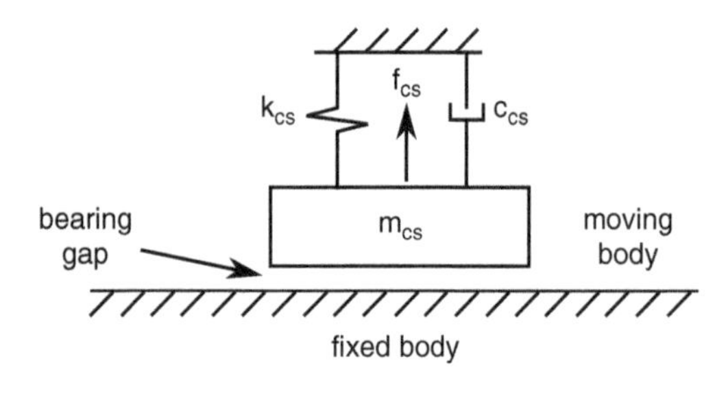

Figure 16.4 Model of the dynamics of the moving system. Source: Republished with permission of Elsevier, from Aguirre G, Al-Bender F and Van Brussel H, 2010; permission conveyed through Copyright Clearance Center, Inc.

constant electric field and d_{33} is the piezoelectric constant (m V^{-1}) for an electric field parallel to the poling direction of the material.

Assuming the actuators are driven by voltage, the second equation in (16.15) is sufficient in order to represent the electromechanical coupling in the model. This equation can be rewritten in terms of voltage, elongation and force, which are the degrees of freedom of the model

$$f_z = \frac{A_p}{s^E L}(\delta h_s - n d_{33} \delta v) \tag{16.16}$$

where f_z is the force taken by the piezo-actuator, A_p is the area of the piezo in contact with the structure, L is the length of the actuator and n is the number of disks in the stack. This force is linearly dependent on the degrees of freedom of the problem, and can therefore be considered as an internal force, and represented in FE formulation as contributions to K, to be added to the equations of the structural degrees of freedom. This contribution is divided among the four nodes (see further) that represent each piezo-actuator.

The 2D mesh of the model is adapted so that the contact between each actuator and the bearing plate is represented by four nodes at the vertexes, and the contribution of the piezo-actuator element is distributed among them

$$f_z = [K_z] \cdot \{\delta h_s\} + [K_{zv}] \cdot \{\delta v\} \tag{16.17}$$

16.2.6 Controller

The active compensation principle is based on changing the voltage applied to the piezo-actuators as a function of the change of some measured state, normally the relative displacement between the bearing surfaces. As the bearing is assumed to be fixed in this model, δh_{cs} represents this relative displacement. Therefore, the element representing the controller couples the displacement of the counter surface node to the change in the electric potential applied to the piezo-actuators. For a controller to be represented in the general formulation of the active bearing model in Eq. 16.1, it must be possible to represent it with the following structure

$$\delta v + m_c \cdot \delta \ddot{h}_{cs} + c_c \cdot \delta \dot{h}_{cs} + k_c \cdot \delta h_{cs} = 0 \tag{16.18}$$

For harmonic analysis, where the signals are represented by complex numbers (in general $xe^{(j\omega t+\beta)}$, with x the amplitude, ω the frequency, t the time and β the phase for zero time), and by virtue of the relationship

$$\frac{\partial}{\partial t}(xe^{(j\omega t+\beta)}) = j\omega x e^{(j\omega t+\beta)}, \tag{16.19}$$

Equation (16.18) can then be simplified to:

$$\delta v + c'_c \cdot \delta \dot{h}_{cs} + k'_c \cdot \delta h_{cs} = 0 \tag{16.20}$$

As an example, a PID controller with position as feedback signal is represented as:

$$\delta v + \left(k_D - \frac{k_I}{\omega^2}\right) \cdot \delta \dot{h}_{cs} + k_P \cdot \delta h_{cs} = 0 \tag{16.21}$$

with k_P, k_I and k_D the proportional, integral and derivative gains of the controller and ω is the frequency.

16.2.7 Coupled Formulation of the Model

We have defined above the finite element implementation of the simulation models of the relevant effects for the design of active aerostatic bearings. Combining their contributions in the general formulation of the model (see

Eq. 16.1), the following representation is obtained for the system matrices

$$M = \begin{bmatrix} 0 & 0 & 0 & 0 \\ 0 & M_s & 0 & 0 \\ 0 & 0 & m_{cs} & 0 \\ 0 & 0 & m_c & 0 \end{bmatrix} \qquad C = \begin{bmatrix} C_p & -c_{ps} & -c_{ps} & 0 \\ 0 & C_s & 0 & 0 \\ 0 & 0 & c_{cs} & 0 \\ 0 & 0 & c_c & 0 \end{bmatrix}$$

$$K = \begin{bmatrix} K_p & -k_{ps} & -k_{ps} & 0 \\ K_{sp} & K_s + K_z & 0 & K_{zv} \\ 0 & 0 & k_{cs} & 0 \\ 0 & 0 & k_c & 1 \end{bmatrix} \qquad f = \begin{bmatrix} f_p \\ f_{sp} \\ f_{cs} \\ 0 \end{bmatrix} \tag{16.22}$$

16.3 Bearing Performance and Model Validation

A detailed validation of the static and dynamic characteristics of active air bearings requires a specialised test setup, where a number of factors need to be considered. The design of such a test setup is described next in this section. Then, the performance of active air bearings is characterised and the validity of the simulation model presented above is discussed.

16.3.1 Test Setup for Active Aerostatic Bearings

The test setup (see Figure 16.5) is built around the air film, which is formed between the flexible surface of the active bearing and a counter surface that is assumed to be rigid. The bearing stands on top of the rigid surface, minimising the moving mass to achieve higher dynamic performance.

The static pre-load applied to the bearing defines the thickness of the air film and has a great influence on the dynamic performance of the bearing. A pneumatic actuator is used, which can be regulated up to 1 kN pre-load adding almost no inertia to the system.

An electromagnetic shaker is used to apply axial disturbance forces to the bearing, in order to assess its dynamic stiffness. It is placed between the pneumatic actuator and the bearing, and it can provide forces up to 20 N peak-to-valley with frequencies between 1 and 2000 Hz.

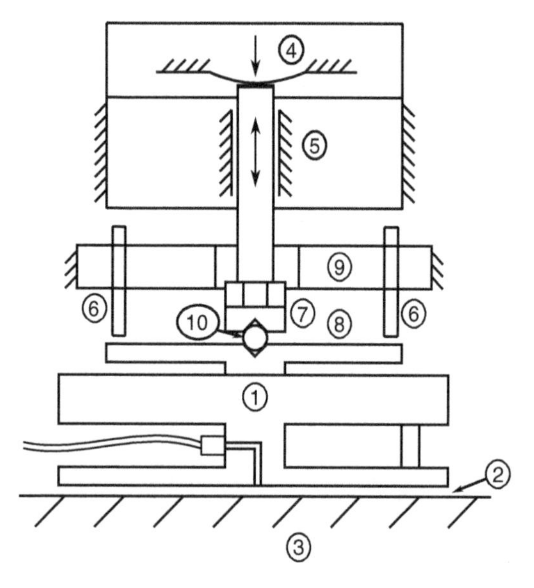

Figure 16.5 Schematic of the test setup. (1) active air bearing, (2) air film, (3) rigid counter surface, (4) pneumatic actuator for pre-load, (5) shaker, (6) displacement sensors, (7) force sensor, (8) measuring target, (9) displacement sensor holder, (10) ball joint connection. Source: Republished with permission of Elsevier, from Aguirre G, Al-Bender F and Van Brussel H, 2010; permission conveyed through Copyright Clearance Center, Inc.

The position of the bearing needs to be measured, both for evaluation of its performance and for use as feedback signal. Three capacitive sensors are used in order to detect both axial displacement and tilt in the bearing. The sensors are mounted equidistant from the center of the bearing, on a frame isolated from the shaker and from the bearing deformation.

Finally, a piezoelectric force cell is placed in series between the shaker and the bearing. Using different sensitivity levels it can measure both the static pre-load applied by the pneumatic actuator (over 1000 N) and the disturbance force applied by the shaker (below 100 N) with high resolution. The measured force is used for analysing the load carrying capacity and the stiffness of the bearing.

16.3.2 Active Bearing Performance and Model Validation

The test setup presented above has been used to perform some experimental tests in order to demonstrate the performance of the active air bearings and to validate the simulation model. The need for a multiphysics coupled model has been motivated by the relevant influence of the flexibility of the bearing plate on the behaviour of the system. Therefore, the experiments have been performed on two bearing prototypes with different plate thicknesses.

First the quasi-static properties of the bearings have been tested, looking at the air gap thickness as a function of the applied load. Then the dynamic performance up to 1500 Hz is analysed (enough to include the first resonance frequency), first in passive operation (no control), looking at the bearing displacement produced by either a change in the voltage applied to the piezo-actuators or in the load applied to the bearing. Finally the active operation is demonstrated by applying feedback control to reduce the vibration produced by the variation of the external load. In all cases simulation and experimental results are presented, compared and discussed.

Bearing Prototypes

A sketch of the design of the two bearings is presented in Figure 16.6. One of the bearings has a constant plate thickness of 10 mm, while for the other one the thickness is reduced to 6 mm in the central portion in order to reduce the bending stiffness, and thus increase the deformation produced by the piezo-actuators. The thickness is not reduced at the periphery of the bearing in order to limit the deformation produced by the air pressure in between the piezo-actuators (see Section 16.3.4).

Both bearings are made of brass, in order to obtain a very low bearing surface roughness, with a pad diameter of 80 mm and a restrictor diameter of 1 mm. The bearing surface is machined with a micro-conical profile, or conicity of five micrometer variation to produce a convergent air gap. Three piezoelectric stack d_{33} actuators of $5 \times 5 \times 18$ mm are used.

Load Carrying Capacity

Figure 16.7 shows the load carried by the thin and thick plate bearings respectively, as a function of the air gap thickness, with 0 and 100 V, respectively, applied to the piezo-actuators and an air supply pressure of 6 bar. In order

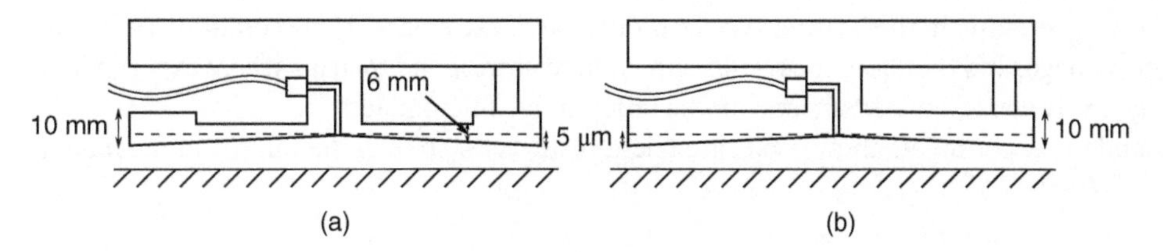

Figure 16.6 Schematic of thin (a) and thick (b) bearing plates with conical surface (not to scale). Source: Republished with permission of Elsevier, from Aguirre G, Al-Bender F and Van Brussel H, 2010; permission conveyed through Copyright Clearance Center, Inc.

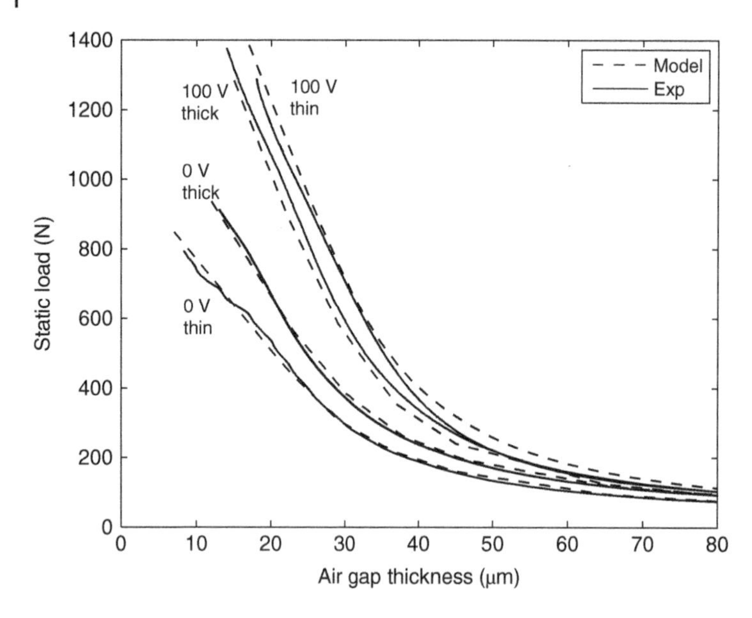

Figure 16.7 Load-gap relation for thin and thick bearing with 6 bar supply pressure. Source: Republished with permission of Elsevier, from Aguirre G, Al-Bender F and Van Brussel H, 2010; permission conveyed through Copyright Clearance Center, Inc.

to have a common reference in all cases, the gap height, as the nominal value, being defined here as the distance between the centre of the bearing surface and the counter surface. The shape of the bearing surface depends on the initial profile and on the deformation produced by the piezo-actuators and the air pressure in the gap, and thus the first contact between the bearing and the counter surface happens at different air gap heights for each case. Due to the lack of axial symmetry of the plate stiffness, the deformed bearing surface does not have a conical profile, but the term "conicity" is still used hereafter as a measure of the average height difference between the centre and the periphery of the bearing pad.

The load carrying capacity of air bearings depends on the geometry of the bearing gap: a positive gap conicity (gap height decreasing with radius) increases significantly the load that can be carried. This is confirmed in Figure 16.7, where the tests with 100 V, due to the conicity generated by the elongation of the piezo-actuators, show the highest load capacity.

The flexibility of the bearing plate has counteracting effects on the load capacity. On the one hand, it affects the elongation of the piezo-actuators, and thus the lower stiffness of the thin plate enables a higher conicity and load carrying capacity at 100 V, compared to the thick bearing. On the other hand, it also leads to a deformation of the bearing surface due to the air pressure, which reduces the conicity, and thus the thick bearing can carry higher loads with 0 V applied to the piezo-actuators (see 16.3.4 for more detail).

When higher static loads are applied, the bearing surface and the counter surface get in contact at some points of the periphery, while air can still flow out at other points. This effect shows low repeatability due to the freedom of the bearing to tilt, and the uncertainty in the bearing surface profile. Such extreme working conditions must be avoided to prevent surface damage, and therefore the results shown here correspond to loads where experiments show good repeatability, typically above a minimum gap on the order of five micrometers.

The simulation results and the experiments show good agreement in all cases proving the validity of the model for static analysis.

Passive Dynamic Stiffness

Figures 16.8 and 16.9 show the passive dynamic stiffness of the bearing, i.e. the ratio between the axial force and the the axial displacement of the bearing as a function of frequency, with a constant voltage (here 50 V) applied

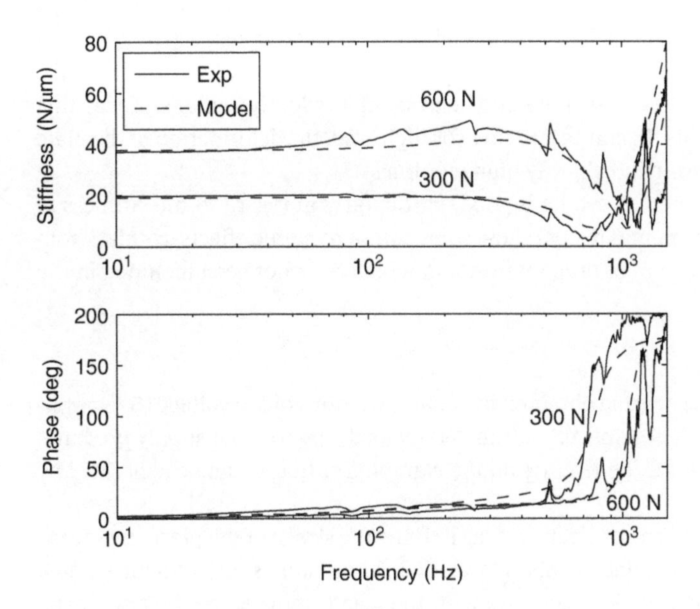

Figure 16.8 Dynamic stiffness of thin plate bearing. Source: Republished with permission of Elsevier, from Aguirre G, Al-Bender F and Van Brussel H, 2010; permission conveyed through Copyright Clearance Center, Inc.

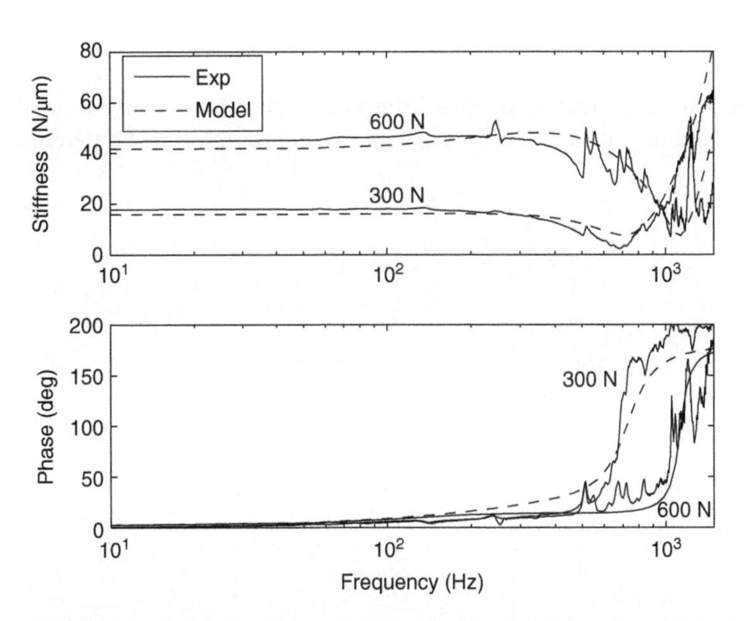

Figure 16.9 Dynamic stiffness of thick plate bearing. Source: Republished with permission of Elsevier, from Aguirre G, Al-Bender F and Van Brussel H, 2010; permission conveyed through Copyright Clearance Center, Inc.

to the piezo-actuators, and an air supply pressure of 6 bar, for the thin and the thick bearing respectively. In both cases the response under 300 and 600 N static loads is analysed.

Comparing the four cases presented here, for low static load the thin bearing offers slightly higher stiffness, due to the higher conicity induced by the piezo-actuators. In contrast to this, for high load, the deformation of the plate under the pressure is dominant and the thick bearing has slightly higher stiffness.

The agreement between simulation and experimental results is good, particularly in regard to the stiffness at low frequencies and around resonance. The experimental results show some minor dynamic effects, coming from resonances or tilting effects in the test setup, which are of little interest and which have not been included in the model.

Dynamic Displacement Generation

In Figures 16.10 and 16.11, the displacement of the bearing obtained for small variations of the voltage (5 V amplitude) is represented, again for the thin and the thick bearing, for 300 and 600 N load, and 6 bar air supply pressure. This result indicates how much displacement or force, depending on the clamping stiffness, can be generated by the active bearing.

The influence of the stiffness of the bearing plate is most clear here, as it limits the stroke of the piezo-actuators, and therefore the thin bearing can produce higher displacements. The static load has almost no influence at low frequencies, but it does on the resonance frequency, which increases with the load (produced by the pneumatic shaker, thus not increasing the inertia) due to the higher support stiffness.

The agreement between simulation and experimental results is again good at low frequencies. At higher frequencies, the resonance frequency is well predicted, but the amplitude of vibration is not properly estimated in some cases.

Active Disturbance Rejection

In a further analysis of the active bearing, the active performance of the bearing is evaluated. Figure 16.12 shows the dynamic stiffness of the active bearing using integral feedback control with two arbitrarily chosen different

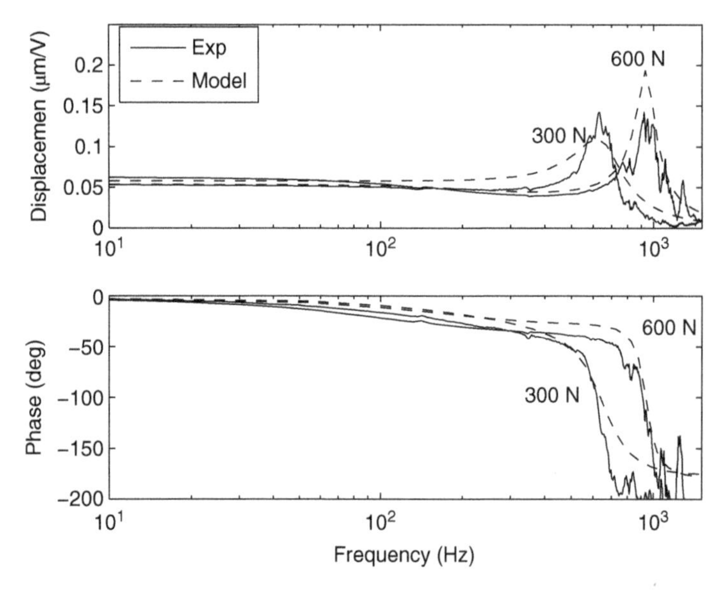

Figure 16.10 Dynamic displacement of thin plate bearing. Source: Republished with permission of Elsevier, from Aguirre G, Al-Bender F and Van Brussel H, 2010; permission conveyed through Copyright Clearance Center, Inc.

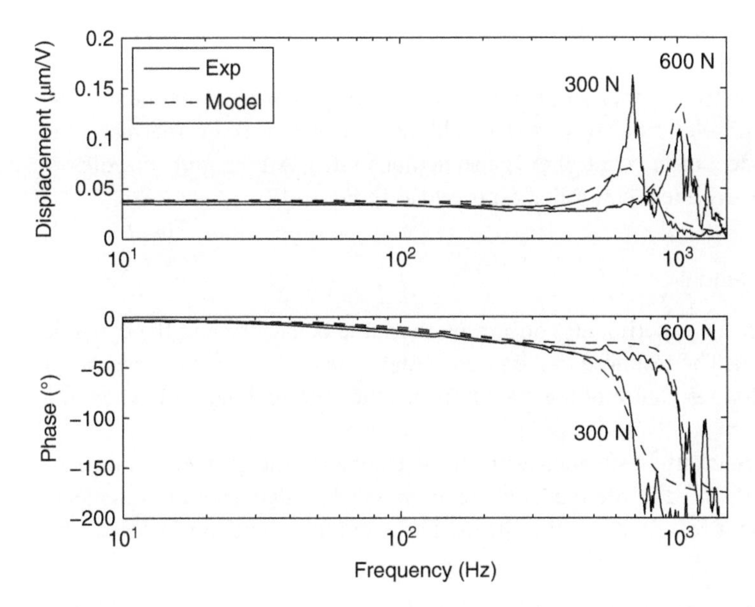

Figure 16.11 Dynamic displacement of thick plate bearing. Source: Republished with permission of Elsevier, from Aguirre G, Al-Bender F and Van Brussel H, 2010; permission conveyed through Copyright Clearance Center, Inc.

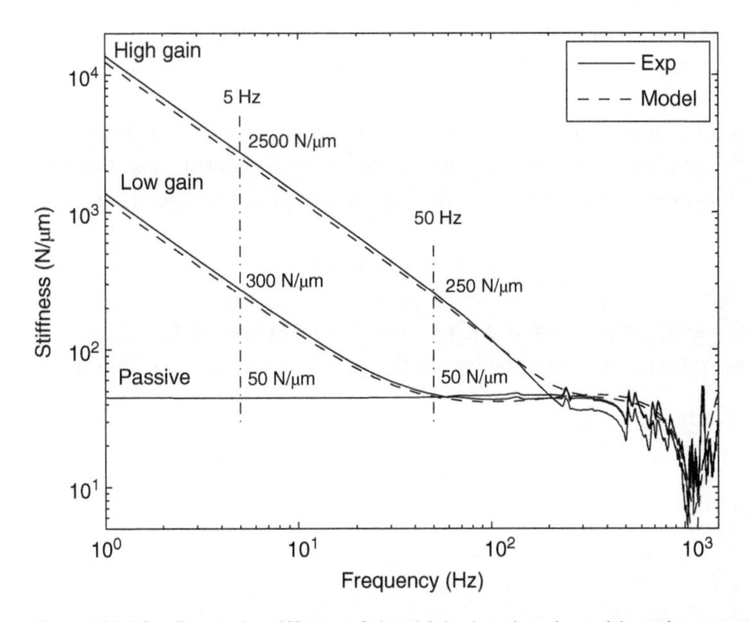

Figure 16.12 Dynamic stiffness of the thick plate bearing with active control. Source: Republished with permission of Elsevier, from Aguirre G, Al-Bender F and Van Brussel H, 2010; permission conveyed through Copyright Clearance Center, Inc.

gains, the highest of the two being close to the stability limit. The thick bearing is used here, with static loads of 300 N and 600 N, respectively, and 6 bar air supply pressure.

The results show, on the one hand, that the active compensation method works, increasing the dynamic stiffness of the bearing in a high bandwidth (quasi-infinite static stiffness, five-fold increase at 50 Hz). Furthermore, the validity of the model for representing the effect of the controller is also demonstrated. An integral controller is used here, but the model is able to deal with any linear controller, expressed as in Eq. 16.18.

16.3.3 Discussion on the Validity of the Model

The simulation results agree fairly well with the experiments concerning the static behaviour and the dynamic stiffness, both in passive and active operation. The response to the piezo-actuators is also well estimated, except for the amplitude of vibration around the first resonance of the system (bad estimation of damping), where the agreement is not so good.

Despite the difficulty to model accurately relevant effects such as the imperfections in the geometry of the surface and restrictor, and the simplifications assumed in the model to get affordable simulation times, the model agrees sufficiently well with the real performance, and it can thus be used as a tool for aiding in the design and optimisation of active air bearings.

16.3.4 Analysis of the Relevance of Model Coupling

The development of the model presented here has been motivated by the need for a tool to assist in the design of active aerostatic thrust bearings, due to the relevance of the interaction between the pressure in the air gap and the deformation of the bearing surface, which to our knowledge has not been considered in existing models for air bearing design.

In this section, some simulation results are presented to demonstrate this claim, by analysing the static and dynamic properties of an active bearing for different plate thicknesses. One of the active air bearings used in the previous section has been taken as reference for these simulations, considering bearing plate thicknesses of 2, 5, 10 and 15 mm, with 500 N static load.

Static Analysis

The coupling between the structural flexibility and the pressure is clearly observed in Figure 16.13, where the deformation of the periphery of the bearing plate is shown. The waviness in the profile is due to the variation in

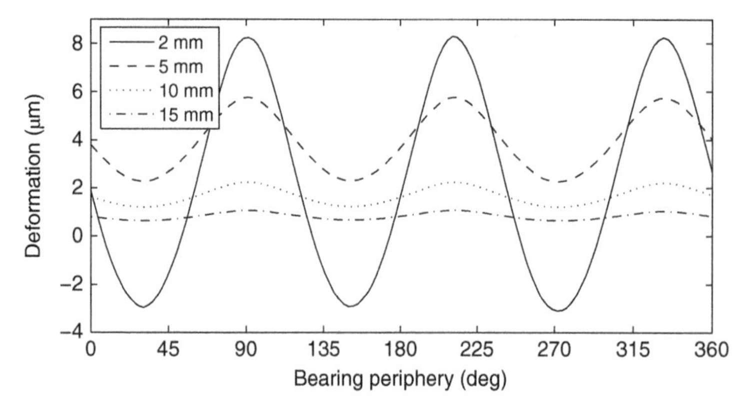

Figure 16.13 Deformation of bearing periphery under influence of air pressure and piezo-elongation, for different plate thicknesses, 50 V applied to the piezos and 6 bar air supply pressure. Source: From Aguirre, G. 2010.

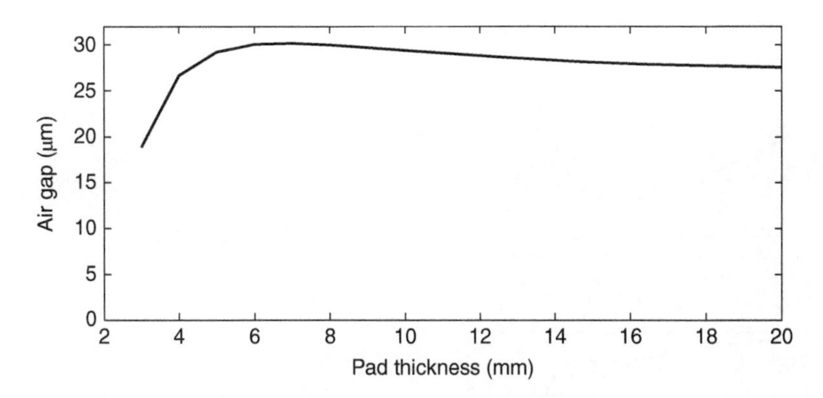

Figure 16.14 Air gap height as a function of bearing plate thickness. Source: From Aguirre, G. 2010.

the stiffness produced by the three piezoelectric actuators distributed in the periphery. The thinnest bearing plate presents a profile height variation, due to the combined action of the elongation of the piezo-actuators and the pressure on the surface, of 11 μm, while for the thickest case the profile variation is 0.4 μm. The high stiffness of the 15 mm plate limits the mean value of the deformation in the plate periphery to 0.8 μm, while for the 5 mm bearing this value increases until 4 μm. For the 2 mm bearing the deformation due to the pressure is dominating, and the average deformation is reduced to 2 μm.

In a similar analysis, Figure 16.14 shows the air gap height, defined as the distance between the centre of the bearing and the counter surface, as a function of the bearing plate thickness. The gap height varies between 18 and 30 μm for the thickness varying from 3 to 7 mm, and decreases slowly for higher plate thickness.

Dynamic Analysis

The dynamic characteristics of active air bearings are also strongly dependent on the thickness of the bearing plate. The stiffness of the aerostatic bearing support as a function of the plate thickness is evaluated in Figure 16.15.

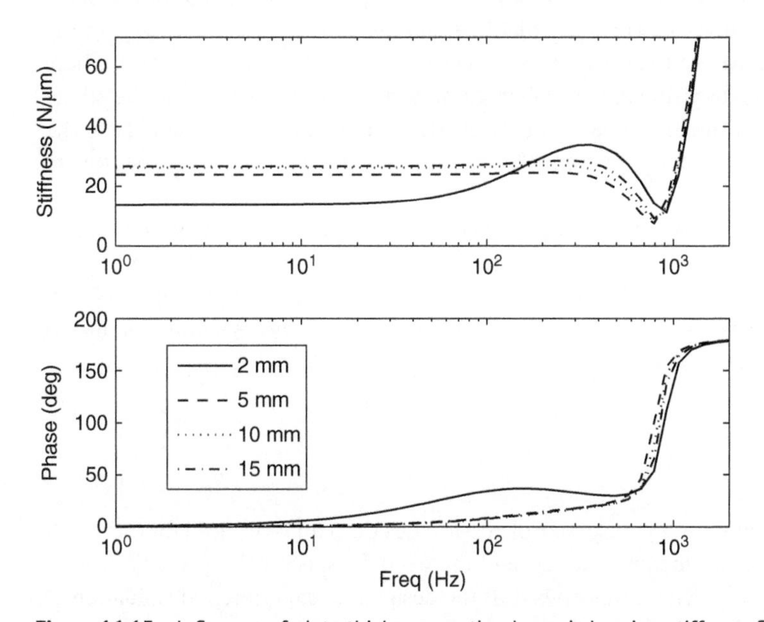

Figure 16.15 Influence of plate thickness on the dynamic bearing stiffness. Source: From Aguirre, G. 2010.

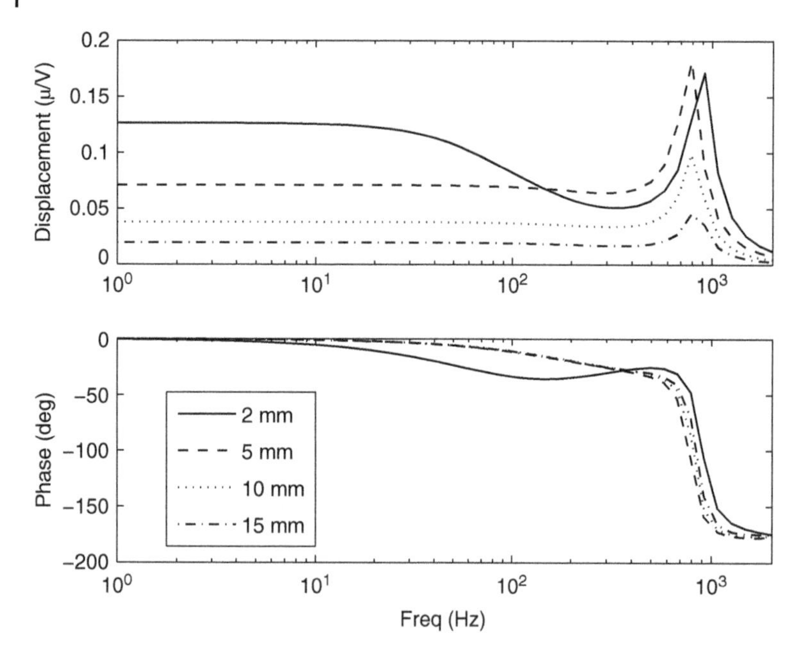

Figure 16.16 Influence of plate thickness on the dynamic bearing displacement generated by piezo-actuators. Source: From Aguirre, G. 2010.

The static stiffness increases with the plate thickness, from 14 N/μm for the 2 mm plate to 27 N/μm for the 15 mm plate. The thinnest bearing shows an increase of the stiffness, exceeding the others above 200 Hz, due to the squeeze-film effect (see Chapters 7 and 15), before dropping at the first mechanical resonance. The other bearings do not show this increase in stiffness, and keep constant stiffness until resonance.

In Figure 16.16 the dynamic displacement of the active air bearing produced by a variation in the voltage of the piezo-actuators is shown. As expected, the thinnest bearing produces the highest deformation of the plate, and thus the highest displacement of the bearing. At 1 Hz, the actuator sensitivity varies from 130 nm V^{-1} for the 2 mm plate to 20 nm V^{-1} for the 15 mm plate. Above 100 Hz, a similar effect as with the stiffness is observed, where the displacement of the thinnest bearing decreases due to the higher stiffness of the air support, while the others remain constant. The resonance frequency of the 5, 10 and 15 mm bearings is approximately the same, 800 Hz, while for the 2 mm bearing, the resonance frequency is higher, 900 Hz, even with a lower static stiffness, due to the squeeze film effect.

The static and dynamic simulations presented above demonstrate that the coupled interaction between the pressure distribution in the bearing gap and the deformation of the plate, produced by the air pressure and by the piezo-actuators, is very relevant to the performance of the active bearing. This presents a complex optimisation problem,which has to be solved, for the design of active air bearings, in which the multiphysics simulation model presented above proves to be an essential tool.

16.4 Active Aerostatic Slide

We have seen that active aerostatic bearings offer interesting control possibilities that can be utilised for the development of active aerostatic slides with sub-micrometer accuracy, and the need for specific design optimisation tools has been addressed in the previous section. The next step is thus the design and experimental validation of an active aerostatic slide prototype, which is discussed hereafter.

16.4.1 Design of the Active Slide Prototype

The design of the active aerostatic slide prototype (see Figure 16.17) is discussed here. The main components of the system are represented schematically in Figure 16.18.

A granite block, 1 m long and 0.4 m wide is used as support and guiding element for the carriage, built in aluminium with 0.5 × 0.5 m top surface. The linear motion is controlled by an ironless linear motor with a high resolution encoder.

Eight aerostatic bearings, six of them active, are used to constrain and control the motion of the slide in the other dof, following the four guiding surfaces on the granite block.

Four active bearings are used for vertical support, even if three are in principle enough to constrain the three dof (vertical displacement, pitch and roll), to increase the stiffness of the system, since they provide a higher bearing surface and better distribution in the carriage. In order to take advantage of actuation symmetry and allow higher compensation forces, all four bearings are active.

Figure 16.17 Active aerostatic slide prototype. Source: Aguirre, G. 2010.

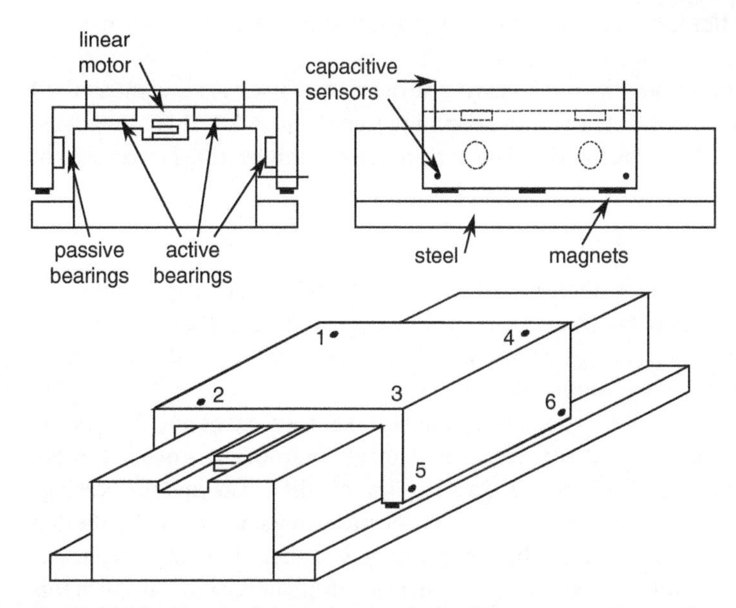

Figure 16.18 Sketch of the active aerostatic slide prototype. Source: From Aguirre, G. 2010.

Since the weight of the slide (50 kg) only provides 125 N pre-load force per bearing, a magnetic pre-load system has been implemented, with 20 disk magnets placed on the lower part of the carriage, facing a steel edge rigidly connected to the granite block, with a magnetic gap that can be regulated up to 5 mm in order to adjust the pre-load force.

Lateral displacement and yaw rotation are constrained by four air bearings, two on each side of the granite block. The bearings on one side are active (positions 5 and 6), while the two on the other side are passive. The role of the passive bearings is to pre-load the active bearings in order to achieve an optimal working gap, and their pre-load can be regulated by means of a screw that connects the carriage to the passive bearing, with a flexible interface element placed between them to reduce the sensitivity of the generated force to the rotation of the screw and enable a finer tuning of the pre-load.

The pre-load applied to the bearings has been tuned in order to have a nominal air gap between 10 and 15 μm. According to the results for single bearings from Section 16.3.2, this corresponds approximately to a total load of 3000 N in vertical direction: combined magnetic attraction and weight of the carriage (500 N).

Five capacitive sensors embedded in the carriage, at the points indicated by a dot on the figure, are used to measure the relative motion between the slide and the granite block. The main design aspects are discussed in more detail next.

Active Air Bearings

The active air bearings in the slide have been designed using the tools presented in the previous section. The most relevant design requirement to be considered is the stroke needed in order to fulfill the following tasks:

- Compensate the flatness error of the guiding surfaces, typically below 5 μm m^{-1}.
- Compensate the effect of disturbances such as external or inertial forces. The stiffness of the aerostatic bearings gives a good estimation of the displacement that needs to be compensated as a function of the applied force.
- The functionality of the slide can be extended with short stroke (up to 10-20 μm) position control in any of the degrees of freedom controlled by the bearings.

The stroke should be well adapted to the real needs, since a high stroke requires larger piezos, requiring more supply current for dynamic operation. For this prototype a bearing stroke of approximately ten micrometers has been chosen, sufficient for compensating the flatness error and for rejecting disturbance forces or for short stroke position control.

The stiffness of the aerostatic film support increases with the bearing surface area, and thus it is maximised within the space available. Bearing design corresponds to the thin bearing presented in 16.3.2. It has a variable plate thickness, 10 mm in the periphery and 6 mm in the interior. The performance of this bearing is discussed in Section 16.3.2.

Measurement

Compensation of measured errors offers the possibility to increase the accuracy of mechanical systems, but when aiming at sub-micrometer applications, measurement becomes the main obstacle to overcome. Position measurement is always relative, with no absolute reference such as gravity that can be taken as reference, and smart calibration strategies are needed to obtain an absolute position reference.

Capacitive distance sensors are used to measure the relative displacement between the carriage and the guiding surfaces of the granite block. Since granite is not conductive, thin steel plates (0.1 mm thick) are glued to the guiding surfaces under the path followed by the capacitive sensors. This implies that the sensor and the bearings must follow separate paths. Since the sensors move along these steel targets, the measurement $s(x)$ will be affected by the straightness error of the targets $t(x)$ (see Figure 16.19), which is typically well above the micrometer. $s(x)$ cannot be used directly as feedback signal to control the position of the carriage during linear motion, since this would transmit the error $t(x)$ to the motion of the slide.

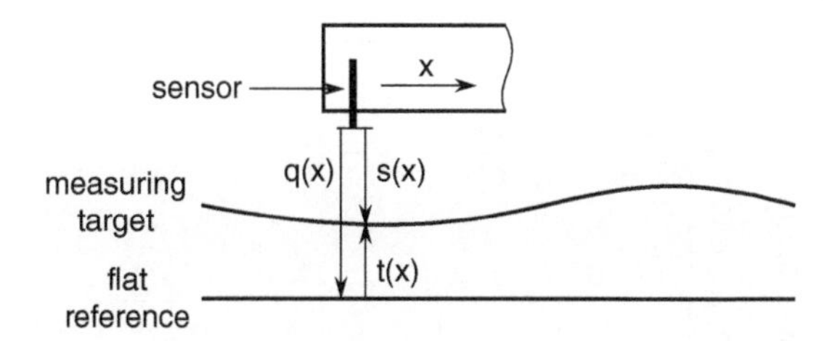

Figure 16.19 Influence of target straightness error $t(x)$ on sensor measurement $s(x)$. Source: From Aguirre, G. 2010.

In order to obtain an absolute measurement $q(x)$ (i.e. relative to a common reference during linear motion) of the position of the slide, a calibration of the profile of the targets is required to obtain $t(x)$. A software reference is thus obtained that can be added to the measurement provided by the sensors during operation to eliminate the effect of the straightness error of the targets

$$q(x) = s(x) - t(x) \tag{16.23}$$

The accuracy of the calibration of the straightness of the measuring targets must be better than the expected accuracy in the motion of the slide. Several methods have been proposed for the straightness calibration, notably the reversal method (Evans et al. 1996), and in (Park et al. 2006; Ro et al. 2010) some of these methods are applied to linear slides.

16.4.2 Identification of Active Slide Characteristics

Before implementing active control strategies, the main characteristics of the active aerostatic slide prototype presented above are identified here. First the passive stiffness of the slide for vertical loads applied on the carriage is presented. Then, the dynamic response of the system to the voltage applied to the bearings is discussed.

Passive Stiffness

Since the stiffness is the derivative of the applied force with respect to the measured displacement, the multiple possible combinations of points where the force can be applied and the displacement measured lead to many possible definitions of stiffnesses (or cross-coupled stiffnesses when the measurement and force positions are different), which can be misleading for understanding the behaviour of the system.

In order to avoid this, the response of the slide to external disturbance forces is illustrated here with two experiments, in which a force normal to the top surface (7.5 N amplitude) of the slide is applied at different frequencies consecutively in a wide frequency range, and the displacement at different points is measured. In the first case, the force is applied at the centre of the carriage, so that the effect of the force is distributed more or less equally over the four bearings. In the second, the force is applied at one corner (position 1), in order to analyse the case in which most of the effect of the force is taken by one of the bearings (see Figure 16.20). The displacement measured

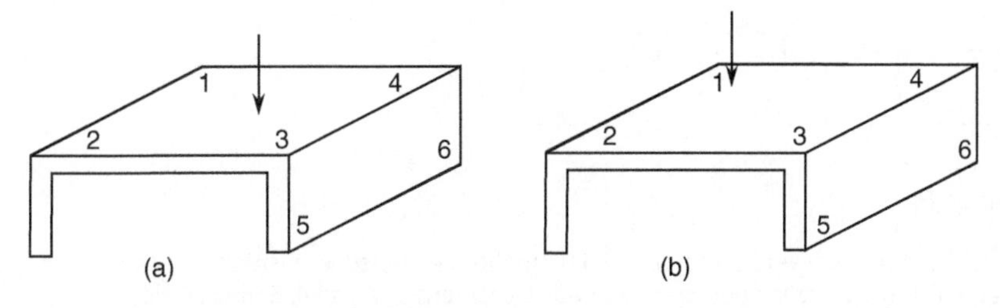

Figure 16.20 Positions at which the excitation force is applied to the slide, (a) centre of the slide, (b) a corner, position 1. Source: From Aguirre, G. 2010.

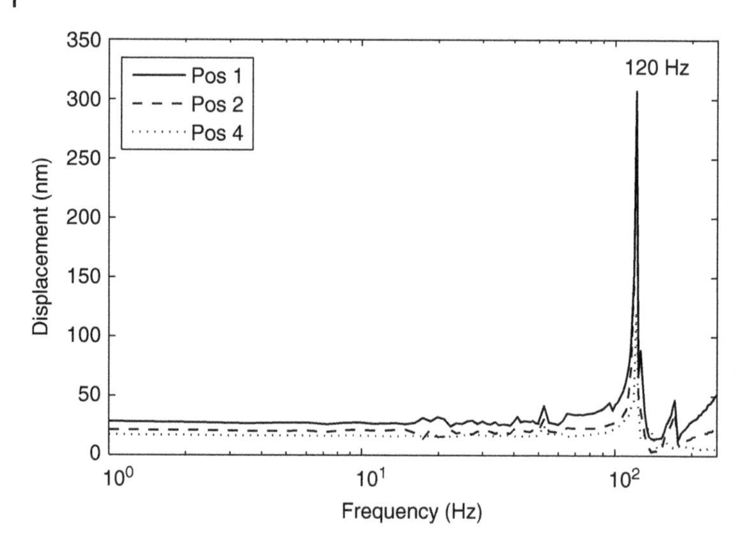

Figure 16.21 Displacement amplitude of the slide for a vertical force of 7.5 N amplitude applied at the centre of the carriage, as a function of the frequency at which the force is applied, measured at positions 1, 2 and 4. Source: From Aguirre, G. 2010.

by the three vertical sensors is shown as a result, in Figure 16.21 for the case of the force applied at the centre and in Figure 16.22 for when the force is applied at position 1.

When the force is applied at the centre of the carriage, the three measurements show similar displacement. At low frequencies, the displacement is between 20 and 30 nm, and when the force is applied at higher frequencies, the first main resonance of the system is observed at 120 Hz. This resonance corresponds to the first vertical vibration mode of the slide on the aerostatic support.

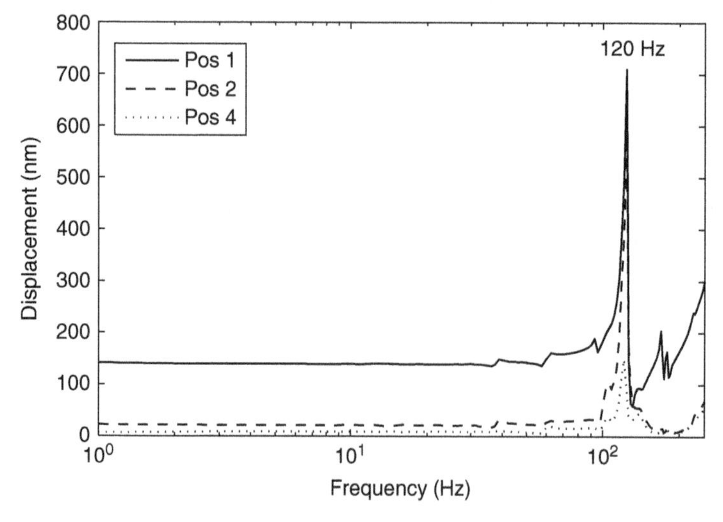

Figure 16.22 Displacement amplitude of the slide for a vertical force of 7.5 N amplitude applied at position 1 of the carriage, as a function of the frequency at which the force is applied, measured at positions 1, 2 and 4. Source: From Aguirre, G. 2010.

When the load is applied at position 1, the difference between the displacements measured at each point increases. The applied force affects mainly the bearing closest to it, and therefore, at low frequencies, the measurement at position 1 shows the highest deflection and lowest stiffness (140 nm, 50 N μm^{-1}), with a much lower displacement at the other points, 20 nm and 375 N/μm at position 2 and 10 nm and 750 N μm^{-1} at position 4. As in the centred force case, the first resonance is observed at 120 Hz.

Dynamic Actuation

The dynamic response of the slide to a low amplitude action on the piezoelectric actuators is analysed here. The dynamic sensitivity (nm V^{-1}) of the slide to the actuation on the bearing in position 2 is shown in Figure 16.23 for the vertical motion (positions 1, 2, 3 and 4). A 50 V offset and a 5 V amplitude sinusoidal input is applied to the active bearing.

At low frequencies, the sensor next to the actuated bearing (position 2) measures the highest displacement (61 nm V^{-1}). At positions 1 and 3, a displacement in phase with the actuation is measured, but of lower amplitude (8 and 2 nm V^{-1}). At position 4, on the opposite corner to the actuation, the displacement of 15 nm V^{-1} is measured, but with opposite direction to the actuation.

Due to the cross coupling between different actuators and sensors, the optimal performance of the slide will be achieved by means of Multiple-Input-Multiple-Output (MIMO) control. However, in the following results, Single-Input-Single-Output control (SISO) is used (i.e. each sensor–actuator pair is controlled independently), which proves to be sufficient for demonstrating the validity of the compensation strategy.

16.4.3 Active Performance

Once the dynamic characteristics of the system have been identified, the next step is to analyse its active performance. The goal here is to demonstrate the capability of the active system to control the position of the carriage with high accuracy and bandwidth. The linear motion, provided by the linear motor, is out of the scope of this book.

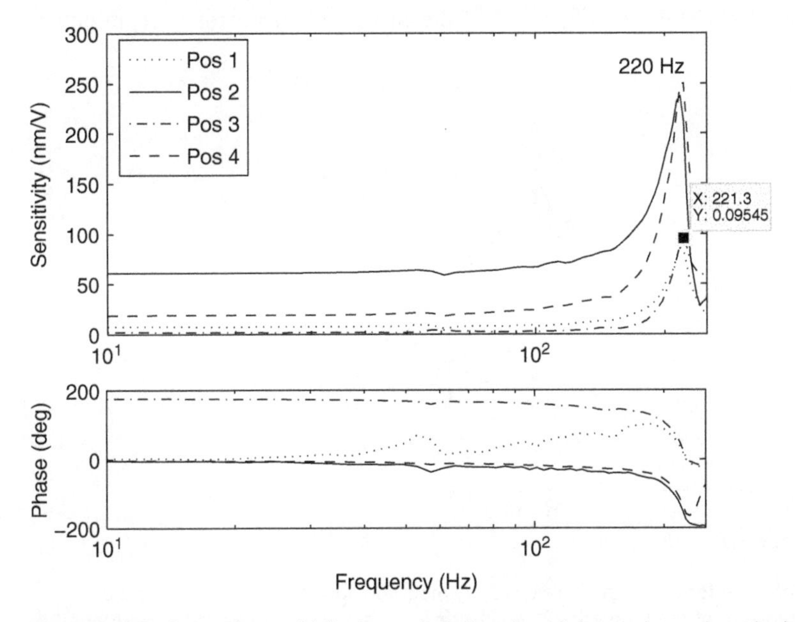

Figure 16.23 Dynamic sensitivity in vertical direction of the carriage when the bearing in position 2 is actuated. Source: From Aguirre, G. 2010.

The vertical performance of the slide is demonstrated here. Since SISO control is used, this limits the actuation to three active bearings and the control to three dof. Three vertical bearings are used here for controlling the vertical displacement and pitch and roll rotations.

A SISO proportional-integral controller has been implemented for each active bearing-sensor pair. For trajectory tracking tests, the hysteresis of the piezoelectric actuators is compensated for by means of a feedforward model (Aguirre et al. 2012).

The active stiffness and the tracking accuracy are tested in the following.

Active Stiffness

The effect of the active control on the stiffness of the slide is evaluated here with the same tests performed to identify the passive stiffness of the system, i.e. a dynamic vertical force (7.5 N amplitude) applied at the centre and at one corner of the top surface of the carriage, and the displacement measured by the three vertical sensors.

Both results, with the load applied at the centre (Figure 16.24) and at position 1 (Figure 16.25), indicate that the control system increases the stiffness at frequencies below 60 Hz. The integral control leads to quasi-infinite stiffness at low frequencies, and the improvement of the stiffness depends on the frequency of the disturbance, e.g. at 10 Hz the stiffness is increased ten-fold approximately.

In order to illustrate the effect of the active control, in Figure 16.26 the measurement of one of the vertical sensors is shown, when a 7.5 N amplitude and 10 Hz force is applied to the slide, with and without active control. Without control, a vibration of 70 nm peak-to-valley is measured, while with active control the effect of the applied force is not observed because it is of lower amplitude than the positioning resolution of the slide.

Given that only a 70 nm stroke is required for compensating the 7.5 amplitude force, the active system should be able to compensate much higher forces. This has not been proved experimentally, however, due to the force limitation of the shaker.

Tracking Control

The tracking control capabilities of the slide are evaluated by applying a reference signal representing a vertical motion of the slide, at frequencies from 1 to 50 Hz. In Figure 16.27 the measured rms error is represented

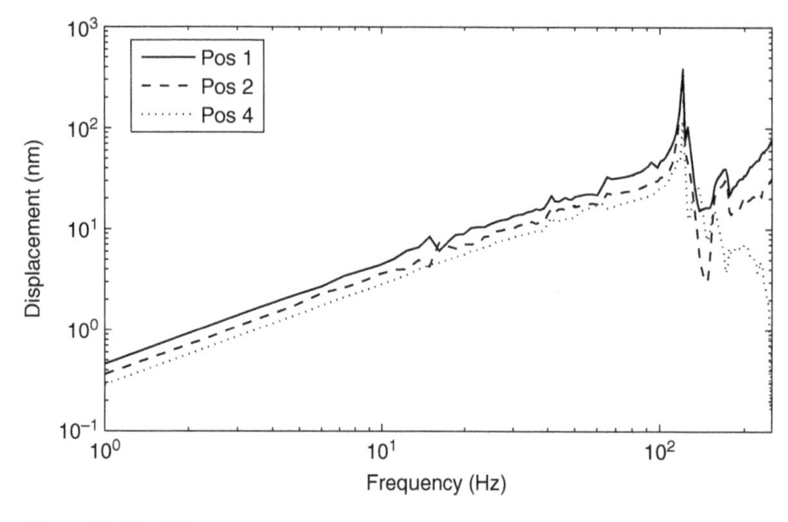

Figure 16.24 Displacement amplitude of the slide for a vertical force of 7.5 N amplitude applied at the centre of the carriage when the active control is on, as a function of the frequency at which the force is applied, measured at positions 1, 2 and 4. Source: From Aguirre, G. 2010.

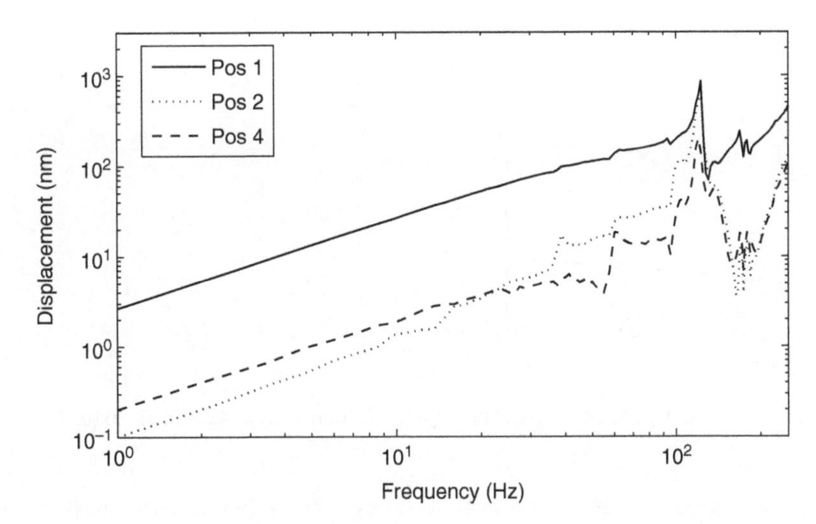

Figure 16.25 Displacement amplitude of the slide for a vertical force of 7.5 N amplitude applied at position 1 of the carriage when the active control is on, as a function of the frequency at which the force is applied, measured at positions 1, 2 and 4. Source: From Aguirre, G. 2010.

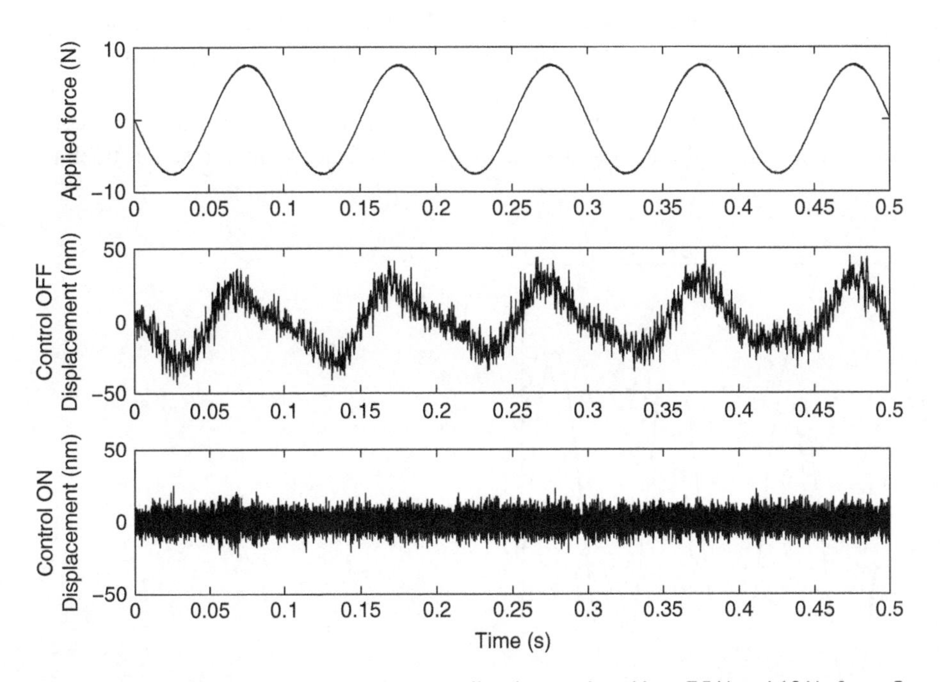

Figure 16.26 Effect of active control on the vibration produced by a 7.5 N and 10 Hz force. Source: From Aguirre, G. 2010.

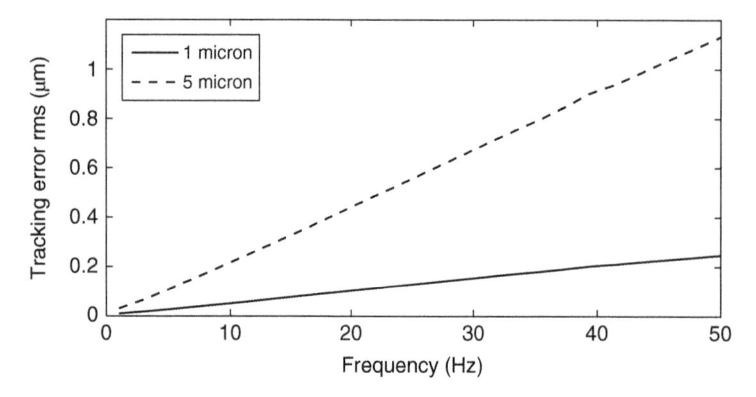

Figure 16.27 Tracking error for vertical motion of 5 and 1 μm peak-to-valley amplitude. Source: From Aguirre, G. 2010.

as a function of frequency, for a reference signal of 5 and 1 μm peak-to-valley amplitude (i.e. 1.76 and 0.35 μm rms respectively). The error increases with the frequency of the tracking signal, reaching 10% error at 8 Hz.

This relatively big error at high frequencies is due to the phase delay in the tracking because of the large inertia of the carriage. As shown in Figure 16.29 for the case of 5 μm motion, the system follows the input signal quite accurately in amplitude, but with increasing phase lag. In Figure 16.28 the reference and measured signals at 50 Hz are represented, which show a *clean* tracking (no other relevant frequency components) of the input signal.

The combination of this feedback strategy with feedforward control, as indicated at the start of Section 16.4.3 for single air bearings, adapted to multi-dof control, is expected to lead to much better tracking results.

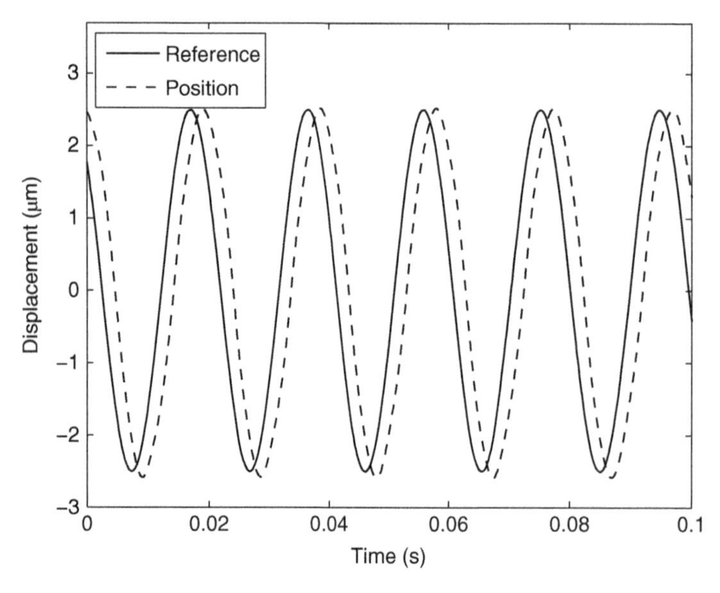

Figure 16.28 Reference and measured signals at 50 Hz and 5 μm peak-to-valley. Source: From Aguirre, G. 2010.

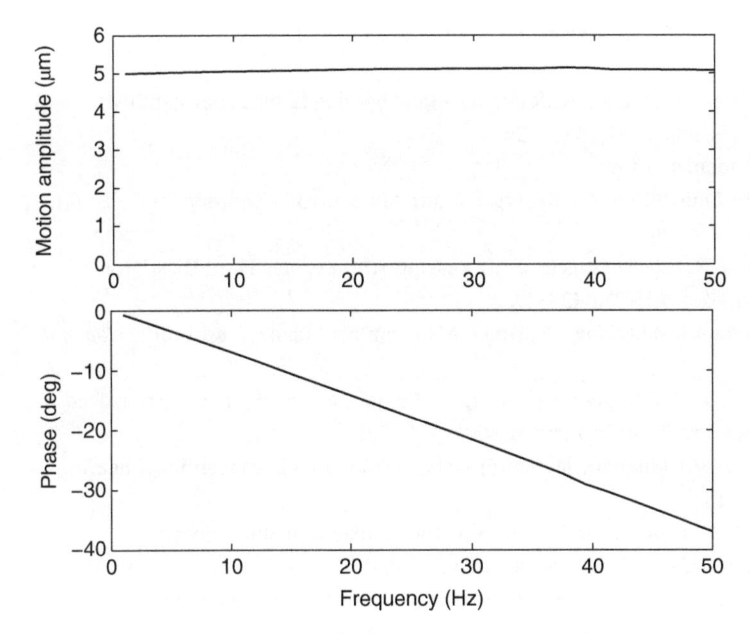

Figure 16.29 Frequency response for tracking of 5 μm peak-to-valley reference. Source: From Aguirre, G. 2010.

16.5 Conclusions

In this chapter, a linear slide with sub-micrometer accuracy requirement in six degrees of freedom has been chosen as a plausible industrial application for active air bearings described in Chapter 15, and a first prototype has been developed and principally tested. The first challenge to achieve this prototype is the optimisation of the design of the active air bearings, considering the coupled interaction between fluid dynamics, structural flexibility, piezoelectricity and control. As a tool to perform this optimisation, we have proposed and validated a multiphysics simulation model that can form the basis for carrying out this task.

The next step has been the design and construction of the active slide, analysing the interaction of several active bearings and their integration into a more complex system, where other effects, such as manufacturing limitations or calibration can determine the design requirements.

The basic functionality of the active slide prototype has been demonstrated in that it could maintain sub-micrometer tracking error in the presence of disturbances. The implementation of calibration strategies and advanced control techniques should allow for the achievement of such levels of accuracy during operation.

References

Aguirre, G., (2010). Optimized Design of Active Air Bearings for Ultra-Precision Positioning Slides PhD thesis KU Leuven – Dept. Mechanical Engineering.

Aguirre, G., Al-Bender, F. and Van Brussel, H., (2010). A multiphysics model for optimizing the design of active aerostatic thrust bearings. *Precision Engineering* 34(3): 507–515.

Aguirre, G., Janssens, T., Van Brussel, H. and Al-Bender, F., (2012). Asymmetric-hysteresis compensation in piezoelectric actuators. *Mechanical Systems and Signal Processing* 30(0): 218–231.

Al-Bender, F., (1992). Contributions to the Design Theory of Circular Centrally Fed Aerostatic Bearings PhD thesis Leuven, Belgium.

Al-Bender, F., (2009). On the modelling of the dynamic characteristics of aerostatic bearing films: From stability analysis to active compensation. *Precision Engineering* 33(2): 117–126.

CSC - IT Center for Science n.d. Elmer finite element software.

Evans, C.J., Hocken, R.J. and Estler, W.T., (1996). Self-calibration: Reversal, redundancy, error separation, and absolute testing. *CIRP Annals* 45(2): 617–634.

Jayson, E.M., Murphy, J., Smith, P.W. and Talke, F.E., (2003). Effects of Air Bearing Stiffness on a Hard Disk Drive Subject to Shock and Vibration. *Journal of Tribology* 125(2): 343–349.

Ono, K., (1975). Dynamic Characteristics of Air-Lubricated Slider Bearing for Noncontact Magnetic Recording. *Journal of Lubrication Technology* 97(2): 250–258.

Park, C.H., Oh, Y.J., Shamoto, E. and Lee, D.W., (2006). Compensation for five DOF motion errors of hydrostatic feed table by utilizing actively controlled capillaries. *Precision Engineering* 30(3): 299–305.

Peijnenburg, A., Vermeulen, J. and van Eijk, J., (2006). Magnetic levitation systems compared to conventional bearing systems. *Microelectronic Engineering* 83, 1372–1375.

Peng, J.P. and Hardie, C.E., (1995). A Finite Element Scheme for Determining the Shaped Rail Slider Flying Characteristics With Experimental Confirmation. *Journal of Tribology* 117(1): 136–142.

Piefort, V. (2001). Finite Element Modelling of Piezoelectric Active Structures PhD thesis Universite Libre de Bruxelles.

Preumont, A., (2006). *Mechatronics: Dynamics of Electromechanical and Piezoelectric Systems.* Springer.

Ro, S.K., Kim, S., Kwak, Y. and Park, C.H., (2010). A linear air bearing stage with active magnetic preloads for ultraprecise straight motion. *Precision Engineering* 34(1): 186–194.

Tang, T., (1971). Dynamics of Air-Lubricated Slider Bearings for Noncontact Magnetic Recording. *Journal of Lubrication Technology* 93(2): 272–278.

Zienkiewicz, O., Taylor, R. and Zhu, J., (2005). *The Finite Element Method. Its Basis & Fundamentals* 5 edn. Elsevier Butterworth-Heinemann.

17

On the Thermal Characteristics of the Film Flow

17.1 Introduction

The solutions to the Reynolds equation, pertaining to externally-pressurised as well as self-acting bearings, which have been provided throughout this book, have been restricted to the isothermal flow case. We shall see later that this assumption does not depart appreciably from most actual situations encountered in practice, especially if one employs a realistic mean temperature. Moreover, the solution method applies also to the more general case of polytropic flow, or more precisely, to the case when the density is a given function of the pressure. However, we cannot know beforehand which case to choose since that depends on the heat transfer to and from the film. To establish this, we need to solve the (thermal) energy equation simultaneously with the continuity and momentum equations, which adds another dimension of complexity to the modelling problem and to its solutions. A more pertinent issue is not so much that of selecting the right polytropic exponent but rather establishing the amount of thermal power generated in the aerodynamic-bearing gap, how it flows through the bearing walls and along the film, and what consequences this has on the bearing characteristics.

The thermal behaviour of the flow in gas bearings and the associated heat transfer have not enjoyed as much or as thorough treatment as other problems such as load, stiffness, dynamics, etc.

Schlichting (1968) presents a systematic treatment of the thermal behaviour of open boundary-layer flows, which provides a very good theoretical foundation, in regard to general equations, properties and considerations as well as intricacies of the problem, to enable one to deal with those aspects.

One of the first treatments of thermal aspects in gas bearings was perhaps that of Hughes and Osterle (1957) who reasoned that the flow state will lie somewhere between the isothermal and adiabatic solutions, which describe the two limiting flow conditions, but thought that in many practical situations the flow is adiabatic. Consequently, they formulated the adiabatic flow situation between two parallel plates, i.e. the Couette problem, and provided a solution for it, which exhibits the *thermal* wedge phenomenon, namely that heat generation in the parallel gap gives rise to a positive pressure build-up (see later in this chapter).

Perhaps the first fairly complete and comprehensive treatment of the thermal behaviour of gas bearings was given by Constantinescu (1969), using simplified theory and considering both adiabatic and (mostly) isothermal walls, to cover the most essential aspects of topic and provide useful qualitative and quantitative results. Constantinescu re-derived the theory of Hughes and Osterle (1957) and concluded that the thermal wedge has only limited relevance when the flow is not wholly adiabatic and when the sliding number is small. Pinkus (1990) devoted a book to the thermal aspects of fluid-film bearings in which heat dissipation is calculated while neglecting possible pressure gradients. This approach then substantially simplifies the models as it uncouples the energy and Reynolds equations from one another. In this chapter, we shall also begin the treatment by a similar approach, where we shall see that the thermal problem is relevant only when the heat flow from the bearing surfaces is

Air Bearings: Theory, Design and Applications, First Edition. Farid Al-Bender.

Companion website: www.wiley.com/go/AlBender/AirBearings

severely restricted, such as would be the case in bump-foil bearings and hinge-supported tilting-pad bearings, in particular when the viscous friction is large, namely, at high bearing numbers. Thus, Salehi and Heshmat (2000) set up such a Pinkus-type model to evaluate thermal aspects in the steady-state operation of bump-foil bearings and seals, where they reported good agreement with experimental results. Radil and Zeszotek (2004) conducted experiments on compliant foil bearing at speeds from 20 to 50 krpm and loads ranged from 9 to 222 N to show that both load and speed were responsible for heat generation, though with a more pronounced role of the speed. Peng and Khonsari (2006) presented a comprehensive thermohydrodynamic (THD) analysis applied foil bearings, which considers a simplified 3D energy equation. A THD analysis of compliant-flexure pivot tilting pad bearings has been presented by Sim and Kim (2008) who treat the Reynolds and 3D energy equations simultaneously with boundary conditions being established through global energy balance considerations. San Andrés and Kim (2010) present a THD that is based on thermal energy transport model, which includes convection and conduction through the bearing surfaces. They reported good agreement between model and experimental data and showed that the gas film temperature increases rapidly due to the absence of a forced cooling air that could carry away the recirculation gas flow and thermal energy drawn by the spinning rotor. Paulsen et al. (2011) present detailed mathematical modelling for non-isothermal lubrication of a compressible fluid film journal bearing, in order to gauge whether and when this type of analysis should be of concern. They apply the analysis to circular, elliptical and three-lobed journal bearings comparing the THD results with isothermal results from the literature. Load capacity, stiffness, and damping coefficients are determined by the solution of the standard Reynolds equation coupled to the energy equation. They conclude that their numerical findings support the common literature assumption of the bearing being isothermal when operating in the low bearing number regime, regardless of the bearing geometry, and therefore suggest that thermohydrodynamic effects in air-lubricated journal bearings should be considered in cases where the bearing is heavily loaded and static and dynamic properties or the temperature itself are of special concern.

In this chapter, we shall propose approximate solutions to the energy equation in two cases: (i) the case of an aerodynamic film, where we first decouple the energy equation from the momentum equation, by means of some simplifying assumptions, and then consider the thermal-wedge, adiabatic flow problem; (ii) the case of an EP centrally-fed bearing, where we solve the film entrance problem by supplementing the energy equation to the momentum equation in a somewhat simplified, though effective manner. In particular, we shall be interested in how far the flow departs from isothermal assumption, when the walls are assumed to be either wholly isothermal, wholly adiabatic or a combination of both. Nevertheless, other aspects of the problem, such as the heat transfer characteristics of the bearing arrangement and the possible thermal expansion of the bearing surfaces, may be also of academic as well as practical interest, but are not in the scope of this treatment.

We shall see that the choice of the polytropic exponent does not have a drastic effect on the pressure distribution, especially for EP bearings. Rather, there are two more important issues that warrant considering thermal effects in a gas film:

- High sliding velocities, such as those obtained in high-speed journal and thrust bearings, cause large heat generation as a result of viscous-friction heat dissipation. As an example consider $V = 100\ \mathrm{m\ s^{-1}}$, $h = 10^{-5}$ m, $\mu = 2\ 10^{-5}$ Pa s. Then by application of Petrov's formula for viscous friction, we obtain a heat generation rate of

$$q = \frac{\mu V^2}{h} = 20\,\mathrm{kW\ m^{-2}}.$$

  This is a significant amount of heat that needs to be evacuated from the bearing film, not least in order to avoid undesired or even catastrophic deformations of the bearing surfaces. (Note also that q varies quadratically with V, so that at a speed of $300\ \mathrm{m\ s^{-1}}$, which is not unheard of in present ultra-high speed applications, we have $q = 180\ \mathrm{kW\ m^{-2}}$!). Two pertinent questions arise here: (i) how much of this heat could be evacuated by convection through the gas film? (ii) how would possible temperature rise in the film affect the pressure directly or indirectly through affecting gas properties and thus the general functioning of the gas film?

- For EP bearings, often used as components (slideways, radial bearings, etc.) in precision machines, gas expansion at inlet to the film (particularly in orifice-compensated bearings) could result in local cooling of the bearing counter surfaces, which in its turn could generate undesirable thermal gradients, which could unfavourably influence the behaviour of ultra-precision machines. It would therefore be interesting to formulate some qualitative and quantitative understanding of this type of situation.

The objective of this chapter is to formulate suitable answers to those two concerns.

17.2 Basic Considerations

Let us begin by restating the steady-state energy equation, from Chapter 2, for the plain (2D) case, with x being the flow direction and z the gap direction:

$$\rho c_p u \frac{\partial T^*}{\partial x} = u\frac{\mathrm{d}p}{\mathrm{d}x} + k\frac{\partial^2 T^*}{\partial z^2} + \mu\left(\frac{\partial u}{\partial z}\right)^2 \tag{17.1}$$

where, c_p is the specific heat capacity at constant pressure, T^* is the absolute temperature and k is the thermal conductivity of the gas. The term on the L.H.S. of the equation accounts for convection. On the R.H.S., we have first the term representing the material pressure derivative, which is peculiar to compressible flow, followed by the conduction term and finally the dissipation term.

Making use of the (viscous momentum equation)

$$\frac{\mathrm{d}p}{\mathrm{d}x} = \mu\frac{\partial^2 u}{\partial z^2},$$

the first and third terms on the R.H.S. of Eq. 17.1 can be coalesced, so as to eliminate the pressure, to yield

$$\underbrace{\rho c_p u \frac{\partial T^*}{\partial x}}_{\text{convection}} = \underbrace{k\frac{\partial^2 T^*}{\partial z^2}}_{\text{conduction}} + \underbrace{\mu\frac{\partial}{\partial z}\left(u\frac{\partial u}{\partial z}\right)}_{\text{modified dissipation}} \tag{17.2}$$

(17.3)

In regard to the modified dissipation term, we note that

$$\int_0^h \frac{\partial}{\partial z}\left(u\frac{\partial u}{\partial z}\right)\mathrm{d}z = u\frac{\partial u}{\partial z}\bigg|_0^h = 0,$$

whenever either or both of the velocity or its derivative vanish at both of the walls. In other words, Poiseuille flow on its own will cause no dissipation of energy; but Couette flow will. Their combination should be assessed according to the situation at hand: (i) if the Poiseuille flow is imposed by external pressurisation in the same direction as the Couette flow, then it will reduce dissipation; (ii) if the Poiseuille component arises as a consequence of the aerodynamic action, then its contribution depends on the pressure distribution, which could be either to increase dissipation or to decrease it; but it will generally be in much smaller magnitude as compared to the Couette contribution.

To begin with, we normalise the equation by putting:
$H = h/h_0, Z = z/h, X = x/L, \wp = \rho/\rho_0, U = u/V\ T = (T^* - T_0)/T_0,$
where T_0 is, for now, an arbitrarily chosen reference temperature, to obtain:

$$\wp \mathrm{Re}^* U\frac{\partial T}{\partial X} = \frac{1}{\mathrm{Pr}}\frac{\partial^2 T}{\partial Z^2} + \mathrm{Ec}\,\frac{\partial}{\partial Z}\left(U\frac{\partial U}{\partial Z}\right) \tag{17.4}$$

where,

- $\mathrm{Re}^* = \dfrac{\rho_o V h_o^{\,2}}{\mu L}$ is the reduced Reynolds number
- $\mathrm{Pr} = \frac{c_p \mu}{k}$ is the Prandtl number
- $\mathrm{Ec} = \frac{V^2}{c_p T_o}$ is the Eckert number based on the reference temperature T_o

We shall see that the subsequent solutions to Eq. 17.4 depend on three dimensionless parameters, which are combinations of the three numbers above:

- $\mathrm{Pr}\ \mathrm{Ec} = \frac{\text{dissipated power}}{\text{conducted power}}$
- $\mathrm{Ec}/\mathrm{Re}^* = \frac{\text{dissipated power}}{\text{convected power}}$
- $\mathrm{Pr}\ \mathrm{Re}^* = \frac{\text{convected power}}{\text{conducted power}}$

Note also that by virtue of the equation of state $\rho_o \mathfrak{R} T_o = p_o$, we have

$$\frac{\mathrm{Ec}}{\mathrm{Re}^*} = \frac{\mathfrak{R}}{6c_p}\Lambda = \frac{1}{6}\left(\frac{\kappa - 1}{\kappa}\right)\Lambda,$$

where,

$$\Lambda = \frac{6\mu L V}{p_o h_o^{\,2}}$$

is the bearing number.

Firstly, in order to see the problem in its simplest guise and gain some idea about basic behaviour, we consider Couette flow in a parallel channel with a stationary lower surface and with the upper surface moving at a constant velocity V. We further assume that the pressure, and consequently also the density, are constant, i.e. $\wp = 1$, (except for the effect of changing temperature, which will be neglected for the time being). This amounts to decoupling the energy equation from the Reynolds equation and the equation of state. Furthermore, for small changes in the temperature, it would be justifiable to assume c_p, μ, k to be constant. We shall consider three cases: (i) isothermal walls; (ii) adiabatic walls; (iii) one adiabatic wall and one isothermal wall. In all cases, we neglect entrance transients and consider only the asymptotic solution.

For initial conditions on the temperature, we assume, without loss of generality, that the entrance air temperature T_a is uniform and equal to the wall temperature T_w and choose $T_o = T_w = T_a$.

17.2.1 Isothermal Walls

First, we begin by neglecting convection entirely in Eq. 17.4, with $\wp = 1$ and $U = Z$, and solving the conduction problem:

$$0 = \frac{1}{\mathrm{Pr}}\frac{\partial^2 T}{\partial Z^2} + \mathrm{Ec}\,\frac{\partial}{\partial Z}\left(U\frac{\partial U}{\partial Z}\right) \tag{17.5}$$

with boundary conditions: $T = 0$ at $Z = 0, 1$.

Integrating twice w.r.t. Z and substituting the b.c., yields

$$T = \frac{\mathrm{PrEc}}{2}Z(1 - Z) \tag{17.6}$$

That is, the temperature profile across the gap will be parabolic with mean value being equal to $\frac{\mathrm{PrEc}}{12}$.

N.B. When the wall temperatures are not equal, one can use the same method of solution to show that, in this case, there will be an additional linear temperature gradient across the gap between the two walls, see Figure 17.1(a).

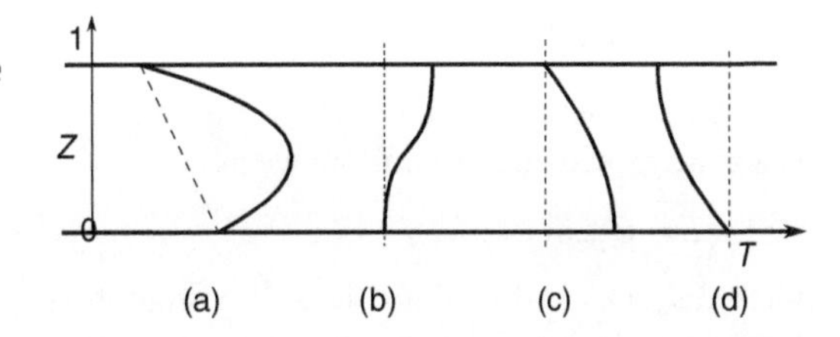

Figure 17.1 Temperature profiles across the gap. (a) Isothermal walls at different temperatures, (b) adiabatic walls, (c) lower surface adiabatic, upper surface isothermal, (d) lower surface isothermal, upper surface adiabatic.

Subsequently, we utilise an approximate integral method based on this solution to obtain the general solution to Eq. 17.4, by writing

$$T = \Theta(X)6Z(1 - Z),$$

where, $\Theta(X)$ may be taken as the mean gas temperature across the gap (N.B. the mean of $6Z(1 - Z)$ is equal to unity), to solve

$$\mathrm{Re}^* \frac{\partial}{\partial X} \int_0^1 UT\mathrm{d}Z = \frac{1}{\mathrm{Pr}} \int_0^1 \frac{\partial^2 T}{\partial Z^2}\mathrm{d}Z + \mathrm{Ec} \int_0^1 \frac{\partial}{\partial Z}\left(U\frac{\partial U}{\partial Z}\right)\mathrm{d}Z \tag{17.7}$$

This results in

$$\frac{\mathrm{Re}^*}{2}\frac{d}{\mathrm{d}X}\Theta(X) = -\frac{12}{\mathrm{Pr}}\Theta(X) + \mathrm{Ec} \tag{17.8}$$

which upon integration w.r.t. X yields:

$$\Theta(X) = \left(\Theta(0) - \frac{\mathrm{PrEc}}{12}\right)\exp\left(-\frac{24}{\mathrm{PrRe}^*}X\right) + \frac{\mathrm{PrEc}}{12} \tag{17.9}$$

or, with $\Theta(0) = 0$ (= the entrance normalised temperature), we have:

$$\Theta(X) = \frac{\mathrm{PrEc}}{12}\left(1 - \exp\left(-\frac{24}{\mathrm{PrRe}^*}X\right)\right) \tag{17.10}$$

Thus, we see that the convective part, being represented by the exponential function, decays at the rate of $\frac{24}{\mathrm{PrRe}^*}$, which is quite fast when PrRe^* is small enough, as most often is the case ($\mathrm{PrRe}^* < 1$), so that we are left with the conduction part of Eq. 17.6 as the dominant part. As an example, we take the same value from the introductory section, viz. $V = 100$ m s^{-1}, $h = 10^{-5}$ m, $\mu = 2\ 10^{-5}$ Pa s, Pr $= 0.72$ (air) and $L = 25$ mm to get:

$$\frac{24}{\mathrm{PrRe}^*} \approx 1700,$$

so that the exponential term will decay completely a few micrometers downstream of entrance.

Furthermore, the steady-state mean film temperature is equal to:

$$\frac{\mathrm{PrEc}}{12},$$

which is independent of c_p and ρ_0 since it depends only on conduction.

Thus, we conclude that, except at the entrance, the mean of the temperature in the gap will be constant along the flow. That is, the flow will likewise be isothermal, with mean temperature generally not equal to the wall temperature, though not appreciably larger:

$$\frac{T^* - T_0}{T_0} = \mathrm{Pr}\ \frac{V^2}{12c_pT_0},$$

where, $T^* - T_0 = \Delta T$ is the difference between mean gas temperature and wall temperature (which we have assumed also to be the ambient temperature).

Substituting the values of the above example, yields:

$$T^* - T_0 = 0.6^\circ\mathrm{K}.$$

17.2.2 Adiabatic Walls

In this case there is no heat transfer between the walls and the gas so that $\partial T/\partial Z = 0$ at the walls $Z = 0, 1$. To establish a steady-state solution, we assume:

$$T = \Theta_{\rm aw}(X) + f(Z),$$

where, $\Theta_{\rm aw}(X)$ is the adiabatic wall temperature, i.e. the temperature of the gas at the walls, while $f(Z)$ is the difference of the gas temperature with the wall temperature. Thus, f satisfies the following b.c.:

$$f(0) = 0, \qquad f'(0) = f'(1) = 0.$$

Inserting this in Eq. 17.4 (with $\wp = 1$) yields:

$$\text{Re}^* Z \frac{\text{d}}{\text{d}X}\Theta_{\rm aw}(X) = \frac{1}{\text{Pr}}\frac{\text{d}^2 f(Z)}{\text{d}Z^2} + \text{Ec}, \tag{17.11}$$

which after rearrangement gives:

$$f''(Z) = \text{PrRe}^*\Theta'_{\rm aw}(X)Z - \text{EcPr}, \tag{17.12}$$

where the primes denote ordinary derivatives w.r.t. to the argument of the function. Integrating once w.r.t. Z and applying the b.c. on the derivatives $f'(Z)$ at the walls, yields

$$f'(Z) = \frac{\text{PrRe}^*\Theta'_{\rm aw}(X)}{2}Z^2 - \text{EcPrZ}, \tag{17.13}$$

and ($f'(1) = 0$)

$$\Theta'_{\rm aw}(X) = 2\frac{\text{Ec}}{\text{Re}^*}. \tag{17.14}$$

Integrating the previous two equations and assembling, we obtain

$$T = \Theta_{\rm aw}(0) + 2\frac{\text{Ec}}{\text{Re}^*}X + \text{EcPr}Z^2\left(\frac{Z}{3} - \frac{1}{2}\right), \tag{17.15}$$

(Obviously, our assumptions on the initial conditions $T_{\rm w} = T_{\rm a} = T_{\rm o}$ imply that $\Theta_{\rm aw}(0) = 0$.)

From this solution, we see that:

- The wall temperatures increase steadily at the rate $2\text{Ec}/\text{Re}^*X$. As an example consider $V = 100$ m s^{-1}, $h = 10^{-5}$ m, $\mu = 2\ 10^{-5}$ Pas, $c_p = 1$ kJ kg^{-1} K^{-1}, $\rho_{\rm o} = 1$ kg m^{-3}, then $\Delta T/x = 4\ 10^4$ K m^{-1} (or 400 K cm^{-1}!). This very high value points to the ineffectiveness of convection to evict the dissipated energy out of the film and the need to have sufficient conductivity of the bearing surfaces.
- The lower surface is hotter than the upper (sliding) surface with the difference between them being equal to the function $\text{EcPr}Z^2\left(\frac{Z}{3} - \frac{1}{2}\right)|_{Z=1} = -\text{EcPr}/6$ (which is approximately equal to 1.2 K) in the above example.
- We can verify that the temperature distribution between lower and upper surface, i.e. $\text{EcPr}Z^2\left(\frac{Z}{3} - \frac{1}{2}\right)$, has zero derivatives at $Z = 0, 1$.

From these results, we conclude that the convection rate is much smaller than the dissipation rate, so that convection (by Couette flow only) is very inefficient in evacuating the heat generated by viscous dissipation. Inspection of the equations and the parameters will reveal that the main culprit in this shortcoming is the low density of air. If air were to have the density of water (with everything else being the same), most of the heat would have been evacuated by convection.

Adding Poiseuille flow, via a pressure gradient along the bearing, will increase convection without adding extra dissipation, by increasing the air flow. However, a small calculation with realistic values will show that even then the convection will still fall short of evacuating the dissipated heat effectively.

Thus, conduction through the bearing surfaces is essential, if one wishes to keep the temperature rise in the bearing at bay. For this purpose, one can calculate the viscous heat dissipation using Petrov's law, define an allowable bearing surface temperature and then determine the temperature gradients in the bearing and housing. From this, one can then determine the amount of cooling needed to maintain thermal equilibrium in the bearing gap.

17.2.3 One Adiabatic Wall and One Isothermal Wall

This situation occurs when it is impossible to effectively cool the moving part of a bearing (such as the shaft, rotor, ... in a JB) while being able to maintain the stationary part (e.g. the bearing housing) at a constant temperature, e.g. by forced cooling. In this case, we use the same temperature form as in the previous case, and with the lower wall being the adiabatic wall, we have b.c.:

$$f(0) = f'(0) = 0, \qquad f(1) = -\Theta_{\text{aw}}(X).$$

Solving as before, we obtain:

$$T = \Theta_{\text{aw}}(X) + \frac{\text{EcPr}}{2}\left[\left(1 - \frac{2\Theta_{\text{aw}}(X)}{\text{EcPr}}\right) Z^3 - Z^2\right], \tag{17.16}$$

with,

$$\Theta_{\text{aw}}(X) = \frac{\text{PrEc}}{2}\left(1 - \exp\left(-\frac{6}{\text{PrRe}^*}X\right)\right) \tag{17.17}$$

So,

$$\Theta_{\text{aw}}(X) \to \frac{\text{PrEc}}{2}$$

i.e. a constant temperature wall, which is six times larger than in the case of both walls being isothermal.

N.B. It is not important which surface is moving since the amount of heat flux remains the same in both cases.

As an example, consider as before $V = 100 \text{ m s}^{-1}$, $h = 10^{-5}$ m, $\mu = 2\ 10^{-5}$ Pa s, then the excess temperature of the adiabatic wall $\Delta T \approx 3.5$K.

Finally, the different distributions are depicted qualitatively in Figure 17.1.

17.3 Adiabatic-Wall Reynolds Equation and the Thermal Wedge

In the previous treatment, we assumed the pressure gradients, density and other fluid-property variations to be negligible. That is often referred to as the "Couette approximation". In the following, we shall present a solution that considers the simultaneous interaction between the energy, the continuity, the momentum and the state equations and thereby derive a Reynolds equation pertaining to flow between adiabatic walls. The basic principle employed in deriving this equation is to employ mean values of the temperature T_{m} and the velocity U_{m} across the gap. This is justified owing to the viscous-flow assumption, which assumes the pressure and the density to be constant across the gap. However, the fluid properties c_p, μ are assumed, as yet, to be constant.

We first state the basic equations:

The velocity and its derivative: From Chapter 4, we have

$$U = \frac{u}{V} = Z - \frac{3H^2}{\Lambda}\frac{\partial P}{\partial X}Z(1 - Z),$$

or

$$U = \frac{u}{V} = Z - \alpha Z(1 - Z),$$

with

$$\alpha = -\frac{3H^2}{\Lambda}\frac{\partial P}{\partial X}.$$

Thus, the mean velocity is:

$$U_{\text{m}} = \frac{1}{2} + \frac{\alpha}{6}$$

and

$$\frac{\partial U}{\partial Z} = \alpha(1 - 2Z)$$

so that

$$U\frac{\partial U}{\partial Z}\bigg|_0^1 = U\frac{\partial U}{\partial Z}\bigg)_{Z=1} = 1 - \alpha\,.$$

The equation of state using mean values across the gap:

$$P = \wp T_{\mathrm{m}}\,.$$

The continuity equation

$$\wp U_{\mathrm{m}} H = \bar{m} = \wp H\left(1 - \frac{H^2}{\Lambda}\frac{\mathrm{d}P}{\mathrm{d}X}\right),$$

where $\bar{m}$ is the dimensionless mass flow rate, being constant along the flow.

The energy equation using mean values across the gap:

$$\wp U_{\mathrm{m}}\frac{\mathrm{d}T_{\mathrm{m}}}{\mathrm{d}X} = \frac{\mathrm{Ec}}{\mathrm{Re}^*}(1 - \alpha)\,.$$

Derivation

We begin from the state equation:

$$T_{\mathrm{m}} = \frac{P}{\wp} = \frac{PU_{\mathrm{m}}H}{\wp U_{\mathrm{m}}H} = \frac{PU_{\mathrm{m}}H}{\bar{m}} = \frac{PH(\frac{1}{2} + \frac{\alpha}{6})}{\bar{m}}\,.$$

Substituting for T_{m} in the energy equation yields:

$$\wp U_{\mathrm{m}}\frac{\mathrm{d}}{\mathrm{d}X}\left(\frac{PH\left(\frac{1}{2} + \frac{\alpha}{6}\right)}{\bar{m}}\right) = \frac{\wp U_{\mathrm{m}}}{\bar{m}}\frac{\mathrm{d}}{\mathrm{d}X}\left(PH\left(\frac{1}{2} + \frac{\alpha}{6}\right)\right)$$

$$= \frac{1}{H}\frac{\mathrm{d}}{\mathrm{d}X}\left(PH\left(\frac{1}{2} + \frac{\alpha}{6}\right)\right) = \frac{\mathrm{Ec}}{\mathrm{Re}^*}(1 - 2\alpha)$$

After rearrangement and substituting for α, we obtain:

$$\frac{\mathrm{d}}{\mathrm{d}X}\left[PH\left(H^2\frac{\mathrm{d}P}{\mathrm{d}X} - \Lambda\right)\right] = -\frac{2\mathrm{Ec}}{\mathrm{Re}^*}\left(3H^3\frac{\mathrm{d}P}{\mathrm{d}X} + \Lambda H\right) = -\frac{(\kappa - 1)}{\kappa}\frac{\Lambda}{3}\left(3H^3\frac{\mathrm{d}P}{\mathrm{d}X} + \Lambda H\right) \tag{17.18}$$

This is the adiabatic-flow Reynolds equation. Note that the L.H.S. is the same as that of the isothermal Reynolds equation. However, the R.H.S. is now different from zero: it has in fact the form of the EP, isothermal porous bearing (see Chapter 13), but with pressure-gradient and bearing-number dependent feed flow.

17.3.1 Results and Discussion

Eq. 17.18 depends on the gap height function $H(X)$ in addition to two parameters: Λ and κ. In order to examine the thermal-wedge effect, we eliminate the physical (or rather geometrical) wedge by making $H = 1$, i.e. uniform gap. Figures 17.2 through 17.5 plot the pressure and temperature distributions for for increasing values of $\Lambda = 1, 3, 5, 10$.

We note that:

- For small values of Λ, the pressure build-up is very modest and can be neglected in favour of the approximate "Couette flow" solution. We note also that this latter, i.e. $T = 2\mathrm{Ec}/\mathrm{Re}^* X$ produces a temperature rise that is close to the exact, Reynolds equation solution.

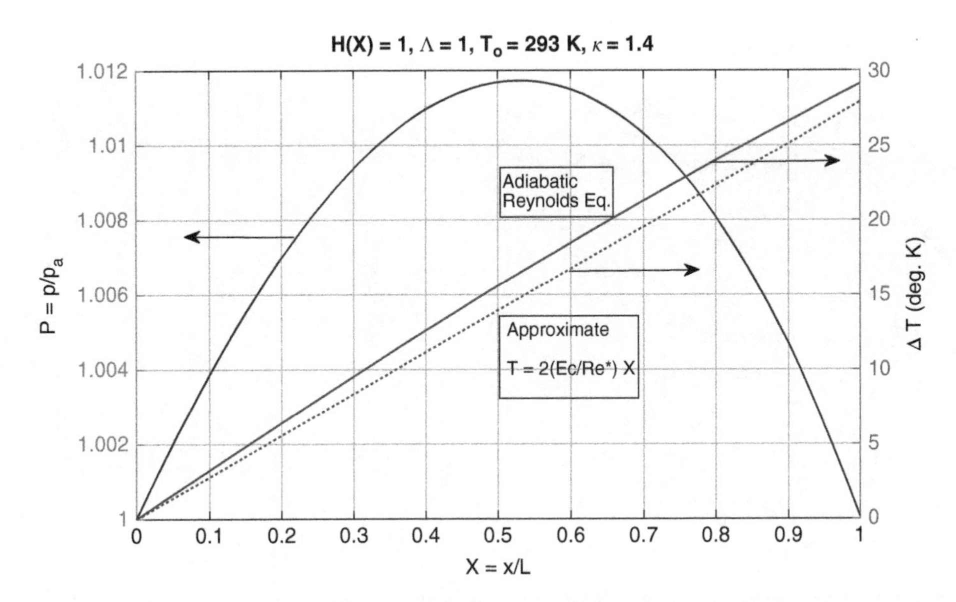

Figure 17.2 The pressure and temperature distributions for adiabatic flow between two parallel surfaces. $\Lambda = 1$, $T_o = 293$, $\kappa = 1.4$. Solid line temperature curve is obtained from adiabatic-wall Reynolds equation; dashed line is from approximate solution $T = 2\text{Ec}/\text{Re}^* X$.

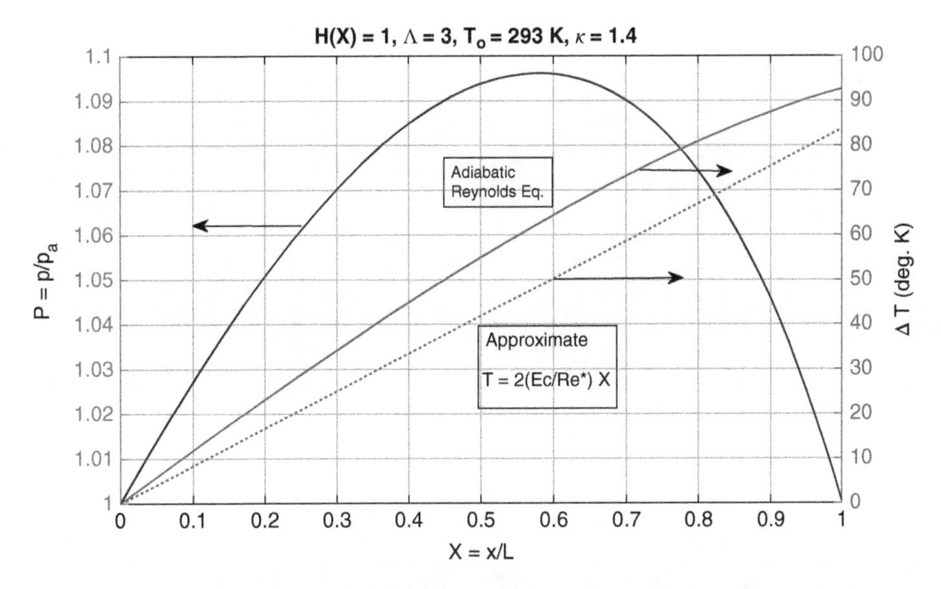

Figure 17.3 The pressure and temperature distributions for adiabatic flow between two parallel surfaces. $\Lambda = 3$, $T_o = 293$, $\kappa = 1.4$. Solid line temperature curve is obtained from adiabatic-wall Reynolds equation; dashed line is from approximate solution $T = 2\text{Ec}/\text{Re}^* X$.

- For higher values of Λ, the pressure build-up becomes appreciable and the temperature curves (approximate and exact) deviate likewise appreciably from each other.
- For this latter situation, the temperature rise is extremely high for the solution to be of practical value.

Thus, we conclude that although the thermal wedge will occur, its relevance will be restricted to special situations in which sliding speeds are extremely high while heat evacuation is severely limited, which are very rare in present-day applications.

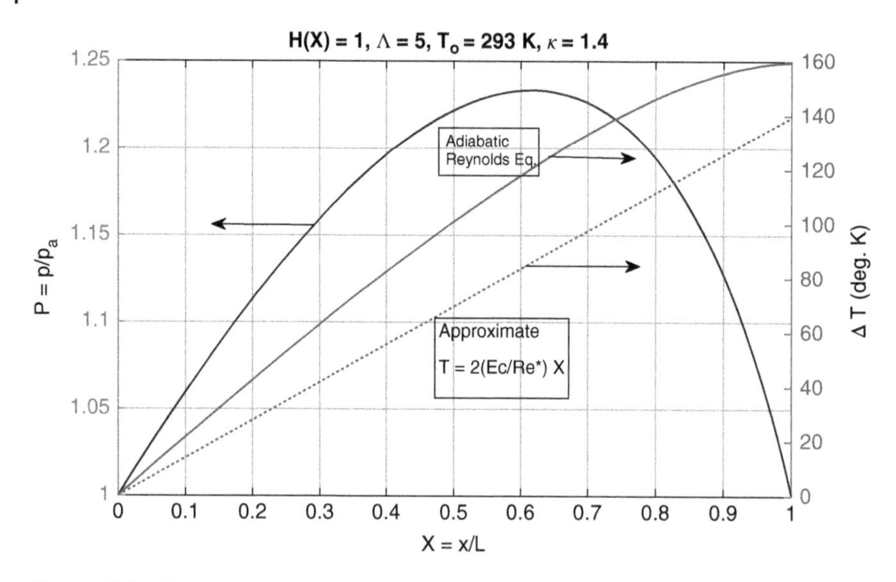

Figure 17.4 The pressure and temperature distributions for adiabatic flow between two parallel surfaces. $\Lambda = 5$, $T_o = 293$, $\kappa = 1.4$. Solid line temperature curve is obtained from adiabatic-wall Reynolds equation; dashed line is from approximate solution $T = 2\text{Ec}/\text{Re}^* X$.

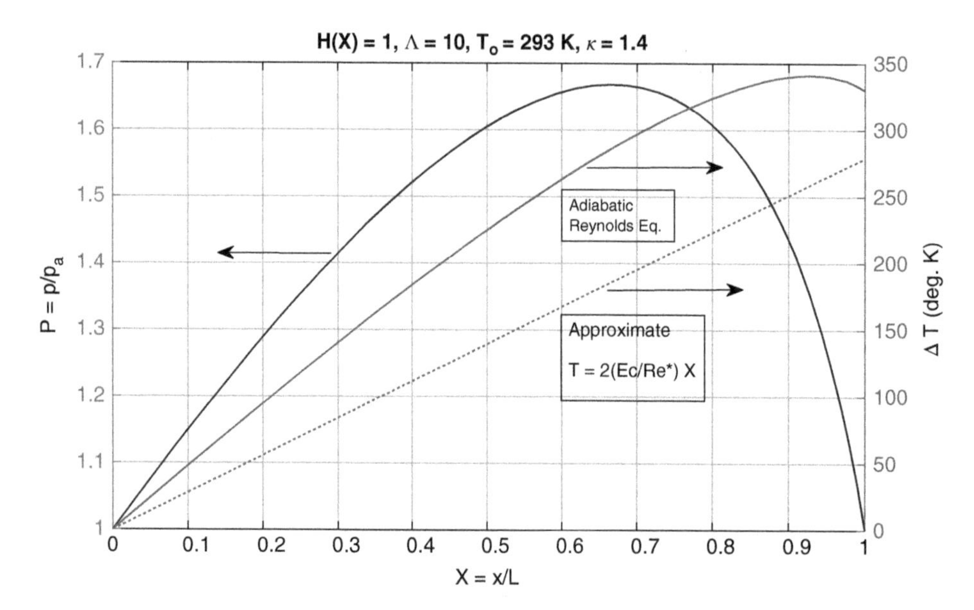

Figure 17.5 The pressure and temperature distributions for adiabatic flow between two parallel surfaces. $\Lambda = 10$, $T_o = 293$, $\kappa = 1.4$. Solid line temperature curve is obtained from adiabatic-wall Reynolds equation; dashed line is from approximate solution $T = 2\text{Ec}/\text{Re}^* X$.

17.3.2 Effect of Temperature on Gas Properties

Generally, except for the viscosity, the other gas properties, such as specific heats, thermal conductivity and κ (the ratio of the specific heats) are almost insensitive to temperature change, for moderate temperature differences (on the order of 100 ° C). The viscosity obeys the approximate formula Constantinescu (1969):

$$\frac{\mu}{\mu_0} = \left(\frac{T}{T_0}\right)^n, \qquad n = 0.76 \text{ (air)},$$

where the temperatures are absolute.

This yields a change of $\sim 10\%$ for a temperature rise of 50 K above 300 K atmosphere. Viscosity values are listed in Table 17.1 for a variety of gases at 1 bar, from Huber and Harvey (2011).

Effect of Increased Viscosity on Bearing Film Behaviour

In self-acting bearings experiencing large viscous warming, the viscosity will rise, almost proportionally with the temperature rise, as we see from the formula and Table above. This will:

- Increase load capacity, especially for small to moderate sliding numbers. In that case, the increase will be approximately in the same proportion. For large sliding numbers, the load becomes almost independent of its value, and thus of the value of viscosity.
- Obviously, and more importantly, also increase the friction torque, which in its turn will further increase dissipation, until thermal equilibrium is attained. Care must be taken, therefore, to ensure that the bearing system does not enter into an unstable thermal cycle.
- Since the sliding number and the squeeze number will both increase in the same proportion, we might intuitively expect that the dynamic film characteristics will not change appreciably. However, for the same bearing load, an increased viscosity leads to smaller eccentricity ratio and that might lead to reduced whirl resistance.

17.3.3 Conclusions on the Aerodynamic Case

From the above solutions and results we can draw the following conclusions:

- a simplified solution, employing only Couette heating (Petrov's law, neglecting pressure flow) and mean temperatures across the gap, is sufficient to describe the problem adequately when bearing numbers are moderate or walls sufficiently isothermal;
- when at least one of the bearing walls is isothermal, conduction across the gap is sufficient to keep the film temperature reasonably small;
- convection has negligible effect w.r.t. conduction; thus, the film temperature will increase appreciably when both walls are adiabatic;
- except for the viscosity, other gas properties are not so sensitive to temperature variation; an increase in viscosity owing to temperature rise leads to increased load capacity but also to increased friction losses.

17.4 Flow Through Centrally Fed Bearing: Formulation of the Problem

Our main concern here is not the thermal management of the bearing in the presence of high rates of viscous dissipation, but rather to discover the evolution of the temperature distribution over the bearing film and walls. This might be of paramount importance in precision and ultra-precision motion control systems. Furthermore, we would like to find out whether the isothermal assumption, which we always used in the Reynolds equation is justified.

Table 17.1 Viscosity of gases, at pressure 1 bar, as function of temperature. Source: Huber M and Harvey A n.d. Viscosity of gases. CRC Handbook of CHemistry and Physics 92nd Ed., 6–229–6–230.

	(Viscosity in μPa s at 1 bar)	**100 K**	**200 K**	**300 K**	**400 K**	**500 K**	**600 K**
	Air	7.1	13.3	18.5	23.1	27.1	
Ar	Argon	8.1	15.9	22.7	28.6	33.9	38.8
BF_3	Boron trifluoride		12.3	17.1	21.7	26.1	30.2
ClH	Hydrogen chloride			14.6	19.7	24.3	
F_6S	Sulfur hexafluoride	15.3			19.7	23.8	27.6
H_2	Normal hydrogen	4.1	6.8	8.9	10.9	12.8	14.5
D_2	Deuterium	5.9	9.6	12.6	15.4	17.9	20.3
H_2O	Water			9.8	13.4	17.3	21.4
D_2O	Deuterium oxide			10.2	13.7	17.8	22.0
H_2S	Hydrogen sulfide			12.5	16.9	21.2	25.4
H_3N	Ammonia			10.2	14.0	17.9	21.7
He	Helium	9.6	15.1	19.9	24.3	28.3	32.2
Kr	Krypton		17.4	25.5	32.9	39.6	45.8
NO	Nitric oxide		13.8	19.2	23.8	28.0	31.9
N_2	Nitrogen	7.0	12.9	17.9	22.2	26.1	29.6
N_2O	Nitrous oxide		10.0	15.0	19.8	24.1	27.9
Ne	Neon	14.4	24.1	31.9	38.6	44.8	50.6
O_2	Oxygen	7.7	14.7	20.7	25.8	30.5	34.7
O_2S	Sulfur dioxide		8.6	12.9	17.5	21.7	
Xe	Xenon		15.7	23.2	30.5	37.2	43.5
CO	Carbon monoxide	6.7	12.9	17.8	22.1	25.8	29.1
CO_2	Carbon dioxide		10.1	15.0	19.7	24.0	28.0
$CHCl_3$3	Chloroform			10.2	13.7	16.9	20.1
CH_4	Methane	3.9	7.7	11.1	14.2	17.0	19.5
CH_4O	Methanol		6.6	9.7	13.0	16.4	19.8
C_2H_2	Acetylene			10.4	13.5	16.5	
C_2H_4	Ethylene		7.0	10.4	13.6	16.5	19.2
C_2H_6	Ethane		6.4	9.4	12.2	14.8	17.1
C_2H_6O	Ethanol				11.6	14.5	17.0
C_3H_8	Propane			8.2	10.8	13.3	15.6
C_4H_{10}	Butane			7.5	9.9	12.2	14.5
C_4H_{10}	Isobutane			7.5	9.9	12.2	14.4
$C_4H_{10}O$	Diethyl ether			7.6	10.1	12.4	
C_5H_{12}	Pentane			6.7	9.2	11.4	13.4
C_6H_{14}	Hexane				8.6	10.8	12.8

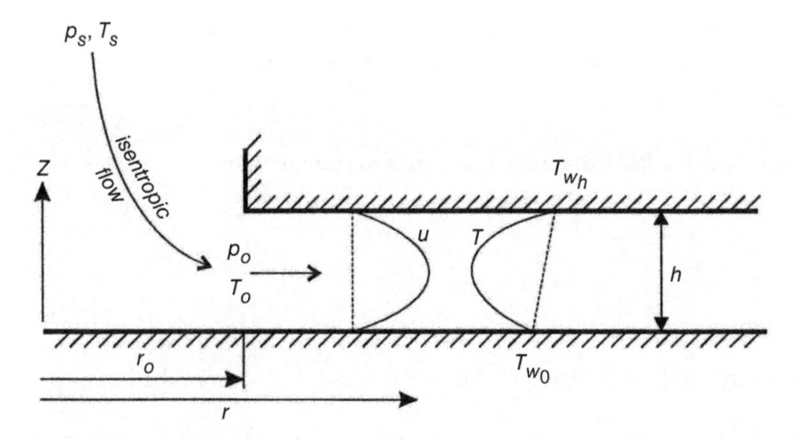

Figure 17.6 Flow configuration showing the temperature distribution. Source: Al-Bender F 1992.

The flow configuration is depicted in Figure 17.6 (Al-Bender 1992), where, in addition to the description given in Chapter 3, a typical temperature distribution in the film is sketched. The assumptions relating to the velocity field, given in Chapters 3 and 5, are taken to hold true. In addition, we make the following simplifying assumptions regarding the temperature and the viscosity:

- The temperature distribution is symmetric about the mid-plane of the flow, i.e. the wall temperatures are equal (but not necessarily constant):

  $$T_{\mathrm{w}0}(r) = T_{\mathrm{w}h}(r) = T_{\mathrm{w}}(r)$$

  This assumption is made in order to ensure consistency with the assumption of symmetric velocity distribution: it does not affect the basic nature of the problem. Furthermore, in the special case when the walls are assumed to be isothermal, their temperature will be (realistically) assumed equal to the stagnation temperature of the gas, i.e.

  $$T_{\mathrm{w}} = T_s \qquad \text{(isothermal walls)}$$

  This is seen to be consistent with the fact that, at stagnation, the gas temperature is equal to that of its surroundings, viz. the bearing body.
- The viscosity is everywhere constant. A solution, (that would be more complicated,) can also be obtained with this condition removed, i.e. with the assumption of temperature dependent viscosity. However, owing to the limited temperature variation of the flow, the viscosity variations will play only a minor role. In other words, we will be mainly interested in the thermal influence on the *state* of the gas; more precisely, the density variation.
- The temperature (and therefore also the density) variation across the gap is small in comparison to the wall temperature (and the density at the wall). Let us note that, in the extreme case of sonic entrance, the ratio of the wall temperature to the mid-plane temperature is only 1.2.

Further, we shall, for simplicity, consider only the case of uniform bearing gap, i.e. $h \equiv$ constant, since this plays an unessential role in the problem.

The flow equations, normalised w.r.t. gap entrance values, are recalled from Chapter 2:

Equation of State

$$P = \wp T \tag{17.19}$$

Note that, by virtue of the BL assumptions, the pressure is independent of Z, i.e. $P = P(R)$. Since the temperature varies, in the general case, across the gap, i.e. $T = T(R, Z)$, it follows that the density must likewise vary across the gap, or $\wp = \wp(R, Z)$. As we shall be interested in the mean temperature and the mean density, we may, by virtue of the third assumption above, express the pressure as a product of these two. Thus, we have:

$$P(R) = \wp(R, Z)T(R, Z) \approx \bar{\wp}(R)\bar{T}(R) \tag{17.20}$$

where the bars indicate gap-averaged values, and the last expression is accurate to second order.

Feed Flow

$$P_s = \left(1 + \frac{(\kappa - 1)}{2}\mathrm{M_o}^2\right)^{\kappa/(\kappa-1)} \tag{17.21}$$

The corresponding equation for the temperature in this region is obtained by substituting from the isentropy and the state, to give:

$$T_s^* = 1 + \frac{(\kappa - 1)}{2}\mathrm{M_o}^2 \tag{17.22}$$

Film Flow

The continuity, momentum and energy equations are respectively:

$$\frac{1}{R}\frac{\partial}{\partial R}(\wp RU) + \frac{\partial}{\partial Z}(\wp W) = 0, \tag{17.23}$$

$$\wp \mathrm{Re}_o^* \left(U\frac{\partial U}{\partial R} + W\frac{\partial U}{\partial Z}\right) = -\frac{\mathrm{Re}_o^*}{\kappa \mathrm{M_o}^2}\frac{\mathrm{d}P}{\mathrm{d}R} + \frac{\partial^2 U}{\partial Z^2} \tag{17.24}$$

$$\wp \mathrm{Re}_o^* \left(U\frac{\partial T^*}{\partial R} + W\frac{\partial T^*}{\partial Z}\right) = \frac{\mathrm{Ec_o Re}_o^*}{\kappa \mathrm{M_o}^2}U\frac{\mathrm{d}P}{\mathrm{d}R} + \frac{1}{\mathrm{Pr}}\frac{\partial^2 T^*}{\partial Z^2} + \mathrm{Ec_o}\left(\frac{\partial U}{\partial Z}\right)^2 \tag{17.25}$$

Eliminating the pressure gradient between the last two equations, and rearranging, yields a more compact version of the energy equation, thus:

$$\begin{aligned}\wp \mathrm{Re}_o^* \left(U\frac{\partial T^*}{\partial R} + W\frac{\partial T^*}{\partial Z}\right) &= -\wp \mathrm{Ec_o Re}_o^* U\left(U\frac{\partial U}{\partial R} + W\frac{\partial U}{\partial Z}\right) \\ &\quad + \frac{1}{\mathrm{Pr}}\frac{\partial^2 T^*}{\partial Z^2} + \mathrm{Ec_o}\frac{\partial}{\partial Z}\left(U\frac{\partial U}{\partial Z}\right)\end{aligned} \tag{17.26}$$

Remark

Note that the last equation may also be expressed as follows:

$$\wp \mathrm{Re}_o^* \left(U\frac{\partial \mathcal{H}}{\partial R} + W\frac{\partial \mathcal{H}}{\partial Z}\right) = +\left(\frac{1 - \mathrm{Pr}}{\mathrm{Pr}}\right)\frac{\partial^2 T^*}{\partial Z^2} + \frac{\partial^2 \mathcal{H}}{\partial Z^2} \tag{17.27}$$

where,

$$\mathcal{H} = T^* + \frac{\mathrm{Ec_o}}{2}U^2 = \frac{c_p T + u^2/2}{c_p T_o}$$

is the the *normalised enthalpy*.

If the Prandtl number Pr is set equal to unity, (as is often done in approximate analysis,) the resulting energy equation will be in the enthalpy alone, and *similar* in form to the momentum equation with zero pressure gradient. Consequently, *similarity* integrals may be found analytically in some limiting cases, see e.g. (Fainzil'ber 1962; Schlichting 1968), in the form:

$$\mathcal{H} = A_0 + A_1 U + A_2 \partial U/\partial Z, \qquad (\mathrm{d}P/\mathrm{d}R \equiv 0).$$

17.5 Method of Solution

With the mean density given as a function of the pressure, a solution to the momentum equation may be obtained, as has been shown in Chapters 3 and 5, briefly restated as follows:

$$U = Q(R)G(Z;\ R) \Rightarrow \partial^2 G/\partial Z^2 = -n + mG^2;\ I = \int_0^1 G\mathrm{d}Z$$

with, (note that $H \equiv 1$)

$$m = \mathrm{Re}_0^* \bar{\wp} \mathrm{d}Q/\mathrm{d}R, \quad n = -(\mathrm{Re}_0^*/\kappa \mathrm{M}_0{}^2)(\mathrm{d}P/\mathrm{d}R)/Q, \quad R\bar{\wp}QI = 1$$

Our problem consists now of solving the energy equation simultaneously with this problem, in order to determine the average density. First, we transform the convective terms in equation 17.26 using the von Mises variable $\xi = R$, to obtain:

$$\wp \mathrm{Re}_0^* \left(U\frac{\partial T^*}{\partial \xi}\right) = -\wp \mathrm{Ec}_0 \mathrm{Re}_0^* U\left(U\frac{\partial U}{\partial \xi}\right) + \frac{1}{\mathrm{Pr}}\frac{\partial^2 T^*}{\partial Z^2} + \mathrm{Ec}_0\frac{\partial}{\partial Z}\left(U\frac{\partial U}{\partial Z}\right) \tag{17.28}$$

where the transverse velocity components have been eliminated. Recalling that the assumption of separability of the momentum equation was equivalent to putting:

$$\partial U/\partial \xi = \mathrm{d}Q/\mathrm{d}R\ G(Z;\ R),$$

we write the temperature function in a similar form:

$$T^* = T_\mathrm{w}(R) + T_\mathrm{f}\Theta(Z;\ R)$$

For simplicity of subsequent calculation, we set, in addition:

$$\int_0^1 \Theta(Z;\ R)\mathrm{d}Z = 1$$

such that

$$\bar{T}^* = T_\mathrm{w} + T_\mathrm{f}$$

Now, equation 17.28 may be integrated across the gap, if we replace the density by its mean value and in doing so make the following approximation:

$$\wp \partial T^*/\partial \xi \approx \bar{\wp}\mathrm{d}\bar{T}^*/\mathrm{d}R$$

Finally, equation 17.28 becomes:

$$\bar{\wp}\mathrm{Re}_0^*\left(U\frac{\mathrm{d}\bar{T}^*}{\mathrm{d}R}\right) = -\bar{\wp}\mathrm{Ec}_0\mathrm{Re}_0^* Q^2 \mathrm{d}Q/\mathrm{d}R\ G^3 + \frac{1}{\mathrm{Pr}}T_f\frac{\partial^2\Theta}{\partial Z^2} + \mathrm{Ec}_0\ Q^2\frac{\partial}{\partial Z}\left(G\frac{\partial G}{\partial Z}\right) \tag{17.29}$$

which is easily integrated w.r.t. Z once G and the boundary conditions on Θ are known. The required integrals of G, G^3, etc. are obtained from the differential equation for G. It is easily shown, e.g., that:

$$\int_0^Z G^3\mathrm{d}Z = \frac{3}{5m}\left(3n\int_0^Z G\mathrm{d}Z + 2(m/3 - n)Z + G\frac{\partial G}{\partial Z}\right)$$

Recurrence relations of this type reduce the requirement to the primitives of G only. These we denote by γ and Γ, such that:

$$\partial\Gamma/\partial Z = \gamma, \qquad \partial\gamma/\partial Z = G$$

with,

$$Z = 0\ :\ G = \gamma = \Gamma = 0$$

(Note that $I = \gamma(1;\ R)$.)

17.5.1 Solutions

The boundary conditions on Θ, in equation 17.29 are:

$$Z = 0,1 \ : \Theta = 0, \quad Z = 1/2 \ : \ \Theta_Z = 0$$

In the case of an adiabatic wall, we have the extra condition:

$$Z = 0,1 \ : \Theta_Z = 0. \quad \text{(adiabatic walls.)}$$

Integrating equation 17.29, with the above boundary conditions, three times w.r.t. Z, we obtain a differential equation for $\bar{T}^*$ in the form:

$$\frac{\mathrm{d}\bar{T}^*}{\mathrm{d}R} = f(\text{flow parameters})$$

This equation may now be solved simultaneously with the entrance problem, determining at each step the average temperature which is the required proportionality factor between the pressure and the density.

17.6 Results and Discussion

Before presenting sample results, let us first examine the following limiting cases pertaining to an adiabatic wall.

The Case of Adiabatic Wall
leads to the following analytic result:

$$\bar{T}^{*\prime} = -\frac{3}{5}\frac{\mathrm{Ec}_o Q}{\mathrm{Re}_o^* \wp}\left(\frac{2}{I}(m/3 - n) + 3n\right)$$

(Note that we will use primes to indicate ordinary differentiation w.r.t. R.)

The two extreme limits of uniform entrance velocity and of viscous flow may now be examined:

1. $I \to 1$: $n \to m \to \infty$ yields the result:

   $$\frac{\bar{T}^{*\prime}}{\bar{T}^*} = \frac{(\kappa - 1)}{\kappa}\frac{P'}{P}$$

   which is recognised as the *isentropy* relation.
2. $I \to 2/3$: $n \to 8$; $m \to 0$ yields the result:

   $$\bar{T}^{*\prime} \to 0$$

   which means that the flow becomes *isothermal* at this limit. This somewhat curious result was arrived at also by Hughes (cited in Constantinescu (1969)) using approximate analysis. It should be noted, however, that although the flow becomes isothermal, its mean temperature will, in the general case, be different from the stagnation temperature. Conversely, it will be apparent from the subsequent results that in the limit of viscous flow with isothermal walls, no heat transfer would take place between the walls and the fluid.

In the following, typical results obtained by this method will be given and discussed. Note, however, that the entrance velocity distributions, i.e. the values of I_o, in these examples, have been chosen in such a way that the mean flow be decelerating downstream of the entrance. This was necessary in order to ensure the stability of the flow at high entrance Mach numbers coupled with small Reynolds numbers. The values of I_o are indicated for each case.

Figure 17.7(a) shows the mean temperature distribution along the bearing land in the case of isothermal wall (with temperature equal to the stagnation temperature). Figure 17.7(b) shows the rate and direction of heat trans-

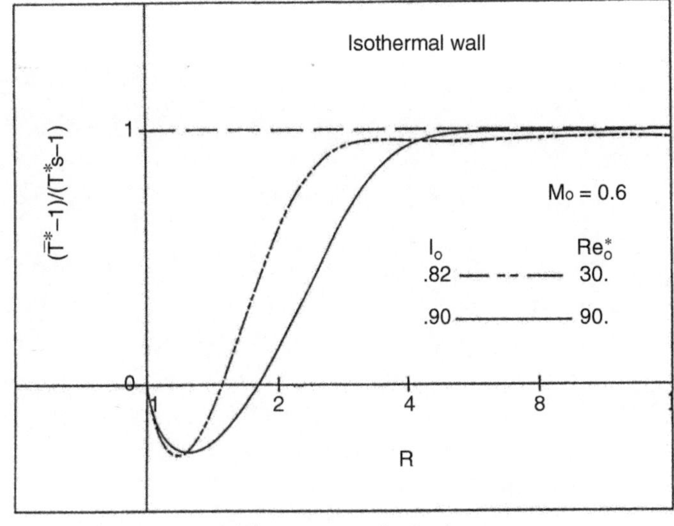

(a) Temperature distribution

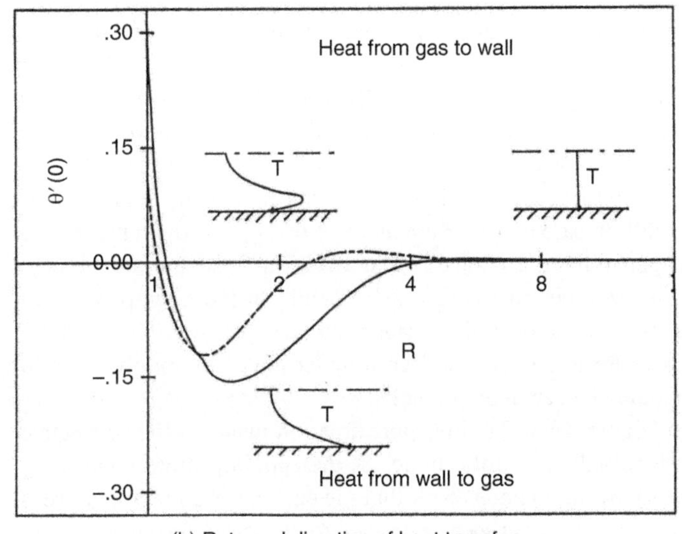

(b) Rate and direction of heat transfer

Figure 17.7 Thermal behaviour in the case of isothermal walls. Source: Al-Bender F 1992.

fer taking place between the gas and the walls, (being proportional to $\Theta_Z(0;\ R)$ denoted for simplicity by $\Theta'(0)$). It is noteworthy that, just downstream of the entrance, the fluid initially looses heat to the wall despite the fact that its average temperature is less than that of the wall. This is owing to the peculiar "reverse" temperature distribution across the gap, as sketched on the figure. Shortly thereafter, the heat flow begins to take place from the wall to the gas until the gas temperature asymptotically approaches the wall temperature. The temperature distribution is evidently dependent on the velocity distribution. Consequently, if the initial velocity distribution is not uniform, we have an *anomaly* at the entrance in that the gas is too *cool* and would, therefore, never regain its stagnation temperature. (This is evidenced more clearly for the case of $Re^*_o = 30.$, where there is a second region of heat transfer from the gas to the wall.) If, on the other hand, the initial velocity distribution *is* uniform, the gas temperature will approach the stagnation (= wall) temperature in the limit of viscous flow. In any case, at the limit

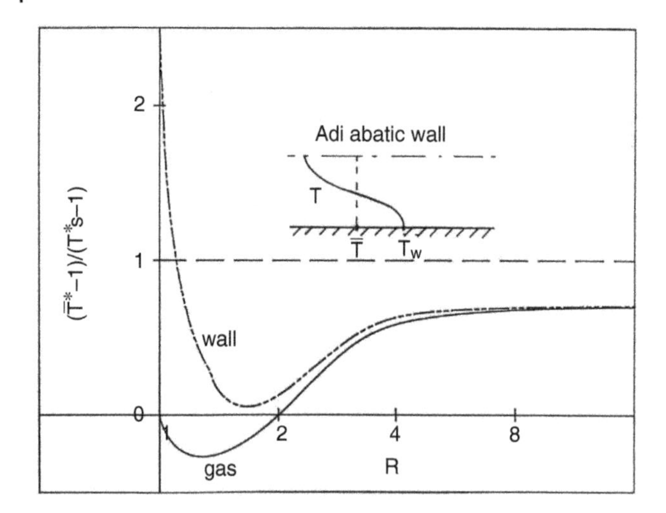

Figure 17.8 Temperature distributions in the case of adiabatic walls. Source: Al-Bender F 1992.

of viscous flow, the gas attains thermal equilibrium, whereby no heat transfer takes place between it and the wall, i.e. it approaches the adiabatic wall case discussed above.

Figure 17.8 shows the temperature distributions of the gas and the wall for the case of an adiabatic wall. The gas mean temperature initially follows the isentropic relation discussed in the first limiting case above. It begins thus by decreasing and increasing with the pressure (not shown). However, viscous friction will cause it to depart gradually from the isentropic behaviour until, reaching the viscous region, it tends asymptotically to an isothermal regime, as has been predicted in the second limiting case above. From about $R = r/r_o = 5$, the temperature quotient plotted in the ordinate $(\bar{T}^* - 1)/(T_s^* - 1)$ approaches the value 0.7. If, as an example, we assume the stagnation temperature $T_s = 300\text{K}$, then the mean entrance temperature is $\bar{T}_0 \approx 280$ K, while the mean temperature in the viscous region is $\bar{T}_{\text{visc}} \to 294$ K; a difference of only 6 degrees from the stagnation temperature. This example serves to show that, even when there is no heat transfer to the gas, the isentropy formula (commonly taken to represent *adiabatic* flow Constantinescu (1969)) would yield erroneous results.

The wall temperature is also plotted in the same figure. Owing to the approximation made on the convective temperature terms, the temperature distribution is initially not uniform across the gap; but rather completely determined by the velocity distribution. This leads to the anomalous result that the entrance wall temperature is appreciably higher than the stagnation temperature. This does not seem, however, to influence the overall result, since the wall temperature quickly drops, and begins qualitatively to follow the gas mean temperature. Eventually, the difference between the two temperatures closes down when they begin to approach the same limit. This means then that the temperature distribution across the gap becomes nearly uniform in the viscous region.

Remark

The case of an adiabatic wall has been treated merely for its academic interest. In practice, the bearing parts are mostly metallic. Noting that metals have thermal conduction coefficients that are several orders of magnitude higher than those of gases, it would be reasonable to assume the bearing surfaces to act like heat sinks. That is to say that the isothermal wall assumption is the one that is more applicable to the bearing problem. The remainder of this section will, therefore, be concerned with that assumption.

The thermal behaviour of the flow, in the case of isothermal walls, may be summarised by examining the influence of the entrance reduced Reynolds number Re_0^* on the polytropic exponent χ. However, we have to give an appropriate definition for χ, when it be a variable. Let us, first, recall the general definition from Chapter 2:

$$\frac{p}{\rho^{\chi}} = \text{const.}, \qquad 1 \leq \chi \leq \kappa$$

We note that this equation becomes dimensionally unsound when χ is a variable; i.e. χ would be not only a variable but also of varying physical dimensions. To overcome this, we may define χ w.r.t. a fixed reference $()_\mathrm{r}$, such that:

$$\frac{(p/p_\mathrm{r})}{(\rho/\rho_\mathrm{r})^{\chi}} = \text{const.},$$

In this case, values of χ lying between 1 and κ may be obtained only when stagnation is used as the reference. Thus, for our problem, we shall define:

$$\chi = \frac{\log(p/p_\mathrm{s})}{\log(\rho/\rho_\mathrm{s})}$$

Figure 17.9 plots χ in function of the radius for various values of Re_0^*, with $\mathrm{M}_0 = 0.3$, (the results are not significantly affected by the value of M_0.) By virtue of the isentropic assumption in the feed region, the value of χ at entrance is always equal to κ. From then on, χ begins to drop (or decay,) approaching the value of unity asymptotically. The rate of decay depends on Re_0^*; the higher the latter the lower the former. These values may provide a rough guide for the thermal behaviour of the flow in the entrance region.

Finally, a pressure distribution example is given to show the thermal influence on the load capacity and the mass flow rate. Figure 17.10 shows the pressure distributions in a bearing film with isothermal walls, obtained by an exact solution, using this method, and by the isothermal state assumption (Chapter 5) respectively. It is seen that the difference is small and confined to the entrance region. The exact solution results in a comparatively smaller pressure depression owing to the fact that the fluid is "cooler", and consequently, more dense than in the isothermal case. The relative error incurred on the load capacity and the mass flow rate by adopting one or the other assumption remains less than 1 percent. Therefore, it would be justifiable to use the isothermal assumption for general design purposes since it also greatly simplifies the problem.

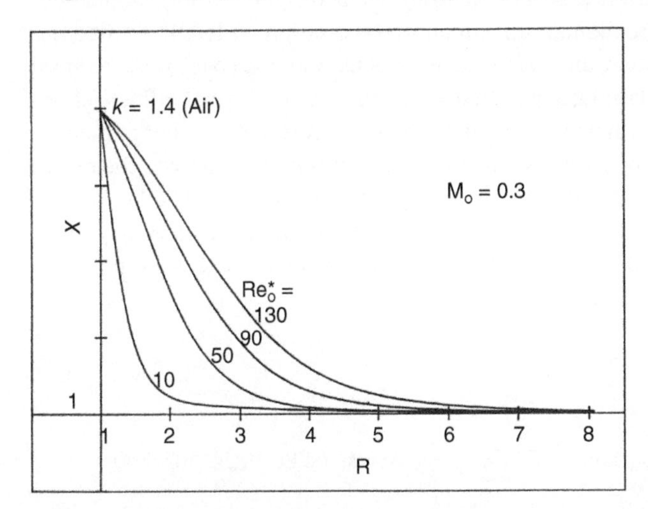

Figure 17.9 Behaviour of the polytropic exponent χ. Source: Al-Bender F 1992 Contributions to the Design Theory of Circular Centrally Fed Aerostatic Bearings PhD thesis Leuven, Belgium.

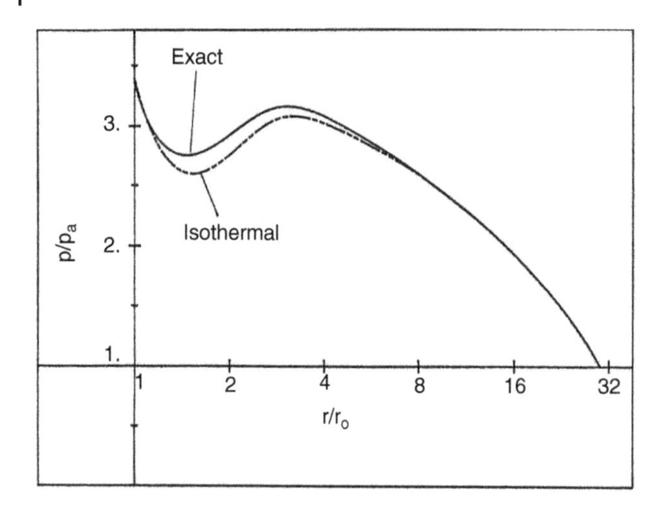

Figure 17.10 Pressure distributions in the cases of isothermal walls ("exact") and isothermal state ("isothermal") for a bearing with: $p_s/p_a = 5$, $r_a/r_o = 30$, $\Lambda_f = 59$ ($Re_o^* = 45.8$, $M_o = 0.76$). Source: Al-Bender F 1992.

17.7 Conclusions

In the first part of this chapter, we showed that viscous friction dissipation can reach high proportions when the relative speed between the bearing surfaces increases. The heat generated by this friction can hardly be evacuated by convection of the flowing air so that conduction through the bearing surfaces will be the only effective mechanism to accomplish this. It might suffice in certain situations to cool only one bearing member, e.g. the housing, which might simplify the task greatly. Formulas have been provided to quantify the various aspects of the thermal problem based on dimensionless parameters.

In the second part, attention was focused on the thermal aspects of centrally fed aerostatic bearings. A method of solving the energy equation simultaneously with the momentum equation has been presented. It is based on approximating the convective terms in the energy equation, and using the mean density and viscosity values across the gap, all of which being considered sufficient for obtaining a quantitative idea about the thermal influences on the flow. The results show that the flow becomes isothermal in the viscous region regardless of whether or not the walls are thermally conductive. If the walls are isothermal, as is usually the case in gas bearing surfaces, the mean gas temperature will approach the wall temperature (assumed equal to the stagnation temperature) in the viscous region. For this case, typical behaviour of the polytropic exponent, in relation to the entrance reduced Reynolds number, has been shown. A bearing example shows that the thermal influence on the pressure distribution, and consequently the load capacity and the flow rate, is negligible.

References

Al-Bender, F. (1992 Contributions to the Design Theory of Circular Centrally Fed Aerostatic Bearings PhD thesis Leuven, Belgium.

Constantinescu, V.N., (1969). *Gas Lubrication*. ASME.

Fainzil'ber, A., (1962). New similarity integrals in heat and mass transfer processes. *International Journal of Heat and Mass Transfer* 5, 1069–1080.

Huber, M. and Harvey, A.H., (2011). Viscosity of gases. *CRC Handbook of CHemistry and Physics* 92nd Ed., 6-229–6-230.

Hughes, W.F. and Osterle, J.F., (1957). On the adiabatic couette flow of a compressible fluid. *Zeitschrift für angewandte Mathematik und Physik*, 8(2): 89–96.

Paulsen, B., Morosi, S. and Santos, I., (2011). Static, dynamic, and thermal properties of compressible fluid film journal bearings. *Tribology Transactions* 54(2): 282–299.

Peng, Z.C. and Khonsari, M.M., (2006). A thermohydrodynamic analysis of foil journal bearings. *Journal of Tribology* 128(3): 534.

Pinkus, O., (1990). *Thermal Aspects of Fluid Film Tribology*. ASME Press: New York.

Radil, K. and Zeszotek, M., (2004). An experimental investigation into the temperature profile of a compliant foil air bearing. *Tribology Transactions* 47(4): 470–479.

Salehi, M. and Heshmat, H., (2000). On the fluid flow and thermal analysis of a compliant surface foil bearing and seal. *Tribology Transactions* 43(2): 318–324.

San Andrés, L. and Kim, T., (2010). Thermohydrodynamic analysis of bump type gas foil bearings: A model anchored to test data. *Journal of Engineering for Gas Turbines and Power*, 132 p 042504., 042504.

Schlichting, H., (1968). *Boundary-Layer Theory* 6 edn. McGraw-Hill, New York.

Sim, K. and Kim, D., (2008). Thermohydrodynamic analysis of compliant flexure pivot tilting pad gas bearings. *Journal of Engineering for Gas Turbines and Power* 130(3): p 032502.

Index

Air Bearings: Theory, Design and Applications, First Edition. Farid Al-Bender.

Companion website: www.wiley.com/go/AlBender/AirBearings

g

i

j

k

t